EURAM 88

Volume 3: Biology

EUREM 88

Proceedings of the 9th European Congress on Electron Microscopy held in York, England, 4–9 September 1988

Volume 3: Biology

Edited by H G Dickinson and P J Goodhew

Institute of Physics Conference Series Number 93
Institute of Physics, Bristol and Philadelphia

CODEN IPHSAC 93 1 – 616 (1988)

British Library Cataloguing in Publication Data

European Congress on Electron Microscopy
(*9th: 1988: York, England*)
EUREM 88.
1. Electron microscopy
I. Title II. Goodhew, Peter J. (Peter John), *1943–* III. Dickinson, H.G. IV. Institute of Physics V. Series
502'.8'25

ISBN 0-85498-184-5 (vol 1)
0-85498-185-3 (vol 2)
0-85498-186-1 (vol 3)
0-85498-187-X (Set—vols 1, 2 and 3)

Library of Congress Cataloguing-in-Publication Data are available

Published under The Institute of Physics imprint by IOP Publishing Ltd
Techno House, Redcliffe Way, Bristol BS1 6NX, England
242 Cherry Street, Philadelphia, PA 19106, USA

Printed in Great Britain at The Bath Press, Avon

Congress organisation

Executive Committee

Secretary General
A W ROBARDS, Institute of Applied Biology, University of York.
J N CHAPMAN, Department of Physics and Astronomy, University of Glasgow.
B C COWEN, Micro Instruments Ltd, Oxford.
H G DICKINSON, Department of Botany, University of Reading.
P J GOODHEW, Department of Materials Science and Engineering, University of Surrey.

Scientific Organisation Committee

Chairman
J N CHAPMAN, Department of Physics and Astronomy, University of Glasgow.
P D AUGUSTUS, Plessey Research (Caswell) Ltd, Towcester, Northants.
G R BULLOCK, Ciba-Geigy Ltd, Basle, Switzerland.
M J CULLEN, Muscular Dystrophy Research Laboratory, Newcastle General Hospital.
H Y ELDER, Institute of Physiology, University of Glasgow.
H M FLOWER, Department of Metallurgy and Materials Science, Imperial College, London.
P J GOODHEW, Department of Materials Science and Engineering, University of Surrey.
C J HUMPHREYS, Department of Materials Science and Engineering, University of Liverpool.
M H LORETTO, Department of Metallurgy and Materials, University of Birmingham.
M OSBORNE, Department of Physiology, University of Birmingham.
T F PAGE, Department of Metallurgy and Engineering Materials, University of Newcastle-upon-Tyne.
A W ROBARDS, Institute of Applied Biology, University of York.
P SHAW, Department of Cell Biology, John Innes Institute, Norwich.

International Scientific Advisory Board

M BORGERS, Belgium.
A BOURRET, France.
A CSANADY, Hungary.
A DELONG, Czechoslovakia.
J GJÖNNES, Scandinavia.
L H MARGARITIS, Greece.
A B MAUNSBACH, Denmark (Secretary, IFSEM).
N PIPAN, Yugoslavia.
P G PLOAIE, Rumania.
G POZZI, Italy.
A RAMBOURG, France.
M C RISUENO, Spain.
B RODKIEWCZ, Poland.
H ROSE, German Federal Republic.
P SKALICKY, Austria.
U B SLEYTR, Austria.
B UHRIK, Czechoslovakia.
K D VAN DER MAST, Netherlands.
E WISSE, Belgium (Secretary, CESEM)

Trade Advisory Committee of the Royal Microscopical Society (Exhibition):

Chairman
R SETTERINGTON, Carl Zeiss (Oberkochen) Ltd.
Vice-Chairman
M J CAPERS, Hitachi Scientific Instruments Ltd.
A H CLEAVER, JEOL UK Ltd.
P CRAWLEY, Cambridge Instruments Ltd.
T R JOHNSON, ISI Europe.
R S PADEN, Cambridge Scanning Co. Ltd
A W ROBARDS, University of York.
D J ROEBUCK, Philips.
E W SAMUEL, Link Analytical Ltd.
R K STACEY, Olympus Optical Co. (UK).
A TODD, Prior Scientific Instruments Ltd.
F WILKINSON, Carl Zeiss (Jena).
J P ZEIDLER, E Leitz (Instruments) Ltd.

Social Programme Committee

Chairman
A J WILSON, Centre for Cell and Tissue Research, University of York.
A W ROBARDS, University of York.

British Joint Committee for Electron Microscopy

Chairman
P J GOODHEW, Department of Materials Science and Engineering, University of Surrey.

Secretary/Treasurer
G R BULLOCK, Ciba-Geigy Ltd, Basle, Switzerland.

Committee of European Societies for Electron Microscopy

Chairman
A W ROBARDS, UK.
Secretary/Treasurer
E WISSE, Belgium.
Committee Members
I M HÖRL, Austria.
J HEYDENREICH, DDR.
F LENZ, German Federal Republic.
S VIRAGH, Hungary.

International Federation of Societies for Electron Microscopy

President
G THOMAS, USA.
General Secretary
A B MAUNSBACH, Denmark.
Vice-President
H HASHIMOTO, Japan.
Committee Members
D J H COCKAYNE, Australia.
A DELONG, CSSR.
A HOWIE, UK.
K OGAWA, Japan.
E F J VAN BRUGGEN, Netherlands.
E ZEITLER, German Federal Republic.

Previous European Congresses:

1956	Stockholm
1960	Delft
1964	Prague
1968	Rome
1972	Manchester
1976	Jerusalem
1980	The Hague
1984	Budapest

Preface

These three volumes contain the tutorials, invited presentations, and extended abstracts of the posters and oral contributions to the 9th European Congress on Electron Microscopy. We have included all the tutorial topics, and most of the papers relating to instrumentation, in volume 1, while the remaining material is divided almost equally between volume 2 (physical and materials applications) and volume 3 (biological applications). The session numbers of the papers are provided in the Contents List for the benefit of conference participants.

The submitted papers reflect the diverse range of applications of electron microscopy in the late 1980s and also the number of complementary techniques which have developed alongside the established SEM and TEM approaches. A glance at the subject index shows the vast scope of the conference, from electron optics to altarpieces, mouse knees and mudcake. A further look shows that the preoccupation with high resolution is still very much with us, but that an increasing number of workers are now using image analysis and surface science techniques.

Associated with the congress is a large exhibition of instruments and accessories. This shows the diversity of our subject in a different way, and provides a welcome contrast to the scientific sessions. It is very important that we retain these links with the experimental basis of our subject.

It is our pleasure to thank the many people who have given their time and effort to make this congress successful. We hope that our introduction of tutorials will prove popular, and we thank the presenters who gave us more extensive written papers for the benefit of those microscopists who could not attend the congress. Our invited speakers contribute greatly to the quality and timeliness of the oral sessions and we thank them warmly. Behind the scenes of any large conference there are many who slave unseen and our thanks extend most sincerely to them, including our scientific committee and its helpers, and the staff of the Royal Microscopical Society, who undertook the administrative arrangements.

J N Chapman,
B C Cowen,
H G Dickinson,
P J Goodhew,
A W Robards

Contents

Chapter 3: Plants

Chapter 5: Muscle

Chapter 6: Invertebrates

Chapter 7: Vertebrates

Chapter 9: Pathology

Chapter 14: Image Processing

Chapter 15: Biological Macromolecules

Chapter 17: Membranes

Chapter 18: Cytoskeleton

Chapter 19: Histochemistry, Cytochemistry and Immunocytochemistry

Chapter 20: Histochemistry, Cytochemistry and Morphometry

Scientific sessions
Biology

Inst. Phys. Conf. Ser. No. 93: Volume 3, Chapter 1
Paper presented at EUREM 88, York, England, 1988

The cryosectioned replica technique

Amanda E Semper * and David M Shotton

The Department of Zoology, University of Oxford, South Parks Road, Oxford OX1 3PS, England (* Present address: CIBA-GEIGY AG, K-125.5.14, Postfach, CH-4002 Basel, Switzerland)

ABSTRACT: By combining rapid freezing, freeze-fracture, cryoultramicrotomy and immunocytochemical labelling procedures, we have developed a new specimen preparative method, the cryosectioned replica technique, which permits the simultaneous viewing of labelled cytoplasmic antigens and freeze-fracture replicas.

1. INTRODUCTION

To date, five main approaches have been developed for combining cytochemistry with the freeze-fracture technique:

(a) Label-etch Widely used since the earliest days of freeze-etching (Pinto da Silva and Branton, 1970), this procedure involves the labelling of antigens on dissociated cells, and the subsequent correlation of their distribution on the etched outer cell surfaces, after conventional freeze-etching and replication, with that of the intramembrane particles (IMPs) on the P fracture faces. Its application is limited to those cells which can survive freezing in the absence of cryoprotectants.

(b) Label-fracture Developed by Pinto da Silva and Kan (1984), in this technique cells in suspension are specifically labelled with colloidal gold probes before being freeze-fractured and replicated. Then, rather than using bleach to digest away the biological material, as is conventional, the replicas are floated on distilled water, allowing free cellular components and unbound gold on the undersurface to be gently washed away. When viewed in the electron microscope, electron dense gold labels attached to the outer half of the plasma membrane, which remains associated with the replica after the gentle washing procedure, can be readily distinguished through the EF replica. Although only labelled extracellular half-membranes are recovered for observation, the surface distribution of label can also be correlated with P face morphology if complementary replicas are studied. As with label-etch, however, label-fracture is restricted to use with surface antigens on cell suspensions or readily-dissociable tissue.

(c) Fracture-label In the fracture-label procedure, the material is bulk fractured by grinding under liquid nitrogen, thawed without replication, and then exposed to a specific cytochemical reagent conjugated to an electron dense marker such as ferritin (Pinto da Silva *et al.*, 1981a). In initial applications the material was then processed for resin embedding, thin sections being cut perpendicular to the labelled fracture faces, which were then viewed edge-on (Pinto da Silva *et al.*, 1981a, 1981b). In a later development, the labelled specimen was critical point dried and then replicated with platinum/carbon before digestion of the tissue, thus permitting the *en face* observation of expanses of labelled fracture faces (Pinto da Silva *et al.*, 1981c, 1981d). However, these replicas showed bad shrinkage and were formed over fractured surfaces in which, upon exposure to aqueous environments, had undergone reorganisation into interrupted bilayers. Their structure was thus not comparable with conventional freeze-fracture replicas.

(d) The sectioned replica technique In this technique (Rash, 1979), tissue is conventionally freeze-fractured and then lightly shadowed with platinum, but stabilising deposits of carbon are omitted. Thus in the lee of IMPs, tissue is still accessible to the immunoreagents to which the thawed tissue and its attached replica are then exposed. Subsequently the immunolabelled tissue and replica are dehydrated and embedded, and thin sections are then cut at, and parallel to, the plane of the replica/tissue interface. When viewed in the EM, electron dense markers can be seen associated with certain IMPs within fragments of the replica, which in turn can be related to underlying structures within the sectioned tissue. Thus, labelled replicas of conventional appearance are achieved, but only of low labelling density and small replica areas (Rash *et al.*, 1982).

(e) The labelled replica technique In an attempt to obtain larger expanses of labelled replica, Rash *et al.* (1981, 1982) modified the sectioned-replica technique to give the labelled-replica technique. In this, after labelling a previously fractured and replicated specimen, the specimen is not embedded for plastic sectioning but rather is either freeze dried or critical point dried, stabilised with carbon, and the tissue then digested before viewing the double-thickness replica. The advantage of this procedure over that of fracture-label is that membrane rearrangement does not occur.

These five techniques for freeze-fracture immunocytochemistry are concerned with the localization of antigens either at the cell surface or within the membrane itself. Their potential for use in the identification of cytoplasmic antigens is either intrinsically impossible or has apparently not been investigated.

We wish now to report the development and initial results of a new procedure, the cryosectioned replica technique, in which a cryoultramicrotome is used to cut frozen semi-thin sections from the surface of a conventionally freeze-fractured and replicated specimen, exposing cytoplasmic antigens which may then be immunolabelled.

As a model system for the development of this technique, we have employed normal human skeletal muscle, in cryosections of which fast and slow muscle fibres may be distinguished immunocytochemically using anti-myosin isoform specific monoclonal antibodies (Semper *et al.*,1988).

2. MATERIALS AND METHODS

(a) Fixation and freezing Samples of healthy human muscle were fixed in an isotonic PIPES buffer containing 2% formaldehyde and 0.1% glutaraldehyde, quenched with 1% sodium borohydride, and infiltrated at $4^{o}C$ for 24h with 0.6M buffered sucrose. Infiltrated specimens were mounted on aluminium supports in preparation for freezing by impact against a liquid helium-cooled copper block. Samples were then stored under liquid nitrogen until use.

(b) Freeze-fracture Samples were fractured conventionally using a Balzers BAF 301 freeze-etching apparatus. To maximise the possibility of obtaining plasma membrane fractures, the muscle fibres were fractured at $-115^{o}C$ with the edge of the knife parallel with, and the sweep of the knife normal to their longitudinal axis. The fractured specimen surface was immediately replicated, with deposition of conventional thicknesses of platinum/carbon and of carbon.

(c) Transfer to the cryoultramicrotome The aluminium planchettes used as specimen supports were not compatible with the conventional specimen chuck of the Reichert Ultracut E/FC 4D cryoultramicrotome. Therefore, a modified specimen chuck was designed, incorporating an old Balzers freeze-fracture four-specimen table, in which could be fitted an aluminium specimen support, attached to a specially-made base which fitted the specimen chuck holder of the Reichert ultramicrotome.

This chuck was first pre-cooled by submersion in liquid nitrogen. The fractured, replicated specimen, still at -115°C, was quickly removed from the Balzers 301 into liquid nitrogen, loaded onto the chuck, and the loaded specimen chuck then rapidly transferred to the specimen arm of the cryoultramicrotome, which had been pre-cooled to approximately -120°C. The temperature of the specimen, knife and chamber was then raised towards the sectioning temperature of -60 to -70°C.

(d) Cutting cryosectioned replicas The replicated specimen was aligned such that the direction of sectioning ran parallel with the fibre longitudinal axis (i.e. at right angles to the direction of freeze-fracturing). Semi-thin (300-500 nm) cryosections were then cut on 45° tungsten-coated glass knives, using a 3° clearance angle.

(e) Section recovery and storage To permit subsequent immunolabelling, the sections were initially inverted, while still resting on the cold knife face, by careful manipulation with a pair of eyelash probes. Sections were then recovered conventionally on droplets of 2.3M sucrose in phosphate-buffered saline (PBS), the replicated surface of the section thus facing away from the droplet, and were transferred to the hydrophillic surface of 100-mesh nickel EM grids bearing a polylysine-coated formvar film. These grids, with sections uppermost, were stored on multi-well glass microscope slides under sucrose/PBS, containing 0.5% formaldehyde to promote section attachment, and examined by phase contrast microscopy to enabled grids bearing potentially useful cryosectioned replicas to be identified. Prior to subsequent processing, the sucrose was then removed by transfer of the grid though several droplets of PBS on Nescofilm.

(f) Immunolabelling When desired, the cryosectioned replicas, supported replica-side down on the formvar coated nickel EM grids, where at this stage immunolabelled with a humam skeletal myosin isoform-specific monoclonal antibody (or an irrelevant control) and 15 nm colloidal gold, using the three-step technique described by Semper *et al.* (1988) with slight modifications to allow staining by immersion of the grids.

(g) Embedding, adsorption staining and electron microscopy Following removal of PBS by further washes in distilled water, the cryosectioned replicas were stained in 2% uranyl acetate oxalate and finally embedded in 2% methyl cellulose. Once dry, grids were viewed immediately using a Philips 400T EM at 80 or 100 KV.

Papers giving full technical details of these procedures are currently in preparation.

3. RESULTS

(a) Appearance of cryosectioned replicas This procedure yielded semi-thin cryosections which incorporated expanses of replica of 15 to 20 um^2 in area. Ultrastructural details of the cryosectioned muscle tissue, such as Z lines, M lines and mitochondria, could be distinguished both adjacent to and also through the replicas (Fig. 1). The contiguity of structural features between replica and tissue shows that the replicas had not detached from the tissue upon which they were formed during subsequent processing. Cryo-sectioned replicas containing replicated areas of both fractured muscle cytoplasm and of fractured P face and E face muscle plasma membrane were obtained. As expected, whereas P face replicas were associated with underlying sectioned muscle cytoplasm in which elements of the sarcomere, such as Z lines, could be identified, the material incorporated with E face replicas was that of the extracellular matrix, for example collagen bundles, and sometimes cytoplasm from the adjacent muscle fibre.

(b) Immunolabelling of cryosectioned replicas Observation of stereo pairs of cryosectioned replicas, immunolabelled to reveal muscle fibre types, showed gold particles to be clearly visible, both adjacent to and beneath areas of replica, attached to the cryosectioned muscle tissue in a fibre type-specific manner. The level of labelling of fast fibres with anti-slow myosin monoclonal N0Q7.5.4D (Semper *et al.*, 1988) was equivalent to, or even less than, that of a negative control section, prepared during the same immunolabelling experiment. However, the specific localization of N0Q7.5.4D to A band areas of the sarcomere was less apparent in these labelled cryo-sectioned replica specimens than in either conventional labelled thin cryosections (Semper *et al.*, 1988) or unreplicated semi-thin cryosections mounted on EM grids and labelled conventionally on the surface of droplets of immunoreagents.

4. DISCUSSION

A method has been developed which enables 300-500nm cryosections to be cut from the fractured and replicated well-frozen surface of muscle. These sections incorporate extensive areas of replica (typically 15-20 um^2, but occasionally up to 75-100 um^2), together with a longitudinal section of the corresponding region of the underlying muscle tissue. Although employing tissue only weakly fixed with 2% formaldehyde/0.1% glutaraldehyde, the section thickness ensures good section integrity. Inherent contrast within these 300-500nm sections of muscle was high, but structural definition was found to be enhanced by brief exposure to uranyl acetate oxalate prior to embedding in methyl cellulose.

Thus, this new method for preparing sectioned replicas allows much greater expanses of replica to be recovered than was possible by previous methods (Rash, 1979). The immunolabelled cryosectioned replica technique, application of which is not limited to muscle tissue, produces sections in which, for the first time, cytoplasmic antigens may be localised with specific antibodies and visualized using gold-antibody conjugates that can be readily observed through the replicas, thus permitting correlations to be made between replica structure and antigen distribution. However, for accurate interpretation it must be ensured that such sarcolemmal replica areas are derived from the same cell as the underlying tissue. Thus, only for cryosectioned replicas containing P face plasmalemmal fractures is the immunohistochemical identification of a cytoplasmic antigen and its correlation with plasma membrane morphology possible.

The immunolabelled cryosectioned replica technique at its present state of development will now make it possible to unambiguously correlate the morphology of sarcoplasmic fractures of normal human muscle fibres with a known muscle fibre type, a comparison that could not previously reliably be made (Shafiq *et al.*, 1979), and should permit study of the fibre-type specificity of sarcolemmal changes induced by neuromuscular diseases such as Duchenne muscular dystrophy (Schotland *et al.*, 1981: Appleyard *et al.*, 1982; McLean *et al.*, 1986), and thus may increase the usefulness of electron microscopy in early diagnosis.

5. ACKNOWLEDGEMENTS

This work has been made possible by research grants to DMS by The Muscular Dystrophy Group of Great Britain and by an SERC Research Studentship to AES.

6. REFERENCES

Appleyard S T, Witkowski J A, Shotton D M and Dubowitz V 1982 *Biol. Cell* 45 273
McLean B, McLean G A and Shotton D M 1986 *Muscle and Nerve* **9** 501
Pinto da Silva P and Branton D 1970 *J. Cell Biol.* **45** 598
Pinto da Silva P and Kan F W K 1984 *J. Cell Biol.* **99** 1156
Pinto da Silva P, Parkinson C and Dwyer N 1981a *Proc. Natl. Acad. Sci. USA* **78** 343
Pinto da Silva P, Parkinson C and Dwyer N 1981b *J. Histochem. Cytochem.* **29** 917
Pinto da Silva P, Kachar B, Torrisi M R, Brown C and Parkinson C 1981c *Science* **213** 230
Pinto da Silva P, Torrisi M R and Kachar B 1981d *J. Cell Biol.* **91** 361

Rash J 1979 In *Freeze Fracture: Methods, Artifacts and Interpretations* ed J E Rash and C S Hudson (New York: Raven Press) pp 153-160
Rash J E, Johnson T J A and Giddings F D 1981 *J. Cell Biol.* **91** 275a
Rash J E, Hudson C S, Albuquerque E X, Edelfrawi M E, Mayer R F, Graham W F and Johnson T J A 1982 *Ann. N. Y. Acad. Sci.* **337** 38
Schotland D L, Bonilla E and Wakayama Y 1981 *Acta Neuropathol.* **54** 181
Semper A E, Fitzsimons R B and Shotton D M 1988 *J. Neurol. Sci.* **83** 93
Shafiq S A, Leung B and Schutta H S 1979 *J. Neurol. Sci.* **42** 129

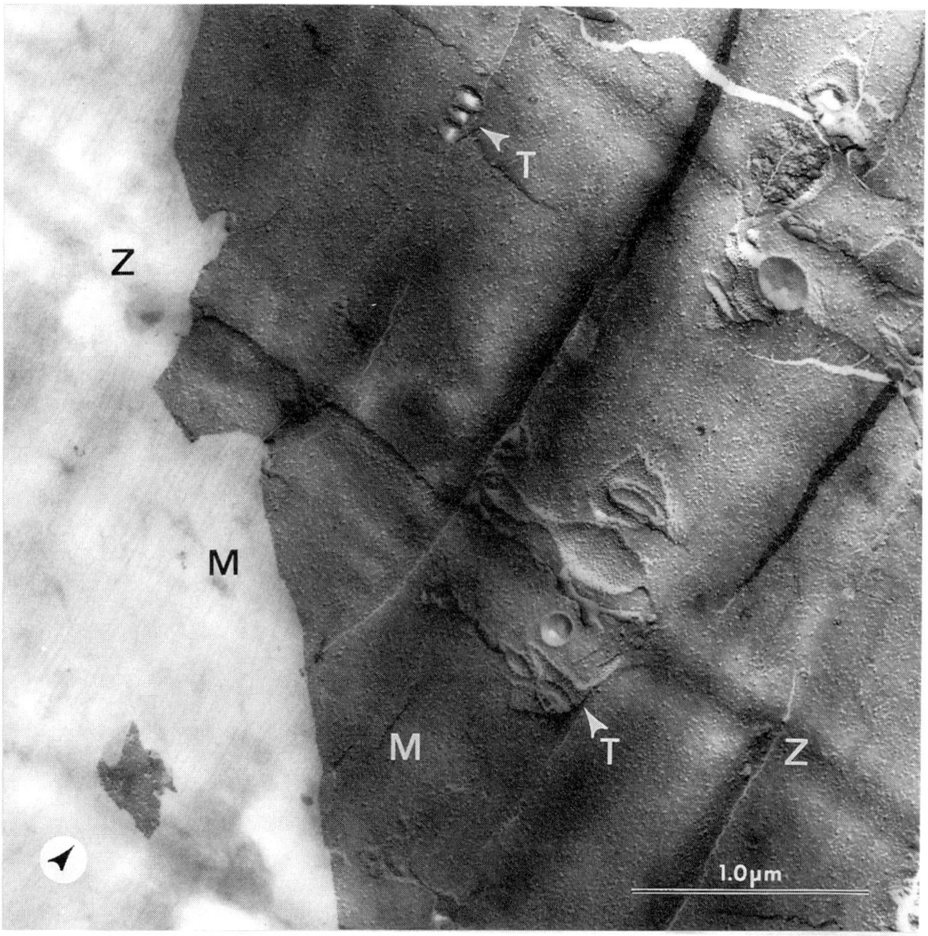

Figure 1. An unlabelled cryosectioned replica prepared from human *temporalis* muscle which incorporates an area of longitudinally-fractured sarcoplasm. Details of the ultrastructure of the underlying tissue, such as Z and M lines (Z and M) and even the thick filaments, can clearly be distinguished through the freeze-fracture replica. The replica shows features typical of a sarcoplasmic fracture, such as fractured triad junctions (T) and elements of the network of sarcoplasmic reticulum tubules.

Inst. Phys. Conf. Ser. No. 93: Volume 3, Chapter 1
Paper presented at EUREM 88, York, England, 1988

Theory of rapid freezing

W B Bald

Wolfson Unit for Applied Cryobiology, Institute for Applied Biology, University of York, York YO1 5DD

ABSTRACT

The quantities which characterise the rapid freezing of materials are shown to be the Biot modulus Bi and the Stefan number St. The value of Bi is the quantity which distinguishes between valid solutions for the rapid solidification of metals and alloys and the rapid freezing of biological materials. Supercooling of the specimen to an initial uniform nucleation temperature is valid for metals and alloys but is inapplicable to biological materials. Comparisons between approximate analytical solutions and numerical solutions in the cryofixation of biological specimens suggest that the pseudo steady state solution where the Jakob number $Ja \ll 1$ is reasonably valid.

1. INTRODUCTION

The rapid freezing or solidification of materials from the liquid state is of great interest to both the material scientist and the low temperature microscopist.

The commercial development of ferromagnetic metallic glasses for transformer design and the development of new hard magnets are but two examples of the use of rapidly solidified alloys in the field of materials science. Some of the techniques and methods used in this field have been discussed recently by Jones (1986) and by Adam (1986). In low temperature microscopy the initial specimen preparation stage is known as Cryofixation. This is the process whereby the specimen is rapidly frozen to minimise the growth of ice crystals which might otherwise obscure the underlying ultrastructure. Detailed discussions on the methods used for cryofixation can be found in the books by Robards and Sleytr (1985), Steinbrecht and Zierold (1987) and Bald (1987).

Irrespective of which of the two foregoing areas of rapid solidification are being used, it is exceedingly difficult, if not impossible, to measure the cooling rate distribution within a sample. Thus the emphasis for quantifying the resulting material structure after freezing must rely largely on theoretical solutions.

In this paper the basic field equations which define the rapid freezing problem will be discussed and which can be adequately solved using numerical

boundary element method. However, these methods of solution give little insight into the various physical parameters which affect any particular solution to the rapid freezing problem.

The one-dimensional analytical solution to the temperature field equations will therefore be examined. This solution highlights the dependency of specimen temperature and cooling rate on two dimensionless quantities, namely, the Biot modulus Bi and the Stefan number St. It is the value of the Biot modulus which separates the types of solution applicable to either the field of materials science or the life sciences.

In rapid freezing problems where $Bi \ll 1$ temperature gradients throughout the specimen cross-section are small and consequeently approximate solutions using the methods described by Gill et al (1981)(1984) are valid for metallic drops.

Cryofixation of biological samples, however, usually involves specimens where $Bi > 1$ and where the initial conditions assumed for metallic drops are no longer valid. The remainder of the paper will therefore examine possible solutions to cryofixation problems and compare these with the more exact finite element numerical methods.

2. THE BASIC TEMPERATURE FIELD EQUATIONS

If we neglect natural convection in the liquid phase and any internal heat generation, such as metabolism in biological systems, the basic equations which describe the temperature field at any instant in a specimen are given by Fourier's heat conduction equation as

for the liquid phase,

$$\alpha_\ell \nabla^2 T_\ell = \frac{\partial T_\ell}{\partial t} \qquad (1)$$

and for the solid phase,

$$\alpha_s \nabla^2 T_s = \frac{\partial T_s}{\partial t} \qquad (2)$$

where α is the thermal diffusivity, T the temperature, t time and subscripts and s refer to the liquid and solid phases respectively.

Equations (1) and (2) are coupled together by the heat energy balance at the moving interface where the latent heat of fusion L is released. At the liquid-solid interface we therefore have

$$\rho_s L \frac{dn_i}{dt} = K_s \frac{\partial T_s}{\partial n}\bigg|_{n=n_i} - K_\ell \frac{\partial T_\ell}{\partial n}\bigg|_{n=n_i} \qquad (3)$$

where ρ is the density, K the thermal conductivity and n is the vector normal to the moving interface.

Equations (1)-(3) together with the initial and boundary conditions define a class of mathematical problems known as Stefan problems or moving boundary problems. In the majority of cases there are no exact analytical solutions to these types of problem and therefore sophisticated numerical techniques are usually employed.

3. THE SEMI-INFINITE PLANAR FREEZING PROBLEM

Numerical solutions to equations (1)-(3) subject to the appropriate boundary conditions give little insight into the various quantities which influence the results in specific cases. It is therefore necessary to apply approximate methods of analysis and we will begin by considering the freezing of a semi-infinite planar region exchanging heat at its surface with a coolant at some temperature T . (Fig.1)

The solution to this one-dimensional

problem is further simplified if the liquid phase temperature is initially assumed to be constant at the freezing temperature Tf. Equations (1)-(3) therefore simplify in this case to

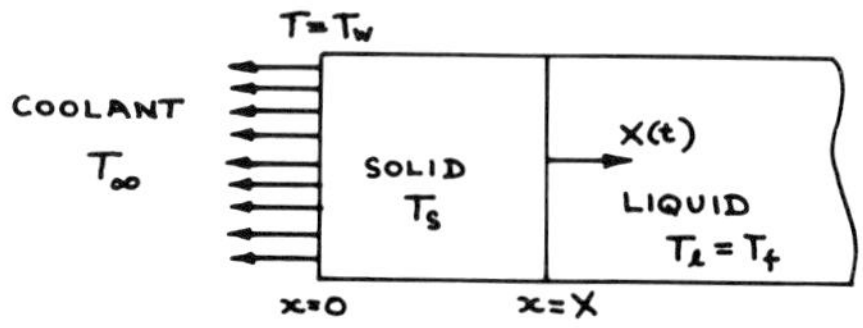

FIG.1. FREEZING OF A SEMI-INFINITE PLANAR REGION IN A COOLANT AT TEMPERATURE T_∞

for the solid phase,

$$\alpha s \frac{\partial^2 Ts}{\partial x^2} = \frac{\partial Ts}{\partial t} \tag{4}$$

at the moving interface x=X,

$$\rho_s L \frac{dX}{dt} = K_s \frac{\partial Ts}{\partial x} \tag{5}$$

If heat is extracted from the surface of the specimen x=o of fig 1 by Newton cooling then the following boundary condition applies to the problem.

at x=o,

$$K_s \frac{\partial Ts}{\partial x} = h(T_w - T_\infty) \tag{6}$$

where T_W is the temperature on the surface of the solid at x=o and h is the surface heat transfer coefficient assumed as constant.

Equations (4) (5) and (6) can be written in dimensionless form as fully discussed by Hill and Dewynne (1987) where it is shown that the solutions are dependent on the dimensionless quantities

$$\begin{aligned} Bi &= hR/Ks \\ St &= L/c_s(T_f - T_\infty) \end{aligned} \tag{7}$$

where Bi is the Biot modulus, S the Stefan number, c the specific heat and R is a convenient length dimension.

Re-arranging the last of expressions (7) the Stefan number can be written

$$St = \left[\frac{c_s(T_w - T_\infty)}{L} + Ja\right]^{-1} \tag{8}$$

where

$$Ja = c_s(T_f - T_w)/L \tag{9}$$

is known as the Jacob number. This number is the ratio of the heat absorbed by the solid phase to the latent heat of fusion at any instant during rapid freezing.

In the case of very rapid cooling where the surface heat transfer coefficient h is large then the value of $T_w \rightarrow T_\infty$, St = 1/Ja and the solutions are equally dependent on the Jakob number Ja.
Table 1 lists some typical property values for metals and ice which approximately represents most biological materials. Also shown in this

table are the calculated values for the Biot modulus Bi and the Stefan number St assuming a droplet size R = 50 μm h = 4w/cm K and T = 100K.

TABLE 1

MATERIAL	Tf (K)	L (J/g)	Cs (J/gK)	Ks (w/cmk)	αs (cm^2/s)	Bi	St
ALUMINIUM	933	370	0.9	2.2	0.94	0.009	0.49
COPPER	1356	204	0.39	4.0	1.16	0.005	0.42
GOLD	1336	65	0.13	3.1	1.25	0.0064	0.4
IRON	1813	251	0.45	0.8	0.23	0.025	0.32
LEAD	600	23	0.13	0.35	0.24	0.057	0.35
NICKEL	1723	307	0.44	0.2	0.055	0.1	0.43
SILVER	1336	118	0.23	3.8	1.55	0.0053	0.41
TIN	505	60	0.23	1.8	1.07	0.011	0.64
ICE	273	334	2.1	0.022	0.0114	0.9	0.92

From the values listed in Table 1 for Bi and St it is obvious that the major difference between the metals and frozen biological material is the value of the Biot modulus Bi. It is therefore the Biot modulus which determines the type of solutions which are valid for rapid freezing in either the materials sciences or the life sciences. Consequently, prior to nucleation and solidification of any specimen the temperature gradients will be negligible in most metals but could be very significant in biological samples.

Thus the basic assumption used by Gill et al (1981)(1984) of an initial uniform nucleation temperature T* within a spherical drop prior to rapid solidification is justifiable for metals but is invalid for biological specimens. It is also worth noting that because of the structure of most biological specimens the degree of supercooling (T_f -T*) attained during rapid freezing is small compared to most metals.

4. RAPID FREEZING OF BIOLOGICAL SPECIMENS

Most biological specimens which undergo cryofixation prior to microscopical analysis can be defined in geometrical terms as either finite thin flat specimens, cylindrical, or spherically shaped specimens. Irrespective of the specimen shape, exact analytical solutions to the one-dimensional problem defined by equations (4)-(6), are not available.

In any specific cryofixation problem the surface heat transfer coefficient h, and therefore Bi, is governed by the particular experiment. In each case the solution is therefore determined by the value of St or, by virtue of (8) and (9), by the value of the Jakob number Ja. If Ja>>1 the amount of heat stored in the solid phase will be large compared to the latent heat L and the specimen will cool in the same way as a homogenous solid with negligible freezing effects. On the other hand, for Ja<<1, the latent heat effect will predominate and the pseudo steady state solution first derived by London and Seban (1943) is valid. It is therefore reasonable to assume that the true temperature distribution and cooling rate within a biological specimen during rapid freezing must lie somewhere between these two extreme assumptions.

Consider for example the rapid cooling of a spherical specimen of radius R subjected to constant Newtonian cooling at its surface.

For the case where Ja>>1, the solution for the temperature distribution at time t and location r within the spherical specimen is given by Carslaw and

Jaeger (1959) as

$$\frac{T-T_\infty}{T_s-T_\infty} = \frac{2Bi}{R}\sum_{1}^{\infty}\left(\frac{\sin r\beta n}{r\beta n}\right)\frac{\left[R^2\beta_n^2+(Bi-1)^2\right]\sin R\beta_n}{\beta_n\left[R^2\beta^2 n+B_i(B_i-1)\right]}\exp\left(-\alpha_s\beta_n^2 t\right) \quad - (10)$$

where the βn's are the roots of the equation

$$R\beta_n \operatorname{Cot} R\beta_n + Bi = 1 \quad (11)$$

In the cryofixation of biological specimens we are mainly interested in the cooling rate distribution within the sample and how this affects the resulting ice crystal size. Thus the cooling rate of significant interest is the rate when T = T* at same time t = t* at any location r = R - X.

Denoting the cooling rate at the instant of freezing by B_f then by differentiating (10)

$$B_f = -\frac{2(T_s-T_\infty)\alpha_s Bi}{R}\sum_{1}^{\infty}\left(\frac{\sin r\beta n}{r\beta n}\right)\frac{\left[R^2\beta_n^2+(Bi-1)^2\right]}{\left[R^2\beta_n^2+Bi(Bi-1)\right]}\beta_n \sin R\beta_n \exp\left(-\alpha_s\beta_n^2 t\right) \quad (12)$$

Using equation (10) to find the nucleation time t = t* when T = T*, this value can be used in expression (12) to determine the cooling rate distribution within a spherical specimen for the case where Ja>>1.
In cases where the latent heat of fusion predominates Ja<<1 and the cooling rate distribution for a spherical specimen has been derived by Bald (1987) as

$$B_f = -\frac{h^2(T^*-T_\infty)R^4}{\rho_s K_s L(R-X)^4\left[1+BiX/(R-X)^2\right]} \quad (13)$$

Equations (12) and (13) have been plotted in figure 2 to illustrate the variations in cooling rate as the freezing front advances into the spherical specimen. Also shown in this figure are some numerically computed values at different locations obtained using finite element methods. The numerical quantities used in this figure were R = 0.5 mm, Bi = 6.6, T_s = T* = 258K, T = 78K which are typical of the values experienced in cryofixation. Figure 2 clearly shows that for the rapid freezing of biological specimens the latent heat effect predominates and the pseudo steady state solution where Ja<<1 is reasonably valid.

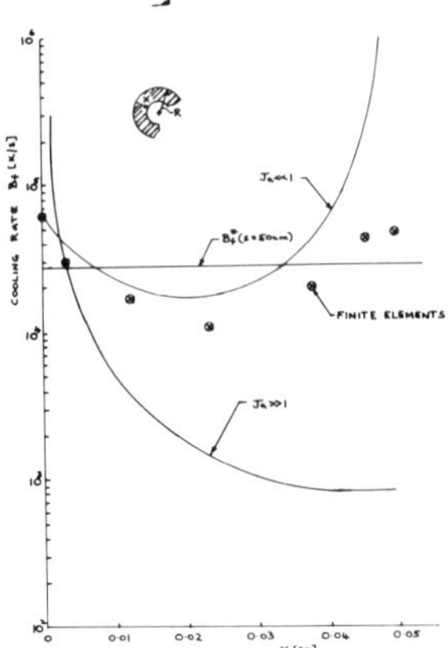

FIG.2. COMPARISON BETWEEN THEORETICAL SOLUTIONS FOR THE RAPID FREEZING OF A SPHERICAL BIOLOGICAL SPECIMEN AND NUMERICALLY COMPUTED VALUES

5. ICE CRYSTAL SIZE

The ultimate aim of rapid freezing or solidification is to minimise the dendritic cell size in materials science or the ice crystal size in the case of cryofixation.

According to Jones (1986) the dendrite cell size s is related to the cooling rate B by the empirical equation

$$s = A\, B_f^{-n} \qquad (14)$$

where A and n are constants and typically $0.25<n<0.5$.

Bald (1986)(1987) has proposed a theoretical model for the ice crystal size s during the cryofixation process. For the case of constant Newtonian cooling at the specimen surface the critical cooling rate B * for a particular ice crystal size s is given by the expression

$$B_f^* = \left\{\frac{\pi}{128}\right\}^2 \frac{U^2 \rho_s L}{K_s\left(\frac{s}{2r^*}-1\right)^4} \left\{1+\sqrt{1+32\left(\frac{s}{2r^*}-1\right)}\right\}^4 \qquad (15)$$

where r is the radius of the critical nucleus and U is the growth velocity of the dendritic ice crystal.

The value of B * obtained from equ (15) for an ice crystal size s = 50 nm is also shown in fig 2 assuming L = 220J/g, r* = 4 nm, U = 10 cm/s. In regions of the specimen cross-section where $B_f < B_f^*$ the resulting ice crystal sizes would be expected to be greater than the chosen value S = 50 nm.

6. CONCLUSIONS

Theoretical analysis shows that for any given specimen the cooling rate and resulting dendritic crystal size are dependent on the Biot modulus Bi and the Stefan number St or Jakob number Ja. Low values for Bi characterise the type of solution which is valid for the rapid solidification of metals. In the case of cryofixation of biological specimens numerical analysis suggests that the pseudo steady state solution where Ja<<1 is reasonably valid.

7. REFERENCES

Adam C.M. 1986 Science and Technology of the Undercooled melt. NATO ASI series (Lancaster : Martinus Nijhoff) pp. 186-207.

Bald W.B. 1986 J. Microscopy 143, 1, 89

Bald W.B. 1987 Quantitative Cryofixation (Bristol : Adam Hilger)

Carslaw H.S. and Jaeger J.C. 1959 Conduction of heat in solids (Oxford : Oxford University Press)

Gill W.N., Jang J.Y. and Mollendorf J.C. 1981 Chem. Eng. Comm. 12, 3

Gill W.N., Jang J.Y. and Mollendorf J.C. 1984 J. Crystal Growth 66, 351

Hill J.M. and Dewynne J.N. 1987 Heat conduction (Oxford: Blackwell Scientific)

Jones H. 1986 Science and Technology of the undercooled melt. NATO ASI series (Lancaster : Martinus Nijhoff) pp. 156-185

London A.L. and Seban R.A. 1943 Trans. ASME 65, 771

Robards A.W.and Sleytr U.B. 1985 Practical methods in Electron Microscopy (Oxford : Elsevier) Vol. 10.

Steinbrecht R.A. and Zierold K. 1987 Cryotechniques in Biological Electron Microscopy (London : Springer-Verlag)

Freeze etching and cryo scanning electron microscopy (CSEM) of plant and animal tissues with SCU 020

T. Müller*, R. Guggenheim[1], M. Düggelin[1], P. Mestres[2], A.C.Van Aelst[3], W. Heyser[4], W. Kumpfer[4]

* Balzers Union AG, Balzers, Principality of Liechtenstein
1 SEM-Laboratory, University of Basle, Switzerland
2 Dept. 3 of Saarland University - Anatomy, Homburg, FRG
3 Dept. of Plant Cytol. and Morph., Agricultural Univ. Wageningen,NL
4 Dept. 2 of Bremen University - Biology, Bremen, FRG

Introduced more than two decades before, freeze etching has become established as a cryopreparation technique for transmission electron microscopy (TEM). For some years now the advantages of CSEM have also been well known: The specimens are preserved in their natural state, can be observed and recorded either uncoated or coated, and are damaged less by the electron beam due to the freeze stabilisation.

In order to combine these two techniques, the BALZERS SCU 020 cryo-preparation unit was developed (Müller et al., 1986). This high vacuum system allows suspensions, as well as plant and animal specimens to be processed. First results have been obtained on surfaces of plant leaves infected with parasitic fungi (Grabski et al., 1987).

In this paper we will present results of freeze etched suspensions and plant and animal tissues at a higher magnification level. The specimens were freeze immobilised on gold carrier plates or glass cover slips by plunging them into liquid propane, or by high pressure freezing them (Moor, 1987) with liquid nitrogen (HPM 010, BALZERS). Fracturing was performed at -130° C or -90° C with a cold, motor driven knife. The specimens were kept frozen hydrated (-130° C) in the preparation chamber or deep etched, i.e. partially freeze dried at -90° C to -60° C on the SEM cold stage of the SCU 020 (CSEM-controlled). After sputter coating at -130° C or -90° C, the specimens were inserted on-line, under high vacuum conditions into the SEM-cold stage for recording at -130° C.

References:
Müller T, Guggenheim R, Düggelin M, Lüönd G, 1986
Proc. XIth Int. Cong. on Electron Microscopy, Kyoto 3: 2233-2234
Grabski Ch, Guggenheim R, Düggelin M, Lüönd G, Müller T, 1987
Proc. DGE Tagung Bremen in Eur. J. of Cell Biol. Suppl. 19, 44:19
Moor H, 1987 in: Cryotechniques in Biological Electron Microscopy, ed. by R. A. Steinbrecht and K. Zierold, part III pp 175 - 191
Van Aelst A C, Müller T, Düggelin M and Guggenheim R, in press

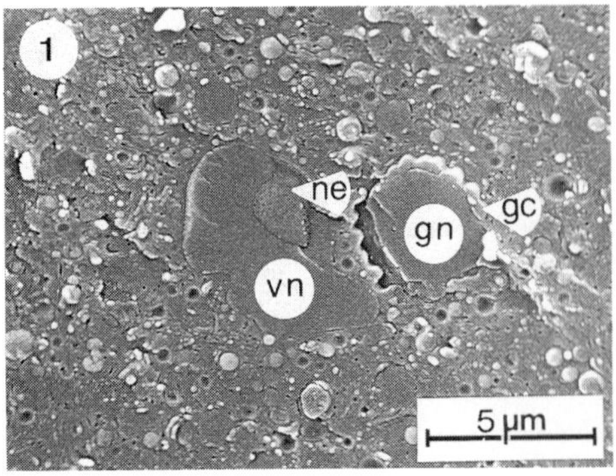

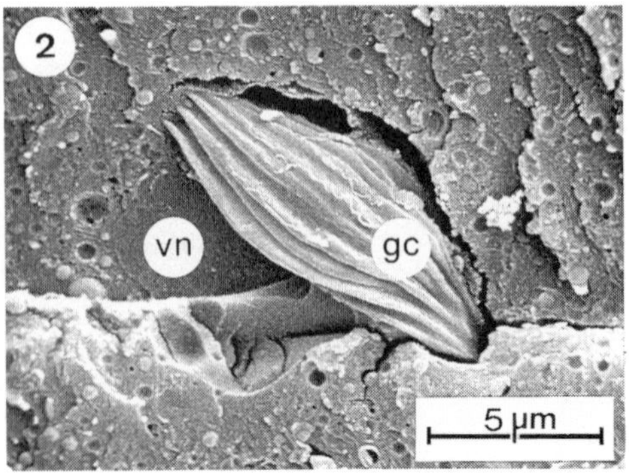

Figs. 1 and 2: Plunge frozen, freeze fractured (frozen hydrated) pollen of Papaver dubium L. Note the pores in the envelope (ne) of the vegetative nucleus (vn). The generative cell (gc) contains a nucleus (gn) and shows wrinkles in longitudinal direction (Van Aelst et al.).

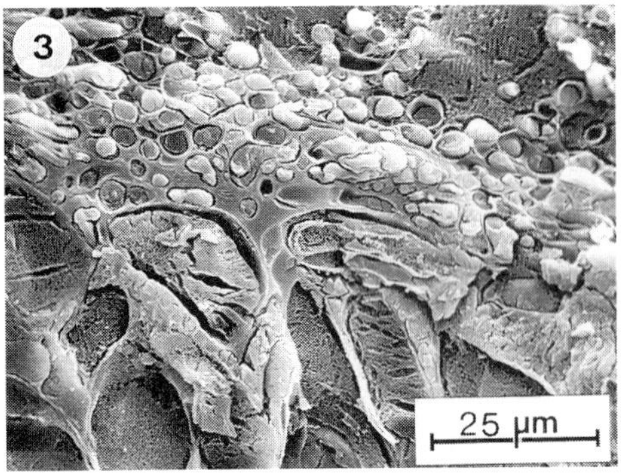

Fig. 3: High pressure frozen, freeze fractured (partially freeze dried) fine root of Fagus silvatica with white mycorrhiza.

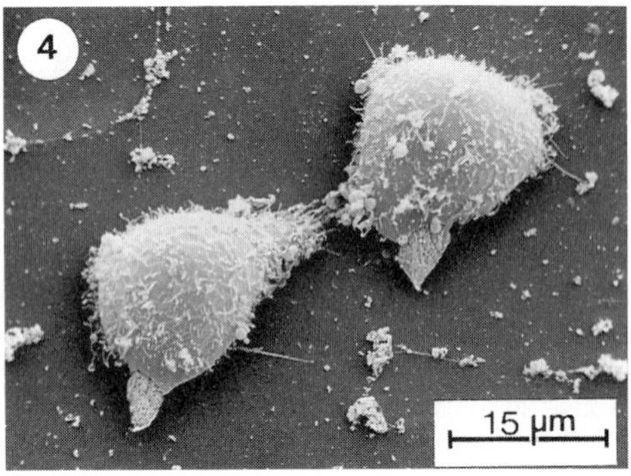

Fig. 4: Plunge frozen, partially freeze dried culture (48 h) of human amnion cells. The height of the cell bodies is remarkable.

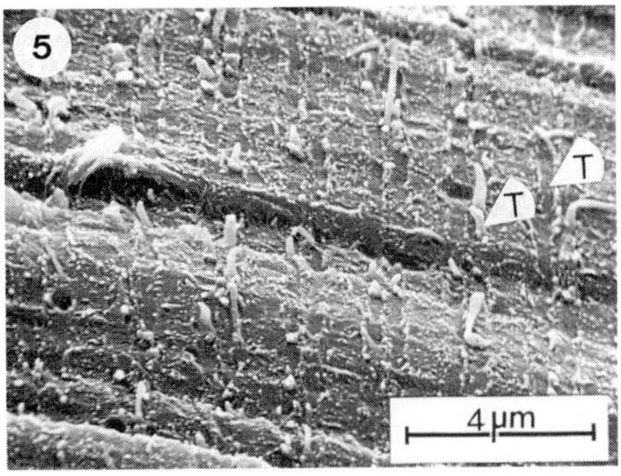

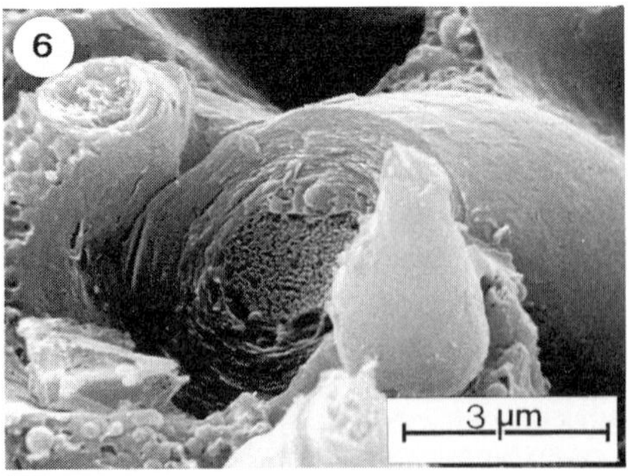

Figs. 5 and 6: Plunge frozen, freeze fractured (partially freeze dried) rat tissue. Striated intercostal muscle exhibit the typical two triads per sarcomer (Fig. 5). In the cerebellum white substance a cross fractured nerve fiber with axoplasm and myelin sheat is visible (Fig. 6).

Common artifacts associated with biological material examined by low-temperature scanning electron microscopy

N D Read and C E Jeffree

Department of Botany, University of Edinburgh, King's Buildings, Mayfield Road, Edinburgh EH9 3JH, UK

ABSTRACT: Most artifacts associated with biological material examined by low-temperature scanning electron microscopy arise because frozen-hydrated specimens contain water. Common artifacts originate during cryofixation, partial freeze-drying, freeze-fracturing, specimen transfer and electron beam irradiation

1. INTRODUCTION

An accurate interpretation of microscopical images requires an understanding of the artifacts produced during specimen preparation and observation. Most of the common artifacts arising in ambient-temperature scanning electron microscopy are avoided by examining frozen-hydrated material. Low-temperature scanning electron microscopy (LTSEM), however, produces its own characteristic artifacts and most arise because frozen-hydrated specimens contain water.

2. ARTIFACTS ASSOCIATED WITH CRYOFIXATION

(a) Expansion during conversion of water to ice. This can cause slight, overall increases in specimen volume. It often results in gross rupturing of cells, tissues and extracellular material due to differential expansion during freezing (Read *et al* 1983; Beckett and Read 1985). Precise measurements of dimensional changes which occur during cryofixation for LTSEM are lacking in the literature.

(b) Ice crystal damage. Freeze-fractured, partially freeze-dried materials typically contain segregation zones (Fig. 1). These predominantly result from ice crystal growth during freezing. The spaces between the segregation zones represent ice crystals subsequently removed during partial freeze-drying. The segregation zones contain the "biological matrix" (i.e. organelles, macromolecular structures, solutes and some hydration water) (Beckett and Read 1985). The extent of the ice crystal damage determines the spatial resolution achievable with LTSEM.

(c) Redistribution of water and solutes. This has been demonstrated in leaves and results in the formation of extracellular ice deposits. The types of deposit vary depending on cooling rates and the direction of thermal gradients during cooling. Whether it is possible to prevent water and solute displacement during cryofixation in bulk plant tissue remains to be determined (Jeffree *et al* 1987).

3. ARTIFACTS ASSOCIATED WITH FREEZE-FRACTURING

(a) Mechanical damage. This commonly results in rupturing which may be difficult to distinguish from rupturing induced by expansion during freezing.

(b) Smearing of fracture faces. This occurs if the fracture knife is not sufficiently precooled prior to fracturing or has insufficient clearance angle.

4. ARTIFACTS ASSOCIATED WITH PARTIAL FREEZE-DRYING

Shrinkage during partial freeze-drying can be substantial but precise measurements of shrinkage are lacking in the literature. Materials of high water content (e.g. mucilages and agar) often become considerably ruptured due to differential shrinkage during partial freeze-drying (Fig. 2) (Beckett and Read 1986).

5. ARTIFACTS ASSOCIATED WITH SPECIMEN TRANSFER

The frozen specimen acts as a cold trap and can become contaminated with condensed, frozen water of varied morphology. The specimen surface can also become contaminated with ice present in the coolant.

6. BEAM DAMAGE

Frozen-hydrated specimens are very sensitive to beam damage. This is usually only a problem at magnifications higher than x1000 and seems to result mainly from the radiolysis of ice by the electron beam (Read *et al*. 1983; Beckett and Read 1986).

7. REFERENCES

Beckett A and Read N D 1986 In *Ultrastructure techniques for Microorganisms* eds Aldrich H C and Todd W J (New York:Plenum) pp 45-86
Jeffree C E, Read N D, Smith J A C and Dale J E 1987 *Planta* **172** 20
Read N D, Porter R and Beckett A 1983 *Can J. Bot*. **61** 2059

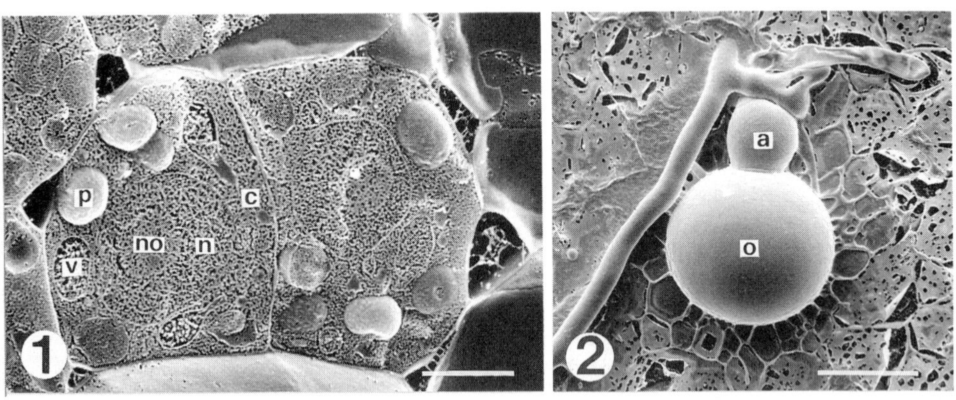

Fig. 1. Freeze-fractured, partially freeze-dried mesophyll cells of a developing leaf of *Phaseolus vulgaris*. Note segregation zones and nuclei (n), Nucleoli (no), plastids (p) and vacuoles (v) within the cytoplasm (c). Bar = 4 μm

Fig. 2. Partially freeze-dried oogonium (o) and antheridium (a) of *Phytophthora infestans* in culture. Rupturing of the underlying agar was due to differential shrinkage during partial freeze-drying. Bar = 20 μm

Cryofixation of plant tissue by high pressure freezing

D. Studer*, M. Michel+, and M. Müller+
Labor für EMI, +Institut für Zellbiologie; *Mikrobiologisches Institut, LFV, ETH-Zürich, Universitätsstr. 2, 8092 Zürich.

Introduction: The currently accepted rapid freezing techniques permit cryofixation of 10 - 20 µm deep superficial layer of an untreated biological sample. This is often an insufficient depth when one wishes to analyse complex botanical systems such as fungus/host interactions, or root nodules, by low temperature procedures (freeze-fracturing or freeze-substitution). High pressure freezing (Riehle and Höchli, 1973) is at present the only method which permits freezing of samples up to a thickness of 500 µm in the microcrystalline state (Müller and Moor, 1983; Hunziker, 1984). This report explains the feasibility of combining high pressure freezing and freeze substitution for the ultrastructural analysis of plant tissues. Freeze substitution followed with low temperature embedding additionally facilitates immunocytochemical experimentation.

Methods: 2 mm disks of "Golden delicious" apple leaves, infected by the fungus *Venturia inaequalis* are immersed in tap-water containing 5% methanol and evacuated for 5 minutes to fill intercellular spaces. The disks are placed in the cavity (ϕ 2 mm, depth 0,2 mm) of a cylindrical aluminium platelet of ϕ 3 mm and a thickness of 0,5 mm and sandwiched with a second platelet of the same dimensions without a cavity. This sandwich is then introduced to the HPM 010 machine (Balzers Union FL) and frozen by high pressure. After freezing the sandwich is immediatly transferred to liquid nitrogen. The sample is then substituted at -90°, -60° and -30°C (kept at each step for 8 hours) in acetone containing 2% OsO_4 and to 0°C for an hour. The samples are then washed in acetone and embedded stepwise in Epon-Araldite. Root nodules of soybeans infected with the bacterium *Bradyrhizobium japonicum* strain 110 are sliced to a thickness of 200 µm and prefixed for 15 minutes in tapwater containing 0,2% glutaraldehyde and 0,3% formaldehyde. Disks of 2 mm are punched out of the slices and further processed as described above for apple leaves. The substitution medium consisted of 2% uranylacetate and 0,2% glutaraldehyde in acetone. After completion of the cryosubstitution the samples were washed with waterfree acetone and, still at -30°C, embedded and polymerised in Lowicryl HM 20.

Results: Apple leaves and fungus show well preserved cells and cell organelles. The cytoplasma is well stained and the membranes show no undulation. The nodule exhibit well preserved plant cells and bacteroids. Most of the peribacteroid membranes are intact. Segregation patterns due to excessive ice crystal formation are rearly visible in both samples.

Discussion: High pressure freezing is a valuable tool for cryofixation of plant material up to a thickness of about 0,2 mm. The mild cryoprotection with 5% methanol improved the yield in adequately frozen specimens from 30% to almost 80% without causing detectable structural alterations. Substitution in acetone/2% OsO_4, and embedding in Epon/Araldite results in better structural preservation than substitution in uranylacetate/glutaraldehyde followed by low temperature embedding in HM20. The latter protocol however is suited for immunocytochemical studies.

Literature:

Hunziker E.B., Herrmann W., Schenk R.K., Müller M., Moor, H. (1984). Cartilage ultrastructure after high pressure freezing, freeze-substitution and low temperature embedding. J. Cell Biol., 98, 267-282.

Müller M., Moor H. (1983). Cryofixation of thick specimens by high pressure freezing. Science of Biological Specimen preparation, SEM Inc., AMF O'Hare (Chicago), 131-138.

Robards A.W., Sleytr U.B. (1985). Low Temperature Methods in Biological Electron Microscopy. Practical Methods In Electron Microscopy, Vol. 10.

Riehle U., Höchli M. (1973). The theorie and technique of high pressure freezing. In: Benedetti E.L., Favard P. (eds.). Freeze-etching Technique and Application. Société Française de Microscopie Eléctronique, Paris, 31-60.

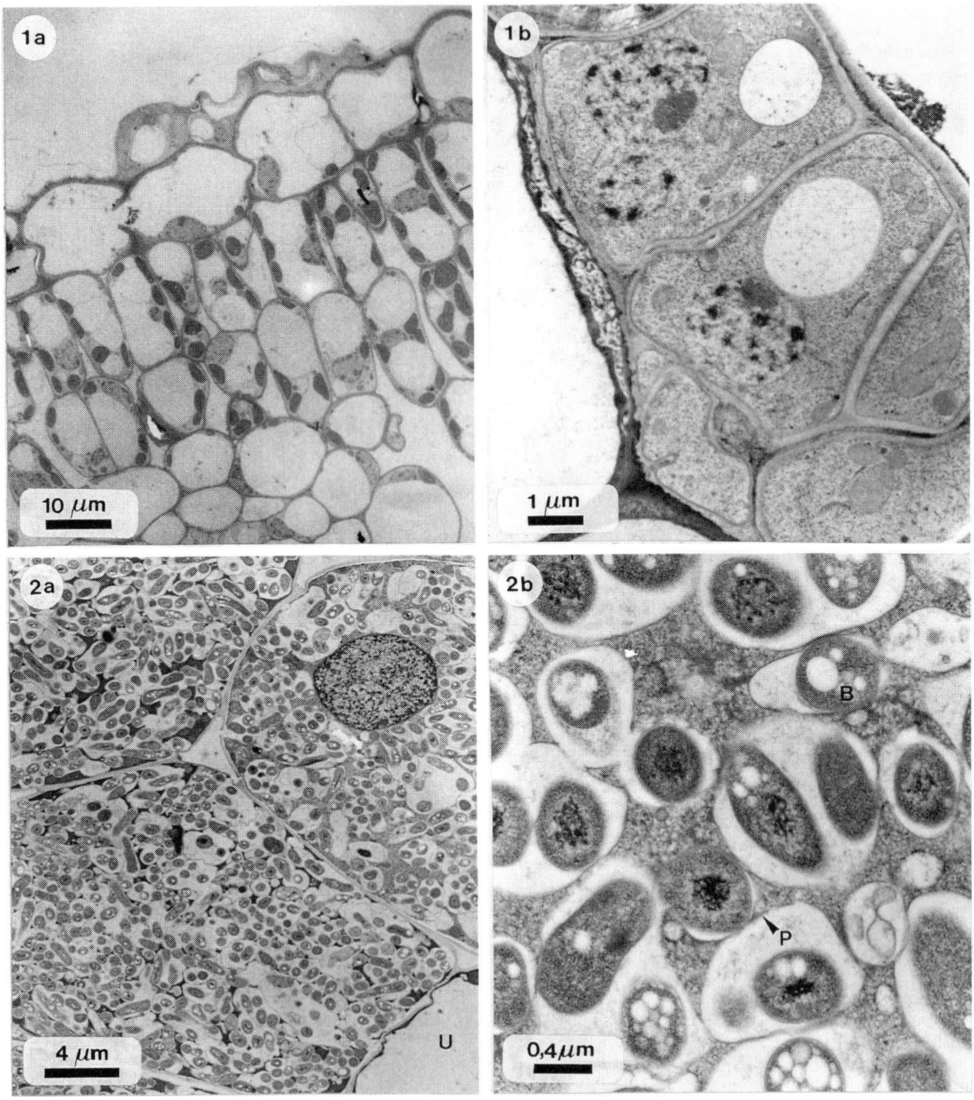

Fig. 1a: Cross-section of a well frozen "Golden delicious" apple leave infected by the fungus *V. inaequalis*

Fig. 1b: Fungus/host interaction site.

Fig. 2a: An uninfected (U) nodule cell and three cells containing bacteroids.

Fig. 2b: The bacteroids (B) are located in the peribacteroid membran (P).

Cryo-electron microscopy of vitrified specimens: study of rat tail tendon

Jean Lepault

Centre de Génétique Moléculaire, C.N.R.S., F-91190 Gif sur Yvette

ABSTRACT: Cryo-electron microscopy of vitrified specimens allows the static and dynamic study of biological specimens in their native aqueous conditions. Although vitrification represents a physical change for biological specimens, their structure, and in particular that of rat tail tendon, is similar in both the vitrified and the liquid state. In these two states biological specimens display similar X-ray diffraction patterns. Cryo-electron microscopy therefore gives a faithful representation of the structure of biological specimens. Vitrification requires the use of cooling methods having a fast freezing rate. Such methods allow one to "freeze" the structure of biological objects, and then to assess the dynamic structure of biological objects.

1. INTRODUCTION

Cryo-electron microscopy arose from two fundamental events. The first one was the demonstration by Taylor and Glaeser (1974) that cryo-electron microscopy is feasible and is a potential method. The second one was the discovery by Dubochet and McDowall (1981) that thin layers of water can be vitrified. Biological objects can then be imaged by electron microscopy under defined conditions (hydration, ionic strength,...) (Lepault et al., 1983), mainly using phase contrast (Lepault and Pitt, 1984).

Since cryo-electron microscopy was developed as a routine technique (Adrian et al. 1984), it has been used in numerous studies. Most of these studies show that cryo-electron microscopy gives more information regarding biological objects than any other electron microscopy method. This is due to two main advantages of cryo-electron microscopy. The first one is the fact that cryo-electron microscopy visualizes hydrated objects and thus reduces the preparation artifact. For example, enveloped viruses are structurally preserved when vitrified (Fuller, 1986). The second advantage of cryo-electron microscopy comes from the study of unstained specimens, which allows the visualization of all the specimen components. For example, internal features could be observed for the acetyl-choline receptor (Brisson and Unwin, 1985), for coated vesicles (Vigers et al., 1986) and for viruses and bacteriophages (Fuller, 1986; Lepault et al., 1987).

The great potential of cryo-electron microscopy to study time dependant phenomenon has also been explored. Cryo-electron microscopy of vitrified specimens calls for the use of a cooling method having a fast freezing rate. Such a method allows the "freezing" of conformational intermediates. For example, intermediates in microtubule depolymerization have been observed (Mandelkow and Mandelkow, 1985). More recently, it has been demonstrated that insect flight muscle thick filaments undergo a polar, reversible rearrangement when they are changed from the relaxed to the rigor state (Menetret et al., J.M.B., in press).

All the interesting results obtained with cryo-electron microscopy of vitrified specimens call for a better characterization of the vitrified state, and in particular its possible effects on the structure of biological objects. To quantify the structural changes of biological objects during vitrification, we studied rat tail tendons.

2. CRYO-ELECTRON MICROSCOPY OF RAT TAIL TENDONS

Rat tail tendon is an ideal specimen to test the effects of vitrification on the structure of biological objects. It is well known that collagen fibers have a structure dependent upon their hydration. Indeed, dried and hydrated collagen fibers have characteritic X-ray diffraction patterns. The effect of vitrification on rat tail tendon can be quantified by both X-ray diffraction and cry-electron microscopy.

X-ray diffraction patterns of vitrified rat tail tendon were performed with the vitrified tendons maintained at c.a.-160°C by a cold stream of nitrogen gas (Gulik and Costello, 1978). X-ray patterns of rat tail tendons are similar, as far as the meridional data are concerned, in both vitrified and liquid states. Precise measurements of the pattern intensities are currently being performed.

Isolated collagen fibrils and tendons were studied by cryo-electron microscopy. Thin fibrils were directly observed in the electron microscope (Plate 1a). Tendons were cryo-sectionned (Plate 1b). Optical diffraction patterns of fibrils are characterized by a strong eighth meridional order (Insert plate 1a) in contrast to X-ray diffraction patterns of hydrated tendons. We then must conclude that the release of thin collagen fibers from the the tendon changes the packing of the collagen molecules.

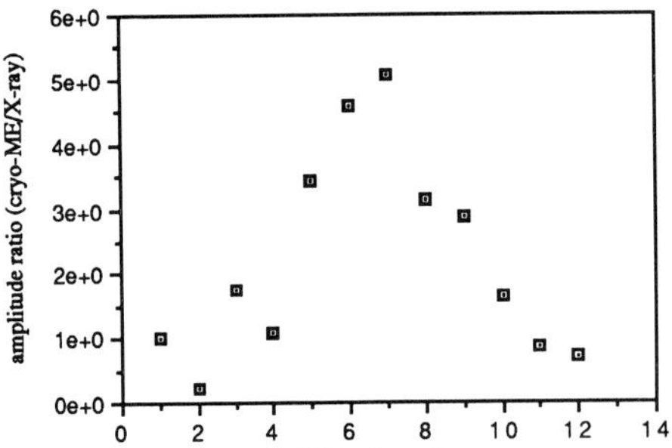

Figure 1: Ratio between the amplitude of the Fourier components deduced from cryo-electron microscopy and those calculated from X-ray diffraction patterns. These data show that cryo-electron microscopy is transferred through a contrast transfer function which can be approximated to a pure phase contrast transfer function.

Cryo-sections of vitrified rat tail tendons display optical diffraction patterns having meridional intensities in agreement with those of X-ray diffraction patterns of hydrated tendons (Insert plate 1b). However, a quantitative analysis (Figure 1) shows that the image Fourier components are modulated by a contrast transfer function. Although the section is c.a. 200nm thick, the contrast transfer function can be considered as being a pure phase contrast function. Effects of the phase contrast transfer function on the image can thus be corrected by a Wiener filter as has been suggested for thin specimens (Lepault and Pitt, 1984). After such a correction, a faithfull image of rat tail tendon can be observed. Such an image is in agreement with the image deduced from X-ray diffraction data and sequence analysis (Hulmes et al., 1977).

3. CONCLUSIONS

This study shows that the structure of vitrified rat tail tendons is similar to that of tendons under physiological conditions. However, since the release of collagen fibrils from the tendon affects the collagen molecule packing, cryo-electron microscopy of collagen has to be done by cryo-sectionning. Although cryo-sections are thick (100-200 nm), information is mainly transfered through phase contrast. Images can thus be corrected for the effects of the contrast transfer function which leads to the "true" image of the tendon. Generalisation of these results to most biological entities seems to be reasonable.

REFERENCES

Brisson, A. and Unwin, P.N.T., Nature, 315, 474-477 (1985).
Dubochet, J. and McDowall, A.W., J. Microsc., 124, RP3-4 (1981).
Fuller, S.D., Cell, 48, 923-937 (1987).
Gulik-Krzywicki, T. and Costello, M. J., J. Microsc., 112, 103-113 (1978).
Hulmes, D.J.S., Miller, A., White, S.W. and Doyle, B.B., J. Mol. Biol., 110, 643-666 (1977).
Lepault, J., Booy, F.P. and Dubochet, J., J. Microsc., 129, 89-102 (1983).
Lepault, J., Dubochet, J. Baschong, W. and Kellenberger, E., EMBO J., 6, 1507-1512 (1987).
Lepault, J. and Pitt, T., EMBO J., 3, 101-105 (1983).
Mandelkow, E.M. and Mandelkow, E., J. Mol. Biol., 181, 123-135 (1985).
Taylor, K.A. and Glaeser, R.M., Sciences, 186, 1036-1037 (1974).
Vigers, G.P.A., Crowther, R.A. and Pearse, B.M.F., EMBO J., 5, 2079-2085 (1986).

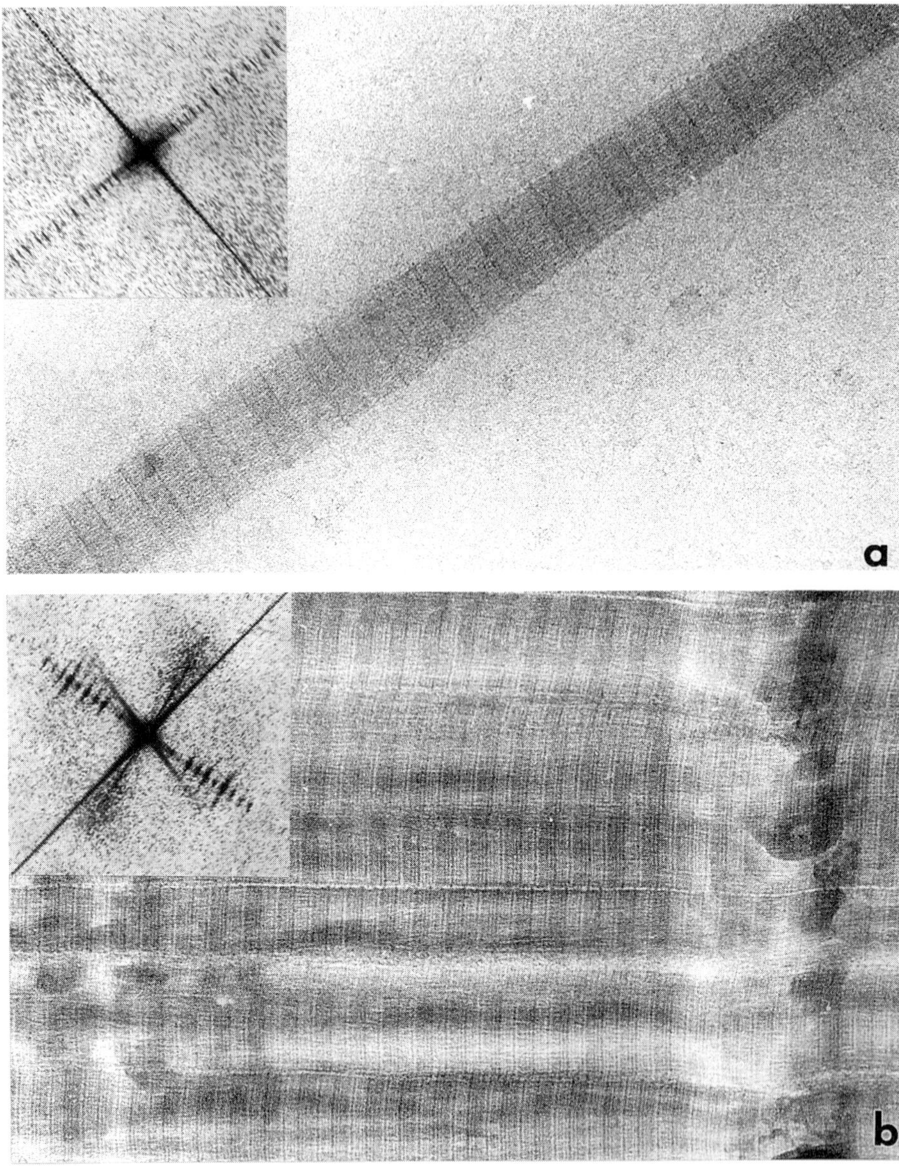

Plate 1 a: Image of a vitrified collagen fibril. Inserted is the correspondant optical diffraction pattern.

b: Image of a cryo-sectionned vitrified rat tail tendon (Courtesy of A. McDowall). Inserted is the correspondant optical diffraction pattern. In a and b, the axial periodicity is equal to c.a. 67 nm.

Inst. Phys. Conf. Ser. No. 93: Volume 3, Chapter 2
Paper presented at EUREM 88, York, England, 1988

Cryo electron microscopy

E. Zeitler

Fritz-Haber-Institut der Max-Planck-Gesellschaft
Faradayweg 4-6, D-1000 Berlin 33 (West) Germany

1. INTRODUCTION

Cryo--a form meaning cold or freezing--combined with electron microscopy implies that the specimens are investigated below ambient temperature. It has, however, become customary to restrict the term to electron microscopical studies carried out at or below the boiling point of nitrogen--that is, below 78K. Only in a few instances are the investigations concerned with the variations of the physical properties of the specimen under the influence of the low temperature, as exemplified in phase transitions /1/ or cathodoluminescence /2/. This would be cryomicroscopy in the true sense. Today cryomicroscopy is practiced as a means of removing the two fundamental shortcomings, mainly in biological research. One is the mismatch between the radiation sensitivity of the electron recording system and that of the organic specimen, and the other is the necessity to dehydrate and thus denature the biological specimen. The radiation damage problem was with us from the early days when in 1934 L. Marton observed the first biological object. With regard to the second problem, progress is also slow and expensive. But it is demonstrated in numerous workshops and monographs /3,4/ and also generally accepted that lowering the temperature of the specimen reduces the two major limitations of electron microscopy. The low temperature, however, must be applied not only during observation but also outside of the microscope during the preparation of the specimen. Thus cryomicroscopy relies primarily on preparation techniques and on instrumentation. Whereas the preparation technique must take into account the particularities of the specimen, the instrumental requirements can be generalized. This has been recognized by the manufacturers who offer devices that can be attached as auxiliaries to the main instrument. Dedicated cryomicroscopes are still rare and expensive. In this survey we shall deal only with the instrumental aspects of cryomicroscopy since the preparation techniques are too specific and numerous and present a field in their own right.

2. THE CRYOTRANSFER SYSTEMS

One approach to avoid the denaturing process of the usual specimen preparation is quick-freezing of the specimen, and if necessary sectioning it at low temperatures (less than 190 K). Another "natural" approach, namely the embedding in ice, has become common practice /5/. In any case, the frozen-hydrated (or freeze-dried) and the ice-embedded objects must be brought cold and uncontaminated into the microscope. Transfer systems for the various microscopes are commercially available /6/. Since, however, the precautions and steps required for the transfer are much the same, we choose from the description of the first device by Heide and Grund /7/ the sketches shown in Fig. 1.

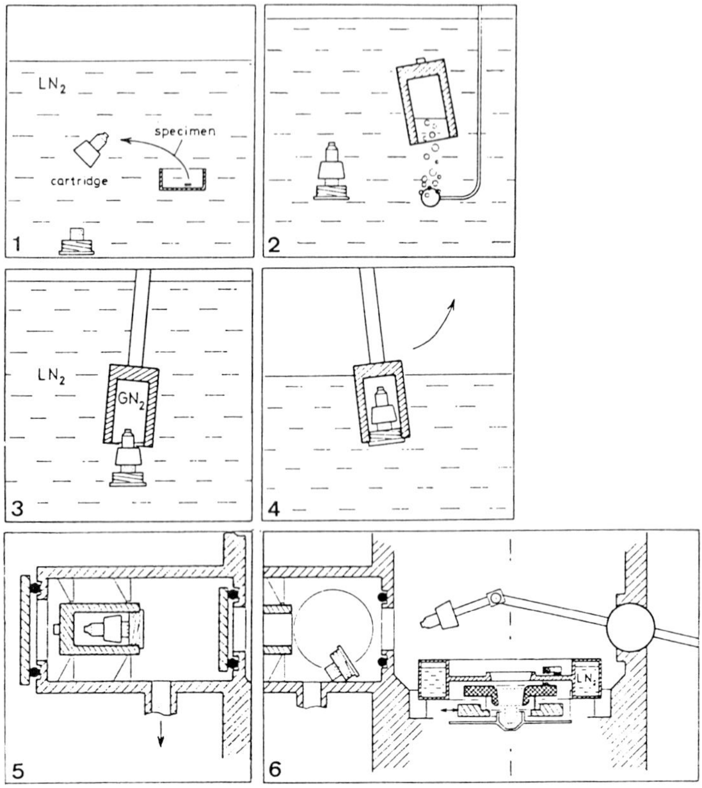

Fig. 1

The various steps are:

1) Introduce the frozen specimen into the cartridge under liquid nitrogen.
2) Cool a container of high heat capacity and remove LN_2 from its inside.
3-5) Transfer the container with cartridge into the airlock chamber of the microscope and pump.
6) Transfer cartridge with specimen to the cold stage of the microscope.

Using a very similar principle, Valdrè and Horne /8/ designed and constructed a freeze chamber directly attached to the microscope with a low-temperature stage. (Both systems, which worked very well, used top-entry objective lenses.)

When a commercial microscope is already equipped with a satisfactory cold stage, only a transfer mechanism is required to image frozen specimens. In times of scarce funds it might be worthwhile to look at the design of Taylor and Glaeser /9/. By modifying the door of the airlock of their JEOL 100B microscope such that cooled containers with the cold specimen cartridge inside could be transferred, they were able to achieve the same results as Heide (by probably simpler means). It must be emphasized that all these solutions are microscope-specific.

3. CRYO HOLDERS

In contrast to the transfer system which is merely used to transfer the frozen specimen to the cold stage of the microscope, cryo holders are both transfer mechanism and cold stage in one. For side-entry objective lens this elegant arrangement comes readily to mind, and indeed up to now only for this type of objective lens are cryo holders available. The long rod of the specimen holder, which hardly differs from the regular one, can readily receive the frozen specimen under LN_2, and the specimen can be protected and cooled constantly during transfer and observation (cooling is achieved either by gaseous coolants or by conduction from a Dewar often directly attached to the rod). The problem of cold stages in general is that the mechanical stability requires firm connections whereas the thermal stability relies on weak thermal couplings. Selection of suitable materials only helps ameliorate this contradiction, but the right compromise is the real crux of the matter. The side-entry solution faces additional problems caused by the long distances from the fulcrum and also by the non-uniformity of the temperature distribution within the specimen chamber. Drift and vibration sometimes from the boiling liquid in the Dewar challenge the designer's ingenuity. The advantage of the side-entry holders is that they can be "retro-fitted" without any alteration of the microscope. The high quality of the cryo holders and their reasonable price certainly engender the growing activity in biological cryomicroscopy.

4. CRYO STAGES

A home-built cryo stage requires a well equipped shop, excellent machinists and a good understanding of cryo technology. An excellent example of a successful design, which entails minimum alteration of a Philips 300 EM, is provided by Taylor et al. /10/. The side-entry stage is cooled by either LN_2 or nitrogen-cooled vapor to an operating temperature of 150 K. Jeng and Chiu /11/ describe a high resolution (3.5 Å) cold stage for a top-entry JEM 100CX according to Hayward and Glaeser /12/, working at 123 K. Special attention has been paid to the reduction of specimen contamination (with ice).

The general user will have to use commercially available stages. For selecting and evaluating these devices some of the considerations which enter the design and determine the philosophy might be helpful. We draw these considerations from the papers of Heide /13/, who has spent a great deal of his professional life with instrumentation for cryo electron microscopy. One might think that the requirements are determined by the lowest temperature desired, but this is not the case. The cold stage must provide the best thermal isolation together with an optimum mechanical stability. The various parts should compensate for temperature-induced expansions. These requirements are more easily fulfilled by nesting two separately cooled shells, the innermost surrounding the specimen and the outer one effecting the thermally isolated connection to the outer world, i.e., to the warm support. If this scheme is observed the question of whether a top-entry stage is superior to a side-entry stage becomes irrelevant. As long as the "principle of concentric nesting" is observed and temperature gradients are absent a thermal drift of the specimen does not occur in either device. For reliable operation of such a stage the cooled diaphragms which permit passage of the beam through the cold specimen space should be exchangeable with ease, because it is there that gases condense and contamination settles. Another "must" is the possibility to measure the temperature of the specimen holder in the immediate vicinity of the specimen. The mere reliance on the boiling points of the coolants is foolhardy, as this temperature has nothing in common with the temperature of the specimen. Wrong conclusions can result from the ignorance of the temperature.

In order to observe all these precautions a fairly involved design is necessary. This system, then, has to be incorporated into a special objective lens, a dedicated objective lens, because modification of a given objective lens is no longer feasible. The question of whether one has to go to liquid helium temperature or only to liquid nitrogen temperature becomes immaterial in the face of the fact that the efforts to have a perfect stage are so immense that the additional costs from LN_2 to LH_2 are marginal. In other words, we might as well use helium and have a stage that can be used (e.g. for Arrhenius plots) in the temperature range from ambient to 6 K!

5. SUPERCONDUCTING LENSES

Superconducting lenses are older than most of today's electron microscopists, namely 47 years /14/. Their origin is tightly connected to the progress in materials science but also to the requirement of high current stability in the lenses. Thanks to the rapid development of electronics, circuitry whose stability is better than one part per million is available today, fulfilling all the requirements of electron microscopy without the use of superconductivity. Historical accounts by pioneers such as H. Fernandez-Moran /15/ and I. Dietrich /16/ attest to the heroic endeavors on the way to today's successful superconducting microscopes. There are two types of lenses using different properties of the superconducting state. The one employs a regular iron shrouding without which the magnetic field distribution would be too weak and hence the aberrations too great. But the solenoids of the lenses superconduct the current to produce fields which readily could surpass the saturation magnetization of the pole pieces. Here only magnetic stability was the real gain. The second type of lens is constructed without any ferromagnetic material. Superconducting could produce a magnetic field which is shaped by a superconducting sinter material, very similar in shape to the iron shroud of the regular lenses. Because of the Meissner-Ochsenfeld effect, the magnetic field cannot penetrate the shielding mantle and is thus concentrated in the gap region. By the same token, disturbing external fields are also rendered ineffective.

Today only a few representatives of these two types are in operation. One is the result of an activity that started in the mid-sixties at Oak Ridge and is at present located at (and known as) the Duke University Cryo Microscope. Construction details and performance at 4 K can be found in an account by Lamvik et al. /17/. Two prototypes of the Dietrich-design shielding lenses work successfully--one at the Max Planck Institute for Biochemistry in Munich and the other in Berlin at the Fritz Haber Institute. Our experience shows that the extraordinarily good mechanical stabilities can be maintained and that the cryo vacuum is an additional bonus of the low temperature, eliminating all specimen contamination /18/. The handling of the microscope is hardly more difficult than that of a commercial one, although a well coordinated team is a prerequisite. Optical properties are superior to a conventional microscope of the same voltage. At the International Congress in Kyoto in 1986, JEOL presented a prototype of a superconducting objective lens which is based entirely on the Dietrich Siemens design /19/. For science, it is lamentable that so few machines are operational.

6. OUTLOOK

Vitreous ice has proven to be an ideal medium for embedding living materials. Sufficient contrast can be achieved only by tremendous underfocus. This, in turn, leads to very ambiguous phase contrast which is either of low resolution or hard to interpret in terms of structure. More distinct is the difference between the surrounding water and the biological specimen, namely the difference in their atomic numbers. It seems that imaging energy filters in combination with cryo microscopes will become the method of choice.

It is obvious that biochemistry must play a greater role in the development of cryo preparation techniques so that they can develop beyond mere accidental recipes. Also an activation of true cryomicroscopy would be desirable. It would be fascinating, for example, if the actual structural changes in the newly discovered high Tc superconductors could be revealed in a cryo microscope. (Actually, these new superconductors could eventually become an essential component in cryomicroscopy.) Apropos, during a recent workshop these thoughts were expressed by Karel van der Mast of Delft: he foresees that a combination of modern lithography techniques customary in small-device production, combined with the molecular beam deposition of the new superconducting oxides, may lead to very small and truly inexpensive superconducting electron optical elements. We will see.

REFERENCES

/1/ J. Mahy, J. Van Landuyt, S. Amelinckx, Y. Uchida, K. D. Bronsema and S. Van Smaalen, "Direct observation of discommensuration arrays in $NbTe_4$ by means of low-temperature electron microscopy", Phys. Rev. Lett. 55, No. 11 (September 1985), 1188.

/2/ J. C. Walmsley and A. R. Lang, "Cathodoluminescence micrography of polycrystalline synthetic diamond", Inst. Phys. Conf. Ser. No. 90, Chapter 9, 1987.

/3/ Second International Low Temperature Biological Microscopy and Analysis Meeting, Cambridge, 1981, J. Microscopy 125, part 2, 1982; J. Microscopy 126, part 1, 1982;
Third International Low Temperature Biological Microscopy and Analysis Meeting, Cambridge 1985, J. Microscopy 140, part 1, 1985; J. Microscopy 141, part 3, 1986;
Cryomicroscopy and Radiation Damage, International Study Group for Cryo Electron Microscopy Workshop, November 1981, Ultramicroscopy 10 (1982) 1;
Cryomicroscopy and Radiation Damage II, International Study Group for Cryo Electron Microscopy Workshop, October 1983, Ultramicroscopy 14 (1984) 169.

/4/ A. W. Robards and U. B. Sleytr, "Low temperature methods in biological electron microscopy", in: Practical Methods in Electron Microscopy, Ed. A. M. Glauert, vol. 10 (North-Holland, Amsterdam, 1985);
R. A. Steinbrecht and K. Zierold, eds., Cryotechniques in Biological Electron Microscopy (Springer Verlag, Berlin, Heidelberg, 1987).

/5/ J. Dubochet and A. W. McDowall, "Vitrification of pure water for electron microscopy", J. Microscopy 124 (1981) 3;
H. G. Heide and E. Zeitler, "The physical behavior of solid water at low temperatures and the embedding of electron microscopical specimens", Ultramicroscopy 16 (1985) 151; R. Rachel, U. Jakubowski and W. Baumeister, "Electron microscopy of unstained, freeze-dried macromolecular assemblies", J. Microscopy 141 (1985) 179.

/6/ W. A. M. Hax and S. Lichtenegger, "Transfer, observation and analysis of frozen-hydrated specimens", J. Microscopy 126 (1982) 275; besides the microscope manufacturers, the following firms--among others--offer cryo transfers: Gatan Inc., Pleasanton, CA 94566 USA; Hexland Ltd, Wantage, Oxon, OX12 9TF, England.

/7/ H. G. Heide and S. Grund, "Eine Tiefkühlkette zum Überführen von wasserhaltigen biologischen Objekten ins Elektronenmikroskop", J. Ultrastructure Research 48 (1974) 259.

/8/ U. Valdrê and R. W. Horne, "A combined freeze-chamber and low temperature stage for an electron microscope", J. Microscopy 103 (1975) 305.

/9/ K. A. Taylor and R. M. Glaeser, "Modified airlock door for the introduction of frozen specimens into the JEM 100B electron microscope", Rev. Sci. Inst. 46 (1975) 985.

/10/ K. A. Taylor, R. A. Milligan, C. Raeburn and P.N.T. Unwin, "A cold stage for the Philips EM300 electron microscope", Ultramicroscopy 13 (1984) 185.

/11/ T.-W. Jeng and Wah Chiu, "High resolution cryo system designed for JEM 100CX electron microscope", Ultramicroscopy 23 (1987) 61.

/12/ S. B. Hayward and R. M. Glaeser, "High resolution cold stage for the JEOL 100B and 100C electron microscopes", Ultramicroscopy 5 (1980) 3.

/13/ H. G. Heide, "Design and operation of cold stages", Ultramicroscopy 10 (1982) 125.

/14/ H. Diepold and J. Dosse, German Patent No. 298216, 1941.

/15/ H. Fernandez-Moran, "Cryomicroscopy. History and outlook", in: Proceedings of the Tenth International Congress on Electron Microscopy, Hamburg, 1982 (Deutsche Gesellschaft für Elektronenmikroskopie eV., Frankfurt/Main 90, 1982), p. 751.

/16/ I. Dietrich, Superconducting Electron-Optic Devices (Plenum Press, New York, 1976).

/17/ M. Lamvik, R. Worsham, D. Kopf and J. D. Robertson, "Construction details of a liquid helium cryostat for a superconducting objective and cold stage", Ultramicroscopy 12 (1983) 79.

/18/ R. Henderson, J. M. Baldwin, K. H. Downing, J. Lepault and F. Zemlin, "Structure of purple membrane from halobacterium halobium: recording, measurement and evaluation of electron micrographs at 3.5 Å resolution", Ultramicroscopy 19 (1986) 147.

/19/ M. Iwatsuki, H. Kihara, K. Nakanishi and Y. Harada, "Cryo electron microscope with superconducting lens", in: Proc. XIth Int. Cong. on Electron Microscopy, Kyoto, 1986, vol. 1, p. 251.

Adaptation of an annular dark field detector capable of single-electron counting to a high resolution field emission scanning electron microscopy

R Reichelt, A Engel and U Aebi

M.E.-Müller-Institute for High Resolution Electron Microscopy at the Biocenter, University of Basel, Klingelbergstrasse 70, CH - 4056 Basel, Switzerland

ABSTRACT: The application of a high resolution scanning electron microscope can be extended towards quantitative studies of unstained biological samples by incorporating an annular dark field detector capable of single-electron counting. Due to the high efficiency and sensitivity of this detector low-dose images of unstained matter can be recorded.

In recent years high resolution scanning electron microscopy (HRSEM) has challenged transmission electron microscopy (TEM) with high resolution surface images of biological samples using secondary electrons (SE) or backscattered electrons (BSE) (Peters 1984; Tanaka et al 1986). However, the yield of these electrons carrying high resolution information is rather small, e.g. the SE-I yield at 20 keV amounts to <1% for the major elements (H,C,N,O,P) constituting biological matter (Joy 1984). Hence, special conditions such as coating of the specimen with a thin film of high atomic number elements (Joy 1984; Peters 1986), imaging at lower acceleration energy (Joy 1984; Crewe 1986) and high recording doses are required to generate a sufficient signal-to-noise ratio (S/N). In case of thin samples 10% or even more of the incident electrons are scattered elastically or inelastically (Fig. 1) and are typically collected with an efficiency of 70% in transmission mode thus providing both high resolution structural and chemical information (Crewe 1970). In particular, an annular dark field (AD) detector capable of single-electron counting is a prerequisite for quantitative analyses such as mass determination of unstained biological specimens at lowest possible electron dose (Engel 1978) exploiting the almost linear relationship between collected electrons and mass thickness (Fig. 1). The AD detector designed for our S-800 (Hitachi /Japan) is shown schematically in Fig. 2. The plastic scintillator glued onto a suitably shaped quartz glass (QG) generates light pulses (LP) transferred by the QG light pipe to the photomultiplier (PM). The peak wavelength of the aluminum coated scintillator (~423 nm) closely matches the maximum PM response (~390 nm). LP are converted to electrical pulses having a height linearly related to the electron energy (Fig. 3). Using a discriminator with a suitably positioned threshold, pulses generated from ≥ 10 keV electrons can then be separated from the noise pulses thus allowing single-electron counting. Under control of an external digital scan generator (via Universal Interface S-6517) the S-800 may serve as a tool not only for imaging unstained biological samples but also for determining their mass or mass distribution. Last but not least, with the above described instrumental extension of a SEM an expensive scanning attachment as needed for TEM becomes rather obsolete.

The performance of the AD detector is demonstrated in Fig. 4 with negatively stained tobacco mosaic virus (TMV) and T4 tubes (a), and with unstained TMV (b). Due to the excellent S/N small details such as the central channel and striation (marked by arrows) are clearly visible even at a beam current as low as ~10^{-11} A (i.e. ~10^3 e/nm^2 at 60,000x). Imaging with low probe current generally yields less contamination and lowers beam damage (Hren 1979).

This work was supported by the M.E.- Müller - Foundation of Switzerland.

References:

Crewe A V 1970 Quart. Rev. Biophys. **3** 137

Crewe A V 1986 Proc. XIth Int. Congr. on Electr. Microsc. eds T Imura , S Maruse and T Suzuki (Tokyo:Japanese Soc. of Electr. Microsc.) pp 2105 -8

Engel A 1978 Ultramicroscopy **3** 273

Hren J J 1979 Introduction to Analytical Electron Microscopy eds Hren J J, Goldstein J I, Joy D C (New York: Plenum Press) pp 481 -505

Joy D C 1984 Microbeam Analysis eds Romig A D, Goldstein J I (San Francisco:San Francisco Press,Inc.) pp 81 -5

Peters K-R 1984 Electron Beam Interactions with Solids for Microscopy, Microanalysis and Microlithography eds Kayser D F, Niedrig H, Newbury D E, Shimizu R (AMF O`Hare: SEM Inc.) pp 363 -72

Tanaka K, Mitsushima A, Kashima Y, Osatake H 1986 Proc. XIth Int. Congr. on Electr. Microsc. eds T Imura, S Maruse and T Suzuki (Tokyo:Japanese Soc. of Electr. Microsc.) pp 2097 -100

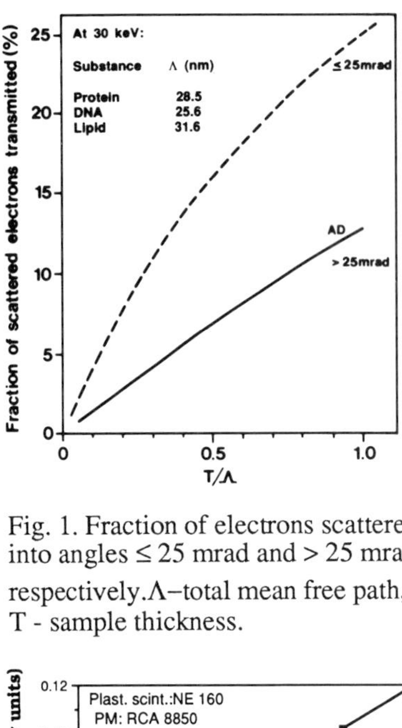

Fig. 1. Fraction of electrons scattered into angles ≤ 25 mrad and > 25 mrad, respectively.Λ–total mean free path, T - sample thickness.

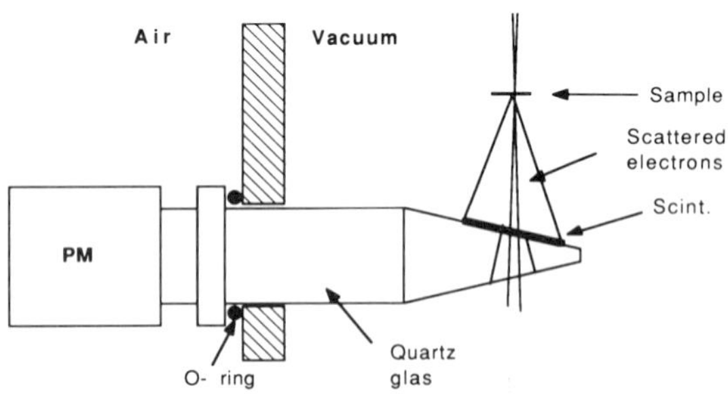

Fig. 2. AD detector with plastic scintillator.Angular collection up to 190 mrad.Max. count rate ~10 MHz.Single-electron counting for E ≥ 10 keV.

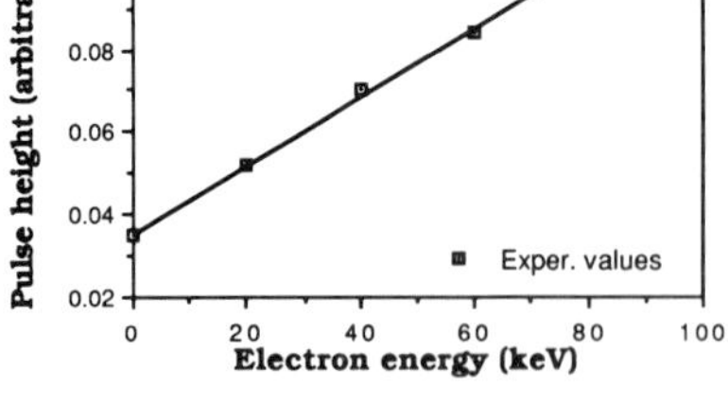

Fig. 3. Pulse height vs.electron energy. Pulse height at 0 keV corresponds to 1 photoelectron-equivalent pulses, i.e. noise pulses.

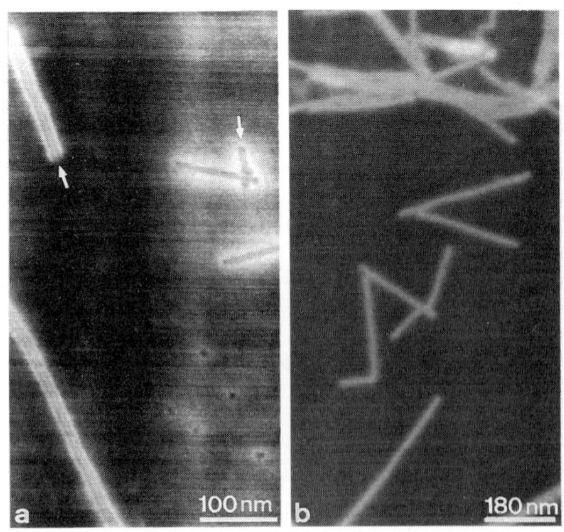

Fig. 4. STEM AD micrographs. Negatively stained TMV and T4-tail tubes (a) and unstained TMV (b). E=30 keV. Dose: ~10^3 e/nm^2.

High resolution shadowing and coating for biological transmission electron microscopy (TEM) and scanning tunnelling microscopy (STM)

H.Gross, M.Amrein, H.Winkler
Institute for Cell Biology, ETH-Hönggerberg
8093 Zürich, Switzerland

G.Travaglini
IBM Research Division Zürich, Research Laboratory
8803 Rüschlikon, Switzerland

Freeze-drying followed by heavy-metal shadowing is a long established approach for routine structural studies in biological electron microscopy. After careful freeze-drying ("maintaining hydration shells"), shadowing at -250°C under ultrahigh vacuum conditions and the application of computer-based averaging technique we were able to visualize on periodic test specimens structural surface features smaller than 2 nm (Fig.1b').
Based on the assumption that after averaging the evaporated metal forms a continuous film with constant thickness in the direction of metal deposition (ideal shadowing) the surface reliefs (Figs.1c,c') were reconstructed from unidirectional shadowing data (Guckenberger 1985). The sixfold internal symmetry of the HPI-layer allowed the reconstruction from one single micrograph. Computer simulation of shadowing (Fig. 1d,d') applied to the reconstructed reliefs shows that for unidirectional shadowing the model assumptions and hence the relief data are correct for the resolution attained. This is further supported by the good agreement of the reconstructed external morphology with the projected structure obtained from the unstained, frozen-hydrated HPI-layer.

We are currently engaged in the application of high resolution rotary-shadowing at -250°C an image averaging to single freeze-dried macromolecules (mitochondrial creatine kinase, recA-DNA complexes, myosin, collagen, proteoglycan) and to freeze-fractured membrane structures.
For rotary shadowed specimens the contrast mechanism is complex. The contrast is inherently low and larger amounts of shadowing material, especially at low elevation angle produce artificial surface structures. Computer simulation is used to differentiate shadowing information from artificial effects (decoration, self shadowing).

Guckenberger R. 1985, Ultramicroscopy 16:357

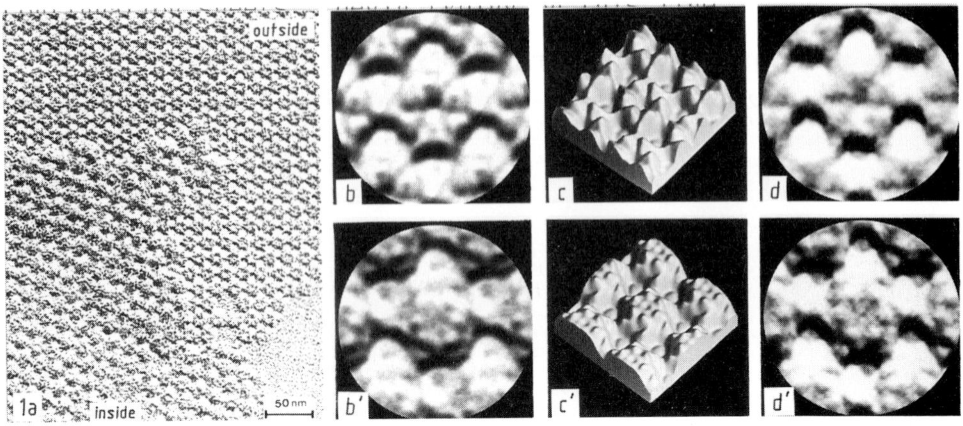

Fig. 1: HPI-layer of the cell wall of the bacterium Deinococcus radiodurans after freeze-drying at -80°C and unidirectional shadowing (45°) with Ta/W at -250°C.
a: micrograph, b,b': averaged micrographs, c,c': reconstructed surface reliefs, d,d': simulations of shadowing (compare with b,b').

In the TEM-approach the shadowing films are routinely stabilized with a carbon-backing layer to minimize post-shadowing artifacts. For obvious reasons such C-coat would blurr fine surface details, when investigated with the tunneling microscope. We found that at the right composition Pt/Ir/C films remain threedimensionally stable after transferring to atmospheric conditions. In addition such films have a small granularity and allow stable tunneling under atmospheric conditions. Fig.2 shows freeze-dried and Pt/Ir/C-coated recA-DNA complexes immaged in the TEM (Fig.2a) and in the STM (Fig.2b) respectively. The STM-image obtained in the constant current mode makes evident that after coating with a thin conducting film relief data at macromolecular level can directly be achieved with the tunneling microscope.

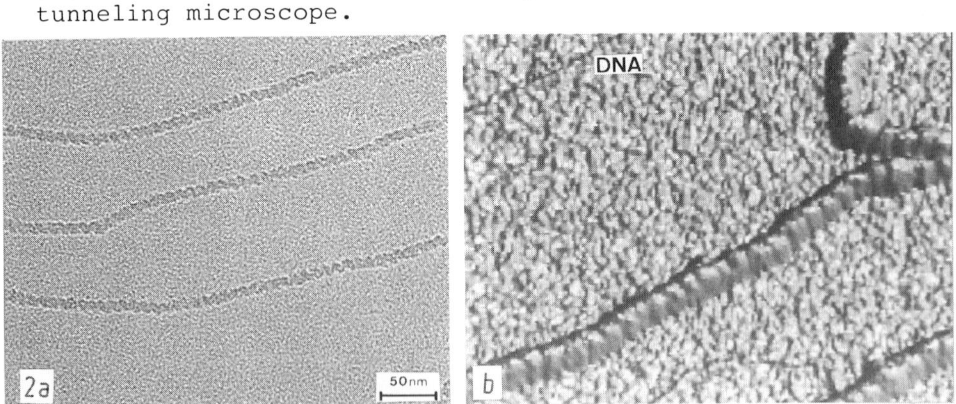

Fig.2: recA-DNA complexes after freeze-drying and rotary-coating (60° elevation) at -80°C with Pt/Ir/C.
a: TEM micrograph
b: processed STM image with simulated shading.

Thin free spanning films of phospholipids for the temperature observation of hydrated proteins and fusion of lipid vesicles

PM Frederik, MCA Stuart, PHH Bomans.
EM Unit, Dept. of Pathology, Univ. Limburg, P.O. Box 616, 6200 MD Maastricht, The Netherlands

Thin and free spanning films of surface active compounds can be obtained by immersing and withdrawing a bare specimen-grid from a solution or suspension of the surface active compound. By adding particulate material to the suspension a free-spanning composite film can be formed. It can be argued that the presence of surface active molecules in thin films is essential to increase the lifetime of a thin film long enough to enable vitrification in coolant. The use of phospholipids to form free spanning films seems to be fascinating since it is a class of naturally occurring surface active compounds of great biological significance.
A suspension of small unilamellar vesicles was prepared by sonication of dimyristoyl phosphatidyl choline (DMPC 16 mM) in water. A free spanning film was prepared and after blotting the film was vitrified within one second by plunging in ethane. In such thin films prepared from DMPC and observed by cryo-electronmicroscopy at 80 K small unilamellar vesicles could be observed (Lepault et al, 1985), as well as the merging of small vesicles in extended leaflets (Frederik et al, submitted). In the latter case thinning of the free spanning film seems to be involved in evoking the membrane rearrangement ("leaky fusion"). The thickness of the free spanning films was estimated with densitometry on the micrographs. The response of the film to electrons was estimated (photographic density D vs electron dose at the specimen site). This relationship was used to relate photographic contrast as estimated by densitometry, to local currents at the specimen level. Latex spheres of various sizes were used as thickness standards to determine the microscope constants involved in contrasttransmission (see Zeitler & Bahr, 1962). The thin rim of a free spanning film of DMPC was usually devoid of vesicles and was found to have a thickness of ca 20 nm. From these thin parts the film is gradually increasing in thickness towards the grid bars and typically some 50 nm thick in the thin area close to the rim where vesicles can be observed. After prolonged draining of a DMPC film at room conditions and observation at ambient temperature in the electronmicroscope free spanning films could still be observed. Thin parts of these films were found to have a thickness of 14 nm and a stepwise increase in thickness towards the grid-bars was observed. The stepheight was determined with densitometry on the micrographs and with quantitation of the darkfield signal in STEM (latex spheres and carbon support films were used for callibration) and found to be 4.3 $\pm$ 1.0 and 4.5 $\pm$ 0.2 respectively. These observations indicate a structure composed of two-bilayers of DMPC (each 4.4 nm thick covered by two monolayers of DMPC at the interfaces for the thinnest part of the film. Each step in the film ("Grandjean terrace") corresponds to the increment of the thickness with one bilayer.
With small unilamellar vesicles prepared from a mixture of cardiolipin and DMPC we could induce fusion of vesicles by adding Ca^{2+} and visualize the fusion products in a free spanning film (fig. 2.). In this type of membrane fusion the involvement of contact areas in the form of an inverted hexagonal phase is inferred on basis of freeze-fracture data (Verkley et al, 1979).
The suspension of small unilamellar vesicles of DMPC can also be used to assist the formation of thin films of particulate material. This has been tested with catalase crystals which are known to be sensitive to dehydration (Taylor & Glaeser, 1975; Dorset & Parsons,

1975; Unwin and Henderson, 1975) as well as to beam induced structural damage (e.g. Chiu et al, 1986). In figure 1 a micrograph is shown from catalase surrounded by vesicles of DMPC and the corresponding electron diffraction pattern shows diffraction spots from catalase beyond 0.42 nm^{-1}. When such a thin film is allowed to drain (for ca 10 seconds instead of less than 1 second) prior to vitrification the configuration of the phospholipid has changed. The vesicles have vanished into extended bilayer sheets (characteristic for DMPC concentrations in excess of 70% w/w) and the catalase crystals still show diffraction spots beyond 0.42 nm^{-1}, the diffraction ring from DMPC. These results illustrate that phospholipid can be used to form a free spanning film of particulate material. The arrangement of the lipid forms an structural indication of the water content of such a film at the moment of its vitrification.

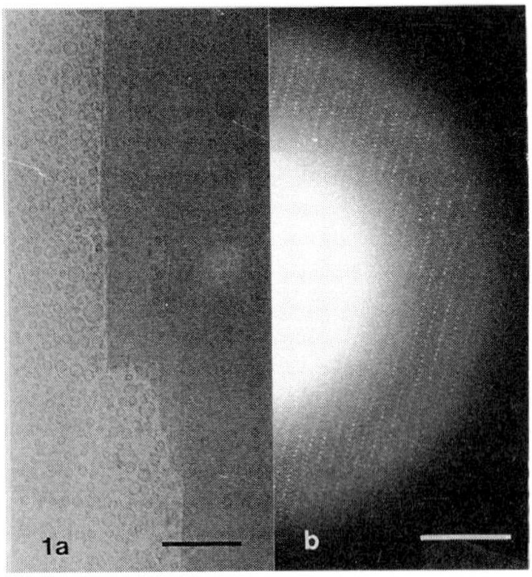

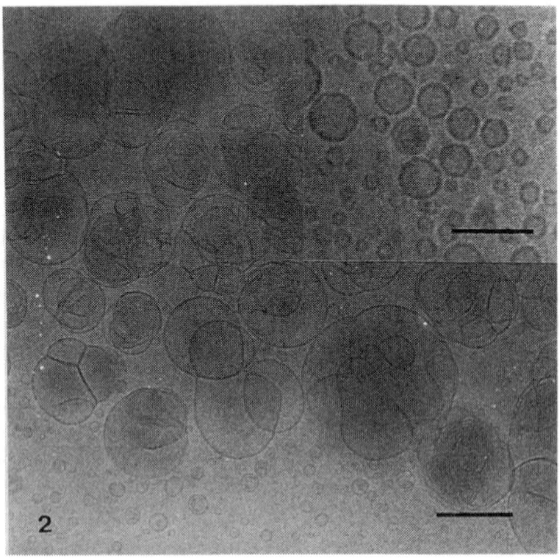

Fig. 1. Catalase crystals in a suspension of DMPC vesicles. The micrograph (a) shows the catalase crystal surrounded by small unilamellar vesicles. The electron diffraction pattern (b) shows diffraction spots beyond the diffraction ring from DMPC (0.42 nm^{-1}), recorded with 75 $e.nm^{-2}$. Bars: 200 nm (a) and 1.0 nm^{-1} (b), temperature 80 K.

Fig. 2. Calcium induced fusion of cardiolipin/DMPC vesicles (3 mM each, 2 mM calcium chloride), vitrified and observed at 80 K. Inset: vesicles without Calcium observed at 80 K. Bar: 200 nm, inset 100 nm.

REFERENCES

Chiu W, Downing K H, Dubochet J, Glaeser R M, Heide H G, Knapek E, Kopf D A, Lamvik M K, Lepault J, Robertson J D, Zeitler E, Zemlin F 1986 *J. Microsc.* **141** 385

Dorset D L, Parsons D F 1975 *Acta Cryst.* **A 31** 210

Frederik P M, Stuart M C A, Bomans P H H, Busing W M 1988 *J. Microsc.* **submitted**

Lepault J, Pattus F, Martin N 1985 *Biochim.Biochys. Acta* **820** 315

Taylor K A, Glaeser R M 1975 *Science* **186** 1036

Unwin P N T, Henderson R 1975 *J. Molec. Biol.* **94** 425

Verkleij A J, Mombers C, Gerritsen W J, Leunissen-Bijvelt L, Cullis P R 1979 *Biochim.Biophys.Acta* **555** 358

Zeitler E, Bahr G F 1962 *J. Appl. Physics* **33** 847

Erythrocyte skeleton replicas obtained with a new freeze drying device

F Lupu, E Constantinescu

Institute of Cellular Biology and Pathology, Bucharest 79691, Romania

ABSTRACT: A device which ensure the protection of frozen specimens against moisture and other contaminants during their transfer from liquid nitrogen in the vacuum chamber and freeze drying process,was designed and built. This device was used to study the ultrastructural features of erythrocyte membrane skeleton and the immunocytochemical localization of spectrin in an "in situ" approach.

1. INTRODUCTION

The contamination of specimens during their mounting on the cold stage of freeze fracture machine, transferring in the vacuum chamber and in vacuum sublimation process, represents an important limitation for deep etching studies of erythrocyte membrane surface and its associated skeleton. We surmounted this restriction by constructing a freeze drying device which shield the specimen against moisture and other contaminants during the whole process.

2. MATERIALS AND METHODS

2.1 Constructive Details

The freeze drying device consists of two parts: a cold trap and a thermic connection (Fig. 1). The cold trap consists of a pure copper cap fixed on a Teflon ring. The thermic connection is made of: a pure copper piece, fixed with a screw set on the microtome arm of the freeze fracture apparatus; a flexible connection made of closely apposited pure copper wires; an assembley made of two copper plates having inbetween the flexible joint, all brought together by four rivets.

2.2 Specimen Preparation

The human erythrocytes were adhered on polylysine coated coverslips and broken by a stream of buffer that mimics the intracellular ionic environment ("inside buffer", IB). The samples were prefixed in periodate-lysine-paraformaldehyde fixative in IB, labeled with antispectrin -5 nm gold particles fixed in glutaraldehyde, mordanted in Formic acid, postfixed in OsO_4, repeatedly washed in water and shortly rinsed in 30% ethanol. The fixed samples are mounted on aluminium quick freezing carriers and frozen by immersion in nitrogen slush (-210°C). The carrier is mounted under liquid nitrogen on the specimen stage and covered with the Teflon-copper made cap which acts as a cold trap. The Teflon ring assures a good thermic isolation,working as a seal against the atmospheric contaminants. The whole assembly is transferred into vacuum chamber of a Balzers BAF 301 apparatus. The moisture on the copper cap was removed by blowing Freon 22. The cold microtome arm is brought into contact with the copper cap by means of the flexible thermic connection through a thin layer or thermal conductive silicone paste. After cooling the microtome arm at -196°C, the samples were deep-etched at -95°C for 30 min in a vacuum of 2×10^{-6} torr or better. Replicas were obtained by rotary shadowing with platinum-carbon at an angle of 23°. Before exposure to the atmospheric pressure, the replicas samples were warmed at room temperature under vacuum. The glass coverslips were removed with hydrofluoric acid and the replicas were cleaned, mounted on grids and examined with a Philips EM 400 electron microscope.

3. RESULTS

Freeze drying of broken erythrocytes (Fig. 2) expose large areas of the true inner surface of the plasmalemma which appears covered by a filamentous meshwork (arrows). The immunocytochemical localization of spectrin using gold labeled specific antibodies (Fig. 3) shows that an important component of the filaments is represented by spectrin (arrowhead).

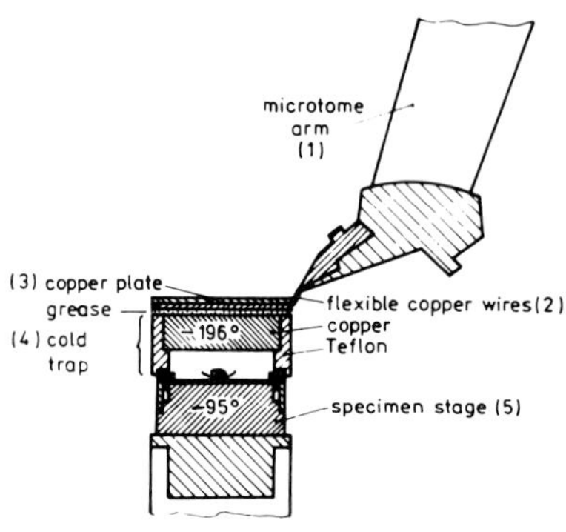

1

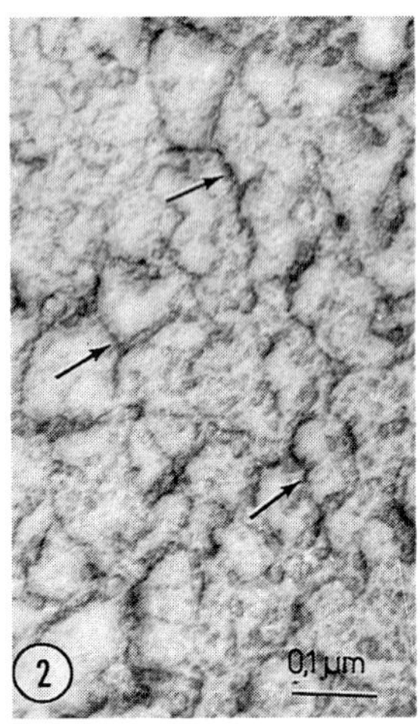

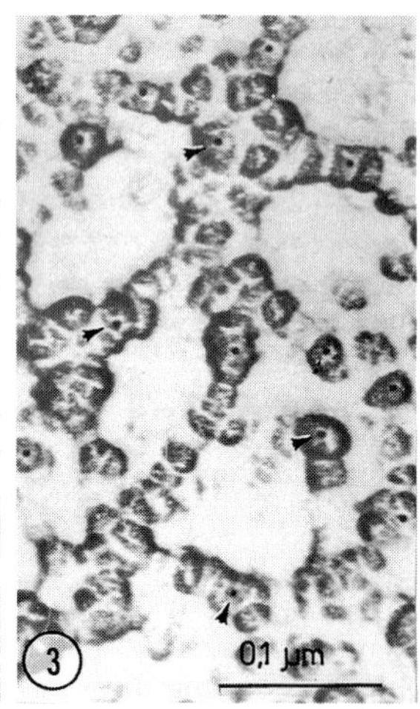

4. CONCLUSION

Our data show that the deep etching and rotary shadowing of immunogold labeled samples represent a valuable approach to the study of the membrane associated erythrocyte skeleton. The technique and the device which we developed may lead to a deeper understanding of the spatial organization of cytoskeletal network of more complex cells.

Supported by Ministry of Education, Romania and NIH (USA) Grant HL-26343 awarded to Drs. Nicolae and Maya Simionescu.

Radiation damage of unstained vitrified biological specimens in STEM

W Tichelaar

European Molecular Biology Laboratory, Postfach 102209, 6900 Heidelberg, FRG

ABSTRACT: The radiation damage of vitrified specimens in STEM has been studied and compared with that in TEM. Despite a ~10^5x higher dose rate in STEM, bubbling appears at the same dose and decay of resolution proceeds with the same rate as in TEM. The usability of STEM regarding vitrified specimens implied by these results is discussed.

We are investigating the applicability of STEM to vitrified specimens. In this context, we first studied radiation damage. Beam damage of vitrified specimens is characterized *i.* by decay in specimen spatial order at low doses, and *ii.* by bubbling at higher doses (>5×10^3 e nm^{-2}). These phenomena are well investigated in TEM. In STEM, several parameters for imaging are different from those in TEM: *i.* the dose rate is about 10^5 times higher; *ii.* the spot size is more than 10^3 times smaller; and, *iii.* the image acquisition is sequential as opposed to parallel. These differences could have an influence on the process of radiation damage.

The samples used were catalase crystals, T4 bacteriophages and tails, and TMV. Specimens were prepared by the perforated foil method (Adrian *et al* 1984). The STEM used was a Vacuum Generators HB5 STEM equipped with a cryo-stage (Homo 1980), operated at ~130K. Specimen transfer to the STEM occurred essentially according to Freeman *et al* (1980). The STEM was operated at 100 kV, and with an objective aperture of 50 µm, and a beam current of ~10^{-11} A. Images were recorded at a magnification of 100,000x or 200,000x, using a 1024^2 raster with 4 to 32 µs/pixel, and using the (BF-DF)/(BF+cDF) signal (where c is a constant). Photographs were taken directly from the monitor screen. Transmission electron microscopy was carried out according to Adrian *et al* (1984), at 25,000x or 43,000x. A Faraday cage and a Keithley solid state electrometer were used to check the speed of the photographic material in TEM, and the beam current in STEM. In order to monitor the decay in specimen spatial order and the onset of bubbling, successive images were taken of a particular specimen area.

Figure 1 shows pictures from series of consecutive exposures of vitrified catalase crystals in STEM in images (a) to (d), and in TEM in (e) to (f). The doses per picture are 1.7×10^3 , and 9×10^2 e nm^{-2}, respectively. Bubbling can be observed at 6.6×10^3 e nm^{-2} in both series. The doses at which bubbling appears in all vitrified suspensions studied in STEM and TEM are listed in Table 1. Figure 2 shows optical diffraction patterns of the vitrified catalase images: of Figure 1a in (a), of Figure 1b in (b), of Figure 1e in (c), and of Figure 1h in (d). The number of unit cells included is ~10^3. The highest reflections in these patterns are the (1,1), (2,0), (6,1), and (3,1) ones, respectively. These reflections and the highest ones observed in similar patterns are plotted as a function of dose in Figure 3. Condensation of water onto the specimen during transfer and image acquisition in the case of STEM, was absent (Results not shown.).

The results show that in the dose range studied, bubbling appears at the same dose, and decay of resolution proceeds with about the same rate in STEM and TEM. Thus, radiation damage is dose-rate independent. The results indicate that the STEM is usable in the study of vitrified specimens, for example to determine the contribution of amplitude contrast in their imaging. Phase contrast imaging, simultaneously with amplitude contrast imaging will be possible when the appropriate detector becomes available. Then, the serial imaging may be advantageous regarding the blurring effect of radiation damage (Bullough and Henderson 1987).

REFERENCES
Adrian M, Dubochet J, Lepault J and McDowall A W 1984 *Nature* **308** 32
Boullough P and Henderson R 1987 *Ultramicroscopy* **21** 223
Freeman R, Booy F and Leonard K 1980 *Proc. 7th Eur. Congr. on Electron Microscopy* eds P Brederoo and G Boom (Leiden: Seventh European Congress on Electron Microscopy Foundation) Vol 1, pp 92-3
Homo J-C 1980 *ibid* eds P Brederoo and W de Priester, Vol 2, pp 650-1

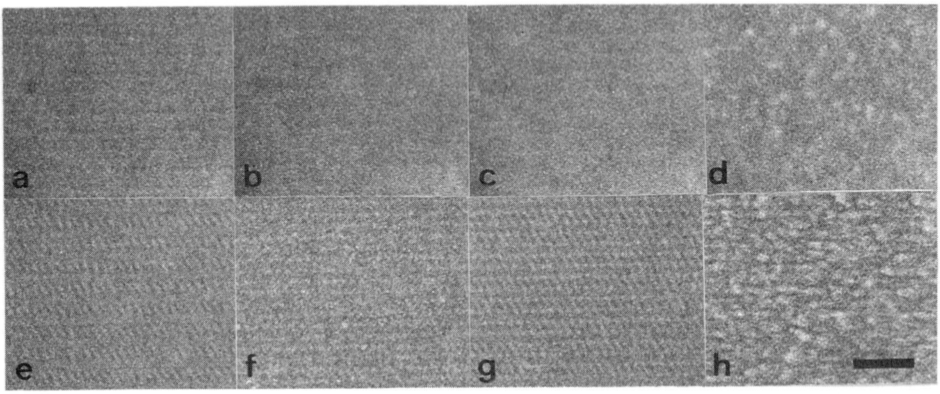

Fig. 1. Bubbling in vitrified catalase crystals. (a) to (d) The first four pictures of a series of STEM images taken of one crystal area with 1.7×10^3 e nm^{-2} per exposure. (e) to (h) The first, second, fifth and sixth image of a series in TEM with 9×10^2 e nm^{-2} per exposure. Bar = 50 nm.

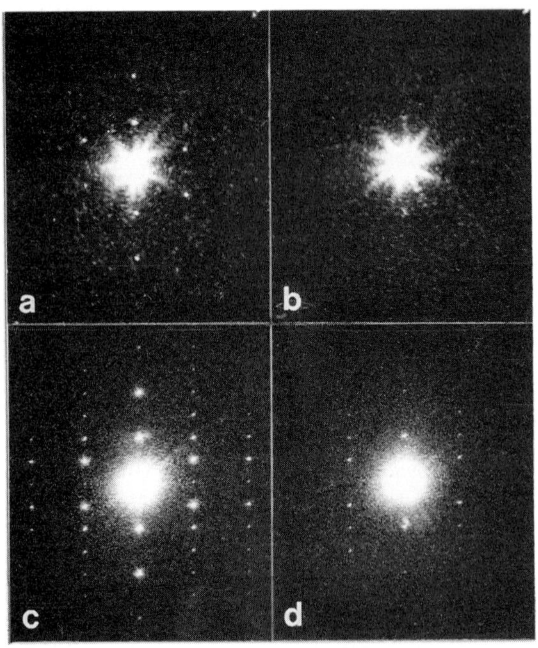

Fig. 2. Optical diffraction patterns of images in Fig. 1. (a) and (b) are from Fig. 1a and b, and (c) and (d) from Fig. 1e and h. (See text for further details.)

Table 1. Electron doses at which bubbling is detected in various vitrified biological suspensions.

Specimen	Electron dose (e nm^{-2}) STEM	TEM
Catalase	7×10^3	5×10^3
T4 (heads)	9×10^3	4×10^3
T4 tails	2×10^4	2×10^4
TMV	2×10^4	3×10^4

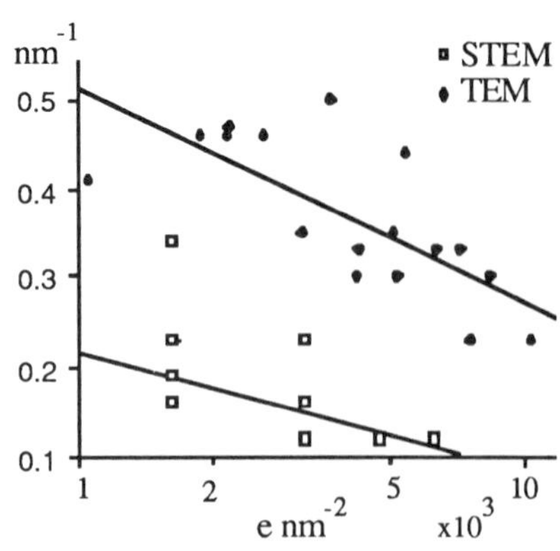

Fig. 3. Decay of spatial order in vitrified catalase crystals with dose.

Inst. Phys. Conf. Ser. No. 93: Volume 3, Chapter 2
Paper presented at EUREM 88, York, England, 1988

Beam induced contrast in vitrified cryo-sections

P.H.H. Bomans, P.M. Frederik
EM-unit, Dept. of Pathology, Univ. Limburg, P.O. Box 616, 6200 MD Maastricht, The Netherlands.

There are conflicting reports about the contrast to be expected in thin cryo-sections prepared from biological material and observed at low temperature in conventional transmission electron microscopy (CTEM) or scanning transmission electron microscopy (STEM). Initially Frederik et al. (1982) reported the absence of contrast in fully hydrated cryo-sections from fresh frozen rat kidney and this observation has since been confirmed by others (e.g. Zierold 1987). A different observation has been reported by McDowall et al. (1983). Fresh liver rapidly cooled and sectioned at low temperature (110 K) showed low but conspicious contrast when observed in CTEM well under focus and under low dose conditions. The repeated and negative attempts to verify this observation has prompted us to concentrate on the following aspects:

- The collection of vitrified water on the cryo-section through vapour condensation can be a serious problem but can be prevented by using correct procedures (Frederik & Busing, 1986).
- Is the section quality (thickness, absence of sectioning artifacts) and contrast in CTEM influenced by the nature of the specimen (e.g. electron scattering properties, vitrified or vitrified/crystalline?

When a fresh tissue sample is rapidly cooled only a small part can be vitrified. If such a sample is sectioned at low temperature it is likely that non-vitrified parts are sectioned as well and contributed to a poor section quality. Therefore we decided to turn to fixed tissues (glutaraldehyde 0.2% + formaldehyde 2% w/w in 0.1M phosphate buffer) treated with cryoprotectants. Fahy et al. (1984) demonstrated that large pieces of tissue and even whole organs can be vitrified upon cooling. At certain (high) solute concentrations the vitrified sample will not recrystallize upon warming. Various solutes can be used to this end and we tested glycerol (67% w/w), propylene glycol (80% w/w) and methanol (80% w/w) to infiltrate aldehyde fixed tissues for subsequent vitrification. This was done with the expectation that within a thin section contrast is depending on the difference in electron scattering properties of cellular components and its surrounding (aqueous) embedding medium (see Carlemalm et al 1985 for review on contrast formation). The solutes employed cover a range of scattering properties; methanol being relatively translucent to the electron beam, glycerol being relatively electron opaque and propylene glycol taking an intermediate position. From samples vitrified in these solutes cryo-sections could be prepared and after cryo-transfer and low temperature observation (80 K) they were found to have a thickness usually between 200 and 300 nm. With low-dose recording it was observed that contrast is induced in these sections during electron-beam irradiation (see figure). Ultrastructural detail became visible at an irradiation dose between 1000 and 2000 e/nm^2, irrespective of the cryoprotectant used. A remarkable feature of an irradiated area is observed when the cryo-section is heated inside the microscope column and subsequently viewed

at room temperature. Unexposed areas are now "collapsed" (upon warming the viscosity drops) and have a smeared appearance without any detail. The beam exposed areas however display still ultrastructural detail.
The induction of contrast by electron beam irradiation of a cryo-section is probably depending on the dose rate (beam current) as well as on the thickness of the irradiated section. It is likely that this form of contrast development is caused by the energy transfer of inelastically scattered electrons, an interaction that will be diminished when using thinner sections. The addition of polyvinyl pyrrolidone (20% w/w) to methanol used for infiltration of tissue samples enabled us to prepare thinner cryo-sections from the vitrified material. Preliminary results suggest that also in cryo-sections thinner than 200 nm contrast is enhanced by beam irradiation. These results stress the importance of low dose observation and thickness determination when contrast in "thin" cryo-sections is discussed.

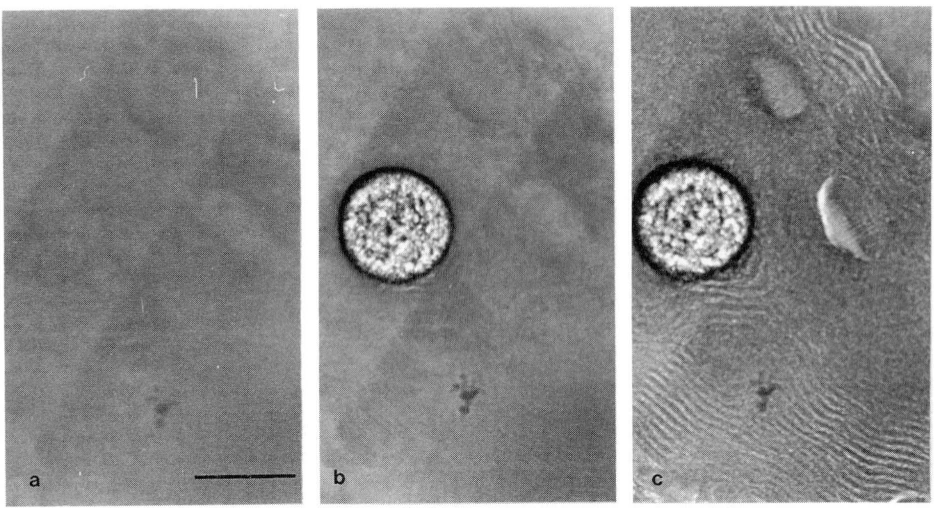

Pancreas fixed in 0.2% glutaraldehyde, 2% formaldehyde and infiltrated with methanol 80% (w/w) and subsequently vitrified in ethane. Cryo-section obtained at 153 K and transferred to the microscope for observation at 80 K. Low-dose image (ca 1000 e/nm^2) shown in (a) displays hardly any ultrastructural detail. In the micrograph (b) the beam was focussed once and immediately defocussed. Some structural detail becomes discernable (ER, nucleus). After prolonged irradiation (c) cellular detail is developed. Bar represents 1 μm.

REFERENCES

Carlemalm E, Colliex C and Kellenberger E 1985 *Adv. in Electronics and Electr. Physics* **63** 269

Fahy G M, MacFarlane D R, Angell C A and Meryman H T 1984 *Cryobiology* **21** 407

Frederik P M, Busing W M and Hax W M A 1982 *J. Microsc.* **126** RP1

Frederik P M, Busing W M 1986 *J. Microsc.* **144** 215

Zierold K 1987 *Cryotechniques in Biological Electron Microscopy* (Berlin: Springer) pp. 132-146

Inst. Phys. Conf. Ser. No. 93: Volume 3, Chapter 2
Paper presented at EUREM 88, York, England, 1988

Countercurrent plunging: improved reproducibility of ultra-rapid cooling

A W Robards[1], P W le R Murray[2] and P R Waites[1]

[1]Institute for Applied Biology, University of York, York YO1 5DD, England; [2]Research Institute for Environmental Diseases, Medical Research Council, P.O. Box 70, Tygerberg 7505, Republic of South Africa.

A novel device has been designed and built to improve cooling rates when plunging into liquid cryogens [Fig. 1 - Patents registered preparatory to commercial production]. A nylon cryogen bath is partially suspended in boiling liquid nitrogen. A magnetically rotating impeller drives a continuous flow of the cryogen up through the central cylinder from which it flows radially outwards in a symmetrical fashion to overflow into the peripheral groove. At the centre of the column, cryogen flow is essentially vertical, thus samples are plunged down through a smoothly rising cryogen stream that can be maintained within 0.5 K of a preselected temperature over the whole 60 mm depth by the device previously described by Murray (1987). *[As usual, stringent safety precautions must be taken when using potentially explosive cryogens such as propane and ethane.]* The opening of the cryogen bath is covered by a centrally perforated lid which provides a chamber above the cryogen surface into which the exhaust, warmed nitrogen gas passes. This avoids any possibility of pre-cooling of specimens in an overlying cold gas layer, avoids frost build-up around the cooling bath, and reduces the likelihood of the production of high concentrations of evaporated cryogen gas.

A copper-constantan thermocouple was prepared by capacitor discharge welding (Robards and Crosby, 1983; Robards and Sleytr, 1985) from 25 µm wires, giving a measured bead size of 70 µm. This thermocouple was plunged at 6.6 m s^{-1} into Freon 12, 25% isopentane in propane, or ethane using the device described by Robards and Crosby (1983). It was also plunged into a horizontally stirred pool of propane, again as described by Robards and Crosby (1983). Cooling rates were determined by using a purpose-built cooling rate meter (Robards, 1981), the output of which was dumped to a BBC Microcomputer for data analysis and cooling rate graph plotting.

The cooling rates illustrated in Fig. 2 demonstrate the substantial superiority of the vertical flow system over a conventional cryogen bath, both in terms of absolute cooling rates [compare curves (1) and (3)] as well as in reproducibility: the small variations in the curves and rates for each cryogen actually reflect differences of merely 1-2 K in the cryogen temperature.

REFERENCES

Murray, P.W. (1987) *Cryo-Letters* 8, 21-24

Robards, A.W. (1981) *Cryo-Letters* 1, 384-391
Robards, A.W. and Crosby, P. (1983) *Cryo-Letters* 4, 23-32
Robards, A.W. and Sleytr, U.B. (1985) *Low Temperature Methods in Biological Electron Microscopy* (ed. A.M. Glauert). Elsevier, Amsterdam.

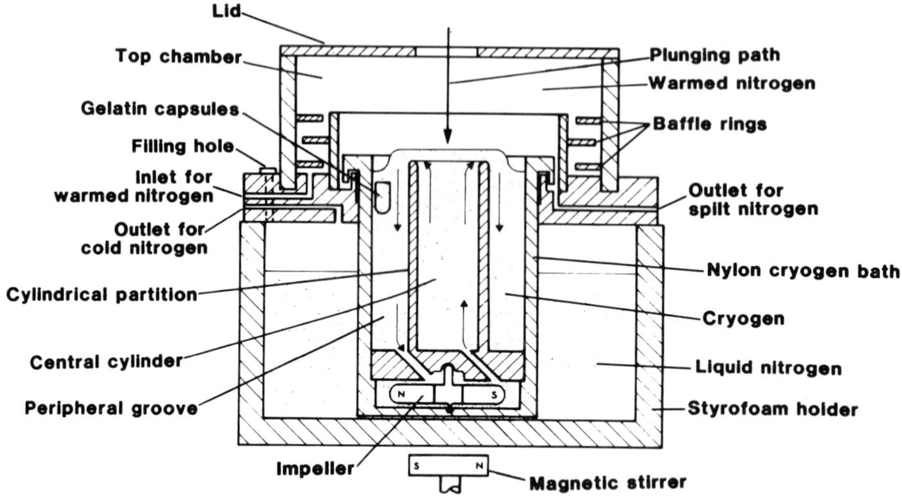

Fig. 1. Sectional diagram of the vertical freezing device

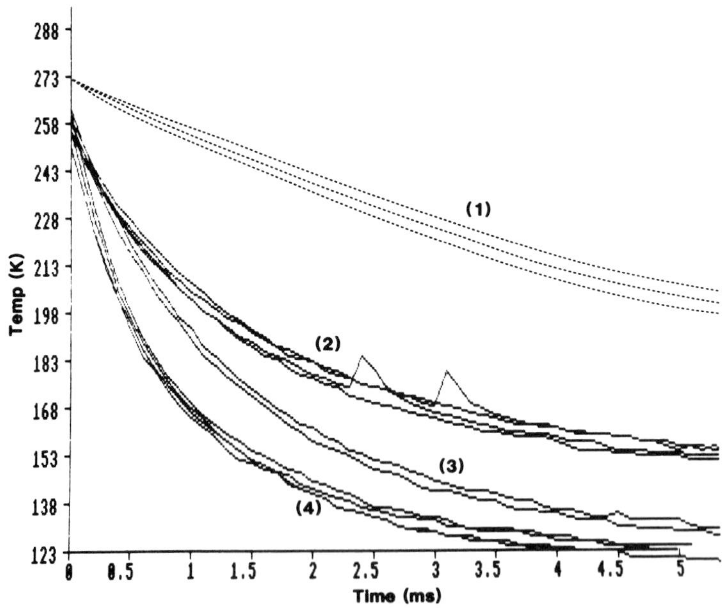

Fig.2. Cooling curves obtained using the identical 70 µm diameter thermocouple. All entry rates = 6.6 m s^{-1}. Cooling rates cited as means of illustrated 273-173 K rates

(1) Horizontally stirred propane (82-88 K). Cooling rate = 12550 K s^{-1}
(2) Vertical freezing device (VFD). Freon 12 (117-126 K) = 34000 K s^{-1}
(3) VFD. 25% isopentane in propane (88 K). = 57550 K s^{-1}
(4) VFD. Ethane (91-97 K). = 91000 K s^{-1}

Morphometric analysis of chromatin arrangement of *in situ* freeze-fractured nuclei

F. Marinelli, S. Squarzoni, E. Falcieri, R. Del Coco, A. Valmori, A. Fantazzini and N.M. Maraldi

Inst. of Citomorfologia C.N.R., c/o Inst. Codivilla Putti, Bologna and Inst. of Anatomia Umana Normale, University of Bologna, Italy.

ABSTRACT: The complex arrangement of interphase nuclei has been studied in freeze-fractured cells by means of image analysis, mainly in the attempt of identifying with morphometric parameters the heterochromatin-interchromatin transitions.

1. INTRODUCTION

Freeze fracturing has been demonstrated to be a powerful technique for describing the native state of isolated nuclei especially when associated with quantitative analysis (Maraldi 1984). Here we report data on the application of freeze fracturing and image analysis to the study of the topological organization of nuclei in situ.

2. MATERIALS AND METHODS

Rat liver fragments and human peripheral blood lymphocytes, cryoprotected in 30% glycerol, frozen in melting Freon 22 and liquid nitrogen, were fractured in a Balzers BAF 400 D at -115°C. Samples were then shadowed with carbon-platinum at 45° and the replicas were cleaned with commercial bleach. Quantitative analyses were performed with a Quantimet 970 image analyser from Cambridge Instruments.

3. RESULTS AND DISCUSSION

The replicas of fractured nuclei of cell types which present different condensation patterns of the chromatin, such as hepatocytes and lymphocytes, have been compared. While in lymphocytes the different organization of heterochromatin and interchromatin areas is easily detectable, in hepatocytes this is much less evident being the nucleoplasm almost filled with particles (Figs. 1,2). The ratio between the number of particles having the size of nucleosome and solenoid fibers, respectively, has been calculated in heterochromatin and interchromatin areas of both cell types. The obtained results (Table I) show a prevalence of nucleosomes in interchromatin in lymphocytes while in hepatocytes this prevalence is slight. By utilizing suitable softwares for determining the topological localization of a given size class of particles, it has been possible to identify perichromatin granules (Fig. 3 encircled). Ribonucleoprotein structures could also be identified by a particular freeze-embedding procedure which allows the observation of contiguous structures (Rash 1979) both replicated in freeze-fracturing and stained in section (Fig. 4).

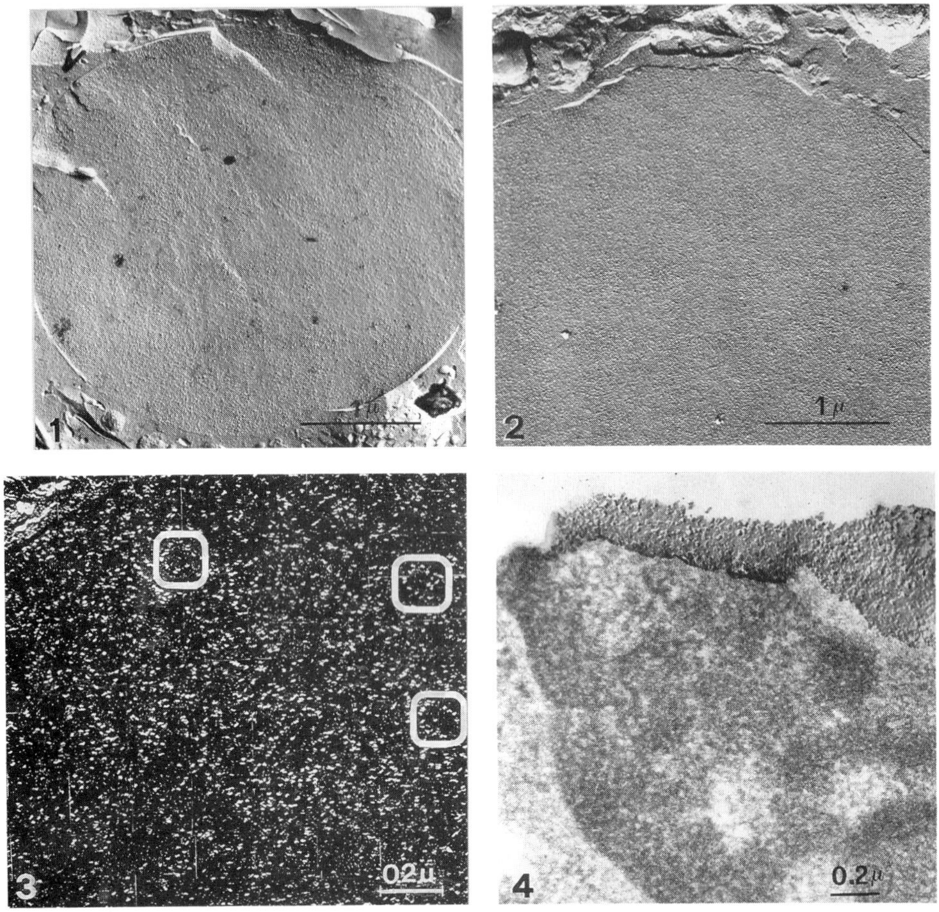

TABLE I	HETEROCHROMATIN		INTERCHROMATIN	
	% OF NUCLEOSOMES	NUMBER PER UNIT AREA	% OF NUCLEOSOMES	NUMBER PER UNIT AREA
HEPATOCYTES	46.2	412±58.7	53.7	378.8±83.9
LYMPHOCYTES	33.8	464±43.9	65.2	346.7±51.2

This comparison could allow to find morphometric parameters of particles such as the interchromatin granules which, being particularly numerous in hepatocytes nuclei, prevent the identification of the nucleosome fibers in interchromatin spaces.

4. REFERENCES

Maraldi N M 1984 *Ultramicroscopy* 12 101

Rash J 1979 *Freeze fracture, methods, artifacts and interpretation* (N Y: Raven Press) pp 153-60

Inst. Phys. Conf. Ser. No. 93: Volume 3, Chapter 2
Paper presented at EUREM 88, York, England, 1988

Improved method for the identification of cell surface structures in field emission SEM: a combination of fast freezing, freeze substitution and critical point drying

P Walther, J Hentschel & D Scott

MPI für Systemphysiologie, Rheinlandamm 201, D-4600 Dortmund 1,

The increased resolving power of field emission SEM's requires improved specimen preparation techniques. From thin sections it is known that artifacts produced by chemical fixation and dehydration are reduced by fast freezing and freeze substitution methods. In order to study the cell surface, freeze substitution can be combined with critical point drying (Steinbrecht & Müller 1987). We tested the feasibility of this approach on kidney epithelial cells in culture.

LLC-PK_1 cells in contact with culture medium were directly fast frozen via propane jet, freeze substituted in methanol containing 1% OsO_4 and 3% glutaraldehyde for 8 hours at -90° C, kept for 6 hours at -30° C, warmed up to room temperature, and were critical point dried in CO_2. Finally the samples were coated by electron gun evaporation with 5 to 10 nm of carbon and by 3 nm of platinum-carbon.

Fig. A represents an overview of a LLC-PK_1 cell culture. Notice that the surface is essentially free of large precipitated material and no ruptures are to be seen. Fig. B shows the cell surface at a higher magnification. Microvilli and small particles on the apical membrane are visible. To ensure that these particles are not precipitates from the culture medium, cells were washed with PBS or isotonic ammonium acetate before freezing. No differences in the surface structures between washed and unwashed cells could be observed. The relation of these particles to membraneous particles visible in freeze etch replicas is currently under investigation by improving the coating and by immunolabelling. Fig. C shows a similar part of a freeze substituted and Lowicryl embedded sample in the TEM. The microvilli have a similar diameter as in the SEM sample and cytoplasmic structures are well preserved.

The technique presented above has the advantage that cell surfaces can be studied without chemical pretreatment before freezing and thus might be suitable for the investigation of dynamic processes in surface composition.

Reference: Steinbrecht R A, Müller M 1987 Cryotechniques in Biological Electron Microscopy eds R A Steinbrecht & K Zierold (Berlin: Springer) pp 149-172

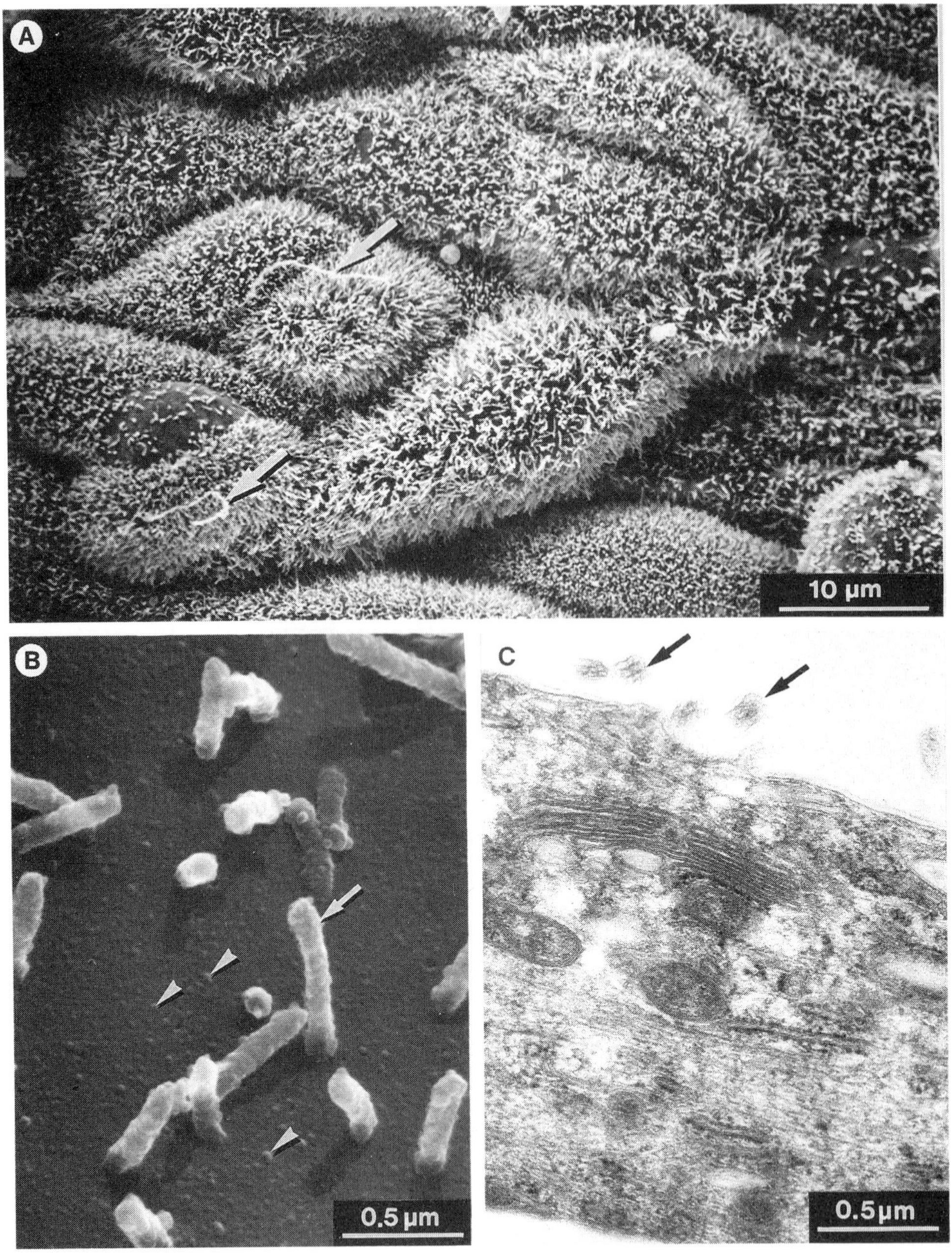

Fig. A: Overview of freeze substituted and critical point dried LLC-PK_1 cells, bearing numerous microvilli and some cilia (large arrows). Fig. B: High magnification showing microvilli (small arrows and small particles on the surface (arrowheads). Fig. C: Cross section of a similar region as in Fig. B, freeze substituted and Lowicryl embedded (small arrows point to microvilli)

Inst. Phys. Conf. Ser. No. 93: Volume 3, Chapter 2
Paper presented at EUREM 88, York, England, 1988

New technique for spray freezing of single cells and macromolecular suspensions

Wiktor Djaczenko* Ambra Canavese* and Antoni Feltynowski**
* Institute of Experimental Medicine, CNR, Rome, Italy
** Am Wurmufer 15,D-8035 Gauting, FRG.

ABSTRACT. We describe simple technique consisting in high pressure injection of cellular or macromolecular suspensions through the layer of LN_2 on top of frozen freeze substitution fixative. A set of Ertalon containers permits easy collection and handling of frozen material without its exposure to the ambient air.

1.INTRODUCTION

Smashing of tissue fragments against cooled mirror metal plate is a commonly used technique for cryofixation of solid biological material (Heuser et al 1979). Plattner et al (1973) proposed a technique of spray freezing using spray gun operating at a pressure of 1 bar. We describe here spray freezing technique based on use of high pressure gun and a special set of containers.

2. MATERIAL AND METHODS

Inner container is removed from the set (fig.1) and filled with any one of the following fixative at room temperature: 1% glutaraldehyde, paraformaldehyde, mixture of 1% acrolein and 1% tris (1-aziridinyl) phosphine oxide (TAPO) dissolved methyl or ethyl alcohol or acetone. Inner container was then frozen in LN_2. The spray gun was filled with 1.5 ml of cellular or macromolecular suspension at room temperature. Friend leukemia (10^7 cells/ml of culture medium) or (Oligochaets) erythrocruorin (15 mg/ml in assembly or dissociation buffer, Chiancone et al 1984). When boiling of LN_2 ceased the spray gun was put in vertical downward position and discharged at intervals, each discharge injecting 0.04 ml of suspension. The total of 10-15 discharges were performed for each sample. Inner container was immediately inserted into dry ice-

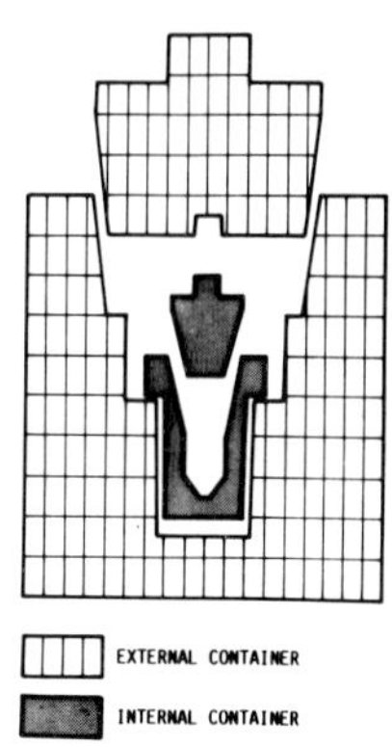

Fig.1 Assembled set of containers made of Ertalon .(Not to scale)

acetone mixture precooled outer container. Freeze substitution lasted for 48-72 h, then the temperature of container was gradually raised to 0°C. Contents of the inner container were centrifuged at proper speed and the supernatant substituted for 1% OsO_4 in organic solvent of the type used for the freeze substitution. Some macromolecules were postfixed in 1% uranyl acetate in organic solvent. After 24 h at 0°C the material was gradually brought to room temperature. The cells were embedded in plastic while macromolecular material was mostly observed untreated and unembedded.TAPO prefixed macromolecules were sufficiently resistant to air drying.

3. RESULTS and DISCUSSION

The quality of freezing offered by our experimental set was homogenously good for macromolecular suspensions. Erythrocruorin both in its integral state and after dissociation did not display any detectable degree of mechanical damage. Spray freezing of erythrocruorin permitted to obtain high resolution electron microscope images of its molecular structure (Fig.2). The quality of freezing of our cellular material was variable and a fraction of cells displayed mechanical and freezing damage. Single cells, in contrast to cell aggregates were best preserved. Mechanical damage to the cells was limited probably because of the high rates of freezing offered by our technique due to high speed of entrance of cells into LN_2 and to high degree of dispersal due to special design of spray gun. In general our set offers several advantages such as high degree of dispersal of injected material, the ease with which frozen material may be handled, minimal exposure to ambient air,minimal handling artifacts.

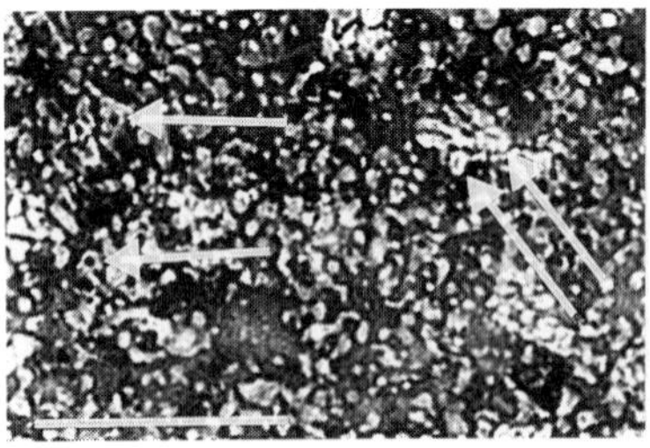

Fig.2 Dissociated erythrocruorin. Arrows show similar subunits. Bar 0.05 μm.

Chiancone E, Vecchini P, Verzilli D and Ascoli F 1984 J Mol Biol **172** 545.
Heuser J E, Rees T S, Dennis M J, Jan Y, Jan L and Evans L 1979 J Cell Biol **81** 275.
Plattner H, Schmitt-Fumian W W and Bachmann L 1973 Freeze-etching techniques and applications ed E N Benedetti and Favard pp 81-100.

The application of cryo-microscopical techniques to the investigation of protein-stabilized foams

A.J. Wilson & A.W. Robards

Centre for Cell & Tissue Research, University of York, York. YO1 5DD

ABSTRACT: Cryo-microscopical methods have been used to monitor changes in protein-stabilised foams during the processes of foam formation and degradation.

1. INTRODUCTION

A wide variety of industrial processes involve the generation of foams. Many of these are of a type in which the main stabilizing factor is protein.

Foams are essentially colloids and contain at least two distinct phases: normally a liquid continuous phase which is often aqueous containing a surfactant, and a gaseous disperse phase which is generally immiscible with the continuous phase.

The volume fraction of the gas in the system is termed the 'disperse phase volume fraction' (ϕ). When this value is large the interfacial area between phases is correspondingly large. The nature of this interface, and its ultrastructural and physical charcteristics, therefore has a direct bearing on the overall morphology of the foam and its ultimate stability.

Structural investigations of foams to a level beyond the resolving power of the light microscope are fraught with problems. Protein-stabilised foams with a large ϕ have no solid architectural component to provide rigidity to their structure. The whole micro-structure of foams is a delicate balance between the pressure inside the bubble and the surface tension forces at its surface. Since the major component of the bubble wall is water, exposure of the foam to a reduced pressure environment would result in vapourisation. Clearly then, foams in their native condition can never be exposed to the high vacuum environment of the electron microscope.

The only method of conferring rigidity to the foam structure is freezing, thereby solidifying the liquid structural component. Such stabilized structures can then be studied by, for example, low-temperature scanning electron microscopy or freeze-fracture replication.

Two general areas in which our understanding of protein-stabilised foams has been increased by the application of such cryo-microscopical methods are illustrated below.

2. RESULTS

i) Investigations of Foam Stability.

These have been performed by monitoring the rates of degradation of foams by various processes

The Polaron E7400 Cryotrans low-temperature SEM system has been used to monitor changes in bubble shape and size distribution during drainage of foamed protein solutions.

Fig. 1. illustrates the appearance of a foamed protein solution which was rapidly frozen to a temperature of 113K, fractured open, then sputter-coated with gold. When the foam is allowed to stand, drainage of the continuous phase occurs. Small bubbles coalesce and the shape of the larger bubbles produced changes from spherical to polyhedral.

ii) Investigations of Foamability.

These have been performed by examining the migration of proteins within the bubble lamellae to the air/liquid interface.

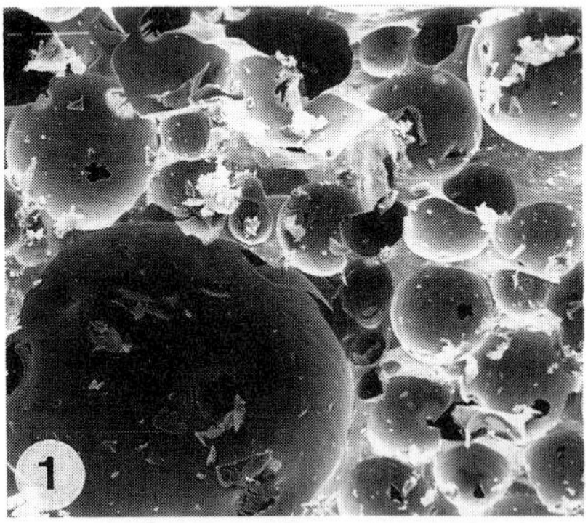

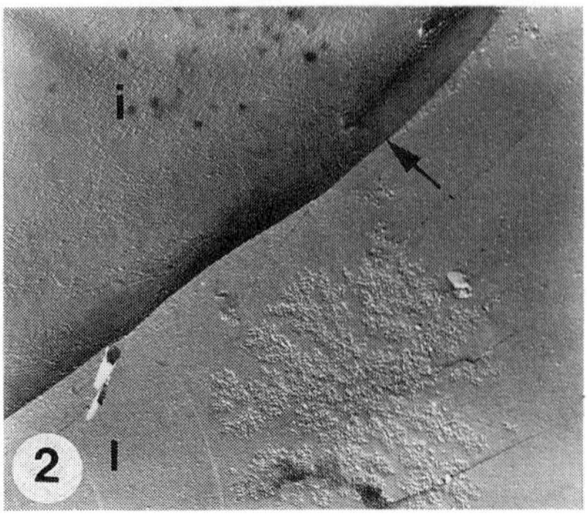

The Leybold Heraeus Bioetch 2005 freeze-fracture unit was used to prepare replicas from small volumes of rapidly frozen foamed milk whey. This contains a mixture of whey proteins and casein.

Fig 2 illustrates the edge of a bubble, where the internal surface of the bubble-wall can be seen (i) together with the gas/serum interface (arrow). Within the lamella between bubbles (l) casein can be resolved in the form of small clusters (micelles) of particles. In addition, individual whey protein particles can be seen scattered throughout the lamella. Some single protein particles appear to be associated with the gas/serum interface and it is felt that these contribute to the pattern on the internal bubble wall surface (i).

Inst. Phys. Conf. Ser. No. 93: Volume 3, Chapter 2
Paper presented at EUREM 88, York, England, 1988

Cryotechniques in the examination of food

Jane F. Heathcock

Structural Studies Unit, Reading Scientific Services Ltd. Lord Zuckerman Research Centre, The University, Whiteknights, Reading, Berkshire, RG6 2LA

ABSTRACT:Cryo-SEM and freeze-etching techniques for TEM have been used to examine and to compare the structures of a range of different foods. Adjustments to the temperature of the sample prior to freezing allowed the examination of different surfaces created from fracturing through different planes of weakness. Complementary data on the distribution of elements across the surface was obtained by X-ray microanalysis within the SEM.

1. INTRODUCTION

Manufactured food products frequently contain either fat, sugar or water. In addition, these components can be associated with each other and with other ingredients to form relatively complex structures. As a consequence, food products can be difficult to handle and to prepare for electron microscopy. Low temperature techniques, such as Cryo-SEM and freeze-etching, however, provide a means of examining food which involves only a minimal of pretreatment or change to the specimen. Further information can be obtained by use of the X-ray microanalytical facility of the SEM to generate elemental distribution maps of the low temperature preparations.

2. METHODS

Samples examined included material taken from stages through food manufacturing processes and the finished products including confectionery, emulsions and bakery products. Small samples taken at room temperature (20°C) were supported onto brass rivets (for SEM) or freeze-etching specimen holders and frozen in liquid nitrogen slush (-210°C). Alternatively, the temperature of the samples was adjusted in 5°C steps up to 40°C or down to 5°C prior to freezing. For SEM, the samples were fractured and coated in a Hexland Cryo Chamber before examination on the cold stage within the JEOL JSM35CF SEM or freeze-etched within the Balzers 360 unit prior to examination using the JEOL 1200EX TEM.

3. RESULTS

3.1. Fat in Food

Fat is often shown to constitute the continuous phase in food products with a proportion of the fat being crystalline at least at room temperature. Cryo-SEM shows the structure of many of these crystals present both on surfaces and within products. Greater detail is revealed by TEM in terms of the size, shape and packing characteristics of the individual crystals. The development of fat crystals, for example, and the changes in fat crystallinity as a result of processing or storage can be observed by the use of freeze-etching preparations. The first signs of order in fat systems are frequently of lamellar structures, typical of liquid crystals (Figure 1). More defined crystal shapes then develop with size being dependent on composition and on the conditions of crystallisation. Freezing from room temperature in these crystalline systems reveals only the fat and little of the dispersed phase. Once the crystals have melted out, fractures are created through fat/particle interfaces and through the particles themselves and thus the fine structure of this second phase is observed.

Fat in many other products is the dispersed phase creating different forms of fat-in-water emulsions. Cryo-techniques have made it possible to observe the structure of these emulsions and how the structure develops in the production of foods such as creams, desserts, confectionery and ice cream. Products can be compared based on the size and shape of fat droplets, how droplets associate together and how they associate with other components. X-ray microanalysis in the SEM then enables the particles to be characterised further with, for example, calcium and phosphorous being associated with protein-rich regions.

3.2 Sugar in Food.

Sugar in food can be either crystalline of non-crystalline and there are often examples of foods where both forms are present within the same product. As with fats, sugars can also constitute both continuous and dispersed phases. Cryo-techniques characterise the structure of many of these products providing information, for example, on size and shape of individual crystals within a matrix, the nature of the matrix and the interrelationships of the remaining components. The presence of high levels of amorphous sugar often masks the components present within it. Fondant confectionery, for example, is essentially sugar crystals dispersed in a sugar syrup. Its surface structure shows a thin coating over many regular shaped sugar crystals. This is illustrated in Figure 2 together with an adjacent region of the sample which has been Cryo-fractured. This reveals another aspect of the structure with the dominant feature being high numbers of air bubbles within the amorphous matrix. Adjustments in temperature of these type of products prior to freezing again changes the way in which the material fractures and reveals more of the dispersed phase.

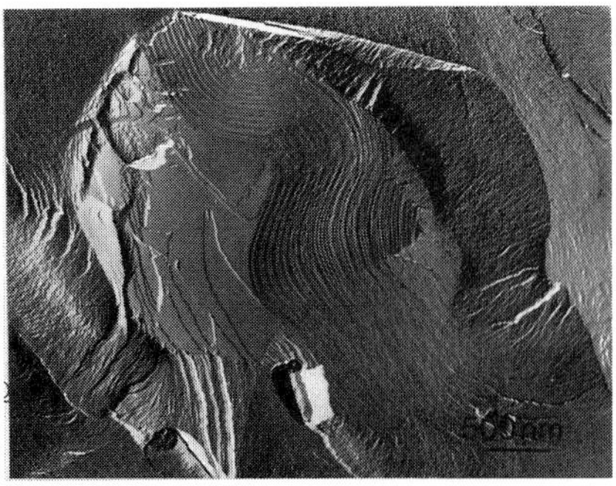

Fig. 1. Freeze-etching preparation of liquid crystalline fat.

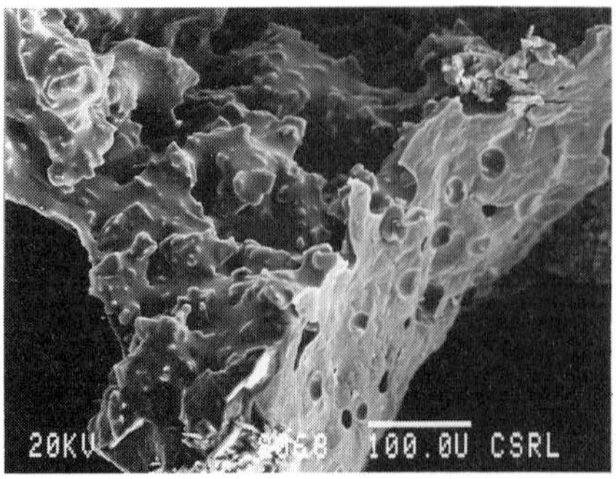

Fig. 2. Cryo-SEM preparation of a sugar fondant.

Inst. Phys. Conf. Ser. No. 93: Volume 3, Chapter 2
Paper presented at EUREM 88, York, England, 1988

Ultrastructural observations in *Papaver dubium* L pollen

A C van Aelst
Dept.of Plant Cytology and Morphology, Agricultural University, Arboretumlaan 4, 6703 BD Wageningen, The Netherlands.

ABSTRACT

The ultrastructure of Papaver dubium L. pollen has been investigated with transmission electron microscopy (chemical fixed and embedded) and low-temperature scanning electron microscopy (freeze fixed and frozen-hydrated at -130° C). In the frozen hydrated state the cytoplasm is well fixed without crystal damage and a three dimensional view of the organelles and membrane structures is visible.

1. INTRODUCTION

Morphological support is needed in order to manipulate gametes for plant breeding and biotechnological purposes. At maturity pollen grains of Papaver dubium are bicellular (Van Aelst et al. 1988), contain 17% sucrose and are relatively dry. In order to collect fast and good preserved electron microscopical images low temperature scanning electron microscopy (LTSEM) is used. In this paper chemical fixation and transmission electron microscopy (TEM) is compared with freeze fixation, freeze fracturing and LTSEM of Papaver dubium L. pollen.

2. MATERIAL AND METHODS

Dried and freeze-stored Papaver pollen was rehydrated in saturated moisture at 20^{0}C for 30 min. For TEM pollen were fixed in: 5% glutaraldehyde and 0.44 M sucrose in 0.1 M phosphate buffer followed by 1%OsO_4 and 0.44 M sucrose in the same buffer, dehydrated in ethanol and embedded in Epon 812. For LTSEM specimens were plunged into liquid propane, mounted on the cryo-stage of a Balzers SCU 020, fractured, sputtered and frozen-hydrated observed at -130^0 C.

3. RESULTS

In the cytoplasm of the vegetative cell different organelles are visible (Figs. 1,2). The plastids (p) do not contain starch, a circular thylakoid membrane is present in the plastid matrix. Mitochondria (m) with cristae are clearly visible (Fig.1). Lipid bodies (l) are frequently present in cross section (Fig.1). In LTSEM cross fractured and unfractured mitochondria (m) and lipids (l) are visible (Fig.2). The lipid bodies are surrounded by stacked rough endoplasmatic reticulum

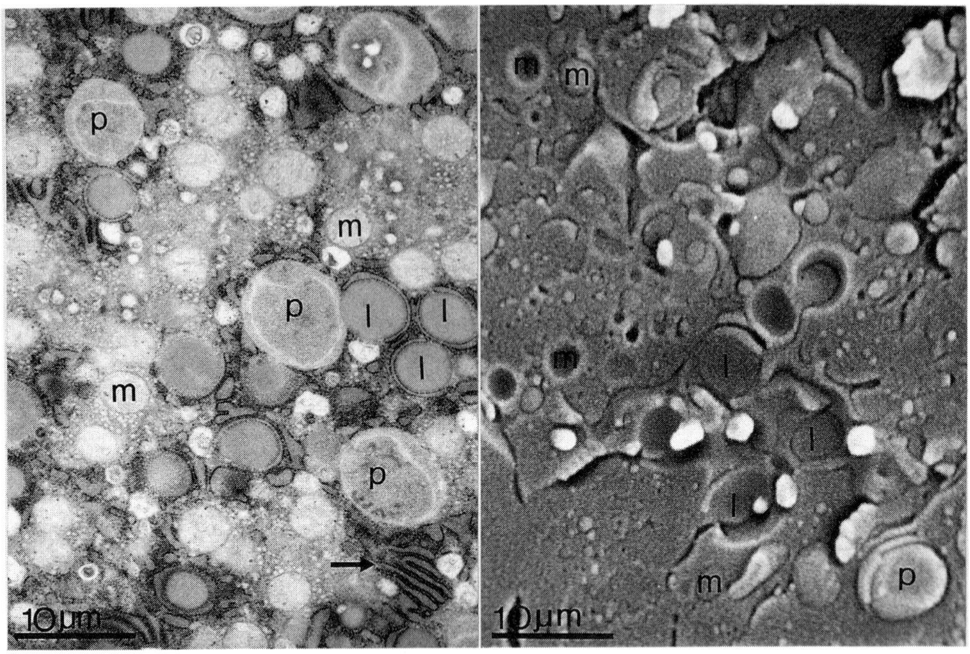

Figures 1,2 Detail of vegetative cell of Papaver pollen grain, Fig.1 TEM micrograph, Fig.2 LTSEM micrograph.

(RER) (→) with large amounts of ribosomes (Fig.1). In Fig.2 is shown that this RER is a large continuous lamellar network between the lipid bodies. Between conglomerates of the lipid bodies and RER association, areas with small vesicles are visible, related to there size, two types are present (Figs.1,2). In fig.2 the dark centre of the smallest structures in this area indicates that the little vesicles are tubes.

4. CONCLUSIONS

Frozen-hydrated freeze-fractured SEM images of Papaver dubium pollen produce important contribution in ultrastructural research. Low-temperature SEM enables us to produce images very quickly. Fast occurring processes as harmomegathy (Pacini 1987) in pollen are recordable with LTSEM. High resolution SEM with low-temperature stage is necessary to visualize small structures like ribosomes and microtubules.

Acknowledgements
The author is much indebted to Dr. T Mueller, M Dueggelin and Prof. R Guggenheim for preparing the images with the LTSEM.

5. REFERENCE

PACINI E (1987) In: Biology of Reproduction and Cell Motility in Plant and Animals, M CRESTI, R DALLAI (eds). University of Siena, Italy.

VAN AELST AC, T MUELLER, M DUEGGELIN and R. GUGGENHEIM. (1988) (submitted to Acta Botanica Neerlandica)

Inst. Phys. Conf. Ser. No. 93: Volume 3, Chapter 2
Paper presented at EUREM 88, York, England, 1988

Low temperature scanning electron microscopy of actinomycetes

J K Warrack and R M Banks

Beecham Pharmaceuticals Research Division, Chemotherapeutic Research Centre, Brockham Park, Betchworth, Surrey RH3 7AJ

ABSTRACT: Actinomycetes are routinely screened by the pharmaceutical industry in the search for novel antibiotics. Scanning electron microscopy is widely used for examining taxonomically important features of these organisms. Low temperature methods allow rapid preparation of specimens and eliminate artifacts introduced by alternative methods. Features of colonies such as liquid droplets are preserved which would otherwise be lost.

1. INTRODUCTION

The sporophore and spore morphology of the bacteria which comprise the order Actinomycetales are of major taxonomic importance. The potential of scanning electron microscopy (SEM) in studying these structures was realised at an early stage (Williams and Davies 1967) and extended the work already done by transmission electron microscopy of whole cells (Tresner et al 1961) and replicas (Dietz and Mathews 1962).

The early SEM work was done on air dried unfixed or osmium vapour fixed material, and these procedures are still advocated (Dietz 1986). The artifacts produced by air drying can cause misinterpretation of specimen morphology. Such artifacts are evident in much published work, and a new approach to specimen preparation is overdue.

In recent years commercial low temperature preparation units and cold stages have become available. These have been used to examine many types of biological specimens (Bastacky et al 1987).

This paper presents results obtained by the use of a Hexland CT1000 cryotransfer system and Philips SEM 501 for the routine investigation of actinomycete morphology. These results are compared with those obtained using alternative specimen preparation methods.

2. METHODS

Actinomycetes were grown on solid media. Samples for low temperature SEM (LTSEM) were prepared by removing a $5mm^2$ block of agar supporting suitable growth and trimming agar from below the colony to leave a specimen approximately 2mm thick. This was mounted onto a specimen stub/holder assembly with carbon cement and immediately frozen in subcooled liquid nitrogen. The specimen was transferred under vacuum to the cold stage of the SEM, held at -75°C, and the sublimation of surface frost observed. When sublimation was almost complete, the specimen was withdrawn to the prechamber and sputter coated with gold, for three minutes at 1.5mA. The microscope stage was cooled to <-170°C and the specimen returned for examination at 15 or 30kV.

For comparison, critical point dried (CPD) preparations were made according to King and Brown (1983) and air dried (AD) material was prepared according to Locci (1976).

3. RESULTS

Streptomyces hygrosopicus var aureolacrimosus

Fig.1. AD. Specimen collapsed

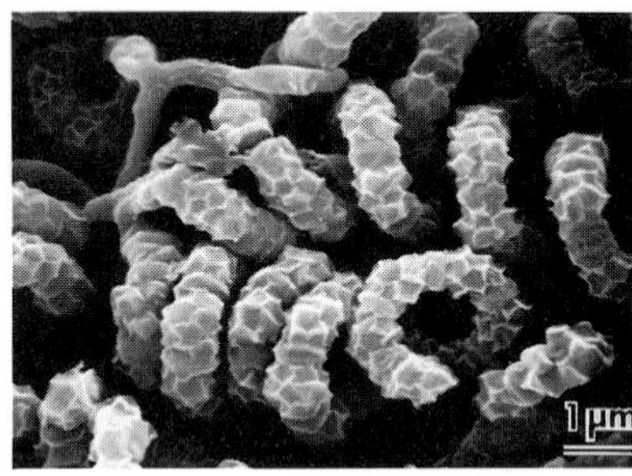

Fig.2. CPD. Specimen distorted

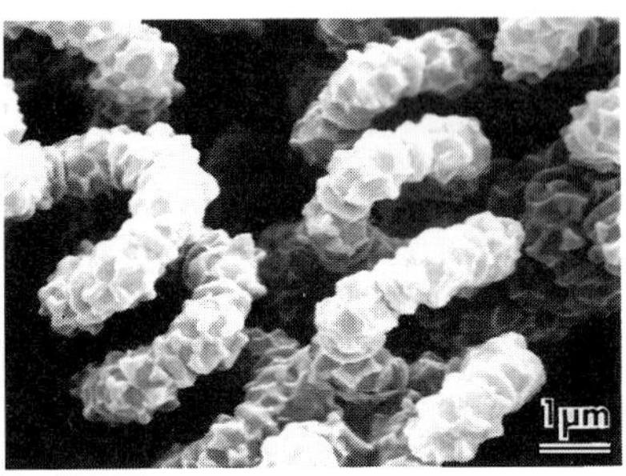

Fig.3. LTSEM. Specimen turgid

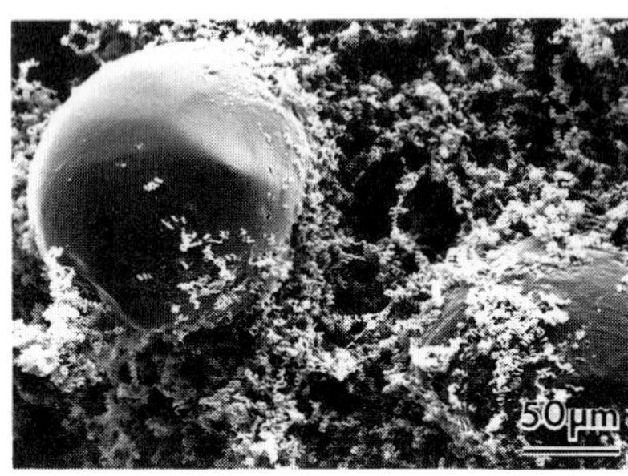

Fig.4. LTSEM. Droplets on colony

4. CONCLUSIONS

Low temperature SEM is an ideal technique for the rapid examination of actinomycetes, eliminating the artifacts induced by air drying and the lengthy preparation procedure required for critical point drying. In addition, volatile components of colonies, such as liquid droplets, can be preserved for SEM only by low temperature methods.

5. REFERENCES

Bastacky J et al 1987 *Scanning* **9** 219

Dietz A 1986 *The Bacteria: A Treatise on Structure and Function Vol. IX* ed. S W Queener and L E Dat (Orlando:Academic Press) pp 1-25

Dietz A and Mathews J 1962 *Appl. Microbiol.* **10** 258

King E J and Brown M F 1983 *Can. J. Microbiol.* **29** 653

Locci R 1976 *Actinomycetes - The Boundary Microorganisms* ed. T Arai (Tokyo:Toppan) pp 249-297

Tresner H D et al 1961 *J. Bacteriol.* **81** 70

Williams S T and Davies F L 1967 *J. Gen. Microbiol.* **48** 171

Inst. Phys. Conf. Ser. No. 93: Volume 3, Chapter 2
Paper presented at EUREM 88, York, England, 1988

Equipment and conditions for freeze substitution with osmium tetroxide fixation of cold treated leaf tissues

DMR Harvey, K Pihakaski

Department of Biology, University of Turku, SF-20500 Turku, Finland

ABSTRACT: The ultrastructure of cold hardened Secale cereale cells from leaves maintained at 278 K or slowly frozen to 257 K was examined by TEM. Preparation without desiccation or thawing was achieved by rapid freezing to 80 K using an electro-magnetically controlled plunging device and freeze-substitution in acetone at 193 K with fixation by osmium tetroxide at 253 K. Slow freezing to 257 K often resulted in severe cell shrinkage and collapse of the cytoplasm; mitochondrial structure was disrupted, but chloroplasts were much less affected.

1. INTRODUCTION

Research into the structural basis of cold hardening and freezing injury in plants requires examination of slowly frozen tissues prepared for TEM without thawing, which might cause additional ultrastructural changes. It is also necessary to produce TEM sections in which the major organelles and membranes are clearly distinguishable. We therefore investigated the use of freeze-substitution in acetone combined with osmium tetroxide fixation.

2. MATERIALS AND METHODS

Plants of Secale cereale c.v. Voima were cold hardened by growth at 278 K for 4 wk. 6 cm long leaf pieces were ice nucleated at 270 K and slowly frozen to 257 K for 1h. These pieces or leaves at 278 K were cut transversely into $\leq$ 1 mm long segments, loaded into Cu mesh baskets and rapidly frozen (Fig. 1). Freeze-substitution and osmium fixation were carried out:

1. 3.3% (w/v) OsO_4 in anhydrous acetone, 193 K, 2 d + 253 K, 5 d*
2. anhydrous acetone rinse, 2 changes, 253 K, 8 h + 16 h
3. infiltration with ERL 4206, Spurr's (1969) resin similar to Harvey et al (1976) except start at 253 K.

* Experiments were carried out in which the temperature was 223 K or 193 K for 5 d.

3. RESULTS AND DISCUSSION

Osmium fixation did not occur to significant extents in 5 d at 223 K or 193 K. The given schedule allows removal of ice by acetone before the temperature is raised for osmium fixation and therefore minimizes the expansion of small ice crystals formed during rapid freezing. Only those cells lacking visible ice crystal damage arising from fast freezing were

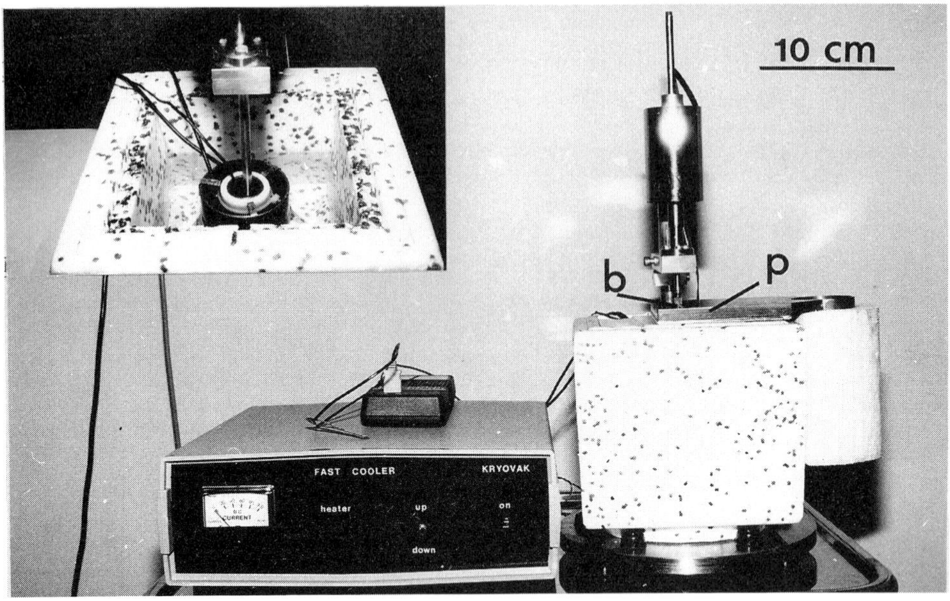

(Fig. 1) Rapid freezer (Kryovak, Finland): taking basket (b) from Cu cold plate (p) at 257 K. Inset: after plunging at 3 ms^{-1} into methylcyclohexane/2-methylbutane mixture in two Cu cylinders cooled by liq. N_2. Shown are attachments to thermocouple and heater of inner cylinder.

examined (Fig. 2a). Cells which were slowly frozen to 257 K were often distorted and collapsed, although the plasmalemma and protoplast may remain in contact with the cell wall. The cytoplasm and mitochondria are disorganized, although there is much less damage to the chloroplasts (Fig. 2b).

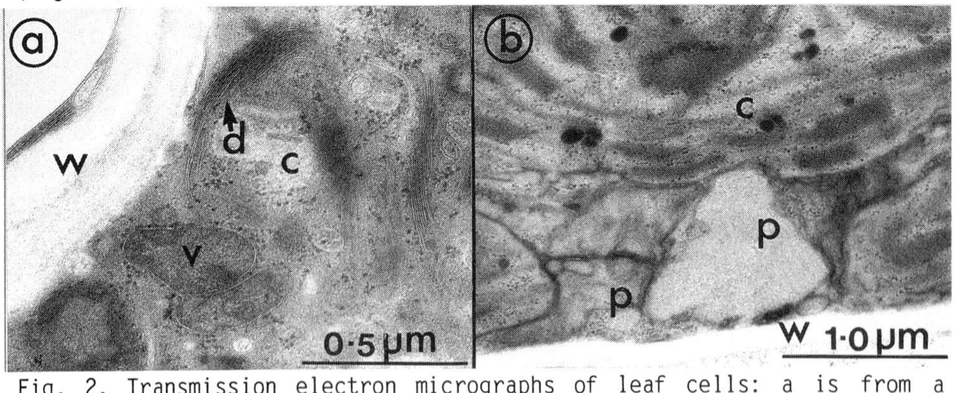

Fig. 2. Transmission electron micrographs of leaf cells: a is from a plant kept at 278 K, b from leaf pieces frozen to 257 K. c=chloroplast, p=cytoplasm, w=cell wall, v=vesicle, d=dictyosome.

4. REFERENCES

Harvey D M R, Hall J L and Flowers T J 1976 J. Microsc. 107 189
Spurr AR 1969 J. Ultrastr. Res. 26 31

Inst. Phys. Conf. Ser. No. 93: Volume 3, Chapter 2
Paper presented at EUREM 88, York, England, 1988

Advantages of low temperature scanning electron microscopy for investigating experimental vaginal candidosis

M J Wilkinson[1], J K Warrack[1], S A Smith[1], A H Everett[2] and P A Hunter[2]

Beecham Pharmaceuticals Research Division, Chemotherapeutic Research Centre[1], Brockham Park, Betchworth, Surrey RH3 7AJ and Biosciences Research Centre[2], Yew Tree Bottom Road, Epsom, Surrey K18 5XQ, UK.

ABSTRACT: By preserving vaginal epithelium in its hydrated state cryoSEM avoided the pitfalls of conventional SEM and histology and permitted a more reliable and comprehensive appreciation of the morphology of the stratum corneum and the distribution of fungal hyphae in a model of chronic vaginal candidosis.

1. INTRODUCTION

We aimed to assess in histological and ultrastructural terms a commonly used model of chronic vaginal candisosis (Sobel et al 1984) to elucidate fungal distribution, morphology and host response. Conventional SEM and wax histology were somewhat inadequate since the organism resides in a loosely associated surface layer (stratum corneum) which proved vulnerable to processing. CryoSEM was applied in an attempt to avoid this problem.

MATERIALS AND METHODS

Ovariectomised, pseudo-oestrus rats with chronic vaginitis following inoculation with C.albicans yeast (B2630, ATCC28366 supplied by Dr.J. Ryley, ICI) were killed by pentabarbital overdose. Approximately 10mm^2 of vaginal epithelium was mounted on a stub, quenched in sub-cooled liquid nitrogen and examined in a Philips 501 SEM equipped with a Hexland cryochamber and cold stage operated at -170°C. Conventional SEM was carried out on glutaraldehyde fixed specimens which had been dehydrated in ethanol and critical point dried from liquid CO_2. Specimens were sputter coated with gold. Wax sections of formalin fixed tissue were obtained by conventional means and stained with Periodic acid - Schiff's reagent.

3 RESULTS

Vaginal epithelium examined by conventional SEM exhibited virtually no mucus (Fig.1). The distribution of squamous epithelial cells was patchy, suggesting that some of this loosely associated material had been lost during processing. Areas depleted of squamous cells showed a smoother layer of non-squamous epithelium interpreted as the surface of the stratum malpighii. Candida hyphae were only associated with the patches of squamous epithelium. A full and reliable assessment of the distribution and morphology of the organism was therefore difficult using conventional SEM. In contrast, the stratum corneum of quenched specimens appeared to remain intact. Only slight etching at -80°C to remove surface ice was needed to reveal numerous fungal hyphae associated with a mixture of mucus and squamous cells (Fig.2). Mechanically induced fractures through the stratum corneum showed that this was a surprisingly deep layer (up to 150µm) which fungal hyphae had permeated fairly uniformly. Penetration of squamous keratinised cells by hyphae was evident. Fungal cells were generally not apparent in the underlying nucleated stratum malpighii.

Wax sections of the defrosted cryoSEM specimen showed a stratum corneum of considerably reduced thickness (typically 25µm). Discontinuity of the metallic surface coating, which appeared as a black line in the sections, indicated that a considerable proportion of surface material was displaced or lost completely during histological processing. The shortcomings of conventional LM and SEM reported here have not been fully appreciated in previous studies of rodent vaginal epithelium (Khosho et al 1987).

In summary, cryoSEM of vaginal epithelium avoids artefacts associated with conventional methods. Instead of the stratum corneum in this infection model being represented as a thin and sometimes sparse layer of cornified cells it should be appreciated as a relatively thick coating that constitutes a quite complex environment throughout which the fungal pathogen is able to thrive.

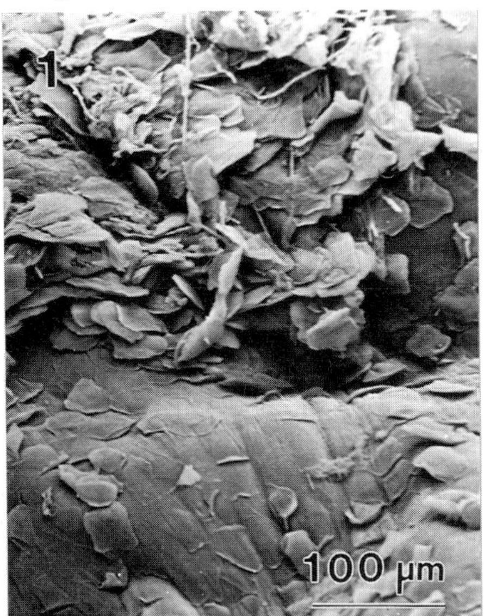

Fig.1 Vaginal epithelium (conventional SEM) showing depleted stratum corneum and smoother underlying stratum malpighii.

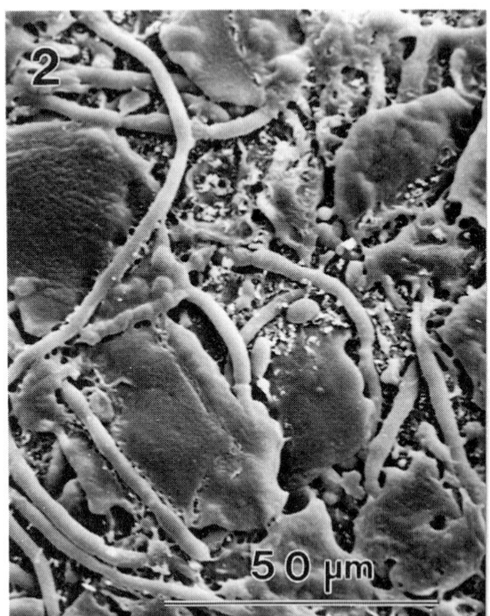

Fig.2 Vaginal epithelium (cryoSEM) showing fungal hyphae, mucus and squamous epithelial cells.

REFERENCES

Kosho F K, Kaufman R C and Amankwah K S 1987 *J.Elect.Micros.Techniques* **5** 117

Sobel J D, Muller G and Buckley H R 1984 *Infection and Immunity* **44** 576

Ultrastructural and cytochemical observations on intranuclear inclusions in *Olea europaea*

J.D. Alché and M.I. Rodríguez-García

Estación Experimental del Zaidín CSIC. Prof Albareda, 1, 18008 Granada, SPAIN

ABSTRACT: The occurrence of intranuclear inclusions in *Olea europaea* has been observed, their paracrystalline structure being appreciated by electron microscopy, their protein nature has been established by protein specific stains under light microscope and preferential stain for ribonucleoproteins under electron microscope.

1. INTRODUCTION

The presence of several types of intranuclear inclusions in plant cells and their proteinaceus composition has been demonstrated by different authors (Fabbri & Menicanti 1970; Perrin 1970; Weintraub et al. 1971; Unzelman & Healey 1972; Vintejoux 1984). In light of their constant presence in some families of Angiosperms, this feature has been proposed as a useful characteristic for taxonomic purposes (Speta 1979, Bigazzi 1984). However, no information has appeared regarding the functional significance of these inclusions in plant physiology. This underlines the possible importance of the presence of such intranuclear inclusions in other species, along with information regarding their structure, chemical composition and characteristics of the tissue in which they appear.

2. RESULTS AND DISCUSSION

The nucleus of mesophyll cells from *Olea europaea* leaves presented crytalloids similar in size or larger than the nucleoli (Fig. 1). These highly electron dense bodies were more or less regularly polyhedral in shape. Compound forms were also occasionally seen which could not immediately be identified as polyhedral-shaped. At high magnification a clearly periodic substructure was visible (period 150-160Å) (Fig. 2). There is evidence of an electron transparent region of uncertain composition surrounding the globoid crystal (soft globoid). Within the electron dense material small rounded electron light areas are also seen (Fig. 3). The inclusions have been shown with protein specific stains to represent crystalloid proteins (Figs. 5 and 6), and also stain positively, although less intensely than nucleolus, with the EDTA-technique, a preferential stain for ribonucleoproteins (Fig. 4).

The notion that these intranuclear inclusions represent stored proteins as a consequence of the high rate of protein synthesis is attractive, although the nucleus has not been considered a reserve substance storing organelle. Current knowledge of protein synthesis favors the idea that the crystalloid body proteins are synthesized in the cytoplasm, to migrate subsequently toward the nucleus (Thomas et al. 1977). In contrast to other species (Morassi & Bigazzi 1980, Bigazzi 1984), no direct relationship has yet been found between crystalloid body proteins and the nucleolus, although they have been observed in close proximity owing to the large area occupied by the inclusions within the nucleus. An investigation of the possible variations in these crystalloid bodies in relation with changing metabolic states in the plant is urgently needed to clarify their physiological role.

3. REFERENCES

Bigazzi M 1984, Caryologia **37**: 269-292
Fabbri F & Menicanti F 1970, Caryologia **23**: 729-761
Morassi B & Bigazzi M 1980, Caryologia **33**: 321-337
Perrin A 1970, Protoplasma **45**: 131-134

Speta F 1979, Pl. Syst. Evol. **132**: 1-26
Thomas D, **Gouranton** I & **Wroblewski** H 1977, Biol. Cell **28**: 195-206
Unzelman J M & **Healey** P L 1972, J. Ultr. Res. **39**: 301-309
Vintejoux C 1984, Ann. Sci. Nat. Bot., Paris **6**: 203-205
Weintraub M, **Ragetli** H W & **Schroder** B 1971, Am. J. Bot. **58**: 182-190

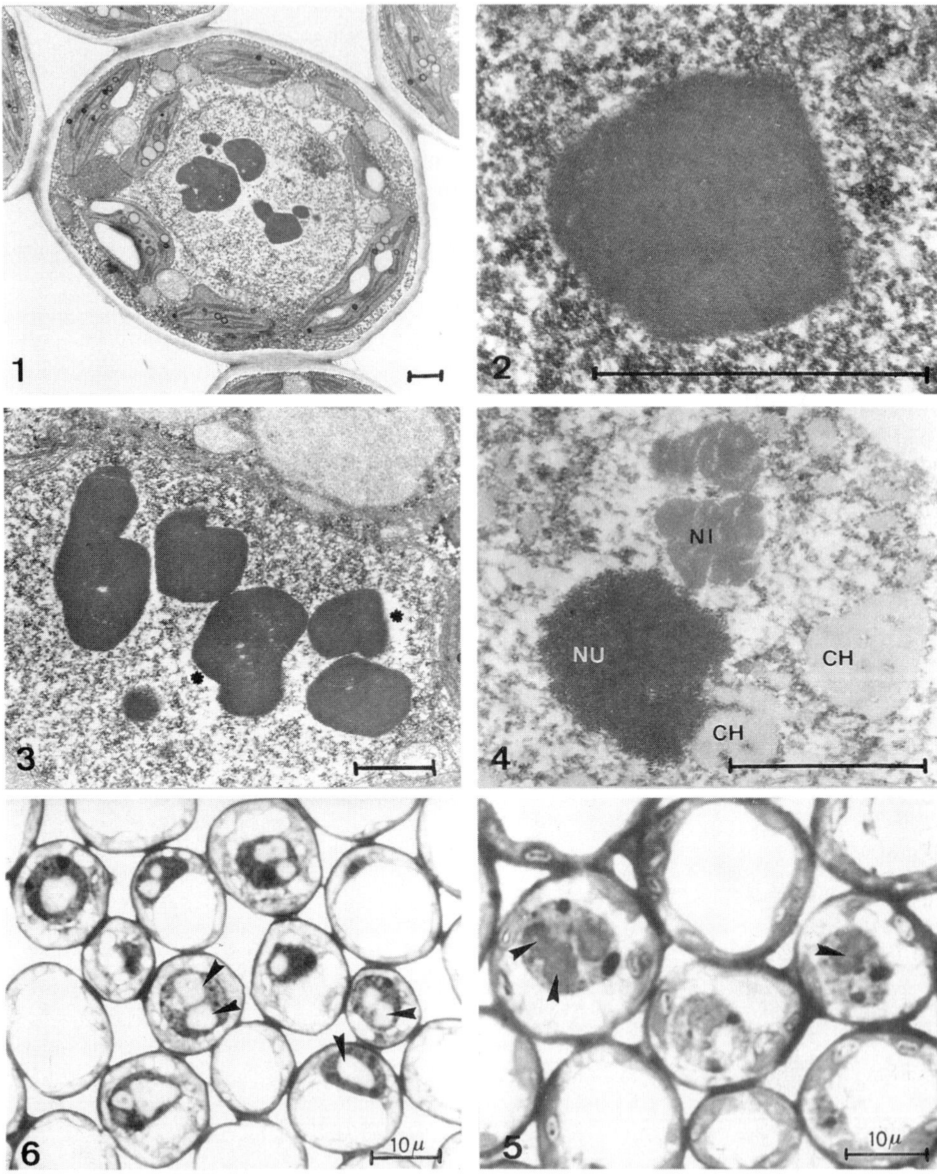

Fig.1 *Olea europaea* mesophyll cell showing crystalline inclusions
Fig.2 High power view of a crystalloid inclusion showing periodic lattice
Fig.3 Group of intranuclear inclusions. Soft globoid area (✱)
Fig.4 EDTA technique. Note differential staining of the nucleolus (NU), Chromosomes (CH) and intranuclear inclusions (NI)
Fig.5 and 6 Light micrographs of cells containing crystalline intranuclear inclusions. Protein specific stains reveal different patterns of positivity. Fig.5: Potassium ferrocyanide/Cl_3Fe; Fig.6: N-PAS stain.

Nuclear inclusions in *Lavandula latifolia* medicus: ultrastructural and cytochemical study and relationships with nucleoli

J.F. MESQUITA and J.D. SANTOS DIAS

Depart. of Botany (Lab. of E.M. and Phycol.), Center for Plant Physiol. and Cytol. (INIC), Univ. of Coimbra, Coimbra, Portugal

ABSTRACT: In parenchymatic cells of different organs of Lavandula latifolia proteinaceous nuclear inclusions are described. Morphometric studies point out to significant and inverse changes in the relative volumes of these inclusions and nucleoli, during cell differentiation.

1. INTRODUCTION

Nuclear inclusions have been frequently observed in plant cells (for Bibliography see Wergin et al. 1970; Thomas and Gouranton, 1979; Esau and Magyarosy, 1979). This study in L. latifolia includes, apart ultrastructural and cytochemical aspects, a stereologic analysis in order to correlate eventual changes in the size of both nucleoli and similar inclusions at different stages of leaf development.

2. MATERIAL AND METHODS

Samples of several organs of L. latifolia were prepared for E.M. according to the current technique. For cytochemical purposes, staining of semithin sections (O.M.) and digestion tests with proteolytic enzymes (E.M.) were carried out. For morphometric studies, volume fractions (Vv) of nucleoli and inclusions (relative to cytoplasm and/or nucleus) were calculated by the formulae of Weibel (1973). Statistical analysis was accomplished by the Student t test and dimensions of structural components were estimated from densitometric traces.

TABLE I

LEAVES	NUCLEOLUS	NUCLEAR INCLUSION
Young leaves (5-6 mm long)	7.74 ± 1.14	3.46 ± 0.68
Mature leaves (40-50 mm long)	2.58 ± 0.78	30.60 ± 2.21
P values	$<$ 0.01	$<$ 0.001

Volume fractions relatively to nucleus (%) of nucleoli and nuclear inclusions at two stages of leaf development in Lavandula latifolia.

3. RESULTS AND DISCUSSION

All organs and tissues of L. latifolia we have studied (root tips, shoot apex, stem, leaves, cotyledons, bracts and glandular trichomes) showed to contain nuclear inclusions, in parenchymatic cells, except roots and glandular tissues (figs. 1, 2). It is clear that the most complicated structures (paracrystalline or crystalline) are made up through packing of elementar long filaments, composed by globular subunits. Then, a square mesh or a transversal striated structure, the striae of which have a mean value of 13nm in width, can be seen (fig. 3). When globular subunits are not evident, the inclusion seems to have a fibrous structure the striations of which are about 17nm in thickness (fig. 4). The proteinaceous nature of the inclusions was demonstrated through their staining with mercuric bromophenol blue and enzymatic degradation by a solution of pronase (0,5%, pH 7.5; 3-12 hours) (fig. 5). In L. latifolia the nuclear inclusions appear frequently in association with nucleoli (figs.1,2). Morphometric and statistical studies which have been carried out in two stages of leaf development showed that during this differentiation there is a significant decrease of the relative volume of nucleoli which is concomitant with an accentuated increase of the volume of nuclear inclusions (Table I). Physical association between nuclear inclusions and nucleoli has been referred in plant cells as either an occasional occurence (Unzelman and Healey, 1972) or a metabolic relationship involving (Wergin et al., 1970; Ciampolini et al. 1980) or not (Bonzi and Bigazzi, 1980) interconvertibility between those structures. The results we present now are favourable to the existence of an inverse relationship between the sizes of both nucleoli and nuclear inclusions.

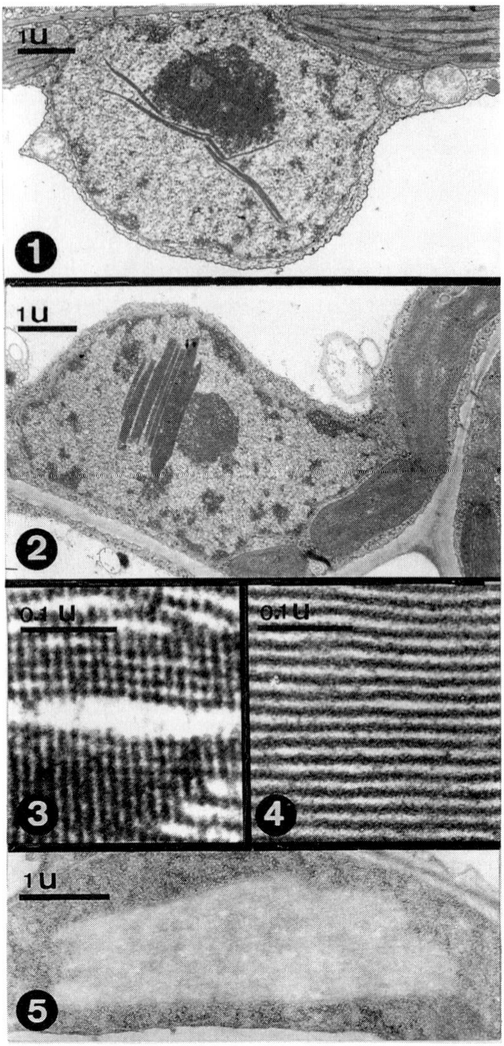

NUCLEAR INCLUSIONS (arrows): close association with the nucleolus (figs. 1,2), structural details (figs. 3,4) and partial pronase digestion (fig.5). (See the text).

BIBLIOGRAPHY

Wergin W P, Gruber P J and Newcomb E H 1970 J. Ult. Res. **30** 533
Thomas D and Gouvanton J 1979 Planta **145** 89
Esau K and Magyarosy A C 1979 J. Cell Sci. **38** 1
Weibel E R 1973 In Hayat, M A Ed. Princ. and Tech. of E.M. (Reinhold) **3** pp 237-296
Unzelman J M and Healey P L 1972 J. Ult. Res. **33** 301
Ciampolini F, Cresti M, Dominicis V D, Garavito R M and Sarfatti G J. Ult. Res. **71** 14

Inst. Phys. Conf. Ser. No. 93: Volume 3, Chapter 3
Paper presented at EUREM 88, York, England, 1988

Immunoelectron microscopy localization of the nucleolar protein B-36 in plant nucleoli

MC Risueño, C López-Iglesias[1], PS Testillano, A Olmedilla and ME Christensen[2]
Centro de Investigaciones Biológicas. CSIC. Velázquez 144. 28006 Madrid. SPAIN
1. Institut des Recherches Scientifiques sur le Cancer. CNRS. Villejuif. FRANCE
2. Dep. of Biology. Texas A. & M. University, College Station. TX 77843. USA

ABSTRACT: Immunoelectron microscopy localization of B 36 nucleolar protein is made in plant cells. Results show that this protein is exclusively located in the dense fibrillar component, where ribosomal transcription takes place.

INTRODUCTION

The nucleolus has a well known role in the eukariotic cell: the production of pre ribosomal particles. In this process rDNA, rRNA and proteins are involved. A rapid association of newly synthesized rRNA with ribosomal proteins takes place to form pre ribosomes. Non ribosomal proteins are also necessary in the processing of pre-ribosomal particles.

In situ localization of several nucleolar proteins has been performed by cytochemical, i.e. Ag-NOR, and bismuth tartrate techniques, and immunocytochemical procedures. The 34K nucleolar protein, B-36, purified from the Physarum polycephalum, has been characterized by Christensen et al. (1986) using monoclonal antibodies. Some of these antibodies have been used in immunocytochemical assays at the electron microscopy, localizing this protein exclusively in the dense fibrillar component (DFC) of Physarum nucleoli (Pierron, personal communication) and of Herpes simplex infected nucleoli (López-Iglesias, personal communication).

In the present study we determined by immunoelectron microscopy technique the distribution of B-36 protein in plant nucleoli.

MATERIALS AND METHODS

Onion proliferating roots were fixed in 4% formaldehyde in phosphate buffer for 1 hour at 4°C and washed with the same buffer. Ultrathin sections of K4M Lowicryl embedded cells were placed on copper grids coated with pyoloform and carbon, and preincubated for 30 min with BSA. Then, grids were incubated in B 36 monoclonal antibody, P2C3 (diluted 1:10) for 1 hour, and after PBS washing, incubated in goat anti-mouse IgG coupled to colloidal gold (GAMIgG G10 Janssen) diluted 1:10 for 30 min. All the incubations were made at room temperature. Sections were stained with uranyl acetate. Controls were performed by incubating sections with PBS in place of the first antibody. Nucleolar ultrastructure was compared in Lowicryl sections stained with uranyl acetate and Pb citrate.

RESULTS AND DISCUSSION

Proliferating roots show different nucleolar morphologies located at different distances from the apex. Depending on nucleolar activity different distribution of the nucleolar components can be found.

Inactive nucleoli (Figs. 1 and 2) are mainly composed by DFC with large heterogeneous fibrillar centres (Fcs) containing great condensed chromatin inclusions. Low-active nucleoli (Fig. 3 and 4) show a segregated structure with the granular component (GC) surrounding the DFC in which Fcs of intermediate type (smaller and more numerous) are located (Risueño and Medina, 1986).

After immunogold labeling gold particles are distributed over the DFC in both types of nucleoli. No-labeling was obtained over Fcs (Figs. 2 and 4) as well as on the GC (Fig. 4).

The nucleolar protein B 36 is highly conserved in eukaryotes: Physarum, rat liver, and in plants (Christensen. personal communication). Several antibodies against nucleolar proteins have detected proteins placed in the DFC such as anti C 23 (Spector et al, 1984), and anti 100K (Escande et al. 1985). but they also show immunoreaction over other nucleolar components. Fcs and/or GC. Another 34K protein similar to the B 36 is the fibrillarin, which has been located in the DFC but also in the Fcs (Ochs et al. 1985: Reimer et al. 1987).

In this way. B-36 antibody can be considered as an important nuclear marker which detects a particular 34K protein exclusively located in the DFC of the nucleolus where the rDNA transcription takes place. The presence of this protein has been related to intact RNA by Christensen et al. (1986). Therefore B 36 could play a role in the transcription and/or processing of the rRNA into the nucleolus.

REFERENCES

Christensen ME, Moloo J, Swischuk JL and Schelling ME 1986 Exp. Cell Res. 166, 77.
Escande ML. Gas N and Stevens BJ 1985 Biol. Cell 53.99.
Ochs RL, Lischwe MA and Spohn WH 1985 Biol. Cell 54,123.
Reimer G. Raska I. Tang EM. Scheer U 1987 Virchows Archives B. Cell Pathol. 54. 131.
Risueño MC and Medina FJ 1986 Nucleolar Structure in Plant Cells (RBC. Springer Verlag) 7, 1 163.
Spector DL, Ochs RL and Busch H 1984 Chromosoma 90,139.

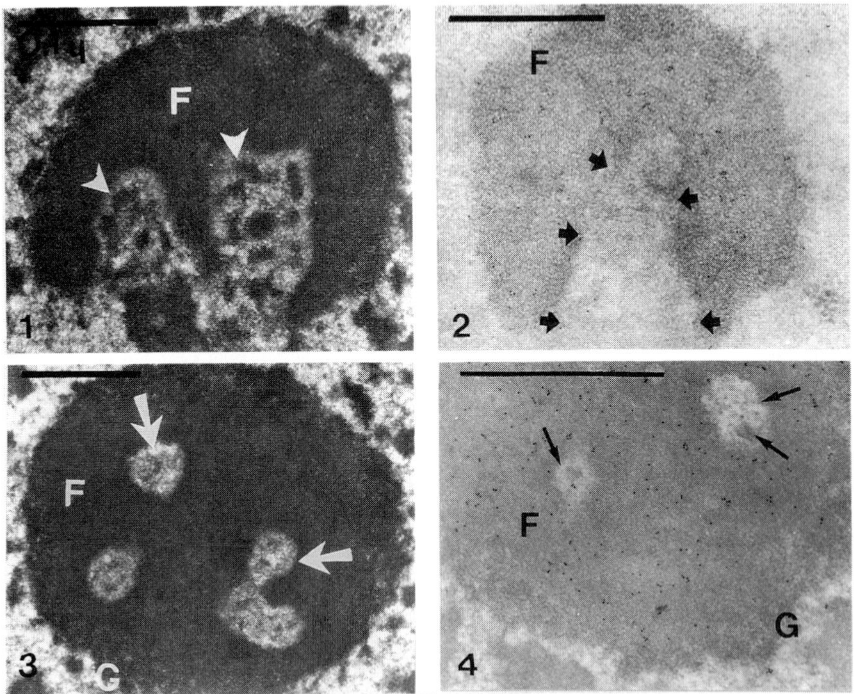

Figs 1 to 4: K4M Lowicryl embedded sections. Figs. 1 and 2: Inactive nucleoli exclusively formed by dense fibrillar component (F) with heterogeneous fibrillar centres (arrowheads). Fig. 1: Uranyl-lead staining. Fig. 2: Immunogold label is placed over the DFC. Figs. 3 and 4: Low active nucleoli. Granular component (G) surrounding the dense fibrillar component (F) with intermediate Fcs (arrows). Fig. 3: Uranyl lead staining. Fig. 4: Immunogold label placed over the DFC, Fcs are free of gold grains.

Inst. Phys. Conf. Ser. No. 93: Volume 3, Chapter 3
Paper presented at EUREM 88, York, England, 1988

Ultrastructural and biochemical changes in the lipid content of Scots pine (*Pinus sylvestris* L) needles in relation to the effects of air pollutants

S Anttonen and L Kärenlampi

University of Kuopio, Ecological Laboratory, Department of Environmental Hygiene, P.O.B. 6, 70 211 Kuopio, Finland

ABSTRACT: The ultrastructural and biochemical content of pine needle lipids were studied on current and previous year needles from two industrial areas and one control area. Ultrastructural observations showed that ageing caused the lipid material of the needle cells to increase. In industrial areas the increase was more pronounced; already the one-year-old needles contained more lipid material than the controls. This increase was also detected biochemically and was highest in the triglyceride and sterol or wax ester lipid fractions of the polluted needles as compared to the controls.

1. INTRODUCTION

Atmospheric pollutants seem to cause changes in the lipid metabolism of conifer needles e.g. on the basis of ultrastructural studies. The increase in lipid material in the cytoplasm and changes in chloroplast internal membrane system, e.g. swelling of thylakoids or a decreased number of granal thylakoids have been reported (e.g. Soikkeli 1981), but biochemical studies on the lipids of needles are rare (e.g. Malhotra and Khan 1978). This paper describes the changes observed in the ultrastructure and in the neutral lipid content of current and previous year pine needles in industrial and background areas.

2. MATERIAL AND METHODS

Current and previous year pine needles from about ten-year-old cultivated trees were collected from two industrial areas, one mainly polluted by SO_2 and the other by SO_2, NO_x and fluoride and one background area in October. Samples for electron microscopy were cut from the middle of the needle and fixed in 2 % glutaraldehyde in 0,1 M phosphate buffer and postfixed in 1 % OsO_4. Thin sections were stained with uranyl acetate and lead citrate. A Jeol Jem 100 B transmission electron microscope operating at 80 kV was used for observations. Needle samples for lipid analysis from about ten trees were also collected, combined and put immediately on dry ice prior to storage in sealed glass containers at -20 ^oC in an atmosphere of nitrogen before lipid extraction. Lipids were extracted from 5 g of needle material by the method of Kates (1972) and were kept frozen until hot isopropanol treatment. Neutral lipids were analysed on TLC-plates (0,5 mm silica gel) with hexane-diethylether-acetic acid (80:20:2 v/v) as a developing solvent. Plates were sprayed with 3 % cupric acetate in 8 % phosphoric acid and the amount of charred material obtained after heating the plate at 180 ^{o}C for 30 minutes was measured by means of a scanning photodensitometer (Shimadzu CS-930 TLC scanner) with authentic standards.

3. RESULTS AND DISCUSSION

The ultrastructural observations showed that the second year needles contained more lipid material than the current year needles. In industrial areas the increase in lipid was higher than in control areas. Natural ageing of pine has been reported to cause a clear increase in the amount of lipid material in the mesophyll (Saastamoinen et al. 1987) of the oldest needles.

Particularly in the current year needles the lipid bodies were usually associated with plastids or mitochondria, so these may be involved in their synthesis. In this same needle class the lipid bodies were occasionally partially dissolved, which probably indicates lipid metabolism. In industrial areas the needles also contained lipid material in the central vacuoles.

Chloroplast stroma and grana thylakoid swelling could occasionally be seen in the industrial area with SO_2, NO_x and fluoride emissions.

The biochemical analysis indicated an increase in neutral lipids during ageing, especially in the amounts of sterol or wax esters and triglycerides. No great differences between needle classes in the amount of free sterols, diglycerides and free fatty acids were detected. In these samples monoglycerides were undetectable. In the second year needles of the industrial areas the sterol or wax ester and triglyceride levels were higher than in controls. Probably the lipid material in the mesophyll is mainly composed of these neutral lipids. Conifers are known to store lipids (Glerum 1980) and propably the lipid material seen in the mesophyll acts partly as an energy reservoir (Selstam and Öquist 1985),but why the lipid content increases during ageing and in the needles exposed to air pollutants should be explained by other processes.

REFERENCES:
Glerum C 1980 N. Z. J. For. Sci. **10** 176
Kates M 1972 Laboratory Techniques in Biochemistry and Molecular Biology eds T S Work and E Work (Amsterdam:North Holland) pp 269
Malhotra S S and Khan A A 1978 Phytochem. **17** 241
Saastamoinen T, Anttonen S and Kärenlampi L 1987 Aquilo Ser. Bot. **25** 153
Selstam E and Öquist G 1985 Plant Sci. **42** 41
Soikkeli S 1981 Ann. Bot. Fenn. **18** 47

Inst. Phys. Conf. Ser. No. 93: Volume 3, Chapter 3
Paper presented at EUREM 88, York, England, 1988

Ultrastructure of maize root cells under high nitrate supply

M Čiamporová

The Institute of Experimental Biology and Ecology, Slovak Academy of Sciences, 81434 Bratislava, Czechoslovakia

ABSTRACT: Root cells of Zea mays grown under high supply of KNO_3 have been studied. Mitochondria with ring-shaped cristae were ubiquitous in the root tissues. Cytoplasm of young hypodermal cells contained fibrillar protein inclusions. Electron transparent cisternae were found in plastids. Contact of ER with plasmodesmata might reveal that transport of nitrate occurs in the root cells.

1. INTRODUCTION

Structural changes of leaf epidermal cells under high KNO_3 concentrations were found (Sitte 1963, Weidinger 1983). Under the excess of nitrate, cell organelles are involved in the processes of accumulation, metabolism and transport of nitrate in root cells (Oji et al 1985, Pan et al 1985, Rufty et al 1986). Ultrastructure of root cells of maize plants that had grown in aerated nutrient solution with control 4 $mmol.l^{-1}$ and high 40 $mmol.l^{-1}$ KNO_3 concentrations for 21 days was investigated. Various tissues and cell types at different stages of their development were examined with TEM Tesla BS 500.

2. RESULTS AND DISCUSSION

Ultrastructure of control cells is adequate to the stage of their differentiation and specialization in the root. Mitochondria with ring-shaped cristae similar to those found in Elodea leaves (Sitte 1963) were observed under high nitrate supply in various types of root cells from the very young meristematic cells (Fig 1) up to mature cells at the root base. This response corresponds to decreased activity of glutamate dehydrogenase in the same roots (Luxová, personal commun.). Numerous fibrillar inclusions were found in the cytoplasm of

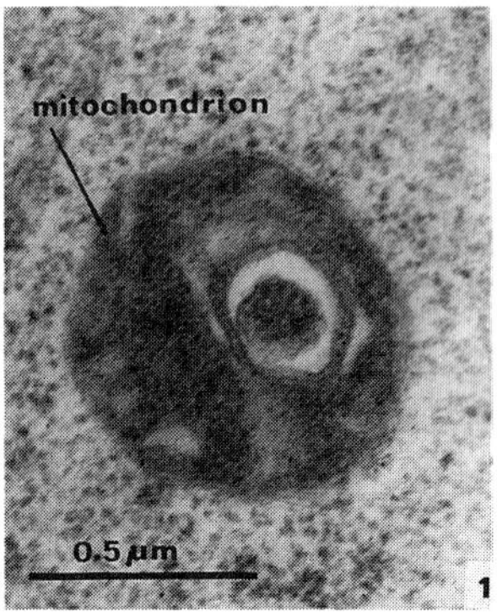

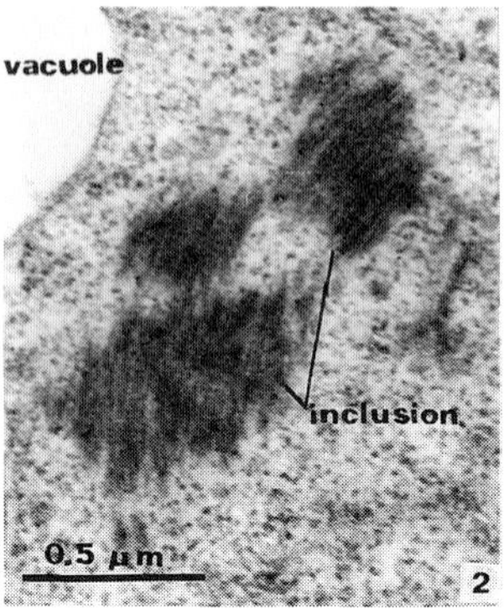

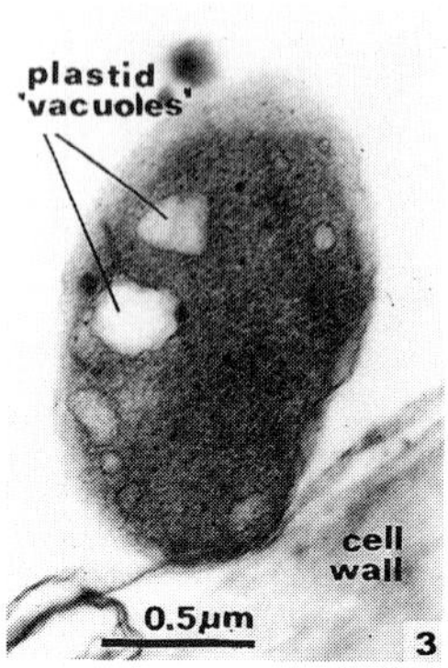

young hypodermal cells (Fig 2). They obviously represent deposits of the excess protein synthesized under the high nitrate supply. When the utilization of protein was slowed down, similar inclusions occurred in root cells of embryo (Yoo 1970) or after chemical stress (Bobák 1980). Electron transparent cisternae found in the stroma of plastids in the more differentiated cortical cells (Fig 3) remind of "plastid vacuoles" that resulted from ion uptake into the organelle (Neumann and Jánossy 1981). Elements of ER are not damaged as in the case of leaf epidermal cells (Sitte 1963, Weidinger 1983). They are often close to plasmodesmata in the more differentiated root cells (Fig 4) creating thus structural pathway for nitrate transport as suggested by Rufty et al (1986).

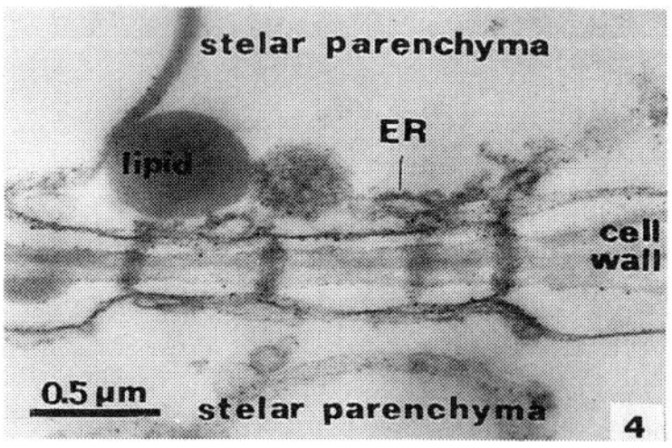

3. REFERENCES

Bobák M 1980 Biológia /Bratislava/ **35** 3

Neumann D and Jánossy A G S 1981 Physiol. Plant. **52** 320

Oji Y, Watanabe M, Wakiuchi N and Okamoto S 1985 Planta **165** 85

Pan W L, Jackson W A and Moll R E 1985 J. Exp. Bot. **36** 1341

Rufty T W Jr., Thomas J F, Remmler J L, Campbell W H and Volk R J 1986 Plan Physiol. **82** 675

Sitte P 1963 Protoplasma **57** 304

Weidinger M 1983 Mikroskopie /Wien/ **40** 210

Yoo B Y 1970 J. Cell Biol. **45** 158

Structural effects of ozone on leaf cells of spinach (*Spinacia oleracea*)

Monica Johansson, Department of Botany, University of Stockholm, S-106 91 Stockholm, Sweden.

Ozone (O_3) is a photochemically produced air pollutant of considerable economical importance because of its severe effects on sensitive agricultural plants, e.g. tobacco and spinach. This report describes ultrastructural injuries in spinach leaves. Greenhouse cultivated plants were used at a stage with 8-12 fully expanded leaves. The treatment was performed in fumigated chambers (Continuous Stirred Tank Reactor) where the plants were kept until sampled. Fumigation was given for 4 h. at three successive days. The O_3 concentrations were 0.05, 0.1 and 0.2 ppm. The controls were also kept in such a chamber and exposed to the background level (0.03 ppm) of O_3.

Palisade parenchyma cells, particularly cells close to the substomatal chamber, were studied. The first signs of injury was the appearance of large vesicles in the cytoplasm and between the cell wall and plasmalemma. Small osmiophilic droplets formed on the chloroplast envelope, and in the cytoplasm occured large, generally compact droplets. The tonoplast dissolved gradually and an aggregation of the cell organelles and cytoplasm toward the center of the cell occurred. Eventually there was a conspicuous degeneration of the cell structures, which were seen dispersed in a strongly electron scattering mass. Characteristic changes in exposed cells were found in the chloroplast. In the stroma large crystalline proteinaceous bodies were formed and this indicated an increase in permeability of the chloroplast envelope (Figure 1). These envelopes remained structurally intact, however. Grana changed their regular stacked appearance and thylakoids became somewhat swollen. In other chloroplasts the thylakoids were little affected but there occurred local swellings of the space between the two membranes of the envelope (Figure 2). Numerous vesicles containing stroma material appeared at such places and probably represent sections through protrusions from the plastid. In these chloroplasts also a number of crystalline bodies developed in the stroma. The envelope of mitochondria occasionally formed similar protrusions as those of affected chloroplasts (Figure 3). Some chloroplasts were swollen and the organization of the thylakoid system was irregular. Chloroplasts with a strongly electron scattering stroma was another kind of injury noticed. Such chloroplasts were sometimes found together with swollen ones.

This study confirms that chloroplasts are conspicuously sensitive to O_3, also in low concentrations, and that their envelopes and thylakoids are damaged.

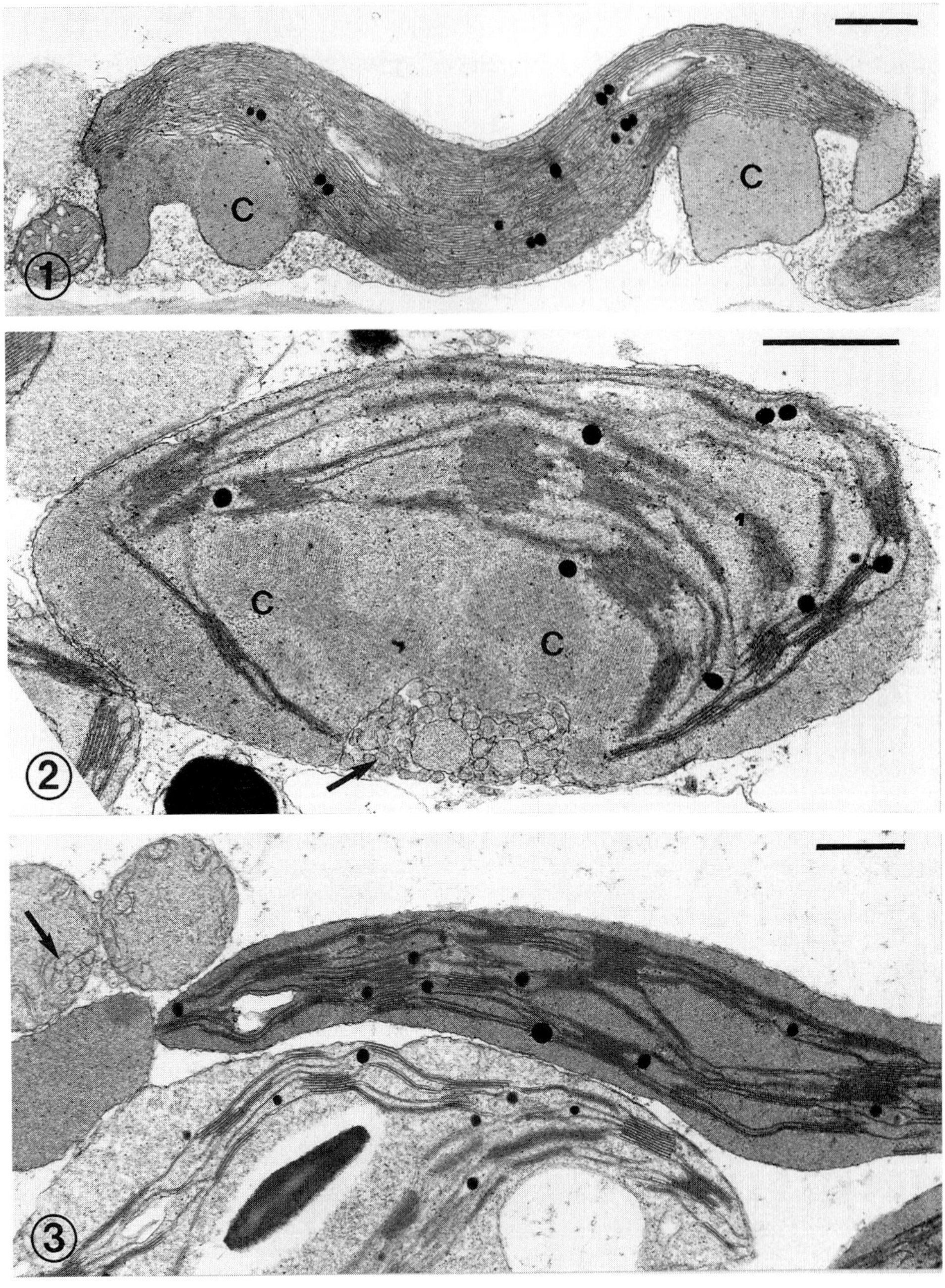

Fig. 1-3. *Spinacia oleracea* var. rebris treated with ozone (0.2 ppm). Arrows indicate protrusion of the chloroplast or mitochondrial envelope. C = Crystalline bodies. Bar = 0.5 µm.

Inst. Phys. Conf. Ser. No. 93: Volume 3, Chapter 3
Paper presented at EUREM 88, York, England, 1988

Effects on leaves of modulating CAM in *Kalanchöe zimbabwensis* through photoperiod

Isabel Santos and R. Salema
Institute of Botany and Centre for Experimental Cytology, University of Porto - Portugal

ABSTRACT:CAM can be modulated in K. zimbabwensis by short days. Comparing leaves grown under short days and long days showed differences in both cell layers and cell size with lower surface/volume cell ratio on the former plants. Protein stroma inclusion linked to CAM was found to fluctuate diurnaly when CO_2 uptake was via CAM.

1. INTRODUCTION

CAM (Crassulacean acid metabolism) can be controlled by photoperiod in certain plants.After 14 days under short days K. zimbabwensis switches from C_3 to CAM. Chloroplasts of some CAM plants have stroma located inclusions build up from tubular elements arranged in parallel hexagonal packing (Santos & Salema,1981) so far found only in members of the "malic enzyme" group and never observed in the "phosphoenolpyruvate carboxykinase" group (Santos & Salema,1981), K. zimbabwensis belonging to the former. A comparative study of K. zimbabwensis leaves developed under short and long days was done and is here reported.

MATERIAL AND METHODS

Plants of K. zimbabwensis were grown under short days (SD, 8h light) and long days (LD, 15h light)in chambers with light of 25 W m^{-2} at the level of foliage. Stereological study was done using at least 60 chloroplasts profiles for each case. Gross anatomic features and fine structural differences at chloroplast level were gathered by both light microscope and electron microscope morphometric analysis of samples from SD and LD plants.

RESULTS

K. zimbabwensis showed a daily acidity variation of 68.2 μeq/g fresh weight when uptaking CO_2 via CAM. Macroscopically comparing both types of leaves SD leaves appeared fatter than LD leaves, in correspondence with water contents (96.72%, 94.12%) and succulence index (2.45, 1.31) respectively. Transections of similar leaves showed that the mesophyll was bifacial, however of low level of differentiation. On the adaxial face cells were more regularly arranged than in the abaxial side, the order being disturbed in places subjacent to the stomatal complexes(Fig.1). Differences in thickness were translated,as an example,in a correspondence of 1.0 mm for SD and 0.8 mm for LD leaves.The thicker leaves had smaller cells and more layers than the thinner ones.Typically average cell size in a in a SD leaf was 74 um whereas in LD leaf was 77 µm with respectively S/V values of 0.91 $\mu m^2/\mu m^3$ and 0.94 $\mu m^2/\mu m^3$. When studying the various mesophyll cell layers it was found that the stroma inclusion showed a gradient with larger aggregates in cells close to veins. Plants following CAM showed a fractional volume Vv (inclusion /plastid) of 8.48 (± 0.25 SEM;$P<0.05$) at subepidermal cells and 11.32 (± 0.41 $P<0.05$),at central cells at 17.00h, time of the day that corresponds to the larger inclusions. Samples collected at the end of night revealed that only very small inclusions were present in the stroma of chloroplasts with peripheric location; chloroplasts near veins showed larger protein deposits with Vv= 7.19(± 0.36;$P<0.05$). Material from LD grown plants showed no protein inclusion in the stroma of chloroplasts either at the end of the light or at the end of the night period,with the exception of the chloroplasts located near veins which by 9.00h showed inclusions with a Vv of 9.22(± 0.82;$P<0.05$).

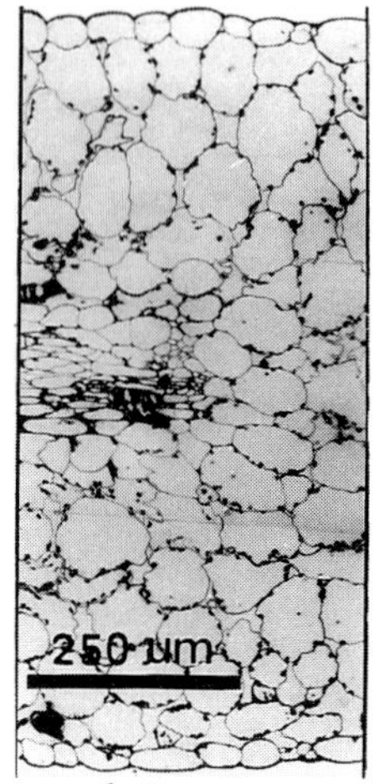

Fig.1 Transection of leaf from SD plant.

4. DISCUSSION

The data obtained showed that in plants following CAM the stroma inclusion had a diurnal variation already described for other CAM plant(Santos & Salema,1983).Interestingly cells near veins showed smaller diurnal variation, possibly due to their different metabolic conditions as deep location and proximity to tissues related to import or export of photosynthates.Preliminary results from *K. zimbabwensis* plants using C_3 photosynthesis(LD)revealed no stroma inclusion on chloroplasts of mesophyll cells in general but curiously enough, inclusions were present in chloroplasts from cells near the veins.This does not contradicts the idea that the protein which follows a diurnal cycle in certain CAM plants is one enzyme somehow related to the PEP-cycle; the above referred cells would have the same genetic composition irrespectivelly of being from SD or LD plants;it is conceivable that being in a situation in which the cycle is not active they would tend to deposit in the stroma of chloroplasts some unnecessary enzyme.

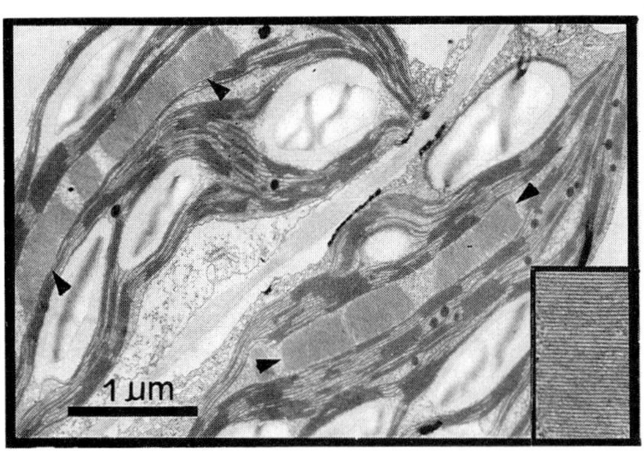

Fig.2 Chloroplasts from subepidermal mesophyll cells,at 17.00 h; SD-plant.Inclusions between arrowheads.

5. REFERENCES

Santos I and Salema R 1981 *Bol. Soc. Broteriana* **53** 1115
Santos I and Salema R 1983 *Z. Pflanzenphysiol.* **113** 29

Tree and needle age related ultrastructural changes in the mesophyll cells of Norway spruce needles from background and industrial areas in Finland

A Wulff, J Ahonen, L Kärenlampi and L Ropponen

University of Kuopio, Ecological Laboratory, Department of Environmental Hygiene, P.O.B. 6, 70 211 Kuopio, Finland

ABSTRACT: Ultrastructure of ageing needles from different aged Norway spruce (Picea abies) trees were studied in two industrial environments and in one background area. Natural needle senescence caused increased number of plastoglobuli and lipid accumulations, light swelling of thylakoids and changes in the appearance of tannin. In industrial environments similar changes could also be detected in younger needles. Other injury types were also observed, obviously caused by air pollutants. The changes observed in the younger trees seemed to be less pronounced in the older trees.

1. INTRODUCTION

A clear sign of the damaging effects of air pollution on coniferous trees is the short needle retention time. There are plenty of articles concerning plant senescence but very little information on conifer needles and senescence (e.g. Sutinen 1987). Our approach in this work has been to reveal how senescence developes in air polluted needles and how the age of the tree affects the injury development. This is important when applying results obtained from young trees to older trees.

2. MATERIAL AND METHODS

Needle samples were collected in the autumn from two industrial areas, one with SO_2 and another with SO_2, NO_x and fluoride emissions, and from one background area. The samples were taken from first, second, fourth, sixth and the oldest living needle class of both young (about 20 year) and old (>50 year) spruce trees. The trees were growing in natural stands and the distance between these two age groups of trees was less than 200 meters so that the trees were exposed to similar ambient air pollutant concentrations.

Cross sections were cut from the middle of the green needles and fixed in 2 % glutaraldehyde in 0.075 M phosphate buffer and postfixed in 1 % OsO_4. Thin sections were stained with uranyl acetate and lead citrate. A JEOL JEM 100 B transmission electron microscope was used for observations.

3. RESULTS AND DISCUSSION

In the background area the chloroplasts remained intact except for slight swelling of the thylakoids in the older needles. The amount of thylakoids remained costant during ageing except in the chloroplasts containing large number of plastoglobuli. The number of plastoglobuli had increased slightly in second-year needles but a clear increase could not be seen until in the sixth-year needles. Only in the oldest living (9th-year) needles were the roundished chloroplasts totally filled with plastoglobuli. In cells where the number and the whitening of plastoglobuli had increased, resolution of thylakoids was often poor and lipid material of irregular shape had increased in the cytoplasm. Contrary to Sutinen's (1987) observations the increase in plastoglobuli number was also observed in cell layers other than the sub-epidermal layer. The amount of cytoplasmic lipids increased during senescence, though in the background area this occurred slowly and only in the oldest living needles were there remarkable lipid accumulations. Penetration of small lipid bodies in to the central vacuole could be seen in the fourth-year and older needles. Surprisingly, Sutinen (1987) had only observed lipid accumulations in old needles from trees in Germany and not in old needles collected in Finland. In the young needles tannin usually appeared as a thin ribbon along the margins of the central vacuole, changing gradually in to a granular form and in the oldest needles the whole central vacuole was often densely filled or had a roughly granulated form of tannin.

In the industrial areas the same changes were stronger and observed earlier in the younger needles. A clear increase in plastoglobuli number could be seen in the second-year needles and the plastoglobuli had filled the whole chloroplast in sixth-year needles. The SO_2 polluted area showed the largest increases of whitened plastoglobuli and irregular shaped lipid which gives support to the earlier findings from SO_2 polluted areas (e.g. Soikkeli 1981). In addition, there were remnants of starch in the chloroplasts, which may indicate disturbances in carbohydrate metabolism. The slight swelling of thylakoids which increased during ageing also occurred most frequently in the SO_2 polluted area. In the area polluted by fluoride, NO_x and SO_2 the curling of thylakoids could be occasionally seen.

The young trees developed the changes slightly earlier than the older trees. This may be due to a more active uptake of air pollutants in young trees. The results support the idea that the effect of air pollution can partly be interpreted as accelerated ageing.

REFERENCES:

Soikkeli S 1981 Ann. Bot. Fennici **18** 47
Sutinen S 1987 Eur. J. For. Path. **17** 65

Inst. Phys. Conf. Ser. No. 93: Volume 3, Chapter 3
Paper presented at EUREM 88, York, England, 1988

Glandular trichomes of *Teucrium Scorodonia* L ultrastructure and secretion

T Antunes and I Sevinate-Pinto

Departamento de Biologia Vegetal, Faculdade de Ciências de Lisboa, Bloco-C_2, Campo Grande 1700 Lisboa

<u>Teucrium scorodonia</u> L. (Labiatae) is a spontaneous plant of our flora rich in essencial oils produced in glandular trichomes. The chemical composition has been investigated by Marco et al (1983). In a previous work (Antunes and Sevinate-Pinto 1987), scanning electron microscopical investigations were carried out to study distribution and morphology of glandular trichomes during leaf development. The young leaves are covered by a dense indumentum of very long non glandular trichomes and a larger number of glandular ones which occur as population of mixed developmental stages. Our results revealed two different types of glandular trichomes (Type I and Type II). Additional epidermal trichomes, projecting over the leaf surface with a visible stalk and involved probably in secretion, have also been observed. The results we present now, emphasize the trichomes Type I - They develop from a single epidermal cell, and at maturity contain six cells including one basal epidermal cell, a very short stalk cell and a multicellular head with four secretory cells (Figure 1). The results obtained by histochemical methods reveal the occurrence of a lipophilic material, but according to our staining reactions it also includes traces of polysaccharides. At stage of active secretion, the glandular head cells are characterized by a dense cytoplasm, a proliferated network of smooth endoplasmic reticulum, dictyosomes few and small vacuoles and plastids (Figure 2). The plastids accumulating osmiophilic material, which is eventually extruded from them, have an amoeboid appearance and are sheathed with tubules of smooth endoplasmic reticulum. Electron dense deposits of similar density also occur between the plasmalemma and the cell wall. The cell wall of the glandular heads shows a separation into two layers forming an annular space filled with secreted material. The results obtained with scanning electron microscope suggest that the trichomes possess a dehiscence mechanism whereby stored material is released by bursting of the cuticule probably caused by the pressure of the secreted material.

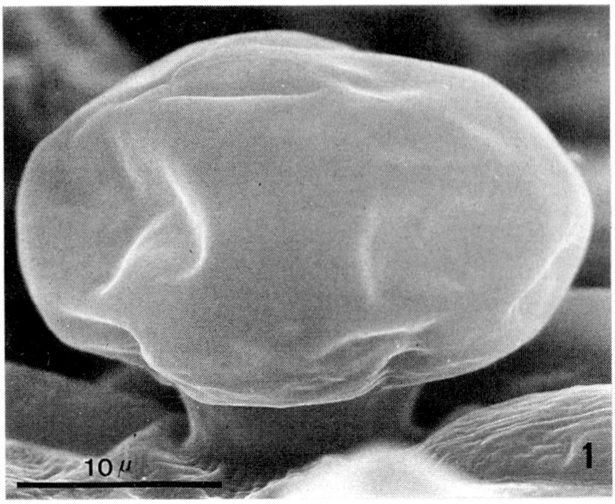

Fig. 1. SEM micrograph showing a trichome

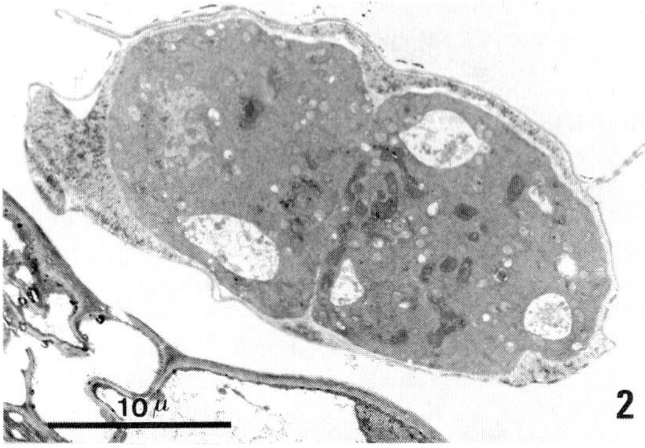

Fig. 2. TEM micrograph from glandular head cells of trichome

References

Antunes T and Sevinate-Pinto I 1987 *XXII Reunião da Sociedade Portuguesa de Microscopia Electrónica*-Portugal

Marco J L, Rodriguez B, Pascual C, Savona G, Piozzi F 1983 *Phytochemistry* 22 727

Paper presented at EUREM 88, York, England, 1988

Histochemical localization of pectin in the stylar transmitting tissue of *Trimezia fosteriana* (Iridaceae)

P-A Bystedt
Department of Botany, University of Stockholm, S-106 91 Stockholm, Sweden

The tissue supporting pollen tube growth in pistils is called transmitting tissue. Most studies of the transmitting tissue have been concerned with the stigma, while less information is available from the stylar and ovarian parts. Monocotyledons usually have a hollow style lined with transmitting tissue while in most dicotyledons the tissue is solid with intercellulars. The transmitting cells are secretory. The secretion contains various nutrients. Pectin has been found as a major component in the secretion of the style in both monocotyledons and dicotyledons (Dashek *et al* 1971, Cresti *et al* 1976). Pollen tubes digest pectin in solid styles and canal secretion in the open styles on their way to the ovary. Convenient ways of determining the presence of pectin is digestion with pectinase and histochemical reaction with iron (cf. Vennigerholz and Walles 1987).

Work presented here is part of a detailed study of the structure and function of the transmitting tissue in *Trimezia fosteriana*. The style of this plant is hollow and the basal part (0.5 mm) is divided into three separate canals which are semi-closed. Mature, unpollinated styles were specifically stained for pectine according to the hydroxylamine-iron method of Albersheim (1965). A positive reaction was obtained for material in the stylar canal and in the intercellulars, Figure 1. The major part of the canal was filled with an electron lucent material. Small amounts of pectin was scattered in it. The cell walls facing the canal were covered with an about 0.2 µm thick layer containing pectin, Figure 2. The intercellulars showed a strong reaction for pectin. The control without hydroxylamine lacked iron precipitation, Figure 3.

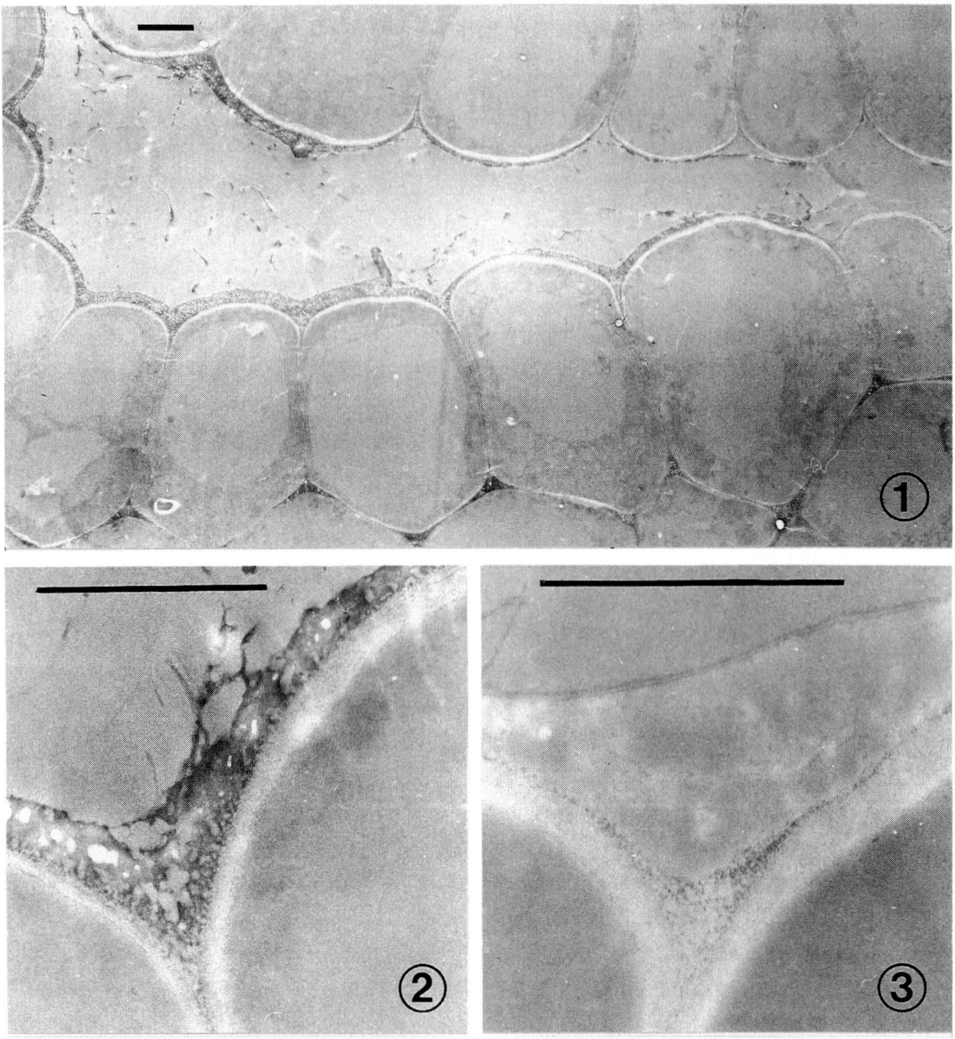

Fig. 1 and Fig. 2. Stylar canal cells. Positive pectin reaction in the secretion and in the intercellulars. Bar = 1 μm

Fig. 3. Control to Fig. 1 and Fig. 2; $FeCl_3$ without hydroxylamine gives no reaction for pectin. Bar = 1 μm

References

Albersheim P 1965 *Protoplasma* 60: 131-135

Cresti M, van Went JL, Pacini E and Willemse MTM 1976 *Planta* (Berl) 132: 305-312

Dashek WV, Thomas HR and Rosen WG 1971 *Amer. J. Bot.* 58: 909-920

Vennigerholz F and Walles B 1987 *Protoplasma* 140: 110-117

Inst. Phys. Conf. Ser. No. 93: Volume 3, Chapter 3
Paper presented at EUREM 88, York, England, 1988

Mature megagametrophyte, nucellus and adventive embryony in *Euphorbia dulcis*

P Gori

Dipartimento di Biologia Ambientale sez. Botanica, Università di Siena, Via Mattioli 4, 53100 Siena, Italy.

ABSTRACT: The mature megagametophyte of Euphorbia dulcis biotypes which develop apomictic embryos has cells, egg included, of apparently normal ultrastructure. At the onset of embryogenesis, the micropylar, middle and chalazal cells of the nucellus differ in degree of cytoplasmic vacuolization, protein synthetic activity and storage and plastid type. The embryos develop from nucellar cells, laterally or apically adjacent the egg apparatus. Up to 8 embryos, each of 2-6 cells, can be seen forming during the degeneration and initial reabsorption of the egg.

1. INTRODUCTION

The present study is part of a research programme on the ovule of different species of the genus Euphorbia (Gori 1987 and references).

2. MATERIALS AND METHODS

Ovules collected from naturally growing Euphorbia dulcis plants were fixed in glutaraldehyde-osmium tetroxide, dehydrated and embedded in an Epon 812-Araldite A/M mixture as recently described (Gori 1987). Thin sections were stained with uranyl acetate and lead citrate.

3. OBSERVATIONS AND COMMENTS

The egg of E. dulcis plants with adventive embryony has a vacuolated micropylar cytoplasm with very few membrane organelles. In the chalazal cell region, besides the nucleus and many mitochondria, plastids with big granules of starch and long ergastoplasm profiles are visible. Here there is no cell wall. The synergids have the usual filiform apparatus. The three antipodals show immature plastids and enlarged rough endoplasmic reticulum cisterns. Their walls are sculptured with transfer cell-like ingrowths as reported in Zea mays (Diboll & Larson 1966) and Gasteria verrucosa (Willemse & Kapil 1981). The central cell of the embryo sac is highly vacuolate; its cytoplasm is limited to a peripheral layer in contact with the nucellus and a reduced central mass which envelops the deeply lobed polar nuclei and is particularly rich in membrane organelles.

There are often large amyloplasts whereas elsewhere there are generally proplastids. The central cell is differentially provided with wall projections but is completely naked where it contacts the egg so that the plasmalemmas of the two cells are in direct contact. At the mature megagametophyte stage, the micropylar, middle and chalazal cells of the nucellus differ in degree of cytoplasmic vacuolization, protein body formation (which is greatest in the micropyle cells) and plastid type (photosynthetically active chloroplasts or amyloplasts carrying different quantities of starch). The nucellus cells above and beside the egg apparatus are involved in embryogenesis, which begins before signs of degeneration are visible in the egg. By the time its ultrastructural details are no longer resolvible, up to 6 or 8 embryos, each of 2-6 cells are forming. The initial cells are characterized by a diffusely electrondense cytoplasm with few vacuoles, small plastids in a great variety of forms only occasionally containing starch, reduced endoplasmic reticulum and possibly autophagic vacuoles (Figure 1).

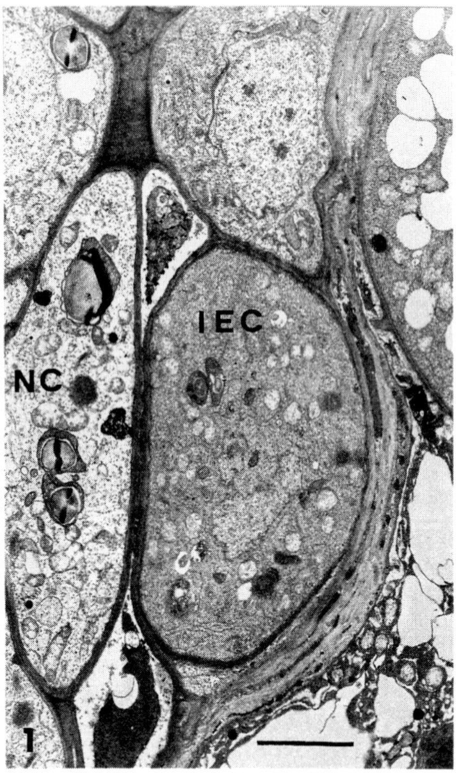

4. REFERENCES

Diboll A G and Larson 1966 Am.J.Bot. 53 pp391-402
Gori P 1987 Ann.Bot. 60 pp563-9
Willemse M T M and Kapil R N 1981 pp25-32

Figure 1. Portions of nucellus (NC) cells and of an initial embryo cell (IEC). Scale mark = 1 micron.

Form and function of the pollen walls in *Stereosandra javanica* (Epidendroideae, orchidaceae)

M.HESSE

Institute of Botany,The University of Vienna,Rennweg 14,Vienna,Austria

ABSTRACT: In Stereosandra javanica,but also in many other orchids,the pollen wall structure is different in the outer and in the inner grains of the pollinia. Only the outermost walls of the outermost grains have massive exines. All other pollen faces have very fragile exines with tiny,incoherent sporopollenin granules. The functional aspects of this uncommon,interesting feature is discussed.

Most spermatophyta produce single pollen grains or pollen tetrads. In all taxa there is no difference in sporoderm structure between the respective pollen grains (with the exception of the apertures,of course). But not so in many orchids: their pollinia consist of a lot of pollen grains,all with strikingly different formed exine structures. Because the sporopollenin content of all non-outermost grains is extremely meager,most orchid pollen cannot withstand either acetolysis or fossilization. In Stereosandra javanica only the distal face of the outermost pollen grains (Fig.1,2) is made up by tooth-like,probably isolated sporopollenin elements,which form a rather complex,thick,massive exine. The exine-"teeth" are fixed within the intine matrix. Already the radial pollen walls of the outermost grains reduce more and more their exine consistency: complex exine elements are lacking (Fig.4). This is especially true for all walls of all the inner pollen grains: their sporoderm consists exclusively of a rather thick,uniform intine with few small,non-coherent sporopollenin exine granules. Such exines are extremely fragile (Fig.3),and their ontogenesis is unknown.
The meaning of these uncommon,highly significant and interesting feature seems to be evident. The pollinating insect of course transports the pollinium as a whole. If there would be a too firm connection between individual pollen grains,at least the central ones would hardly germinate on the stigma surface,and the chance for successful pollen tubes would be at least diminished. Therefore the transfer of all the inner pollen grains would waste time and energy both for the orchid and the flower-visiting insect. Such a nonsense,or better,detrimental actions would be contraproductive for the evolution of most Orchidaceae.
Very interestingly we can find a quite similar reduction of the pollen wall consistency also in other pollinia,e.g. within the Asclepiadaceae ! SCHILL and JÄKEL (1978) published micrographs and notes on an identical configuration of the pollen walls inside the pollinia,while only the outermost grains show massive,thick,coherent exines. Further investigations on this topic is necessary.
Acknowledgements: The orchid material was provided by Dr.Pamela BURNS-BALOGH (Wiesbaden).The investigations was supported by Projekt 5812 B (Fonds zur Förderung der wissenschaftlichen Forschung in Österreich).
SCHILL R,JÄKEL U 1978 Tropische und subtropische Pflanzenwelt **22**, Franz Steiner Verlag,Wiesbaden.

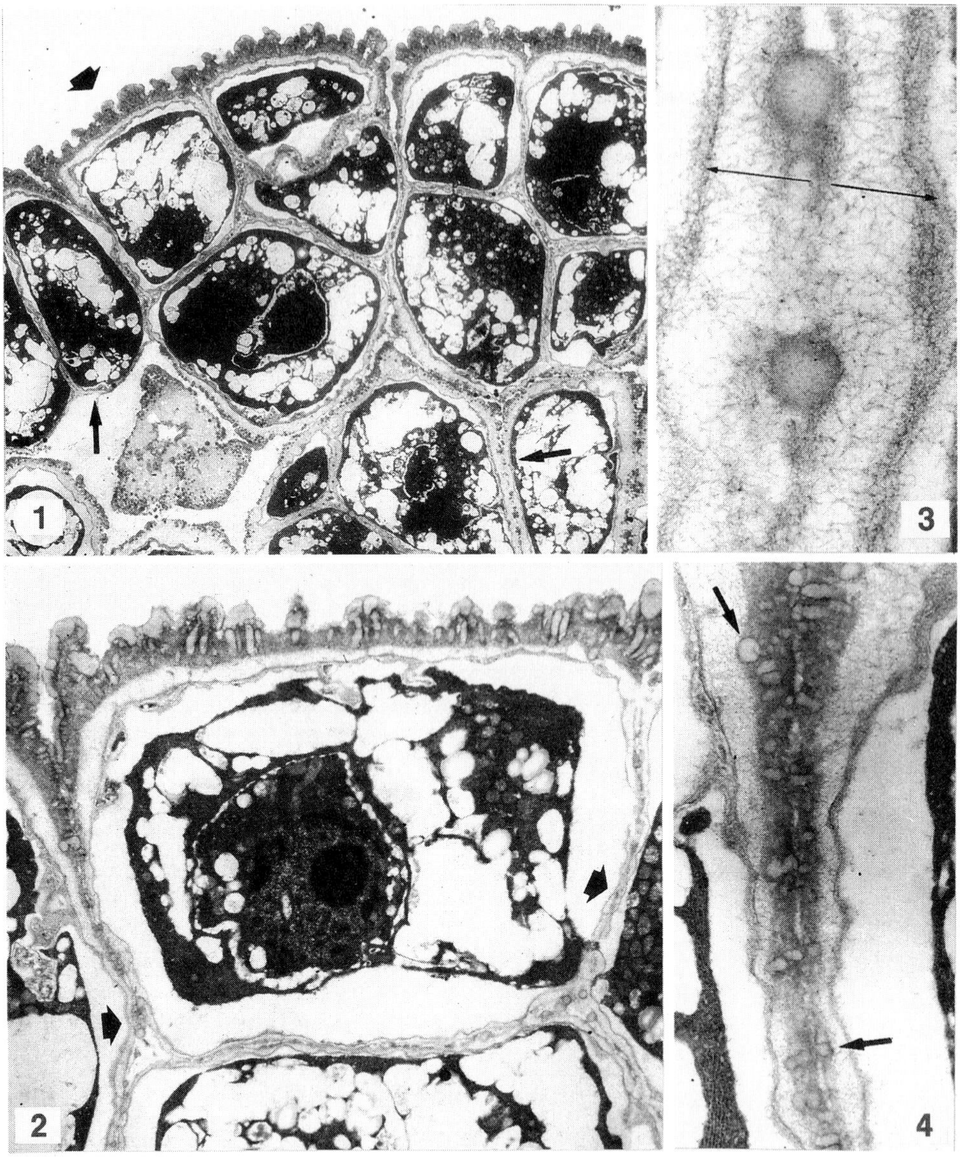

Fig.1. Detail of a Stereosandra javanica pollinium. The outermost pollen grains show an elaborated,thick exine on their distal face exclusively (thick arrow). All other pollen walls have thin,fragile exines embedded within the intine (long arrows).

Fig.2. Peripheral pollen grain with tooth-like exine elements (above) and thin,fragile exines inside the pollinium (arrows).

Fig.3. Small sporopollenin exine elements embedded within the intine (see the arrows) of central pollen grains. The sporopollenin globules are not connected,but most probably completely isolated.

Fig.4. Gradual reduction of sporopollenin exine elements (arrows).

Changes in calcium distribution in nuclei from cultured maize embryos detected with antimonate

H. Kieft and J.H.N. Schel

Dept.of Plant Cytology and Morphology, Agricultural University, Arboretumlaan 4, NL 6703 BD, Wageningen, The Netherlands

ABSTRACT: We have used the antimonate precipitation technique to detect calcium in cultured immature embryos from Zea mays L. Most calcium was present in the euchromatic regions of the nuclei. After prolonged culture the total amount decreased. Because of the relationship of decondensed chromatin and transcriptional activity it is suggested that this method in general might be used to monitor gene expression during plant cell culture.

1.INTRODUCTION

Calcium is known to have several functions during plant development. Amongst those, modulations of the phosphorylation of nuclear proteins has been suggested, resulting in varying gene expression (see e.g. Datta et al. 1985). In this study we localized calcium in meristematic cells of the root apex from immature maize embryos, cultured in vitro. Calcium distribution was registered using the antimonate precipitation method (Slocum and Roux 1982).

2.MATERIALS AND METHODS

Immature embryos from Zea mays L., strain A-188, approx. 12 days after pollination were cultured on a solid medium essentially as described elsewhere, but with the scutellum side down (Fransz and Schel 1987). After different periods in culture the embryos were fixed in 3% paraformaldehyde + 0.25% glutaraldehyde in antimonate buffer. The samples were washed and postfixed in 1% OsO_4 using the same buffer. Final washing was performed in phosphate buffer. After Epon embedding, ultrathin sections were poststained with uranyl acetate and lead citrate.

3.RESULTS

If cultured as described the embryo develops an axis in a short time. We restricted our observations to the root apex. Root apex cells, 8 hours in culture (8 hic),show a high amount of antimonate precipitation in the nucleus, mainly in the euchromatic regions of the nucleoplasm (Fig 1a). There is no precipitation in the nucleolus, but in the nucleolar vacuole

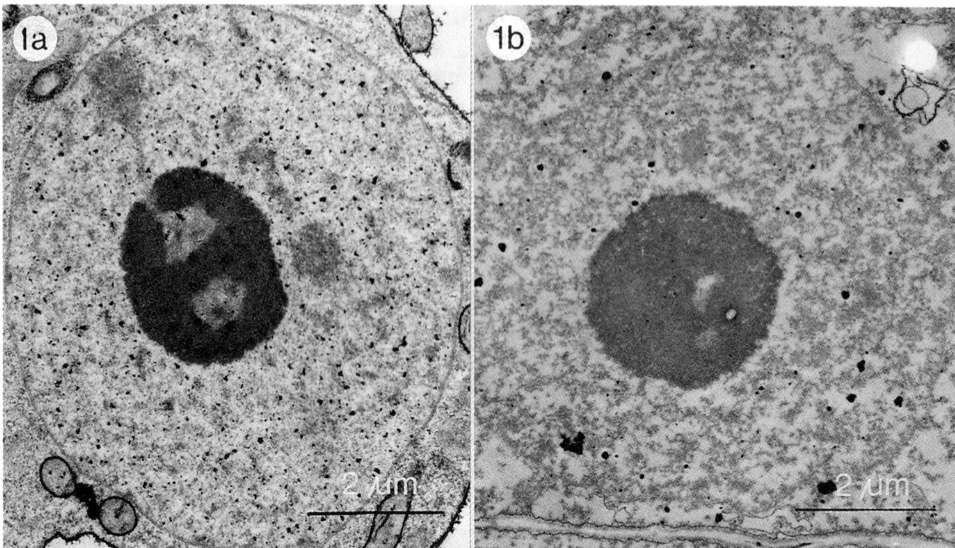

Fig.1:**a.** Nucleus after 8 hic. The euchromatin regions are preferentially labelled. The nucleolus is negative except the nucleolar vacuoles. **b.** Nucleus after 28 hic. Less precipitate is present, but still over the decondensed chromatin.

some antimonate is detected. The cytoplasm of the cell is hardly labelled; only membrane-bound calcium can be detected in the organelles, especially the mitochondria and vacuoles. Apex cells, 28 hic, show a less dense antimonate precipitation in the nucleus and organelles (Fig.1b). Nevertheless, again mainly the euchromatic regions of the nucleus are labelled.

4.DISCUSSION

Obviously, there is a change in calcium distribution when immature embryos are on culture medium for a longer period of time. At the onset of culture there is a high rate of cell formation in the root apex and nuclei show high concentrations of calcium. The lower content of calcium at 28 hic probably indicate a lowered nuclear activity during that period. The fact that precipitates were mainly found in the euchromatic parts of the nucleus supports the evidence that calcium plays an important role in the regulation of chromatin condensation (Datta et al. 1985). Therefore, this technique might be used for monitoring changes in nuclear activity during in vitro plant cell culture.

5.REFERENCES

Datta N, Chen Y and Roux S J 1985 In: Molecular and cellular aspects of calcium in plant development. ed. A J Trewavas (New York: Plenum) pp 115-122

Fransz P and Schel J H N 1987 Acta Bot. Neerl. **36** 247-260

Slocum R D and Roux S J 1982 J. Histochem. Cytochem. **30** 1982 617-629.

Paper presented at EUREM 88, York, England, 1988

Some first observations on calmodulin distribution in embryogenic carrot cell cultures obtained by immuno-electron microscopy

M Kreuger[1)], A A M van Lammeren[1)], S C de Vries[2)] & J H N Schel[1)]

[1)] Department of Plant Cytology and Morphology and [2)] Department of Molecular Biology, Agricultural University, Arboretumlaan 4, 6703 BD Wageningen, The Netherlands

ABSTRACT: Early stages of somatic embryo formation in carrot suspension cultures were examined for calmodulin distribution after embedment in London Resin White and immunogold labelling. The results indicate a higher level of cytoplasmic calmodulin near the edges of the embryogenic cell clumps and an increase of the total amount of calmodulin after prolonged culture.

1. INTRODUCTION

The process of somatic embryogenesis in plants has been studied in detail, being of high importance for commercial plant breeding. Most reports describe biochemical and molecular biological events; a smaller number deals with ultrastructural data (for a survey, see e.g. Ammirato 1983). At the intracellular level, evidence is scarce about the events which trigger a plant cell to pursue an embryogenic pathway. We think that, possibly, changes in calmodulin distribution might be related to that. The present study, therefore, is a first attempt to localize calmodulin in embryogenic cell cultures at the subcellular level, although we are quite aware from the pitfalls which might occur, as pointed out clearly by Roberts *et al.* (1986).

2. MATERIALS AND METHODS

Suspension cultures of *Daucus carota* L., cv Trophy, were used. Culture conditions and induction of embryogenesis were essentially as described elsewhere (Giuliano *et al.* 1983). At various days after embryo induction (d.a.i.) small pro-embryogenic masses (p.e.m.) were sampled, fixed in 3% p-formaldehyde and 0.25% glutaraldehyde in MSB buffer and embedded in London Resin White. No OsO_4 was used. Ultrathin sections were labelled for 1-2 hrs at 37 °C with 15 μl anti-calmodulin (dilution 1:500, raised in rabbit against spinach, gift of dr. S. Wick, St. Paul, Minn.) dissolved in 0.2 M Tris/HCl, 0.4% BSA, 0.04% gelatin and 0.4% Tween-20, pH 7.4. After rinsing, the sections were incubated with the second antibody (GAR-colloidal gold (7 nm), Janssen Pharmaceutica, 10 μl, diluted 1:25 in PBS, 1-2 hrs at 37 °C). The sections were rinsed and stained with uranyl acetate (20 min, 20 °C) and lead citrate (40 sec, 20 °C).

3. RESULTS AND CONCLUSIONS

Some representative micrographs of developmental stages are given in Fig. 1. The distribution of gold particles/μm^2 is given in Fig. 2.

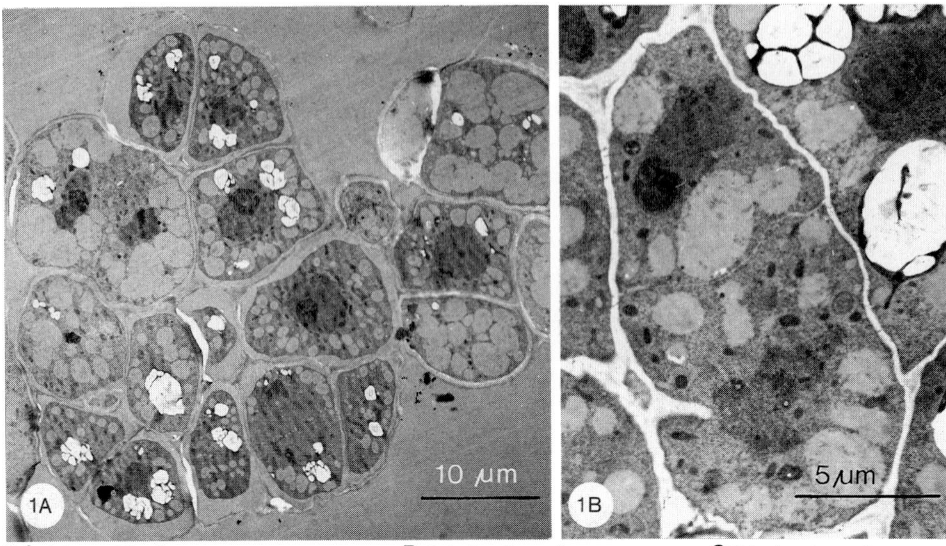

2 A 1 d.a.i.; mid cell B 1 d.a.i.; edge cells C 2 d.a.i.; edge cells

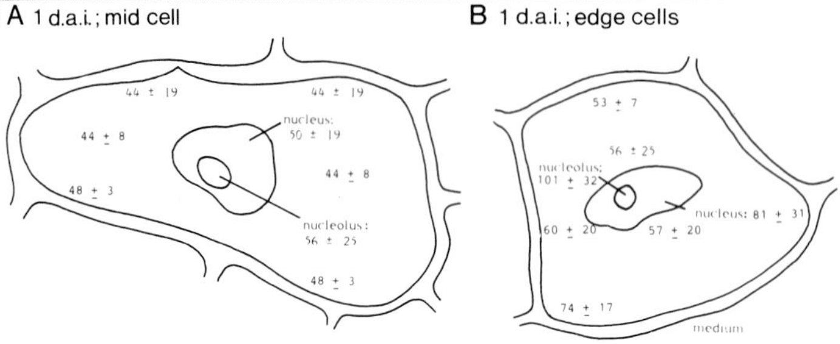

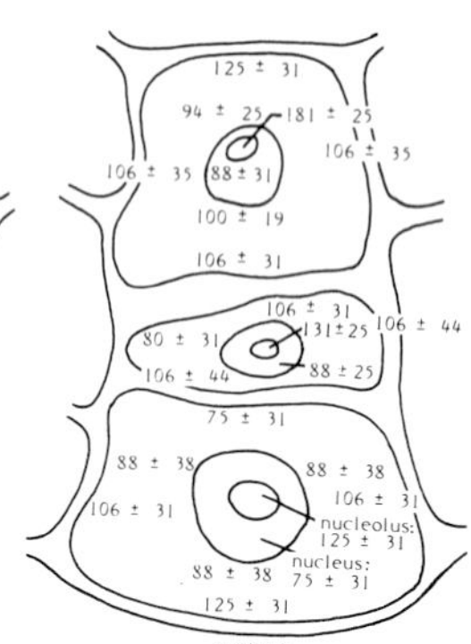

Fig. 1A. Electron micrograph of a p.e.m. at 1 d.a.i. The cells are thin-walled, slightly plasmolyzed, with small vacuoles and electron-dense cytoplasm. B. Higher magnification of a cell from the mid of a p.e.m. at 1 d.a.i. which just had divided. Fig. 2. Schematic representation indicating the number of gold particles/µm^2 for various stages.

Two main effects can be recognized: a. The cells in the mid of the p.e.m. show a rather evenly distribution of gold particles/µm^2 cytoplasm (Fig. 2A) while the cytoplasm of the cells at the edges of the p.e.m. shows a gradient in gold particle distribution; most labelling was found in the region which was oriented to the medium (Fig. 2B). b. After prolonged culture the gradient effect diminished but the overall amount of label has increased (Fig. 2C). In all cases, the highest amounts of label were found in the mitochondria or plastids (difficult to discriminate in these young cells), followed by the nucleolus and the nucleoplasm. Cell walls, starch grains and vacuoles were negative.

4. REFERENCES

Ammirato P V 1983 *Handbook of Plant Cell Culture* ed D A Evans, W R Sharp, P V Ammirato, Y Yamada (New York: McMillan) vol **1**. pp 82-123

Giuliano G, Rosellini G D and Terzi M 1983 *Plant Cell Rep.* **2** 216-218

Roberts D M, Lukas T J and Watterson D M 1986 *Crit. Rev. Plant Sci.* **4** 311-339

Differentiation of the seed coat in *Magnolia grandiflora* L: some histological, histochemical and ultrastructural aspects

J.D. SANTOS DIAS and J.F. MESQUITA

Depart. of Botany (Lab. of E.M. and Phycol.), Center for Plant Physiol. and Cytol. (INIC), Univ. of Coimbra, Coimbra, Portugal

ABSTRACT: In the course of seed differentiation in Magnolia grandiflora L. the colour changes are concomitant with two essential cellular modifications in the sarcotesta: plastidal evolution towards chromoplasts and accumulation of lipid globules in cytoplasm.

1. INTRODUCTION

It is known that the seed coat (testa) shows a great structural diversity in plants, what is considered a characteristic with an important taxonomic significance. Sometimes (zoochoric plants) the testa includes pigment-containing cells which are responsible for the colour of the seeds (see Fahn, 1974; Esau, 1977; Boesewinkel and Bouman, 1984). In Magnolia grandiflora, this colour changes during seed development, from green-whitish, in young or imature seed, to deep red in mature seed. The aim of this work was to investigate the cell modifications which could explain this process.

2. MATERIAL AND METHODS

For this study, samples of seeds of Magnolia grandiflora were harvested at five maturation stages identified by their colour and mean weight (from 16 to 180 mgs). In light microscopy, observations of both fresh material and one-micron sections stained by toluidine blue were carried out. For electron microscopy, the pieces were prepared according to the current technique. Cytochemical studies were performed on semithin sections stained with black Sudan or PAS method (M.O.) and on ultrathin sections treated by Thiery's technique (E.M.).

3. RESULTS AND DISCUSSION

During the maturation of the seed-coat in Magnolia grandiflora, the following zones are differentiated: an exotesta three to four cells in thickness, a well developed and parenchymatous mesotesta and a three-layered endotesta the innermost layer of which contacts with the endosperm (fig. 1). The pigments responsible for the colour of the seeds are exclusively localized in both exotesta and mesotesta (sarcotesta) with great predominance to the latter. The endotesta is completely devoid of pigments (figs. 1,2). So, the cytological changes we have studied during seed maturation, essentially concern the sarcotesta. Two aspects stand out in this differentiation process: the plastidal evolution and the accumulation of lipid globules in the groundplasm. As to the plastids, they begin to present membrane-bound inclusions, with an electrodense contents, similar to those which have been described in other plant tissues (Salema et al. 1972).

Then, they evolve progressively in amyloplasts and/or amylochloroplasts with very reduced grana, and finally, they become membranous-globular chromoplasts (Sitte, 1974; Mesquita and Santos Dias, 1988) in stroma of which starch grains can be seen, too (figs. 4-7).

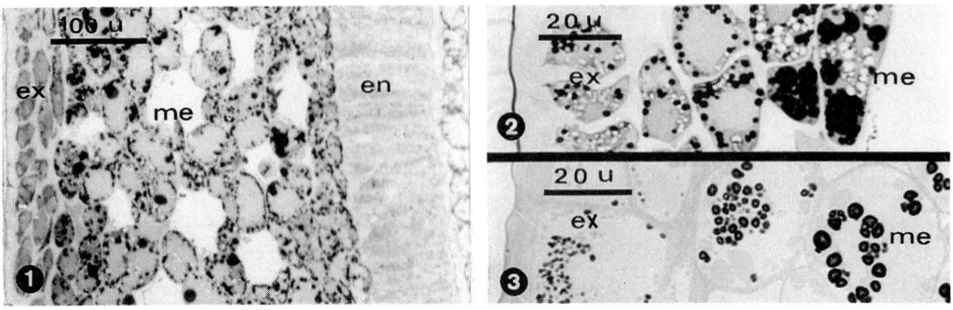

Lipid globules (figs. 1,2) and starch grains (fig. 3) are present in the exotesta (ex) and mesotesta (me), but not in the endotesta (en). Staining with black Sudan (figs. 1,2) and PAS (fig. 3).

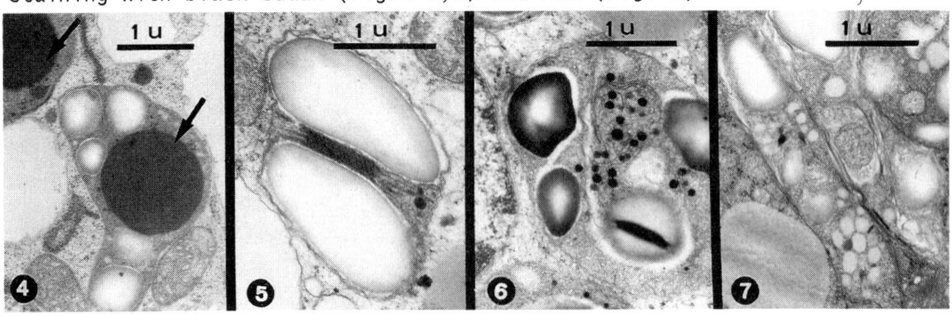

Ultrastructural evolution of plastids: organelles with electrondense membrane bound inclusions (fig. 4 arrows), amylochloroplasts (fig. 5) and chromoplasts (figs. 6,7), can be seen.

Concerning cytoplasmic globules, they are intensely sudanophilic and increase, in number and dimensions, during sarcoteca development this taking place from outer to inner layers of this region of the seed coat (figs. 1,2). Their progressive fusion leads to the appearance of voluminous sudanophilic inclusions which constitute the greatest part of the cellular contents in final stages of differentiation (fig. 2). Although, isolation of plastids and pigment dosage have not been carried out yet, it is not probable that the deep red colour of the mature seed is due to plastidal carotenoids, only. So, cytoplasmic sudanophilic inclusions can to represent lipochromes (secondary carotenoids) accumulated through a mechanism similar to that described in some algae (Lang, 1968; Fátima Santos and Mesquita, 1984).

BIBLIOGRAPHY

Fahn A 1974 Plant Anatomy (Pergamon Press)
Esau K 1977 Anatomy of seed plants (John Wiley and Sons)
Boesewinkel F D, Bouman F In Embriology of Angiosperms B.M. Johri Edit. Springer-Verlag
Salema R, Mesquita J F and Abreu I 1972 J. Submicr. Cytol. **4** 161
Sitte P 1974 Z. Pflanzenphysiol. **73** 243
Mesquita J F, Santos Dias 1988 (in press)
Lang N J 1968 J. Phycol. **4** 12
Fátima Santos M and Mesquita J F 1984 Cytologia **49** 215

Paper presented at EUREM 88, York, England, 1988

Ultrastructural and cytochemical changes in aleurone cells of oat during germination

I Sevinate-Pinto

Departamento de Biologia Vegetal, Faculdade de Ciências, Bloco C_2, Campo Grande, 1700 Lisboa

The aleurone layer of oat (*Avena sativa* L.) and its role in endosperm mobilization was investigated.

To understand the ultrastructural and cytochemical changes which occur during the germination process, the aleurone layer was studied in dry seeds, imbibed seeds, germinated seeds and seedlings at early stages of growth. Several fixation procedures were used to determine which were the most suitable for seed preservation.

The cells contained protein and lipid bodies (Fig. 1) which were mobilized during germination. The matrix of the protein bodies contained two types of inclusions with different histochemical properties. The lipid bodies differed in composition and function. During germination, the cells became vacuolated, rough endoplasmic reticulum proliferated (Fig. 2), the outer layer of the bilayered cell wall was degraded (Fig. 3), and the protein bodies were quickly degraded and mobilized.

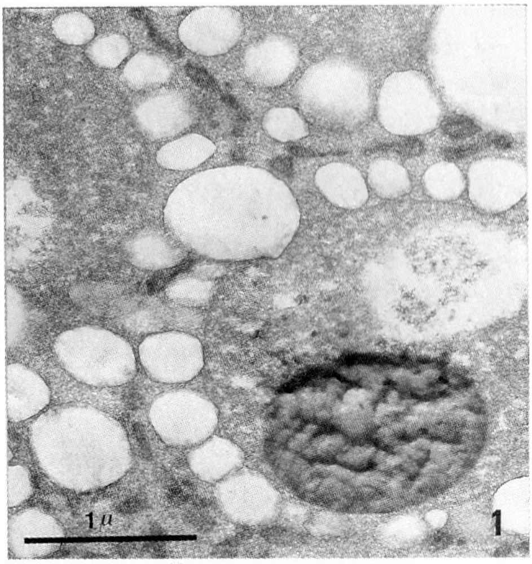

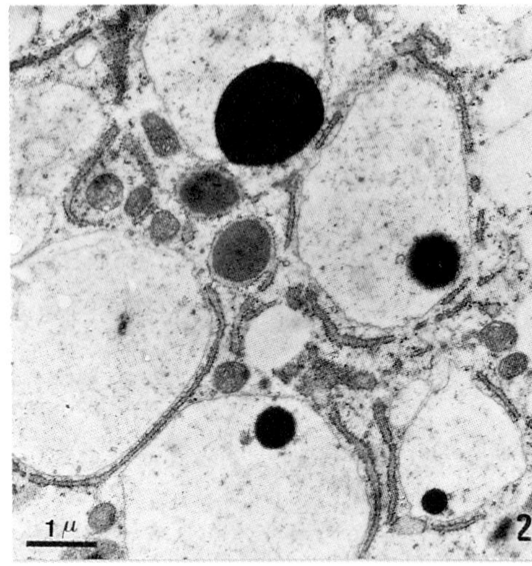

Fig. 1. Section through part of an aleurone cell. The protein bodies are surrounded by lipid bodies

Fig. 2. In this stage the vacuoles are surrounded by RER

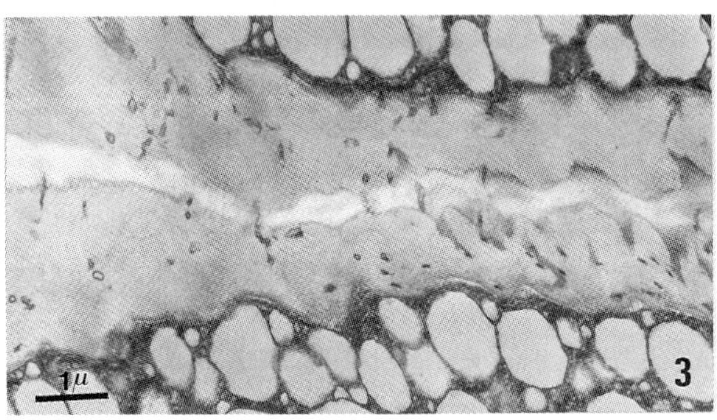

Fig. 3. The cell walls are not imune to degradation. The outer layer of the wall is digested during the germination process

Inst. Phys. Conf. Ser. No. 93: Volume 3, Chapter 3
Paper presented at EUREM 88, York, England, 1988

X-ray microanalysis of the vacuolar content in pear fruits under experimental conditions

Á Keresztes[1], E Bácsy[2], E Kovács[3] and K Bóka[1]

[1]Department of Plant Anatomy, Eötvös Loránd University, Múzeum krt. 4/A, 1088 Budapest, Hungary
[2]Institute of Experimental Medicine, Hungarian Academy of Sciences, Szigony u. 43, 1083 Budapest, Hungary
[3]Department of Microbiology, Central Food Research Institute, Herman Ottó u. 15, 1022 Budapest, Hungary

ABSTRACT: $CaCl_2$ treatment of Hardenpont pears changed the structure of the vacuolar inclusions of the hypodermis without accumulating Ca in them. In the flesh, however, Ca did accumulate in the vacuolar precipitation, accompanied by other elements.

1. INTRODUCTION, MATERIAL AND METHODS

We have been investigating the ultrastructural and chemical effects of different treatments for shelf life extension on mushrooms and fruits (Keresztes et al 1985, Keresztes and Kovács 1987, Kovács et al - in press). As a part of this project Hardenpont pears were treated by dipping in $CaCl_2$ solution immediately after picking, then stored for 2 months together with the control fruits. Ultrathin sections were prepared for conventional electron microscopy (TESLA BS 500) and semithin sections for X-ray microanalysis (JEOL TEMSCAN 100 CX operated at 80 kV in STEM mode, by tilting the specimen with 37°).

2. RESULTS AND DISCUSSION

In the hypodermal cells of the control pears dense vacuolar inclusions (presumably consisting of polyphenols, Williams 1960) developed during storage. With the application of $CaCl_2$ the inclusions became of reticular structure but did not differ compositionally from the control ones when probed by X-ray microanalysis. In the vacuoles of the control fruit flesh no characteristic peaks could be identified in the spectra of the faint precipitate (Fig. 1). $CaCl_2$ caused here a coarse granulation which contained Ca abundantly (Ca peak:background ratio being 5:1), accompanied by other elements (Fig. 2). It is an open question, how Ca accumulation could lead to concomitant sequestering of Na, K, Fe and S in the vacuole.

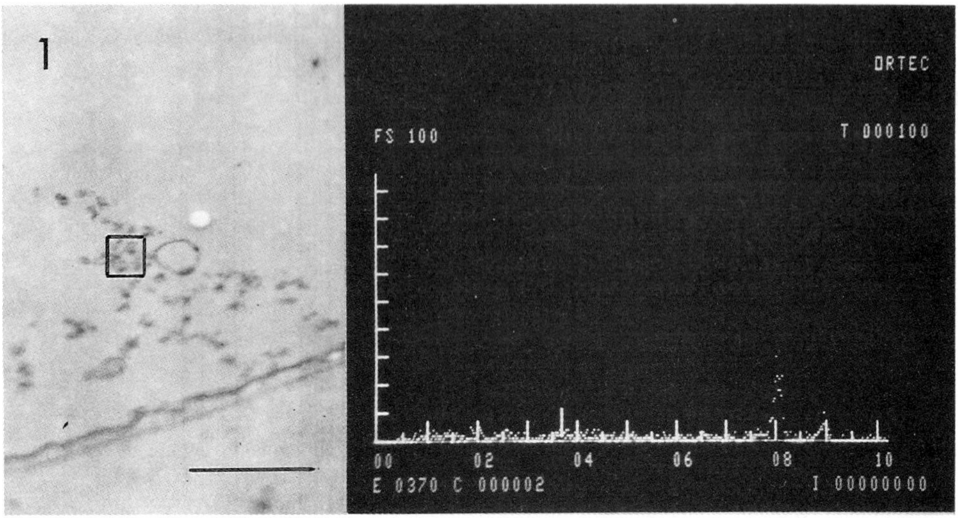

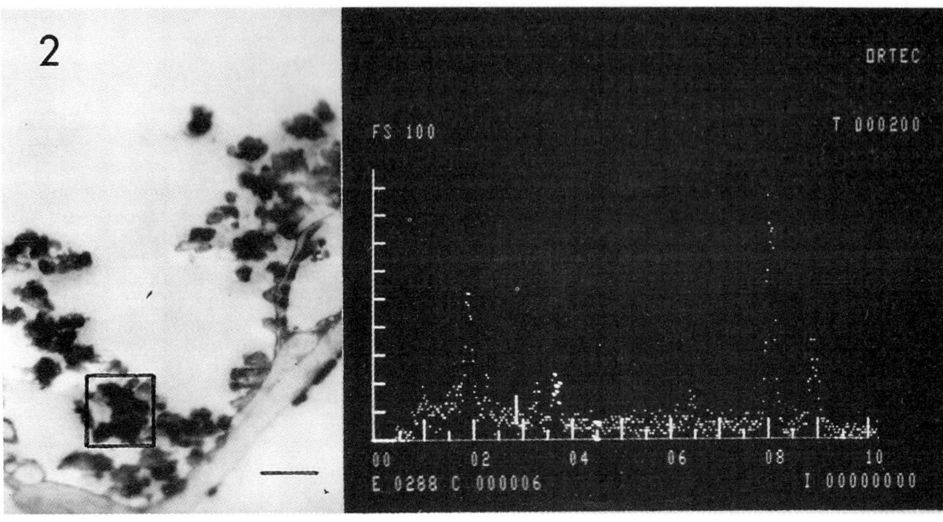

Figs. 1 and 2. Probed areas (marked) and their X-ray spectra in vacuoles of the stored control (Fig. 1) and $CaCl_2$ treated (Fig. 2) fruit flesh. Characteristic peaks in the spectrum of Fig. 2 are Na (1.0 keV), S (2.3 keV), K (3.3 keV) and Fe (6.4 keV). Counts for Ca and background were collected at 3.64 - 3.76 or 4.22 - 4.34 keV, respectively. Bars equal 1 μm.

REFERENCES

Keresztes Á, Kovács J and Kovács E 1985 Food Microstructure 4 349

Keresztes Á, Kovács E 1987 Food Microstructure 6 75

Kovács E, Keresztes Á and Kovács J 1988 Food Microstructure, in press

Williams A H 1960 Phenolics in plants in health and disease ed J B Pridham (Oxford: Pergamon Press) pp 3-7

Structural insensitivity of the fungus *Sclerotinia minor* to mercaptoethanol, an inhibitor of sclerotium formation

C Christias

Department of Biology, National Research Center "DEMOCRITOS", 153 10 Aghia Paraskevi Attikis, Greece.

ABSTRACT: In S. minor mercaptoethanol did not inhibit sclerotium formation and had no effect on hyphal ultrastructure. This contrasts with the fungus Sclerotium rolfsii in which this compound inhibited the formation of sclerotia completely and induced extensive changes in hyphal membranes. The inhibition of sclerotium formation in certain fungi may thus be related to induced ultrastructural alterations in hyphae.

1. INTRODUCTION

In the fungus Sclerotium rolfsii mercaptoethanol inhibited sclerotium formation completely at 2-4 mM without any effect on mycelial growth (Christias 1975). It was also found that mercaptoethanol induced a number of ultrastructural alterations in hyphal membranes (Daskaloyanni and Christias 1980). The fungus S. minor is nearly insensitive to mercaptoethanol. It was of interest, therefore, to investigate whether any effect of the inhibitor on fine structure analogous to that in S. rolfsii existed in S. minor. Such data may help to understand better the mechanism of sclerotium formation.

2. RESULTS

The typical growth and sclerotium formation of S. minor are shown in Fig. 1 (control) 10 days after inoculation. Abundant sclerotia were produced on the mycelium. The inhibitor had no effect on sclerotium formation at 3mM. Mycelium grown at 3mM mercaptoethanol was indistinguishable from the control. At higher concentrations the mycelial growth was inhibited. Fig. 2 shows mycelium grown at 6mM. Abundant sclerotia were produced on the mycelium. Complete inhibition of sclerotium formation was not possible in this fungus even at concentrations at which the mycelial growth was greatly inhibited.

Electron microscopy showed a hyphal ultrastructure typical of fungi, i.e. nuclei limited by a double nuclear membrane bearing numerous nuclear pores, mitochondria with well developed cristae, extensive endoplasmic reticulum, fat globules (Fig. 3). No structural alterations due to the use of the inhibitor were observed. It appeared though that the cytoplasm retained the electron stains less strongly as compared to the control (Fig. 4).

3. DISCUSSION

The mechanism of sclerotium formation in fungi is not well understood in spite of extensive research efforts (Chet and Henis 1968, Chet et al 1966). It is obvious that any compound which may interfere with normal sclerotium development is interesting since it can be used as a research tool for the study of the mechanism of sclerotium formation.

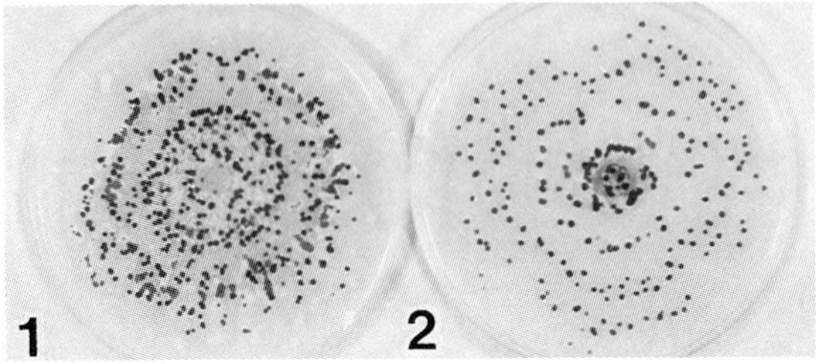

Fig. 1. Sclerotium formation in S. minor (Control).
Fig. 2. Mycelium grown at 6mM mercaptoethanol

In this fungus mercaptoethanol did not inhibit sclerotium formation and had no effect on the ultrastructure of the fungal hyphae. On the contrary, in the fungus S. rolfsii this same compound inhibited the formation of sclerotia completely and induced many ultrastructural alterations in the hyphal membranes. It appears, therefore, that the inhibition of sclerotium formation is related to the induction of changes in the membrane system of hyphae. The mechanism by which these changes are effected is not yet understood but preliminary data indicated that such changes may affect the energy yield in hyphae of sensitive fungi in such a way that they become unable to produce sclerotia.

4. REFERENCES

Chet I and Henis Y 1968 J. Gen. Microbiol. 54 231
Chet I, Henis Y and Mitchell R 1966 J. Gen. Microbiol. 45 541
Christias C 1975 Can. J. Microbiol. 21 1541
Daskaloyanni M and Christias C 1980 Electron Microscopy 1980 Vol. 2 466

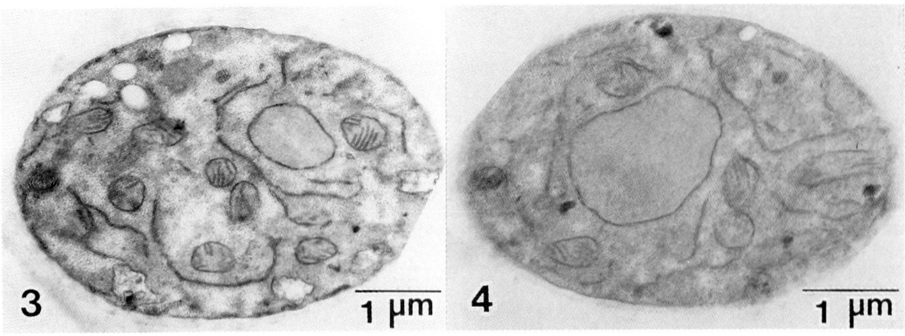

Fig. 3. Typical oblique hyphal section of S. minor
Fig. 4. Oblique section of hyphae grown in 6mM mercaptoethanol

Scanning electron microscopy of calcium-rich crystals at apices of cystidia on gills of the basidiomycete, *Inocybe asterospora*

D. Jones[1], W. H. McHardy[2], M. J. Wilson[2] and Clare Alexander[1], Departments of Microbiology[1] and Mineral Soils[2], The Macaulay Land Use Research Institute, Craigiebuckler, Aberdeen, AB9 2QJ, Scotland, United Kingdom.

ABSTRACT: Crystals at the apices of cystidia on gills of the Basidiomycete fungus Inocybe asterospora have been examined in a scanning electron microscope. X-ray micro-analysis revealed calcium as the major component and chemical tests coupled with light microscope observations strongly suggest that the crystals are calcium oxalate.

1. INTRODUCTION

Cystidia (sterile cells) found on the spore-bearing regions of Inocybe spp. are of diagnostic value and are of two main types (Dennis 1981). They are either 'smooth, hair-like or club shaped', on the gill-edges, or spindle-shaped (or flask-shaped) and crowned at their apices with a 'crest of small crystalloid bodies'. The latter cystidia are found on the gill-face as well as the gill-edge (Dennis 1981). Since there is no published chemical data or ultrastructural details on the 'crystalloid bodies', we present such information on crystal clusters located at the apices of cystidia of the Basidiomycete, Inocybe asterospora.

2. MATERIAL AND METHODS

Fruiting bodies of the mycorrhizal fungus Inocybe asterospora Quel were found in trays and pots containing peat and sand and planted with Sitka spruce [Picea sichensis (Bong.) Carr] and Scots pine (Pinus sylvestris L.) mixtures or with Scots pine seedlings only. The tree seeds were germinated in a forest soil-peat-sand mixture and, since the trays were in close proximity in the glasshouse, there was a spread of fungi from one tray to another and it is likely that I. asterospora came from the forest soil.

Fruiting bodies of I. asterospora were fixed in glutaraldehyde and osmium tetroxide (Jones et al 1987), critical point-dried and sputter-coated with gold before being examined in a scanning electron microscope equipped with an energy dispersive X-ray analyser. For electron probe micro-analysis, unfixed specimens were freeze-dried, mounted onto aluminium stubs with conductive carbon cement and coated with carbon.

3. RESULTS AND DISCUSSION

A typical crystal cluster at the apex of a cystidium from a chemically fixed and critical point-dried gill is shown in Figure 1 and an X-ray distribution map for calcium of this area is illustrated in Figure 2. There is a concentration of calcium in the cystidium itself as well as in the crystals and this effect can be more clearly demonstrated by back-scattered electron imagery where contrast is influenced by atomic number differences.

Fig. 1. SEM of crystals at apex of cystidium of I. asterospora.

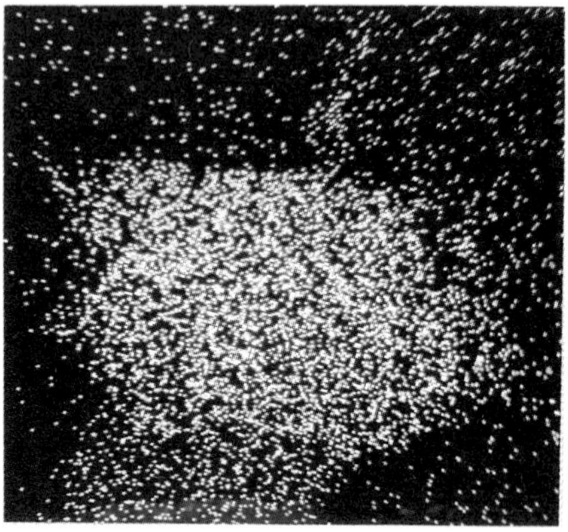

Fig. 2. X-ray distribution of calcium in crystals shown in Fig. 1.

Energy dispersive spectra of the crystals show only Ca. Some 15 µm from the tip there is still mainly Ca with a small amount of K. Near the base of the cystidium there is no Ca; here the main inorganic component is K with small amounts of P, S and Cl.

Attempts to isolate the crystals from cystidia, or record an interpretable X-ray powder pattern from separated gill material, failed. However, the crystals were observed to be soluble in 0.01 M HCl but insoluble in 0.1 M acetic acid proving that they are not $CaCO_3$. This information, together with the observation that their birefringence was high 3rd order and refractive index greater than 1.535, strongly suggests that the crystals are calcium oxalate.

These results are in keeping with those reported by Van Beeck and Scheuemann (1982) on a scanning electron microscopy and X-ray microprobe study of encrustations on cystidia of the basidiomycete Peniophora incarnata.

4. REFERENCES

Van Beeck M and Scheuemann D W 1982 Proc. 10th Int. Congress on Electron Microscopy. Hamburg, pp 407-408.
Dennis R W G 1981 Common British Fungi (Surrey, England: Saiga Publishing Co. Ltd.).
Jones D McHardy W J and Alexander Clare 1987 Scanning Microscopy Vol 1 Number 2, 1423-1429.

Vacuolar activity and phagocytosis as responses to allelopathic stress

LOVETT, J.V., and M.Y. RYUNTYU, Department of Agronomy and Soil Science, University of New England, Armidale, Australia.

Tropane alkaloids, scopolamine and hysocyamine, are liberated from seeds of thornapple (*Datura stramonium* L.) an important weed, and interfere with the growth of seedling roots of some crop plants, including sunflower (*Helianthus annuus* L.). TEM of damaged sunflower root tip cells exposed to thornapple allelochemicals at 0.5% and 0.05% concentrations shows cell division apparently disrupted at metaphase and anaphase.There is evidence also of damage to mitochondria, increasing with concentration of allelochemicals (Fig. 1). Inhibition of mineral uptake in root tip cells treated with allelochemicals has been attributed to alteration of cellular membrane function (1) and to inhibition of electron transport in the mitochondria (2).

Our observations of disordered cellular development, damage to mitochondrial matrices (Exp. 1) and membranes (Exp. 2) and accumulation of food materials may be interpreted as a failure of energy transfer, with the inevitable consequence of reduced seedling root growth.

Sequence A (Figure 1) shows the association of endoplasmic reticulum and Golgi complex with the formation of phagocytic vacuoles (pv). Vacuolar development becomes more pronounced as concentration of allelochemicals increases. Sequestration of cytoplasm and invaginations containing damaged mitochondria are observed in Exp. 1. Phagocytic vacuoles are involved in the breakdown of cell components (cw, n, m).

Sequence B (Figure 1) shows changes in cells at 0.05% (Exp. 1) and 0.5% alkaloid concentration (Exp. 2) as compared with sterile water control. Changes to cell walls and nuclei and an accumulation of food materials (A), particularly around the nucleus (Exp. 2), become more apparent as concentration of allelochemicals increases (Exps. 1, 2).

1. Balke, N.E. 1985. "Effects of allelochemicals on mineral uptake and associated physiological processes". In: The Chemistry of Allelopathy, (Ed. A.C. Thompson). American Chemical Society Symposium, Series 286, Washington, D.C.: 161.

2. Moreland, D.E. and Novitzky, W.P. 1987. "Effects of phenolic acids, coumarins and flavonoids on isolated chloroplasts and mitochondria." In: Allelochemicals: Role in Agriculture and Forestry (Ed. G.R. Waller). American Chemical Society Symposium, Series 330, Washington, D.C.: 247.

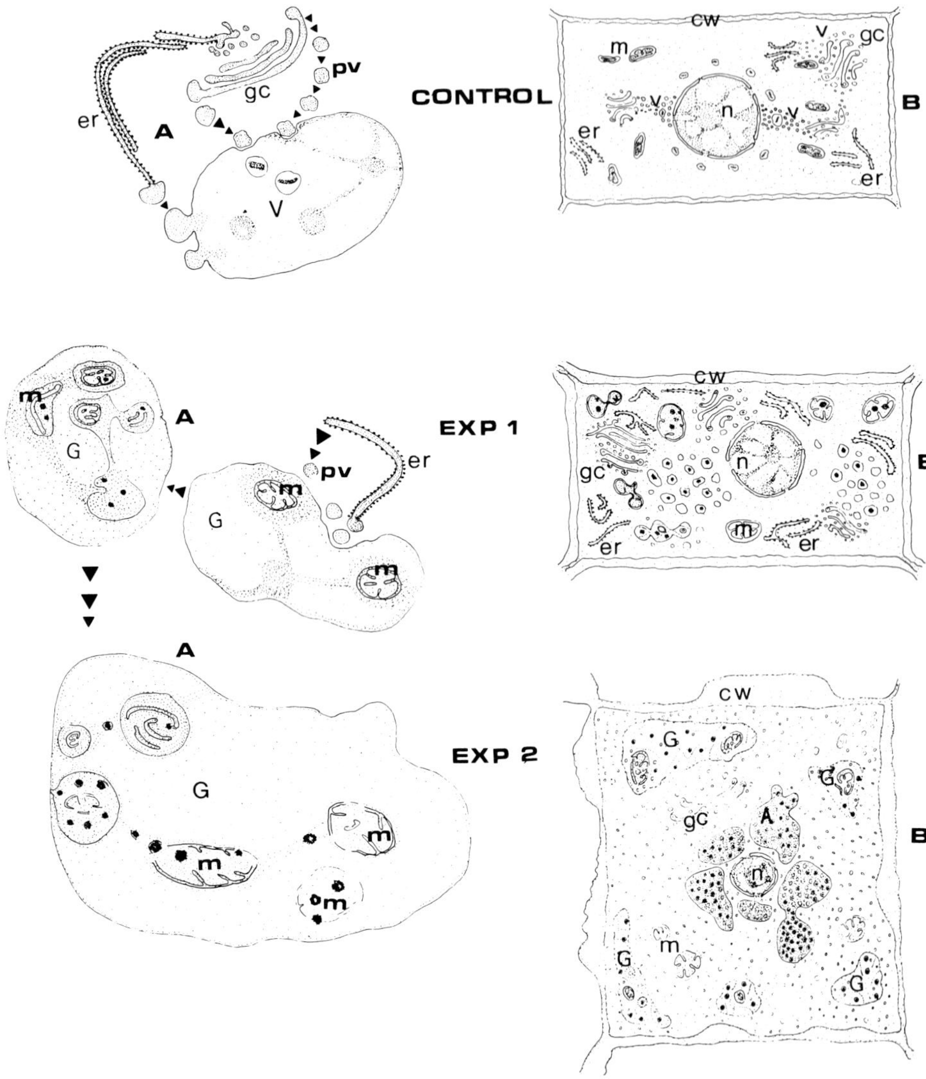

Figure 1.
Morphological features of sunflower root tip cells after 48 h germination at constant temperature.

C = control, sterile water.
Exp. 1 = 0.05% alkaloid treatment
Exp. 2 = 0.5% alkaloid treatment
Abbreviations: A = amyloplast; CW = cell wall; er = endoplasmic reticulum; gc = Golgi complex; pv = phagocytic vacuole; G = giant phagocytic vacuole; m = mitochondrion; n = nucleus; v = vacuole.

Inst. Phys. Conf. Ser. No. 93: Volume 3, Chapter 3
Paper presented at EUREM 88, York, England, 1988

Anabaena azollae: an autotrophic or heterotrophic organism?

R Tavares, J Regalado and F Carrapiço

Departamento de Biologia Vegetal. Faculdade de Ciências de Lisboa, Bloco C2, Campo Grande, 1700 Lisboa

The free-floating aquatic pteridophytes in the genus Azolla contain an N2-fixing cyanobacterium, Anabaena azollae, as a symbiont. Over the last few years, several studies have been made to elucidate the functional relationship between these two partners. One of the most interesting features is the role played by the blue-green algae, Anabaena azollae, in the association. Filaments of A. azollae are localized in a cavity of the dorsal lobe of the fern's leaves, where special ecological conditions stimulate high heterocyst frequency and a vegetative cell differentiation during leaf development. The Anabaena colony associated with the shoot apex comprises generative filaments without heterocysts and, therefore, unable to fix nitrogen. In matures leaves, the Anabaena filaments cease to grow and differentiate heterocysts which are the sites of N2 fixation. Thus the endophyte can supply the association with its total N requirement. In other plant-cyanobacterial symbiosis, the capacity to differenciate heterocysts is identical, but the endophyte loses its capability to fix CO2 and becomes dependent upon the plant for a source of fixed carbon. It means that the photosynthetic carbon reduction pathway is not working in the cyanobacteria.

One important enzyme of this pathway is ribulose-1,5-bisphosphate carboxylase/oxygenase (RuBisCO) and its immunocytochemical detection in the chloroplasts of the Azolla mesophyll cells (Carrapiço and Tavares 1988) confirms the capacity of these organelles to fix CO2. The immunocytochemical localization of RuBisCO in the blue-green algae vegetative cells shows, either in the generative or mature filaments, that the enzyme is present in special polyhedral structures identified as carboxisomes.

Previous data presented by several authors, namely Peters e Calvert (1983), suggested that the cyanobacterium could have a transition mode from photoautrotophic, in generative filaments, to photoheterotrophic or mixotrophic metabolism in mature filaments. This transition was followed by the increasing of differentiation of heterocysts. Our results suggest that Anabaena azollae vegetative cells conserve RuBisCO during the symbiosis development and, probably, are able to fix CO2 during the different stages of the association. It means that the autotrophic condition of the Anabaena cells is maintained during the process of symbiosis and is not inhibited by the photosynthate sugars produced by the host and translocated to the cavity, where the endophyte lives. However, it can not be excluded the possibility that RuBisCO in this symbiotic blue-gree algae can act mainly as an oxygenase rather than a carboxylase if one admits that in cavity, as a consequence of its inner leaf localization, the CO2 partial pressure is low. This can have an important physiological consequence in lowering the O2 concentration in the cavity, which can affect positively nitrogenase activity.

References

Carrapiço F and Tavares R 1988 Plant and soil, in press

Peters G A and Calvert H E 1983 The Azolla-Anabaena Symbiosis. Algal Symbiosis: a Continuum of Interaction Strategies (New York: Cambridge University Press) pp 109-145

Determination of the accessibility of lignocellulosic substrates to enzymatic degradation by immunoelectron microscopy

Ewald Srebotnik and Kurt Messner

Institut für Biochemische Technologie und Mikrobiologie, Abteilung Mykologie, TU-Wien, Getreidemarkt 9, A-1060 Wien, Austria

ABSTRACT: Pine decayed by wood rotting fungi was infiltrated with a concentrated culture filtrate of the white rot fungus Phanerochaete chrysosporium and labelled for lignin peroxidase by post-embedding immunoelectron microscopy. This method enabled us to show that enzymatic attack on pine by lignin peroxidase and presumably also other enzymes of the same size is restricted to the surface of the wood cell wall.

1. INTRODUCTION

The microbial degradation of wood is carried out predominantly by wood rotting fungi. They possess a system which is capable of completely or partly degradating the components of wood. In addition to the biochemistry of the fungi, the mechanism of the microbial degradation of wood fibers largely depends on ultrastructural features of the wood cell wall that limit its accessibility to enzymes. These complicated biochemical ultrastructural correlations have not been clarified so far. The pore size of wood fibers as one of the limiting factors to enzymatic attack and the size of common enzyme molecules are within the same orders of magnitude, that is why penetrability (Cowling & Brown 1969) has always been a matter of discussion. By post-embedding immunoelectron microscopy developed in the past years it has become possible to demonstrate the diffusion and distribution of enzymes in the wood cell wall directly.

2. MATERIALS AND METHODS

Pine wood samples decayed by P. chrysosporium BKM-F1767 (Kirk et al 1986) and the brown rot fungus Fomitopsis pinicola D 34 under conditions described by Messner & Stachelberger (1984) were infiltrated with a 300-fold concentrated culture filtrate (ultrafiltration) of P. chrysosporium grown in liquid culture optimized for lignin peroxidase production (Kirk et al 1986), transferred directly to 70% ethanol, dehydrated up to 90% ethanol and embedded in LR White (Polaron, UK). Thin sections were labelled with a polyclonal rabbit anti-lignin peroxidase serum (Kirk et al 1986) and goat anti-rabbit IgG, conjugated to 5 nm colloidal gold (Janssen, Belgium) as described by Srebotnik et al (1988) and examined in a JEOL 100C TEM.

3. RESULTS AND DISCUSSION

In post-embedding immunoelectron microscopy, the antibodies bind only to the antigens exposed on the surface of the thin sections so that the

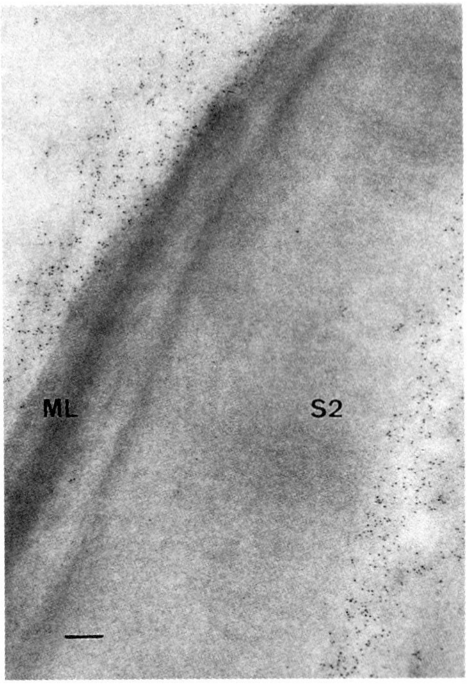

Fig. 1. P. chrysosporium on pine. Cultivation time 86 days. 25% weight loss. The upper half of the wood cell wall has been completely degraded down to the middle lamella (ML). Nevertheless, lignin peroxidase remains only on the surface. Note the border between labelled and unlabelled areas of the wood cell wall (S2 = S-2 layer). Bar = 0.2 µm.

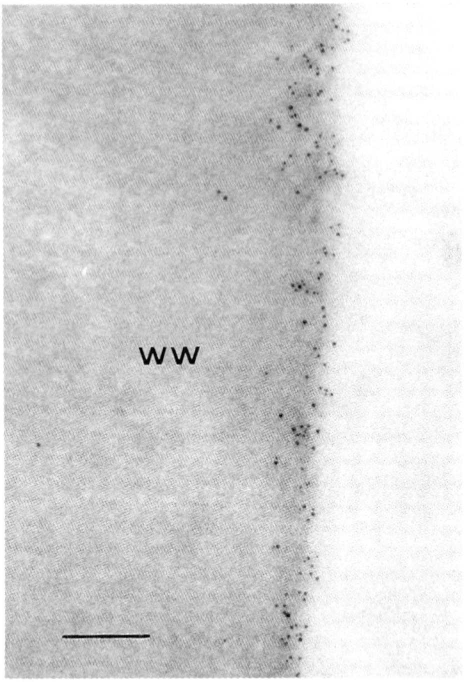

Fig. 2. F. pinicola on pine. Cultivation time 34 days. 50% weight loss. Although brown rot causes the depolymerisation of cellulose over the entire cell wall lignin peroxidase only penetrates superficial areas of the brown rotted wood cell wall (WW) indicating a non-enzymatic depolymerisation mechanism (Cowling & Brown 1969). Bar = 0.2 µm.

antibodies themselves need not diffuse into the wood cell wall when demonstrating enzymes in it. The post-embedding localisation of lignin peroxidase, which is regarded as one of the key enzymes in enzymatic lignin degradation (Kirk et al 1986), therefore clearly indicated that enzymatic wood degradation in vivo takes place only starting from the lumen surface and not in the entire wood cell wall (Figures 1 and 2). It can be assumed that this is true also for other enzymes of a similar size (42 kD), such as cellulases. Continuing these first investigations, the same method may be applied also for other proteins as well as to fundamental studies with respect to a possible use of enzymes in paper industry.

4. REFERENCES

Cowling EB, Brown W 1969 In: Hajny GJ, Reese ET (eds) Cellulases and Their Applications. Washington D. C.: American Chemical Society, pp 152-187

Kirk TK, Croan S, Tien M, Murtaugh K, Farrell R 1986 Enzyme Microb Technol 8 27

Messner K, Stachelberger H 1984 Trans Br Mycol Soc 83 113

Srebotnik E, Messner K, Foisner R, Pettersson B 1988 Curr Microbiol 16 221

Inst. Phys. Conf. Ser. No. 93: Volume 3, Chapter 3
Paper presented at EUREM 88, York, England, 1988

A comparison of ultrastructure in freeze-substituted OsO_4-fixed and chemically fixed rye leaf cells

K Pihakaski, D M R Harvey and U-M Suoranta

Department of Biology, University of Turku, SF-20500 Turku, Finland

ABSTRACT: Non-acclimated and cold acclimated Secale cereale leaf pieces were fixed in i) glutaraldehyde/formaldehyde followed by osmium tetroxide, ii) these aldehyde fixatives with tannic acid, iii) lysine, or iv) freeze-substituted with acetone/osmium. The best overall contrast was with tannic acid fixation; after lysine fixation electron dense granular bodies were observed particularly in acclimated tissue. After tannic acid fixation electron dense lamellar bodies were observed close to the plasmalemma and in the vacuoles; these were not observed in the freeze-substituted specimens, but there were arrays of membranes parallel to the plasmalemma.

1. INTRODUCTION

Previous literature on plant cell ultrastructure concerns mostly material chemically fixed for TEM; however, conventional chemical fixation does not adequately preserve many structural components. We have therefore used tannic acid or diamine (lysine) with glutaraldehyde/formaldehyde to improve the contrast of delicate cell components. We also compared chemically fixed material with the same material prepared by freeze-substitution. Rapid freezing and freeze-substitution allow fixation of tissues at low temperature and should produce less chemical artefacts than conventional fixation. The present study shows results from both of these techniques and by using non-acclimated and cold acclimated material sheds some light on the structural basis of cold hardening.

2. MATERIALS AND METHODS

Seedlings of winter rye (Secale cereale cv Voima) were germinated at 295 K and grown for 10 d, with 16 hrs light period (non-acclimated, NA) or transferred after one week to 278 K (cold acclimated, A). 0.5 x 1.0 mm pieces from the centre of leaf blades were fixed chemically in:

i) 2.5% glutaraldehyde(GA)/1.0% formaldehyde(FA) for 4 hrs at 295 K or
ii) 2.5% GA/1.0% FA + 0.1% tannic acid(TA) (Tiwari & Gunning 1986) or
iii) 2.5% GA/1.0% FA + 0.05 M lysine for 10 or 30 min (Boyles 1984)

made in 0.05 M Na-cacodylate buffer + 5 mM $MgCl_2$ for non-acclimated or 0.1 M Na-cacodylate buffer + 5 mM $MgCl_2$ for acclimated specimens. The specimens were washed in buffer, dehydrated in graded acetone and embedded in Spurr's (1969) resin. Alternatively specimens were prepared by freeze-substitution with osmium tetroxide (Harvey & Pihakaski, in this volume). Thin sections were stained with uranyl acetate and lead citrate, and examined with an JEM-100CX electron microscope (Jeol).

3. RESULTS AND DISCUSSION

The ultrastructure of chemically fixed mesophyll cells did not differ strikingly. There were, however, slight differences between different fixations. Addition of tannic acid fixative produced generally better contrast of both the ground cytoplasm and membranes, than lysine or aldehydes alone. After all three different fixations various unusual membrane configurations were observed in the mesophyll cells either in the cytoplasm close to the plasmalemma or, frequently, also in the vacuoles (Figures 1 and 2). It seemed that they were more frequent in

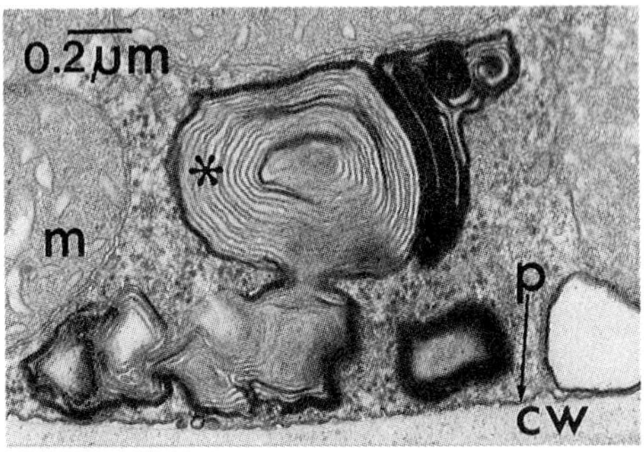

Fig. 1. Membrane whorls(*) in a mesophyll cell of non-acclimated rye fixed in GA/FA/TA. (cw, cell wall; m, mitochondrion; p, plasmalemma).

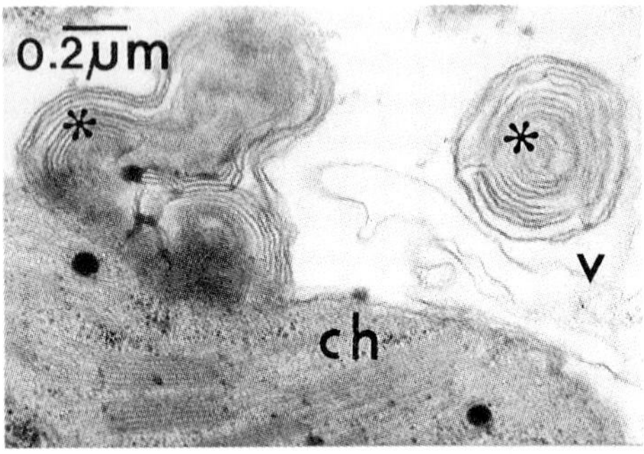

Fig. 2. A membrane whorl (*) in a vacuole and a protrusion of cytoplasm in a lysine-fixed NA-leaf tissue (ch, chloroplast).

non-acclimated than in cold acclimated tissues. In the freeze-substituted material we did not observe these membrane whorls but stacks of membrane strands parallel to the plasmalemma were observed (Figure 3).

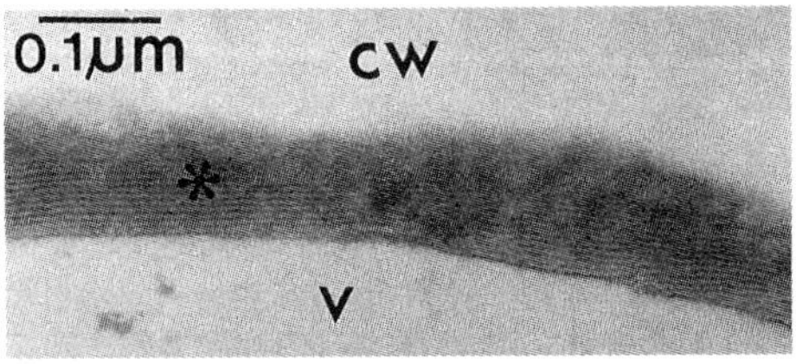

Fig. 3. Stack of membrane strands (*) in freeze-substituted non-acclimated mesophyll cell (v, vacuole).

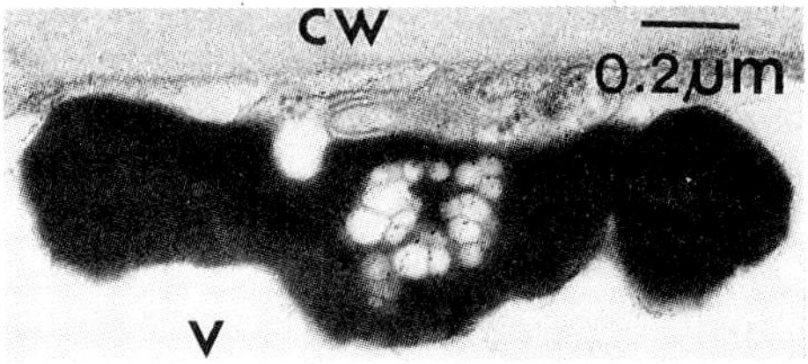

Fig. 4. An electron dense granular body attached to a thin strand of cytoplasm in a lysine-fixed A-cell.

Electron dense bodies, which at high magnification were seen to be composed of tightly packed membrane whorls were present after fixation with tannic acid. The electron dense bodies in the specimens fixed in lysine were granular in structure (Figure 4). Very seldom were microtubules or microfilaments visible in either non-acclimated or cold acclimated material. Preliminary observations revealed that there were more plasmodesmata in the non-acclimated than in cold acclimated tissue.

4. REFERENCES

Boyles J 1984 The Science of Biolgical Specimen Preparation eds J-P Revel T Barnard G H Haggis (Chicago: SEM, Inc) pp 7-21

Harvey D M R and Pihakaski K 1988 The present volume

Spurr A R 1969 J. Ultrastr. Res. **26** 31

Tiwari S C and Gunning B E S 1986 Protoplasma **134** 1

MEATSEM (moist environment ambient temperature scanning electron microscopy) or the SEM of plant materials without water loss

A N Farley*, A Beckett** and J S Shah

H.H. Wills Physics Laboratory, University of Bristol, Tyndall Ave, Bristol, BS8 1TL.
* Now at North Staffordshire Polytechnic, Department of Physics, College Rd, Stoke -on-Trent, ST4 2DE.
**Department of Botany, Woodland Rd, Bristol, BS8 1UG.

ABSTRACT: Biological materials can suffer significant morphological and/or compositional changes during the various preparative techniques necessary for conventional SEM. A considerably improved and relatively new technique was employed which requires no preparative treatments but allows the SEM of fully hydrated specimens at ambient temperature. Perianth cells of *Narcissus sp.* are sensitive to dehydration damage and were used to evaluate the new techniques with reference to LTSEM where dehydration artifacts are minimal.

1. INTRODUCTION: It is advantageous to view biological materials in a fresh and fully hydrated state (7,4). Detectors for secondary electrons cannot be used under the high electron scattering conditions which are necessary to maintain hydration. Backscattered electron (BSE) detectors have been used for viewing a few materials, which are less prone to dehydration or beam damage, at low to medium magnifications. Delicate structures are difficult to image with currently available detectors.

2. MAINTENANCE OF HYDRATION: Water loss is prevented in MEATSEM by placing the specimen in close thermal contact with a surrounding annular water reservoir supported by an additional, variable temperature, water supply. The specimen is maintained at 100% humidity by isolating it from the high vacuum of the rest of the microscope by a differentially pumped stage (6,2).

3. IMAGE IMPROVEMENT: Instead of using a BSE detector, a novel method of signal extraction was incorporated (3). Detrimental effects of ionization were minimised by extracting low energy secondary electrons and ionization products with a 1000 Vm^{-1} field gradient produced by a biasing electrode positioned above the specimen. The image retrieved by detecting the specimen current signal (2,5) was of high resolution and incorporated both the SE and BSE signals.

4. MATERIALS AND EVALUATION: Untreated, hydrated, perianth segments of *Narcissus sp.* were maintained and examined at a saturated water vapour pressure of 1716 Pa. Fully frozen-hydrated, partially freeze-dried, coated and uncoated specimens were prepared for LTSEM (1). Image evaluation was based upon surface morphology, degree of beam damage, presence or absence of charging, resolution and contrast.

5. RESULTS AND DISCUSSION: Surfaces of fully frozen-hydrated (FFH) biological specimens are often partially obscured by ice (Fig.1). Etching will remove this but such specimens are then partially freeze-dried (PFD) and care is needed to avoid undue morphological and dimensional changes in these samples. Uncoated FFH and PFD specimens tend to charge even at low accelerating voltages (Fig.2) and most FFH

biological samples are susceptible to beam damage (1). In contrast, there is no surface ice in MEATSEM and charging does not occur (fig.3). However, it is necessary to use low beam currents (<100pA); this avoids beam damage which otherwise can be recognised as localised areas of cell collapse. Despite the low emissivity, undistorted images with good contrast are obtainable. MEATSEM has also been used successfully for viewing other botanical specimens such as urediniospores of the bean rust fungus *Uromyces viciae-fabae,* conidiophores and conidia of *Aspergillus niger* and frustules of the diatom *Navicula phyllepta* (2). Further developments of this techique will allow the study of dynamic changes in specimen morphology and properties.

6. REFERENCES: 1. Beckett A and Read N 1986 Low Temperature Scanning Electron Microscopy . In: *Ultrastructure Techniques For Micro-organisms.* edit; Aldrich H C and Todd W J. Plenum Press.
2. Farley A N 1987 Ph.D Thesis, University of Bristol.
3. Farley A N and Shah J S 1988; this proceedings.
4. Robinson V N E 1978 *Scanning,* **1**, 149.
5. Shah J S 1987 *Spectrum* , No. 208, 6.
6. Shah J S 1977 British Patent 1477458.
7. Shah J S and Beckett A 1979 *Micron,* **10,** 13.

Figures 1-3 *Narcissus sp.*
Epidermal cells of perianth segments.

Figure 1. LTSEM; FFH; coated
9 kV; bar = 10 microns.

Figure 2. LTSEM; PFD; uncoated
3kV; bar = 10 microns.

Figure 3. MEATSEM 10 Kv.
1716 Pa water vapour pressure.
+ 16 °C. bar = 10 microns.

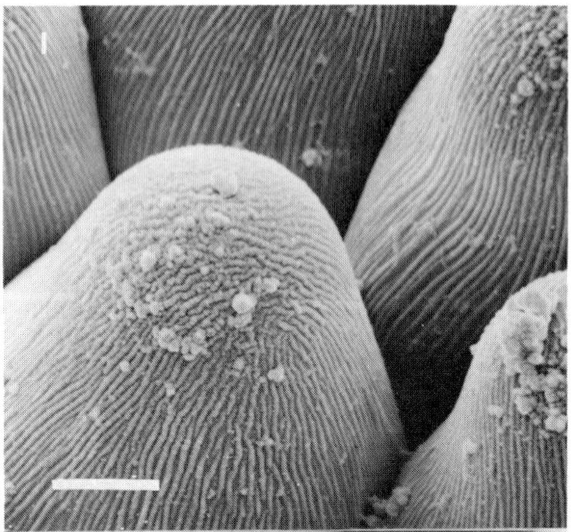

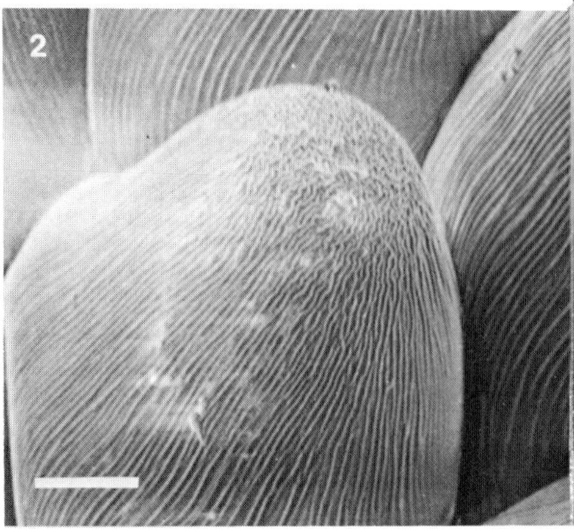

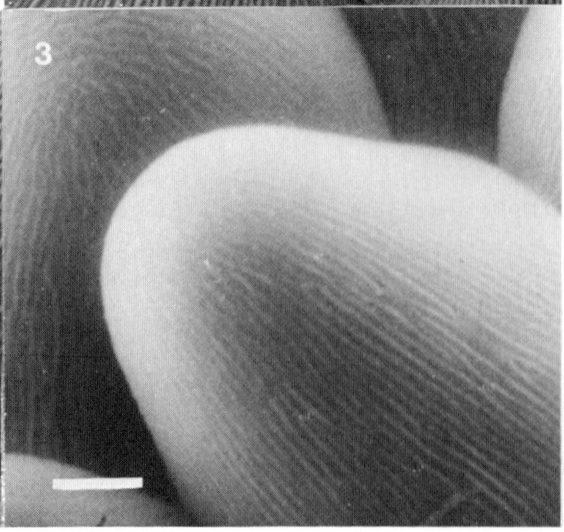

Scanning electron microscopy of skeletal mineralized tissues

A Boyde and SJ Jones

Department of Anatomy and Developmental Biology
University College London, Gower St., London WC1E 6BT

ABSTRACT: We review some of the advances which have been made through the application of SEM to mineralized tissue biology. SEM has played a central role in unravelling problems and improving our understanding of the organisation and development of the mineralized tissues, because these tissues are well suited to the techniques involved and the methods have so much to contribute.

1. INTRODUCTION

The mineralized tissues are distinguished from the soft tissues of the body in being literally hard to prepare for other methods of microscopy. Their matrix and mineral surfaces are the principal regions of activity and interest, yet are simple to prepare for SEM and undergo little deformation in the process. The 3-D structure of a hard tissue is determined at the formative front or interface with the cell type which secretes the organic matrix. By removing the cells from the usually comparatively stable matrix surface, both the mechanisms of organization and the complexities of structure can be understood. The formative (and destructive) cells are arrayed on extensive, rather rigid, surfaces and can be relatively easily displayed. The process of mineralization (which occurs at some depth within, and roughly parallel with, the formative surface) can be witnessed by digesting the organic matrix. Useful information concerning the mineralized matrix structure may be obtained by fracturing the tissue, which will part more cleanly than the soft tissues because it is rigid. Such breaks expose entombed cells and the spaces which they occupy and can also be interpreted to aid our understanding of the fracture propagation process.

From the technical viewpoint it is interesting to note that the earliest recorded SEM studies of hard tissues used backscattered electron (BSE) compositional contrast and x-ray emission imaging and analysis of polished, embedded specimens and topographic contrast BSE imaging of ion beam etched surfaces. After the introduction of commercial SEMs fitted with the Everhart-Thornley detector, there followed a long period in which mainly secondary electron (SE) topographic contrast imaging of specimens with naturally rough surfaces was employed. More recently, there has been a swing back to compositional and topographic imaging using BSE, and some important use of cathodoluminescence (CL) imaging to study discrete aspects of composition.

2. THE SKELETAL TISSUES - SE TOPOGRAPHIC IMAGING

The kinds of surface which can be prepared for examination by SE-SEM are the free external and internal surfaces of bone, cut surfaces which may be generated in the process of exposing the former class, and fractured surfaces which may either be used to expose the free surfaces or to study the nature of bone fracture propagation. The components of bone matrix surfaces are secreted by osteoblasts and, when mineralized, become bone.

They may be added to by deposition of new layers of bone matrix, mineralization usually following in a layer-wise pattern, and they may be resorbed by the influence of special multi-nucleated cells, the osteo-clasts. SEM enables one to differentiate between these forming, resting, and resorbing surfaces. The identification of forming surfaces necessitates the removal of the unmineralized bone matrix to expose the mineralizing front. In adult lamellar bone, this shows mineral replicas of incomplete segments of collagen fibre bundles contrasting with the continuous collagen fibrils and fibre bundles of the matrix itself. If no visible change occurs in the detailed morphology of the bone matrix surface consequent upon an attempt to remove it by dissolution or by oxidation by plasma ashing, then the limits of the bone matrix collagen compartment and the mineralised phase of the bone were confluent and we identify a resting surface. The highest proportion of bone surfaces in mature adult humans is of this resting type with no superficial unmineralized osteoid. A proportion of such resting surfaces, however, show a much smoother texture in which the process of mineralization has spread outside the collagen fibre containing compartment of the bone matrix into a layer of superficial "ground substance": these are the prolonged resting bone surfaces. Resorbing and resorbed surfaces are recognized by the characteristic excavations caused by the osteoclasts. Just the simple ability to examine bone surfaces in this way has contributed much to the understanding of this tissue. For example, we now appreciate the nature of the branching bundle pattern within each successive layer of collagen on the bone matrix surface, and the extent of the domains in which the collagen is roughly parallel on a patch of the bone surface. From studies of isolated bone cells in culture, we can see that this patchwise orientation of groups of osteoblasts which gives rise to the oriented collagen pattern is a property of the behaviour of these cells. We can infer that the orientation of the collagen depends upon the freedom of movement of the cells on the matrix surface: when the cells are no longer able to move, as occurs when an osteoblast is entrapped in the bone matrix to become an osteocyte, the last collagen which it makes may no longer be oriented. We can understand that the process of mineralization in the lamellar bone is one which spreads through the collagen component of the matrix preferentially (Boyde, 1972; Jones and Boyde, 1972, 1977). Maturation, which is an increase in the degree of mineralization of the bone matrix, occurs mainly in the water-filled space of the ground substance compartment of the bone matrix.

An important contribution to our knowledge came from the observation of the differences between the immature, foetal, or woven form of this tissue in comparison with the adult type. SEM study makes obvious the difference in the intermediate level of organisation of the two tissue types. The very cellular woven bone is nearly always present as a fine trabecular form in which the trabeculae occupy roughly half of the available tissue volume. This contrasts with the more usual forms of adult lamellar bone which are either mostly bone (compact bone) or mostly space (spongy or trabecular bone) where 25% of the volume would typically be occupied by bone, the remainder being occupied by haemopoietic tissue or fat storage tissue. SEM allowed the characterization of the organization of the collagen fibres of the woven bone matrix which form a random feltwork of a variety of fibril and fibre bundle sizes. Even more marked are the differences in the morphology of the mineralizing front in woven bone in anorganic or deproteinised specimens: here it is not possible to determine the collagen fibril or fibre orientation, but large numbers of small, separate centres of mineralization are evident. This implies that the mineralization process is spreading through both the ground substance and the collagen compartments of the tissues simultaneously. An analogous spread of mineralization as "calcospherites" occurs in both cartilage and dentine mineralization, but these may reach much larger dimensions than are encountered in foetal bone (Boyde and Jones, 1983; Jones and Boyde, 1984). A remarkable SEM finding stemmed from attempts to clean foetal bone matrix surfaces by removing osteoblasts by trypsinization. It was found that the foetal bone osteoid is largely destroyed by trypsin right down to the level of the mineralizing front (Boyde, 1972). Since it is known that trypsin cannot attack collagen, it has to be supposed that in some way tryspin activates a latent enzyme system built into the bone matrix, secreted by the osteoblasts. Adult osteoid collagen is not affected by trypsin.

It is possible to use SEM observations to characterize the effects of the major, circulating, bone seeking hormones upon the morphology of the free bone surfaces just considered. Parathyroid hormone (PTH, administered as parathyroid extract to rats) caused an increase in the extent of resorption of those parts of flat bone surfaces immediately surrounding the entry points of blood vesssels into the underlying solid bone. However, a high proportion of free bone matrix surfaces acquired a "prolonged resting" morphology, resulting from the mineralization of superficial osteoid under these abnormal circumstances contrasting with the controlled mineralization of deep osteoid at its deep surface, the mineralizing front, in the normal situation (Jones and Boyde, 1970).

3. BONE CELLS

The study of bone cells using SEM was pioneered by Jones (1973). The preservation of bone cells on bone surfaces presents perhaps the most extreme difference in physical properties of two adjacent materials found in nature, the highly hydrated cells contrasting with the extracellular matrix of bone in which the great proportion of the water-filled space of the matrix is occupied by bone mineral. Thus the cells shrink more than the bone matrix on any method of drying. This results in the formation of cracks between the osteoblasts, which delineate the cells. However, it should be borne in mind that the cells are, in vivo, in close contact all around their peripheries. The ability to study osteoblasts as a sheet on tabular bone surfaces such as the endocranial aspect of the rat calvarium allowed the establishment of values for the secretory territories of the individual cells (Jones, 1973). Changes in the morphology of the cells in this sheet could be studied as a consequence of experimental interferences in an organ culture system. In this respect, the effects of calcitonin (CT), PTH and $1,25(OH)_2D_3$ are of greatest interest. CT has no effect upon the morphology of osteoblasts in organ culture, to which these cells adapt by showing an atypical dorsal surface ruffling. However, SEM showed a profound influence of PTH in reducing the extent of this dorsal ruffling and inducing shape changes in the osteoblasts. Clefts of an extent which cannot solely be explained by drying shrinkage artefacts occur between the osteoblasts in the sheet in the earlier time intervals. Later, the osteoblasts spread to produce a multi-layered sheet in which the cells elongate. Removal of PTH by returning to control culture medium was associated with the occurrence of numerous mitoses in the osteoblastic sheet after a few hours. Thus SEM studies provided a clue that the PTH control mechanism for bone resorption lay in an initial control of the osteoblastic sheet. Further, they provided evidence for the mitogenic and bone formation stimulatory activity of PTH (Jones and Boyde, 1976). Recent experiments have shown that the PTH effect on osteoblast shape is not mediated by prostaglandins (unpublished). Active metabolites of vitamin D were also shown to have a direct effect on osteoblast shape.

The characterization of osteoclasts by SEM permitted new insights into the function of these cells. Active osteoclasts may be found to be removing bone in separate sites simultaneously, other non-resorbing portions of the same complex cell being separated at a distance from the bone and sometimes by an underlying layer of osteoblasts. Surfaces where osteoclasts have been active are also characterized by the appearance of another cell type - the released osteocyte (Jones and Boyde, 1977). This cell type, which lies entombed in bone matrix, has been assumed by innumerable authors to participate in calcium homeostasis by itself functioning as a resorptive or demineralizing cell on the surrounding bone matrix. One of the contributions of SEM morphology was to make possible the survey of thousands of osteocyte lacunae, in none of which was any evidence for the attack of the cell on surrounding matrix ever found. The discovery that the released osteocyte was a vital cell which did not fuse with the osteoclasts but which was able to migrate from its previous lacunar bed confirmed the impression that these cells do not participate in bone removal. They almost certainly, however, play an active role in keeping open the canalicular mechanism through bone. Recent studies have shown that osteocytes in bones which are not remodelled, such as those of the auditory ossicular chain, may become mineralized. SEM evidence was not only influential in questioning the significance of the theory of osteocytic osteolysis, but also contributed to understanding how this idea arose. From the examination of the formative surfaces in those bone disease conditions in which osteocytic osteolysis was presumed to have occurred, it could be shown that the mineralization of the matrix surrounding the osteocytes was abnormal in the first instance, remaining incomplete in the adult tissue. Such conditions are frequently associated with a high bone turnover rate as a complicating factor (Boyde, 1980).

4. BSE AND CL COMPOSITIONAL IMAGING

The SE signal from rough biological samples in an SEM does not lend itself to a sensible approach to quantitative image analysis, yet there is much interest centred upon the measurement of histomorphometric parameters to allow a quantitative approach to bone pathology. In this respect the other signals due to electron beam interaction with the specimen in the SEM are more sensibly employed. The data obtained from quantitative analysis of x-rays are considered extensively elsewhere (Nicholson and Dempster, 1980; Roomans, 1980). We have been concerned with assessing the value of the BSE and CL signals. Neither of these signals can be made fully quantitative unless the influence of unknown specimen surface topoography is removed. This is achieved by embedding the bone sample in polymethylmethacrylate (PMMA, stabilised with 5% styrene) so that the whole block can be cut and polished to produce a flat surface. Micromilled surfaces have a much lower relief than those produced by diamond polishing. The large atomic number difference between mineralized bone matrix and PMMA is sufficient to produce strong contrast between bone and non-bone. Thus it is possible to produce reliable data at high acquisition rates to determine the phase volume of bone in bones. However, the newer solid state BSE detection devices have made it possible to differentiate between decimal average atomic number differences and thus between fractions of bone mineralized to different extents. We were able to use our Quantimet 720 image analysing computer to distinguish between non-bone and three mineral density fractions in bone. However, a much cheaper and more versatile system was constructed around an economical micro-computer. This arrangement, perfected for us by PGT Howell, allowed us to set up a system in which we reserved one histogram bin for a measurement of the phase volume of non-bone, another seven being allocated for the density fractions found within the normal range of mineralization in bone tissue. Thus we have established a system which allows the rapid quantitative stereology of bone, even dividing it into a number of fractions which could not previously be studied. The first survey with this system showed a general shift towards more of the bone consisting of more mineralized fractions as a function of increasing age (Howell and Reid, 1986; Reid and Boyde, 1987). Another parameter which changes with age and particularly in the post menopausal female is the phase volume of bone. Thus, in the ageing female, not only is less bone present, but that which is present is more highly mineralized and the tendency of such bones to fracture could be accounted for in this way. However, we would like to be able to study such parameters in three dimensions: the degree of interconnection between different packets of bone having different physical properties must be an important factor determining the functional integrity of this tissue.

CL can also make a contribution to SEM studies of bone. Using a simple, effective and economical CL collector, we can discriminate against the BSE contribution to the so-called CL images produced by previous detection systems (Boyde and Reid, 1983). CL was used to permit the measurement of the phase volume of osteoid (non-mineralized bone matrix). Osteoid has essentially the same density as PMMA and cannot be differentiated from the former on the basis of the strength of the BSE signal. It can be picked out from bone using the the CL signal in different ways. For example, a scintillator can be added to the embedding PMMA so that non-bone gives a high CL signal: the mineralized bone gives a high BSE signal and the osteoid is automatically recognized as the phase which gives neither. Alternatively, the osteoid may be stained by the use of a CL stain which gives rise to a high signal level and automatic recognition. It was also found to be possible to read tetracycline labels in bone. Tetracyclines are antibiotics which bind to the mineralizing front in bone and can be used to study growth in bone. SEM provides the means of reading these growth lines at higher resolution and providing an exact cross correlation with the morphological and density dependent SEM images.

6. TOPOGRAPHIC BSE IMAGING

BSE imaging has valuable uses in studying unembedded specimens of skeletal tissues. It is easy to eliminate charging effects, and difference images from multiple detector segments can be used to enhance topography. The cut surface of a bone sample in a summed image will return the highest signal level which can be thresholded to produce an image from which the stereological phase volume of bone can be measured. On resorption test samples, areas demineralised but not digested by osteoclasts may be distinguished as dark in carbon coated specimens, because of the reduced sub-surface mean atomic number, or bright in gold coated samples because of the local enhancement of surface roughening which causes the accumulation of more gold per unit area locally. Spread osteoblasts and bone lining cells in general, and some osteoclasts, are so thin that the compositional contrast of the underlying mineralized bone phase can be "seen through" the cell in both BSE and BSE dependent SE images.

7. QUANTITATION OF OSTEOCLASTIC FUNCTION - SEM STEREOPHOTOGRAMMETRY

We have used stereophotogrammetric analysis of stereo- pair SE and BSE topographic images in assaying the function of isolated osteoclasts. These cells are seeded on to flat-surfaced slices of bone or dentine. The cells make excavations in the surface of these slices and the resorptive activity can be observed using light microscopic methods. After removing the cells, the substrates are prepared for SEM and stereo pair images taken using a 20° tilt angle difference generated by a high precision goniometer stage. XY and delta X co-ordinates are measured using a specially constructed instrument - the SFS 3 stereo comparator (Ross, 1986; Boyde et al, 1986) - and these co-ordinates transferred to a computer to calculate the depth, area and volume of these features. The establishment of this method as a routine assay for the function of osteoclasts has allowed us to test the influence of drugs and hormones which might affect this process. It has also allowed us to prove that the fully 3-D approach is necessary (Jones et al, 1985; Delaisse et al, 1987). The measurement of the number of resorption pits, for example, only provides data on the settling and adhesion efficiency of these cells. The measurement of area relates to the spreading function. However, the main matter of interest in bone resorption is the volume of tissue digested. It is possible to measure the volume of tissue demineralized - the first step in the resorptive process - separately from the volume of matrix totally digested - the end result of the process. Methods for the completely automated recognition of corresponding feature points in stereo pair images have recently made it possible to conduct these measurements without the intervention of a stereo competent operator, but the operator still has to be trained in the skills of using the SEM and the computer!

8. REFERENCES

Boyde A 1972 The Biochemistry and Physiology of Bone ed G H Bourne (New York: Academic Press) pp 259-310
Boyde A 1980 Metabolic Bone Disease Related Res. **2** Suppl. 239-255
Boyde A, Howell P G T and Franc F 1986 J. Microscopy **145** 257-264
Boyde A and Jones S J 1972 Developmental Aspects of Oral Biology ed H C Slavkin and L A Bavetta (New York: Academic Press) pp 243-274
Boyde A and Jones S J 1979 Scanning Electron Microscopy/1979/**2** 393-402
Boyde A and Jones S J 1983 Cartilage, Vol.1 ed B K Hall (New York: Academic Press) pp 105-148
Boyde A and Reid S A 1983 J. Microscopy **132** 239-242
Delaisse J M, Boyde A, Maconnachie E, Ali N N, Sear C H J, Eeckhout Y, Vaes G and Jones S J 1987 Bone **8** 305-313
Howell P G T and Reid S A 1986 Scanning **8** 139-144
Jones S J 1973 PhD Thesis University of London
Jones S J and Boyde A 1970 Scanning Electron Microscopy/1970 pp 193-200
Jones S J and Boyde A 1976 Cell & Tissue Res. **169** 449-465
Jones S J and Boyde A 1977 Cell & Tissue Res. **185** 387-397
Jones S J and Boyde A 1984 Dentin and Dentinogenesis Vol.I ed A Linde (Boca Raton: CRC Press) pp 81-134
Jones S J, Boyde A, Ali N N and Maconnachie E 1985 Scanning **7** 5-24
Nicholson W A P and Dempster D W 1980 SEM/1980/**2** 517-533
Reid S A and Boyde A 1987 J. Bone & Mineral Res. **2** 13-22
Roomans G M 1980 Scanning Electron Microscopy/1980/**2** 309-320
Ross H F 1986 Scanning **8** 216-220
*** We thank the MRC, the SERC, and Research into Ageing for grant support.**

Inst. Phys. Conf. Ser. No. 93: Volume 3, Chapter 4
Paper presented at EUREM 88, York, England, 1988

Electron microscopic study of muscle cell structure

B Uhrík

Centre of Physiological Sciences, Slovak Academy of Sciences, Vlárska 5, 833 06 Bratislava, Czechoslovakia

There are two fundamental questions associated specifically with muscle physiology: the mechanism of tension generation in contractile units and the mechanism by which the excitation of muscle cell membrane is signalled to contractile proteins, excitation-contraction (E-C) coupling. In the present report three areas of skeletal muscle ultrastructure will be briefly characterised taking examples from recently published papers, (1) myofibrils (the force generating components), (2) cytoskeleton (the network of filaments transmitting force and maintaining cellular integrity), (3) T-system and sarcoplasmic reticulum (SR) (the structures involved in E-C coupling), with the main emphasis on the third area.

1. MYOFIBRILS

Since the advent of electron microscopic techniques we have witnessed a tremendous progress in elucidation of the structural basis of muscle machinery. The demonstration of two sets of interdigitating filaments (Huxley 1957) was a prerequisite for the formulation of sliding filament theory, the dominating theory of muscular contraction. The configuration of myosin cross-bridges and the spatial arrangement of other periodic components of both thick and thin filaments at rest, during activity and rigor have been investigated extensively with different techniques. More conventional preparatory procedures of electron microscopy have been supplemented by elaboration of rapid freezing methods. For electron microscopic study of subcellular organisation of muscle cells in a fully hydrated state two approaches were realised: (1) Frozen hydrated sections of muscle embedded in amorphous ice (McDowall et al 1984). (2) Freezing of isolated filaments in a layer of vitrified ice. Using this method Menetret et al (1987) demonstrated the spacing of 14.5 nm between myosin heads along thick filaments from insect flight muscle, this periodicity being the same as that established with low angle X-ray diffraction.

The promising technique of freeze-fracturing, deep etching and rotary shadowing rendered it possible to see filamentous lattice three-dimensionally in a high contrast, with distinct profiles of cross-bridges. However, the prefixation in glutaraldehyde followed by treatment with neutralised tannic acid before freezing was necessary to make the lattice of myosin filaments and cross-bridges more resistant to the stress exerted by freeze-fracture (Heuser 1987). A noteworthy finding in this study was an absence of C-filaments in the asynchronous flight muscle of a blowfly. Instead, thick filaments appeared to branch at their ends and form lateral contacts with adjacent thin filaments. In other asynchronous flight muscles the C-filaments connect thick filaments directly to Z-lines and may play a role both in maintaining the stiffness of relaxed muscles (White 1983) and in stretch activation during oscillatory movements.

The results of biochemical analyses of muscle constituents are a challenge for electron microscopists to locate different protein species in the ultrastructural scheme of muscle cells. Apart from major proteins involved in contractile activity (myosin, actin, tropomyosin, troponin), there is a number of minor or accessory proteins, some of them still in search of function. In a recent antibody labeling study on three different rabbit muscles Bennett et al (1986) reexamined the location of C-, X- and H-proteins. These proteins are present on one or more of eleven transverse stripes in each half A-band. Differences in labeling patterns were found between psoas, soleus and plantaris muscles. The number of classes of fibre, as determined by the accessory proteins pattern, exceeded the number of fibre types presently recognised. Considerable attention has been paid to the structural organisation of the M-band and to location of M-band proteins. Varriano-Marston et al (1987) have proposed a new model of the M-band with the myomesin forming the cuff along the thick filament, with M-protein being the major component of the M-bridge sheets and MM-creatine kinase constituting nodal points - the thickenings in the centre of the M-bridges.

2. CYTOSKELETON

Following extraction of contractile proteins a network of transverse, longitudinal and circular intermediate filaments was revealed. This network together with other, high molecular weight, proteins titin or connectin and nebulin is assumed to form a cytoplasmic skeleton and be responsible for alignement of myofibrils and filaments, as well as for passive tension and elasticity of muscle cells (Wang 1984, Maruyama 1986). Recently the loss of passive tension after degradation of titin and nebulin by ionising radiation has been reported (Horowits et al 1986). Using immunoelectron microscopy Suzuki et al (1987) demonstrated the localisation of connectin and nebulin in chicken breast muscle: anti-connectin formed three stripes at A-I junction, anti-nebulin gave rise to several stripes near the Z-line. The location of these stripes was reminiscent of N-lines, the elusive structures in I-bands of unknown function (Locker and Wild 1984).

3. T-SYSTEM AND SARCOPLASMIC RETICULUM

The inward spread of excitation and a practically simultaneous activation of all myofibrils is due to the existence of T-system - a network of tubules of predominantly transverse orientation. The T-tubules are invaginations of the cell membrane. Marking of the T-tubule lumen by penetration of extracellular tracers may be necessary for morphometric estimation of relative areas of different intracellular membrane systems. In a crayfish muscle the area of invaginating membranes is about 40 times higher than the area of the peripheral sarcolemma, thus accounting for an unusually high specific electrical capacity (if related to the periphery) in this type of muscle fibre (Uhrík et al 1980). Three-dimensional disposition of the T-system in vertebrates, with its helicoidal arrangement and longitudinal connections between successive networks, have been studied on semi-thin or thick sections in a high-voltage electron microscope (HVEM) after staining of T-tubules (Peachey and Franzini-Armstrong 1983). This method has been especially useful in studies of muscle fibre types with several categories of surface invaginations, such as crustacean short- and long-sarcomere fibres. In long-sarcomere fibres three classes of clefts in addition to T- and Z-tubules were described, whereas in short-sarcomere fibres Z-tubules were absent. The T-tubules originate at sarcolemmal clefts as longitudinally extended sheets and form discoids at places of junctions to SR, the function of Z-tubules remains obscure (Franzini-Armstrong et al 1986, the own unpublished results). An interesting finding is the high density of T-tubules and a reduced amount of non-junctional SR in fast short-sarcomere crustacean muscle fibres, this disproportion being somewhat reminiscent of asynchronous insect flight muscles (for a review on insect muscle structure, see Elder 1975). The participation of T-tubules in E-C coupling may be proved indirectly by destroying their continuity after glycerol wash-out from glycerol preloaded fibres. T-tubules vesiculate and disconnect from sarcolemma with resultant detubulation and E-C decoupling. An intriguing question concerns the differences between reversible and irreversible detubulation. In crayfish both reversible and irreversible detubulation (wash-out of 300 or 400 mmol/l glycerol, respectively) results in a similar, ultrastructurally undifferentiated, detachment of T-tubules from the surface with disintegration of connections between T-system vesicles and the sarcolemma (Zacharová and Uhrík 1978). The underlying mechanisms of reversibility vs. irreversibility remain unknown. If devesiculation, the structural correlate of reversibility, is dependent on ordering influence of cytoskeleton, a disorder in it might be assumed in irreversibly decoupled fibres.

Activation of contractile apparatus is dependent on sudden increase in intracellular concentration of Ca ions. In cross striated skeletal muscles SR is the most important Ca-accumulating and Ca-releasing component. A direct release of Ca from SR by caffeine, eliciting contracture, is accompanied by reversible swelling of SR elements, as demonstrable after conventional double fixation (Uhrík and Zacharová 1976). Electron probe X-ray microanalysis (EPXMA) on freeze-dried cryosections of frog twitch muscle fibres has shown that during 1.2 s tetanus 59% of the Ca content of terminal cisternae

(TC) of SR was released, sufficient to occupy the Ca binding sites on troponin and parvalbumin. At the same time no significant increase in Ca concentration in longitudinal SR (LSR) could be detected (Somlyo et al 1981). In our EPXMA study on cryosections from the same type of muscle after the series of K-contractures with Sr or Ba substituted for Ca, the substituted bivalent cations could be detected only in TC, with Ca occassionally being increased in LSR (Uhrík and Zacharová 1988). An internal cycle of activating Ca during contraction-relaxation process, the release from TC and a return back via LSR, has been observed in an investigation using histochemical precipitation method after tetanus (60 s) or K-contractures (80 s) (Uhrík and Zacharová 1987). Concerning the ability of LSR to accumulate Ca, it is interesting that under influence of hypertonic solution Ca can be translocated into LSR forming granules with Mg and P (Somlyo et al 1977). The dense granular content of TC, assumed to indicate the presence of a Ca accumulating protein, calsequestrin, has been found in LSR as well (Sommer 1982, Franzini-Armstrong et al 1987). The mechanism of Ca transport from LSR to TC could be related to the occlusion ("reversible collapse") of LSR lumen under the influence of cations (Sommer 1982). An intriguing finding on E-faces of freeze-etched TC and free SR is an increase in number of 13 nm pits immediately (1 ms) after electrical stimulation (Nassar et al 1986). Structures of still unknown function on freeze-fracture replicas of TC are indentations and rods. Their number has been related to the speed of contraction and to the amount of asymmetric charge movement (Dulhunty 1987).

EPXMA studies of frog twitch muscle cells are facilitated by the location of triads near to Z-lines and by the dense content of TC imparting sufficient contrast to them even in unstained cryosections. This is not the case of long-sarcomere crayfish muscle fibres: The dyads (TC with apposed T-tubules) are longitudinally oriented, located approximately at the A/I boundary, there are frequent shifts in register between adjacent myofibrils and TC have no comparable densities within their lumen. The lack of densities results probably from the absence of calsequestrin: this protein was not found in sarcotubular fractions of crayfish (Tomková et al 1984). Consequently, it is difficult if not impossible, to identify TC in cryosections of the crayfish muscle (the own unpublished results). Yet the interest to study Ca translocations in this preparation is high, as crayfish (and, generally, invertebrate) muscles are distinguished by Ca electrogenesis, the inward membrane current being carried by Ca ions (for a review, see Zachar 1971). The results of histochemical investigation (Ca precipitated by phosphate and substituted for by lead) have shown that in resting crayfish muscle all SR is evenly "stained", thus making the Ca precipitating technique a suitable pretreatment for three-dimensional examination of the SR net in a HVEM (Uhrík and Zacharová 1979). In a subsequent study (Uhrík and Zacharová 1983) the use of thick sections in HVEM has proved advantageous for a more representative imagining of changes in precipitate distribution upon activation of muscle fibres. Whereas during caffeine contractures an almost complete disappearance of Ca from SR with a moderate increase in myofibrillar precipitates was observed, the potassium contractures resulted only in partial SR depletion yet in much higher density of myofibrillar precipitates. This discrepancy may be explained by an assumption of extracellular Ca contribution to activation in depolarisation-induced contractures in crayfish. A conspicuous finding in this study was the uniform change in precipitate density throughout all SR elements, the SR net behaving as a whole both during Ca-release and subsequent Ca-uptake.

The structural specialisation at the site of apposition of T-system to SR is T-SR junction. The T-SR gap of 10-20 nm is spanned by junctional processes (JP) or "feet" which are not only intercalated between junctional membranes (jSR and jT) but anchored to them: In freeze-fracture replicas the luminal leaflet of jSR has scallops corresponding to JP, the cytoplasmic leaflet of jT is occupied by clusters of 4 particles (Franzini-Armstrong and Nunzi 1983). However, the jT clusters correspond only to alternate jSR scallops. Each foot consists of 4 subunits (Ferguson et al 1984). In grazing sections JP have less dense central core and are arranged tetragonally with a spacing of about 30 nm. Unlike vertebrate T-SR junctions with 2-3 rows of JP, the junctions in crayfish are patchy with 4-8 rows of JP (Uhrík et al 1984). Despite a lower density of JP per unit volume of fibre in crayfish in comparison with frog, a much higher charge is displaced by slow asymmetry currents in crayfish (Henček et al 1984). Hence the slow asymmetry currents in crayfish may be generated in structures different from JP. In spite of general acceptance of the importance of JP in E-C coupling, no specific treatment destroying the feet with the resultant E-C decoupling has been known. In our laboratory a slow development of E-C decoupling in isolated crayfish muscle fibres kept in vitro has been observed. Ultrastructural examination of the fibres did not reveal any changes providing an explanation for the decoupling except for T-SR gaps: Grazing sections of junctional gaps displayed blurred patches without clear-cut dots characteristic of JP in normal junctions (Uhrík et al 1986). An important, still unresolved, problem of muscle physiology is the mechanism of signal transfer across the T-SR gap. One of the suggested mechanisms has been the release of a transmitter substance into the T-SR gap. Besides recently proposed inositol 1,4,5-trisphosphate, Ca has for long been considered a candidate for triggering the Ca release from the SR. We have tried to test histochemically the presence of Ca along junctional T-tubule membrane at rest and its release upon excitation in single frog twitch muscle fibres (Uhrík and Zacharová 1987). The most serious obstacle in this study was the fact that the perfusion of the fibres with the fixing and precipitating solution elicited mechanical activity even in a portion of fibres pretreated by tetrodotoxin and procaine. Nevertheless in fibres without/or with largely reduced activity, triads with precipitates along T-tubule membrane were found in contrast to mechanically active fibres, where the precipitates were lacking. These results are consistent with the trigger calcium hypothesis.

REFERENCES

Bennet P, Craig R, Starr R and Offer G 1986 J. Muscle Res. Cell Motil. **7** 550
Dulhunty A F 1987 Muscle Nerve **10** 783
Elder H Y 1975 Insect Muscle ed P N R Usherwood (London, New York, San Francisco: Academic Press) pp 1-74
Ferguson D G, Schwartz H W and Franzini-Armstrong C 1984 J. Cell Biol. **99** 1735
Franzini-Armstrong C and Nunzi G 1983 J. Muscle Res. Cell Motil. **4** 233
Franzini-Armstrong C, Eastwood A B and Peachey L D 1986 Cell Tissue Res. **244** 9
Franzini-Armstrong C, Kenney L J and Varriano-Marston E 1987 J. Cell Biol. **105** 49
Henček M, Zachar J, Zacharová D, Uhrík B and Novotová M 1984 Neurophysiology (Kiev) **16** 460
Heuser J E 1987 J. Muscle Res. Cell Motil. **8** 303
Horowits R, Kempner E S, Bisher M E and Podolsky R J 1986 Nature **323** 160
Huxley H E 1957 J. Biophys. Biochem. Cytol. **3** 631
Locker R H and Wild D J C 1984 J. Ultrastruct. Res. **88** 207
Maruyama K 1986 Int. Rev. Cytol. **104** 81
McDowall A W, Hofmann W, Lepault J, Adrian M and Dubochet J 1984 J. Mol. Biol. **178** 105
Menetret J F, Hofmann W and Lepault J 1987 J. Muscle Res. Cell Motil. **8** 71
Nassar R, Wallace N R, Taylor I and Sommer J R 1986 Scanning Electron Microsc. **1** 309
Peachey L D and Franzini-Armstrong C 1983 Handbook of Physiology, Sect. 10: Skeletal Muscle eds L D Peachey, R H Adrian and S R Geiger (Bethesda: Amer. Physiol. Soc.) pp 23-71
Somlyo A V, Shuman H and Somlyo A P 1977 J. Cell Biol. **74** 828
Somlyo A V, Gonzales-Serratos H, Shuman H, McClellan G and Somlyo A P 1981 J. Cell Biol. **90** 577
Sommer J R 1982 Z. Naturforsch. **37c** 665
Suzuki T, Sawada H and Maruyama K 1987 Biomed. Res. **8** 285
Tomková Ž, Juhászová M, Zacharová D and Uhrík B 1984 Gen. Physiol. Biophys. **3** 55
Uhrík B and Zacharová D 1976 Pflügers Arch. **364** 183
Uhrík B and Zacharová D 1979 Cell Tissue Res. **202** 343
Uhrík B and Zacharová D 1983 Gen. Physiol. Biophys. **2** 63
Uhrík B and Zacharová D 1987 Histochemistry **86** 305
Uhrík B and Zacharová D 1988 Physiol. Bohemoslov. (in press)
Uhrík B, Novotová M and Zachar J 1980 Pflügers Arch. **387** 281
Uhrík B, Novotová M and Zacharová D 1984 Gen. Physiol. Biophys. **3** 441
Uhrík B, Novotová M, Zacharová D and Rýdlová K 1986 Gen. Physiol. Biophys. **5** 109
Varriano-Marston E, Franzini-Armstrong C and Haselgrove J 1987 J. Electron Microsc. Technique **6** 131
Wang K 1984 Contractile Mechanisms in Muscle eds G H Pollack and H Sugi (New York, London: Plenum Press) pp 285-303
White D C S 1983 J. Physiol. (London) **343** 31
Zachar J 1971 Electrogenesis and Contractility in Skeletal Muscle Cells (Baltimore, London: University Park Press)
Zacharová D and Uhrík B 1979 Cell Tissue Res. **192** 167

Inst. Phys. Conf. Ser. No. 93: Volume 3, Chapter 4
Paper presented at EUREM 88, York, England, 1988

On the mechanism of internalization of asbestos fibers: an *in vitro* study

W Malorni, G Arancia, F Iosi, M Falchi and G Donelli

Department of Ultrastructures, Istituto Superiore di Sanitã, Viale Regina Elena 299, 00161 Rome, Italy

ABSTRACT: Chrysotile fibers represent the main components of asbestos particulates. An in vitro study on the mechanism of cytotoxicity of chrysotile fibers in interphase cells was carried out. A likely involvement of cytoskeletal apparatus in fiber internalization and transport into the nuclear matrix is hypothesized.

1. INTRODUCTION

The generic term asbestos is used to describe the natural fibrous form of chain silicates, of which chrysotile and amphiboles are the two main types. Interesting data have been provided by in vitro studies on the mechanism of interaction of some asbestos fibers with subcellular components. The ability to induce carcinomas and mesoteliomas has been attributed to various physicochemicals properties of these minerals including geometry, size and electrostatic charge of the fibers. Moreover, plasma membrane and cytoskeletal elements seem to be involved in asbestos-induced cytotoxicity. In order to elucidate the mechanisms of internalization of asbestos fibers, an in vitro study on different epithelial cell lines (CG5, HEp-2) exposed to chrysotile was performed by light and electron microscopy.

2. RESULTS AND DISCUSSION

Results obtained can be summarized as follows: a) the mechanism of fiber (< 10 µm) internalization seems to be mediated by some intracytoplasmic components. Fibers are carried through the cytoplasm void of a membrane envelope (Figs. 1a, b) and can also be found in the nuclear matrix (Fig. 1c) and inside the nucleolar electron-dense region (Fig. 1d). b) The transport of chrysotile fibers through the cytoplasm is intimately related to the cytoskeletal apparatus arrangement. In fact, when observed by SEM, the cytoskeletal components appeared to be in close relationship with intracytoplasmic fibers (Figs. 2a, b). Moreover, the microtubular apparatus and microfilament network (Figs. 3a, b, respectively) underwent a rearrangement, as revealed by fluorescence microscopy. The cytoskeletal integrity and function can be soon recovered after fiber penetration. As a general rule, intracellular translocation of chrysotile fibers could be mediated by cytoskeletal elements making interphase cells able to internalize small asbestos fibers. Finally, the observation of chrysotile-induced giant polynucleated cells (Fig. 4) could be related to nuclear internalization and might account for the cytopathogenic effect induced by asbestos fibers.

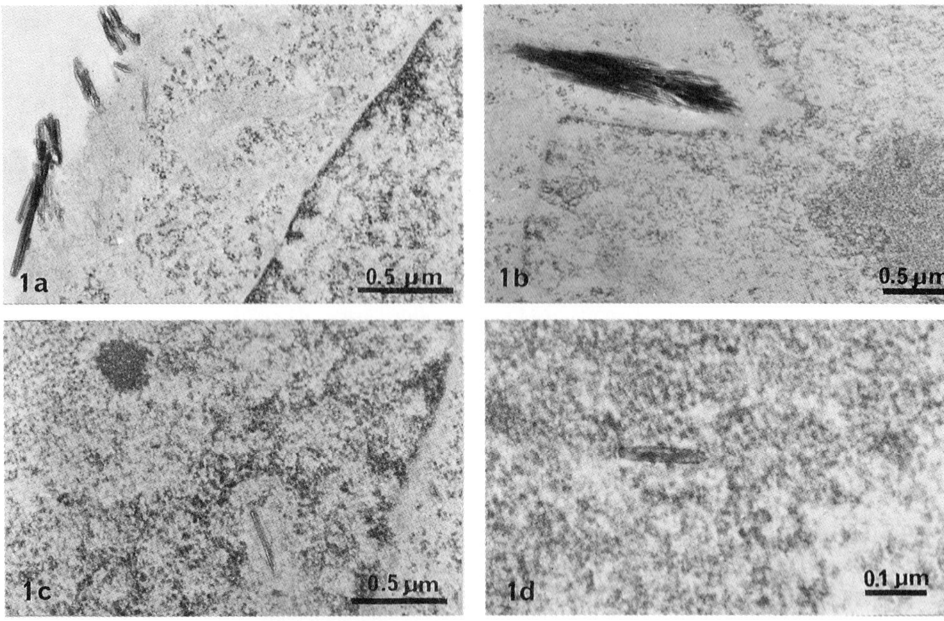
1a
0.5 µm
1b
0.5 µm
1c
0.5 µm
1d
0.1 µm

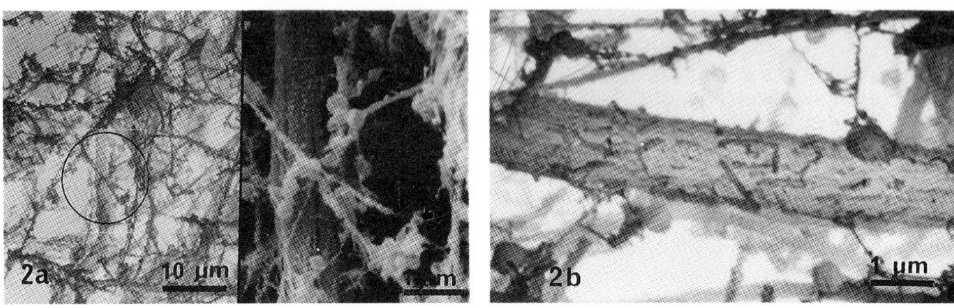
2a
10 µm
2b
1 µm

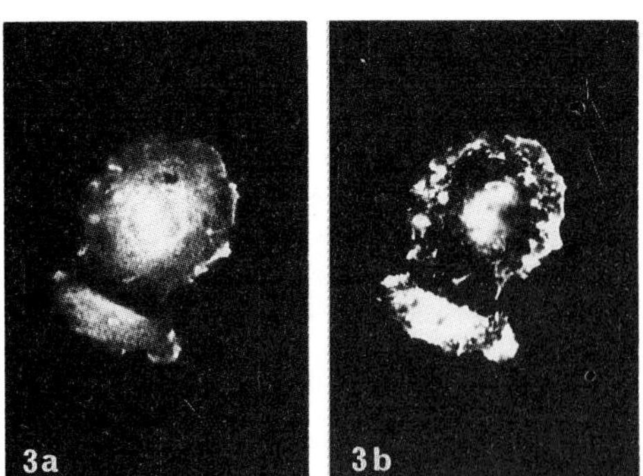
3a
3b

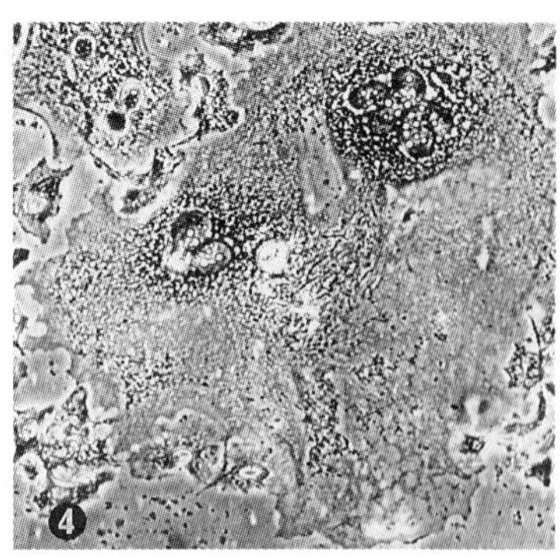
4

Inst. Phys. Conf. Ser. No. 93: Volume 3, Chapter 4
Paper presented at EUREM 88, York, England, 1988

Ultrastructure of human lymphokine activated killer cells

P Groscurth[1], S Diener[1], R Stahel[2] and L Jost[2]

[1] Institute of Anatomy, Div. of Cell Biology, University of Zurich
[2] University Hospital of Zurich, Dept. of Internal Medicine

ABSTRACT: The morphology of LAK cells has been studied by scanning and transmission electron microscopy including immuno-electron microscopy. LAK cells display typical lysosomal granules containing pore forming proteins like other cytotoxic cells. On the other hand the nuclear morphology of LAK cells differs distinctly from that of CTL and NK cells indicating that LAK cells are unique cytotoxic cells of other origin.

1. INTRODUCTION

Lymphokine activated killer (LAK) cells are potent cytotoxic cells which differ from conventional cytotoxic T-lymphocytes (CTL) and natural killer (NK) cells concerning kinetics of activation, specificity of target cells and phenotype of the precursor and effector cells (Grimm et al 1982). Light microscopically LAK cells are classified as large granular lymphocytes since they contain numerous lysosomal granules in the cytoplasm. However no data are available on the ultrastructure of these cells. Therefore we studied the morphologic appearance of LAK cells by scanning (SEM) and transmission electron microscopy (TEM) including immunolabeling by specific antibodies against the pore forming protein "perforin 1 (P1)". The aim of our study was to define differences and / or similarities between LAK and other cytotoxic cells.

2. MATERIAL AND METHODS

LAK cells were obtained from normal donors by continuous in vitro stimulation of peripheral blood mononuclear cells with recombinant interleukin 2 (r-IL2). The cells were harvested at day 5, 10 and 15 after stimulation and subsequently processed for SEM and TEM by routine procedures. Immunostaining of fixed LAK cells was performed on ultrathin cryosections using a monospecific antiserum against P1 and protein A gold complex as previously described (Groscurth et al 1987).

3. RESULTS AND DISCUSSION

SEM morphology of LAK cells was similar to that of other killer cells. By TEM the LAK cells displayed an elongated nucleus with numerous nuclear pockets projecting into the cytoplasm (Figure 1a). In addition peculiar nuclear inclusion bodies (NIB) were detectable in about 5 % of LAK cells obtained 5 days after r-IL2 stimulation (Figure 1c). They were round in shape and contained abundant branched or twisted membrane profiles. Examination of serial sections revealed that the NBI were apparently derived from ER-

profiles which had been trapped in the nucleus by fusion of adjacent nuclear pockets (Figure 1b and c). Number and size of NBI increased in 10 and 15 days old LAK cells indicating that these structures are related to prolonged r-IL2 stimulation.
The cytoplasm of LAK cells contained numerous lysosomal granules which were characterized by an electron dense core and a small halo (Figure 2a). Immuno-electron microscopic studies with anti-P1 antibodies revealed specific staining of the granules whereas other organelles remained unlabeled (Figure 2b). Lysosomes with similar morphology and composition were also present in CTL and NK cells.
Summarizing our study indicates that LAK cells differ from other cytotoxic cells by their nuclear morphology but that they possess similar granules containing cytolytic proteins like CTL and NK cells.

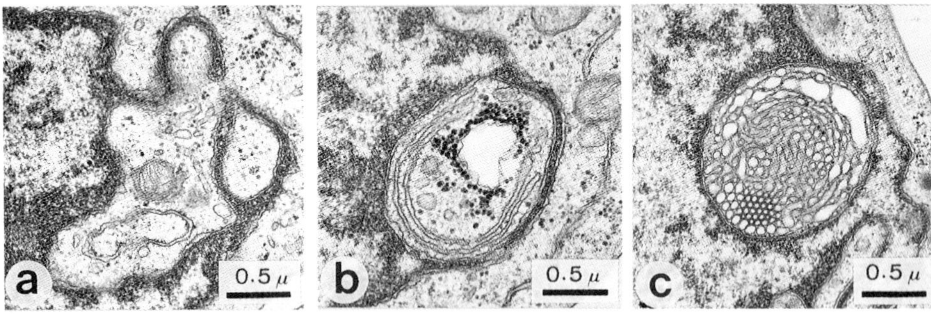

Figure 1: Morphology of LAK cell nucleus, a) nuclear pockets surrounding cytoplasmic areas with membrane profiles, b) ER profiles trapped within the nucleus c) typical nuclear inclusion body

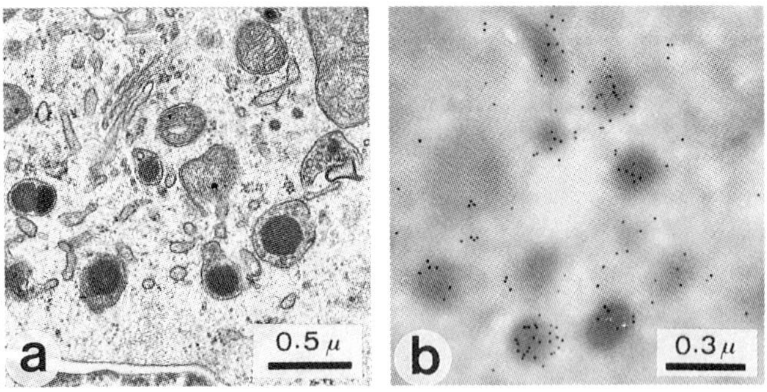

Figure 2: Ultrastructure of LAK cell lysosomes a) characteristic granules with dense core and translucent halo, b) immunostaining of lysosomes by anti-P1 antibody and protein A gold complex

4. REFERENCES

Grimm E A, Mazumder A, Zhang H Z and Rosenberg St A 1982 *J exp Med* **155** 1823
Groscurth P, Qiao B-Y, Podack E R and Hengartner H 1987 *J Immunol* **138** 2749

Inst. Phys. Conf. Ser. No. 93: Volume 3, Chapter 4
Paper presented at EUREM 88, York, England, 1988

Induction of monocytic differentiation of cultured HL60 myeloid leukemia cells as determined by ultrastructural cytochemistry

RHJ Beelen, HJ Bos, GJ Ossenkoppele

Department of Hematology and Cell Biology, Free University Hospital, De Boelelaan 1117, 1081 HV Amsterdam, The Netherlands

ABSTRACT: In this study a clear differentiation was shown of the human promyelocytic cell line HL-60 upon induction by vit D (monocytes) as well as by DMSO (macrophages and granulocytes) as determined by ultrastructural cytochemistry. These results were confirmed by immunological characteristics and function of these cells.

1. INTRODUCTION

Our earlier studies on the ultrastructural peroxydatic activity (PA) pattern have clearly shown that it is an excellent marker for the differentiation of the cells of the mononuclear phagocyte series and granulocyte series (Beelen, 1980, van der Meer et al1982). Also we have shown that the HL-60 promyelocytic cell line can be induced to differentiate into monocytic or granulocytic direction based on quantitative enzyme determination (Wijermans et al 1987). In this study we have now investigated the PA pattern of this cell line HL-60 after induction of different chemicals and have correlated these results with immunological and functional characteristics.

2. RESULTS AND DISCUSSION

The results showed clearly promyelocytic and blastlike cells in the control (no inducer) as well as after the addition of Ara-C (fig 1: PA in the Golgi system, lysosomal granules and RER). Vit. D (1.25 dihydroxy Vit. D3) induced the differentiation into monocytic cells (PA only in lysosomes) as shown in fig 2. Vit. A (retinoic acid) resulted in a differentiation of granulocytes while most remarkably DMSO gave a very pronounced differentiation into both very mature granulocytes as well as mature resident macrophages (PA only in RER). The results, with respect to the ultrastructural cytochemistry, fitted very well in the functional characteristics of these cells, since after induction with Vit. D and also DMSO the cells showed an enhanced Fc receptor activity which correlated with a clear functional antibody dependent phagocytosis (ADP). No ADP was found in the control or after induction with the other 2 chemicals. In agreement with this findings both Vit. D and DMSO resulted in a chemotactic activity to the chemoattractant FMLP comparable to normal human macrophages. Finally Vit. D was found to induce the expression of HLA-DR on this cell-line, which is a marker for the more mature form of the mononuclear phagocyte series, while also antigen presentation could be induced under these circumstances.

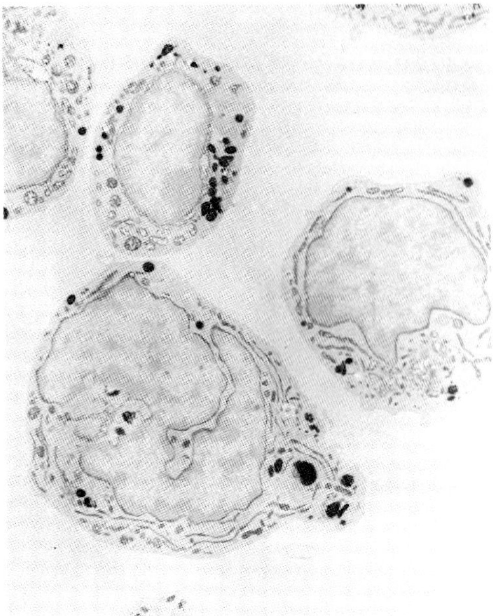

Fig. 1. The cells are promyelocytic with PA in the Golgi system, RER and lysosomal granules.

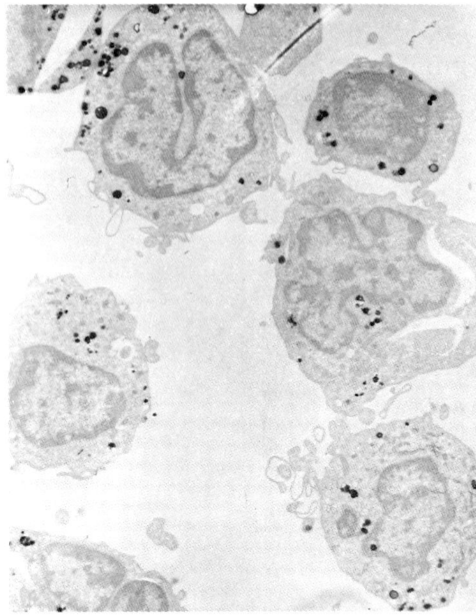

Fig. 2. Showing HL-60 cells after 4 days of incubation with Vit. D. These cells are monocytic with PA only in lysosomal granules.

In conclusion the results are summarized in table 1 which shows that vit. D gave a very pronounced differentiation of the promyelocytic cell line in well differentiated monocytes. The importance of this finding also in the treatment of leukemia with Vit. D in vivo is presently under study.

differ inducer	PA pattern	Fc	ADP	chemotaxis	HLA-DR
-	blasts	-	-	-	-
Ara-C	blasts	-	-	-	-
Vit. A	granulocytic	-	-	-	-
Vit. D	monocytic	+	+	+	±
DMSO	macrophagelike-granulocytic	±	±	±	±

Table 1. Induction of differentiation of HL-60 cells by different-inducers based on ultrastructural cytochemistry, immunological and functional characteristics.

Beelen R H J 1981, Blut **27** 84

Meer J W M van der, Gevel J S, Beelen R H J, Fluitsma D M, Furth R van 1982, J Reticuloendothelial Soc **32** 355.

Wijermans P W, Ossenkoppele G J, Huijgens P C, Imandt L M F M, Waal F C de, Langenhuijsen M M A C 1987, Leuk Res **11** 641.

Remodelling of the intercalated disc in dissociated and cultured adult rabbit ventricular myocytes

K S Shovel, N J Severs and T Powell*.

Department of Cardiac Medicine, Cardiothoracic Institute, Fulham Road, London SW3 and
* University Laboratory of Physiology, Parks Road, Oxford, UK.

ABSTRACT: The internalization of fasciae adherentes with time in culture is described for isolated adult ventricular myocytes.

Isolated adult ventricular myocytes undergo a smoothing-over of the intercalated disc following dissociation and maintenance in culture (Jacobson and Piper, 1986) but the precise mechanism by which this occurs has not been established. The present study set out to elucidate this process.

Suspensions of adult ventricular myocytes were isolated from New Zealand White rabbit hearts by retrograde perfusion via the aorta with crude collagenase in a low calcium Krebs-Ringer bicarbonate buffer as described by Powell et al (1980). Cells from each heart were divided into batches; some were fixed and processed for thin section electron microscopy immediately after isolation, while others were fixed after maintenance in culture medium (Delbecco's MEM in 25mM Hepes containing sodium pyruvate (1g dm^{-3}), glucose (1g dm^{-3}) and horse serum (5% v/v)) at room temperature for periods up to 22h.

Immediately after isolation an average of 71% of the myocytes are rod-shaped, retaining the step-like appearance of the intercalated disc characteristic of intact tissue. During dissociation of the myocytes, the fasciae adherentes junctions are split in half. The convoluted morphology of this adherens region is initially preserved (Figure 1). An average of 54% of myocytes maintained in culture for 22h retain the overall rod-shaped morphology typical of freshly isolated myocytes. The majority of cells, however, show some degree of smoothing-over of the formerly irregular disc elements.

The first stage in this process appears to involve protrusion of cell processes at the borders of the fasciae adherentes junctions (Figure 2). These protrusions subsequently seem to loop over (Figure 3) fusing together to form a continuous sheet over the fasciae adherentes (Figure 4) thereby internalizing them. The new zone of cytoplasm lying between the fasciae adherentes and the surface plasma membrane then appears progressively to expand. These zones typically contain mitochondria and annular gap junctions, but few myofilaments (Figure 5).

The rate at which cells undergo this smoothing-over process is variable, but by 22h most cells have rounded rather than step-like cell ends. The fasciae adherentes remain in their original position at the ends of the myofibrils (Figure 6). This process of internalization appears to differ from that of the zonula adherens in MDBK cells (Kartenbeck et al 1982) where the plaques are released in large aggregates detached from the plasma membrane.

This work was supported by British Heart Foundation grant F100.

REFERENCES

Jacobson SL and Piper HM 1986 J. Mol. Cell. Cardiol. **18** 661
Kartenbeck J, Schmid E, Franke WW and Geiger B 1982 EMBO J. **1** 725
Powell T, Terrar DA, Twist VW 1980 J. Physiol. **302** 131

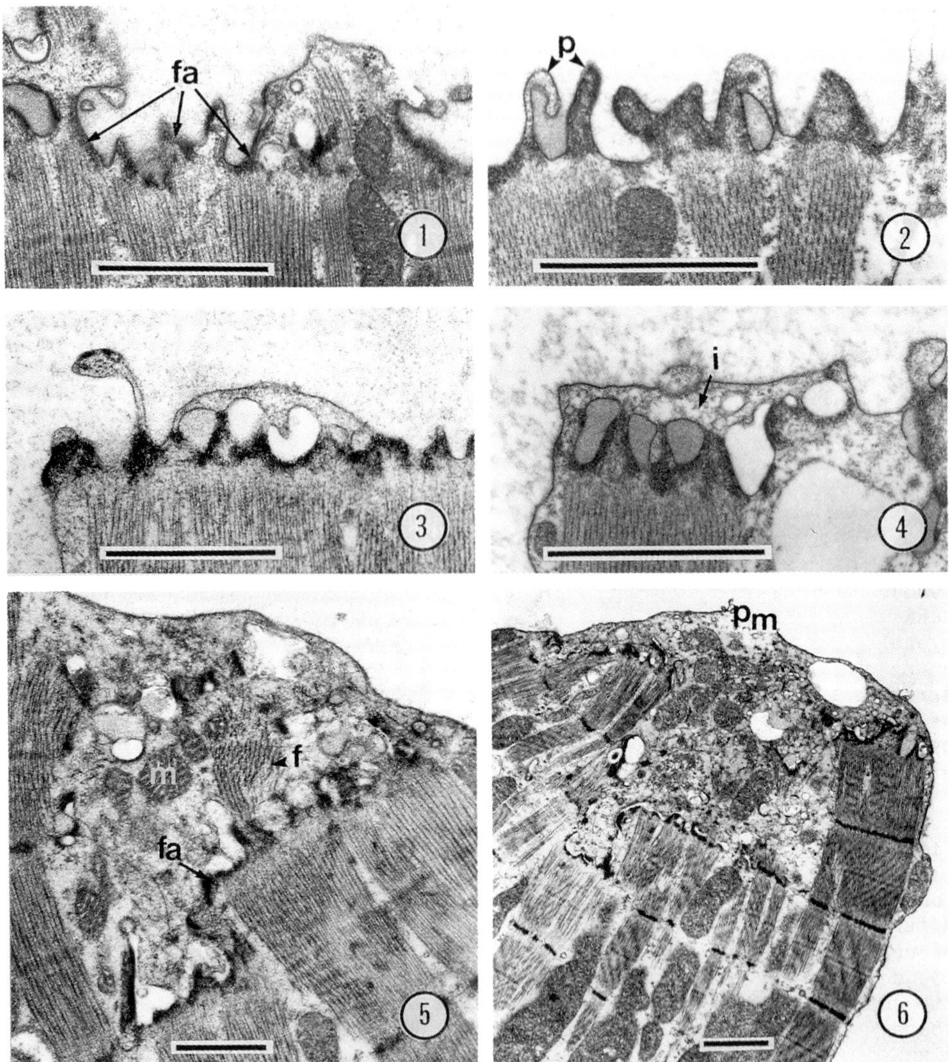

Figures 1-4. Sequence depicting proposed early stages in the smoothing-over process. fa, fascia adherens; p, cytoplasmic protrusions; i, incipient smooth zone (formed by fusions of cell protrusions).
Figures 5-6. Expanded cytoplasmic areas beneath the smoothed membrane in myocytes maintained for 22h. fa, fascia adherens; m, mitochondria; f, myofilaments; pm, plasma membrane.
Magnification bars represent 1 µm.

Inst. Phys. Conf. Ser. No. 93: Volume 3, Chapter 5
Paper presented at EUREM 88, York, England, 1988

Topography and internalization of acetylcholine receptors in cultured myotubes as visualized by surface replication and thin sectioning techniques

D Veltel and H Robenek

University of Münster, Faculty of Medicine, Dept. of Cell Biology, Domagkstr. 3, D-4400 Münster, FRG

We studied the topography and internalization of acetylcholine receptors (AchRs) in cultured rat skeletal myotubes. AchRs were marked with α-bungarotoxin (α-BTX) followed by anti-α-BTX antibodies and FITC-conjugated goat anti-rabbit antibodies or gold-labeled protein A and localized using conventional fluorescence microscopy and thin section electron microscopy. The α-BTX-anti-α-BTX antibody system used detected AchRs with high specificity (Fig 1 a/b). In addition, we combined a postfixation immunogold technique with platinum-carbon surface replication to study the distribution of AchRs on the dorsal surface of cultured myotubes. This new approach allowed us to obtain more detailed information about the distribution of AchR than any other technique used so far (Veltel and Robenek, 1988). Two distinct distributional patterns of AchRs were evident; AchRs appeared dispersed and in clusters on the surface of the myotubes (Fig. 2). Dispersed AchRs were usually distributed diffusely among the clusters. The clusters exhibited a characteristic internal arrangement of AchRs, and patches with tighter arrays of AchRs could be observed within the clusters (rectangles). The precise mechanisms by which the regional differences in the distribution of AchRs in the plasma membrane are generated and maintained are not understood at present, but there is reason to believe that the cytoskeleton is responsible. This interpretation is in agreement with the findings of Bloch and Froehner (1987) who, using an indirect immunofluorescence method, showed that a 43 kD cytoskeletal protein existing concomitantly with AchR clusters may contribute to the unique distribution of this membrane protein. We also studied the mechanism of uptake and the intracellular fate of the internalized AchR. Using pulse-chase experiments and subsequent thin sectioning of the embedded myotubes we could show that the AchR is internalized via coated pits (Fig. 3a) and is transported to endosomes. Later on, a high concentration of gold label (i.e. AchRs) could be observed in cytoplasmic tubular structures (Fig. 3b). At present we have no information about the relationship of these cytoplasmic pools to the intracellular fate of the AchR.

References: Bloch, R.J. and Froehner, S.C., (1987) J. Cell Biol., 104, pp 645-654
Veltel, D. and Robenek, H., (1988) J. Histochem. Cytochem., in press

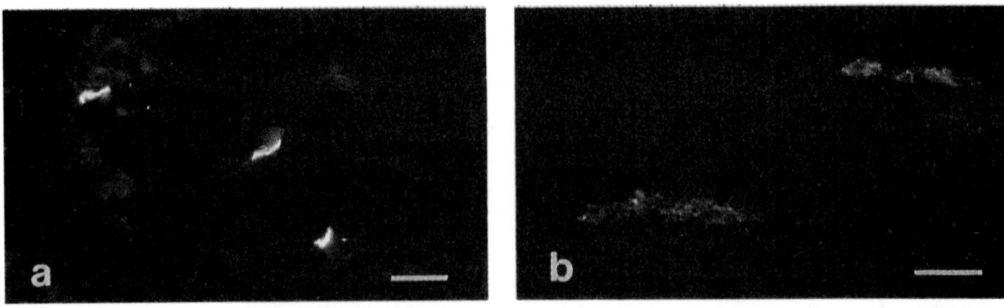

Fig. 1: AchRs labeled with α-BTX-anti-α-BTX antibodies and FITC-conjugated anti-rabbit IgGs. a) Neuromuscular junctions of rat leg muscle. b) AchR clusters on cultured rat myotubes.
Bars: (a) 6 µm, (b) 12 µm

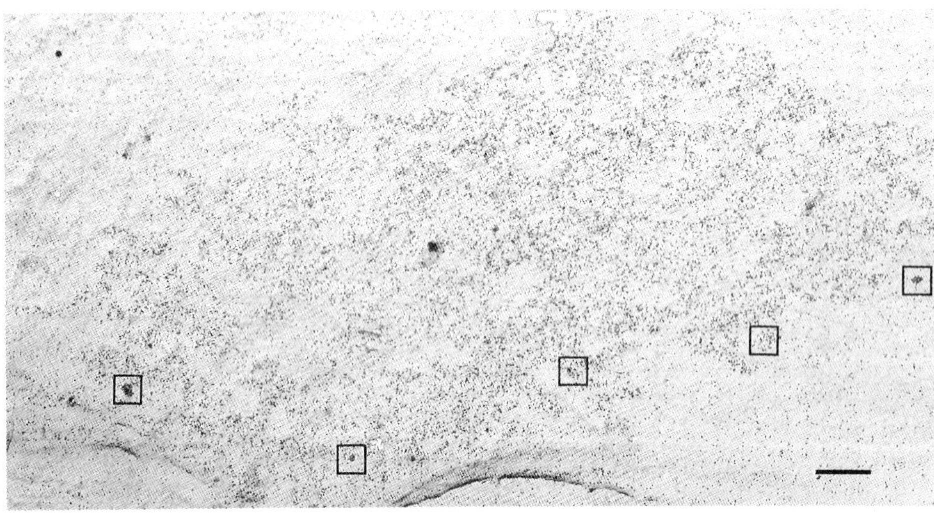

Fig. 2: Surface replica of a myotube dorsal plasma membrane showing the distribution of AchRs. Immunolabeling as described in text. Gold particles at low concentration are more or less randomly distributed over the entire plasma membrane, and at high concentration in clusters. Patches of gold particles clustered into tigher arrays can also be observed (rectangles). Bar: 0.5 µm

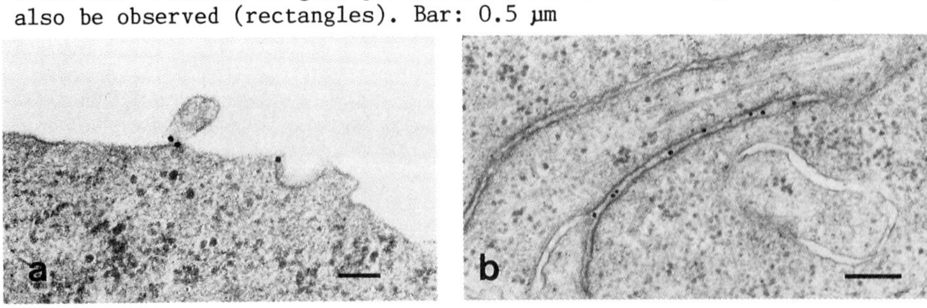

Fig. 3: Sectioned myotube after treatment with α-BTX-anti-α-BTX antibodies and protein A coupled to colloidal gold particles at 4°C. Receptor internalization via coated pits after warming the cells to 37°C (a). Gold label (i.e. AchRs) in cytoplasmic tubular structures 30 min after warming (b). Bars: (a) 0.1 µm, (b) 0.2 µm

Electron microscope revealed heart muscle cell alterations

Sz Virágh

Department of Pathology, Postgraduate Medical School, Budapest, H-1389

ABSTRACT: The fine structure of abnormal non-contractile cytoskeleton filaments, nascent T tubules, unusual couplings of the sarcoplasmic reticulum, giant mitochondria and other mitochondrial alterations occuring in different heart diseases, and an adult-type mitochondrial cardiomyopathy are reported in this publication.

1. INTRODUCTION

This paper deals with TEM revealed human heart muscle cell abormalities observed mostly in biopsy and some in autopsy specimens. The following structures are in the focus of interest: cytoskeleton filaments, sarcoplasmic reticulum (SR) and mitochondria.

2. OBSERVATIONS

The desmin or skeletin intermediate-type filaments are normal constituents of the heart muscle. In patients of an unusual familial cardiomyopathy (CM) these filaments were excessively occumulated in the working (Fig.1) and impulse conducting muscle fibres (Fig.2, see also Stoeckel et al. 1981). However, desmin filament proliferation seems to be related to the destruction of myofibrils in other primary and secondary CM, in which the diameter of the cardiocytes augments and new T tubules form by sarcolemmal invaginations (Fig.3).

The SR demonstrated an extensive proliferation in deseased cardiocytes with intermediate-type filament accumulation, and in the same cells atypical couplings and rod-like bodies were frequently seen (Figs. 4,5).

Mitochondrial alterations, like giant mitochondria formation (Fig.6) are of multiple reasons, we found them in ischemic and alcoholic heart diseases. In other CM-s the rarification of cristae, and matrix calcification indicated the mitochondrial injury (Figs. 7,8). Mithocondrial CM of a 62 year old woman showing abnormal accumulation of mitochondria with dense matrix deposits (Figs. 9,10) is reported here as the first adult case, others were found in child (Hübner and Grantzow 1983).

3. REFERENCES

Hübner G and Grantzow R 1983 Virchows Arch (Pathol Anat) 399 115
Stoeckel M-E, Osborn M, Porte A, Sacrez A, Batzenschlager A and Weber K 1981 Virchows Arch (Pathol Anat) 393 53

See legend for illustration in the text.
The bar indicates one µm unless labeled otherwise.

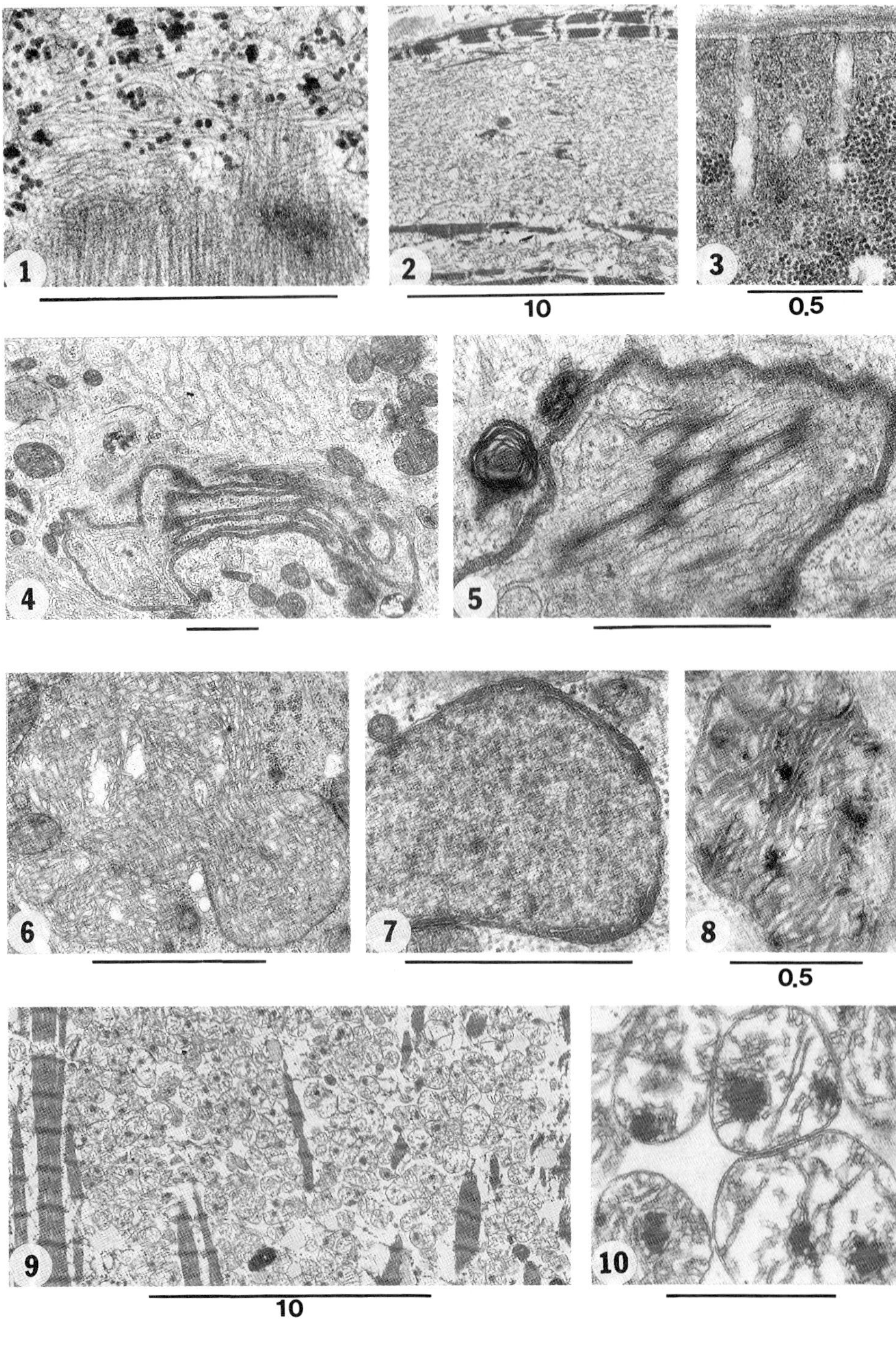

Ultrastructural features of the Golgi apparatus in human cardiocytes

D.Trpinac(1), M,Veg(1), J.B.Vuković(2), D.Avramović(4), K.Avramović(4), G.Teofilovska(3) and B.Stefanović(1) Faculty of Medicine: Institute of Histology (1), Biophysics (2) and Anatomy (3), and Clinic of Cardiology, UKC (4), 11000 Beograd, Višegradska 26, Yugoslavia.

Thin sectioned human heartś right atrial appendiges were interested. The tissue, taken during the cardiac surgery of different patients, was prepared in the routine way. It is know that in man and in some other mammalian species, Golgi apparatus (GA) of the right atria is the dominant site of Atrial Natriuretic Factor (ANF) synthesis. Previously this process had been described mainly on experimental animals and rarely in man (Spasić,1986). So we investigated ultrastructural features of this process in human in different cardiomyopathias. GA is mainly located close around nucleus and consists of a number of stacks, with usualy 4 to 6 flattened cisternae which are filled with osmophilic content.
Around GA cisternae there is a great number of small translucent transport vesicles and bigger vesicles containing electron dense secretory products, presumably ANF containing granules. GA stacks are also frequently found in cytoplasm between two sarcomeras away from the nucleus with the same characteristics as perinuclear GA stacks. In different patients there are a different number of ANF granules. On the serial thin sections it is possible to reconstruct interstacks bridges (Pavelka,1987) which connect distinct GA stacks. Solitary ANF granules are seen between sarcomeres and subsarcolemmaly where the extrusion of granules takes place. Also we found in some cardiocytes GA stacks, sometimes with centrioles subsarcolemmaly located with charasteristics somewhat different from those previously described.

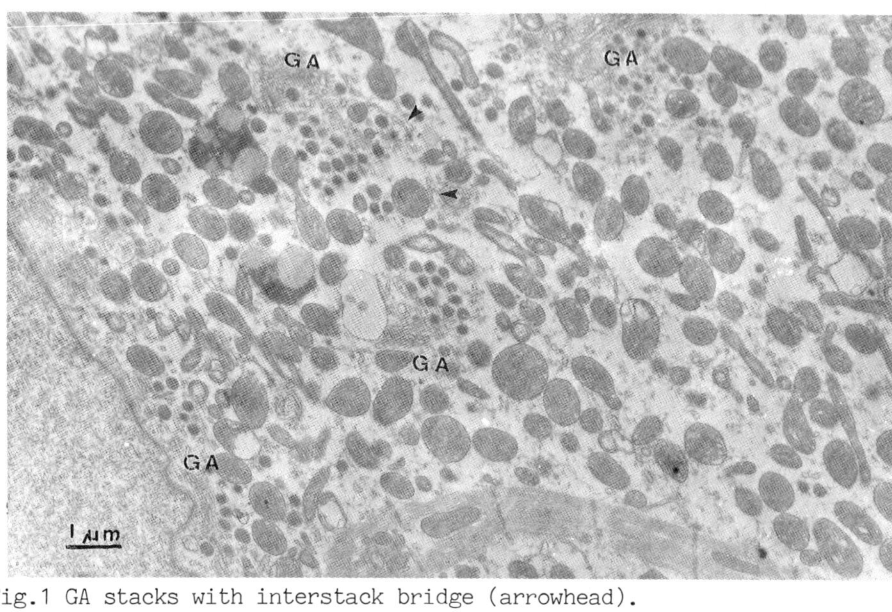

Fig.1 GA stacks with interstack bridge (arrowhead).

There were no visible secretory electron dense product in cisternae and surrounding vesicles, and these small vesicles are similar to endocytotic or exocytotic vesicles on sarcolemma. This finding of two morphologicaly different GA populations in the same cell sugests a possibility of their different functions perinuclear and sarcoplasmatic GA stacks predominantly take place in ANF section, and subsarcolemal stacks might have a role in recycling membrane but not in ANF secretion.

Spasić,P: The Heart as an Endocrine Gland in: 30 years of EM in Serbia Symposium, Ed J.B.Vuković,p.141, Beograd 1986.
Pavelka,M. Functional Morphology of the Golgi Apparatus,p,8, Springer, 1987.

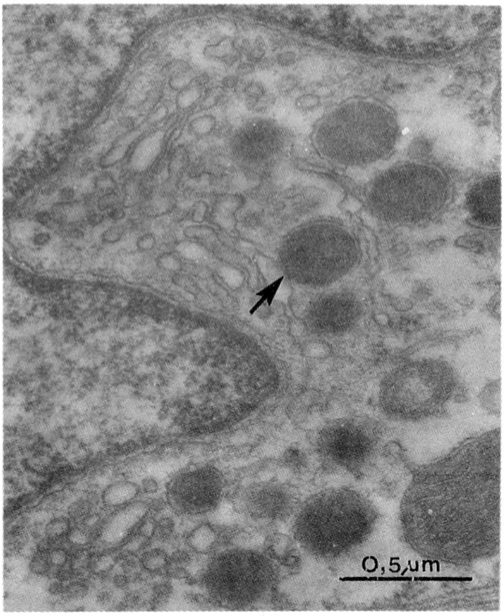

Fig.2. Perinuclear GA stack with ANF granules (arrow)

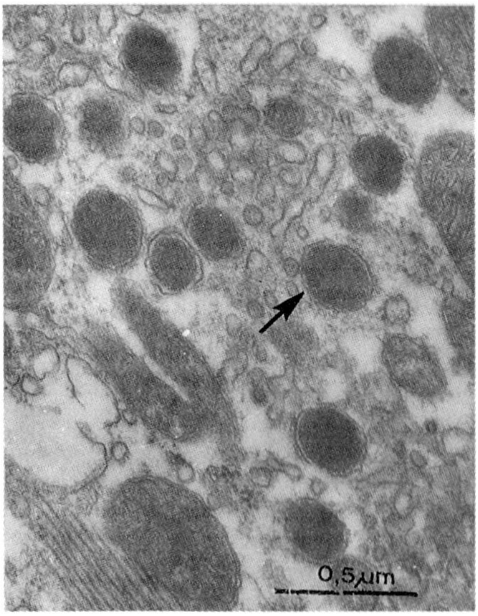

Fig.3. GA stacks between sarcomeras with ANF granules (arrow).

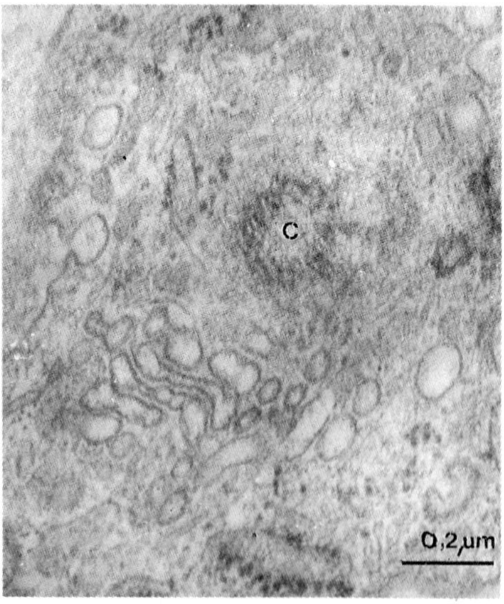

Fig.4. Subsarcolemaly located GA stacks with centriol (C).

The development of detergent-free methods for cardiac gap junction isolation

R G Gourdie, C R Green, N J Severs

Department of Anatomy and Developmental Biology, University College London, Gower Street, London WC1E 6BT and Department of Cardiac Medicine, Cardiothoracic Institute (University of London), Fulham Road, London SW3 6HP

ABSTRACT: The role of electron microscopy (EM) in developing new strategies for detergent-free isolation of cardiac gap junctions is outlined.

The alkali-extraction technique (Hertzberg, 1984) allows a 10-fold increase in the yield of gap junctions from liver plasma membrane preparations compared with detergent-based isolation methods. To date, gap junctions from heart have been extracted only with protocols using detergents (Kensler and Goodenough, 1980; Manjunath et al, 1984). Here, we demonstrate the role of electron microscopy in the development of strategies for detergent-free isolation of rabbit cardiac gap junctions.

Existing methods for isolating gap junctions from heart involve the following steps; (i) homogenisation, (ii) release of membrane from myofibrillar contractile protein by salt-extraction (0.6M KI) and (iii) detergent solubilisation of non-junctional membranes and purification of gap junctions on sucrose gradients.

Monitoring by thin section EM of the sub-cellular components in the homogenate and sucrose gradients prepared under a variety of conditions led to the following observations and modifications in methodology.

Mitochondria are highly susceptible to breakdown, resulting in heavy contamination of the gap junction-containing interfaces on sucrose gradients. This breakdown is minimised by using isotonic low Ca^{2+} buffers (2mM EGTA, 250mM sucrose, 5mM Tris, 5mM $MgCl_2$) during steps (i) and (ii) (Figures 1 and 2). A "lighter" (i.e. lower buoyant density) gap junction-enriched membrane fraction is easily separated on a stepped sucrose gradient from heavier mitochondria (e.g. Figure 3).

Release of membrane from contractile protein (i.e. step (ii)) was achieved using Deoxyribonuclease I (1 mg/ml, 1 hour, 20 C, Sigma, code D5025) rather than extraction with 0.6M KI. DNase I has been suggested to release membrane vesicles from aggregates, increasing the yield of sarcolemma from isolated heart (Philipson et al, 1980). EM indicates that breakdown of the sarcomeric I-band and its probable site of interaction at the intercalated disk may also explain the release of membrane by DNase I.

Treatment of the gap junction-enriched cardiac membrane fraction by sonication in a 20 mM NaOH/1M KCL solution results in further enrichment (30-fold) for gap junctions as measured by EM counts (Figure 4). If this treatment is performed in the absence of KCL (as in the alkali-extraction protocol for liver gap junction isolation (Hertzberg, 1984)), no gap junction structures are detected.

The introduction of these modifications should now make feasible the development of a detergent-free protocol for isolating cardiac gap junctions.

This work was supported by grants from the MRC and BHF (no. 86/39)

REFERENCES

Hertzberg E L 1984 J. Biol. Chem. **259** 9936
Kensler R W, Goodenough D A 1980 J. Cell Bio. **86** 755
Manjunath C K, Goings G E, Page E 1984 Am. J. Physiol. **246** H865
Philipson K D, Bers D M, Nishimoto A Y 1980 J. Mol. Cell. Card. **12** 1159

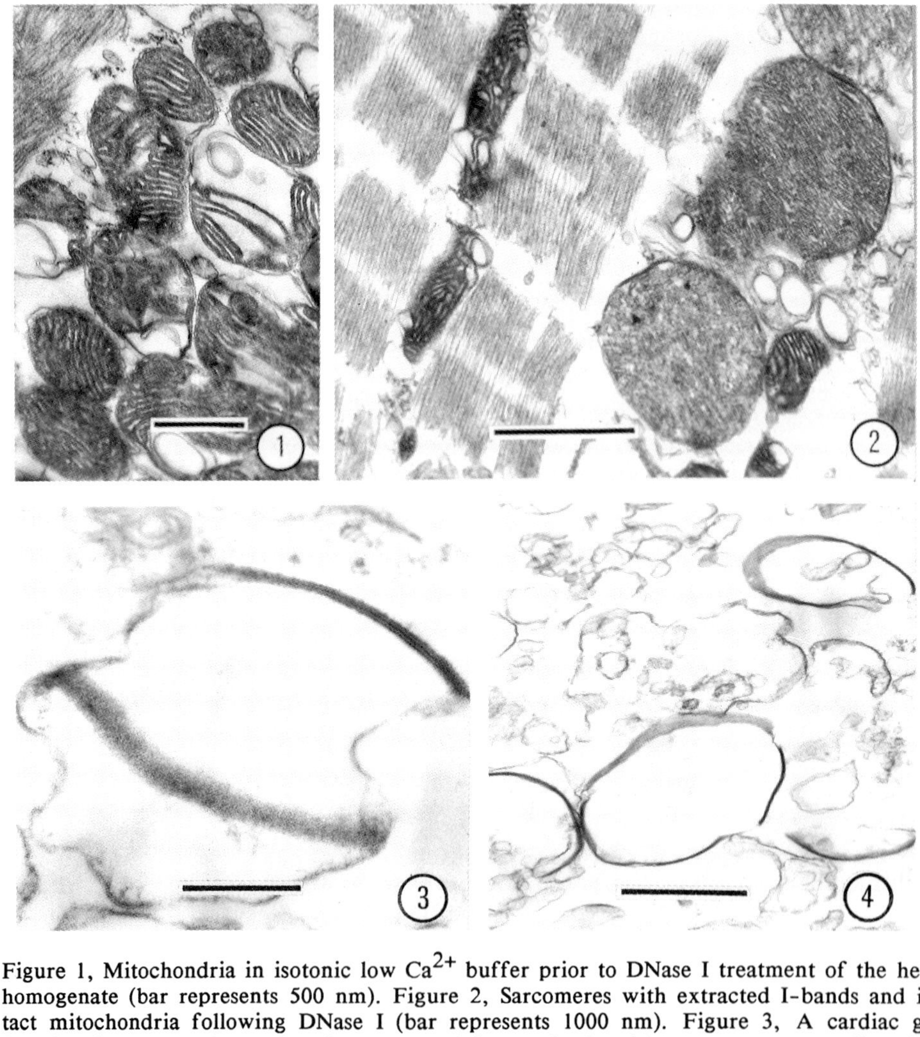

Figure 1, Mitochondria in isotonic low Ca^{2+} buffer prior to DNase I treatment of the heart homogenate (bar represents 500 nm). Figure 2, Sarcomeres with extracted I-bands and intact mitochondria following DNase I (bar represents 1000 nm). Figure 3, A cardiac gap junction in the membrane fraction separated from mitochondria on a sucrose gradient (bar represents 200 nm). Figure 4, Enrichment of gap junctions following the 20mM NaoH/ 1M KCL treatment of the membrane fraction (bar represents 500 nm).

Freeze–fracture demonstration of tight junction abnormalities in airway epithelium of cystic fibrosis (CF) patients

N J Severs and P K Jeffery

Department of Cardiac Medicine & Department of Lung Pathology, Cardiothoracic Institute, Fulham Road, London SW3 6HP

ABSTRACT: Freeze-fracture studies on CF airway epithelium obtained from heart-lung transplant patients and at early post-mortem suggest that tight junction proliferation may be a common pathological feature associated with cystic fibrosis.

The lung is a primary target organ in CF. A functional defect in airway trans-epithelial chloride (Cl^-) transport has been demonstrated (Knowles et al 1983) and attributed to defective control of apical membrane Cl^- channels (Schoumacher et al 1987). In addition, the epithelial tight junctions may act as selective barriers to the paracellular movement of ions (see Welsh, 1987). A unique opportunity to investigate tight junction structure in CF has been provided by the availability of CF lungs from heart-lung transplant patients (n=4) and from early post-mortems (n=4). Comparison was made with grossly normal airways from lungs resected for carcinoma (n=7). Extrapulmonary main and lobar bronchi were dissected and fixed in cacodylate-buffered 3% glutaraldehyde. Fixation was started within 10 minutes of lung removal for CF transplant specimens, ~3 hours after death for the post-mortem CF specimens and within 30 minutes of resection for the controls. Mucosal samples were glycerinated and frozen, and freeze-fracture replicas prepared in a Balzers BAF 400T apparatus.

The tight junction of control specimens typically appeared as a clearly delineated band of 7-14 interconnecting strands circumscribing the apico-lateral plasma membrane of the superficial epithelial cells (Fig 1). In CF specimens, this 'normal' tight junction morphology was retained by some cells; however, a marked proliferation of the tight junction was a consistent feature of others. Aggregates of long, disorganised strands often occurred loosely-associated with the usual tight junction belt (Fig 2) sometimes extending to the base of the lateral membrane (Fig 3). Tight junction material in the form of isolated maculae and fasciae occludentes were also found in deeper layers of the epithelium (Fig 4). These junctional abnormalities were detected in 3/4 transplant patients and 4/4 post-mortem cases of CF. They varied in severity between subjects but were generally more marked in the post-mortem than in the transplant group. The tight junctions of the resection controls were predominantly of 'normal' appearance. Some proliferation/disorganisation was observed in 3/7 control cases, but overall this was minor in extent compared with that observed in transplant patients.

Tight junction proliferation has previously been reported in a range of cell types in response to a variety of experimental conditions (eg Carson et al, 1980; Robenek et al, 1980; Chevalier et al, 1985) and can occur rapidly in post-mortem material (Kachar & Pinto da Silva, 1981). Whatever factors may lie behind this response, the present results from transplant patients and resection controls suggest that the altered tight junction morphology and its possible link with CF-related impermeability deserves further investigation.

We are most grateful to the transplant teams at both Harefield and Papworth Hospitals.

REFERENCES

Carson JL, Collier AM and Hu SS 1980 J.Ultrastruct.Res. **70** 70
Chevalier J, Bourguet J and Pinto da Silva P 1985 J.Cell Biol.**101** 1146a
Kachar B and Pinto da Silva P 1981 Science **213** 541
Knowles M, Stutts M, Spock A, Fischer J, Gatzy J and Boucher R 1983 Science **221** 1067
Robenek H, Doldissen M and Themann H 1980 J.Ultrastruct.Res.**70** 82
Schoumacher RA, Shoemaker RL, Halm DR, Tallant EA, Wallace RW and Frizell RA 1987 Nature **330** 752
Welsh MJ 1987 Physiolog. Rev. **67** 1143

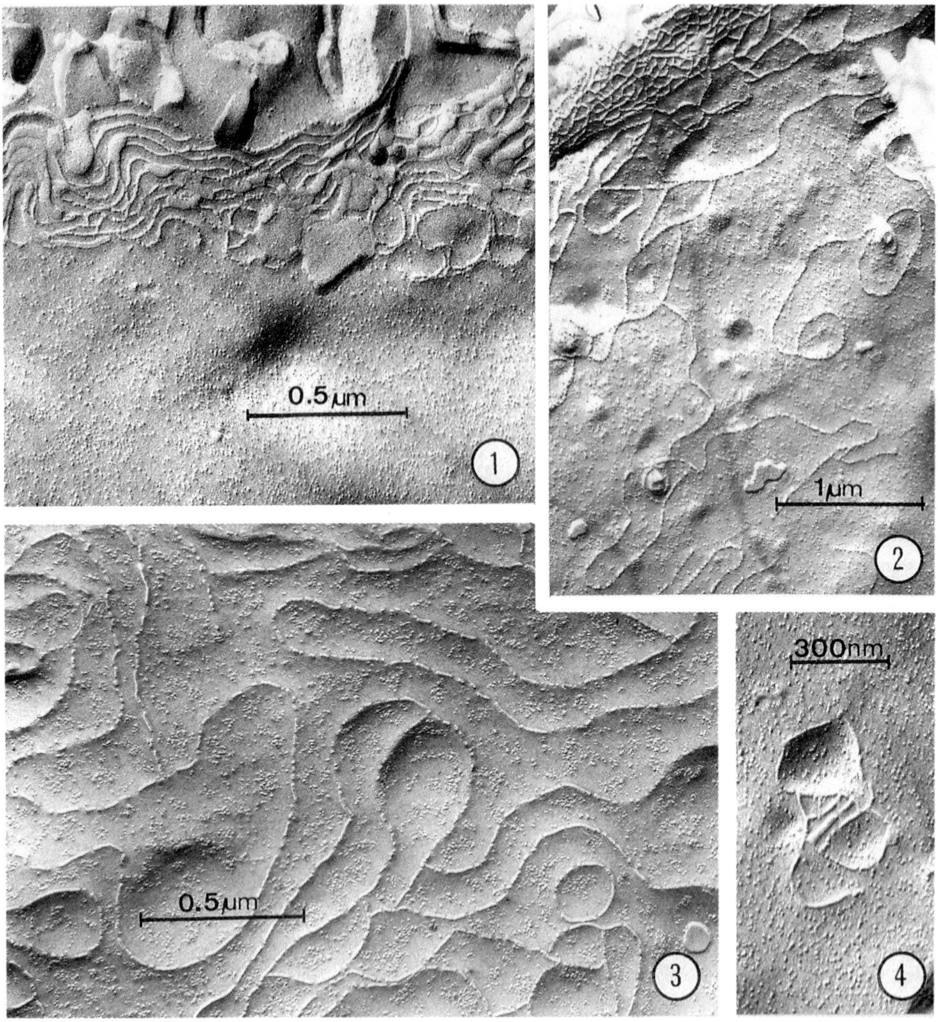

Fig 1. Resection control showing characteristic appearance of the tight junction. Fig 2. Proliferation extending below a region of 'normal' tight junction from a CF transplant patient. Fig 3. Extensive tight junction proliferation at the basolateral surface of the plasma membrane in a CF post-mortem specimen. Fig 4 Macula occludens deep within the epithelium of a CF transplant specimen.

Innervation of afibrillar muscles in the flight apparatus of an insect as seen with scanning electron microscopy

H Bradacs and K Kral

Institute of Zoology, University of Graz, A-8010 Graz, Austria

ABSTRACT: In addition to single innervation, muscle fibers with double innervation can be found in the afibrillar muscles of the honey bee flight apparatus. Two nerve endings running parallel seem to terminate in a common neuromuscular junction (nmj). The morphology of the nmj's varies: there are longish, flat ones as well as compact, elevated nmj's.

1. INTRODUCTION

While the fibrillar flight musculature of Hymenoptera mainly shows single innervation (McCann and Boettiger 1961; Ikeda and Boettiger 1965), there seems to be little information on the innervation of the afibrillar muscle in the flight apparatus of these insects. Scanning electron microscopy (SEM) has been used to study muscle innervation, mainly in vertebrates (Shotton et al 1979; Desaki and Uehara 1981). Difficulties in preparation can, however, prevent the use of this method; for example, the view of the structures under study can be obscured by other tissue. In this work, we have found that the SEM method provides valuable insights into the innervation of the afibrillar muscle of the honey bee flight apparatus.

2. MATERIALS and METHODS

The experimental animals were worker honey bees, Apis mellifera. The afibrillar muscles were exposed by removal of the cuticula on the propodeum and in the pterothorax. The specimens were fixed and dehydrated in the usual way, critical point dried, sputter-coated and viewed with a Zeiss DSM 950.

3. RESULTS

The afibrillar muscle fibers examined have a diameter of 10 - 30 μm and show distinct ribbing. This latter is due to the fact that they have come loose from their attachment at one end and have drawn up. In different places, branches (3 μm) leave the nerve trunk (10 μm). These first run separately, then join together to continue their course to the muscle fiber, which they contact, forming a nmj (Fig. 1). This nmj is longish and narrow (30x5 μm) with a flat form and location parallel to the muscle fiber. Its boundaries are somewhat unclear (Fig. 2). Other nmj's found in this preparation are shorter, wider (20x10 μm) and are distinctly elevated (Fig. 3). The nerve terminals are accompanied by numerous tracheoles (0.5 μm). The tracheoles branch out of tracheas (1 - 2 μm) which sometimes lie against muscle fibers close to nerve endings and send out tracheoles over several muscle fibers (Fig.4).Our observations show double innervation of afibrillar muscle fibers: two nerve endings, each in its own sheath, terminate in a common nmj.

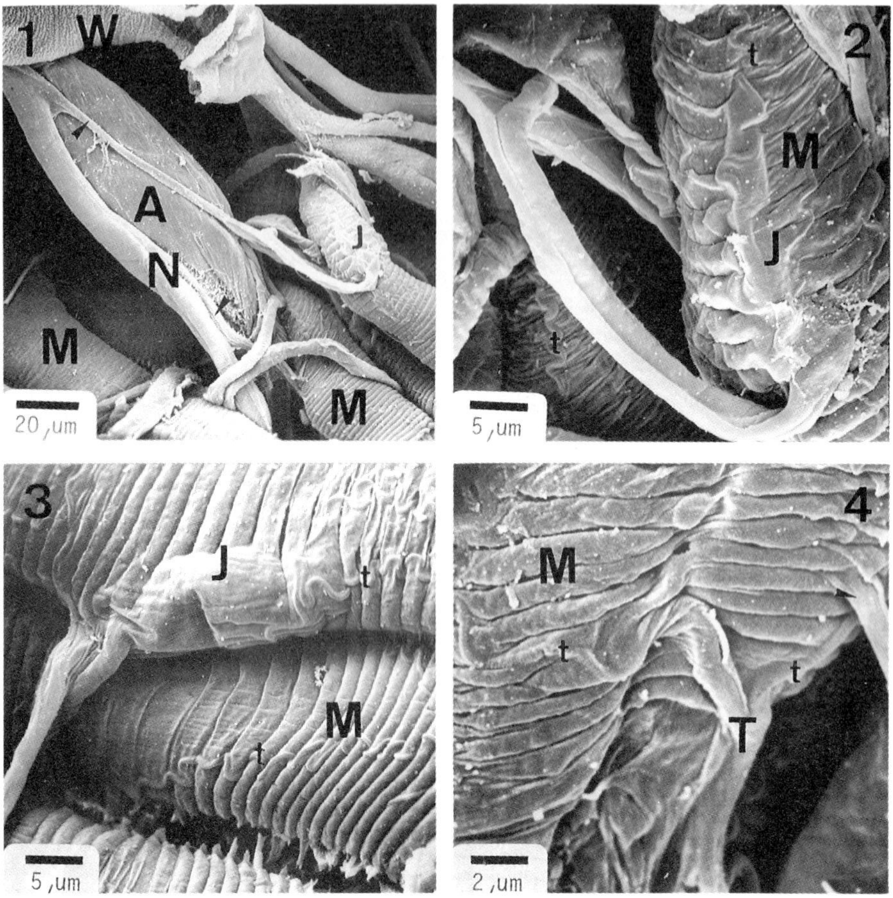

Fig. 1. A nerve-muscle preparation. The nerve trunk (N) sends out nerve branches (arrowheads). J - nmj; M - muscle fibers; A - apodeme; W - air sac wall
Fig. 2. Higher magnification of Fig.1. Two nerve branches form a common nmj. t - tracheoles
Fig. 3. Different form of nmj, which is more compact and elevated
Fig. 4. Trachea (T) close to nerve terminal (arrowhead) branching into tracheoles

4. REFERENCES

Desaki J and Uehara Y 1981 J.Neurocytol. 10 101
Ikeda K and Boettiger E G 1965 J.Ins.Physiol. 11 779
McCann F V and Boettiger E G 1961 J.Gen.Physiol. 45 125
Shotton D M, Heuser J E, Reese B F and Reese T S 1979 Neurosci. 4 427

Inst. Phys. Conf. Ser. No. 93: Volume 3, Chapter 5
Paper presented at EUREM 88, York, England, 1988

Immunolocalization of 215 kDA Mannose 6-phophate receptor in rat heart

V Marjomäki, M Surkka, A Huovila and A Salminen

University of Jyväskylä, Department of Cell Biology, SF-40100 Jyväskylä, Finland

215 kDa cation independent mannose 6-phosphate receptor (MPR^{ci}) was immunolocalized in rat heart muscle in relation to a lysosomal membrane glycoprotein lgp120 using protein A-colloidal gold technique and thin frozen sections. MPR^{ci} was found in distinct reticular vesicular structures, the majority of the label associated with internal membranes. Structures were often colocalized with lgp120-label found especially in the outer limiting membrane. Vesicles labelled mainly with antibodies of either MPR^{ci} or lgp120 were also seen. Reticular vesicular structures possibly represent an intermediate compartment linking the biosynthetic and endocytotic route to lysosomes.

1. INTRODUCTION

Newly synthesized enzymes are routed from the Golgi complex finally to lysosomes by mannose 6-phosphate receptors named after specific recognition of mannose 6-phosphate -residues on oligosaccharide chains of lysosomal enzymes (Sly & Fischer 1982). Ligand-receptor complexes dissociate due to an acid environment that is formed within vesicles budding from Golgi complex. Lysosomal enzymes are then ready for final packaging to lysosomes and receptors for reuse.

So far, two distinct receptors have been identified. One is a cation independent 215 kDa (MPR^{ci}) and the other a 46 kDa cation dependent transmembrane glycoprotein (Lobel et al. 1987 ; Dahms et al. 1987). Several groups have immunolocalized MPR^{ci} in well characterized cell types and tried to elucidate the exact pathway for lysosomal enzyme delivery. Varied methods in immunoelectron microscopy have given slightly different results concerning MPR^{ci} localization and relative amounts in different cell compartments. In general MPR^{ci} is found in Golgi complex, in plasma membrane and in endosomal compartments but not in lysosomes.

Our purpose was to immunolocalize MPR^{ci} in rat heart muscle in relation to a lysosomal membrane marker lgp120 using protein A-gold technique and cryosections.

2. MATERIAL AND METHODS

Thin frozen sections were cut from both tissue blocks prepared from perfusion fixed Wistar rat heart muscle and rat cardiac myocytes isolated by combined collagenase and hyaluronidase perfusion.

Immunolabelling was carried out with protein A- colloidal gold probes as described by Griffiths (1984).
MPR^{ci} was localized using IgG-fraction of anti-rabbit antibody produced against receptor, that had been purified from bovine liver. In double labelling studies we also used a lysosomal membrane marker lgp120 (Lewis et al. 1985), a kind gift of Dr. Gareth Griffiths, EMBL. Protein A- colloidal gold particles of two sizes were used (average diameters 5.6 and 10.7 nm ; Janssen Life Sciences).

3. RESULTS

Intense labelling of MPR^{ci} was found in vesicles with tightly packed tubules inside. MPR^{ci}- antibodies were localized especially in the membranes of the tubules. In double labelling studies lysosomal membrane marker, lgp120, was usually localized in the outer limiting membrane of the same vesicles. Vesicles labelled with both MPR^{ci} and lgp120 were often seen in close proximity to nucleus but also in the vicinity of cell membrane. Also vesicles labelled with only either antibodies were found. Vesicles labelled with only lgp120 were often associated with mitochondria. Those mitochondria showed often also degradative changes, associated vesicles thus probably representing primary lysosomes. Vesicles labelled primarily with MPR^{ci} were often small in size and near the sarcolemma.

4. DISCUSSION

In rat heart muscle the majority of the 215 kDa mannose 6-phosphate receptor was localized in distinct reticular vesicular structures often colocalized with a lysosomal membrane glycoprotein lgp120. High amount of MPR^{ci} labelling was associated with internal tubular membranes and lgp120 label in the outer limiting membrane. Since also vesicles with only lgp120 or MPR^{ci} were found it would seem probable that the vesicles labelled intensively with both antibodies represent an intermediate vesicle linking both the biosynthetic and endocytic pathways to lysosomes in heart muscle as suggested by Griffiths et al. (1988) for normal rat kidney cells. However in order to better characterize the distribution and relative amount of MPR^{ci} in heart muscle a more varied range of spesific markers for different cell compartments have to be used, also because the morphology of heart muscle cells differs greatly from other frequently studied cell types.

5. REFERENCES

Brown W J and Farquhar M G 1984 Cell **36** 295-307

Dahms N M, Lobel P, Breitmeyer J, Chirgwin J M and Kornfeld S 1987 Cell **50** 181-192

Geuze H J, Slot J W, Strous J A M, Hasilik A and von Figura 1984 J Cell Biol. **98** 2047-2054

Griffiths G, McDowall A, Back R and Dubochet J 1984 J Ultrasturcture Res. **89** 65-78

Griffiths G Cell In press.

Lewis V, Green J A, Marsh M, Vihko P, Helenius A and Mellman I 1985 J. Cell Biol. **100** 1839-1847

Lobel P, Dahms N M, Breitmeyer J, Chirgwin J M and Kornfeld S 1987 Proc. Natl. Acad. Sci. USA **84** 2233-2237

Sly W S, Fischer H D 1982 J Cell Biochem **18** 67-85

Inst. Phys. Conf. Ser. No. 93: Volume 3, Chapter 6
Paper presented at EUREM 88, York, England, 1988

Stages in the development of spermatozoa of *Aspidosiphon sp* (Sipunculida)

Waltraud Klepal

Institut für Zoologie der Universität Wien
Althanstraße 14, 1090-Wien, Austria

In **Aspidosiphon sp.** the gonad is situated on the M.retractor, especially where the latter is fixed to the septum. Not much is known about the cytodifferentiation of the male gametes (Rice 1975). A gross outline of gonad development has been given by Gonse (1956) in **Phascolosoma vulgare.**

In **Aspidosiphon sp.** the spermatocytes separate from the testis as clusters of about 20 cells. They swim as a "morula" in the coelomic fluid, where they differentiate to spermatids (Fig.1).

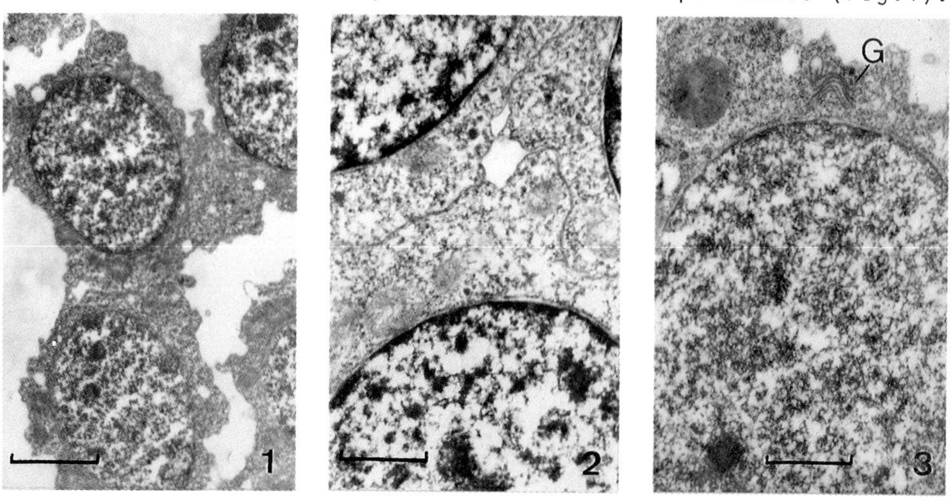

Fig.1. Clusters of spermatocytes. Bar 2μ.
Fig.2. Cells with processes adhering to each other. Bar 1μ.
Fig.3. Nucleus with nucleolus and pores in the nuclear membrane, Golgi within the cytoplasm. Bar 1μ.

In this stage the spermatocytes undergo two meiotic divisions. The cells are of an irregular shape with many processes. In places the cell membranes of neighbouring cells adhere to each other (Fig.2). Their nuclei are large with two or more nucleoli, and pores in their membrane. The cytoplasmic portion of the cell is relatively small, containing RER, Golgi with many vesicles (Fig.3) and a number of mitochondria. The latter may, together with other inclusions, cause the lobed appearance of the cell body. There are only few cristae in the mitochondria. The Golgi is supposedly the place of acrosome formation.

At a later stage the spermatozoa are found within the nephridium. Although not fully mature it may be noted that the spermatozoa are of the primitive type (Franzen 1956). The ovoid nucleus is dense and reticular with vesicles at the acrosomal end. The acrosome is well developed. It has a thin apical portion and thick side-portions around the perforatorium (Fig.4). The flagellum is also developed. It may be wound around the nucleus. At a later stage the nucleus the nuclear material becomes denser and the flagellum of the 9+2 type is free. The centrioles are arranged at right angles to each other. The anchoring fiber apparatus of the distal one is highly complex (Fig.5; Klepal 1987). There are four mitochondria with granular material between them and between them and the centrioles (Fig.6). Only when the spermatozoa are released into the seawater they are fully mature and active.

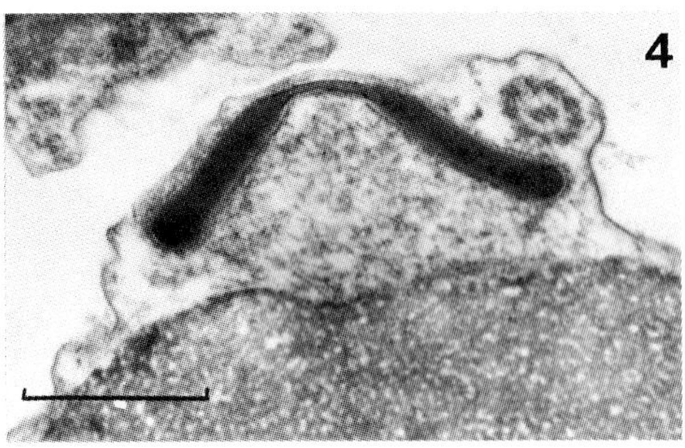

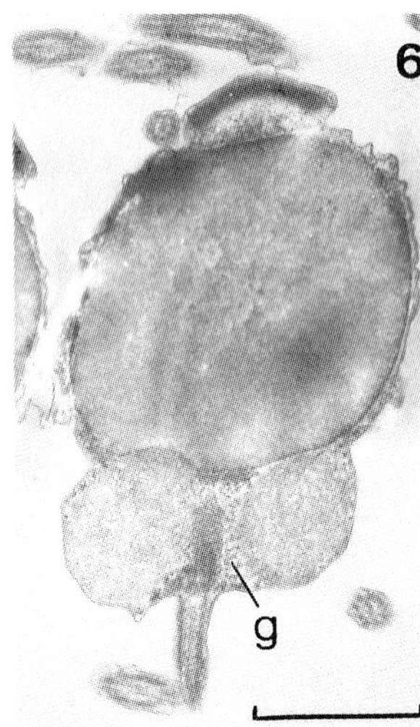

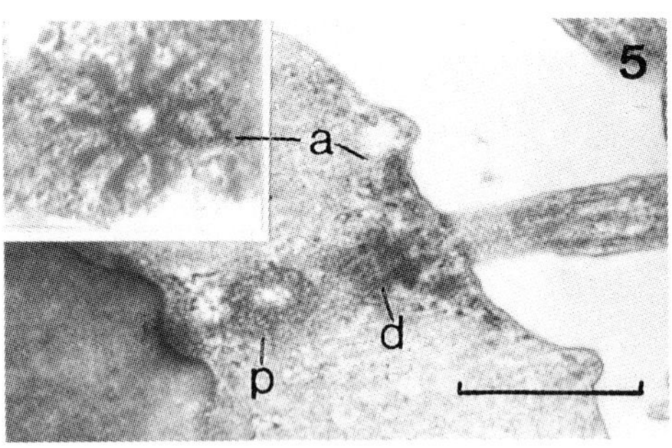

Fig.4. Reticular nucleus with acrosome on apical pole and flagellum wound around the nucleus. Bar 0.5μ.
Fig.5. Flagellum with proximal and distal centriole (p, d) and anchoring fiber apparatus (a). Bar 0.5μ.
Fig.6. Sperm of primitive type with granular material between mitochondria and centrioles (g). Bar 1μ.

References

Franzen A 1956 Zool.Bidr.Uppsala **31** 355
Gonse P 1956 Acta Zool. **37** 193
Klepal W 1987 Eur.J.Cell Biol. Suppl.**18** (43) 43
Rice M E 1975 Reproduction of marine invertebrates **II** pp 67-127

Inst. Phys. Conf. Ser. No. 93: Volume 3, Chapter 6
Paper presented at EUREM 88, York, England, 1988

Ultrastructural evidence of midgut cells in the isopod *Ligia italica* (Isopoda, Crustacea)

J Štrus and K Drašlar
Department of Biology, Biotechnical Faculty, University of Ljubljana, Aškerčeva 12, 61ooo Ljubljana, Yugoslavia

ABSTRACT: The investigation of the foregut-hindgut junction with scanning and transmission electron microscope demonstrated the presence of midgut cells in the digestive tract of Ligia italica. The ultrastructural features of the cells suggest that they are involved in absorption and secretion.

1. INTRODUCTION

The digestive system of isopods consists of a digestive tract lined with chitin and of paired midgut glands-hepatopancreas which are endodermal in origin. The evidence of an endodermal midgut in the digestive tract is controversial (Goodrich 1939, Holdich 1973, Bettica et al 1987). Recently Bettica et al (1987) demonstrated the absence of a true midgut in the digestive tract of different isopods.
The present study was undertaken to find out if the midgut cells are present in the digestive tract of the intertidal isopod Ligia italica.

2. MATERIALS AND METHODS

The digestive tract of intermoult animals was dissected out and fixed in 3,5% glutaraldehyde in o,1 M Sörensen buffer for two hours. The samples were rinsed in buffer, postfixed in 1% osmium tetroxide, dehydrated and embedded either in Paraplast or in Araldite. Thick paraplast sections (3oo µm) were immersed in xylene, critical-point-dried with carbon dioxide and coated with gold on a JFC 11oo sputter coter. The sections and dry-fractured parts of the digestive tract were examined with SEM Jeol 84o A. Ultrathin araldite sections were stained with uranyl acetate and lead citrate and examined with TEM Jeol 8T.

3. RESULTS AND DISCUSSION

A detailed examination of the foregut-hindgut junction revealed a band of cylindrical cells that lack a chitinous intima (Figs. 1,2). The apical surface of the cells is covered with a dense microvillar border (Fig.3). The most characteristic features of the midgut cells are: the absence of a chitinous intima, a dense microvillar border, mitochondria in the subapical region, numerous dictyosomes, secretory vesicles, and multivesicular bodies (Fig. 4). The midgut cells are morphologically similar to the large hepatopancreas cells involved in food absorption and secretion (Štrus 1987). The cells of the hindgut are covered with a chtinous intima (Fig.4).
Our study indicates the presence of an endodermal midgut in the digestive tract of Ligia italica. The ultrastructural characteristics of the midgut cells suggest their absorptive and secretive function.

3. REFERENCES

Bettica A, Witkus R, Vernon G M 1987 J.Crustacean Zool. 7/4
Goodrich A L 1939 J. Morphol. 64 pp4o1-43o
Holdich D M 1973 Crustaceana 29 pp 186-192
Štrus J 1987 Inv. Pesq. 51 pp 5o5-514

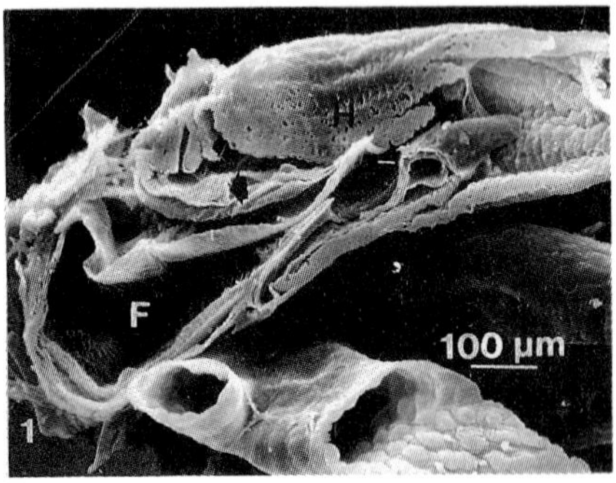

Fig. 1: SEM micrograph of the foregut - hindgut **(F,H)** junction. The arrowhead points to the midgut cells.

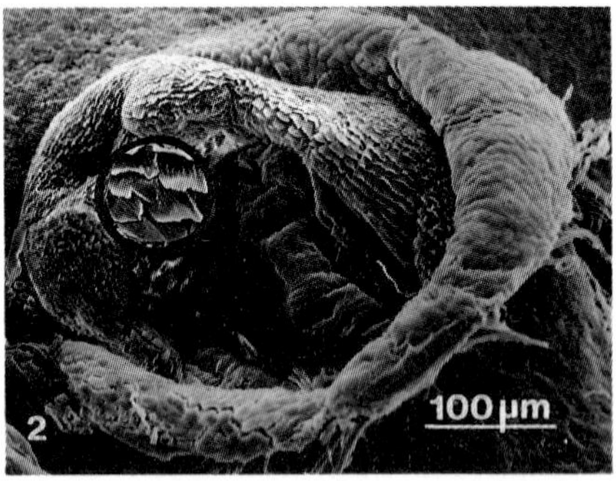

Fig. 2: SEM micrograph of the anterior hindgut with midgut cells and chitinous structure.

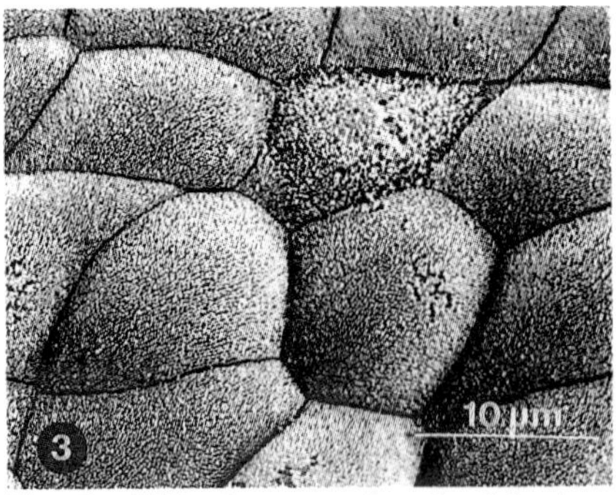

Fig.3: SEM micrograph of midgut cells with dense microvillar border.

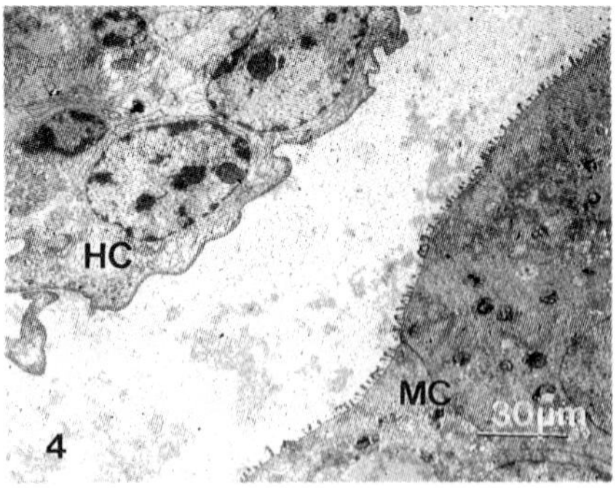

Fig. 4: TEM micrograph of the midgut cell (MC) and hindgut cell (HC).

Inst. Phys. Conf. Ser. No. 93: Volume 3, Chapter 6
Paper presented at EUREM 88, York, England, 1988

Surface ultrastructure of lateral line sensory receptors in *Proteus anguinus* laur (Urodela, Amphibia)

Boris Bulog

Institute of Biology, University of Ljubljana
61001 Ljulbjana, Aškerčeva 12 - POB 141/3, Yugoslavia

ABSTRACT: SEM study of the mechanoreceptive neuromasts and electroreceptive ampullary organs are presented in the blind, neotenous cave salamander Proteus anguinus Laur. These sensory organs are very probably crucial for orientation of Proteus in the dark underground water habitat.

1. INTRODUCTION

The lateral line system is composed of epidermal sense organs distributed over the head and along the body in aquatic amphibians (Russell 1976, Lannoo 1987).

Proteus anguinus, the blind neotenous cave salamander has differentiated lateral line neuromasts and ampullary organs in the integument (Istenič and Bulog 1984). The purpose of our research was to study those sensory organs, which presumably have adequate stimuli in the cave habitat (Istenič and Bulog 1984, Bulog 1986, 1987, 1988).

2. RESULTS AND DISCUSSION

Recent SEM report includes the study of these organs in smaller specimens (37-77 mm). Neuromasts on the snout are surrounded by a groove-like depression and are distributed in a charasteristic pattern (Figs. 1, 3). In the posterior part of the head they are groove less (Fig. 5). The surface portion of the neuromasts is oval, with about 16 sensory cells. These cells are distributed in a longitudinal band and have approximately 60 stereocilia distributed in organ pipe fashion and an eccentrically placed longer kinocilium (Figs. 3, 4, 5, 6). Morphological asimmetry in the ciliary portion is identical with its functional asimmetry. The functional-morphological orientation of the sensory cells is paralell to the longitudinal band of neuromasts (Figs. 3, 4, 5, 6). The sensory cell apical portions in neuromasts are similar to those of morphogenetically earlier P_1 sensory cells in the inner ear (Bulog 1986, 1988). Recent studies of the neuromasts in other urodelans show a similar surface ultrastructure but they differ in the number of sensory cells per neuromast (Lannoo 1987).

Current SEM studies of the ampullary organs on the dorsal part of the head of Proteus support previous results done with light and TEM analysis (Istenič and Bulog 1984). Longitudinal SEM sections of these electroreceptors also give some new information on the threedimensional ultrastructure of the ampullary organs (Figs. 1, 2).

Mechanoreceptive neuromasts and electroreceptive ampullary organs might have a very important role in Proteus orientation during migrations from the central to the peripheral ecotope and back, in communications with its congeners, and in search of prey.

3. REFERENCES: Bulog B 1986 Dissertation
Bulog B 1987 Acta Anat 130/1: 14
Bulog B 1988 J Morphol (in preparation)
Istenič L and Bulog B 1984 Cell Tissue Res 235: 393-402
Lannoo M J 1987 J Morphol 191: 247-263
Russell I J 1976 Lateral line receptors. In: Frog Neurobiology, Springer

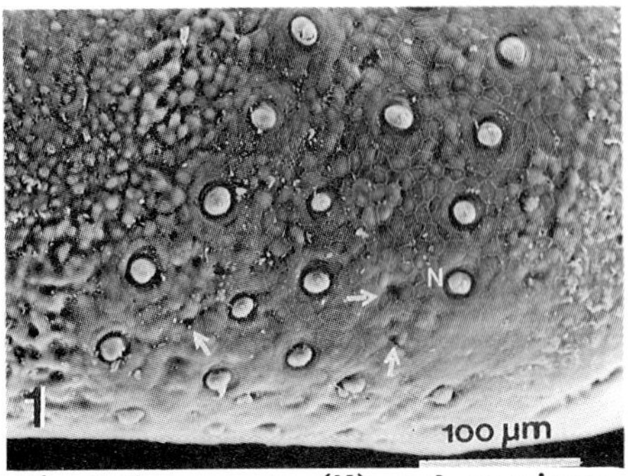

Fig.1 Neuromasts (N) and openings of ampullary organs on snout (▼)

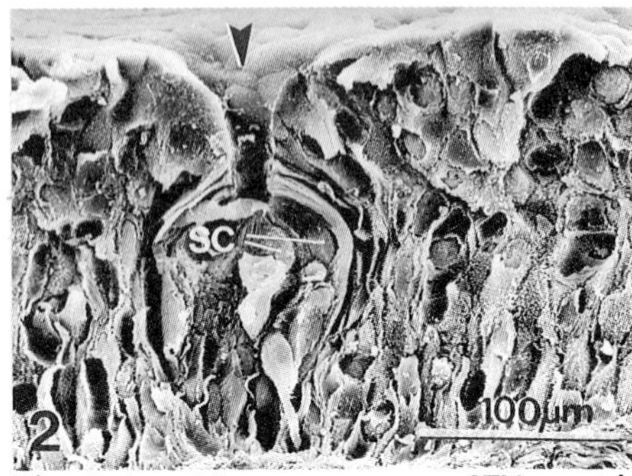

Fig.2 Ampullary organ on SEM section. ▼-opening to canal, SC-sensory cells

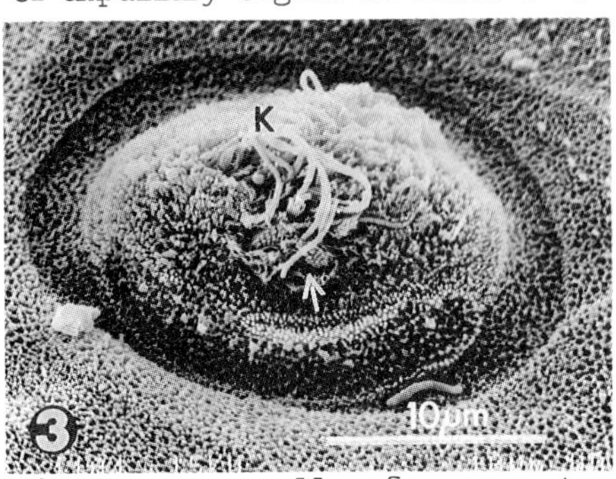

Fig.3 Sensory cells of neuromast. ▼-their orientation, K-kinocilia

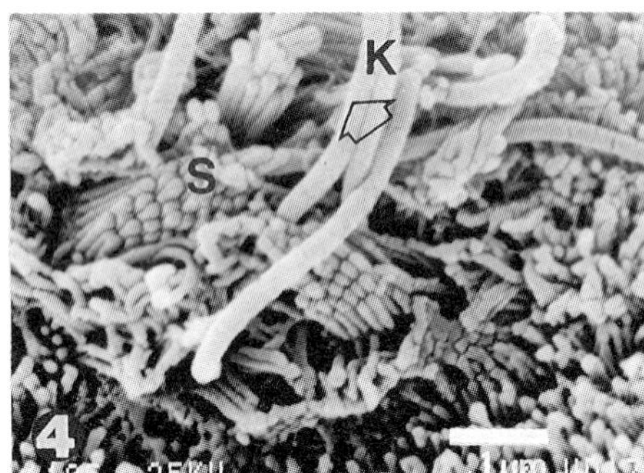

Fig.4 Sensory cells. K-kinocilia, S-stereocilia, ▼-hair cells orientation

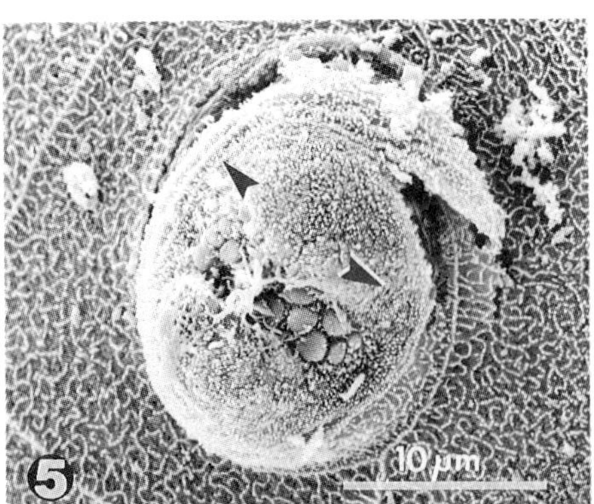

Fig.5 Sensory cells of neuromast have paralell orientation to the longitudinal band and opposite to each other (◄ ►).

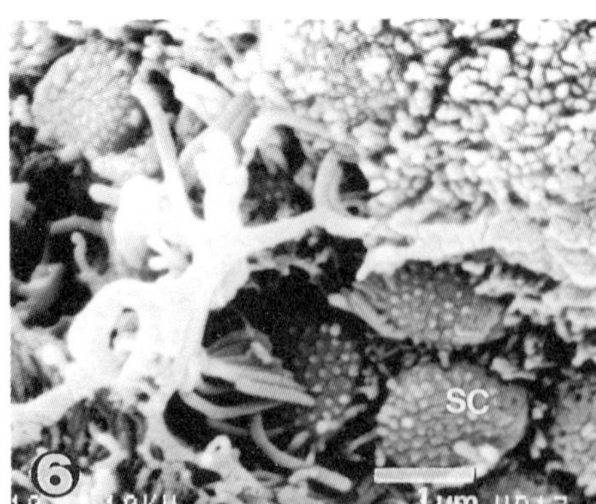

Fig.6 Central part of the band from fig.5 composed of apical portions of sensory cells (SC). During SEM preparation the cilia were lost.

Similarity of roles of nematode epicuticle in intercellular communication with mammalian membranes and root tip epidermis

MATTHEW RYUNTYU

DEPARTMENT OF AGRONOMY AND SOIL SCIENCE, UNIVERSITY OF NEW ENGLAND, ARMIDALE, NSW 2351 AUSTRALIA.

It is important to bear in mind that the successful TEM techniques for nematodes are derived from membrane studies of mammals. This question is of considerable importance when investigating the interactions between the parasite's surface and the components of the host – plant defence systems.

Aporcelaimellus obtusicaudatus, *Mesorhabditis monhystera*, *Clarcus papillatus* and *Pratylenchus brachyurus* (♂♀) were incubated with rat neutrophils in the presence of C3 for 4 hrs. The neutrophils attach and then fall off about 3 hrs later. All species were exposed to neutrophils in the presence of complement, as above, and then stained for ATP–ase. It has been shown that ATP–ase patches on the epicuticle surface are still attached by neutrophils. These studies also reveal that the nematode surface does not contain ATP–ase. No reaction product has been seen in the cuticle layers by TEM, and after examination for the presence or absence of magnesium activated ATP–ase. A cell line RK 13 (Flow Labs. Cat. No 03-550) was included as a control with the same results for the cuticle surface.

All studies indicate that the epicuticle surface can differ and not be similar to mammalian membranes. However, the soil, with dung and decaying plant and animal remains, would be exposed to a wide spectrum of enzymes by the surface of epicuticle. This suggests that the decrease in net negative charge was a function of exposure time of the nematode surface to the rabbit kidney epithelium enzymes (RK 13). It has been shown that despite many basic similarities in morphological features there is nevertheless considerable variation both within and between our species. It would appear that despite similarities, such as semi–permeability to chemical compounds and the presence of certain chemical groups, there are still doubts as to which one can compare these structures.

For example, *A. obtusicaudatus* (free–living species) and *P. brachyurus* (plant parasitic species) were treated with sodium borohydride before: with periodic acid with negative observed by TEM. Sodium borohydride after the periodic acid treatment blocked all staining with negative results too. However, this staining procedure is very specific for carbohydrates: reacts with several groups of polysaccharides, glycoproteins and mucopolysaccharides. The cationized ferritin method enabled us to restrict the labelling only to carbohydrates carrying carboxyl and sulfate groups. At pH 2.5 with cationized ferritin observed the same labelling pattern obtained with the PA–TSC–SP reagents, suggesting that the components being stained were mainly acid mucopolysaccharide residues. As indicated by our results, that the cuticle surface coats (up to 200Å) from the epicuticle appears to contain acid mucopolysaccharides (electron–dense granules).

The ferritin method, as shown above, indicates that the carbohydrates from the epicuticle surface coats, probably, are also secreted from hypodermis cells only.

We do support an hypothesis (see Figure) that the epicuticle surface (e) as well as root (r) tip surface contains a number of highly specific carbohydrate moieties (c) which act as receptors for binding an array of different toxic or non–toxic factors. That polysaccharide–like moieties (pm) composed of secretions from the nematode glands (n), along with various inorganic salts (desquamated cells by root tips (rt)) on the root surface play a decisive role in intercellular communication between plant and nematode (ne) as well as in host parasite recognition. It is well known that pm (a carbohydrate which, on hydrolysis, yields more than 10 monosaccharides), a carbohydrate–derived portion of the structure of a molecule and carbohydrate : aldehyde of ketone derivative of a polyhydric alcohol, particularly of the pentahydric and hexahydric alcohols. Moreover, soluble polysaccharides can be obtained from various soil microorganisms, which in high dilution precipitate specifically the antisera of the corresponding host organisms. Reserve carbohydrates can be stored in the plant (p) or animal epithelium (ae) in the form of high molecular weight hydrolysable compounds, such as starch (s) or glycogen (g).

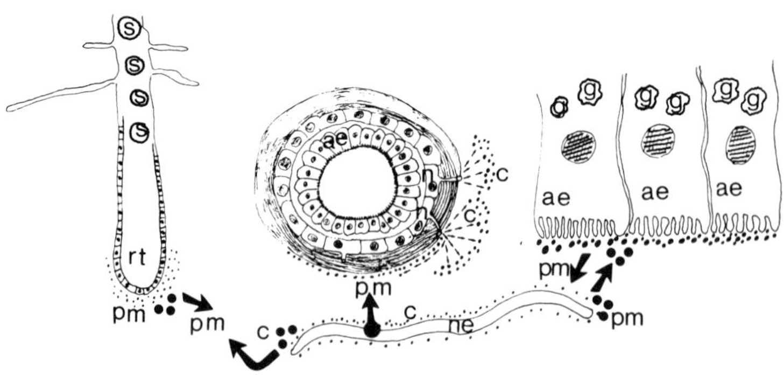

Figure 1: Schematic representation of (pm) in (rt) cells and (ae) and (ne).

Paper presented at EUREM 88, York, England, 1988

Ultrastructure of cells from the adenomera of the digestive gland of *Helix aspersa* M

ALMENDROS, A.; RIOS, A.; BUENO, J.D.; PORCEL, D; ABADIA MOLINA, F.

Dptm. of Cell Biology. Faculty of Sciences. University of Granada. SPAIN.

Introduction

From the beginning of this century several studies have been made on digestive gland of pulmonate gasteropoda using both light microscopy and electron microscopy (Krijgsman 1925, 1928; Mc. Gee-Russell 1955; David y Götze 1963, Sumner 1966, 1969; Abolins-Krogis 1965, 1976). These studies have offered information about the ultrastructure of the gland but several aspects are yet unknown. In this sense, it seems to be interesting to describe ultrastructural characteristics in a gasteropoda poorly studied and of great economical interest, the snail Helix aspersa M. In this paper we describe a study on the ultrastructure of the hepatopancreas of this snail using transmission electron microscopy.

Material and Method

The snails used in this study were not fed for 10 days, in order that the food in the digestive organs be digested, thus aiding the subsequent preparation of specimens in the digestive gland.
The small particles obtained from the digestive gland were fixed in 5% glutaraldehyde and 1% osmium tetroxide, both buffered with 0.2 M sodium cacodylate (pH 7.4). Dehydration was carried out in acetone and propylene oxide, embedding them subsequently in Durcupan ACM. The ultrafine sections obtained in an LKB ultramicrotome Ultrotome III were stained with 5% uranyl acetate and lead citrate. A Philips EM 300 electron microscope was used to make the subsequent observations.

Results

The cells that constitute the adenomera of the digestive gland are situated surrounding the central lumen of ducts and show cytological differences in function of the cellular type observed. Differences are also found for a same cell with respect to the cytoplasmic zone that is observed.
All the descriptions made during the last years about the structure of the digestive gland or hepatopancreas of pulmonate gasteropoda are coincident in that they report several cellular types. Their different denominations were adequately summarized by Sumner in 1966.
Four main types of cells were found: digestive cells, excretory cells, calcic cells and oxyphilic cells. Transversal sections of secretory tubules show that the cells are closely intrused, extending from the basal limit to apical region on a waving course. This cellular distribution difficults to obtain complete sections of the cells in their longitudinal axis, offering partial sections of the different cellular types in an adjacent arrangement. A clasification of these cells has been made on an ultrastructural basis, using as features the characteristics of cytoplasm as well as the presence and distribution of cellular organelles.

a) Digestive cells:

The basal cytoplasm shows abundant organelles, mainly endoplasmic reticulum of small and smooth vesicules, dictiosomes and vacuoles with a variable content. Two different types of vacuoles can be observed in this cytoplasmic region. Type "a" vacuoles are occupied by a clear material whereas type "b" show a more dense substance with a variable granular aspect.

A peculiar feature of these cells is the presence of parallelly associated lengthy cisternae that offer in transversal sections a typical "honeycombed" image. These formations appear sourronded by membrane formations that show some dilatation in several points. The structure organized in this maner probably represents a particular smooth endoplasmic reticulum formation with a ordenate arrangement similar to that observed in other cellular types of invertebrate groups.

In the apical border, the plasma membrane shows a brush border with endocytic formation of vesicles in the bottom region between microvilli.

b) Calcic cells:

This cellular type offers a dense cytoplasm with abundant granular endoplasmic reticulum usually arranged surrounding mitochondria. The most characteristic feature of those cells is the presence of membrane covered granular formation with a concentric lamellar structure.

Such formations correspond to calcium spherules or calcium granules described in other animals and in this model show a different electronic density for one lamella to another. They are distributed all over the cell with a high concentration in the basal region of the cytoplasm.

c) Excretory cells:

They are cells with a clear aspect whose cytoplasm is loose in organelles. The greatest part of the cell is occupied by big vacuoles in a fashion to what has been described for "b" type in digestive cells. A small quantity of type "a" vacuoles is also present in excretory cells.

d) Oxyphilic cells:

This denomination is based on the greater number of mitochondria that appear inside these cells. The abundant mitochondria offer a diversity of size and aspect. There is a scarce presence of other cellular organelles. The abundance of this type of cell in secretory tubules of digestive gland is not important and the cellular size is also small. This situation makes it difficult to obtain oxyphilic cells in sections from adenomera.

Literature

-Abolins-Krogis A 1965. Arkiv För Zoologi, Band 18 nr 7, pg. 85-92.
-Abolins-Krogis A 1976. Cell and Tiss. Res., 172, pg. 455-476.
-David H, Götze J 1963. Z. mikr.-anat. Forsch. Bd 70, pg. 252.
-Krijsman B J 1925. Zeit. vergl. Physiol. Bd 2, pg. 264-296.
-Krijsman B J 1928. Zeit. vergl. Phisiol. Bd 8, pg. 187-280.
-Mc Gee-Russell S M 1955. D. Phil. Oxford.
-Summer A T 1966a. Jour. Royal Micros. Soc. vol. 85, pg. 181-192.
-Summer A T 1969. J. Zool. Lond. v. 158, pg. 277-291.

Inst. Phys. Conf. Ser. No. 93: Volume 3, Chapter 6
Paper presented at EUREM 88, York, England, 1988

The abdominal trichobotria of the first larval stage in the bug *Pyrrhocoris apterus*

K. Drašlar and J. Rode

Department of biology and Institute of biology
Aškerčeva 12, Ljubljana, Yugoslavia

ABSTRACT: Single lateral trichobotria are located on the 3rd, 4th, 5th, 6th and 7th abdominal segments in the bug Pyrrhocoris apterus. The trichobotria are knoblike and surrounded by a smooth cuticle.

1. INTRODUCTION

The sensory hairs - the trichobotria are cuticular mechanoreceptors in insects. Their position and number are constant on abdomen of bugs (Schaeffer, 1975). Trichobotria surrounded by fields of cuticular trichoma (Fig. 1) are located ventrally in the mid of the 3rd and 4th and on the margin of the 5th, 6th and 7th segments. Three different functional types of receptors were described in imago of Pyrrhocoris apterus (Drašlar, 198o). The long, the middle and the short hair receptors differ also in their physiological functions. We investigated the shape and the dsitribution of trichobotria in early larval stages to enable better understanding of their physiological function.

2. MATERIAL AND METHODS

Imagoes collected in nature and laboratory cultured larvae were used for investigations of sensory hairs. Either air dried ufixed animals or G.A.-OsO_4 fixed CP dried animals were coated with gold and examined under JEOL 840A SEM microscope.

3. RESULTS

The distribution and shape of abodominal trichobotria of the first larval stage are different from adults. Grouped lateral and medial trichobotria were described in third and later larval stages and imagoes. All trichobotria of thefirst larval stage are single and located on the margin of the 3rd, 4th, 5th, 6th and 7th abdominal segments. The trichobotrium cup is located within a cuticular knob above the surface of a smooth cuticle (Fig. 2). During later development the cuticular knob sinks slowly to the level of a surroundings cuticle (Fig. 3). The denticles of the outer edge of the trichobotrial cup appear either during the first larval stage or in later stages. The fields of surrounding trichoma are well developed by the third larval stage (Fig. 4).

4. REFERENCES

Drašlar K., 198o: Dissertationes SAZU XXII/5

Schaefer, W. C., 1975: J. Insect Morphol-Embriol., 4(3)

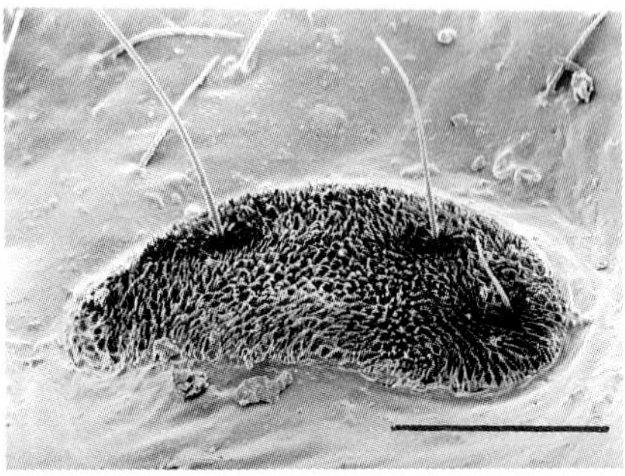

Fig.1.Trichobotria by imago 5th abdominal segment (scale bar 1oo µm)

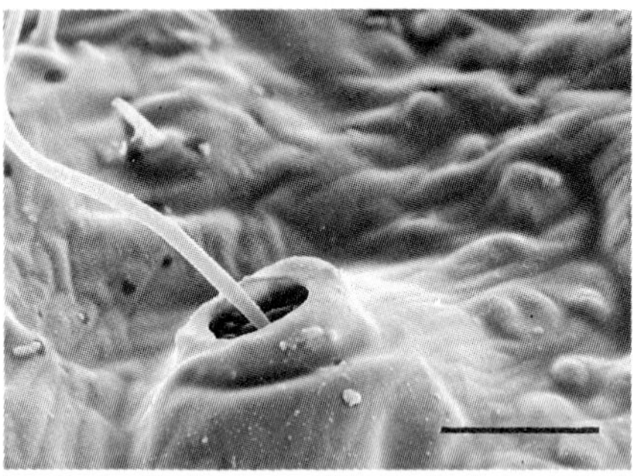

Fig. 2. Single trichobotrium in 1st alrval stage, 5th abdominal segment (scale bar 1o µm)

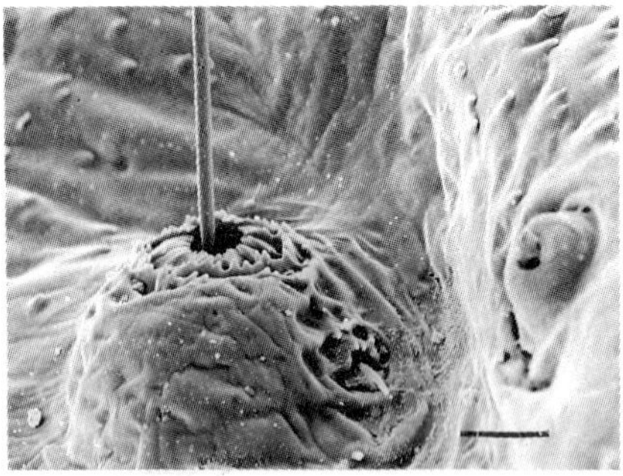

Fig. 3. Single trichobotrium in late state of 1st larval stage 5th abdominal segment (scale bar 1o µm)

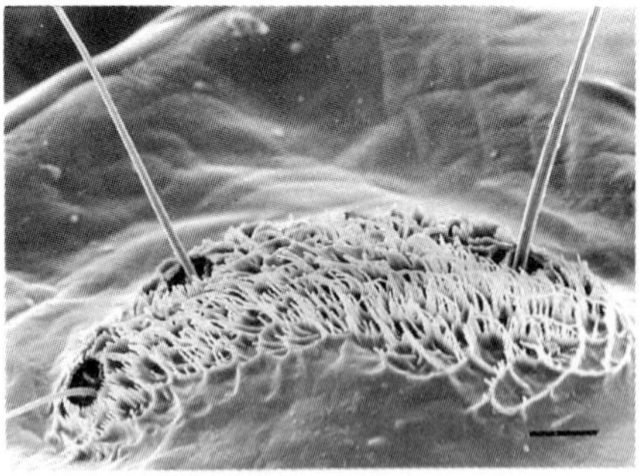

Fig. 4. Trichobotria by 3rd larval stage, 5th abdominal segment (scale bar 1o µm)

Inst. Phys. Conf. Ser. No. 93: Volume 3, Chapter 6
Paper presented at EUREM 88, York, England, 1988

A scanning electron microscopic study of the embryonic development of the house dust mite, dematophagoides farinae (Pyroglyphidae, Actinotrichida) from Blastula to the hatching of the larva

M G Walzl

Department of Comparative Anatomy & Morphology, Institute of Zoology, University of Vienna, Althanstraße 14, A-1090 Vienna, Austria

ABSTRACT: Six stages of the embryonic development of the allergen producing mite Dermatophagoides farinae are represented for the first time by means of scanning micrographs. The results are compared to those obtained from other actinotrichous mites.

1. INTRODUCTION

Embryological investigations of Acari are rare and fragmentary. Only some species of medical and economic importance have been studied in detail by means of light microscopy (Anderson 1973). Because of technical problems in the fixation of the eggs due to an impenetrable chorion, no electron microscopic studies abaut the development of actinotrichous mites have been made. In the course of investigations into the structure and development of house dust mites it became evident that the scanning electron microscope can contribute to clear up the external features of the embryonic development of Dermatophagoides farinae.

2. MATERIAL AND METHODS

Fertilized eggs of D. farinae, kept separate at a temperature of 25°C and 75% relative humidity, were fixed in CARNOY's fluid 20, 36-40, 48, 68-70, 74-76, and 170 hours after oviposition (=hrs.a.o.). Before fixation the eggs were laid in 0,25% sodium-thiosulfate to swell the chorion and pricked with two needles. Preparation for SEM was continued according to the method of Walzl & Waitzbauer (1980). Owing to the small size of the objects, specially constructed tubes were used for the CP-treatment.

3. RESULTS AND DISCUSSION

The embryonic development in the ovoid eggs (fig.1) - 160-180µm and 70-90µm in size- takes 170-180 hours. The formation of the uniform blastoderm is finished 20 hrs.a.o. (fig.2). 36-40 hrs.a.o. the blastoderm is differentiated to the germ band on the ventral side with developing prosomal limb buds and to the dorsal extraembryonic ectoderm (fig.3). 48 hrs.a.o., the dorsal closure of the embryo is completed except for a round plate of extraembryonic ectoderm in the posterior region (fig.4). The cheliceral limb buds (ch), called cephalic lobes by Hughes (1950), are formed before the pedipalpal pair (p) and the three pairs of ambulatory limb buds (1-3) (fig.4 and 5). As in other actinotrichous mites (Hafiz 1935, Hughes 1950, Langenscheidt 1958), there is no evidence for the existence of a fourth pair of ambulatory limb buds. 74-76 hrs.a.o. however, two median bulges become visible behind the third pair of ambulatory limbs, which can be interpreted as the last prosoma- and the first opisthosoma segment (fig.6). About 100 hrs.a.o. two egg teeth become visible at the basis of the chelicerae, sign of the finished chitinisation of the prelarva. The hexapod larvae show a characteristic posure in the egg, with the tips of both anterior leg pairs pointing backwards and those of the third leg pair pointing forwards (fig.7). The hatching larva splits the egg shell longitudinally, leaving the prelarval integument as exuvia (ex) in the egg shell together with the egg teeth (et) (fig.8).

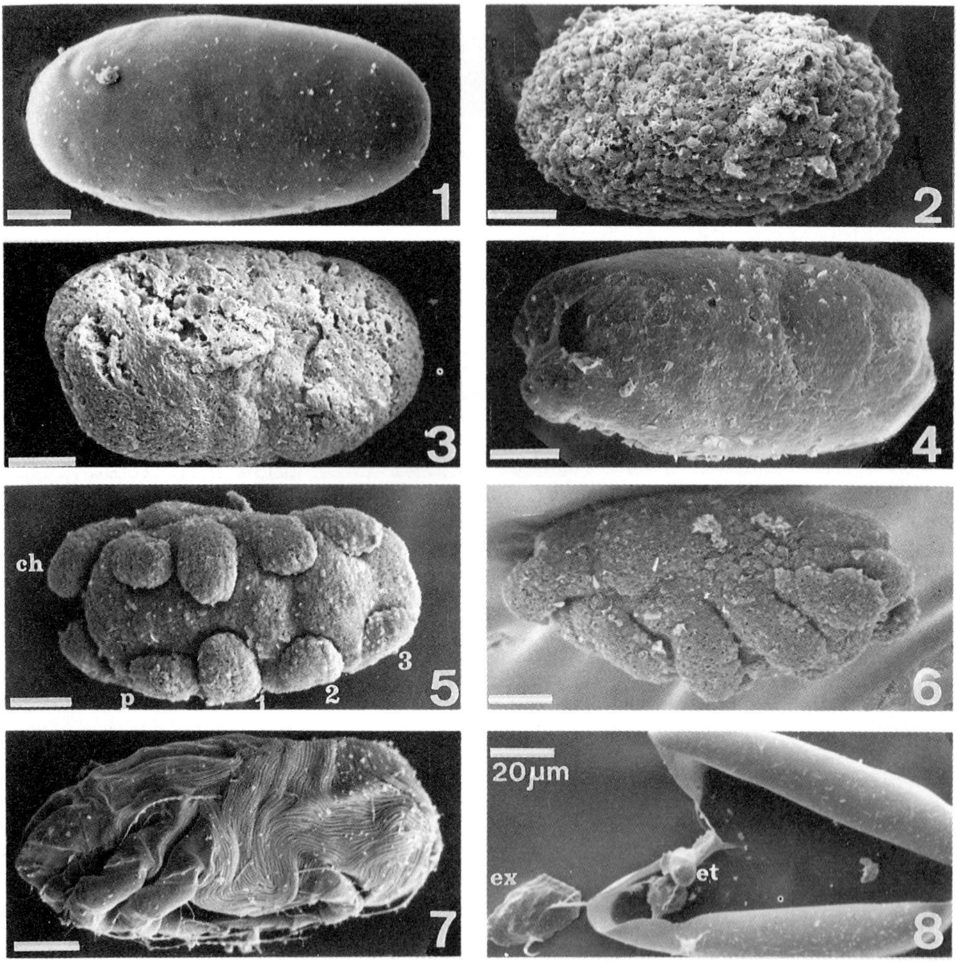

4. REFERENCES

Anderson D T 1973 Embryology and Phylogeny in Annelids and Arthropods (Oxford:Pergamon) pp 400-7
Hafiz A 1935 Proc.Roy.Soc.London Ser.B **7** 174
Hughes T E 1950 Proc.Zool.Soc.London **119** 873
Langenscheidt M 1958 Z.Parasitenkunde **18** 349
Walzl M and Waitzbauer J 1980 Mikroskopie (Wien) **36** 164

Supported by FWF Project 4015

Ultrastructural study of putative endocrine cells in the midgut of the adult worker honeybee

H. Raes*, W. Bohyn**, F. Jacobs* & P.H. De Rycke*

* Laboratory of Zoophysiology, K.L. Ledeganckstraat 35
** Laboratory of Electron Microscopy, Sint-Pietersnieuwstraat 41
State University of Gent, B-9000 Gent (Belgium).

Gut endocrine cells are widely distributed through the animal kingdom. They have also been described in several insect ordines (Nishiitsutsuji-Uwo and Endo 1981). Their occurrence in the midgut of honeybees was therefore to be expected.

We found small endocrine-like cells dispersed between the columnar cells in all but the anterior fourth of the midgut. Like the columnar cells, they seem to stem from the regenerative crypts; per crypt-complex, one, exeptionally two of them, can be found. As yet, we were able to discern two different gut endocrine cell types : a more abundant basal-granulated cell and a basal-vesiculated cell.

Apart from the different morphology of their secretory products, which may correspond to different types of peptide hormones, the ultrastructure and general outlook of the two cells is very similar. Both are relatively electron dense with a broad cell base and a tapering apex which opens into the lumen. The small cell top carries few short microvilli.

The cell is divided in two functional compartments by the nucleus. The supra-nuclear region is the site of secretion production with many well developed cisternae of R.E.R. and Golgi complexes, as well as newly formed secretion granules in different stages of development (Figure 1). This part of the cell also contains free ribosomes, polysomes, some mitochondria and multivesicular bodies. In older cells, particularly of the vesicular type, several large secondary and tertiary lysosomes and some mineralised granules can be found close to the cell apex.

The infra-nuclear region contains many mitochondria, polysomes, some glycogen and short stacks of R.E.R.; this part of the cell is dominated by the abundance of secretory product (Figure 2). In the basal-granular cells this appears as small (100-200 nm), more or less electron dense membrane bound granules (Figure 3). In the basal-vesicular cells, it appears as electron lucent vesicles which may contain a flocculent material; their form as well as their size are variable (Figure 4).

In contrast to the columnar cells, the endocrine cells have no basal labyrinth; the close contact between secretion products and the basal cell membrane suggests secretion towards the haemolymph by exocytosis.

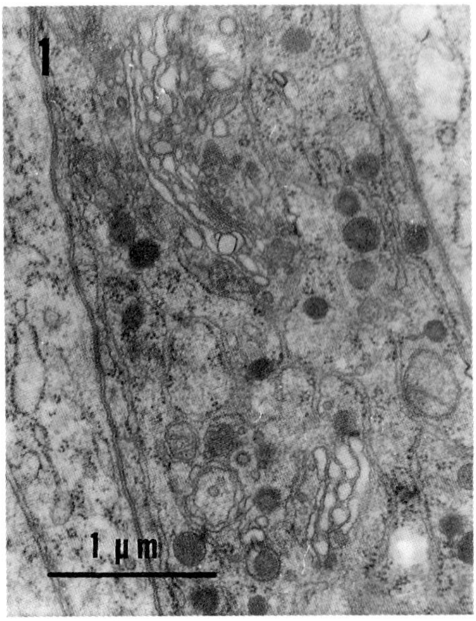

Figure 1 : Golgi complexes with newly formed granules

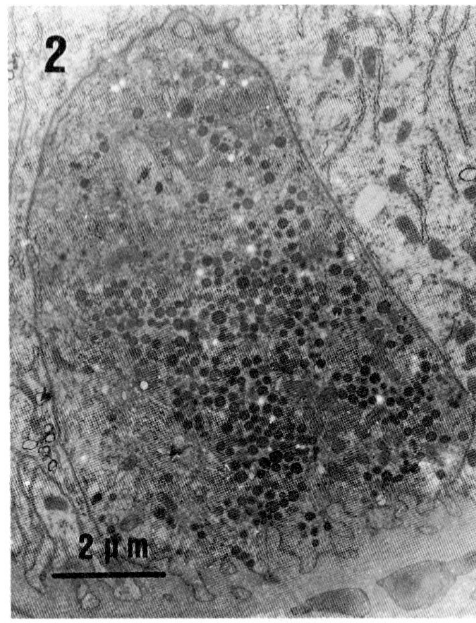

Figure 2 : Infra-nuclear region of basal-granulated cell

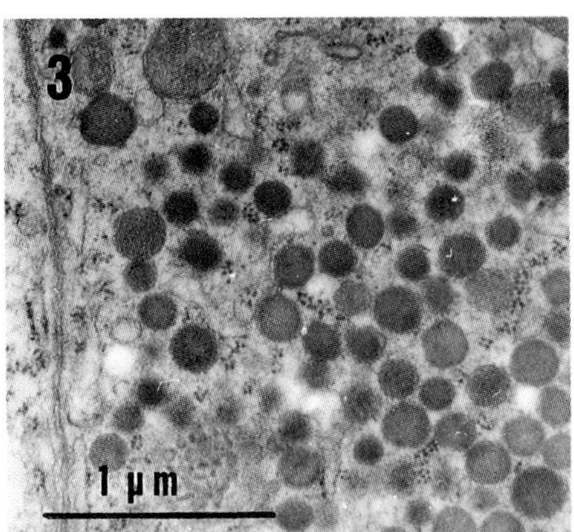

Figure 3 : Secretory granules in basal-granulated cell

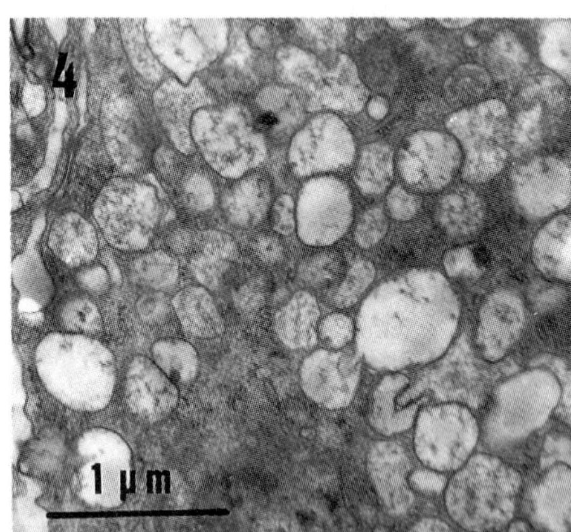

Figure 4 : Secretory vesicles in basal-vesiculated cell

REFERENCES

Nishiitsutsuji-Uwo J, Endo Y 1981 Biomed. Res. 2 (1) 30

Inst. Phys. Conf. Ser. No. 93: Volume 3, Chapter 6
Paper presented at EUREM 88, York, England, 1988

Different types of plasma-membrane-derived coated vesicles in developing honeybee ocelli

M A Pabst

Department of Histology and Embryology, Harrachgasse 21, 8010 Graz, Austria

ABSTRACT: From the beginning of microvilli formation in photoreceptor cells of two day old pupae and further on coated pits pinch off from the microvilli bases indicating photoreceptor membrane degradation. Only two to four day old pupae also show double membraned coated vesicles budding off from membranes of two connecting photoreceptor cells, and their microvilli bases. They may be involved in screening pigment granule formation occuring only at this time. Coated pits and vesicles were also seen in corneagenous cells during lens development, in photoreceptor cells and glial cells.

1. INTRODUCTION

Coated pits and pinched-off coated vesicles forming on plasma membranes are known to be involved in receptor-mediated endocytosis. Additionally in the arthropod retina coated pits budding off from the bases of microvilli are involved in the degradation of photoreceptor membranes (Blest et al 1984). Coated vesicles may also be involved in screening pigment granule formation and in lens development.

2. RESULTS and DISCUSSION

Even in a two day old pupa (Pp-pupa after Rembold et al 1980) when microvilli development just started, as in all later pupal and adult life stages, coated pits and vesicles are seen in photoreceptor cells at the bases of microvilli (Figure 1, 2). Supposed membrane shedding already seems to start during development of microvilli as also seen by Hafner et al (1982) in developing crayfish retina. Only in two to four day old pupae double membraned coated vesicles are also seen forming as invaginations of two neighbouring photoreceptor cells seen from distally to the nucleus, also from bases of already developing microvilli (Figure 2). At this same time and area screening pigment granule formation takes place from vesicles also showing two membranes during first developmental stages. The inner membrane separates the small latticed grey praegranule from the developing electron-dense part (Figure 2). On and near the bases of microvilli of corneagenous cells coated pits and vesicles are found during development of the lens (Figure 3). They always show electron-dense material connected to their inner membrane. Smooth secretory vesicles are also detected there. Locke (1969) finding these two types of vesicles in epithelial cells during development of the protein epicuticle, suggests that macromolecules not incorporated and no longer needed are carried off by receptor-mediated endocytosis. Coated pits are also found in membranes of receptor cells surrounded by glial cells in pupae and adult bees down to their axonic region (Figure 4) and in glial cells, as already found in the intermediate retina of complex eyes (Saint Marie et al 1984), facing hemolymph spaces.

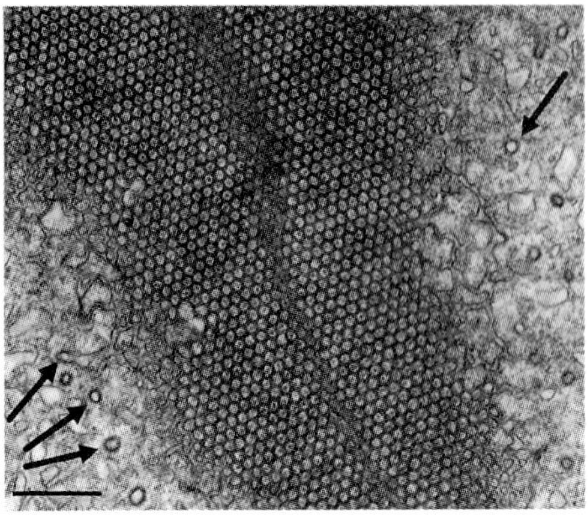

Fig. 1. Just hatched bee. Cross section of a rhabdom. Arrows point to coated pits and vesicles. Scale bar: 0.5 µm.

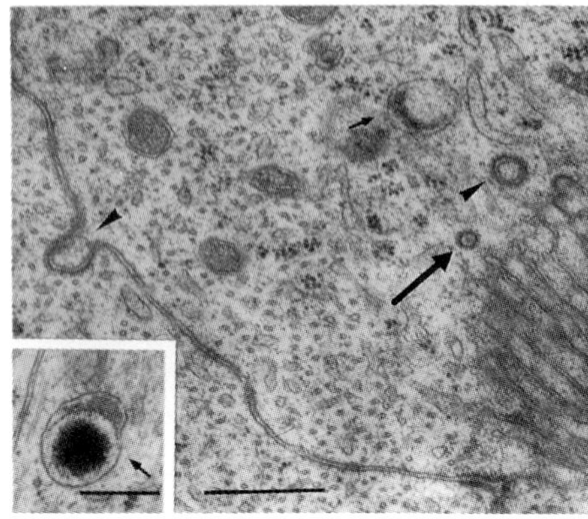

Fig. 2. Two day old pupa: coatea vesicle (arrow) and double membraned coated pit and vesicle (arrowheads). Small arrows point to developing screening pigment granules. Scale bar: 0.5 µm, insert: 0.2 µm.

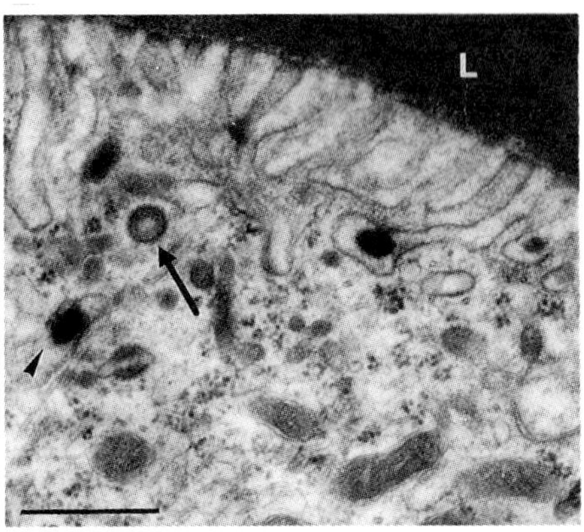

Fig. 3. Just hatched bee: coated vesicle (arrow) and secretory vesicle (arrowhead) near microvilli basis of a corneagenous cell. L=lens Scale bar: 0.5 µm.

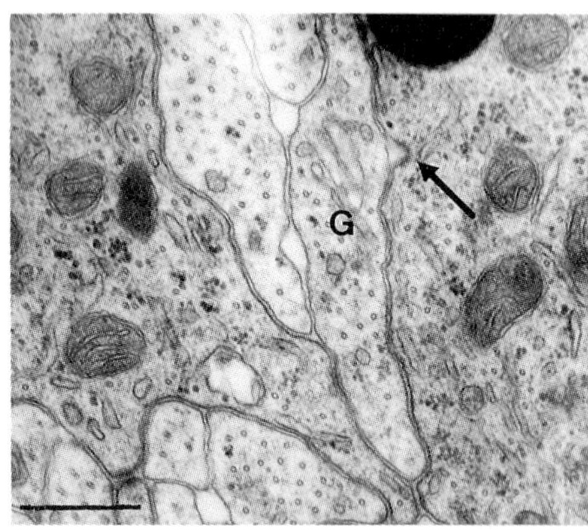

Fig. 4. Five day old pupa: coated pit (arrow) on the membrane of a receptor cell surrounded by a glial cell (G). Scale bar: 0.5 µm.

3. REFERENCES

Blest A D, Stowe S and De Couet H G 1984 Sci.Prog., Oxf. 69 83

Hafner G S, Tokarsky T and Hammond-Soltis G 1982 J. Morph. 173 101

Locke M 1968 J. Morph. 127 7

Rembold H, Kremer J P and Ulrich M 1980 Apidologie 11 29

Saint Marie R L, Carlson S D and Chi C 1984 Insect Ultrastructure 2 (New York, London: Plenum Press) pp 435-75

Inst. Phys. Conf. Ser. No. 93: Volume 3, Chapter 6
Paper presented at EUREM 88, York, England, 1988

Tracheoles in the developing and adult ocelli (*Apis mellifica*: Hymenoptera)

M A Pabst

Department of Histology and Embryology, Harrachgasse 21, 8010 Graz, Austria

ABSTRACT: When ocelli are already largely differentiated tracheoles grow in from proximal between glial cells. So in the 6-7 day old pupae, about one day before hatching, tracheoles can be seen to the rhabdomeric region of photoreceptor cells. The tracheoblasts cover large areas in close contact with neighbouring glial cells, connected by tight junctions. A direct contact to photoreceptor cells was never found. In imagos the cytoplasmic envelope of tracheoles is reduced and only small areas contact glial cells.

1. INTRODUCTION

Tracheoles, the fine endings of ramified tracheae, where the gas exchange takes place, were found in ocelli only in tapetal sheath cells of dragon flies (Ruck and Edwards 1964) and in the tapetal layer of some dipterans (Goodman 1970). The development of tracheoles in ocelli is so far undiscribed.

2. RESULTS and DISCUSSION

Ocelli develop principally during the pupal stage of honeybee. In a two day old pupa (Pp-pupa after Rembold et al 1980), when the lentinogenous and the retinogenous layer are clearly visible and have started their differentiation, tracheal elements can be found close to ocelli but no tracheoles can be detected inside ocelli. Rare first ingrowths of tracheoles are observed three days later (Pdl-pupa) in the axonic region of receptor cells secreting cuticular material. In a 6-7 day old pupa (Pdm-pupa), about one day before hatching, tracheoles are present from the proximal to the rhabdomeric region of receptor cells (Figure 1). Now these tracheoles usually have far fewer fibrils in their lumen than in Pdl-pupae but the outer cuticle still appears pentalaminar. In just hatched bees no more fibrils are found in the tracheolar lumen. The cytoplasm of tracheoblasts in Pdm-pupae shows many microtubuli, arranged parallel to the long axis of the cell, probably important for their ingrowth. Delicate undulations, oriented toward the long axis of the tracheole, are seen in intertaenidial spaces of the tracheolar cuticle of all ages of bees (Figure 2). Ingrowing tracheoblasts have no basement membrane. They cover large areas surrounded by glial cells and are connected to them by tight junctions (Figure 2). They never are in direct contact to photoreceptor cells. In imagos, when the hemolymph spaces get wider the contact to glial cells is reduced to some small areas and also the cytoplasmic envelope itself is reduced (Figure 3) but no naked tracheoles, i.e. without any cell lining, as in the rectum of forager honeybees (Noirot and Noirot-Timothée 1982) are seen.

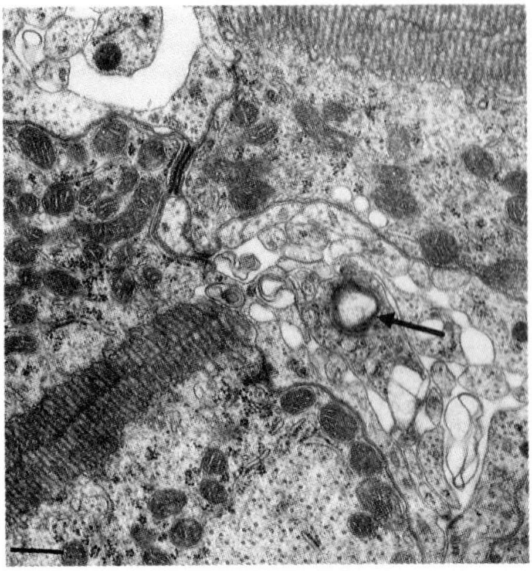

Fig. 1. 6-7 day old pupa: A tracheole (arrow) surrounded by glial cell processes in the rhabdomeric region of photoreceptor cells. Scale bar: 0.5 µm.

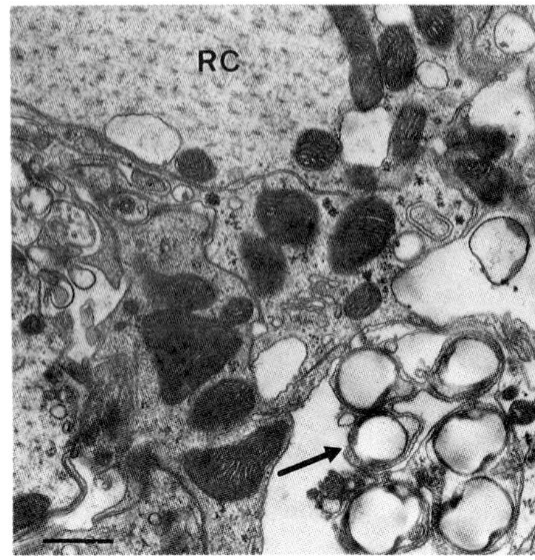

Fig. 3. Forager bee: Ramified tracheoles (arrow) in the axonic region of receptor cells. Note their small cytoplasmic envelope at this age. RC= receptor cell axon. Scale bar: 0.5 µm

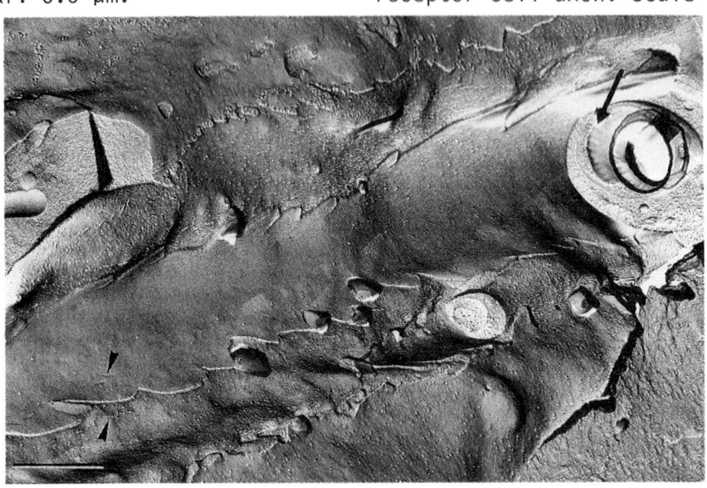

Fig. 2. Freeze-etch replica of a just hatched bee: Arrow points to delicate undulations of the cuticular intima. Tight junctions (arrowheads) between the tracheoblast and a neighbouring glial cell. Scale bar: 0.5 µm.

3. REFERENCES

Goodman L J 1970 Adv. Insect Physiol. 7 97

Noirot C and Noirot-Timothée C 1982 Insect Ultrastructure 1 (New York, London: Plenum Press) pp 351-81

Rembold H, Kremer J P and Ulrich M 1980 Apidologie 11 29

Ruck P and Edwards G A 1964 J. Morph. 115 1

Inst. Phys. Conf. Ser. No. 93: Volume 3, Chapter 6
Paper presented at EUREM 88, York, England, 1988

Evolutionary aspects as deduced by ultrastructural analysis of the chorion in five species of 'picture-winged' Hawaiian Drosophila

Lukas H. Margaritis, Katherine Dellas and Michael Kambysellis
Athens University, Dept.of Biology and New York University, Dept.of Biology

ABSTRACT: Scanning electron micoscopy and gel electrophoresis of the chorion proteins in five Hawaiian Drosophila species has revealed that certain variations occur which are related to the phylogenies of the species as well as to the microenvironment of the egg-laying substrate.

1. INTRODUCTION

The hundreds of Drosophila species endemic to the Hawaiian Islands provide an excellent group of organisms for evolutionary studies (1) especially since frequent volcanic activity has established geographical isolation. The species are well adapted to their diverse ecological niches by modification of their reproductive strategies (2). Thus, the process of choriogenesis which has been studied extensively (3,4) becomes an attractive system for evolutionary studies. Using D.melanogaster as a reference we aim in corelating structure/composition of the chorion with evolutionary and/or ecological characteristics.

2. MATERIALS AND METHODS

We have used scanning electron microscopy of selected follicles collected in the field and polyacrylamide gel electrophoresis as described elsewhere(3)

3. RESULTS AND DISCUSSION

Scanning electron micrographs of the endochorion are shown in figures 1 through 21 as follows: figures 1,6,11 for Drosophila grimshawii, figures 2,7,12 for D.heteroneura, figures 3,8,13,16,19 for D.adiastola, figures 4,9,14,17,20 for D.macrothrix and figures 5,10,15,18,21 for D.formella. Magnification is indicated by a bar equalling 100 microns in figures 1-5, five microns in figures 6 and 8 and ten microns in the remaining figures. These eggs are large compared to D.melanogaster (3), have a dorsal ridge and four rather than two respiratory appendages, the length of which correlates to the ovipositional substrate. Higher magnifications reveal the various structural details of the respiratory appendages (figures 6-10), the operculum (figures 11-15), the follicle cell imprints (figures 16-18) and the posterior pole (figures 19-21)

Following in vitro culturing in the presence of tritiated proline the chorion proteins of the five species are displayed in fluorograms and show distinct differences compared to D.melanogaster (figure 22b): D.grimshawii (fig.22a), D.formella (fig.22c), D.adiastola (fig.22d), D.heteroneura (fig.22e) and D.macrothrix (fig.22f).

Given the known ecological niches of the egg-laying substrate and the phylogenies based on other criteria it appears that the chorion substructure and composition has been evolutionary selected to fullfill the physiological requirements during embryogenesis(5).

REFERENCES

1. Carson H. et al. 1970. In essays in Evolution and Genetics(eds.Hecht and Steere) N.Y.Appleton-Century
2. Kambysellis, M.P. and Heed, W.B. 1971. Amer.Nat. 105,31-49
3. Margaritis L.H. et al.1980. J.Cell Sci.43, 1-35
4. Margaritis L.H. 1986 Can J.Zool. 64,2152-2175
5. Margaritis L.H. 1985 In Comprehensive Insect Physiology, Biochemistry and Pharmacology,Vol1 pp153-230. (eds L.I.Gilbert and G.A.Kerkut) Pergamon Press.

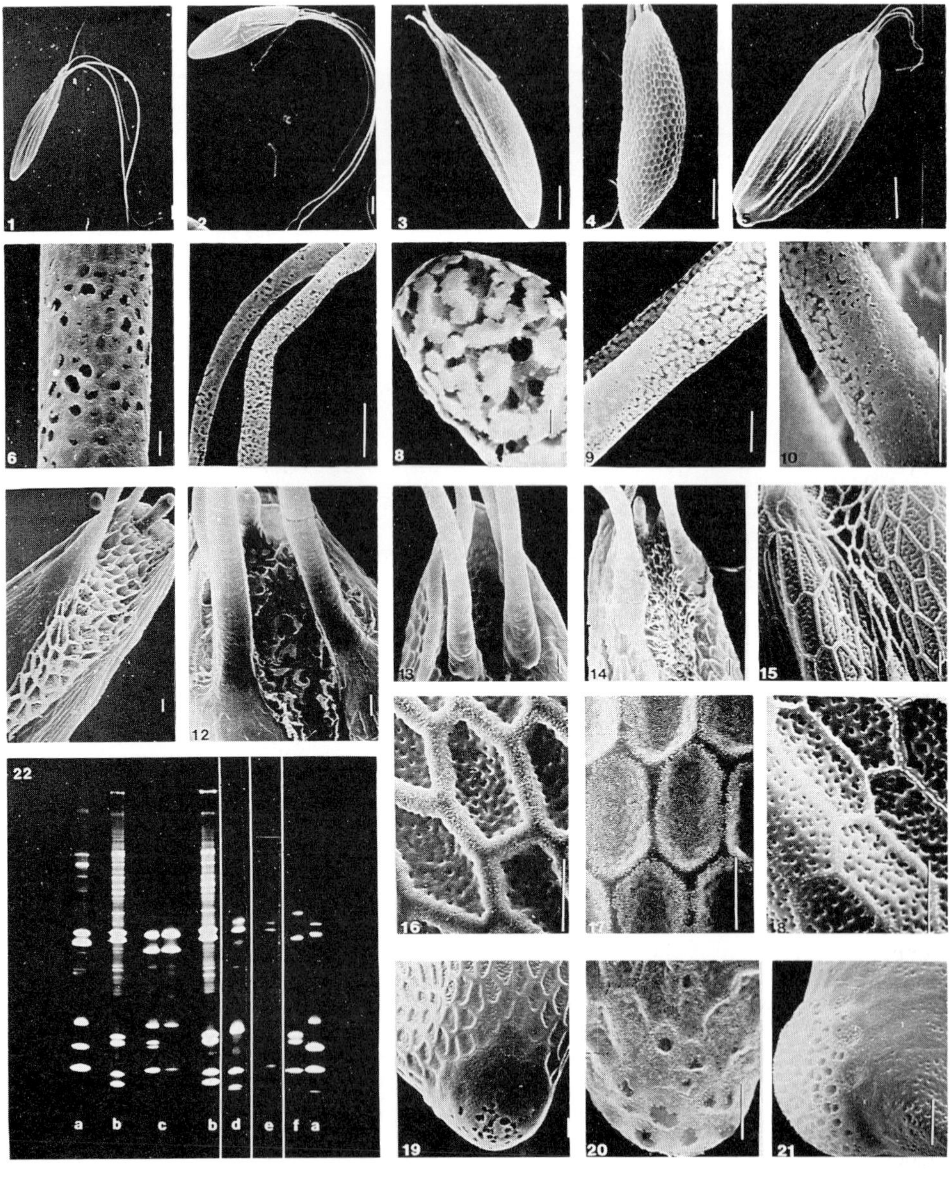

Particular uptake by teleostean thrombocytes

E Bielek

Institute of Histology and Embryology, University of Vienna, Schwarzspanierstr. 17, A - 1090, Vienna, Austria

ABSTRACT: Thrombocytes of trout and carp accumulate particulate material (ferritin, colloidal carbon, latex, cell debris, lipid droplets) regularly in large vacuoles in the cell centre.

1. INTRODUCTION

The phagocytic capacity of teleostean thrombocytes has been the subject of numerous controversial observations (see Suzuki 1984). Its interpretation is complicated by the difficulties of identification of this cell type, which may asssume different shapes and resembles often lymphocytes. As thrombocytes can be unequivocally identified ultrastructurally by characteristic features analogous to mammalian platelets, i.e. marginal band and surface connected system (SCS) (e.g. Bielek 1979, Daimon et al. 1979), the possible uptake of different particles was investigated electron microscopically.

2. MATERIAL AND METHODS

In vivo uptake was studied after injection of colloidal carbon into fingerlings of trout (Salmo gairdneri; by heart puncture, in PBS).
In vitro uptake included incubation of peripheral blood cells of adult carp (Cyprinus carpio L.) with ferritin, colloidal carbon, latex (0.8 µm)(in PBS, at room temperature, for 1 hour). Observations from cytolysis experiments with HeLa tumour cells were also used.
Further handling followed electron microscopical routine methods (fixation with 1-3% glutaraldehyde in cacodylate buffer, 0,1 M, pH 7,4, postfixation after Palade, embedding in Epon 812).

3. RESULTS AND DISCUSSION

Thrombocytes of both species showed regular uptake of all particles tested (ferritin, carbon, latex) and also of cell debris resulting from lysis of tumour cells and autologous lipid droplets in a contaminated blood probe. The particles were found in clear vesicles obviously belonging to the surface connected system, but were often concentrated in several large vacuoles near the nucleus apart from SCS and supporting microtubular marginal band. These accumulations occurred irrelevant of the round, oval or spindle shape of the cell. Degranulation of presumptive lysosomes into the vacuoles indicating digestion was not observed. Active uptake (phago- or endocytosis) has been doubted for thrombocytes in fish as well as in other vertebrates (e.g. Thuvander et al. 1987, Zucker-Franklin 1981). In fish, digestion seems especially improbable, as lysosomal enzymes as peroxidase or acid phosphatase have not been found regularly in thrombocytes below the amphibian class (Daimon et al. 1985, Hine et al. 1987 resp.). Passive uptake by way of the SCS as proposed for teleostean and e.g. mammalian thrombocytes (Ellis 1976, Zucker-Franklin 1981 resp.) might be more plausible. The few reports of particular uptake by teleostean thrombocytes or even negative results in the same species, e.g. recently on pronephric thrombocytes of the carp (Temmink and Bayne 1987) are rather against any important clearance function of this cell type. Nevertheless, the regular large accumulations of quite different particles observed in the present study suggest a greater defensive role than supposed till now, possibly requiring specific conditions or triggering, and with its significance perhaps in passive storage as proposed for avian thrombocytes (Chang and Hamilton 1979).

4. REFERENCES

Bielek E 1979 Zool. Jb. Anat. 101 19
Chang C F and Hamilton P B 1979 J Retic. Soc. 25 585
Daimon T, Gotoh Y, Kawai K and Uchida K 1987 Histochemistry 82 345
Daimon T, Mizuhira U, Takahashi J and Uchida K 1979 Cell Tiss.Res. 203 355
Ellis A E 1976 J Fish Biol 8 143
Hine P M, Wain J M and Boustead N C 1987 New Zealand Fish Res.Bull.28 75p.
Temmink J H M and Bayne C J 1987 Dev. Comp. Imm. 11 125
Thuvander A, Norrgren L and Fossum C 1987 J Fish Biol. 31 197
Suzuki K 1984 Bull. Jap. Soc. Fish. 50 1305
Zucker-Franklin D 1981 J. Cell Biol. 91 706

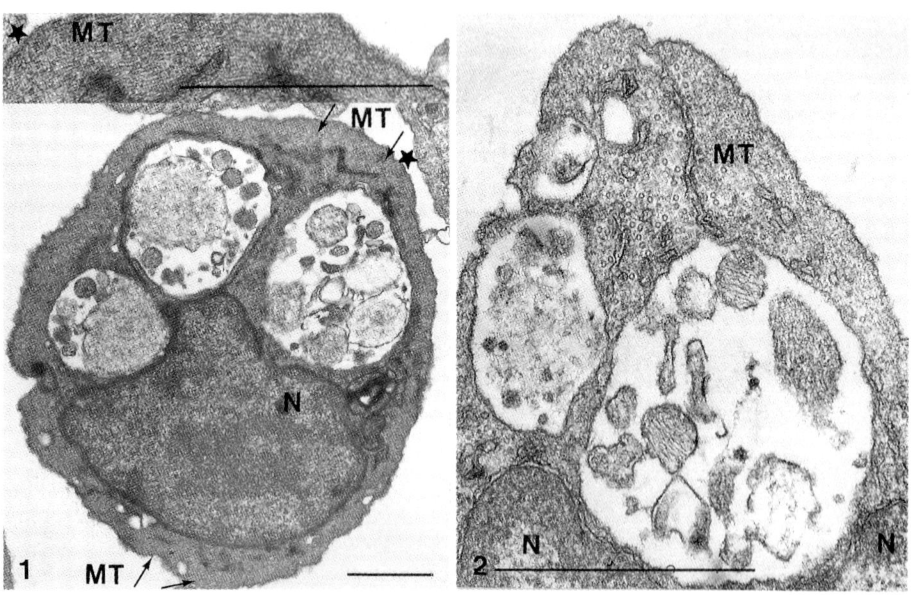

Fig. 1. Thrombocyte of a carp showing large vacuoles containing cell debris. The marginal clear vesicles are part of the apparently reduced surface connected system; MT = marginal microtubules (see also inset ✶); N = nucleus. — = 1 µm.
Fig. 2. Detail of a similar, spindle shaped cell.

Quantitative measures of the distribution of surface antigens on thymocytes during capping

P M Lackie and D M Shotton

Department of Zoology, Oxford University, South Parks Road, Oxford OX1 3PS

ABSTRACT: Lymphocyte sialoglycoprotein on the surface of rat thymocytes was labelled using a monoclonal antibody and immunogold. The outline of thymocytes in sections and the location of gold label was digitized from electron micrographs using a simple personal computer system. Two quantitative indexes of label aggregation and capping were calculated. One took account of cell surface irregularities the other did not, both were independent of cell orientation and magnification. The biological relevance of these two indexes is discussed. This analysis can be extended to light microscopic images of live cells.

1. INTRODUCTION

We have investigated the distribution of surface molecules on lymphoid cells. Cross-linking of these surface molecules using multi-valent ligands such as antibodies, caused patches to form which then aggregated and migrated, in a temperature and energy dependent manner, to one discrete region on the cell. The mechanism by which this "cap" forms is still disputed and in order to investigate the process, we have developed a method to quantify antigen distribution over the cell surface. The antigen molecules can be considered to be on the surface of a sphere around which they are free to move. While this may be an adequate approximation for some purposes, the presence of irregularities and cell processes may severely bias data collected using such estimates. We therefore developed a method of analysis which takes account of such irregularities.

2. METHODOLOGY AND RESULTS

We used thymocytes to develop a system for analysis, looking at the surface distribution of lymphocyte sialoglycoprotein (LSGP). Using fresh rat thymocytes kept at 4°C in cell culture medium (RPMI 1640), LSGP was labelled with a mouse monoclonal antibody (W3/13). After washing, a second antibody layer labelled with either a gold or fluorescent marker was used to visualise the position of the LSGP molecules. This second antibody also provided additional cross-linking. If cells were then warmed to 37 ^{o}C capping occurred over a period of 3-8 minutes and could be seen by fluorescence microscopy. For electron microscopic analysis ultrathin sections of Araldite embedded gold labelled cells were cut and then photographed using a Philips 400 electron microscope.

The outline of the cell and the position of gold label was plotted from the electron micrographs using a digitizing tablet attached to an IBM PC compatible computer. Subsequent data analysis was carried out using a FORTRAN program written and compiled on the personal computer. The centre of mass of the cell was calculated and the position of the gold and cell outline relative to this. The position of each gold particle was then described as a vector giving its distance from the centre and its angle from an arbitrary polar axis. The angle was calculated as (a) simply the angle of the gold from the centre or (b) the distance along the cell perimeter (from the intersection of the polar axis) scaled relative to the total perimeter so that each unit corresponded to one degree. These vectors were then converted to unit vectors on a unit circle centred at the origin. The mean vector for the gold label was calculated for each cell, giving an average direction and a length related to the spread around the perimeter of the cell. The

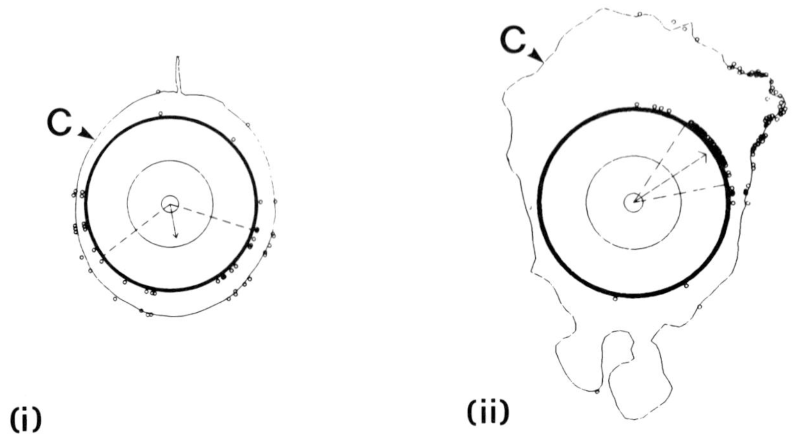

Figure 1. Cells were labelled with W3/13 and immunogold, cell (i) was uncapped cell (ii) capped for 8 minutes at 37 ^{o}C. Cell outline (C) and gold points (small circles) are shown, the heavy circle represents the unit circle, the light lines are 0.5 and 0.1 divisions on the radius. Gold points on the unit circle show the position of the unit vectors. Arrow shows the mean vector dashed lines indicate one mean angular deviation either side of this. Method (a) was used for both these cells.

length of the mean vector could theoretically vary between 0 (perfectly even distribution of the unit vectors around the unit circle ie: gold evenly spread on cell surface) and 1 (unit vectors identical ie: all gold particles at one point). This also allowed us to calculate a measure for the mean angular deviation of the gold label. This was calculated, using the mean vector, for a large number of cells and used as a quantitative index of the degree of aggregation of the surface molecule on cells in the population. The mean vector varied from 0.41 in uncapped cells to 0.94 in fully capped cells with an associated mean angular variation from 62.7 to 19.8 degrees. The second method of calculation gives comparable results but is more sensitive in detecting cells where the label was all at one pole but dispersed over a convoluted cell membrane. The mean angular deviation provides a quantitative index of capping which is independent of the orientation of the cell and the magnification of the micrograph.

3. CONCLUSIONS

The mean angular deviation indicates the degree of surface label aggregation on a cell. It can be used to describe the speed at which this changes for a population of cells under different experimental conditions. This method of analysis will be particularly useful when applied to the distribution of two or more molecular species as it allows statistical testing of association between more than one antigen or one antigen and another cell component. Biologically, the index (a) (-see above) is relevant to initial cell - cell or cell - substrate interaction whereas the angular spread of surface molecule is important in initial binding. A consideration of surface irregularity (index (b) above) is important when looking at the density of molecules on the surface and subsequent surface- cytoskeletal interaction. This may be central to eliciting physiological responses initiated by receptor cross-linking, in such instances analysis assuming a spherical cell shape could be very misleading. Initial studies indicate that similar indexes can be applied to the distribution of gold (seen by epi-polarization microscopy) or fluorescent labels on the surface of living cells during capping.

The olfactory border and its supraepithelial fine structures

U. Heinzmann, G. Grevers* and J. Plendl

Dep. Pathology, Ges. Strahlen- und Umweltforschung mbH München
D-8042 Neuherberg.*HNO-Klinik und Poliklinik der Universität München.

ABSTRACT: The olfactory epithelium with its apical fine structures were examined in perfusion fixed, adult hamsters (Mesocricetus auratus) and NMRI-mice. Both species show a comparable topography and fine structures of the olfactory border. The polymorphic olfactory receptors are hidden in a meshlike plexus of variable densities consisting of long, bifurcated supporting cell microvilli and filamentous structures of unknown origin. The pleomorphisme of the olfactory receptors is also discussed with respect to recent immunhistochemical findings .

1. INTRODUCTION

The nasal mucosa of various species have been examined by means of transmission electron microscopy (Andres 1969; Jafek 1983; Menco et al 1985). Recent immunhistochemical results demonstrate a heterogeneity (Allen and Akeson 1985; Mollicone et al 1985; Plendl and Schmahl 1988) within the receptor cells of the olfactory epithelium. These olfactory receptors, hidden in a regionally rather or fairly dense plexus of various profiles (Heinzmann and Grevers 1987), were observed by scanning electron microscopy. This study was undertaken to elucidate the physical relationships of the olfactory border and its supraepithelial fine structures in adult hamsters and NMRI-mice.

2. MATERIAL AND METHODS

Anaesthetized adult animals of both sexes (10 hamsters-Mesocricetus auratus and 10 NMRI-mice) were first fixed by perfusion with phosphate-buffered 1% glutaraldehyde solution and were subsequently treated by immersion. The nasal septa and the lateral walls with the conchae nasales were dissected and rinsed. All samples were dehydrated through a graded series of ethanol and dried from CO_2 and sputter-coated with gold. A light- and a scanning electron microscope were used to examine the tissues..

3. RESULTS

The nasal cavities in both species were generally coated by the respiratory epithelium concering the two lower parts of the septum nasi, the concha ventralis nasale, the posterior / inferior parts of the concha dorsalis and the rims of the conchae ethmoidales. The olfactory area is located near the most dorsal region of the nasal cavity and extends onto the conchae ethmoidales. Its apical fine structures consisted of microvillous profiles of supporting cells, a loose, meshlike plexus of unknown origin and a variety of polymorphous, ciliated dendrite endings. Topographic variations presented a range of densities in the olfactory border, including a heterogeneity of the microvillar length and degree of arborisation. The ciliated, dendritic endings of the olfactory cells were found to be enmeshed in the overlying blanket of microvilli. Some receptors radiated up to 20 cilia about 1 μm in length. In other receptors, the cilia were abruptly tapered at a rate such that a distal segment has an average diameter smaller than half of the proximal segment. The distal segments were arranged in a parallel fashion occasionally over more than 70 μm into the area between the surface of the epithelium and the dense plexus of the olfactory border.

4. DISCUSSION

Despite several investigations on the olfactory epithelium in quite a number of different species (Andres 1969; Jafek 1983; Menco et al 1985; Heinzmann and Grevers 1987), the origin of the olfactory border with its various profiles remains indefinite. We suppose an additional structure beneath the microvilli and cilia contributes to the distal demarcation of the olfactory border which covers the olfactory receptors. Recently, various immunhistochemical results have to the presumption that a cellular heterogeneity exists amongst olfactory receptors (Allen and Akeson 1985; Fujita et al 1985; Mollicone et al 1985). In fact, we can find polymorphous olfactory receptors in both adult species, but this seems more likely to be a question for the stage of differentiation. Farbman (1986) reported on a non-synchronous development of the olfactory receptor neurons which can be replaced during postnatal life (Graziadei and Monti Graziadei 1983).

6. REFERENCES

- Allen W K and Akenson R 1985 Dev. Biol. 109 393
- Andres K H 1969 Z. Zellforsch. 96 25
- Farbman A I 1986 Chem. senses 11 3
- Fujita S C, Mori K, Imamura K and Obata K 1985 Brain Res. 326 192
- Graziadei P P C and Monti Graziadei G A 1983 Am. J. Otolaryngol. 4 228
- Heinzmann U and Grevers G 1987 Acta Anat. 130 41
- Jafek B W 1983 Laryngoscope 93 1576
- Menco B Ph M and Farbman A I 1985 J. Cell Sci. 78 283
- Mollicone R, Trojen I and Oriol R 1985 Dev. Brain Res. 17 275
- Plendl J, Schmahl W 1988 Anat. Embryol. 177 459

Inst. Phys. Conf. Ser. No. 93: Volume 3, Chapter 7
Paper presented at EUREM 88, York, England, 1988

Combined TEM/SEM study on the conjunctival epithelium of reptiles

U Hiller and H J Dieterich

Institut für Anatomie, D-4400 Münster, W Germany

ABSTRACT:The morphology of the reptilian conjunctiva has been studied by means of TEM and SEM. Both the chameleon and agamid species demonstrate conjunctival lamellae which lack the iguanid lizards. In all cases the conjunctiva is endowed with more or less differentiated protrusions and microvilli.

In contrast to mammalian - including human - conjunctiva the conjunctival epithelium in reptiles is highly differentiated. This applies both to the structure of the epithelial cells themselves and the arrangement of the conjunctiva in general. This will be demonstrated in chameleons, agamids, and iguanids.

The conjunctiva of the chameleon is typically marked by several lamellae which run parallel to the palpebral margin from the inner to the outer canthus of the bulbar face. This prominent lamellated area consists, as the TEM reveals, of a basal stalk of connective tissue covered by a stratified epithelium. To a greater or lesser extent cells with remarkably high electron density are distributed throughout the conjunctiva (Fig.1).

The conjunctival surface pattern of the eyelid appears especially in the lamellated region, and that both inter- and intraspecifically in wide-ranged diversification. Very common are microvillous projections and flattened protrusions of the cells at the edges of the lamellae. Moreover, in Chamaeleo chamaeleon some microprojections are extremely elongated having spoonlike broadenings at their ends. In other chameleon species there is always a lamellated area, but the range of cellular projections may be completely different, which can be demonstrated in the case of Ch. ceylanicus (Fig.2).

Like the chameleons agamid reptiles possess similar lamellated regions, too. But in the group studied so far the projections are only slightly elongated, although similarly variable in shape and distribution. So some prominent epithelial cells are covered by partly branched protrusions as it is known from the avian conjunctiva. Very prominent projections could be documented in some cases only. In the TEM the agamid epithelium resembles that of the chameleon. The surface is covered by microvilli, and here again the epithelial cells are richly endowed with mitochondria, Golgi systems and numerous different small vesicles, suggesting high physiological activities.

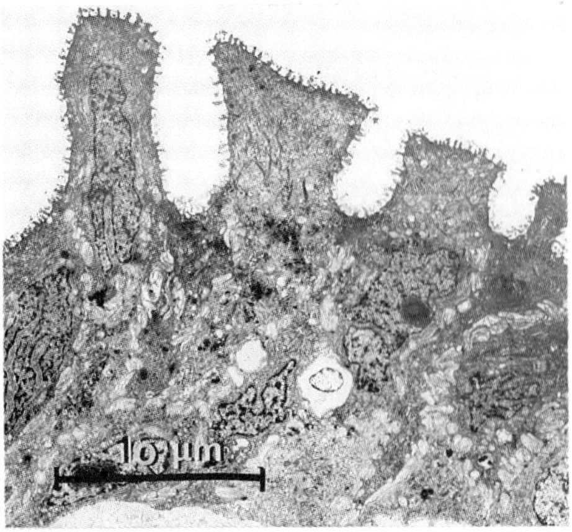

Fig. 1 Ch. ceylanicus

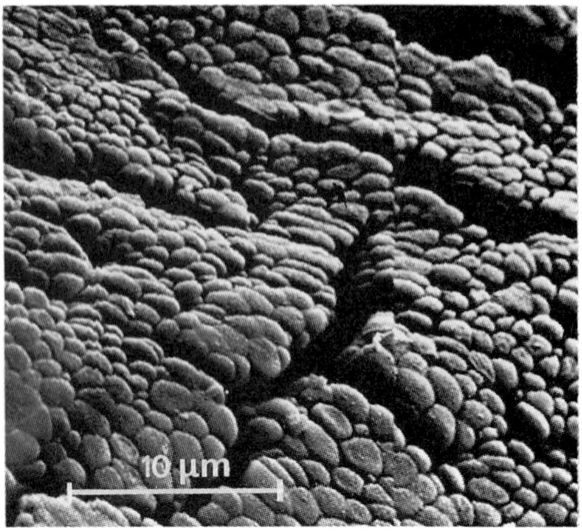

Fig. 2 Ch. ceylanicus

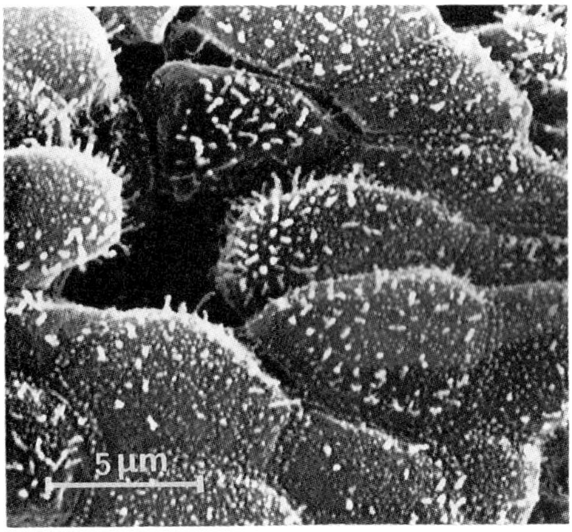

Fig. 3 A. carolinensis

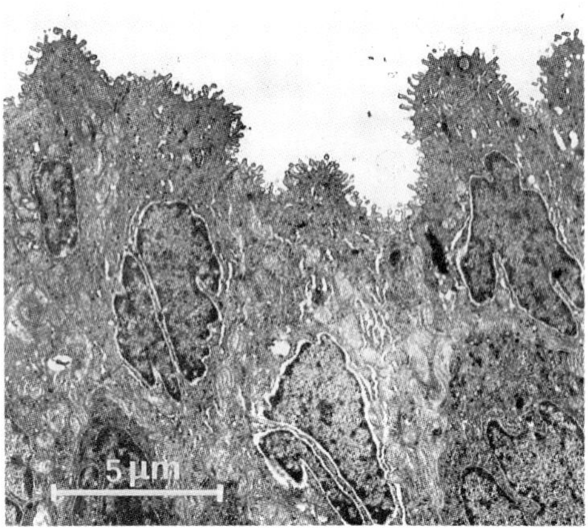

Fig. 4 A. carolinensis

In the iguanid lizard Anolis carolinensis the inner surface of the eyelids lacks a lamellated area. But also in this case there are cellular protrusions of varying sizes, situated more or less clearly in rows parallel to the palpebral margin. So besides microvillous structures, multiply-branched protrusions as well as elongated or spherical projections can be found (Fig.3). In this area the epithelium consists of two layers. The columnar cells of the upper layer have bulged apical regions, covered by blunt small microvilli. Only some of them are notably more elongated than the other ones. The nuclei of all epithelial cells are irregular in shape but somewhat longish and oriented perpendicular to the surface (Fig.4). The cytoplasm demonstrates high electron density and a conspicuous abundance of mitochondria and endoplasmic reticulum. Depending on the eyelid area goblet cells can be found to be more or less prevalent. These findings of Anolis carolinensis do not differ fundamentally from those yielded by the study of the species mentioned above.

Nerve fibres in the intestinal mucosa of a cartilaginous fish

G Tagliafierro, L Borgiani *, M Canepa *, G Faraldi, G G Rossi, T Zanin *

Istituto di Anatomia Comparata, Università di Genova, Genova, Italy
* Istituto di Anatomia Patologica dell'Ente Ospedaliero "Ospedali Galliera", Genova, Italy

ABSTRACT: Numerous peptide-containing nerves were demonstrated in the intestinal mucosa of the cartilaginous fish *Scyliorhinus canicula*. At electron microscope these nerves show dense cored vesicles of different size and morphology.

Numerous regulatory peptides has been immunohistochemically found in gastrointestinal nerves of cartilaginous fishes (Holmgren and Nilsson 1983; El-Salhy 1984; Cimini *et al* 1985; Tagliafierro *et al* 1988). Nevertheless little is still known about their ultrastructural aspects, localization and distribution in the intestinal mucosa. The aim of the present work is to investigate the presence of these peptides in the enteric autonomic nerve system of the cartilaginous fish *Scyliorhinus canicula* L. by immunohistochemistry and ultrastructural methods. Furthermore we want to get some information about the co-localization of different peptides in these enteric nerve fibres.

Small pieces of the internal portion of the spiral valve (which is a particular aspect of the cartilaginous fish gut corresponding to the mammalian small intestine) were fixed by immersion in 1) formalin-calcium (1% $CaCl_2$ in 4% formaldehyde) or 2) Karnovsky fixative and postfixed in 1% osmium tetroxide. The immunoreactivity was detected by the indirect immunofluorescence and PAP methods using antisera against 1) synthetic amphibian bombesin (1:400 CRB, UK); 2) natural porcine VIP (1:400 CRB, UK; 3) natural porcine PHI (1:200 CRB, UK); and 4) serotonin (1:500 INC, USA). Controls were performed by absorption of the primary antiserum with its corresponding antigen. For ultrastructural data, ultrathin sections from Durcupan ACM (Fluka, Switzerland) embedded specimens, were used after uranyl acetate and lead citrate staining. Observations were carried out by Philips M 202 and Zeiss 109 T electron microscopes.

At light microscope nerve fibres showing positive staining to VIP and PHI antiserum were present in the lamina propria and in the muscularis mucosae (Figure 1). Nerve fibres showing VIP-, bombesin-, PHI- and serotonin-like immunoreactivity were detected in the wall of blood vessels round smooth muscular fibres. Some of these peptides seem to be localized in the same nerve terminal (Figure 2). At electron microscope nerve fibres containing electron dense granules of different size and morphology were demonstrated in the same localization of the neuropeptides and serotonin observed by immunohistochemistry (Figure 3). Furthermore much smaller nerve fibres characterized by dense-cored vesicle rich varicosities were seen beneath the intestinal epithelium in close proximity with endocrine or non-endocrine cells. The nerve terminals localized round smooth muscular fibres often showed co-localization of vesicles of different size

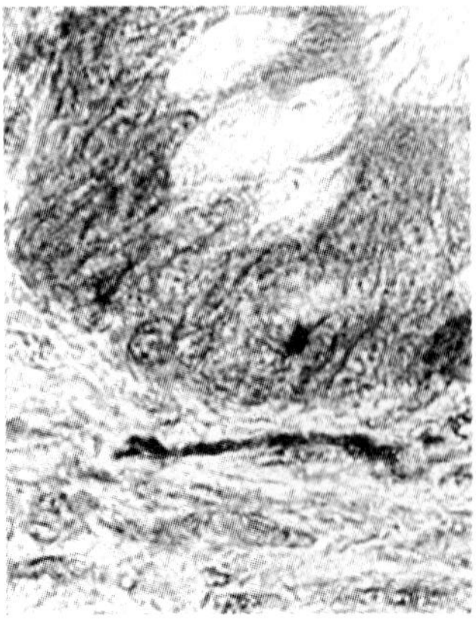

Fig. 1. Nerve fibre immunopositive to VIP antiserum. 20 μ

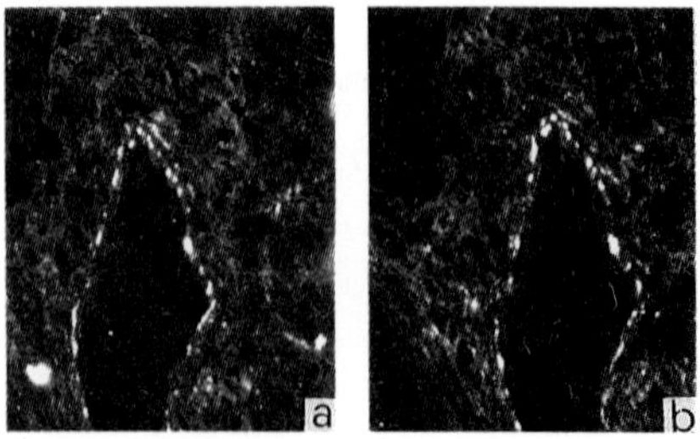

Fig. 2. Adjacent semithin sections: nerve fibres in the blood vessel wall immunopositive to VIP antiserum (a) and to PHI antiserum (b). 30 μ

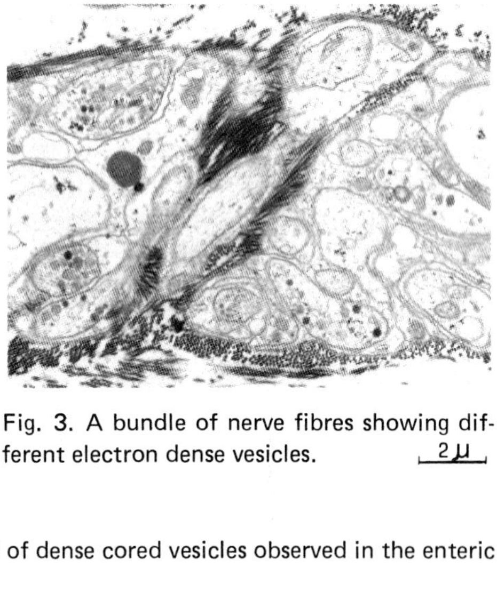

Fig. 3. A bundle of nerve fibres showing different electron dense vesicles. 2 μ

and morphology.

These observations suggest that there is a rich peptidergic innervation in the intestinal mucosa of the cartilaginous fish *Scyliorhinus canicula* as found in higher vertebrates (Costa *et al* 1986). From our ultrastructural and immunohistochemical data, we can also hypothesize a probable co-existence of a classical neurotransmitter and a peptide or of two different peptides in some nerve terminals. A further immunocitochemical study is necessary to identify at electron microscope the different kind of dense cored vesicles observed in the enteric nerve fibres.

REFERENCES

Costa M, Furness J B and Gibbins I L 1986 *Progr. Brain Res.* **68** 217-239

Cimini V, Van Noorden S, Giordano Lanza G, Nardini V, McGregor G P, Bloom S R and Polak J M 1985 *Peptides* **6** (suppl. 3) 373-377

El-Salhy M 1984 *Histochemistry* **80** 193-205

Holmgren S and Nilsson S 1983 *Cell. Tissue Res.* **234** 595-618

Tagliafierro G, Bonini E, Faraldi G, Farina L and Rossi G G 1988 *Cell. Tissue Res.* **(accepted, in press)**

Inst. Phys. Conf. Ser. No. 93: Volume 3, Chapter 7
Paper presented at EUREM 88, York, England, 1988

Diet dependent ultrastructural features in the liver of sea bass (*Dicentrarchus labrax* L)

J Meseguer, M A Esteban, J Cabezas and B Agulleiro

Department of Cell Biology, Faculty of Biology, University of Murcia. Spain.

Adult sea bass specimens were distributed in three groups and kept in a experimental aquaria with running seawater at 17ºC and exposed to natural photoperiod. All fish were fed on commercial diets which contained saturated fatty acids (control group or G0), unsaturated (group 1 or G1) or polyunsaturated (group 2 or G2) fatty acids.

The liver of G0 specimens (Figures 1,2) showed

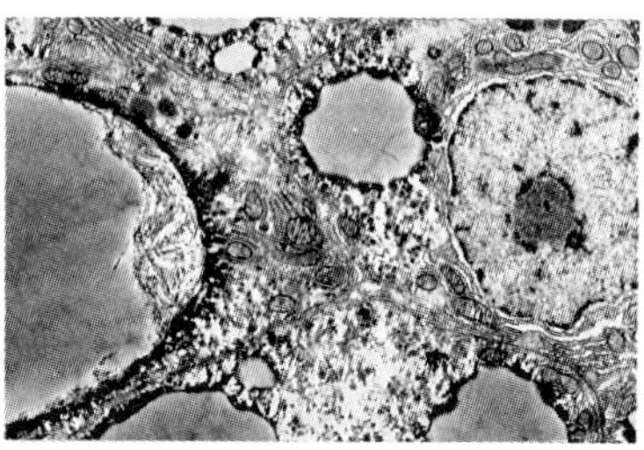

Fig. 1. G0 hepatocyte. x 4750.

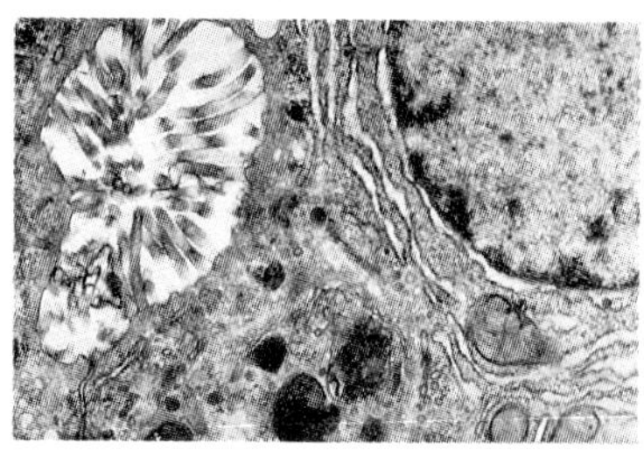

Fig. 2. Cytoplasmic area and bile canaliculus (G0 hepatocyte). x 12000.

ultrastructural features similar to those described in other teleosts (Ferrer et al. 1982, Tanuma 1980, Weis 1972, Welsch and Storch 1973). The hepatocytes of G1 specimens contained abundant glycogen and numerous lipid droplets which were smaller than those observed in the hepatocytes of the G0 fishes (Figure 3). Fat-storing cells in the G1 liver had numerous lipid droplets (Figure 4). G2 hepatocytes had very scarce lipid droplets and more abundant glycogen than G0 and G1 hepatocytes

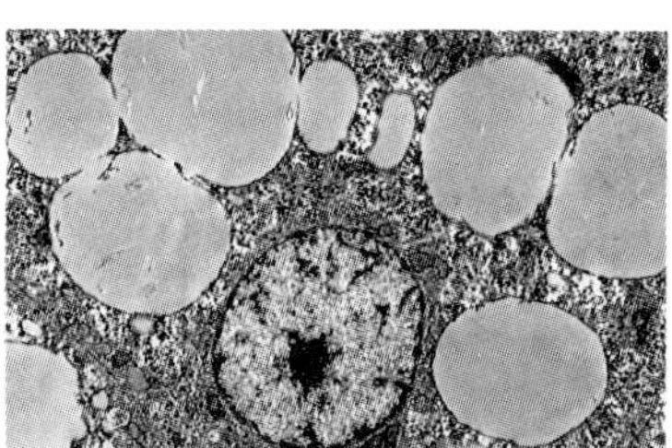

Fig. 3. G1 hepatocyte. x 6300.

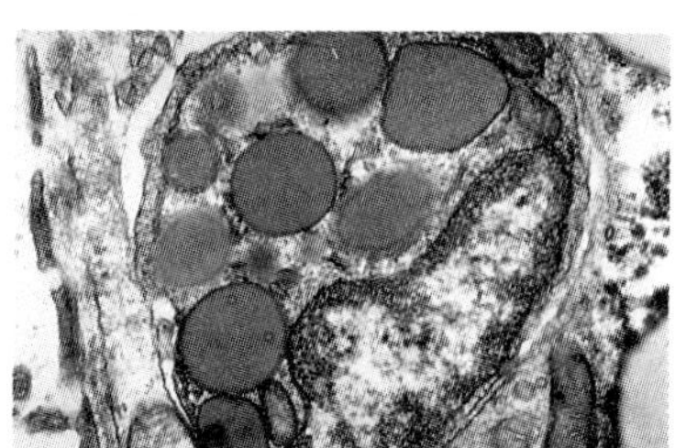

Fig. 4. Fat-Storing cell (G1) x 15000.

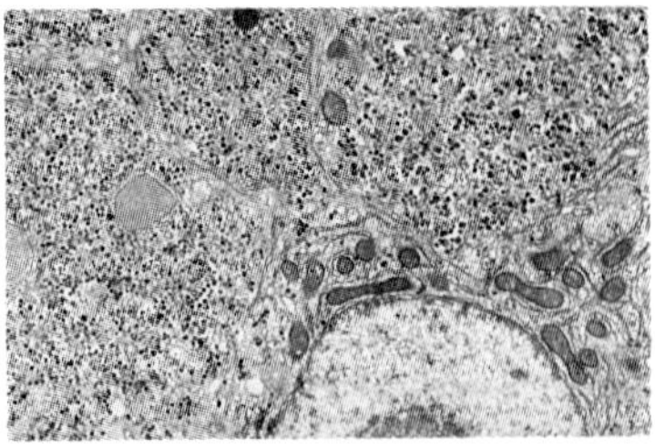

Fig. 5. G2 hepatocyte. x 6000.

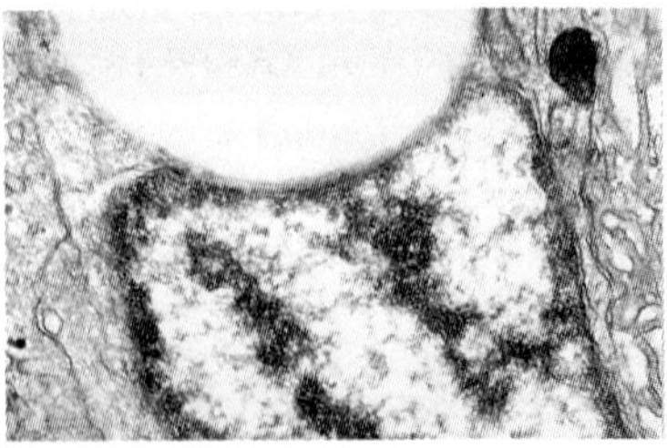

Fig. 6. Fat-Storing cell (G2). x 16000.

(Figure 5). Fat-Storing cells showing a single, big lipid droplet (Figure 6) and some Filament-Rich cells were also seen. Numerous collagen bundles and some fibroblasts were frequently observed surrounding the blood vessels (Figure 7).

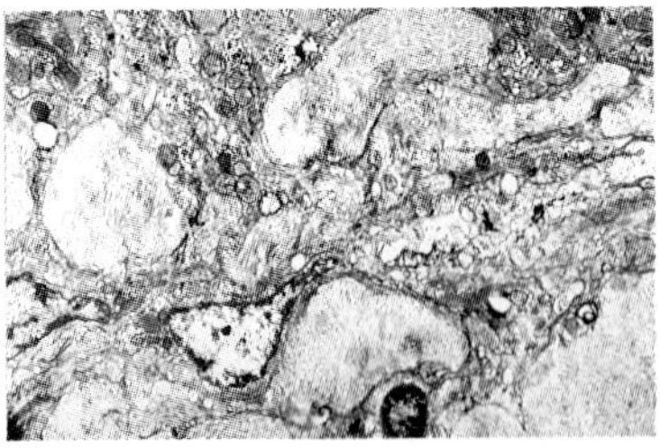

Fig. 7. Collagen bundles in the perivascular zone (G2). x 3250.

Ferrer C, Agulleiro B, Meseguer J and Hernández F 1982 Morfol. Normal Patol. Sec A 6/2 281.

Tanuma Y 1980 Arch. Histol. Jap. 43 1.

Weis P 1972 Amer. J. Anat. 133 317.

Welsh V N and Storch V N 1973 Arch. Histol. Jap. 36 21.

Blood cell formation in the head kidney of the sea bass (*Dicentrarchus labrax* L)

MA Esteban, J Meseguer, J Muñoz and B Agulleiro

Department of Cell Biology, Faculty of Biology, University of Murcia. Spain.

Erythropoiesis, thrombopoiesis and granulopoiesis in head kidney of sea bass were studied. Blood cells formation, unlike mammals, take place intermingledly in the same area outside the sinusoids.

Erythroid serie consisted of candidate erythroid progenitor

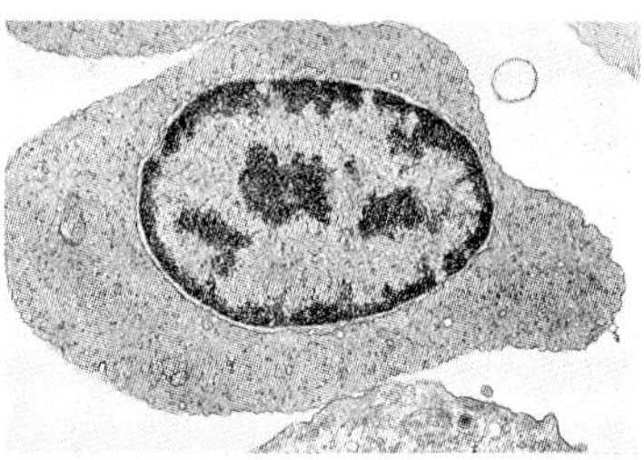

Fig. 1. Polychromatophilic erythroblast. x 10000.

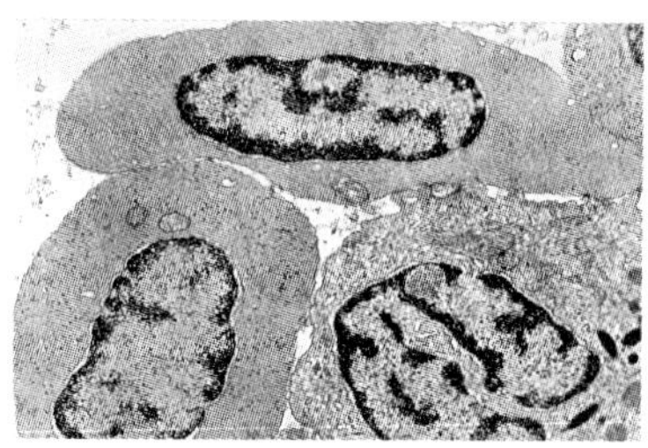

Fig. 2. Young erythrocyte. x 6000.

(proerithroblast), basophilic erythroblast, polichromathophilic erythroblast (Figures 1,2), acidophilic erythroblast, young erythrocyte (Figure 2) and mature (old) erythrocyte, according to the nomenclature proposed by Fänge (1986). Erythropoiesis involves the progressive condensation of the chromatin, the concentration increase of the haemoglobin and the progressive reduction of the ribosome number.

Prothrombocyte (Figure 3) and thrombocyte (Figure 4) stages were observed. Mature thrombocytes were distinguished from lymphocytes by the well-developed surface connected canalicular system

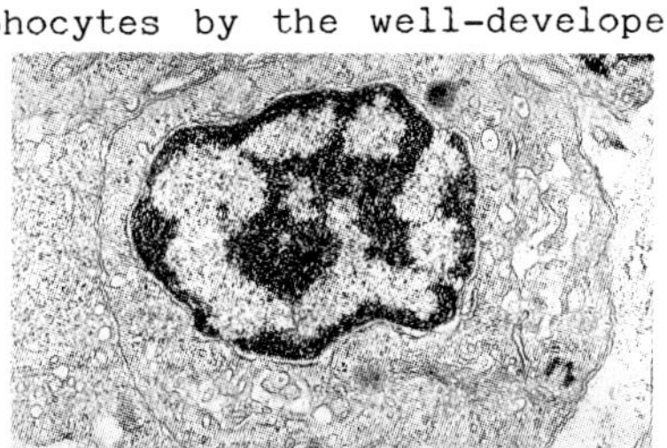

Fig. 3. Prothrombocyte. x 12600.

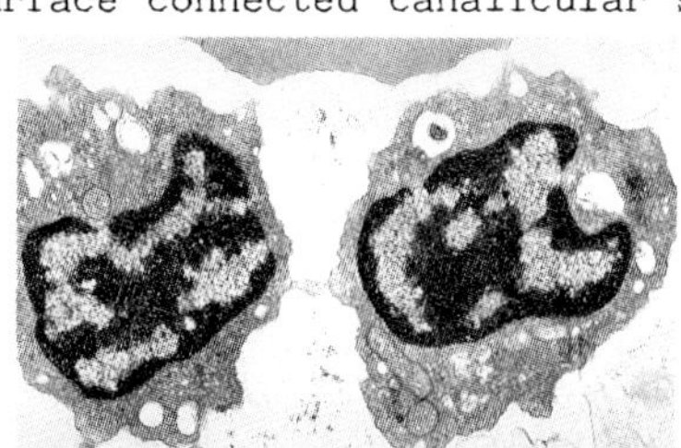

Fig. 4. Thrombocytes. x 10000.

'Fig. 5. Heterophilic promyelocyte. x 9450.

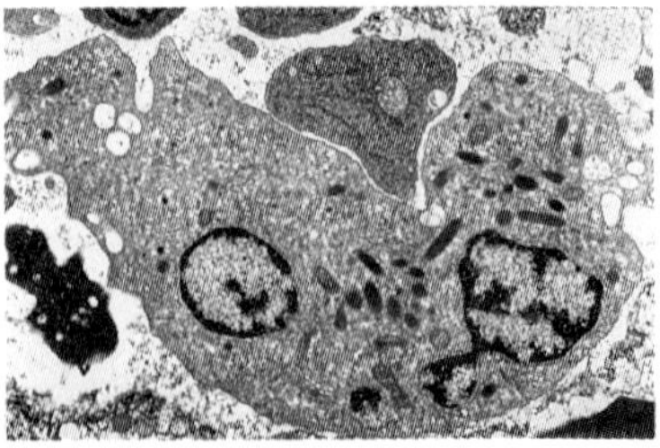

Fig. 6. Mature heterophilic leukocyte. x 6300.

(SCCS) and the marginal bundle of microtubules (Savage 1983).

Granulopoiesis consisted or heterophilic, acidophilic and basophilic series although only one or two series have previously been described in other teleost species (Bayne 1986). Promyelocyte (Figure 5), myelocyte (Figures 7,9), metamyelocyte and mature granulocyte (Figures 6,8) stages were identified as in other vertebrates (Meseguer et al. 1985, Brederoo et al. 1986). The basophilic granulocytes showed some round osmiophilic granules (Figure 9) similar to those found in other teleosts.

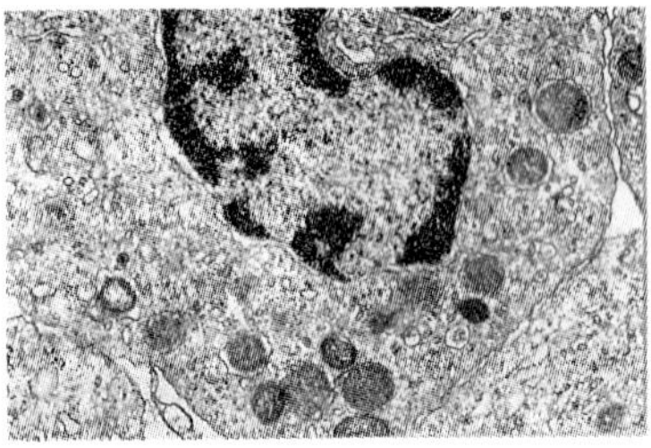

Fig. 7. Acidophilic myelocyte. x 12600.

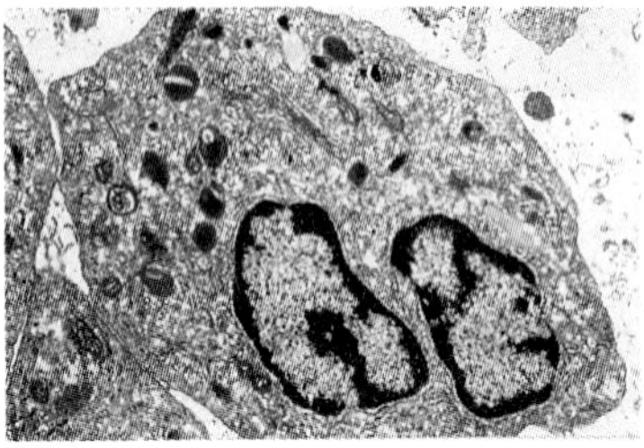

Fig. 8. Mature acidophilic leukocyte. x 5000.

Bayne C J 1986 Veterinary Immunology and Immunopthology (Netherlands) pp 141-52.

Brederoo P, Van der Meulen J and Daems W Th 1986 Histochemistry. 84 445.

Fänge R 1986 Fish Physiology: Recent Advances (London) pp 1-23.

Meseguer J, Lozano M T and Agulleiro B 1985 J. Submicrosc. Cytol. 17 391.

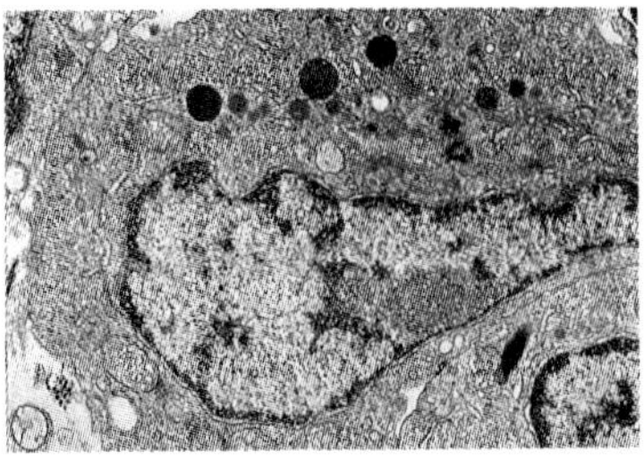

Fig. 9. Basophilic metamyelocyte. x 7500.

Inst. Phys. Conf. Ser. No. 93: Volume 3, Chapter 7
Paper presented at EUREM 88, York, England, 1988

The vascular system of the developing pig liver as shown by SEM of corrosion casts

E Hasselager & R Leiser

Department of Anatomy, Royal Veterinary and Agricultural University, Copenhagen, Denmark and Institut für Tieranatomie der Universität Bern, Bern, Switzerland

ABSTRACT: A modified low viscosity resin shows excellent casts of the vascular system including sinusoids in the pig liver. Sinusoids show poor organization in the fetal liver and lobules can not be defined. Lobulation develops short after birth simultaneously with the connective tissue, characteristic of the pig.

1. INTRODUCTION

The fine structure of the vascular system in the liver has not been studied in the pig. The lobules in pig liver is described by Wünsche (1982, 1985) using india ink for demonstration of venous supply. Scanning electron microscopy of vascular casts gives excellent opportunities for the study of hepatic fine architecture and of the sinusoidal system including the endothelial lining cells.

2. MATERIAL and METHODS

The livers from 5 fetuses from 46-100 days in pregnancy and 6 piglets 5-24 days old were injected with a Batson compound and corroded as described earlier (Leiser and Kohler, 1983). The livers were injected *in situ* via the portal/umbilical vein or the hepatic veins. Natural as well as fractured and cut surfaces were examined.

3. RESULTS

Good vascular fillings were obtained by both routes of injection of the modified resin. The method can be used early in fetal life, where the extracorporal umbilical veins give convenient access to the portal system.

In the fetal liver no lobulation and no preferred orientation of the sinusoids could be observed (fig. 1) The sinusoids are closer together as compared to postnatal liver. Sinusoidal calibre varied being smaller close to portal veins and increasing before joining central veins. Imprints of endothelial cell nuclei can be detected both on venous (fig. 1) and on sinusoidal casts (fig. 2).

The postnatal liver had a marked lobulation given by the vascular architecture and the clefts left by corrosion of interstitial connective tissue (fig. 3). The anastomosing sinusoids together with an overall radial distribution in the lobules are well demonstrated in the scanning electron microscope. The diameter of the sinusoids increases in the central parts of the lobules.

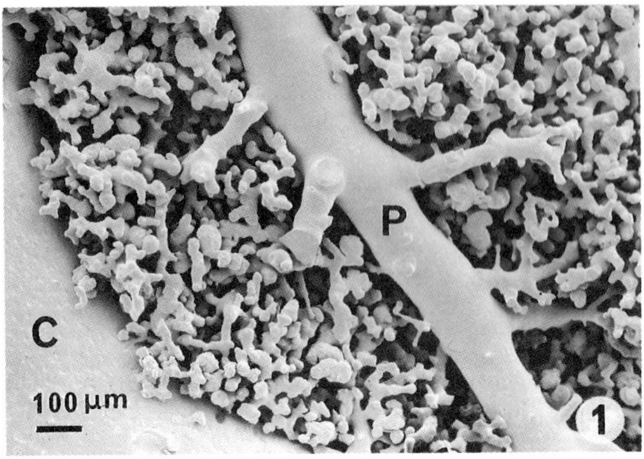

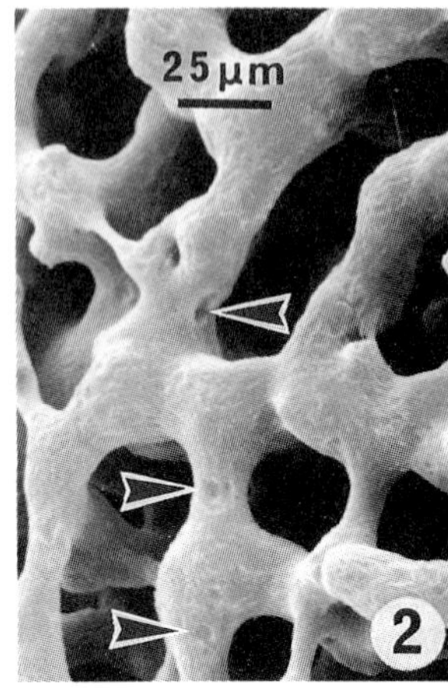

Fig. 1. SEM of corrosion cast of liver from a pig fetus, 46 days in pregnancy. Injection through umbilical vein. Portal vein (P), central vein (C) with nuclear imprints.
Fig. 2. SEM of corrosion cast of liver sinusoids from pig fetus (100 days). Nuclear imprints from endothelial cells ( ). Umbilical injection.

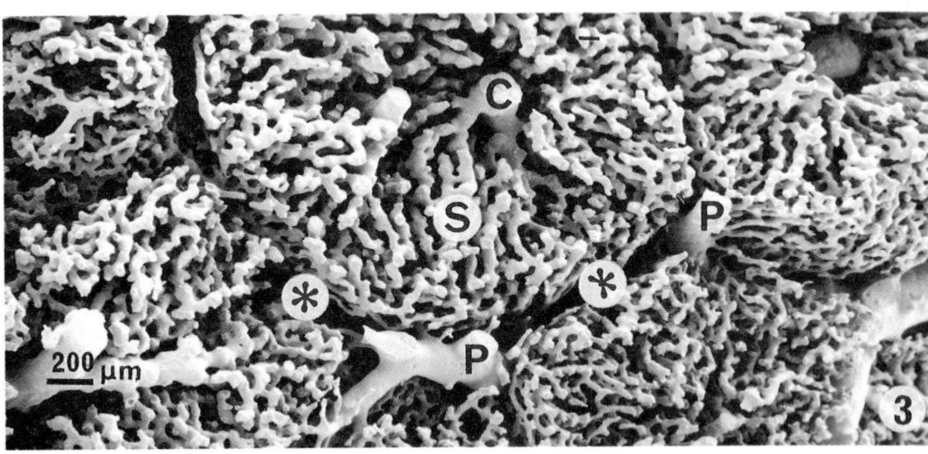

Fig. 3. SEM of corrosion cast of liver from piglet, 24 days old. Injection through portal vein. Lobules separated by clefts (*). Portal veins (P) with radicles entering lobules and providing blood to sinusoids. Sinusoids (S) in radial pattern going to central vein (C).

5. REFERENCES

Leiser R and Kohler T 1983 *Anat. Embryol.* 167 85
Wünsche A 1982 *Zbl. Vet. Med. C. Anat. Histol. Embryol.* 10 342
Wünsche A 1985 *Zbl. Vet. Med. C. Anat. Histol. Embryol.* 14 15

Extracutaneous pigment systems in the *Sparus auratus*

A.Zuasti and C. Ferrer

Department of Cell Biology. School of Medicine.University of Murcia. 30100 Murcia. Spain.

It´s frequent to find large pigmented nodules called melanomacrophage centres distributed at random in the interstitium of the hemopoietic tissues of fishes, especially in the kidney, liver and spleen depending on the species (Roberts,1975; Oguri,1983; Agius and Agbede,1984 and Rowley and Page, 1985).

This study was undertaken in order to establish the ultrastructural characteristics of the numerous pigmented nodules found in the kidney, liver and spleen of the teleost fish Sparus auratus and to verify if these are similar in the different organs.

In the kidney we have observed a great number of melanomacrophage centres formed by rounded electron-dense cells, similar to the melanocytes in which numerous oval melanin granules appear either individuallyor grouped together surrounded by a membrane. Most of them are strongly melanized. The melanomacrophage centres are surrounded by a layer of fibroblast cells (Figure 1). Occasionally we have also observed some isolated melanin-bearing cell (Figure 2).

In the spleen the melanomacrophage centres are numerous and they are formed by pigment cells surrounded by a thin capsule of fibroblast cells. In the cytoplasm of the pigment cells we immediatly notice the great filamentous bodies together with the melanin granules. These cells are similar in appearance to macrophages which phagocyte melanin (Figure 3).

In the liver of the Sparus auratus, appear dark brown pigment centres found in large numbers in the hepatic parenchyma. These pigment-bearing cells are spherical and surrounded by a membrane. The granules are slightly elongated or oval and homogeneously electron-dense, and coexist with granules of irregular shape and variable electron opacity; lamellar structures are frequently noted in larger compound granules (Figure 4).

Considering these results we may conclude that the melanomacrophage centres of the Sparus auratus are morphologically different in those organs studied. In the kidney the cells which form these centres are similar to melanocytes; in the spleen they are macrophages which phagocyte melanin and in the liver the cells are similar to macrophages without melanosomes in the cytoplasm.

Agius C and Agbede S A 1984 J.Fish Biol.24 pp 471-488
Oguri M 1983 Bull.Japan Soc.Sci.Fish 49 pp 1679-1681
Roberts R J The pathology of fishes (Ribelin W E and Wigaki G eds.Univ. Wisconsin Press) pp 399-428
Rowley A F and Page M Fish Immunology (Manning M J and Tatner M F eds. Academic Press London) pp 273-284

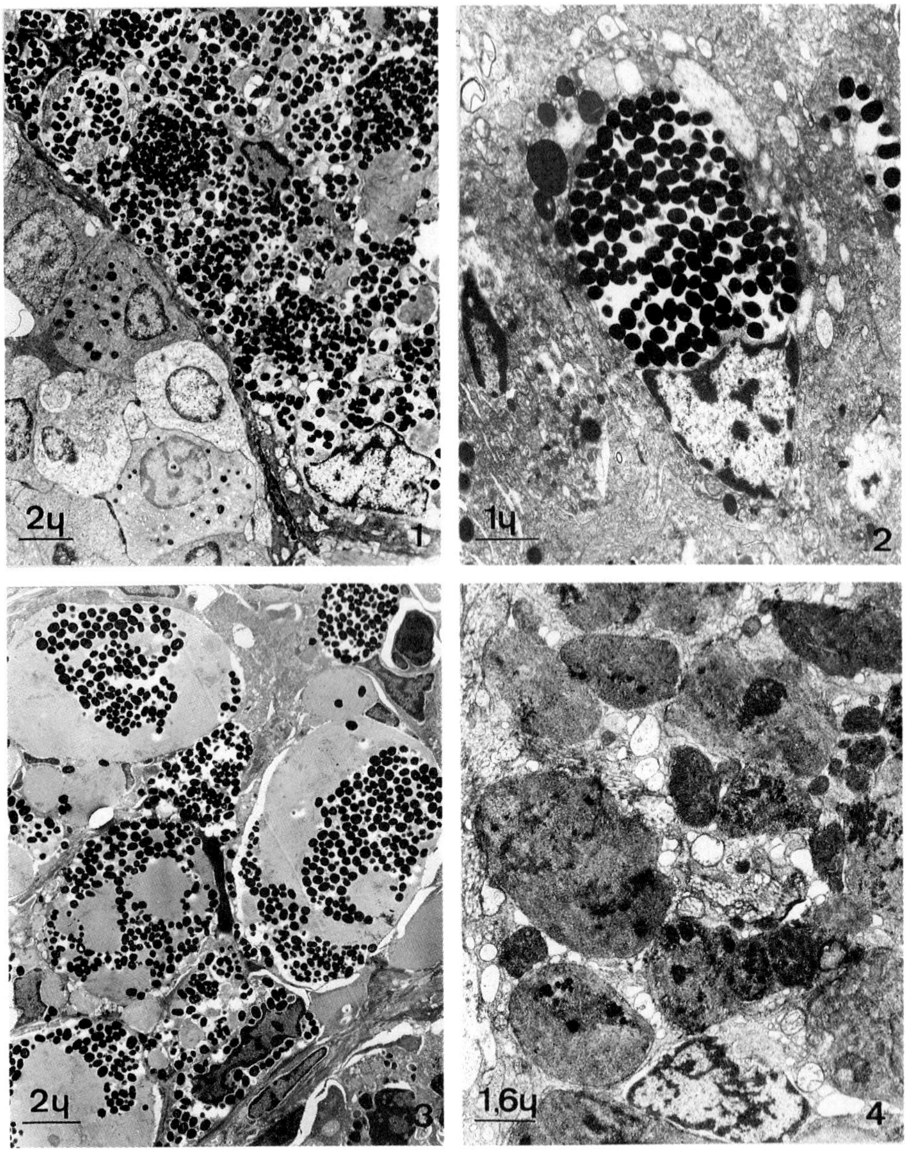

Figure 1. Electron micrograph of the kidney, showing a large aggregate of melanin-bearing cells.

Figure 2. An isolated melanin cell with numerous melanosomes in the kidney of the Sparus auratus.

Figure 3. Melanomacrophage centre in the spleen of the Sparus auratus showing melanosomes and filamentous bodies in the cytoplasm.

Figure 4. Pigment centre located between the hepatocytes of the liver of the Sparus auratus.

POSTER

Inst. Phys. Conf. Ser. No. 93: Volume 3, Chapter 8
Paper presented at EUREM 88, York, England, 1988

Regulation of placental fluid by Hofbauer cells

Ramazan Demir[a] and Türkân Erbengi[b]

[a]Dept. of Histo.Embryo. Medical Faculty, Akdeniz University, Antalya
[b]Dept. of Histo.Embryo. Medical Faculty, İstanbul University, İstanbul

ABSTRACT: The vacuolated Hofbauer cells continued to imbibe fluid by micropinocytosis and phagocytosis, a new equilibrium between extra-and intracellular fluid quantity resulted after a definite level was attained and vacuoles collapsed so that the fluid level of the stroma was balanced.

1. INTRODUCTION

In the course of the development of the human placenta, as pregnancy progresses, structural modification of the chorionic villi become especially within villous stroma. Kaufmann et al (1977) described various aspects these stromal differentiation throughout pregnancy. The present study was undertaken in order to investigate the histofunction of the Hofbauer cell in the immature intermediate villi (Kaufmann, 1982) of human placenta by transmission electron microscopy (TEM).

2. FIGURE PRESENTATION

Hofbauer cells appered in various shapes particularly within the immature intermediate villus, and mostly characterized by numerous long and thin cytoplasmic processes and cytoplasmic vacuoles. Sometimes the cytoplasmic projection anastomose end-to-end or end-to side, in this way the formation of new large vacuoles take place periferally (Fig.1). According to our results a type of Hofbauer cells containing numerous vacuoles was concerned with that function. While the vacuolated Hofbauer cells continued to imbibe fluid by micropinocytosis and phagocytosis, a equilibrium between extra and intra cellular fluid guantity resulted after a definite level was attained. In this case the fluid, that was collected in the vacuoles, was transported outside of the cell by means of cytoplasm, and vacuoles collapsed so that the fluid level of the stroma was blanced (Fig.1). Free water and molecules of soluble substances which are taken up by Hofbauer cells are an important potential. These cells may be responsible for many important processes such as to regulate the level of water in the placenta or to decrease the amount or fetal serum proteins in the villus stroma. This task of Hofbauer cells is very important because of the regulation between mother and fetus. These cells might play a role in controlling the direction of stromal fluid movement, like a system of unfixed valves or sphincters (Castellucci and Kaufmann 1982; Demir and Erbengi 1984,1987a) within the stromal channels. The most original results of this study are tubular structures which extend between the nucleus and the extracellular ground substance crossing the cytoplasm (Fig.2). To our knowledge these findings have not been reported in the literature before. With this special features the Hofbauer cells give the impression of conveying information and controlling complex regulative processes in the villus stroma during the development and the differentiation of the chorionic villi core. These opinions and interpretations are also supported by the second tubular structure that was found in the Hofbauer cells, i.e. the cilium-like structure limited by three membranes and included electron dense substance (Fig.2).

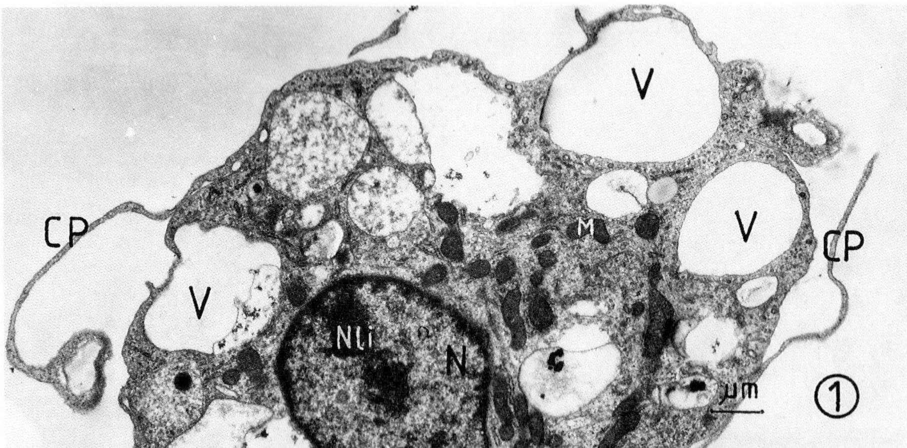

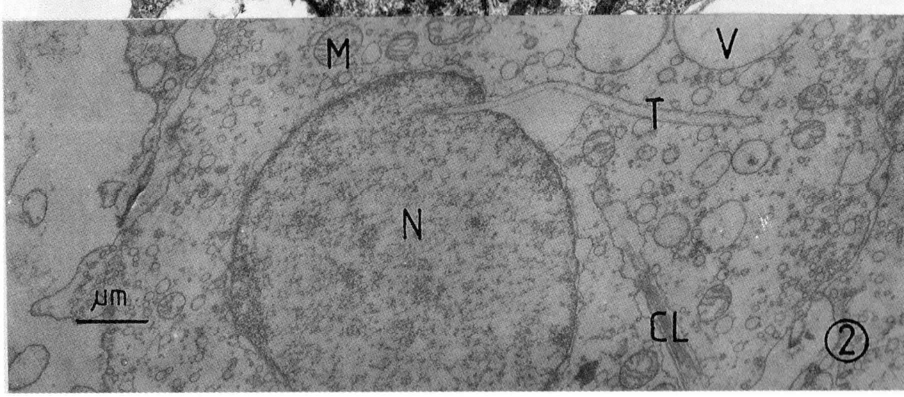

Fig.1 and 2. New vacuoles forming by cytoplasmic projections (CP), collapsed vacuoles (V) and the peculiar tubular structure penetrating into the nucleus (T) and the second structure (CL) limited by three membranes are seen. Normal placenta, 4-6 weeks of pregnancy.

In conclusion, Hofbauer cells appear to play a rol in the development and differentiation of the placental villi core, in the vascularization of villi, in the regulation of the water blance and in the transportation of the various substances; and they may, indirectly, induce the production of collagen fibers. We can consider these cells as intrastromal regulation centers.

Castellucci,M.;Kauffman,P.:(1982).Placenta 3:269-286.
Demir,R.;Erbengi,T.:(1984). Acta Anat. 119:18-26.
Demir,R.;Erbengi,T.:(1987a).Acta Anat. 130(1):24·
Kauffmann,P. et al.:(1977).Cell Tissue Res. 177:105-121
Kauffmann,P.:(1982).Bibliot.Anat. 22:29-39.

Ultrastructure of the human amnion epithelium

Hieronim Bartel*, Wacław Dec**

*Department of Histology and Embryology, **Department of Obstetrics and Gyneacology, Medical Academy of Łódź, Poland.

The development of the technique of amniocentesis, and the importance of this method in prenatal diagnosis has resulted in a greater appreciation of the physiology and biochemistry of amniotic fluid and the changes is undergoes during pregnancy. There are only a few papers on the human amnion ultrastructure. It is mainly connected with the availability of human material, good fixation and selecting the proper sections. On the present stage of studies the ultrastructure of the amnion of at term pregnancy is better recognized than that of early pregnancy. There is no literature available at present on ultrastructure of the amnion membranes after 12 weeks of pregnancy. Whereas it is known from comparing the amnion morphology of the early and at term pregnancy that there are significant differences, and this proves that dynamic changes occur in the amnion during pregnancy.

The aim of this study has been the presentation of submicroscopical morphology of the amnion from the very early periods of pregnancy and comparing it with the amnion at term pregnancy. The observations were concentrated on the epithelium of all three parts of the amnion (amnion reflectum, amniochorion, and cord amnion). Early pregnancy epithelial cells were squamous and had a few apical microvilli (Fig. 2), late in pregnancy cells were cuboidal with numerous microvilli and a well-developed glycocalix. Epithelial cells from early pregnancy had large glycogen stores, well-developed Golgi apparatus RER, and relatively few 10 nm cytoplasmic filaments. Later, cells were similar except for less glycogen, somewhat more lipid droplets, and numerous 10 nm filaments. Lateral cells borders were complexly interdigitiated and desmosomes were the predominant type of cells junction (Figs. 2,3). Occasional tight junctions were observed early in pregnancy. Early in pregnancy the basal cell surface was smooth, and late in pregnancy the basal surface was thrown into numerous folds (Fig. 1) and had hemidesmosome - like junctions.

The results support the view that amniotic epithelial cells remain metabolically active thrughout pregnancy. The glycogen and lipid represent intracellular energy stores, and the 10 nm filaments probably constitute part of cytoskeletal system for the cells. The functional significance of the increasingly complex apical surface may be related to the increased transport of metabolities by the epithelium.

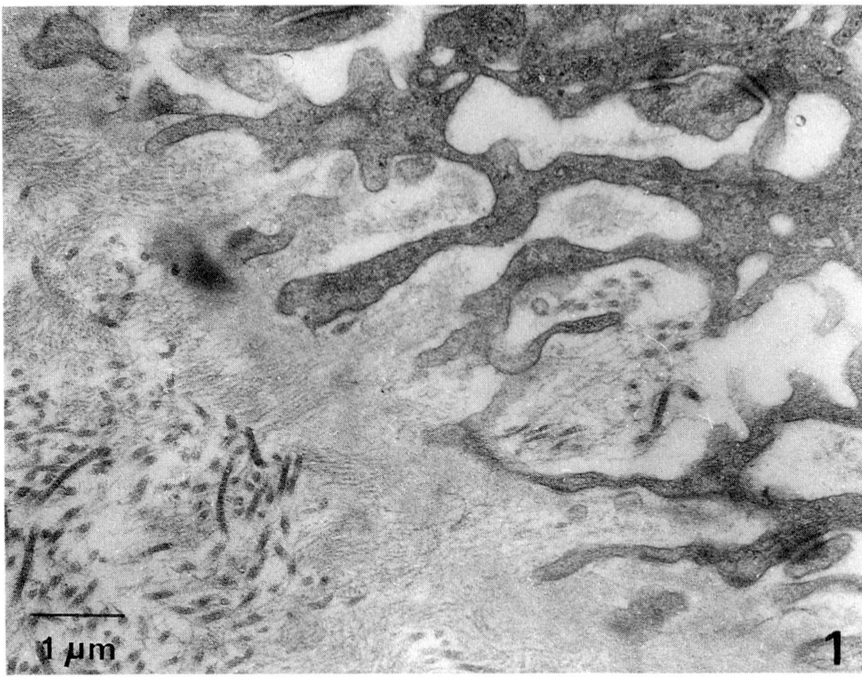
1 µm
1

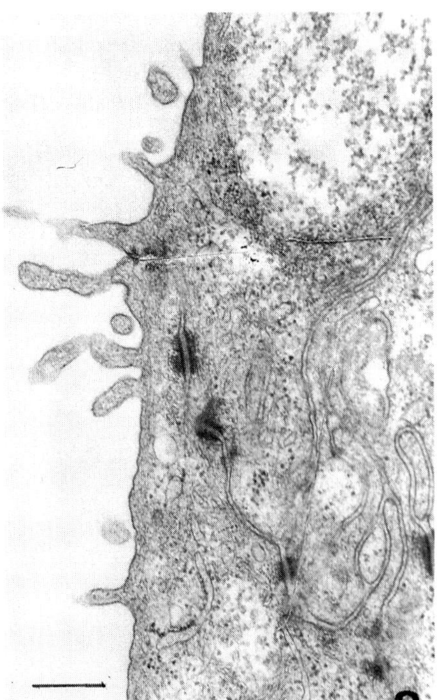
1 µm
2

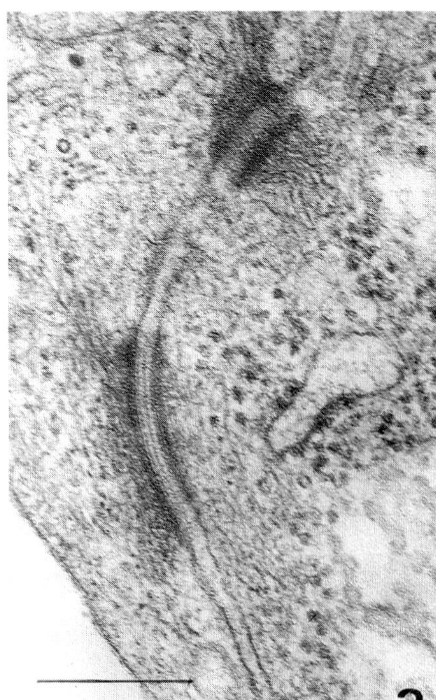
1 µm
3

Early identification of amacrine cells in human foetal retina

D van Driel, JM Provis and FA Billson

Department of Ophthalmology, Sydney University, Sydney, NSW, Australia 2006

ABSTRACT: Developing amacrine cells in the inner nuclear layer (INL) of the retina of a 15 week human foetus have been identified by their ultrastructural characteristics. At this age, cells in the INL range from immature precursor cells to relatively mature cells. Although the amacrine cells have few adult characteristics, they can be positively identified.

1. INTRODUCTION:

Development of the macula and foveal depression of the human retina is not well understood, even though blindness or severe visual impairment occurring in the first year of life is most commonly caused by developmental anomalies. By 15 weeks gestation in the human, the retina of the eye at the macula contains, in immature form, all the cell layers observed in the adult eye. Ganglion cell axons (which comprise the optic nerve) and the ganglion cells themselves are present in higher numbers than are seen in the adult, an example of the overproduction of cells characteristic of the developing foetus (Provis and Penfold, 1988). However amacrine and bipolar cells which synapse with the ganglion cells are still not easy to identify at this age, although synapses from both classes of cells are known to be present (Spira and Hollenberg, 1973). These cells are found within the inner nuclear layer (INL) together with Muller cells and horizontal cells. In this study, amacrine cells were identified from ultrastructural characteristics in serial sections through a part of the macula of a 15 week human foetal retina and cells of a range of maturity were found.

The mature amacrine cell, situated on the innermost margin of the INL, has an indented nucleus with an approximately round profile. The cytoplasm contains numerous mitochondria and polyribosomes, smooth endoplasmic reticulum (sER), and a Golgi apparatus and often a cilium vitreal to the nucleus. The cellular processes, found within the inner plexiform layer (IPL), characteristically contain scattered synaptic vesicles and occasional microtubules, mitochondria and sER. They form conventional synapses onto bipolar cells, ganglion cells and other amacrine cells and are postsynaptic to the ribbon synapses of bipolar cells.

2. RESULTS:

Initially, amacrine cells were identified by tracing amacrine processes, containing conventional synapses, back to the soma. Identification of other amacrine cells was then possible by studying the characteristic ultrastructure. The soma cytoplasm (Figure 1) resembled the adult in the types of cell organelles, although tended to be of darker staining intensity in the more immature cells. Processes within the IPL (Figure 1, inset) contained microtubules and only occasionally other organelles; synaptic vesicles were not seen except in association with synapses.

One of the earliest signs of differentiation detected in the amacrine cell was the migration of the centrioles and Golgi apparatus to the vitreal side of the nucleus. Indentation of the nucleus was also seen, however this appeared to be a feature of ALL the immature cells of the INL at this age, not just the amacrine cells, suggesting that maturity eliminates this characteristic in all but the amacrine cells. Nuclei of younger amacrine cells were greatly elongated in a direction perpendicular to the plane of the retina, similar to the majority of cells within the INL at this age; with increasing maturity the nucleus became more rounded, although no nuclei were seen with an adult-like appearance.

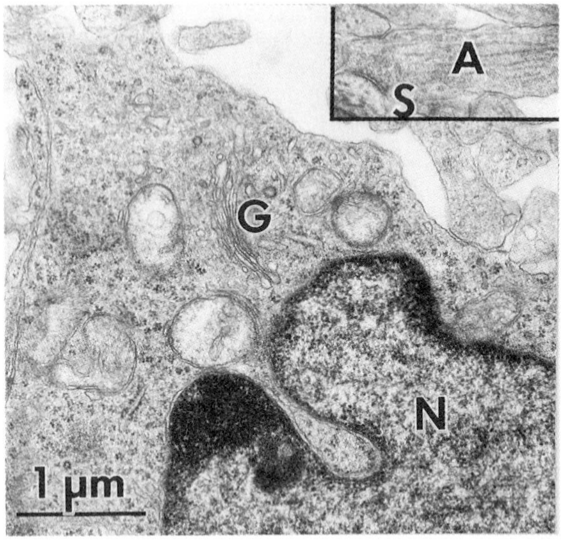

FIGURE I: Amacrine cell and IPL process (inset;A). N:nucleus; G:Golgi; S:synapse

Conventional synapses (from amacrine cells) were not common and some identified amacrine cells were not involved in any synapses, either pre- or post-synaptically, suggesting the formation of synapses to be a late phenomenon in the maturation of these cells. Conventional synapses were immature, with few synaptic vesicles in the presynaptic process and variable membrane thickening in both the pre- and post-synaptic process. Punctum adhaerens junctions were common.

REFERENCES:

Provis J, Penfold P 1988 Progress in Neurobiology (in press)
Spira AW, Hollenberg MJ 1973 Dev Biol 31:1-21

Intercellular junctions and endocytosis are characteristics of hematopoiesis in organ cultures *in vitro*

JG Sharp, SL Mann, G Udeaja and DA Crouse

Department of Anatomy, University of Nebraska Medical Center
Omaha, NE 68105 USA

ABSTRACT: The cells present in the adherent layer of typical murine long term bone marrow cultures (LTMC) can be removed from the flask, reaggregated and grown as three dimensional organ cultures (OC). In addition to stromal cells, assayable stem cells (CFU-s) as well as more differentiated hematopoietic cells are present. Hematopoiesis in these organ cultures is characterized by abundant evidence of receptor mediated endocytosis and intimate membrane associations between hematopoietic and stromal cell populations and between adjacent hematopoietic cells.

1. INTRODUCTION

Hematopoiesis requires the continuous production of mature blood cells from stem and progenitor cells located in bone marrow. Hematopoietic stem cells (HSC) can be maintained long term in vitro in the adherent layer of LTMC (Dexter et al 1984). Although such cultures initially contain many maturing and differentiating hematopoietic cells, after 3-4 months of culture macrophages and non-hematopoietic stromal cells predominate in the adherent layer. Hematopoiesis is largely limited to "cobblestone" areas of the adherent layer (Allen 1981), even though the adherent layer still contains HSC capable of reconstituting lethally irradiated recipients (Sharp et al 1985). To assess if the geometric constraints imposed by the less than three dimensional "monolayer" LTMC system might underlie the loss of differentiating cells, we have examined the morphology of LTMC adherent cells when aggregated into three dimensional OC.

2. MATERIALS AND METHODS

Murine LTMC were established by the method of Dexter et al (1984). The cultures were re-fed with fresh bone marrow after 3 weeks and subsequently demi-depopulated at weekly media changes for a minimum of 12 weeks. Some cultures were sacrificed for in situ morphology. The adherent layer cells of other cultures were made into a slurry using a sterile cell scraper and concentrated by gentle centrifugation. Aliquots were dispensed onto polycarbonate filters resting on Gelfoam platforms in Petri dishes. These slurries rapidly re-aggregated as small OC. After 2 - 9 hours these OC were sacrificed for light and electron microscopy or transplanted under the kidney capsule of intact or lethally irradiated syngenic recipients.

3. RESULTS

Organ cultures of LTMC adherent layers showed active hematopoiesis with many early granulocytes, and some megakarocytes among scattered stromal cells. Primitive hematopoietic cells and mitotic figures were observed frequently. These findings suggest that when LTMC adherent layer cells are aggregated into OC, active hematopoiesis is maintained or even stimulated and hematopoietic cells may be selectively included in the OC. Morphologically atypical junctions were observed between many cells, frequently connecting early hematopoietic cells with macrophage-like cells and sometimes adjacent myeloid cells. These latter cells were often at similar stages of differentiation, suggesting synchronous development of coupled cells. Numerous coated pits and vesicles were observed, particularly in the vicinity of junctional complexes. When these OC were transplanted to lethally irradiated recipients, they established active hematopoietic microenvironments, gave rise to spleen colonies and reconstituted hematopoiesis and immune function (Sharp et al 1985).

4. DISCUSSION

Although the adherent layer of an LTMC is capable of maintaining HSC for several months, differentiating hematopoietic cells initially present are lost from such cultures. This may reflect a loss of the granulocyte adhesion molecule, hemonectin, with time in culture (Campbell et al 1987). Overall, there is minimal proliferation of cells in LTMC unless the cultures are disturbed by feeding or shaking (Dexter et al 1984). Hematopoiesis occurs at discrete sites in the cultures and the overall ratio of stromal cells to differentiating hematopoietic cells is high. When the adherent layer was aggregated into a three-dimensional OC, the ratio of hematopoietic cells to stromal cells at sites of active hematopoiesis was increased. Hematopoietic foci were associated with mitoses, frequent intercellular junctions and evidence of the uptake of materials from the surroundings. These observations suggest that intercellular junctions, potentially indicating cell communication, together with receptor mediated endocytosis, particularly in areas of adjacent macrophage-like cells, are important characteristics of differentiating hematopoietic cells. These features do not appear to be adequately provided by monolayer cultures but can be achieved in OC of adherent cells from such cultures. Organ cultures, which mimic in vivo spatial relationships, provide an excellent means of studying the morphology and cell physiology of the earliest stages of hematopoiesis.

5. ACKNOWLEDGEMENTS

These studies were supported by the Nebraska Department of Health, UNMC Seed Grant funds and the Nebraska Bankers Association.

6. REFERENCES

Allen T D 1981 Microenvironments in hematopoietic and lymphoid differentiation ed R Porter and J Whelan (London: Pitman) pp 38-60

Campbell A D, Long M W and Wicha M S 1987 Nature 329 744-746

Dexter T M, Spooncer E, Simmons P and Allen T D 1984 Long Term Bone Marrow Culture ed D G Wright and J S Greenberger (New York: Liss) pp 57-96

Sharp J G, Crouse D A, Jackson J D, Schmidt C M, Ritter E K, Udeaja G C and Mann S L 1986 Br. J. Cancer 53(Suppl VII):133-136

Inst. Phys. Conf. Ser. No. 93: Volume 3, Chapter 8
Paper presented at EUREM 88, York, England, 1988

Hormone induced changes in the sertoli cell

G Viehberger, G Lunglmayr

The Institute of Histology and Embryology, University of Vienna, Schwarzspanierstr. 17, A-1090 Vienna and the Department of Urology, Mistelbach General Infirmary, A-2130 Mistelbach, Austria

ABSTRACT: GnRH-analogue therapy in the male causes a significant increase of number and size of lysosomes in the Sertoli cell. Acid phosphatase, thiamine pyrophosphatase and aryl sulfatase have been demonstrated. An induction by enhanced germ cell depletion and degeneration is assumed.

1. INTRODUCTION

Highly potent GnRH-analogues are successfully used in the palliative therapy of advanced prostate cancer as an alternative to orchiectomy. Due to hypophyseal down regulation and Leydig cell inactivation they induce serum testosterone levels within castrate ranges after four weeks of application. Maturation arrest and germ cell depletion and degeneration are followed (1-3). Knowing about the close relationship between germ cells and Sertoli cells in the seminiferous epithelium, it was the aim of the study presented now to complete prior reports (1) on Sertoli cell morphology after therapy induced germ cell depletion.

2. MATERIAL AND METHODS

Testis biopsies from prostate cancer patients were taken up to 36 weeks after the onset of therapy (3.6mg ICI 118.630 per month), fixed in 2% glutaraldehyd and 1% osmic acid in 0.1M sodium-cacodylate buffer, dehydrated, embedded in Epon 812, cut and stained with uranyl acetate and lead citrate. Acid phosphatase, thiamine pyrophosphatase and aryl sulfatase have been demonstrated in accordance to methods given by Lewis and Knight (4).

3. RESULTS AND DISCUSSION

Compared to testis biopsies taken before treatment (Fig.1), Sertoli cells from therapy induced germ cell free testis tubules exhibit enlarged smooth endoplasmic reticulum, highly active Golgi fields, bundles of filaments and as most prominent alteration numerous large lysosomes (Fig.2 and inset). Besides the previously described enzymes (acid and alkaline phosphatase, peroxidase and catalase)(1), they contain large amounts of thiamine pyrophosphatase and aryl sulfatase (Figs.3-5). The increase of lysosomes and their high enzymatic activity might be the result of GnRH-analogue induced germ cell depletion and degeneration, because their cell debries seem to be phagocytized by the Sertoli cells similar to residual bodies in normal seminiferous epithelium.

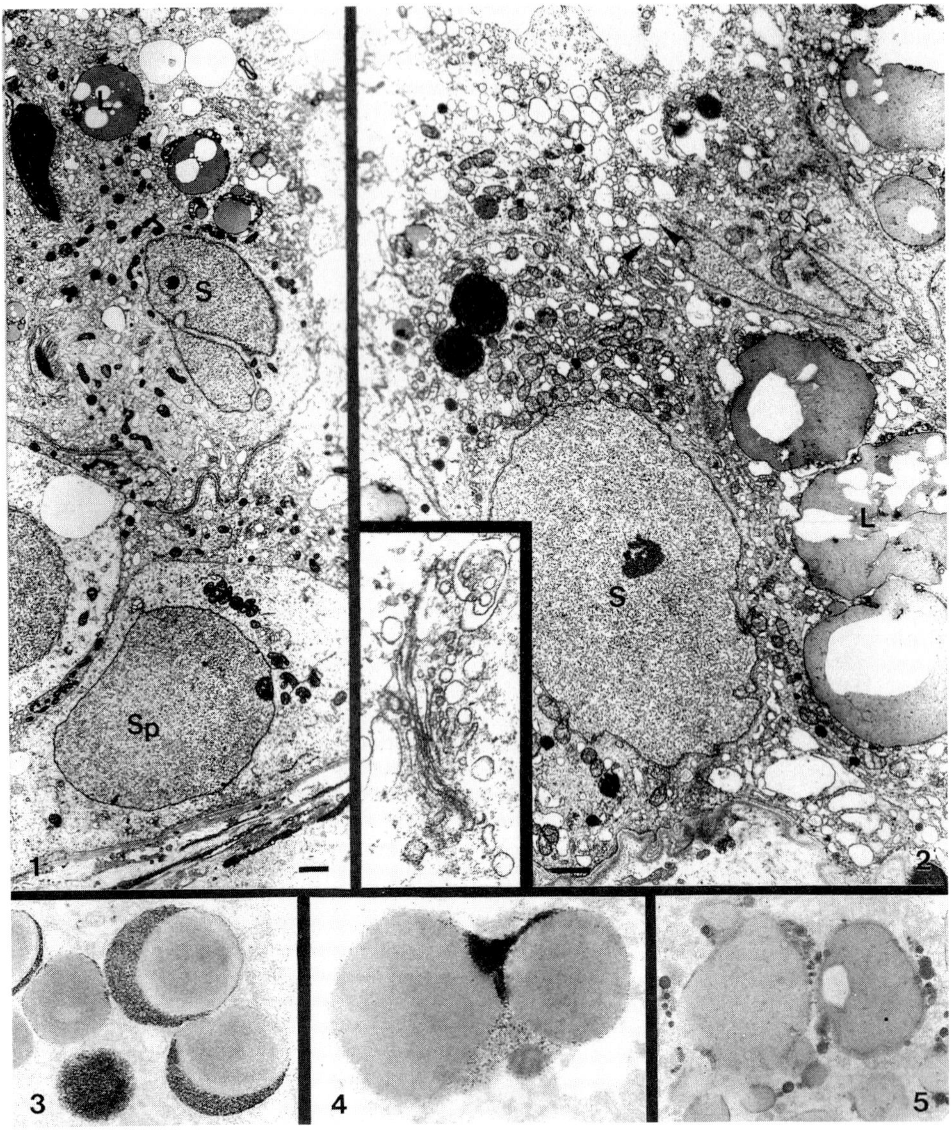

Fig.1: Tubular cross section, untreated (L:lysosome, S:Sertoli cell, Sp: spermatogonia, barr:1µm).
Fig.2: Testis tubule 30 weeks after GnRH-therapy (L:lysosome, S:Sertoli cell, arrowheads:smooth endoplasmic reticulum, barr:1µm).
Figs.3-5: Enzyme reactions (3:acid phosphatase, 4:aryl sulfatase, 5:thiamine pyrophosphatase).

Literature: 1)Viehberger, G. et al.(1985): Verhandlungsber., X.Vet.-Human-med.Gemeinschaftstagung, Berlin, 97-101. 2)Viehberger, G.; Lunglmayr, G. (1985): Europ.J.Cell Biol., Suppl.10, 73. 3)Lunglmayr, G. et al.(1988): Urol.Res., in press. 4) Lewis, P.R.; Knight, D.P.(1977): Staining methods for sectioned material. North-Holland, Amsterdam-New York-Oxford.

Inst. Phys. Conf. Ser. No. 93: Volume 3, Chapter 8
Paper presented at EUREM 88, York, England, 1988

X-ray microanalytical and mapping histology of dental tissues

M.C. SANCHEZ-QUEVEDO, P.V. CRESPO , E. FERNANDEZ-SEGURA, A.CAMPOS
Dept. Biologia Celular. Facultad de Medicina y Odontologia. 18012. Granada SPAIN

ABSTRACT: Elemental mapping distribution of calcium and mineralization index is determined in dental tissues with three different methods. Discrepancies were found between procedures based on Freeze-drying and chemical fixation.

1. INTRODUCTION

The development of SEM and its application to mineralized tissues in the last two decades has made this one of the most useful instruments in bone and teeth research. The use of electron probe-X-Ray microanalysis in association with SEM contributed in this sense to our present understanding of tooth biomineralization (1)(2). The applications of analytical electron microscopy to dental tissues involves serious methodological problems which are mainly related with sample preparation (4). The present study consisted of a microanalytical investigation of tooth mineralization through a comparison of three different methods for sample preparation. Elemental mapping histology and mineralization index were determined in dental enamel and dentine.

2. MATERIAL AND METHODS

Thirty dental specimens from Wistar rats divided into three groups of ten specimens each were used. Dental slices 0.3-0.5 mm thick were obtained as previously described (5). In method I the samples were fixed for six hours in 2% glutaraldehyde in sodium cacodylate buffer at ph 7.4. The material was dehydrated in ascending grades of acetone, critical point dried and coated with carbon. The samples in method II were fixed with 70% ethanol for 24 hours and then dehydrated in a graded ethanol series, continuing as under method I above. The samples in method III were cryofixed, cooled with liquid nitrogen and freeze-dried at -45º C, then coated with carbon (4). The samples were qualitatively analyzed in a Philips SEM 505 equipped with an EDAX system. Data presented are ZAF corrected.

3. RESULTS

Average calcium and phosphorous levels in counts per 50 sec. with the three methods allow us, after ZAF corrections, to calculate the mineralization index. The mineralization index is significantly higher in enamel than in dentine in method I and method II as clearly shown by X-ray microanalytical mapping histology (Fig 1). However method III showed no significant differences in mineralization index despite the fact that X-ray microanalytical mapping histology once again revealed noticeable topographical differences.

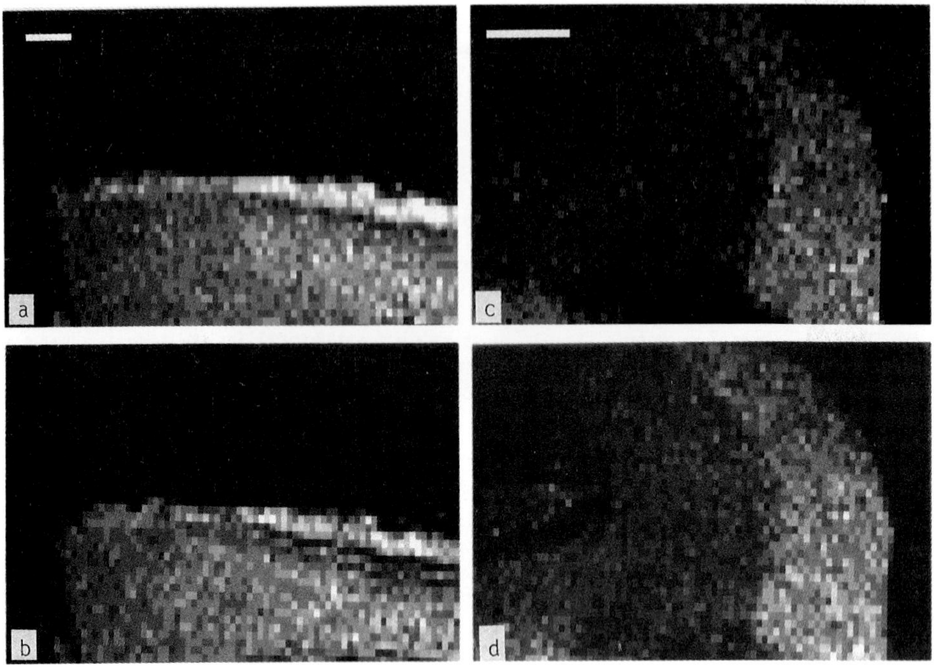

Fig. 1 1a. M-III Ca 78 x 1b. M-III P 78 x
1c. M-II Ca 150 x 1d. M-II P 150 x 0.1 mm

4. DISCUSSION

The present comparative study of the three methods for processing dental tissues for SEM and microanalytical histology throws light on the apparent differences between chemical values of Ca/P and local distribution patterns. It seems clear in M-I and M-II that the elements of interest in the present study are probably associated with macromolecular complexes in dental tissues (4)(5) implying the likelihood of transformation of amorphous into apatitic mineral (3). It has yet to be elucidated whether these phenomena, particularly the latter, are related with topographical distribution, which is prevented by M-III.

5. REFERENCES

1. Boyde, A. 1986 Scan. Electron. Microsc. 4, 1537
2. Dempster ,D.W. 1979 Scan.Electron Microsc 2, 513
3. Morgan, A. 1979 Scan. Electron Microsc. 2, 635
4. Panessa,B. 1983 Scan. Electron Microsc4, 713
5. Sanchez-Quevedo, M.C. 1986 Histol. Med. 2, 149

Paper presented at EUREM 88, York, England, 1988

Determination of the distribution of equivalent diameters of dentinal tubules by combination of scanning electron microscopy and image analyser

G Pavlović and P Jovanić

Faculty of Stomatology, Institute of Chemistry, Belgrade

ABSTRACT: The distribution alteration of the equivalent diameters of the dentinal tubules on the dentine surface, exposed to eugenol for different time intervals (10, 20 or 30 days), has been determined. The analysis of the diameter alterations of the dentinal tubules, before and after the action of eugenol, has been carried out by light microscope and the combination of scanning electron microscopy and image analyser.

1. INTRODUCTION

The dentinal tubules are one of the main structural characteristics of the dentine. Due to the possible changes of the dentinal tubules that may occur as a consequence of application of different medical drugs in dental practice, it is very important to determine the distribution of their equivalent diameters.

2. MATERIALS AND METHODS

We have used intact molars of adults. The teeth have been cut in the longitudinal cuts through the surface, polished with the diamond cream, cleaned with ultrasound and steamed with golden layer of 100 Å.

Analysis of the samples has been done by means of scanning electron microscope, Philips SEM model 515, image analyser Leitz TAS+, and the Leitz light microscope as well.

The teeth treated with eugenol before being coated with the golden layer, have been prepared in the similar way.

3. RESULTS

Twenty segments on the each dentine surface (the number of spots per surface measure unit) have been selected for the determination of the distribution of the equivalent diameters of the dentinal tubules.

The lower values of the average wideness of the dentinal tubules calculated by the combination of the secondary electrons and the image analyser may be explained in terms of the so called "edge effect" (Figure 1).

However, by application of backscattered electrons, more visible pictures of the dentinal tubules depicting the same structural details have been obtained, whereby the equivalent diameters were in accordance with those determined by the light microscope (Figure 2).

The results of the determination of the equivalent diameter alterations of the dentinal tubules in a function of the duration of eugenole, exposure, are shown on the Figure 3.

The slight extension of the dentinal tubules, proportional to the duration of the eugenol exposure, has been observed.

On the basis of these investigations, it was concluded that the transformations of the dentinal tubules may occur under the influence of eugenol, which represents one of the basic components of all dental medicines, containing zinc oxide-eugenol.

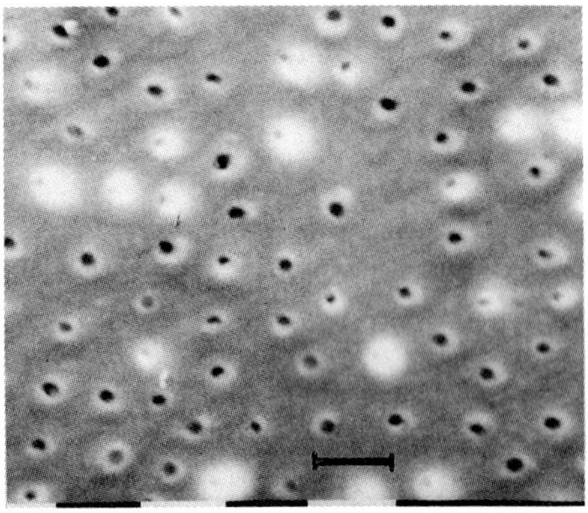

Figure 1. Scanning electron micrograph (secondary electrons-SE)

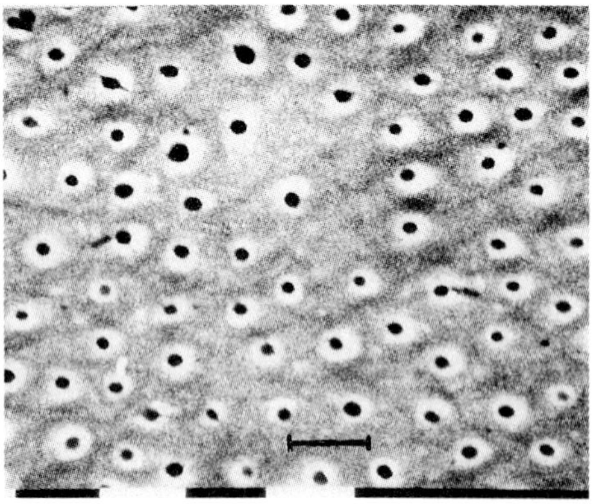

Figure 2. Scanning electron micrograph (backscattered electrons)

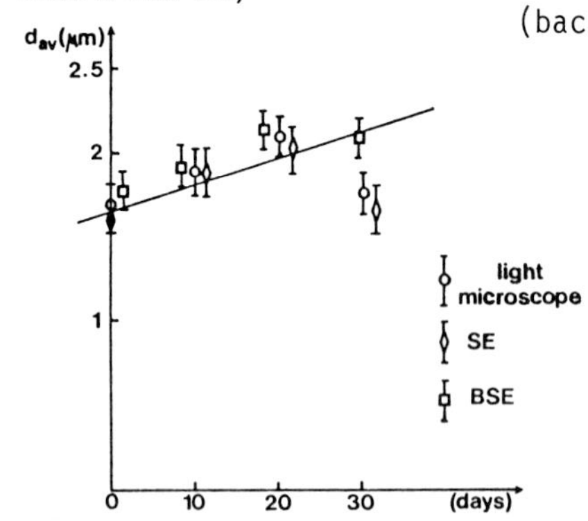

Figure 3. The dependence of the diameter alterations of the dentinal tubules in function of the duration of eugenol exposure

Inst. Phys. Conf. Ser. No. 93: Volume 3, Chapter 9
Paper presented at EUREM 88, York, England, 1988

Electron microscopy in pathology

Arvid B. Maunsbach

Department of Cell Biology, University of Aarhus, DK-8000 Aarhus C, Denmark

ABSTRACT: Electron microscopy has contributed immensely to human pathology at the cellular, subcellular and molecular level and is an indispensable tool in experimental pathology. Diagnostic electron microscopy has for years been much more discussed but consensus seems approaching regarding a few areas where electron microscopy is truly diagnostic and other areas where electron microscopy is an important adjunct to other diagnostic methods.

1. INTRODUCTION

Electron microscopy has contributed fundamentally to the understanding of disease processes at the cellular and subcellular levels and a wealth of information has been gathered about the ultrastructural alterations of diseased cells (Trump and Jones 1978-1983; Johannessen 1978-1985; Erlandson 1981; Ghadially 1982; Carr and Toner 1982; Johannessen 1982; Dolman 1984; Henderson et al. 1986; Moss 1986). Likewise the literature on the applications of electron microscopy in experimental pathology is voluminous and illustrates the usefulness of different types of electron microscope techniques in understanding e.g. pathogenetic mechanisms. Electron microscopy was initially expected to have a great future also in diagnostic pathology but this application has created many controversial opinions, ranging from strong acceptance (Laschi 1974, 1980) to almost complete rejection (Bergstrand 1982). However, there is now some agreement about those areas where different forms of electron microscopy are useful aids in diagnosis (Johannessen 1984; Dardick 1987).

2. TRANSMISSION ELECTRON MICROSCOPY

Transmission electron microscopy is the dominating method in diagnostic electron microscopy and is usually applied in combination with conventional specimen preparation methods. However, several technical modifications have been introduced to improve its usefulness in diagnostic pathology, including rapid embedding procedures (Johannessen 1973), methods for reducing sampling problems (Eyden 1987) and application of low magnification electron microscopy (Rippstein et al. 1987). Kidney tubules, which are extremely sensitive to variations in the preparation procedure (Maunsbach 1979), show improved preservation in surgically removed human kidneys if the extirpated kidneys are fixed by vascular perfusion (Møller et al. 1982) and haematology specimens show excellent preservation upon fixation and embedding directly on glass slides (Yasuda and Toida 1986). Procedures have also been developed for the electron microscope analysis of small samples of cells obtained through fine needle aspiration (Wills et al. 1987; Mackay et al. 1987a).

Transmission electron microscopy has been used for years in the analysis of renal biopsies and the ultrastructural lesions have been well characterised in a large variety of renal diseases (Brun and Olsen 1981; Heptinstall 1981; Zollinger and Mihatsch 1982). Procedures have been developed for systematic semiquantitative analysis of glomerular ultrastructure in human renal biopsies (Bohman et al. 1979; Maunsbach et al. 1981). The results demonstrate that a few well defined lesions show a very high reproducibility of evaluation and that a classification depending largely upon one or more of these lesions can be made with great precision, even if a limited amount of material is analyzed. Electron microscopy aids in the differential diagnosis in particular by defining changes in the glomerular capillary wall (Brun and Olsen 1981; Seymour et al. 1983). Altogether electron microscopy has been estimated to contribute to the diagnosis of renal biopsies in 6-30% of the cases (Collan et al. 1978; Mithatsch and Zollinger 1980; Muehrcke et al. 1979). In a series of 91 consecutive biopsies from patients with glomerulonephritis it was found that the light microscopic diagnosis was revised by electron microscopy in about 13% of the biopsies (Olsen et al. 1983).

Transmission electron microscopy is essential when diagnosing the immotile-cilia syndrome (Afzelius 1976) where several different forms of the syndrome can be distinguished on morphological grounds (Afzelius 1981). Also in inherited storage diseases the characteristic ultrastructural changes of the lysosomes may occasionally give a clue to the diagnosis (Malmqvist et al. 1971). Recently characteristic ultrastructural markers related to the endoplasmic reticulum have been reported in lymphocytes from patients with AIDS or AIDS-related conditions (Orenstein et al. 1987; Hansmann et al. 1987). Future studies will undoubtedly uncover additional ultrastructural markers of diagnostic significance in various diseases.

The overwhelming majority of tumors can be classified without the aid of electron microscopy but for a limited number of tumors electron microscopy is usefull in the diagnosis. In particular the histogenesis of some types of poorly differentiated neoplasms can be determined on the basis of ultrastructurally defined junctions, filaments, granules, microvilli, glycogen inclusions or other ultrastructural features (Sobel and Marquet 1980; Henderson et al. 1986). Electron microscopy has also been helpful when diagnosing some astrocytomas (Scheithauer and Bruner 1987), sarcomas (Erlandson 1987), renal tumors of childhood (Mierau and Beckwith 1987) and metastatic tumors of unknown origin (Hammar et al. 1987) and may serve as an adjunct to light microscopic immunocytochemistry (Mackay et al. 1987b). While electron microscopy is of limited value in routine diagnostic gynecologic pathology or diagnosis of breast tumors it may play an imporant part in selected cases (Ferenczy 1987; Nesland et al. 1987). Transmission electron microscopy has also been used to classify adenomas (McNicol 1987; Holm-Nielsen and Steen Olsen 1988) as well as neoplastic cells in effusion fluids in cases where light microscopic diagnosis alone was unable to give a conclusive answer (Mukherjee 1982).

3. IMMUNOELECTRON MICROSCOPY

With the availability of a large variety of specific antibodies, including monoclonal antibodies, immunocytochemistry at the light microscopic level has rapidly become an indispensable tool in diagnostic pathology. The rapid development of new methods for immunocytochemistry at the electron microscope level (Roth 1986; Silver and Hearn 1987) has opened the possibility to combine immunocytochemistry with electron microscopy for diagnostic purposes. Typical applications include the identification of endocrine tumors on the basis of immunolabeling of secretory granules in the cells (Polak and Varndell 1984) and the identification of intermediate filaments (Moll et al. 1986). It is likely that the number of similar applications will raise rapidly.

Immunoelectron microscopy has also found a useful appplication in diagnostic virology where several methods have been developed for the identification of viruses in clinical specimens (Kjeldsberg 1980, 1986; Doane 1987; Doane and Anderson 1987). In this field immunoelectron microscopy can be used to identify the viruses as well as to enhance the sensitivity of the detection system. Thus, the detection of viruses by immunoelectron microscopy is much more sensitive than direct electron microscopy.

Immunological reactions of the cell surface have also been analyzed by scanning electron microscopy and the results indicate that scanning immunoelectron microscopy may have some useful applications, in particular when analysing human leukaemia and lymphoma cells (Gamliel et al. 1981). A future method for specific localization of antigenic sites on cell surfaces may involve the use of scanning electron microscopy to demonstrate small colloidal gold markers conjugated with protein A (de Harven et al. 1984; de Harven 1987). This marker can be recognized on the basis of its atomic number contrast by using the backscattered electron imaging mode of the SEM.

4. SCANNING ELECTRON MICROSCOPY

Scanning electron microscopy has been widely applied in pathology and has, in conjunction with light microscopy and transmission electron microscopy, contributed information in e.g. gastrointestinal pathology (Carr et al. 1984; Bonvicini et al. 1985, 1986), haematology (Polliack 1981) and cardiology and ophtalmology (Versura et al., 1986; Laschi 1987). However, there is general agreement that scanning electron microscopy only rarely contributes to the diagnosis (Buss and Hollweg 1980; Carr et al. 1980; Carter 1980; Carr and Toner 1981). One main reason is that observations by SEM are not specific enough to provide a safe diagnosis that may have clinically consequences and that the same results can be obtained almost invariably by conventional light microscopy, sometimes in combination with transmission electron microscopy or immunocytochemistry. In addition few pathology laboratories are equipped for routine SEM. In the near future SEM is therefore likely to be primarily an adjunct to other microscopic techniques in diagnosis.

5. X-RAY MICROANALYSIS

X-ray microanalysis has developed rapidly in biomedical research, including diagnostic pathology. Until now the applications are limited in number but the results suggest that X-ray microanalysis has a place in specific fields of diagnostic pathology (Roomans 1983, 1984). The examples of X-ray microanalysis include the analysis of particulates, such as asbestos (Craighead and Mossman 1982), the analysis of pathological accumulations of endogenous or exogenous particulates, including metals (Roomans and Forslind 1980; Parkes 1982; Forslind 1984; Pasquinelli et al. 1987), and precipitates of endogenous ions such as calcium. In addition it has applications in medico-legal diagnoses (Abraham 1980).

6. CONCLUSIONS

- Electron microscopy in various forms has contributed fundamentally to our understanding of disease processes at the cellular and subcellular level but in diagnostic pathology it remains an adjunct to light microscope examination.

- Electron microscopy has diagnostic significance only in a small number of cases, probably less than 1-2% of surgical or biopsy specimens. In many other cases electron microscopy supports the light microscope diagnosis but it is not necessary for the diagnosis.

- The diagnostic value of electron microscopy is very well documented for some renal biopsies, certain types of tumors and a few other diseases such as the immotile-cilia syndrome.

- Light microscopic immunocytochemistry is rapidly developing into a strong competitor to electron microscopy as an adjunct to diagnostic light microscopy. Future developments may include an increasing use of ultra-structural immunocytochemistry for diagnostic purposes.

- Scanning electron microscopy has contributed information about various diseases, but has so far been of diagnostic value only in exceptional cases.

- The practicability of diagnostic electron microscopy must always be related to the high costs of instrument investments and the necessity of maintaining a functioning EM-laboratory and well-trained personel.

5. REFERENCES

Abraham J L 1980 Scanning Electron Microsc. 1980/IV 171

Afzelius B A 1976 Science **193** 317

Afzelius B A 1981 Am. J. Hum. Genet. **33** 852

Bergstrand A 1982 Ultrastruct. Pathol. **3** 397

Bohman S-O, Deguchi N, Gundersen H J G, Hestbech J, Maunsbach A B and Olsen S 1979 Lab. Invest. **40** 433

Bonvicini F, Maltarello M C, Versura P, Bianchi D, Gasbarrini G and Laschi R 1986 Scanning Electron Microsc. II 687

Bonvicini F, Zoli G, Maltarello M C, Bianchi D, Pasquinelli G, Versura P, Gasbarrini G and Laschi R 1985 Scanning Electron Microsc. III 1279

Brun C and Olsen S 1981 Atlas of Renal Biopsy (Copenhagen: Munksgaard)

Buss H and Hollweg H G 1980 Scanning Electron Microsc. III 139

Carr K E, Kamel H M H, Toner P G, McGadey J and Wong A 1984 Scanning Electron Microsc. II 761

Carr K E, McLay A L C, Toner P G, Chung P and Wong A 1980 Scanning Electron Microsc. III 121

Carr K E and Toner P G 1981 J. Microscopy **123** 147

Carr K E and Toner P G eds 1982 Cell Structure - An Introduction to Biomedical Electron Microscopy (Edinburg, London, Melbourne, New York: Churchill Livingstone)

Carter H W 1980 Scanning Electron Microsc. III 115

Churg J and Sobin L H 1982 Renal Disease. Classification and Atlas of Glomerular Diseases (Tokyo: Igaku-Shoin)

Collan Y, Klockars M and Heino M 1978 Nephron **20** 24

Craighead J E and Mossman B T 1982 N. Engl. J. Med. **306** 1446

Dardick I 1987 Ultrastruct. Pathol. **11** v-vi

de Harven E 1984 Ultrastruct. Pathol. **11** 711

de Harven E, Leung R and Christensen H 1984 J. Cell Biol. **99** 53

Doane F W 1987 Ultrastruct. Pathol. **11** 681

Doane F W and Anderson N 1987 Electron Microscopy in Diagnostic Virology: A Practical Guide and Atlas (New York: Cambridge University Press)

Dolman C L 1984 Ultrastructure of Brain Tumours and Biopsies. Methods in Laboratory Medicine vol. 5 (New York: Praeger Publishers)
Erlandson R A 1981 Diagnostic Transmission Electron Microscopy of Human Tumors (New York: Masson Publishing USA)
Erlandson R A 1987 Ultrastruct. Pathol. **11** 83
Eyden B P 1987 Ultrastruct. Pathol. **11** 449
Ferenczy A 1987 Ultrastruct. Pathol. **11** 335
Forslind B 1984 Scanning Electron Microsc. 1984/I 183
Gamliel H, Leizerowitz R, Gurfel D and Polliack A 1981 J. Microscopy **123** 189
Ghadially F N 1982 Ultrastructural Pathology of the Cell and Matrix (London: Butterworths)
Hammar S, Bockus D and Remington F 1987 Ultrastruct. Pathol. **11** 209
Hansmann M-L, Kaiserling E, Müller-Hermelink H K and Hui P-K 1987 Ultrastruct. Pathol. **11** 389
Henderson DW, Papadimitriou J M and Coleman M eds 1986 Ultrastructural Appearances of Tumours 2nd ed (Edinburg: Churchill Livingstone)
Heptinstall R H 1983 Pathology of the Kidney vols 1-3, 3rd ed (Boston: Little, Brown and Company)
Holm-Nielsen P and Olsen T S 1988 Ultrastruct. Pathol. **12** 27
Johannessen J V 1973 Kidney Int. **3** 46
Johannessen J V 1978-1985 Electron Microscopy in Human Medicine vols. 1-12 (New York: McGraw-Hill)
Johannessen J V 1982 Diagnostic Electron Microscopy 2nd ed (New York: McGraw-Hill)
Johannessen J V 1984 Electron Microscopy 1984 vol. 3 Life Sciences II (Budapest: Programme Committee of the Eight European Congress on Electron Microscopy) pp 2149-54
Kjeldsberg E 1980 Path. Res. Pract. **167** 3
Kjeldsberg E 1986 Ultrastruct. Pathol. **10** 553
Laschi R 1974 Gazz. Ingl. **3** 83
Laschi R 1980 Biol. Cell **37** 305
Laschi R, Pasquinello G and Versura P 1987 Scanning Microsc **1** 1771
Mackay B, Fanning T, Bruner J M and Steglich M C 1987a Ultrastruct. Pathol. **11** 659
Mackay B, Ordónez N G and Khoursand J 1987b Ultrastruct. Pathol. **11** 483
Malmqvist E, Ivemark B I, Lindsten J, Maunsbach A B and Mårtensson E 1971 Lab. Invest **25** 1
Maunsbach A B 1979 Electron Microscopy in Human Medicine, Vol. 9 Urogenital System and Brest ed J V Johannessen (London: McGraw-Hill) pp 143-65
Maunsbach A B, Bohman S-O, Gundersen H J G and Olsen T S 1981 Proc. 10. Tagung Elektronenmikroskopie ed H Luppa and J Heydenreich (Leipzig: Gesellschaft für Topochemie und Elektronenmikroskopie der DDR) pp 44-47
McNicol A M 1987 Histopathol. **11** 995
Mierau G W and Beckwith J B 1987 Ultrastruct. Pathol. **11** 313
Mihatsch M J and Zollinger H U 1980 Path. Res. Pract. **167** 88
Moll R, Osborn M, Hartschuh W, Moll I, Mahrle G and Weber K 1986 Ultrastruct. Pathol. **10** 473
Moss T H 1986 Tumours of the Nervous System - an Ultrastructural Atlas (London: Springer-Verlag)
Muehrcke R C, Mandal A K, Gotoff S P, Isaacs E W and Volini F I 1979 Arch. intern. Med. **124** 170.
Mukherjee T M 1982 J. Submicrosc. Cytol. **14** 717
Møller J Chr, Skriver E, Olsen S and Maunsbach A B 1982 Ultrastruct. Pathol. **3** 375
Nesland J M, Holm R, Lunde S and Johannessen J V 1987 Ultrastruct. Pathol. **11** 293
Olsen S, Bohman S-O, Hestbech J, Gundersen H J G, Petersen V P, Deguchi N and

Maunsbach A B 1983 Acta path. microbiol. immunol. scand. Sect. A **91** 53
Orenstein J M, Preble O T, Kind P and Schulof R 1987 Ultrastruct. Pathol. **11** 673
Parkes W R 1982 Occupational Lung Disorders, 2nd ed (London: Butterworths)
Pasquinelli G, Scala C, Borsetti G P, Martegani F and Laschi R 1985 Scanning Electron Microsc. III 1133
Polak J M and Varndell I M 1984 Electron Microscopy 1984 vol. 3 Life Sciences II (Budapest: Programme Committee of the Eight European Congress on Electron Microscopy) pp 2155-2162
Polliack A 1981 J. Microscopy **123** 177
Rippstein P, Cavell S, Boivin M and Dardick I 1987 Ultrastruct. Pathol. **11** 723
Roomans G M 1983 Electron Microscopy in Human Medicine vol. 11B ed J V Johannessen (London: McGrawHill) pp 89-156
Roomans G M 1984 Electron Microscopy 1984 vol. 3 Life Sciences II (Budapest: Programme Committee of the Eight European Congress on Electron Microscopy) pp 2201-2204
Roomans G M and Forslind B 1980 Ultrastruct. Pathol. **1** 301
Roth J 1986 J. Microscopy **143** 125
Scheithauer B W and Bruner J M 1987 Ultrastruct. Pathol. **11** 535
Seymour A E, Canny A and Spargo B H 1983 Ultrastruct. Pathol. **4** 123
Silver M M and Hearn S A 1987 Ultrastruct. Pathol. **11** 693
Sobel H J and Marquet E 1980 Path. Res. Pract. **167** 22
Trump B F and Jones R T eds 1978-1983 Diagnostic Electron Microscopy vols. 1-4 (New York: John Wiley and Sons)
Versura P, Maltarello M C, Bonvicini F, Caramazza R and Laschi R 1986 Scanning Electron Microscopy **III** 1229
Wills E J, Carr S and Philips J 1987 Ultrastruct. Pathol. **11** 361
Yasuda H and Toida S 1986 Ultrastruct. Pathol. **10** 577
Zollinger H U and Mihatsch M J 1978 Renal Pathology in Biopsy. Light Electron and Fluorescent Microscopy and Clinical Aspects (Berlin: Springer Verlag)

Inst. Phys. Conf. Ser. No. 93: Volume 3, Chapter 9
Paper presented at EUREM 88, York, England, 1988

An immuno-electron microscopic study of the tissue cyst of *Toxoplasma gondii* in mouse brain

DJP Ferguson

Nuffield Department of Pathology, University of Oxford, John Radcliffe Hospital, Oxford

ABSTRACT: Ultrastructurally, it was observed that the cysts were within intact host cells. Immunocytochemically, the anti-*Toxoplasma* antibody reacted specifically with the cyst wall, the ground substance of the cyst and certain areas within the parasites predominantly the rhoptries and micronemes. No *Toxoplasma* antigens were present in the host cell cytoplasm or surrounding tissue. This study shows that the intracellular location of the tissue cysts protects the parasites from immune attack by preventing the exposure of *Toxoplasma* antigens.

1. INTRODUCTION

Toxoplasma gondii is a protozoan parasite which infects all warm blooded animals including man. Infection results in an initial proliferation of the parasite which, with the onset of the immune response, enters a chronic phase where it forms tissue cysts predominantly in muscle and the CNS. The chronic infections are asymptomatic but if the host immune system is compromised (tumour therapy, transplantation or AIDS) a recrudescence of the proliferative form from the tissue cyst can cause serious disease. In the brains of immunocompetent hosts, the tissue cysts containing numerous parasites, elicit no immune response. The reason for this is unknown although it has been proposed that the intracellular location of the cyst masks them from the host immune system (Ferguson and Hutchison, 1987). To study this we have used immunoelectron microscopy to identify the distribution of *Toxoplasma* antigens associated with tissue cysts.

2. MATERIALS AND METHODS

Mice were infected with *T. gondii* and examined at intervals between 7 days and 22 months. The brains were either immersion or perfusion fixed in 4% glutaraldehyde in phosphate buffer. Small blocks were dehydrated and embedded in LR White resin. Thin sections were stained by an indirect immunocytochemical technique using a polyclonal anti-*Toxoplasma* IgG fraction obtained from the serum of previously infected rabbits*. The location of the primary antibody was visualised using swine anti-rabbit immunoglobulin conjugated to 15nm colloidal gold particles. The sections were counterstained with uranyl acetate and lead citrate prior to examination.

* Antibody kindly supplied by Dr. E. Pettersen, SSI, Copenhagen, Denmark.

3. RESULTS AND DISCUSSION

The cysts were present in various regions of the brain but predominantly in the cerebral hemispheres. The cysts were of variable size and stained positively for *Toxoplasma* antigen by light microscopy (Fig. 1) but induced no inflammatory response from the host. Ultrastructurally, using conventional processing, it was found that the cysts were present within intact host cells (Fig. 2). By immunoelectron microscopy it was observed that *T. gondii* antigens were present in the cyst wall and ground substance of the cyst (Fig. 3). Within the parasites the antibody reacted predominantly with the rhoptries and micronemes (Fig. 4). There was no specific staining of the host cell or surrounding tissue.

These observation are consistent with the *Toxoplasma* antigens being contained in the cyst within the host cell cytoplasm. Since no extracellular antigen is presented to the host immune system, the parasites are protected from immune attack.

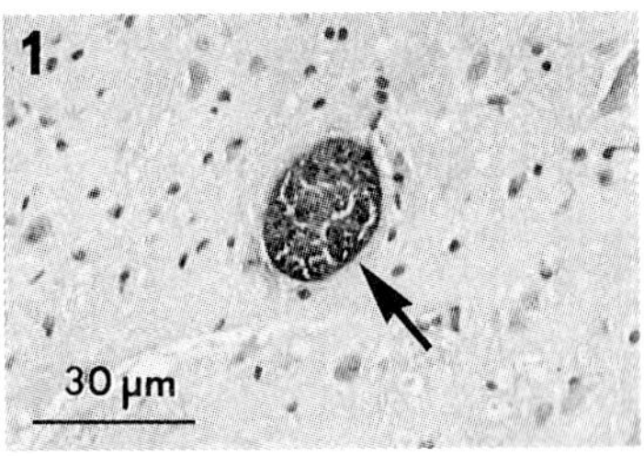

Figure 1. Light micrograph of a tissue cyst (arrow) within the brain. Immunoperoxidase stained.

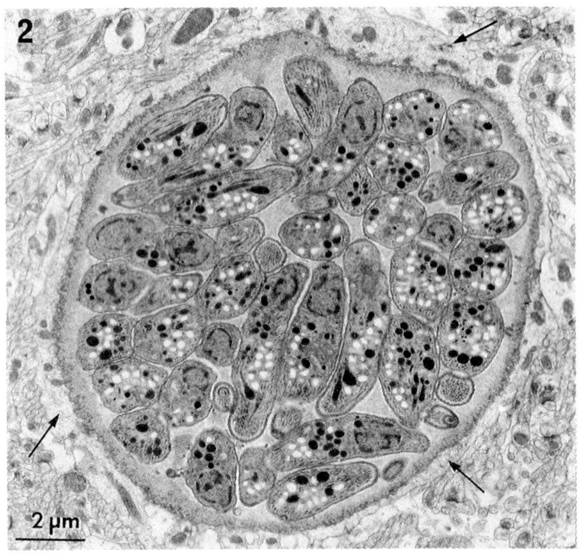

Figure 2. Electron micrograph of a tissue cyst present within host cell cytoplasm (arrows). Conventionally processed.

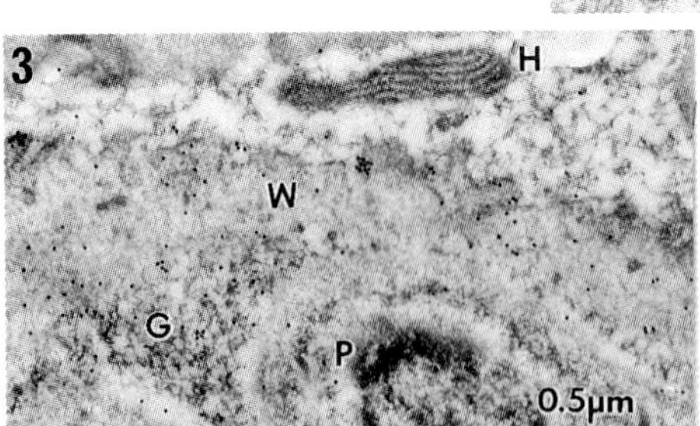

Figure 3. Gold particles attached to the cyst wall (W) and ground substance (G). H - host cell; P - parasite.

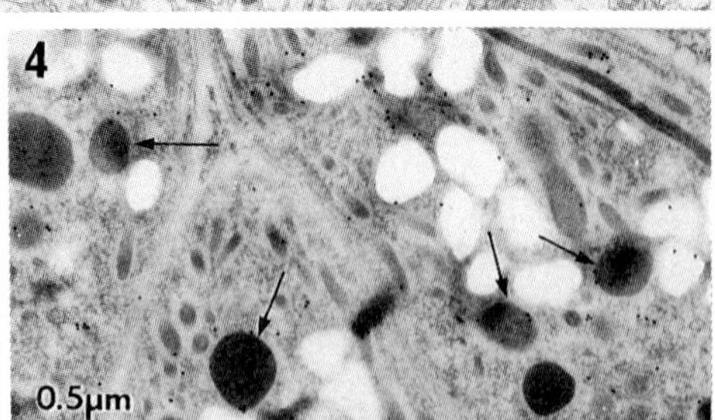

Figure 4. Detail of a parasite showing the gold particles associated with rhoptries (arrows).

REFERENCE

Ferguson DJP and Hutchison WM 1987 Virchows Arch A **411** 39.

Simultaneous triple-immunogold staining of virus and host cell antigens in ultrathin cryosections using monoclonal primary antibodies

L.Bastholm*, M.H.Nielsen*, S.Chatterjee** and B.Norrild***
*Institute of Pathological Anatomy and ***Medical Microbiology, University of Copenhagen, 11 Frederik D. V vej
DK-2100 Copenhagen, Denmark
**Dept. of Pediatrics University of Alabama, Birmingham.

ABSTRACT: A method for simultaneous triple-immunogold staining of antigens in ultrathin cryosections is described. Each antigen is stained by a indirect method using specific mouse monoclonal as primary antibody, rabbit anti mouse IgG as second antibody, and gold-conjugated goat anti rabbit IgG as third antibody. Between each antigen staining cycle sections are covered by methyl cellulose and exposed to formalin vapour.

1. INTRODUCTION:

Simultaneous immunocytochemical staining of multiple antigens in ultrathin sections is the most direct way to visualize the mutual spatial relationship of antigens at the subcellular level.In case of visualization of multiple antigens in ultrathin cryosections it is the only method available. We have invented the use of an immunocytochemical triple-staining method originally developed for resin embedded material (Wang et al, 1985) for triple-staining ultrathin cryosectioned material employing mouse monoclonal antibodies specific for the three antigens.

2. MATERIAL AND METHODS:

Herpes Simplex Virus type I (HSV) infected human embryonic lung cells (HEL) were cultivated and prepared for electron microscopy (Norrild et al. 1988). Mouse monoclonal IgG specific for 1) HSV glycoprotein-C (gC)(Koga et al 1986), 2) HSV glycoprotein-D(gD)(Koga et al 1986), and 3) +ß tubulin (Amersham) were used as primary antibodies.Affinity purified rabbit anti mouse IgG (KAM) as second and goat anti Rabbit (GAR) conjugated to either 5,lo or 15 nm colloidal gold particles (Janssen) as third antibody.
The staining procedure includes: Staining for the first antigen by a three layer indirect method using sufficient concentration of KAM and gold conjugated GAR to saturate available antigenic epitopes on the primary and the secondary antibodies. Cryosections were covered by methylcellulose and exposed to paraformaldehyde vapour at 80 C-for 30 minutes followed by removal of the methylcellulose. Staining for the second and third antigen was carried out by repeating the

three layer staining procedure using GAR conjugated to a second and third size class gold particles.Between the second and the third staining cycle were sections treated with methylcellulose and formalin vapour as described above. Controls included :1) replacement with unspecific mouse monoclonal antibodies.

3. RESULTS AND COMMENTS:

Microtubules were either labelled 1) for gC, gD and tubulin or 2) for one glycoprotein and tubulin or 3) for tubulin. Golgi vesicles and intra- and extracellular virus were labelled for gC and gD. Varying the sequence of primary antibodies used for first, second and third staining cycle resulted in labelling of identical structures. To prevent cross-labelling antigenic epitopes should always be completely saturated with antibody before the formalin vapour treatment after each staining cycle. The formalin vapour destroy all free anti-IgG binding sites on the antibodies.

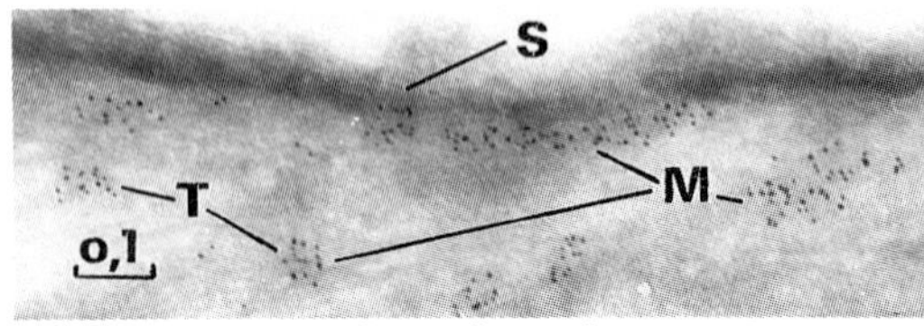

Fig.1. Cytoplasmic microtubules(M) immunostained for tubulin(T)(5nm gold). S marks the cell surface.

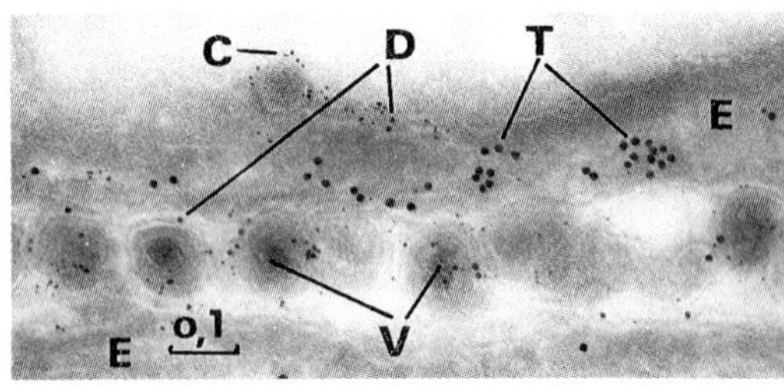

Fig.2. Simultaneous immunostain for tubulin(T)(15nm gold), glycoprotein C(C)(5nm gold) and D (D)(10nm gold). Virus are marked V, cell extensions E.

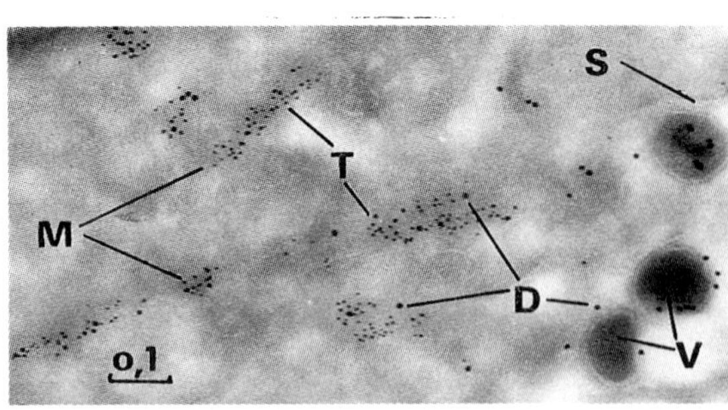

Fig.3. Simultaneous immunostain for tubulin(T) (5nm gold) and glycoprotein D(D)(10nm gold). Microtubules are marked M, virus V, and cell surface S.

REFERENCES:

Bastholm L, Nielsen MH and Larsson L-I. 1987 Histochem.87 229
Koga J , Chatterjee S, Whitley RJ.1986 Virology 151 389
Nielsen MH, Bastholm L, Chatterjee S, Norrild B EUREM 88
Wang B-L, Larsson L-I. 1985 Histochemistry 83 47

Inst. Phys. Conf. Ser. No. 93: Volume 3, Chapter 9
Paper presented at EUREM 88, York, England, 1988

Contribution of scanning/transmission (STEM) immune electron microscopy to the understanding of HIV infection in man

M.I. Herrera [1], I. Santa Maria[1], R. Nauta[2], W. Hax[2], J. González Lahoz[1], and R. Nájera[1]

[1]: Instituto de Salud Carlos III. Majadahonda (Madrid). Spain.

[2]: Philips Applications Laboratory. International Bussines Center for Electron Optics. Building AAE, Eindhoven. Holland.

ABSTRACT: Unconventional electron microscopy , using a TEM/STEM system, has been applied to pinpoint HIV antigens in infected peripheral blood mononuclear cells from patients. An immunogold labelling technique has been used to detect the viral antigens. Sensitive and specific visualization of the gold marker (backscattered electrons image) and correlation in topographical images (secondary electrons detection) have been possible. Quantitative studies reveal a number of infected cells greater than that anticipated by other methods. The results suggest that this technique should help to study HIV replication "in vivo" and other virus-cell systems.

The AIDS virus, HIV, has been asigned to the Lentivirus subfamily of the Retroviridae on the basis of sequence homologies and structural studies combining conventional TEM pre-embedding IEM and immunocryoultramicrotomy (Gelderblom, 1987). These studies have beutifully shown the stages of HIV entry, assembly and budding at the surface of "in vitro" infected cells.

However, little is yet known about the stages of HIV replication "in vivo" as the search for the virus or its antigens in thin sections of clinical specimens encounters the difficulty of localizing the small number of wholly structured particles which might be present in them. We report here the results of applying a different type of E.M. approach. HIV antigens have been localized in peripheral blood mononuclear cells (PBMC) by combining immunogold surface labelling techniques with scanning/transmission electron microscopy.

Monoclonal antibodies against three structural proteins (p24, gp41 and p17) and 40 nm colloidal gold particles linked to goat anti-mouse IgG have been used in the labelling procedure, while the presence of the gold marker has been detected using the backscattered electrons imaging mode of the microscope.

The application of this technique to study PBL from asymptomatic seropositive individuals, AIDS related complex (ARC) patients and full blown AIDS patients reveals that the number of HIV antigen producing cells is greater than it was anticipated from neutralization tests and other biochemical studies. For instance, the estimate of the percentage of infected cells in some AIDS patients is about 30%. Fig. 1 shows the paired electron micrographs corresponding to one of these cases. In patients with ARC, the technique has allowed the observation of several labelled cell membrane protrusions, as the one shown in Fig. 2, similar to the budding process typical in the assembly of retroviruses. In samples from symptomless, seropositive individuals the percentage of labelled cells appears to be as high as 18%. The technique has also allowed us to pick up some labelled cells as those shown in Fig. 3, in PBMC from asymptomatic, yet seronegative, individuals at risk of AIDS.

Though these are just preliminar results, they suggest that this type of approach to the study of AIDS infection should be useful for detecting the virus early in the latency period of the disease and should also help to solve some of the unresolved mysteries of AIDS replication "in vivo".

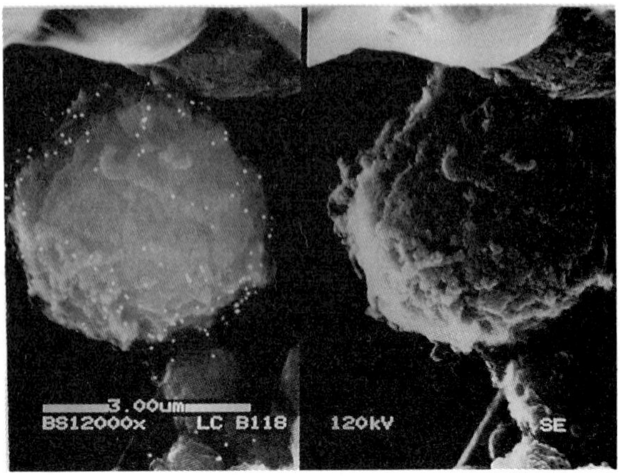

Figure 1

Figure 2

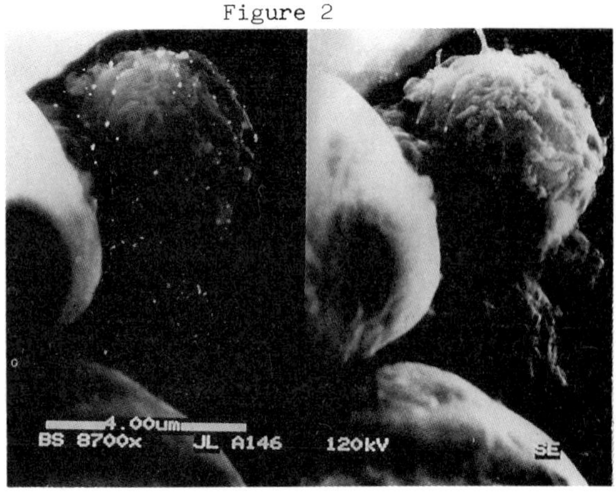

Figure 3

References.-
Gelderblom, H.R.; Hausmann, E.H.; Ozel,-M.; Pauli,G, and Koch,M. "Fine structure of human immunodeficiency virus (HIV) and immunolocalization of structural proteins". 1987. Virology 156. pp 171-176.

Cytoplasmic viral core aggregates and budding of mature virus and sperhules in mouse brain following semliki forest virus infections

S Pathak and H E Webb

Neurovirology Unit, Rayne Institute, St. Thomas' Hospital, The United Medical and Dental Schools of Guy's and St. Thomas' Hospitals.

Electron microscopical (EM) studies of virulent (L10) and avirulent (A774) Semliki Forest virus (SFV) replication in mouse central nervous system (CNS) show budding of mature virus (virions) and spherules from the cell membranes. Cytoplasmic viral core aggregates (VCA) consisting of granular, amorphous and fibrous material (Fig. 1) are rarely reported (Pathak and Webb 1978; 1983). Our aim is to understand the nature of these VCA and their role in the virus replication cycle.

This study shows that VCA are present in the cytoplasm of neurons and oligodendrocytes in all ages of mice infected wth either strain of SFV. 2-7 day old (baby), 12 and 14 day old mice were inoculated intraperitoneally or intracerebrally with either strain. Brains were sampled on post inoculation day (PID) 1-5 and prepared for both routine and immunoelectron microscopy (Evans and Webb, 1984).

VCA were labelled with gold conjugated antibody specific for viral nucleocapsid protein (Fig. 2). Sometimes nucleocapsids were seen in association with VCA. In all L10 infections and A774 infection of baby mice nucleocapsids and VCA were often associated with the cell membranes, at sites where many virions and spherules were budding (Figs. 3,4 and 5). However, in A774 infection of 12 and 14 day old mice there was increased accumulation of VCA in the juxtanuclear region of the cytoplasm (Fig. 1) but reduction in the budding of virions and spherules despite high virus titres.

The spherules are viral in nature as they label for SFV viral glycoprotein E1/E2 (Evans and Webb 1986) (Fig. 6). These spherules contain a dense thread like structure (Figs. 4, 5 and 6) which resembles viral RNA (Hsu et al 1973).

VCA contains viral nucleocapsid protein and possibly viral RNA and appears to be involved in the formation of nucleocapsids and budding of virions and spherules. In older mice VCA accumulates in the cytoplasm when virus budding is restricted.

Evans N R S and Webb H E 1984 J. Histochem. Cytochem. **32** 372
Evans N R S and Webb H E 1986 J. Neurol. Sci. **74** 279
Hsu M T, Kung H J and Davidson N 1973 Cold Spring Harbor Symposia on Quantitative Biology **38** 943
Pathak S and Webb H E 1978 J. Neurol. Sci. **39** 119
Pathak S and Webb H E 1983 Neuropath. Appl. Neurobiol. **9** 313

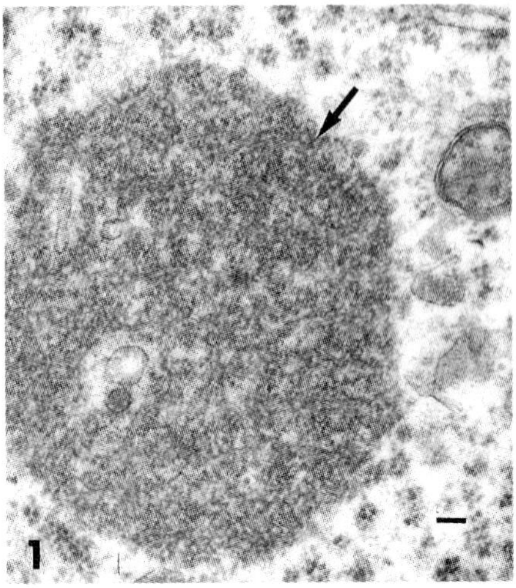

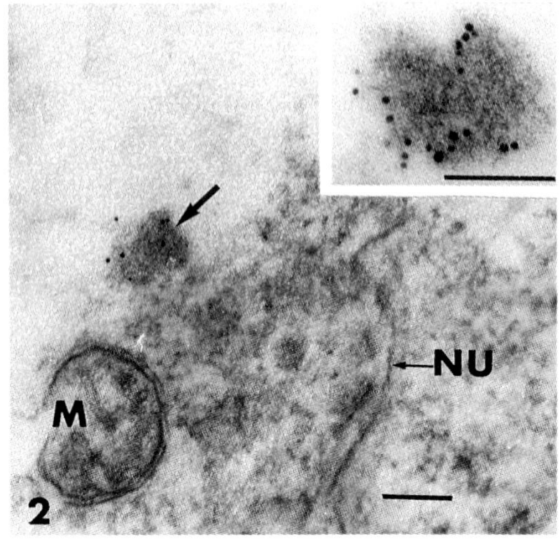

Fig. 1. Large VCA (arrow) in the cytoplasm of a neuron. 12 day-old mouse brain. A774, PID 5.

Fig. 2. Cytoplasmic VCA (arrow) labelled with gold (G10) conjugated antibody against viral nucleocapsid protein. Frozen section. 12 day-old mouse brain. A774, PID 3. Inset: Another VCA labelled.

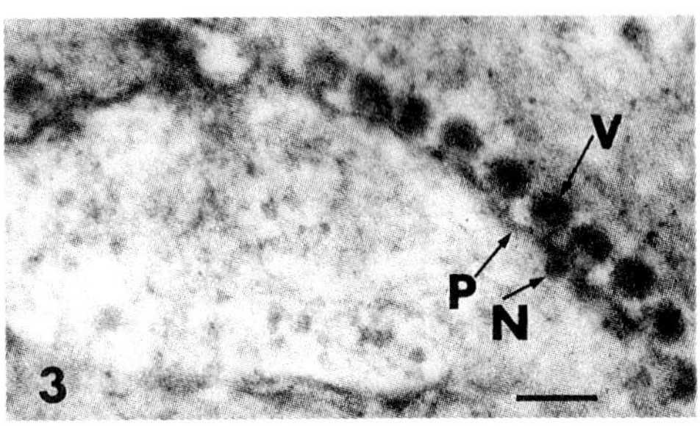

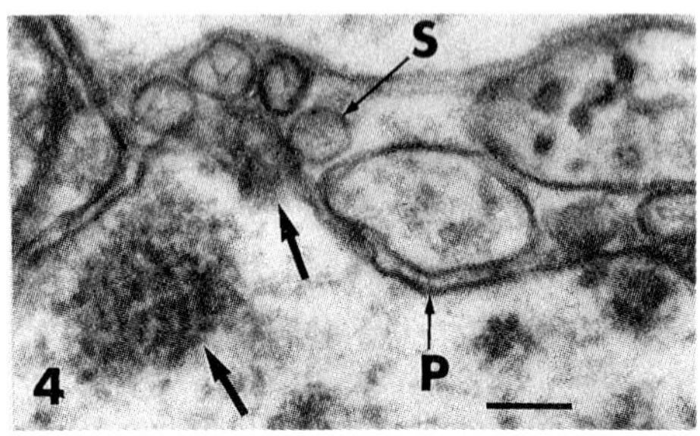

Figs. 3 and 4. Neuronal processes showing budding of virions (V) and spherules (S) from the plasma membrane (P). Nucleocapsids (N) and VCA (arrow) seen underneath. Baby mouse brain. L10, PID 2.

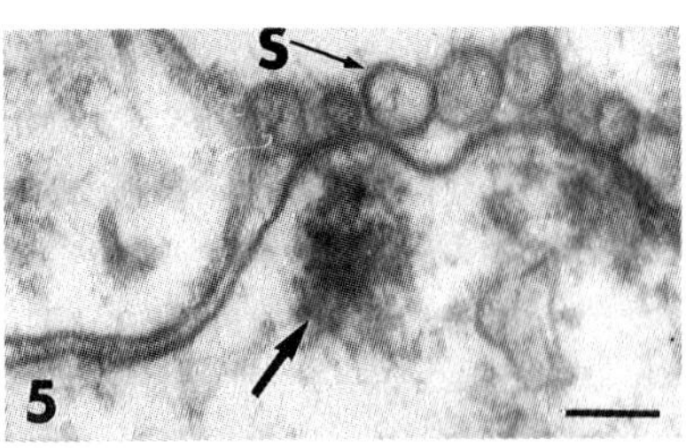

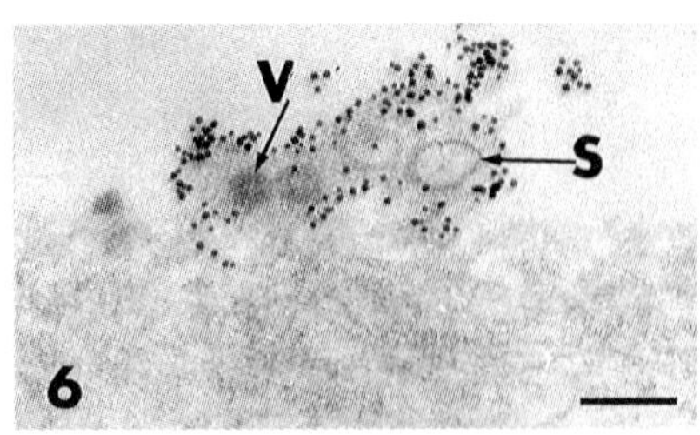

Fig. 5. Another neuronal process showing VCA (arrow) and spherules (S). Baby mouse brain. L10, PID 2.

Fig. 6. Virions (V) and spherules (S) labelled with gold (G5) conjugated antibody against E1/E2. Mouse brain cell culture (Fig. 6 with kind permission of J. Neurol. Sci.).

M = mitochondrion, Nu = Nuclear membrane. Bar = 0.1 µm.

Ultrastructural alterations of small intestine in cases of alpha chain disease

N EROL, R SEZER, T ERBENGİ, Z MUNGAN, G BOZTAŞ

Department of Gastroenterology Internal Medicine and Department of Histology and Embryology, İstanbul Faculty of Mediine Istanbul TURKEY

ABSTRACT: 18 cases of alpha chain disease were studied clinically and ultrastructurally in the small intestinal biopsy specimens. Surface epithelium were found usually normal, but in some cases mildly or patchily or severely altered. Alterations were shortness, irregularity and loss of microvilli, increased lysosomes and lymphocyt infiltration in the epithelium and dense lymphoplasmocytic infiltrations in the lamina propria. Especially diffuse, mature looking plasma cell infiltrations and in some cases of alpha chain disease atypical malign plasma cells were found significantly.

Alpha chain disease (α-CD) is one of the most common form of the primary small intestinal lymphoma presented with severe malabsorption. Since the first demonstration of α-CD (Rambaud JC 1968) more than 170 patients have been reported (Doe WF 1972, Rambaud JC 1976 and Asselah CH 1982) but very few ultrastructural studies were carried out. In the prognosis of α-CD, early diagnosis is very important because remission may be achieved following antibiotic therapy. During the last 15 years we studied 18 of α-CD showing severe malabsorption and multipl per oral small intestinal biopsies were carried out using standard methods. Diagnosis was verified by demonstration of pathological α-CD protein on immune electrophoresis with monospesific antisera to IgA, both in plasma and small intestinal secretions. Light microscopy (LM) and electronmicroscopical (EM) studies of small intestinal biopsy specimens were made according to standard methods. All of the patients were young adults with equal sex distribution. Small intestinal alterations under LM, typical for α-CD were partial or subtotal villous atrophy and dense mononuclear cell infiltrations in lamina propria. EM revealed intact surface epithelium (Figure 1) and no severe nucleocytoplasmic alterations in most of the early cases. Only minor changes such as shortening, irregularity and loss of microvilli (←) were noted. Mildly or patchily moderate (Figure 2) or more severly altered surface epithelium were found from place to place in advanced cases (Figure 3). Increased lysosome (⇞)· and lymphocytic infiltration of surface epithelium were also prominent (Figure 4). Infiltration of Plasmocytic cells (PC) (⇞) especially mature looking and actively secreting PC and in some cases distrophic (Figure 5) or malign PC (Figure 6) with pale and larged nucleus and distended cysternae of rough endoplasmic reticulum were noticed. A good clinical improvement and long lasting remission have been achieved following large spectrum antibiotic therapy but no morphological responce were noted.

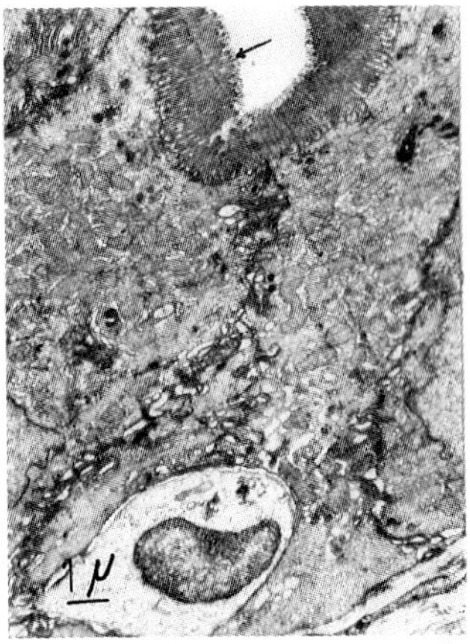

Fig.1. Small intestinal epithelium with intact microvilli

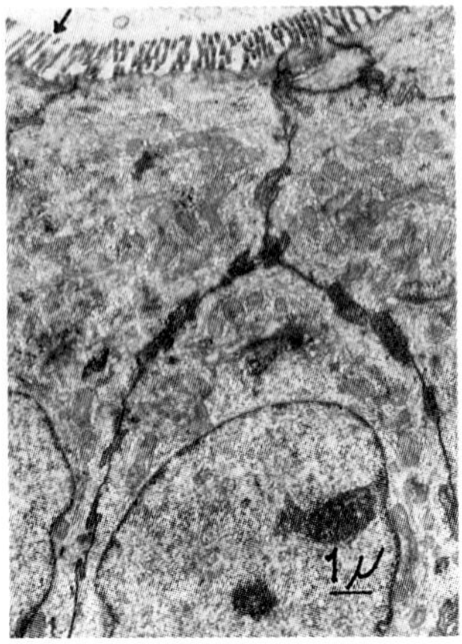

Fig.2. Moderate alteration of microvilli

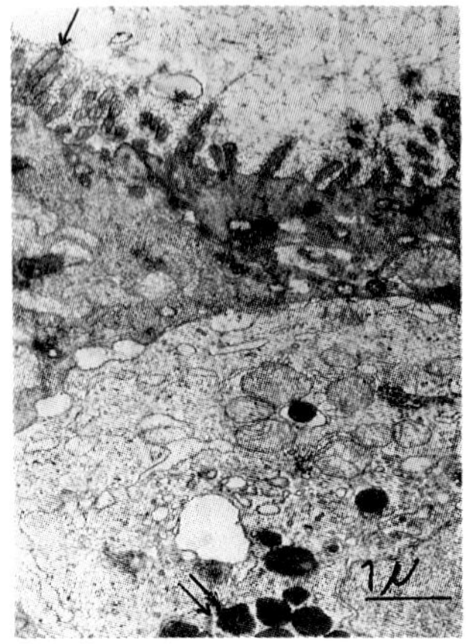

Fig.3. Irregularities and loss of microvilli and increased lysosome

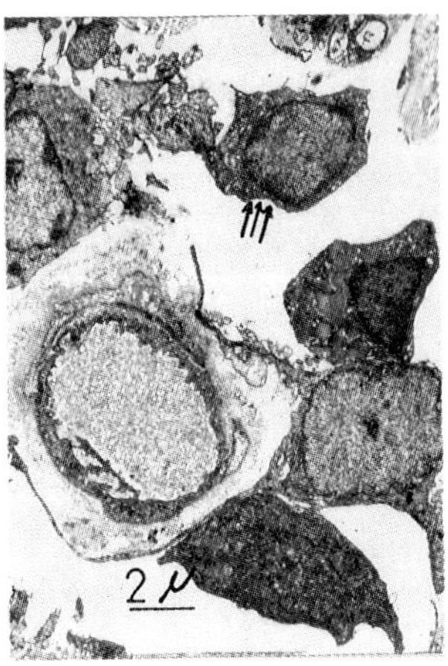

Fig.4. Cellular infiltrations in lamina propria

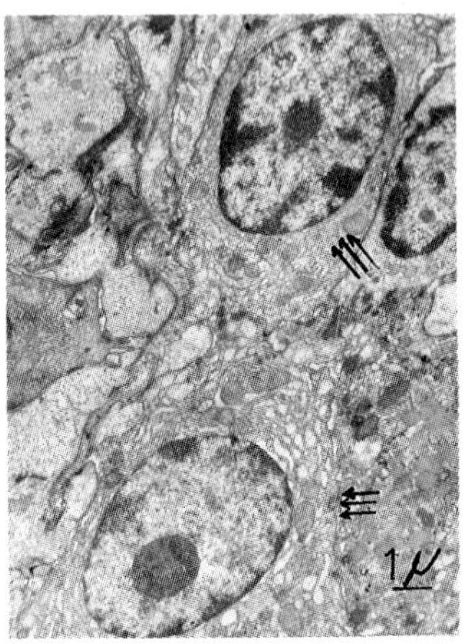

Fig.5. Plasma cell infiltrations in lamina propria

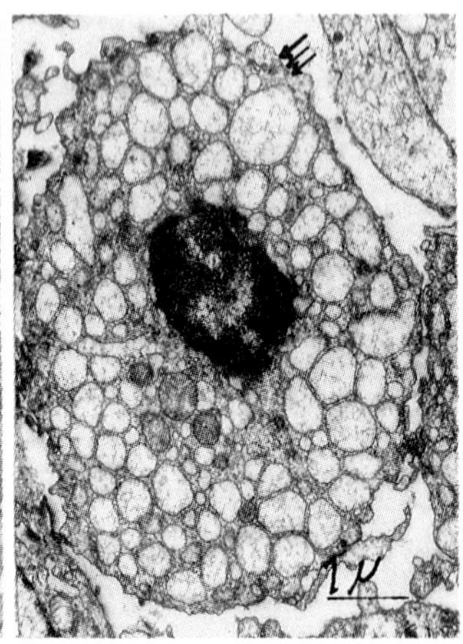

Fig.6. Increased plasma cells showing malignity

References:
- Rambaud J C, Bognel C, Prost A, Bernier J.J, Lequintrec Y, Lambling A, Danon F and seligmann M. 1968. Digestion 1, 321.
- Rambaud J C, Seligmann M. 1976. Clini in Gastroenterelogy 5, 341.
- Asselah C H and Asselah F. 1982 Butterworths. International Medical Review. Gastroenterology. 2, 174.
- Doe W F, Henry H, Hobbs J R, Avery Jones E, Dent C E and Booth C C. 1972. Gut. 13, 945.

Fingerprint and fingerprint like deposits in patients with *Lupus glomerulonephritis*

Hvala A, D Ferluga

Institute of Pathology, Faculty of Medicine, 61105 Ljubljana, Yugoslavia

ABSTRACT: In 102 kidney biopsies performed in 96 patients with lupus glomerulonephritis (L-GN) fingerprint (FP) structures were present in 9 (8,8 %) patients. Most frequently they were found in glomerular transmembranous and mesangial deposits and only in some cases in subepithelial, subendothelial and tubular ones. In 2 cases they persisted in repeated biopsies after 2 or 6 years.

Renal immune deposits in patients with L-GN are electron microscopically homogenous but sometimes they are organized. FP structures found in 6 to 10 % patients with L-GN have been so far according to the literature only rarely studied in details (Grishman 1967, Alpers 1984).

MATERIAL, METHODS, RESULTS: 102 kidney biopsies of 96 patients with L-GN were examined. FP formations have been found in 11 biopsies of 9 patients: in 10 cases of diffuse mixed proliferative and membranous GN with mesangial-transmembranous deposits, and in 1 case of mesangial GN with mesangial deposits only. In 9 biopsies FP formations were found within and on both sides of glomerular basal membrane (fig.1); in 3 cases they were found simultaneously within mesangium (fig.2) and in one case only FP were found simultaneously in the tubular basal membrane. In one case glomerular FP were found only subendothelially while in another one only subepithelially. In two female patients the FP structures were expressed in repeated biopsies after 2 or 6 years. The average diameter of dark lines in FP structures was 12 nm (10 - 15 nm). On both sides toward the light lines projections can be noted (fig.1,2,3). Individual dark lines show a tubular organization (fig. 2). In one patient FP like structures were discovered in the adventitia of a small artery. They showed a periodical crystalline structure (fig.4).

DISCUSSION: In our control group of 465 kidney biopsies with other forms of GN FP structures have not been found. Their finding seems to be rather specific and suggests the diagnosis of L-GN although they were also described in few cases of some other forms of GN (Grishman et al 1967, Alpers et al 1984). In respect to their morphological characteristics - though not to morphometric ones - they remined us on deposits of cryoglobulins in mixed cryoglobulinemy IgG-IgM. Morphologically similar microtubules without adequate FP structure were, apart from this, found in individual cases of diabetic glomerulosclerosis, in one case of glomerular fibrillosis and in even some cases of normal kidneys. Their origin and role are stil unclear.

Grishman E, Porush JC, Rosen SM, Churg J 1967 Lab.Invest. 16 717-25
Alpers CE, Hopper J, Bernstein MJ, Biava CG 1984 Ann.Intern.Med. 100 66-68

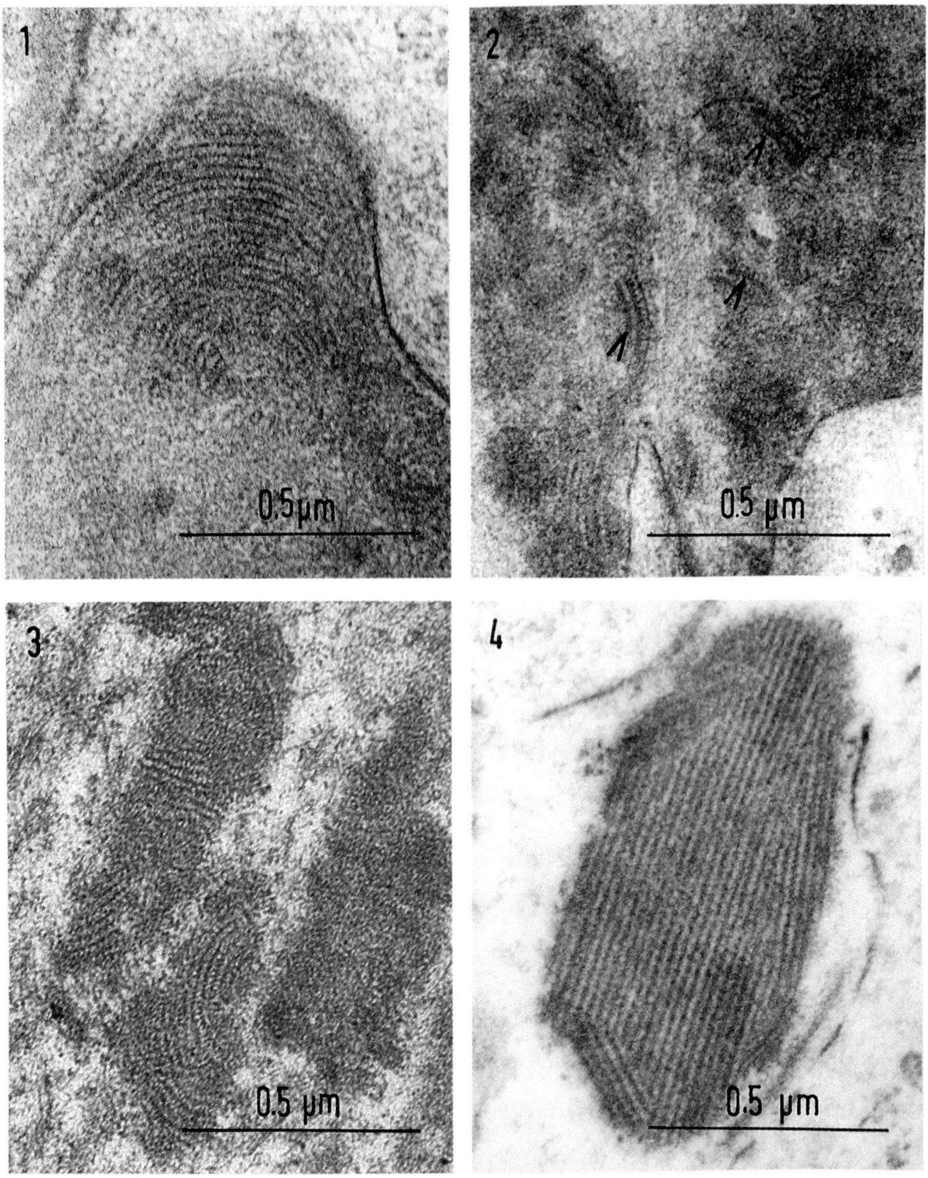

Fig. 1. Fingerprint deposit within glomerular basement membrane. OsO_4 Epon

Fig. 2. Fingerprint deposits within mesangial matrix. Tubular structure of dark bands (arrows). OsO_4, Epon

Fig. 3. Fingerprint deposits within tubular basement membrane. OsO_4, Epon

Fig. 4. Crystalloid deposit in the adventitia of small artery. OsO_4, Epon

Histopathological effects of cadmium on the somatic tissues of the ovary of *Tilapia nilotica*

A.Herrera

Institute of Biology,College of Science,University of the Philippines,Diliman,Quezon City,Philippines

ABSTRACT:Four-day old fry exposed to 0.5 mg.per liter cadmium chloride for eight weeks had damaged somatic tissues of the ovary.

I. INTRODUCTION

Cadmium is an environmental pollutant affecting the biological systems of many organisms.This is the first report on the effect of cadmium on the somatic tissues of the ovary of *Tilapia nilotica* in tropical waters.

Four-day old fry were obtained from the Institute of Fisheries Development and Research of the University of the Philippines.They were distributed to 40-liter tanks at 30 fry per aquarium.Total water replacement was done every 24 hours for the first week then every 48 hours until the termination of the experiment.Fish were fed daily to satiation with rice bran and ground fish.Continuous aeration was applied to the water.

For light and electron microscopy,the ovaries were dissected free from other tissues.They were cut into small cubes ($1mm^3$) then transferred to vials and fixed for two hours in two changes of fresh glutaraldehyde.They were then washed in buffer,postfixed in osmium tetroxide,washed again in buffer and dehydrated in acetone.Infiltration was done in propylene oxide and araldite resin.The tissues were embedded in pure resin and polymerized in 60°C for four to five days.Ultrathin sections were mounted on uncoated grids and stained with uranyl acetate and lead citrate.Observation was done on a JEOL U-100 electron microscope at the Natural Sciences Research Institute of the University of the Philippines.

Ultrastructural changes observed in the somatic tissues of

the ovaries are shown in the figures.

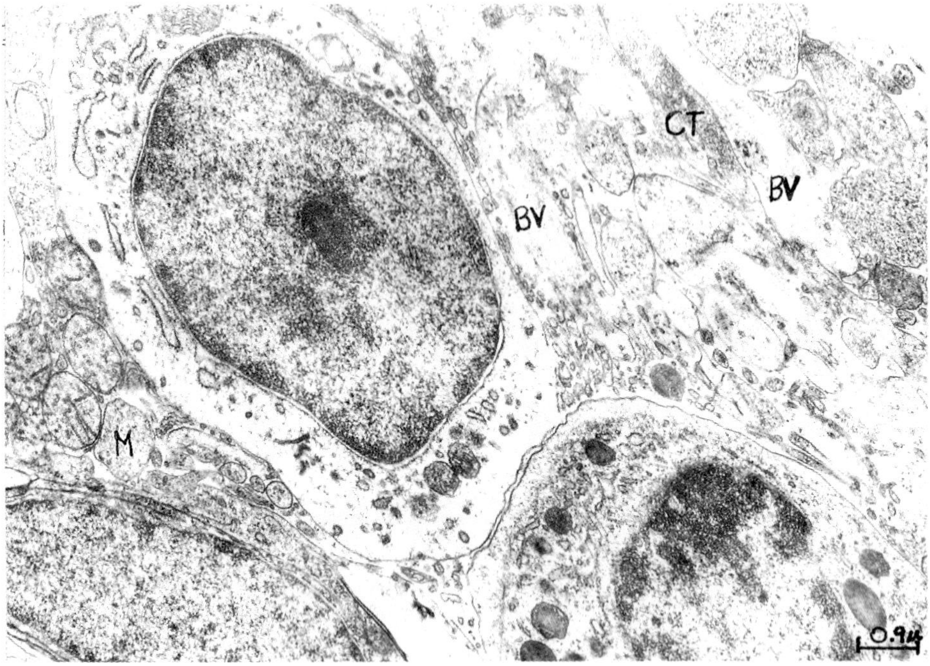

Figure I shows necrosis of the blood vessels,dilation of the mitochondria of granulosa cells and degeneration of stromal connective tissues(CT).

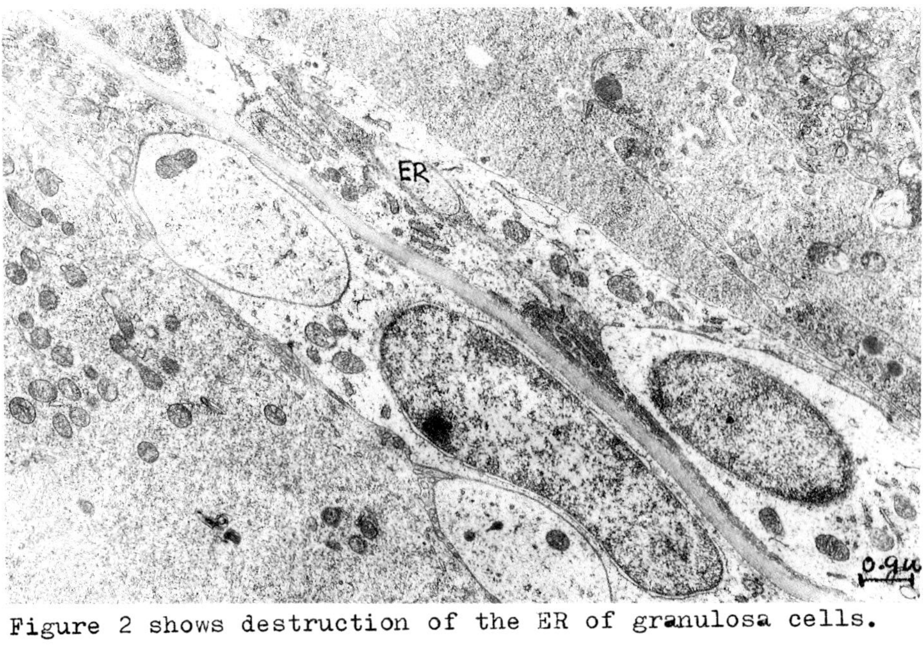

Figure 2 shows destruction of the ER of granulosa cells.

Ultrastructural changes in the chick hepatocyte as a response to ethylic intoxication during the postnatal period

Abaurrea, M.A.; Rios, A.; Ostos, M.V. and Torres, M.I.

Dptm. of Cell Biology. Faculty of Sciences. University of Granada. SPAIN.

Introduction

The effects of ethanol consumption have become the subject of increasing interest in recent years as it has become more and more clear that ethylic intoxication is a frequent cause of disease in industrialized communities. The development of a valid experimental model for studying ethanol-induced organic alterations is a key factor in this research.

Chicks have shown themselves to be useful models in this respect for observing the effects of ethanol in the embryonic and perinatal periods of development, especially as there is no postnatal maternal influence to interfere with an objective analysis of the effects produced in the organism by this drug.

At the organic level the liver, as the main site for the metabolism of ethanol, is the principal subject of research when looking for the primary changes brought about by ethanol consumption (Lelbach 1967; Christoffersen et al. 1971; Rubin and Lieber 1972).

In this work we have analysed the effect of long-term ethanol administration on the cellular structure of the hepatocyte during the postnatal period in chick.

Material and Methods

The experient was made by keeping different groups of newborn chicks for a month on diets containing various concentrations of ethanol. The experimental groups were supplied with 4%, 8%, 12% and 25% concentrations of ethanol in their drinking water and a solid, standard chicken diet. Control animals were fed with a similar solid diet and an isocaloric solution of sacharose as the different experimental groups.

The chicks were killed by decapitation after a month and samples of their livers were processed for observation under an electron microscope The samples were fixed in 4% glutaraldehyde, and 2% osmium tetroxide both tamponed in cacodylate (pH=6,8). Inclusions were made in Spurr resin and ultrathin sections were observed, after contrast in uranyl acetate and lead citrate, in a Zeiss EM 10C electron microscope.

Results and Commentary

The first aim of this work was to determine the minimal dose of ethanol that is able to produce observable ultrastructural alterations in the hepatocyte.

The hepatocytes of the chicks supplied with 4%, 8% or 12% of ethanol showed ultrastructural features similar to those of the control groups. The characteristic pattern of cellular structure of hepatocyte was present in the treated animals. Their liver cells offer the presence of one or two spherical nuclei and regular distribution of cellular organelles. Frequently, rough endoplasmic reticulum cisternae are surronding mitochondria poor in crests. Lysosomes with a variable content were present, as well as multivesicular bodies in which small vesicules are included in a membrane-limited dense matrix. Small lipid droplets were usually present in the hepatocytes of these groups of animals.

When the diet was supplemented with 25% of ethanol, the hepatocytes of the chicks showed marked differences, easily observable on electron microscopic images, compared to the controls. Such cytoplasmic structures as mitochondria, endoplasmic reticulum, peroxysomes and lysosomes were altered to various degrees. Mallory bodies were also present with this treatment.

A great volume of the cytoplasmic area was occupied by lipid vacuoles of different sizes, thus producing displacement of the nucleus and other cytoplasmic organelles. This feature coincided with descriptions reported by other authors in mammalian experimental models (Romert and Matthiessen 1984).

The mitochondria showed a rounded shape and their crests were decreased or even absent. A great number of osmiophilic granules were present in the mitochondrial matrix. These result agree with those of other authors describing mitochondrial alterations (Schaffner et al. 1963; Klion and Schaffner 1968; Romert and Matthiessen 1984).

Megamitochondria as described in literature (Bruguera et al. 1977), with a size up to 10-12 microns, were not to be found in our experiments.

The functional significance of mitochondrial modification is very difficult to evaluate but it may be due to alterations in mitochondrial membrane synthesis.

The lipid droplets lack limiting membrane and so they can join up to form greater drops. Multivesicular bodies are prominent and similar to those occuring in the hepatocytes of animals fed with smaller concentrations of ethanol.

Mallory bodies, widely described in the literature in association with hepatic alterations induced by ethanol (French 1981, 1983), are usually present in our observations. They appear as paracrystalline structures composed of parallely asociated filaments frequently related to small lipid droplets.

Literature

- Rubin, E., Lieber, C.S. 1972. Int. Rev. Exp. Pathol. 11,117.
- Lelbach, W.K. 1967. Acta Hep., 14,9.
- Christoffersen, P. et al. 1971. Acta Pathol. Micr. Scand. 79,150.
- Romert, P. and Matthiessen, M.E. 1984. Acta anat. 120,190.
- Schaffner, F. et al. 1963. J. Amer. Med. Assoc. 183,343.
- Klion, F.M. and Schaffner, F. 1968. Digestion 1,2.
- Romert, P. and Matthiessen, M.E. 1983. Virchows Arch. 399,299.
- Bruguera, M. et al. 1977. Gastroenterology 73,1383.
- French, S.W. 1981. Hepatology 1,76.
- French, S.W. 1983. Arch. Pathol. Lab. Med. 107,445.

Inst. Phys. Conf. Ser. No. 93: Volume 3, Chapter 9
Paper presented at EUREM 88, York, England, 1988

Effects of leucine, histidine and 3-methyladenine on vinblastine-induced autophagic vacuole accumulation in Ehrlich ascites cells

EL Punnonen and H Reunanen

Department of Cell Biology, University of Jyväskylä, Vapaudenkatu 4, SF-40100 Jyväskylä, Finland

ABSTRACT: The effects of leucine and histidine, and 3-methyladenine on vinblastine-induced autophagic vacuole accumulation in Ehrlich ascites cells was studied using electron microscopic morphometry. Vinblastine increased the cytoplasmic volume fraction of autophagic vacuoles significantly in 30 min, and the volume increased slowly for 120 min. 3-Methyladenine prevented this effect almost completely. In amino acid-treated cells the autophagic vacuole accumulation began after a 60-min incubation, and after 120 min the volume fraction of autophagic vacuoles was almost the same as in solely vinblastine-treated cells.

1. INTRODUCTION

The microtubule inhibitor vinblastine (VBL) increases the number of autophagic vacuoles (AV's) in many cell types and tissues. Several mechanisms have been proposed for this effect: 1) The disappearance of intact microtubules inhibits the fusion of autophagosomes and lysosomes leading to the accumulation of the former. 2) VBL inhibits the acidification of autolysosomes, which retards degradation. 3) VBL stimulates the formation of new autophagosomes.

VBL increases the volume fraction of AV's in Ehrlich ascites tumour (EAT) cells (Hirsimäki et al 1984). In these cells the effect of VBL does not seem to be mediated by retarded fusion between AV's and lysosomes. Firstly, there are very few dense lysosomes in EAT cells (unpublished observation). Secondly, VBL accumulates mostly acid phosphatase-positive AV's (Reunanen, unpublished), and thirdly, other microtubule inhibitors do not induce AV accumulation (Reunanen et al 1988). This study was done to further clarify the action of VBL in EAT cells.

Amino acids (Kovács et al 1981) and 3-methyladenine (3MA) (Seglen and Gordon 1982) are specific inhibitors of deprivation-induced autophagosome formation in hepatocytes. Amino acids also inhibit VBL-induced AV accumulation in hepatocytes (Kovács et al 1982). In this study electron microscopic morphometry was used to investigate whether leucine and histidine, the most effective sequestration-inhibitory amino acids in hepatocytes (Seglen and Gordon 1984), and 3-methyladenine prevent the VBL-induced AV accumulation in EAT cells.

2. METHODS

EAT cells were grown in the peritoneal cavities of NMRI mice and suspended in modified Krebs-Ringer phosphate buffer pH 7.4 containing glucose (1 mg/ml). The cells were pre-incubated with leucine (10 mM) and histidine (10 mM), or with 3-methyladenine (20 mM), or without additions for 30 min. The incubation was started by adding VBL (0.1 mM) or physiological saline. Samples for electron microscopy were taken after 1, 30, 60 and 120 min. The morphometric analysis was made by point counting. A total of 48 micrographs from four different experiments were analyzed for each test point.

3. RESULTS

Figure 1. The cytoplasmic volume fraction (mm^3/cm^3) of AV's in different experimental groups. ○,Control; □,leucine+histidine; △,3MA; ◆,VBL; ■,leucine+histidine and VBL; ▲,3MA and VBL. Standard errors are indicated by the bars. a, Statistically significant difference as compared with solely VBL-treated cells ($p<0.05$).

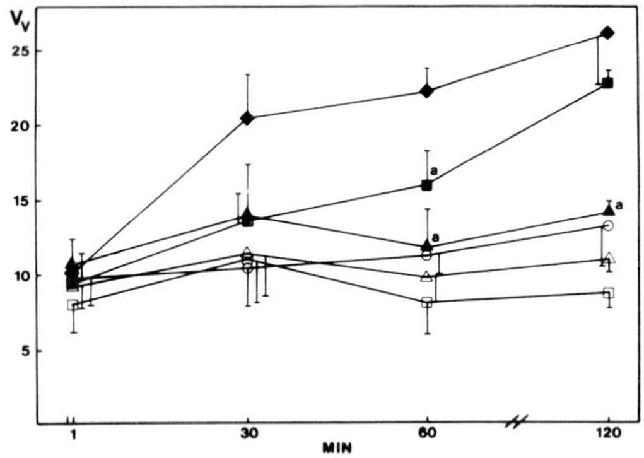

VBL accumulated electron-lucent AV's the contents of which were not recognizable. The amount of autophagosomes remained on the control level (approximately 2.4 percent of total AV volume).

4. DISCUSSION

VBL accumulated autolysosomes in EAT cells. The amount of autophagosomes remained on the control level. This result supports the view that in these cells the effect of VBL is not mediated by the inhibition of fusion between autophagosomes and lysosomes.

Leucine and histidine retarded the VBL-induced accumulation of AV's. The retardation was slighter than that observed in isolated hepatocytes (Kovács et al 1982). Leucine and histidine seem to retard but not prevent the formation of new autophagosomes. 3-Methyladenine prevented the VBL-induced AV accumulation almost completely.

In VBL-induced AV accumulation in EAT cells the autophagosomes obviously form by the same mechanisms as in amino acid deprivation-induced autophagocytosis in isolated hepatocytes.

5. REFERENCES

Hirsimäki P, Marttinen MT and Hirsimäki Y 1984 Stereology and Morphometry in Pathology ed Y Collan et al (Kuopio: Kuopio Univ. Press) pp 223-228

Kovács AL, Grinde B and Seglen PO 1981 Exp. Cell Res. **133** 431

Kovács AL, Reith A and Seglen PO 1982 Exp. Cell Res. **137** 191

Reunanen H, Marttinen M and Hirsimäki P 1988 Exp. Mol. Pathol. **48** 97

Seglen PO and Gordon PB 1982 Proc. Natl. Acad. Sci. USA **79** 1889

Seglen PO and Gordon PB 1984 J. Cell Biol. **99** 435

Monoclonal antibody Ki-M4R detects sinus lining cells and follicular dendritic cells in rat lymph nodes. An immunoelectron microscopic study

Hansmann ML, Wacker HH, Frahm SO, Radzun HJ, Parwaresch MR
Institute of Pathology, University of Kiel, FRG

INTRODUCTION

Reticulum cells are important constituents of lymphoid tissue. They have the function of antigen trapping and presentation to the lymphoid cells. The present immunoelectron microscopic investigation deals with a monoclonal antibody (mAb) reactive with follicular dendritic cells and so-called sinus lining cells in lymphoid tissue of rat.

MATERIALS AND METHODS

The mAb Ki-M4R was produced by immunization of Balb/c mice with peritoneal macrophages of CAP rat as described elsewhere [1]. Immunoreactivity was demonstrated light microscopically using an alkaline-phosphatase technique on frozen sections [1] and immunoelectron microscopically with a preembedding technique [2]. For conventional electron microscopy tissue was fixed in 4% glutaraldehyde and embedded in araldite.

RESULTS

Light microscopy:

The mAb Ki-M4R stained follicular dendritic cells of germinal centers, sinus lining cells, and high endothelial and postcapillary venules.

Electron microscopy:

The marginal and intermediate sinuses were lined by two different cell types. The first type, the so-called sinus lining cells, contained oval, sometimes slightly indented nuclei with coarse heterochromatin (Fig. 1). The cytoplasm exhibited only a few organelles such as mitochondria and endoplasmic reticulum. These cells were usually in close proximity to collagen fibers and resembled follicular dendritic cells although they lacked numerous processes. They were strongly positive with the mAb Ki-M4R in the cytoplasmic membrane (Fig. 2). The cytoplasm, organelles, nuclei and the adjacent collagen fibers were usually negative. The second cell type exhibited the features of macrophages, with round to slightly irregular nuclei and often many lysosomes in the cytoplasm (Fig. 1). Occasional phagolysosomes were found. This cell type was usually negative with the mAb Ki-M4R, as were the lymphocytes in the sinus. In the germinal centers, Ki-M4R reacted with follicular dendritic cells; the reaction product, like that in sinus lining cells, was localized on the surface membrane.

DISCUSSION

Sinuses are an important lymph node structure regulating the traffic of the lymphoid cells and antigenic particles. The sinus lining cells detected by the mAb Ki-M4R may play an important role in antigenic trapping and in transport of the antigen to the germinal centers [3]. Immunoelectron microscopy showed that Ki-M4R stains the sinus lining cells but not the macrophages and lymphocytes in the sinuses. Ultrastructurally,

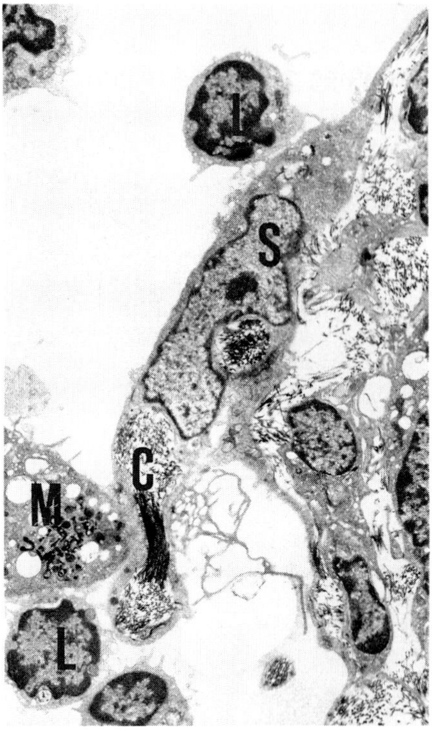

Fig 1: Sinus lining cell (S) in contact with a macrophage (M), lymphocyte (L), and collagen bundle (C). Rat lymph node. Electron micrograph x11,000.

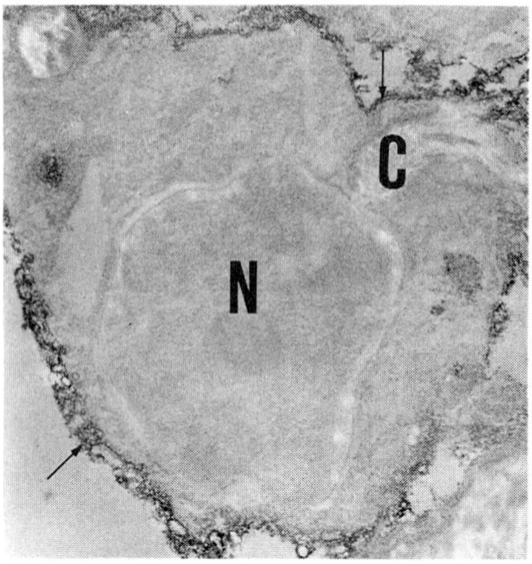

Fig 2: Sinus lining cell. Immunoreaction product on the cell membrane (↑). Nucleus (N), collagen bundle (C). Immunoelectron micrograph x11,000.

sinus lining cells and follicular dendritic cells are quite similar; the ultrastructural and immunoelectron microscopic data speak in favor of a common origin of both cell types. The mAb Ki-M4R allows clear detection of these cell types in rat lymphoid tissue and may be useful in studying antigen trapping under experimental conditions.

REFERENCES

1. Wacker HH, Radzun HJ, Mielke V, Parwaresch MR: Selective recognition of rat follicular dendritic cells (dendritic reticulum cells) by a new monoclonal antibody Ki-M4R in vitro and in vivo. J Leuk Biol 41:70-77, 1987.
2. Hansmann ML, Radzun HJ, Kaiserling E, Parwaresch MR: Immunoelectron microscopic demonstration of tissue antigens with monoclonal antibodies. Virchows Arch [Cell Pathol] 46:1-12, 1984.
3. Szakal AK, Holmes KL, Twe JG: Transport of immune complexes from the subcapsular sinus to lymph node follicles on the surface of nonphagocytic cells, including cells with dendritic morphology. J Immunol 131:1714-1727, 1983

Scanning electron microscopy in the backscattered electron imaging (BEI) mode of human natural killer cells

E. FERNANDEZ-SEGURA, J.M.GARCIA,M.C. SANCHEZ-QUEVEDO, A. CAMPOS

Dept Biologia Celular. Facultad de Medicina y Odontologia 18012 Granada. SPAIN

ABSTRACT: Human natural killer (NK) cells were isolated from PBL, morphologicaly characterized by SEM in SEI mode and cytochemicaly typified in BEI mode. The conjugation assays between NK and K-562 cells showed lysis phenomena in target cells.

INTRODUCTION

Natural killer (NK) cells are a subset of peripheral blood lymphocytes with peculiar morphological, phenotipic, and functional characteristics. NK cells display the structural features of large granular lymphocytes (LGL) (Timonen 1981).and the ability to lyse neoplasmic cells, virus-infected cells and haematopoyetic cell precursors without restriction by the expression of the Major Histocompatibility Antigens on target cells (Young 1987). SEM in SEI and BEI mode have recently been used to determine the nature of the various haematopoyetic series (Soligo 1981). The present paper develops these techniques as applied in the study and characterization of human NK cells.

MATERIALS AND METHODS

Human NK cells were isolated from PBL by fractionation in discontinuous Percoll density gradients (1.053-1.077 g/ml)(Timonen 1981). For cytochemical studies (acid phosphatase) the cells were incubated in modified Gomori medium and processed for SEM in SEI and BEI mode according to Soligo (1983). K-562 line cells were used to investigate cytotoxicity.

RESULTS

Human NK cells present low density (Figure 1) and SEM in the SEI mode

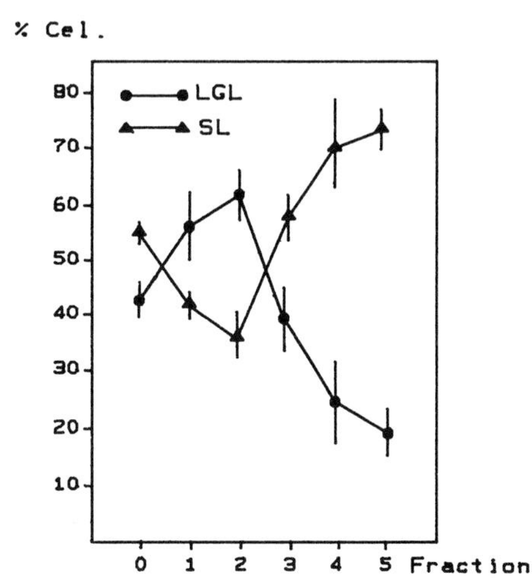

Figure 1.

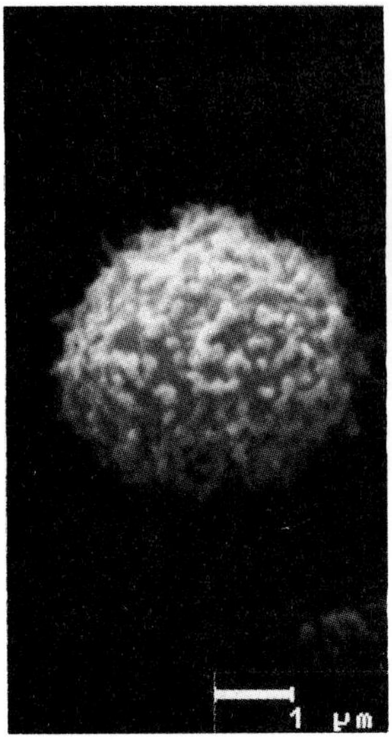

Fig 2

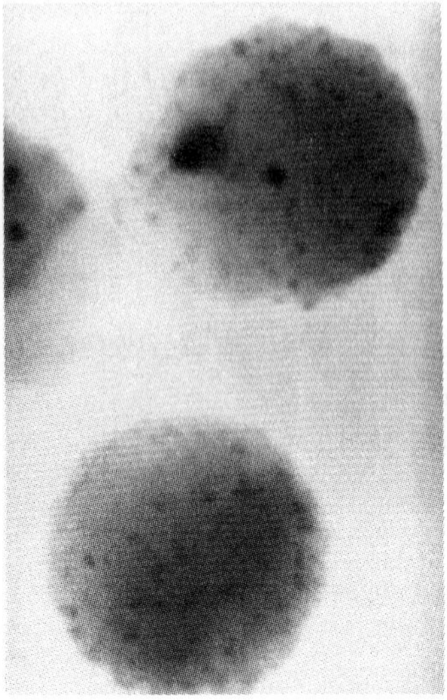

Fig 3

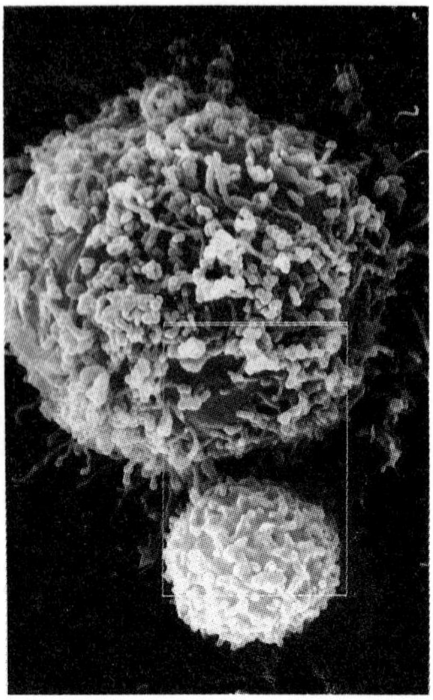

Fig 4

shows them to measure 4-4.5 um in diametre and posses microridges and microvillous surface features (Figure 2.). In BEI mode the cells are characterized by a posiyive granular acid phosphatase distribution (Figure 3.). Monocytes and small lymphocytes in SEI mode present ruffles and smooth surface features while BEI mode reveals positive disperse acid phosphatase reaction in the former and a focal reaction pattern in the latter. Conjugation assays showed cellular contacts between NK and K-562 cells (Figure 4.). Microblebs and holes illustrating cell lysis phenomena were observed as a consequence of such contact on the target cells.

DISCUSSION

The use of SEM in SEI and BEI mode allow investigators to define three-dimensional morphological features in accordance with cytochemical properties. These techniques applied to the study of purified pools of human NK cells isolated from PBLs (de Harven 1987) provides evidence of significant differences between these cells and both monocytes and small lymphocytes. The precise conjugation assay used in the present study confirm the cytolytic nature of human NK cells isolated and characterized by SEM in SEI and BEI mode.

REFERENCES

de Harven E 1987 Ultras Pathol 11 711.
Soligo D, Lampen N and de Harven E 1981 Scann Electron Microsc 1981/II 95.
Soligo D, Pozzoli E, Nava MT Polli N, Lanbertenghi-Deliliers G and de Harven E 1983 Scann Electron Microsc 1983/IV 1795.
Timonen T Ortaldo JR and Herberman RB 1981 J Exp Med 153 569.
Young JD and Cohn ZA 1986 Cell 46 641

Inst. Phys. Conf. Ser. No. 93: Volume 3, Chapter 9
Paper presented at EUREM 88, York, England, 1988

Lymphocyte-hepatocyte interaction in chronic hepatitis

Heinz David, Petra Reinke and Ingrid Uerlings
Humboldt University Berlin, School of Medicine (Charité),
Dept. of Ultrastructural Pathology and Electron Microscopy
Institute of Pathology and Clinic of Internal Medicine

Chronic hepatitis (chronic-persistent, chronic-aggressive hepatitis) is largely caused by hepatitis B virus, though some of its aetiology has not been completely elucidated (Non-A-non-B hepatitis).

In hepatitis B, for example, primary cytotoxicity is associated neither with the existence nor with replication of the virus in hepatocytes, taking place, for example, in a strongly pronounced manner as a consequence of immunosuppression in the wake of kidney transplantation.

Cells are damaged by cytotoxic lymphocytes which enter into the liver lobule to adversely affect, usually up to lytic necrosis, the hepatocytes which contain HBcAG or HBsAG.

Within the liver lobules, growing amounts of lymphocytes are recordable from the sinusoid and migrate through openings in endothelial cells of the sinusoidal wall into the Disse space which is getting enlarged. Some of the sinusoidal microvilli will be flattened.

Small and large lymphocytes (cytotoxic T-cells, killer cells, zero cells), mononuclear phagocytes, and plasma cells are recordable from in between hepatocytes in the Disse and intercellular spaces (Fig. 1,2).

Interaction between T8-lymphocytes (cytotoxic/suppressor T-cells), on the one hand, and hepatocytes, on the other, occurs via HBcAG expressed at their plasma membrane, and to some extent also via HBsAG. Parallel to cytotoxic effects, there are numerous lymphocyte-hepatocyte contact relations without any damage inflicted upon the hepatocytes.

The results obtained have shown that in chronic hepatitis B lymphocytes and mononuclear macrophages may have two possible consequences:

1. Cytotoxic effects, primarily along with the buildup of lytic necroses of hepatocytes;

2. No primary toxic action with compression of adjacent hepatocytes up to emperipolesis of lymphocytes (i.e. complete absorption of a lymphocyte by a hepatocyte) with the possibility of cytotoxic effects in the case of changing immunological situation.

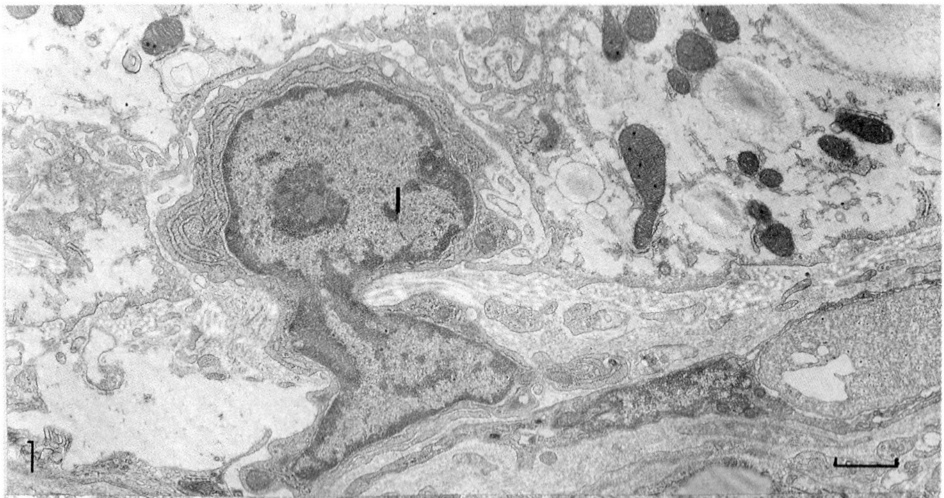

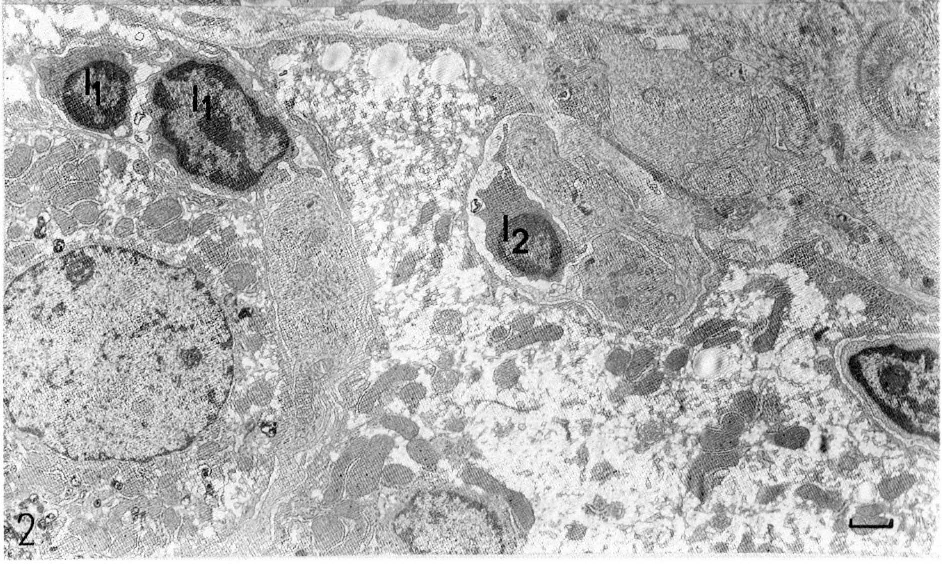

Fig. 1: Invasion of a large lymphocyte (l) through a gap of an endothelial cell into the Disse space.(x 10,000)

Fig. 2: Several lymphocytes (l_1) in vading into the intercellular space between hepatocytes (h) and into hepatocytes (l_2). x 6,000

Inst. Phys. Conf. Ser. No. 93: Volume 3, Chapter 9
Paper presented at EUREM 88, York, England, 1988

Alteration of junction of hepatocyte in chronic liver disease

Hiromoto Yasuda*, Suzuko Toida*, Masami Yamanaka**, Ichiro Takahashi*** and Makoto Kako****
Department of Pathology*, Internal Medicine**, Central Lab. of Electron-microscopy***, Teikyo University School of Medicine, 2-11-1, Kaga, Itabashi-ku, Tokyo, and Dept. of Int.medicine, Teikyo University Hospital Mizonokuchi****, 74, Mizonokuchi, Takatsu-ku, Kawasaki, Japan.

Abstract: Junctions of hepatocyte in human biopsy cases,including several chronic hapatitis, and liver cirrhosis were investigated with thin section electronmicroscopy and especially, with freeze-fracture replica method (FF). In chronic hapatitis with intra-hepatic cholestasis, bile canaliculi were mostly dilated with decrease or lack of microvilli. There were irregularly arrnaged junctional strands of more than four to five layers around the lumen, but, with not so prominent disruption. In contrast to the chronic hepatitis, junctional strands of liver cirrhosis were prominently decrease in number of less than two to three layers with occasional disruption of strands around the narrow lumen.

1.Materials and methodes: A number of the cases were biopsied mostly with needle biopsy and a few cases with cirrhosis were from autopsy materials of relatively fresh ones. Sections were fixed with 1% glutar-aldehyde for 2 hours and 1) post fixed with 1% OsO_4 solution for 2 hours,embedded with Epon 12 and ultra-thin sectioning with Reicherdt om-U3 and 2)the other ones for FF., were treated for 2 hours with 30% glycerol and then frozen for freeze-fracturing in liquid nitrogen. Replicas were made by usual method equipted with Eiko FD-2A type.

2.Results and comments: 1) In thin section electronmicroscopy, liver cells of chronic hepatitis with intra-hepatic cholestasis showed marked dilatation of bile canaliculi with prominent irregularities of cell junction to indicate decrease in number or lack of microvlli and filled with lamellar dense bile substances in their lumina (Fig.1). 2) On the other hand,there were irregularly arranged or elongated junctional strands of more than four to five layers with vertical ridges perpendicular to the pallarel ones, and not so prominent disruption or discontinuation of strands in the domain of bile canaliculi 3) In contrast to those observed in chronic hapatitis, the lumen of bile canaliculi in liver cirrhosis tends to be narrow with increase in number of short micro-villi, and junctional strands were prominently decrease in number of less than two or three layers with occasional disruption or discontinu-ation and sometimes, loss of verically arranged ridges.

There have been several reports to anlyse the junctional complex of hepatocytes in the experimental animals.Spelman et al. described that the strands number ranges from three to seven in normal liver, and also reported reversible alterations in hepatocyte gap and tight junctions occur as a results of administration of a diet including oval cell proliferations. But, little is known to analyse the hepatocyte junction with FF. on human biopsy case.

Our results are suggested that the irregularity or elongation of junctional strands are not always to indicate the chronicity of liver cell injury, but, decrease in number of the layer of strands or especially, disruption or discontinuation, associated with loss of vertical ridges are one of the important structural marker of real chronicity to represent the irreversibility of junction in liver cirrhosis.

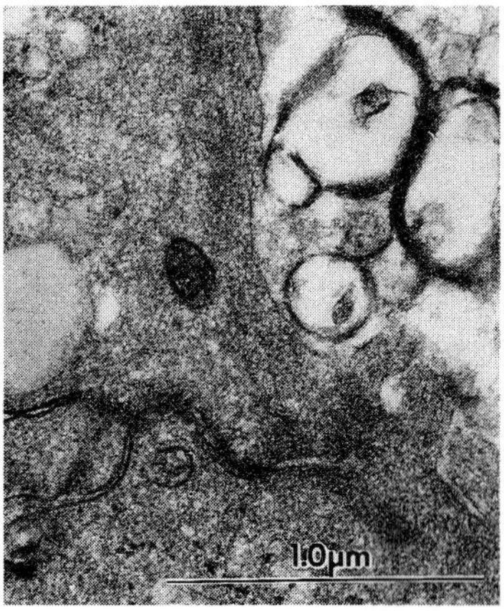

Fig.1. Chronic hepatitis, 46 year old male. Dilatation of bile canaliculi with irregular cell junction.

Fig.2. Chronic hepatitis, FF. image, same case as Fig.1. Irregularly arranged and elongated junctional strands.

Fig.3. Liver cirrhosis, 57 year, male. FF. Irregularly arranged and decrease in number of junctional strands with loss of vertical ridges

Reference

Spelman, L.H. et al.: Am.J. Pathol., 125:379-392, 1986

Inst. Phys. Conf. Ser. No. 93: Volume 3, Chapter 9
Paper presented at EUREM 88, York, England, 1988

Ultrastructural pathology of perisinusoidal functional unit of the liver after preservation and transplantation

Heinz David, Klaus Gellert, Ingrid Uerlings, and Matthias David
Humboldt University Berlin, School of Medicine (Charité), Dept. of Ultrastructural Pathology and Electron Microscopy Institute of Pathology and Clinic of Surgery

The outcome of liver transplantation has proved to depend on the following factors:

1. Interval between death of donor and organ collection;
2. Period and success of preservation;
3. Surgical technique;
4. Postoperative ischaemic processes;
5. Immunological processes.

Qualitative and quantitative ultrastructural studies so far have been primarily conducted into hepatocytes, whereas the authors of this paper have concentrated their investigations on changes to the perisinusoidal functional unit (endothelial cells, Kupffer cells, Ito cells, Disse space, sinusoidal zone of hepatocyte).

The sinusoidal plasma membrane and adjacent cytoplasm are primarily affected by consequences of preservation, perhaps also in combination with postmortal ischaemia. Results include swelling of microvilli, vesiculation with cell components (usually hyaloplasm, rarely cell organelles, such as endoplasmic reticulum, ribosomes, and mitochondria) first of all in the Disse space and to some extent accompanied by protrusion of endothelial cell components reaching into the sinusoidal lumen (Fig. 1,2). The blebs are getting detached in the process and come to lie in the Disse space from where they can escape into the sinusoid through gaps in the sinusoidal wall. Larger vesicles, some of them in conjunction with erythrocytes, may fill up the sinusoidal lumen and may thus cause stagnation of blood flow and ischaemia of hepatocytes.

Morphometric investigations have shown that one hour after reflow the diameter of sinusoidal wall and Disse space is enlarged from 2.33 µm to 2.83 µm. Wall width increases from 3.74 µm three days after transplantation to 6 µm after 76 days due to incorporation of fluid in the Disse space and as a result of an increase in collagen fibres and partial propagation of Ito cells.

Processes of erythrophagocytosis are recordable from Kupffer cells, especially in the wake of transplantation.

The authors' own findings as well as some scattered publications are likely to suggest that a major role is played in the pathogenesis of post-transplantation liver damage by the perisinusoidal functional unit and that this role may be of decisive importance to the positive or negative outcome of liver transplantation.

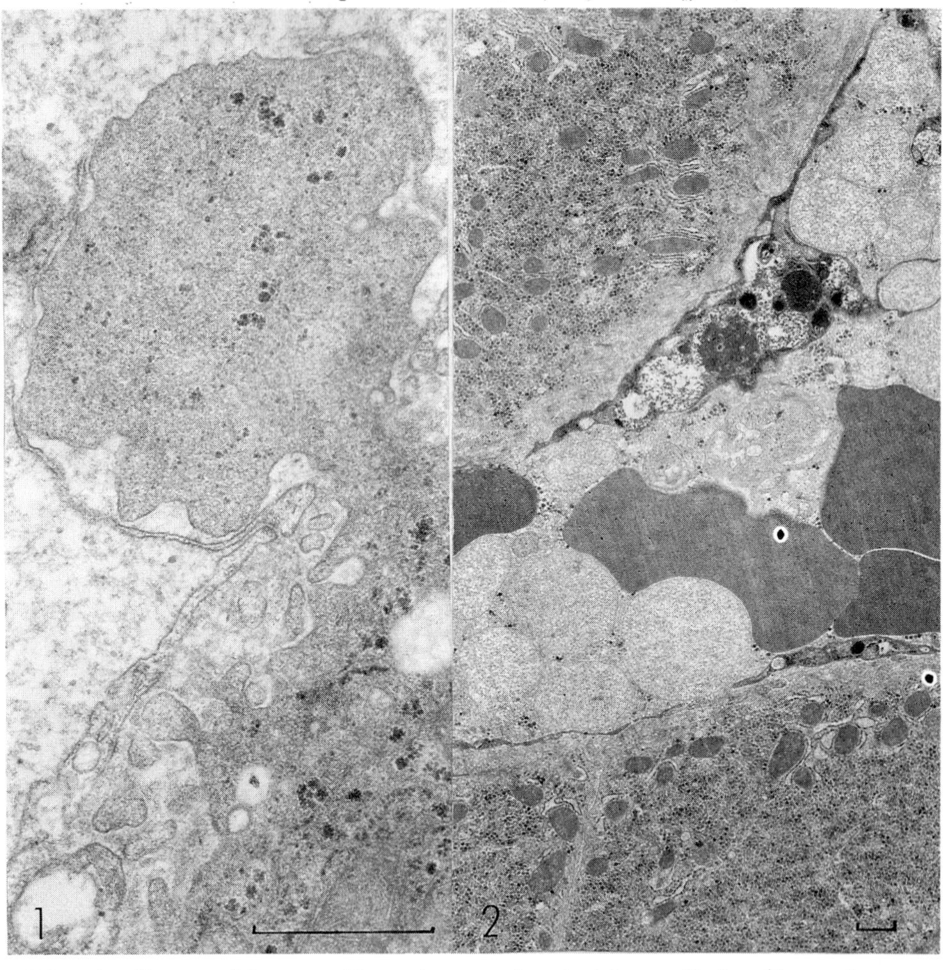

Fig.1: Human liver after transplantation. Into the sinusoidal lumen through a gap of an endothelial cell protruded bleb of a hepatocyte. x 30,000.

Fig.2: Human liver after preservation. Numerous blebs fill the sinusoid compressing the erythrocytes and occluding the lumen. x 6,000.

Inst. Phys. Conf. Ser. No. 93: Volume 3, Chapter 9
Paper presented at EUREM 88, York, England, 1988

The effects of tamoxifen citrate and toremifene citrate on the ultrastructure of rat liver

P Hirsimäki, Y Hirsimäki and L Nieminen
Farmos Group Ltd, Research Center, PO Box 425, SF-20101, Turku Finland

ABSTRACT: The effects of equimolar doses of the antiestrogens tamoxifen citrate and toremifene citrate on rat liver were studied in a 52 weeks toxicity study. Liver tumors were found at the highest dose level of tamoxifen citrate (45 mg/kg/day). Morphometric analysis revealed that the volume densities of peroxisomes, mitochondria and residual bodies were elevated in the hepatocytes of tamoxifen treated rats. In the tumors which appeared to be hepatocarcinomas the volume density of nuclei was slightly elevated. The proliferation of peroxisomes may be connected to tumor development.

1. INTRODUCTION

Tamoxifen citrate and toremifene citrate (manufacturer: Medipolar, Oulu, Finland) are non-steroidal antiestrogens developed against estrogen receptor positive tumors. We compared their effects on the ultrastructure of rat liver.

2. MATERIAL AND METHODS

Dosing was performed orally to female rats (n=5) daily for 52 weeks. Dose levels used were equimolar: toremifene 12 mg/kg and 48 mg/kg, tamoxifen 11.3 mg/kg and 45 mg/kg. Controls received corresponding volume of the suspension medium carboxymethyl cellulose (CMC). A three-month recovery period was included.

Morphometric analysis of hepatocytes was performed on the highest dose levels of toremifene and tamoxifen after 52 weeks dosing. Specimens were also taken from three tumors (tamoxifen 45 mg/kg). The livers were perfused via lower Vena cava using a mixture of 2.5% glutaraldehyde and 2% paraformaldehyde in 0.1 M phosphate buffer (pH 7.4) as fixative. Thin sections were stained on grid with uranyl acetate and lead citrate and examined with a JEM 100 CX (Jeol) electron microscope. Morphometric analysis was performed with point counting using 5000 x primary magnification and 1:9/ 121:1089 double-square lattice (Weibel and Bolender 1973, Rohr et al 1976).

3. RESULTS

After 52 weeks of dosing liver tumors were found (diameter 0.2 mm- 1 cm) in four out of five animals at the highest dose level of tamoxifen citrate. After the 3-month recovery period the tumors were even larger.

In morphometric analysis it appeared that the volume density of peroxisomes, mitochondria and residual bodies was significantly elevated in the hepatocytes of tamoxifen treated rats. In the tumors the volume density of the nuclei was slightly elevated. In microscopical evaluation the tumors appeared to be hepatocarcinomas. In toremifene treated rats only the volume density of mitochondria was slightly elevated.

4. DISCUSSION

It is known that several drugs which lower blood cholesterol level in man increase the number of liver peroxisomes in rodents (Reddy et al 1982, Reddy and Lalwani 1983). They are considered non-mutagenic and non-genotoxic hepatocarcinogenic agents acting probably by means of oxidative stress. Tumors develop only in organs in which peroxisome proliferation occurs, i.e. mainly in liver (Reddy and Rao, 1987).

Both tamoxifen and toremifene besides of their antiestrogenic effect might be "hypolipidemic drugs" in rats because they both lowered serum cholesterol level significantly in a 24-week toxicity study in male and female rats at dose levels of 48 mg/kg (Nieminen, 1983). However, in the present study only tamoxifen induced significant proliferation of peroxisomes. It can be concluded that the slight proliferation of peroxisomes may be connected with the tumor development in tamoxifen treated rats.

5. REFERENCES

Nieminen L 1983 Study report. Farmos Group Ltd, Finland
Reddy J K and Lalwani N D 1983 CRC Crit. Rev. Toxicol. 12 1
Reddy J K and Rao M S 1987 Arch. Toxicol. Suppl. 10 43
Reddy J K Warren JR Reddy M K and Lalwani N D 1982 Ann. NY Acad. Sci. 386 81
Rohr H Oberholzer M Bartsch G and Keller M (1976) Int. Rev. Exp. Pathol. 15 eds G W Richter and M A Epstein (London: Academic Press) pp 233-325
Weibel E R and Bolender R P 1973 Principles and techniques for electron microscopic morphometry 3 ed M. A. Hayat (New York: Van Nostrand Reinhold Co) pp 239-296

Ultrastructural study of intravascular growth of disseminated B16F10 melanoma cells in the liver

E Hilario and SF Aliño .

Department of Cell Biology and Morphologic Sciences, Fac. of Medicine and Dentistry, University of the Basque Country. Leioa. 48940 Vizcaya. SPAIN.

ABSTRACT: B16F10 melanoma tumour cells were intravasculary inoculated into the liver of C57/BL6 mice for arrest and growth study. Animals were sacrified at intervals ranging from 1 minute to 10 days. Tumour cells were arrested preferently in the periportal regions of the liver acinus, showing an intravascular proliferation pattern. Metastatic foci appeared as well-defined nodules flattening the surrounding liver tissue, suggesting an expansive tumour growth.

1. INTRODUCTION

One of the properties of malignant tumour cells is their ability to spread and form secundary tumours elsewhere in the body. The initial steps of the metastatic process involve the arrest and retention of tumour cells in the target organ, and their ability to develope clonogenic foci (Fidler 1978, Poste and Fidler 1980).

2. MATERIALS AND METHODS

Syngenic C57BL/6 mice aged 9-10 weeks were inoculated, via a mesenteric vein or intrasplenic, with a suspension of B16F10 melanoma cells ($4x10^6$ or $5x10^5$, respectively). Cell viability test was assessed by Trypan blue exclusion stain (greater than 95%) and only single-cell suspensions were used. Animals were sacrified, by cervical dislocation, at intervals ranging from 1 minute to 10 days, after tumour cells inoculation. The livers were perfused with 1.5% glutaraldehyde solution and tissue blocks, less than 1 mm^3, were processed for transmision electron microscopy by convencional methods. The semithin section were observed with a Olympus CH2 microscope. Ultrathin sections, stained with uranyl acetate and lead citrate, were observed in a Philips EM 300.

3. RESULTS AND DISCUSSION

In the first moment after cell inoculation, free cells were placed preferently in periportal or subcapsular regions of the liver. The arrested tumour cells were confined to the vascular spaces, and showed direct relation with endothelial cells or hepatocytes. Migration or diapedesis of the tumour cells was not observed in any of the time periods studied. On the other hand, intravascular tumour cell proliferation was associated with hepatocytic deformation in the tumour edges. These results are in agreement with the observations of Crissman

et al. 1985 in the lung, but in contrast, we did not observe tumour cell-endothelial plasma membrane contact.

The edge of the metastatic foci was well-delimited and the surrounding liver structures showed a flatted imagen, suggesting an expansive tumour growth (Figure 1).

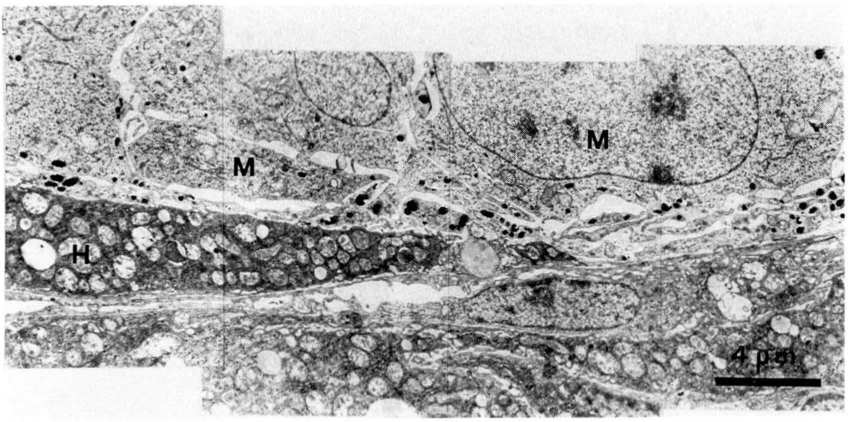

Fig. 1. The margin of the metastatic focus in well-delimited, and compresses the adjacent liver tissue. H: hepatocyte; M: melanoma cell.

In summary, these results suggest that in the liver, the B16F10 melanoma cells are arrested and lodged, preferently, in the sinusoidal lumen of periportal regions where they proliferate and grew by an expansive process, and in any case active extravasation by migration or diapedesis of tumour cells was observed.

4. ACKNOWLEDGMENTS

This work was supported by grants UPV 327.09-33/86 and UPV 075.327-30/87 from the University of the Basque Country.

5. REFERENCES

Crissman J D, Hatfield J, Schaldenbrand M, Sloane B F and Honn K V 1985 Lab. Invest. **53** 470
Fidler I J, Gersten D M and Hart I R 1978 Adv. Cancer Res. **28** 149
Poste G and Fidler I J 1980 Nature **283** 139

Spindle cell carcinoma of oesophagus simulating carcinosarcoma

S.Gazilerli and N.E.Aydın

Department of Histology and Department of Pathology, Medical Faculty, Atatürk University, Erzurum, TURKEY

An oesophagus cancer with biphasic light microscopic appearance of poorly differentiated squamous cell carcinoma and sarcomatous spindle-cell component was detected in a 75 year old man (Figure 1).

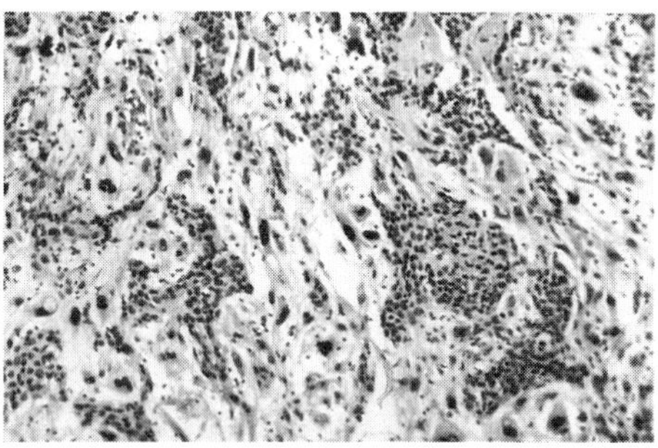

Figure 1.

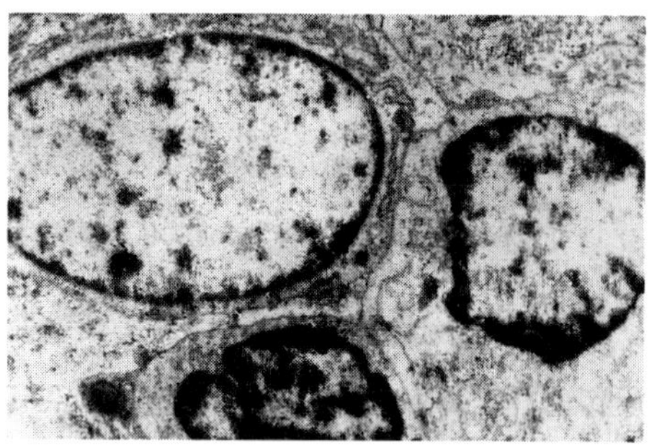

Figure 2.

Electron microscopy revealed epithelial cell groups (Figure 2) and spindle cells around these (Figure 3). Epithelial cells were heterogeneous in respect to their size, nucleus structure and organels. Cells possessing tonofilaments (Battifora 1976) could be discerned among them (Figure 4).

Intercellular spaces widened as the cells gained an irregular and elongated appearance at the periphery of the epithelial cell groups (Figure 5). The spindle cells nearby also showed variable nuclear and cytoplasmic features. Nuclei were markedly interdigitated in all. However, two kinds of cells could be discerned with different nuclear and cytoplasmic density. Big vacuoler structures were seen predominantly in those with dense nucleus and cytoplasm (Figure 4,5).

The ultrastructure of these spindle cells may point to a mesenchymal origin (Lattes 1982). It is also possible that these take origin from epithelial cells (Battifora 1976, Matsusaka et al. 1976, Ming 1973) as seen in similar tumours originating in aerodigestive tract (Zarbe et al.

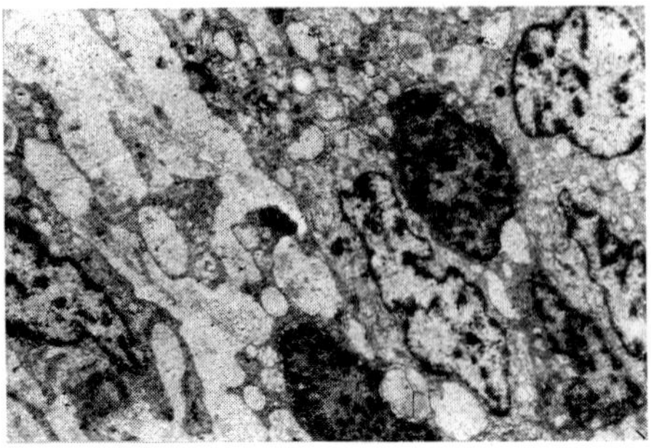

Figure 3.

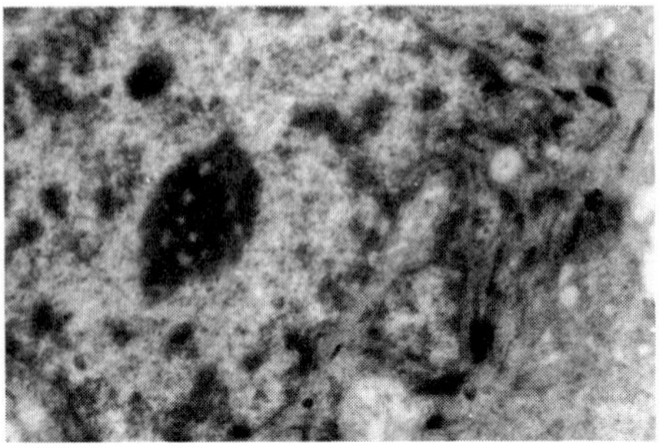

Figure 4.

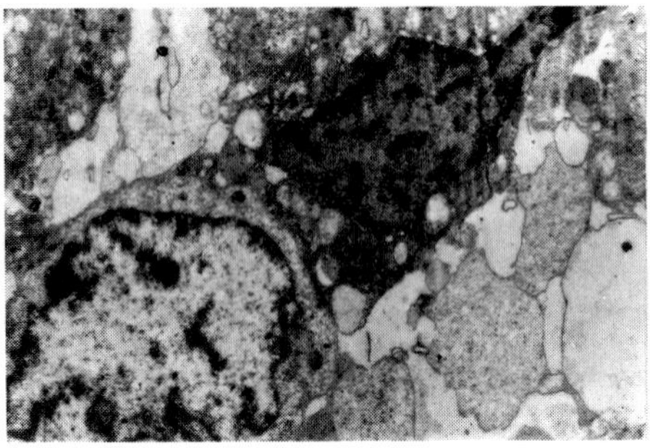

Figure 5.

1986). In our view, this sarcomatous appearing spindle cell component has originated from the epithelium giving a biphasic appearance simulating carcinosarcoma.

References:

Battifora H 1976 *Cancer* **37** 2275

Lattes R 1982 *Tumors of the Soft Tissues, Atlas of Tumor Pathology* Fasc. 1 (Revised) 2nd ed. (Washington DC: AFIP) pp. 123-255

Matsusaka T, Watanabe M, Enjoji M 1976 *Cancer* **37** 1546

Ming S-C 1973 *Tumors of the Esophagus and Stomach, Atlas of Tumor Pathology* Fasc.7 2nd ed. (Washington DC: AFIP) pp.76-80

Zarbo R J , Crissman J D, Venkat H, Weiss M A 1986 *Am. J. Surg. Pathol.* **10** 741

Inst. Phys. Conf. Ser. No. 93: Volume 3, Chapter 9
Paper presented at EUREM 88, York, England, 1988

An ultrastructural study of fine needle aspirates from benign lesions of the breast

DJP Ferguson and CA Wells

Nuffield Department of Pathology, University of Oxford, John Radcliffe Hospital, Oxford

ABSTRACT: The benign breast parenchyma present in fine needle aspirates (FNA) consisted of clumps of epithelial cells many showing luminal differentiation. There was no evidence of underlying myoepithelial cells or basal lamina. The extraction process appeared to cause shearing between the epithelial cells and myoepithelial cells/basal lamina. Certain of the characteristic naked nuclei could result from the lysed myoepithelial cells.

1. INTRODUCTION

There is a rapid expansion in the use of fine needle aspiration (FNA) cytology in the diagnosis of breast disease because it provides a rapid and accurate result and is relatively non-invasive. The procedure involves the removal of a cell sample through a fine needle by applying suction. Although the effect of this process on the characteristic in situ architecture of the breast parenchyma is not clearly understood, the cytological features can accurately differentiate between samples from malignant and benign (normal) lesions. The parenchyma of the breast consists of epithelial cells with underlying myoepithelial cells which are separated from the connective tissue by an intact basal lamina. In benign lesions, the normal architecture is affected but both types of epithelial cells are retained and the basal lamina is intact. The cytology of benign lesions is characterised by clumps of parenchymal cells with a background of naked nuclei. To understand the effect of FNA on breast parenchyma we have tried to correlate the ultrastructural appearances with the cytological characteristics.

2. MATERIALS AND METHODS

FNA were obtained from the breast lesions of women attending the Breast Clinic at the John Radcliffe Hospital. Part of the material was used to make smears for cytological diagnosis. The remainder was placed into 4% glutaraldehyde and processed by routine techniques for electron microscopy. This study is based on the examination of 20 patients with benign lesions.

3. RESULTS AND DISCUSSION

Cytologically, the FNA from the benign lesions contained breast parenchyma which presented as cohesive clumps of cells with no individual cells "falling off" at the periphery (Fig. 1). The cells were of normal size and exhibited morphological variations which were interpreted as representing epithelial and myoepithelial cells. An additional characteristic is the presence of naked nuclei of unknown origin.

The ultrastructural examination of these clumps of parenchyma showed only epithelial cells similar to those observed in situ (Ferguson, 1988). Many of the cells exhibited normal polarisation with luminal formation (Fig. 2). However, although there was some variation in shape and electron density (Fig. 3), there was little evidence for myoepithelial cells which differs from the cytological interpretation. In addition, no remnants of basal lamina or connective tissue were found associated with the clumps (Fig. 4).

From the ultrastructural observations, it would appear that the forceful removal of the breast parenchyma results in a shearing between the epithelial cells and the basal lamina. This shearing could result in the disruption of the myoepithelial cells which are more firmly attached to the basal lamina by hemidesmosomes. Therefore, it is possible that a number of the naked nuclei result from lysis of the myoepithelial cells.

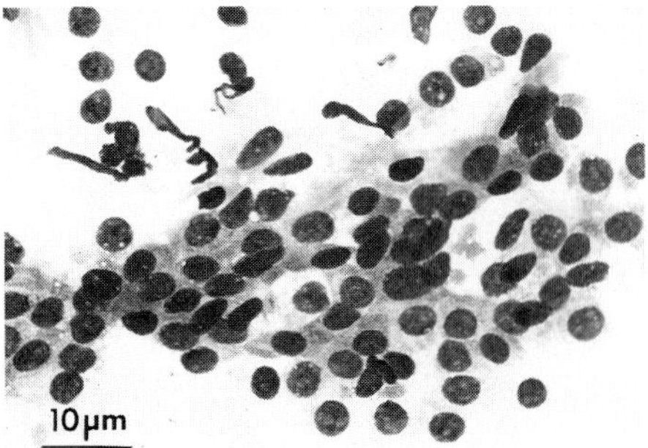

Figure 1. Cytological appearance of a clump of benign parenchymal (epithelial) cells.

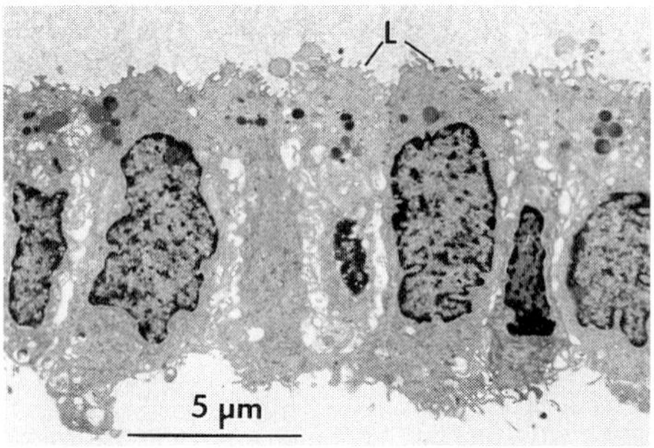

Figure 2. Electron micrograph of a group of polarised epithelial cells showing lumen formation (L).

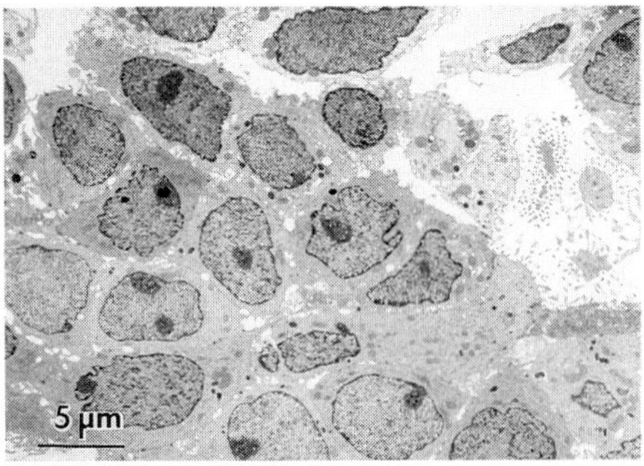

Figure 3. Clump of benign epithelial cells showing variations in shape and electron density.

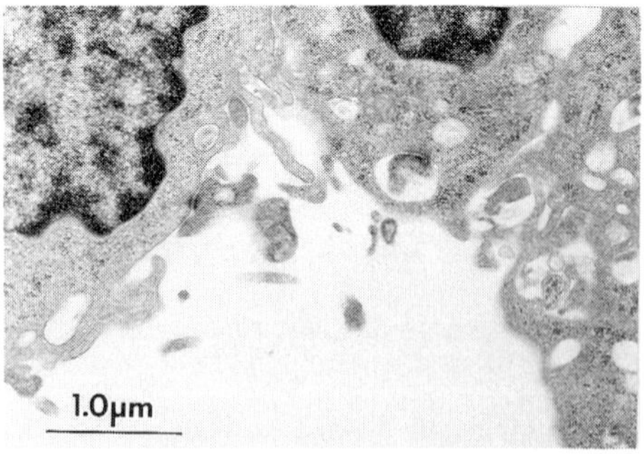

Figure 4. Detail of the basal region of the epithelial cells shown in **Fig. 2.** Note the absence of myoepithelial cells and basal lamina.

REFERENCE

Ferguson DJP (1988) Cell Tissue Res. (In press).

Paper presented at EUREM 88, York, England, 1988

Ultrastructural studies on the testicular microlithiasis

T Erbengi

Department of Histology and Embryology, İstanbul Faculty of Medicine, İstanbul/TURKEY

ABSTRACT: In this study testicular intratubular bodies are described in the seminiferous tubules of a patient with testicular hypoplasia.Our ultrastructural observations have revealed that the vesicles and electron dense granules of the degenerated Sertoli cells and intratubular deposits of collagen fibers are the origin in the formation of the microliths.Electron probe microanalysis of the same concretions demonstrated the presence of calcium and phosphorus peaks.In addition,semiquantitative analysis data of the sections showed that 65 % Calcium and 34 % Phosphorus were present in the concretions.

There are studies in literature which report on radiographically and histologically demonstratable calcification in abnormal gonads and gonadal tumors.However,both the structure and the formation mechanism of these calcifications remain to be elucidated.Oiye (1928) named these structures "calculi". Blumensaat (1929) has proposed that these intratubular structures observed in testes of prepubertal children may arise from degenerated spermatogonia which have fallen into the tubulus lumen.

The only electron-microscopic study on these structures is a report by Vegni-talluri et.al.(1980).Besides,the two studies on the structural analysis of these intratubular structures have yielded conflicting results.Weinberg et.al. (1973) have observed broad and weak diffraction lines at 2.78 Å and 3.45 Å with X ray diffraction studies and have suggested that these lines probably reflect the presence of calcium and phosphate.In 1980, Vegni Talluri et.al. have reported that only calcium is present in the EDAX x-ray spectrum.In order to provide insights into this phenomenon,we decided to investigate the ultrastructure of "testicular microlithiasis" and the steps in the formation of these structures.

Case report: The patient was a 24-year old university student with karyotype XY.During the operation for gynecomastia,biopsies were taken from the left and right testes.Electron-microscopic studies: Biopsies were prefixed in % 2,5 phosphate-buffered glutaraldehyde and postfixed in % 1 osmium tetroxide.Electron probe microanalysis (EPMA): The ultrathin sections prepared for transmission electron-microscope (JEOL 100C) were used and the analysis were conducted with a TRACOR IN-2000 Energy Dispersive Spectrometer (EDS).Furthermore,semiquantitative measurements were done on the EPMA-processed samples.

Based on our observations,we propose the following steps in the formation of microliths (Figure 1). Uniunvolved seminiferous tubule segments were lined by undifferentiated cells (US) and spermatogonia (Sp) Figure 2;a)The formation of numerous vesicles (V) of variable size in degenerating Sertoli cells (SC) in the seminiferous tubule wall (Figure 3);b) The occurence of 1-5 osmium-dense granules (←) and lipid inclusions (L) in these vesicles (Figure 4,5); c) The lysis of the cell membrane and the release of the vesicles into the lumen (Figure 3,5); d) The accumulation of collagen fibers (C) (Figure 3) in the tubules. Finally,the intratubular microlith surrounded by concentric lamellae is formed.

We have observed calcium and phosphorous lines in the EPMA analysis of these intratubular structures (Figure 6).We did not observe the magnesium line

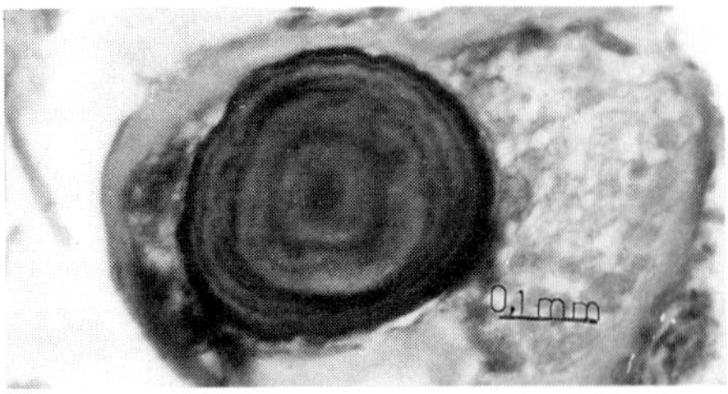

Fig.1. Light micrograph shows Microlith, PAS staining

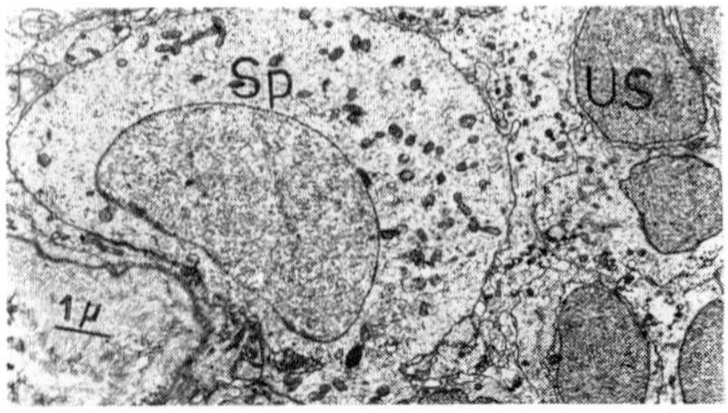

Fig.2.Uninvolved segment of seminiferous tubule

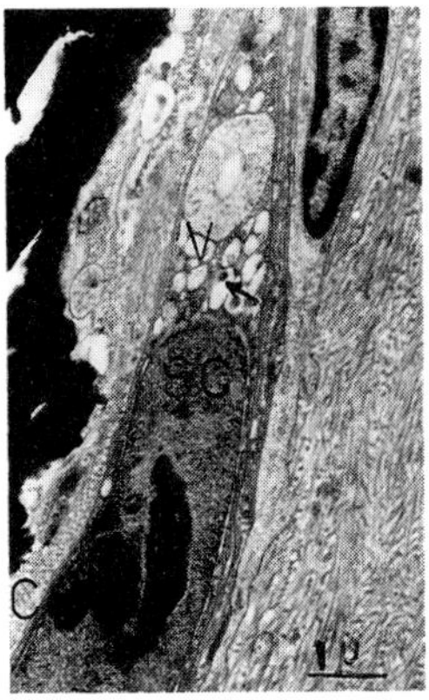

Fig.3.Degenerating Sertoli cell

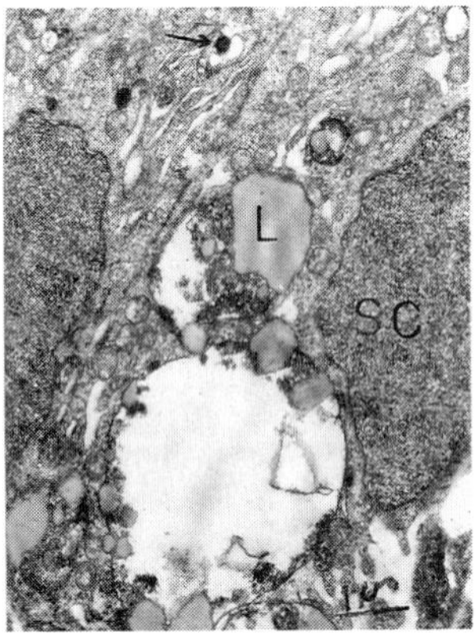

Fig.4.Early degeneration in Sertoli cell

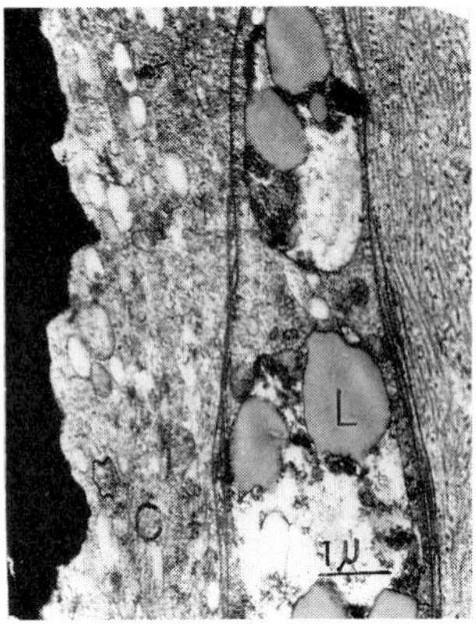

Fig.5.Same appearance as in Fig.4.

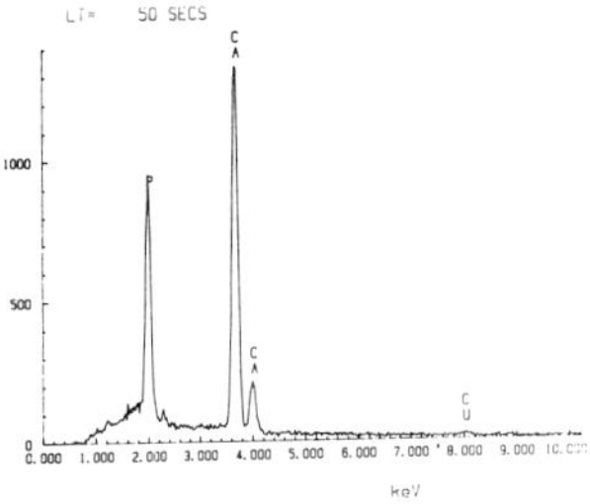

Fig.6.EPMA results

reported by Weinberg et.al. (1973).In contrast to the observations of Vegni-Talluri et.al.(1980) our findings demonstrate the presence of the phosphorus line as well.That the semiquantitative analysis has yielded % 65 calcium and % 34 phosphorus supports the suggestion that the structure of the concretion is calcium phosphate.

References:

Blumensaat C 1929 Virch Arch Path Anat *273* 51

Oiye T 1928 Beitr Path Anat *80* 479

Priebe C J Jr, Garret R 1970 Pediatrics *46* 785

Vegni-Talluri M, Bigliardi E, Vanni M G and Tota G 1980 J Urol *124* 105

Weinberg A G, Currarino G and Stone I C Jr 1973 Arch Path *95* 312

Ultrastructural study of spermiogenesis on nitrofurazone-treated mice

L Pereira*, J F Soares Pessoa**, A Jorge Silva**, E Marques Fontes**

*Departamento de Biologia, Universidade de Aveiro, Aveiro, Portugal
**Centro de Farmacologia e Toxicologia Veterinária, Universidade Técnica de Lisboa, Portugal.

ABSTRACT: The aim of the present work was to study Nitrofurazone experimentally-induced seminiferous tubules mórfological alterations in mice. The ultrastructural changes in germs cells are described, particulary abnormal spermiogenesis.

1. INTRODUCTION

Nitrofurazone is extensively used in the commercial animal exploration due to its antimicrobial properties. Pereira *et al* (1987) have shown that administration of Nitrofurazone to mature mouse caused seminiferous epithelium degeneration and necrosis. In view of the wide use of this Nitrofuran it was thought of interest to study the morphological effects on germinal cells of seminiferous tubules.

2. MATERIALS AND METHODS

Charles River Mice (8 weeks of age) were given subcutaneously Nitrofurazone suspended in parafin (40mg/ml) at doses of 0.2mg/g body Wt, each 72 h during 10 days. Control animals received equivalent volumes of the vehicle. The testis were dissected out and seminiferous tubules were isolated. After glutaraldehyde/osmium tetroxide fixation by imersion, segments of seminiferous tubules were processed routinely for electron microscopy.

3. RESULTS

Seminiferous tubules of control mice showed aparently normal ultrastructural features. The degree of abnormalities is very pronounced between Sertoli cells (Figure 1) and spermatids of treated mice. Desquamation and varying degrees of degenerative changes, including multinucleated cells were noted in spermatid population. Large incidences of spermatid abnormalities were observed: the cytoplasm of round spermatids is filled with peculiar inclusions; invariably, observation of the spermatid showed the acrosome to be abnormal in outline, (Figure 2 and 3) although was highly developed; the presence of apical projections resulted in acrosome asymmetry and most frequently vesicular inclusions were also noted; ultrastructural details of damaged spermatids indicated the perinuclear cisterna to be swollen. Malformed spermatozoa were also documented. (Figure 4).

4. CONCLUSIONS

Ultrastructural observations of seminiferous tubules revealed swollen spaces between Sertoli cell junctions of treated mice, suggesting blood-testis barrier alterations. On the other side, recognition of irregularities during spermiogenesis steps, namely injured acrosome formation and degenerated cytoplasm, suggests that in these experimental conditions, Nitrofurazone causes damage on the fine structure of mice seminiferous tubules, although these conclusions were obtained in some experimental conditions differing from the *in vivo* situation.

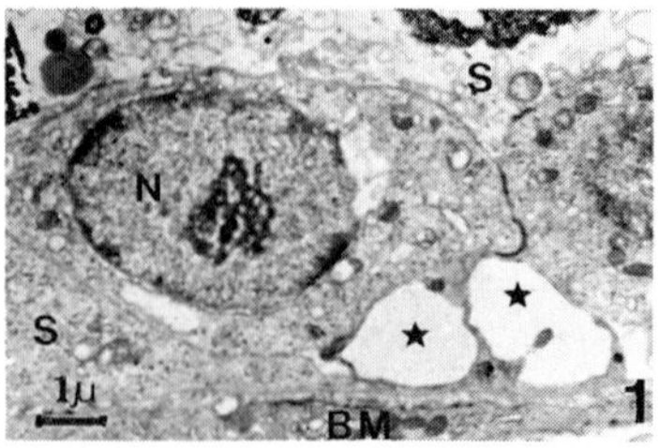

Fig. 1. Note large spaces between Sertoli cell junctions (★)
N = Nucleus; S = Sertoli cell; BM - Basement membrane.

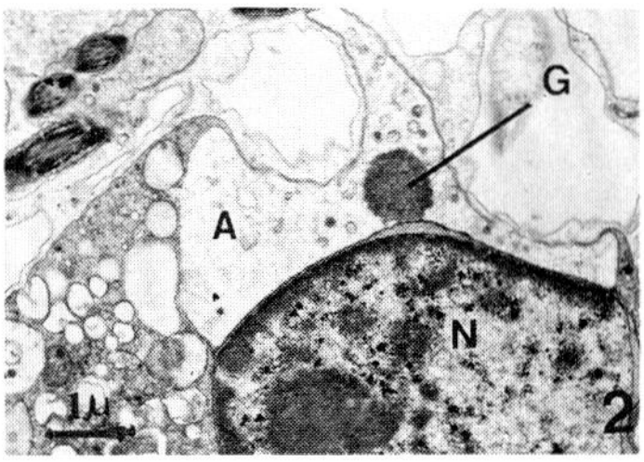

Fig. 2. Asymmetric acrosome in a damaged spermatid.
An irregularity of chromatin condensation is noted.
A= acrosome; G= acrosome granule· N= nucleus.

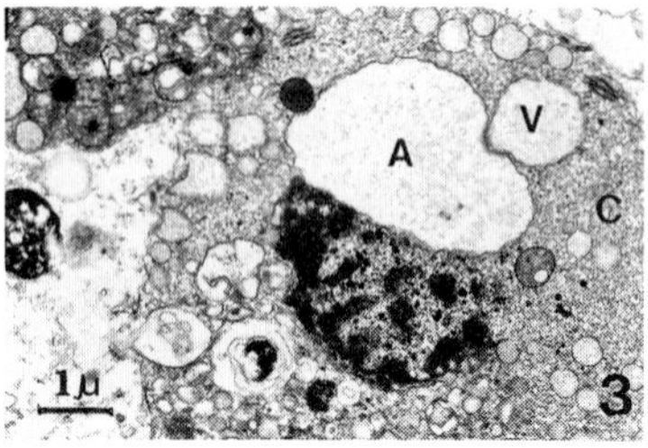

Fig. 3. A large vesicle is apposed to an abnormal acrosome. The cytoplasm of this spermatid is filled with degenerated material.
A= acrosome; V = Vesicle; C= cytoplasm.

5. REFERENCES

Pereira L, Pessoa J F C, Silva A J and Fontes E M 1987 XXII. *Ann.Meeting of Portuguese Society of Electron Microscopy.* (Évora: Portugal) p6

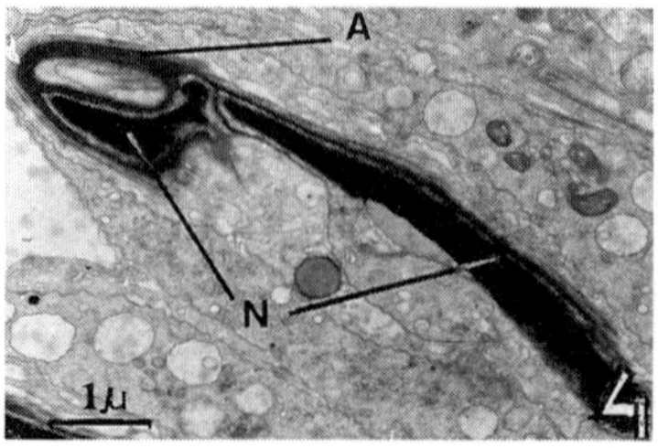

Fig. 4. A malformed and elongated acrosome.
A= acrosome; N= nucleus.

ACKNOWLEDGMENTS

The authors wish to thanks Doutor Helena Ferronha of National Laboratory of Veterinary Research (Lisbon)

Peritoneal dialysis fluid induces an acute elicitation in the rat peritoneal cavity: a study on peroxidatic activity

H J Bos, J C de Veld and R H J Beelen[1]

Departments of Cell Biology and Haematology[1], Medical Faculty and University Hospital[1], Free University, Amsterdam, The Netherlands.

ABSTRACT: Peritoneal macrophages from uninfected Continuous Ambulatory Peritoneal Dialysis (CAPD) patients showed in general two different endogenous peroxidase activity (PA) patterns: exudate and negative. This suggests, in accordance with the animal model, that a local chronic elicitation exists in CAPD-patients. This chronic elicitation might be induced by the volume alone (massage) or the dialysis fluid used. Therefore in the rat both saline and dialysis fluid were injected. Our results do indicate that the dialysis fluid indeed induced an elicitation especially when compared with the minor saline effect.

INTRODUCTION

In vivo in the animal model peritoneal macrophages (mφ) can be devided by their PA pattern in resident (PA-activity in nuclear envelope; NE, and rough endoplasmic reticulum; RER), exudate (PA-activity in granules; GR), exudate-resident (PA-activity in GR, NE and RER) and PA-negative mφ. In the normal steady state 90% of the mφ are resident-cells. After acute elicitation exudate and exudate-resident mφ appear, whereas after a chronic elicitation exudate and PA-negative mφ appear (Beelen et al, 1982).

The kinetics of the different cell types in vivo and in vitro do indicate that in an acute inflammatory state exudate mφ are induced, which transform into resident macrophages via exudate-resident cells, while in a more chronic inflammatory state the recruted exudate mφ loose very quickly the myeloperoxidase and so transform into PA negative mφ. Peritoneal mφ from uninfected CAPD patients showed in general two different PA patterns: exudate and negative. (Bos et al, 1988) This suggests, in accorance with the animal model, that a chronic elicitation exists in the peritoneal cavity of CAPD patients. This chronic elicitation might be induced by the volume alone (massage) or the dialysis fluid used, therefore in the animal model both saline and dialysis fluid were injected.

MATERIALS AND METHODS

Peritoneal cells from male SPF-Wistar rats were harvested as described by Beelen et al (1982) and peritoneal mφ were reacted for PA as described by Bos et al (1988).

* Supported in part by the Kidney Foundation, Bussum and Foundation of "The Three Lights", The Hague, The Netherlands.

RESULTS AND DISCUSSION

Cellular Composition. After injection of dialysis fluid the number of peritoneal mφ and neutrophilic granulocytes show a dramatically increase with a maximum after 24 hr (fig. 1A), whereas saline has only a minor effect (fig. 1B). In both experimentst the initial cellular composition was present 4 days after interperitoneal administration.

Cytochemistry of peritoneal mφ. In the peritoneal cavity of the normal steady state rat, 90% of the mφ are resident. After injection of dialysis fluid the number of exudate mφ rose sharply with a peak at 24 hr and then decreasing, returning to the normal level after 8 days (fig. 2A). When compared with the pronounced effect of dialysis fluid, interperitoneal administration of saline had only a minor effect: A maximum of exudate cells was found after 16 hr and then decreased returning to the normal level after 8 days (fig. 2B). Interperitoneal administration of dialysis fluid resulted, as expected on the cellular composition, in an acute elicitation. This acute elicitation is comparable with that after interperitoneal administration of new born calf serum (Beelen et al, 1982).

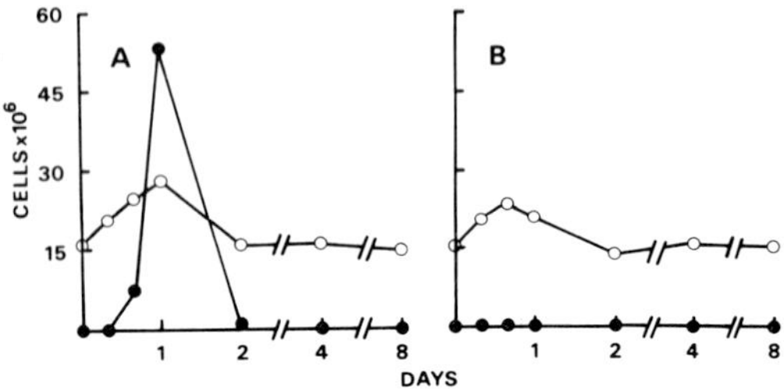

Fig. 1 Changes in the total number of mφ (—○—) and neutrophilic granulocytes (—●—) in peritoneal exudates during the first 8 days after an injection of dialysis fluid (A) or physiological saline (B).

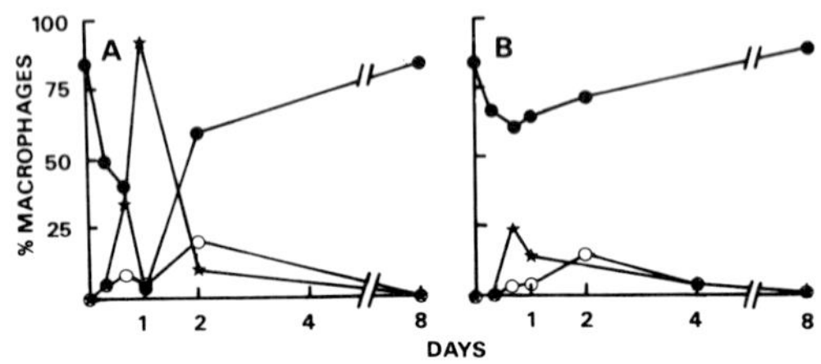

Fig. 2 Percentual course of the resident mφ (—●—), exudate mφ (—★—) and exudate-resident mφ (—○—) in the inflammatory peritoneal exudates elicited with peritoneal dialysis fluid (A) and physiological saline (B).

Our results do indicate that dialysis fluid used in CAPD patients indeed induced a sterile inflammatory state in the rat peritoneal cavity, especially when compared with the minor saline effect. Therefore frequent interperitoneal administration of dialysis fluid in CAPD patients might induce a chronic elicitation, which is in agreement with our previous results (Bos et al, 1988).

The clinical significance of the induction of an elicitation by peritoneal dialysis fluid is unknown. One may claim that the aim should be to have a high percentage of resident mφ during CAPD treatment, because this presence reflects an unstimulated peritoneum. On the other hand, resident mφ represent an end stage of a phagocytic cell line , less able to attack invading microorganism.

In conclusion: In accordance with our results in the animal model, dialysis fluid might be the inducer of the chronic elicitation present in CAPD patients (Bos et al, 1988), which may lead to peritoneal fibrosis.

REFERENCES

Beelen R H J and Fluitsma D M 1982 Immunobiol. 161 pp 266-73

Bos H J, Bronswijk H van, Helmerhorst T J M, Oe P L, Hoefsmit E C M and Beelen R H J 1988 J. Leuk. Biol. 43 pp 172-8

Inst. Phys. Conf. Ser. No. 93: Volume 3, Chapter 9
Paper presented at EUREM 88, York, England, 1988

A quantitative study of rabbit peroneal nerve by means of electron microscopy

G Germanà, G P Germanà, M Gugliotta*, V Micale•, A Toscano*, E Ciriaco and U Muglia

Institutes of Veterinary Anatomy and Neurology*, University of Messina (Italy) - Istituto Sperimentale Talassografico C.N.R•., Messina (Italy)

ABSTRACT: The Authors calculated the area, diameter, perimeter and total number of myelinated and unmyelinated axons of rabbit peroneal nerve.

I. INTRODUCTION

The relationship between the conduction speed and the diameter of peripheral nerve axons has led many researchers to study the morphometric typization of various nerves in many Mammals. Previous observations carried out in our laboratory on the regeneration speed of rabbit sciatic nerve subjected to crushing and treated with $ACTH_{4-10}$ (Muglia et al 1985, Girlanda et al in press) have shown a regenerative response different from that one reported by other AA., in the same experimental conditions, in the rat (Bijlisma et al 1983). The lack, in the literature, of morphometric observations on rabbit sciatic nerve and the difference in the regenerative behaviour of this nerve compared with that one of rat have led us to undertake a morphometric study of the peroneal nerve, a branch of the sciatic nerve previously considered for our regenerative observations, in normal physiological conditions.

2. MATERIALS AND METHODS

Peroneal nerve samples were taken from 10 New Zealand rabbits 4 cm above the knee joint, processed for E.M. and cut at the ultramicrotome transversely to their course. The ultrathin sections were photographed at the E.M. and the obtained prints examined by means of a computerized semi-automatic scanner, Mop Videoplan Zeiss. The area, diameter and perimeter of axons, both myelinated and unmyelinated, were calculated. Histograms and statistics were automatically worked out by the scanner computer from the obtained data.

3. RESULTS AND CONCLUSIONS

The mean diameter of peroneal nerve was 754.650±22.84 μm; 4I08±I04.344 myelinated fibers were counted,representing 22.82% of the entire axon population of each nerve and I4I47 ± ±437.428 unmyelinated fibers,representing 77.I8%. The diameter of myelinated fibers ranged between I.92 and I6.I5 μm, with an average of 6.76 ± 0.88 μm; the area between 2.90 and 204.90 μm^2, with an average of 43.I6 ± I.I2 μm^2; the perimeter between 6.29 and 53.59 μm,with an average of 22.60 ± 0.29 μm. The histogram showing the relative frequencies of myelinated fibers vs fiber diameter presents a bimodal distribution.The first frequency peak,between 4 and 5 μm of diameter,represents about 24% of fibers;the second peak,between I2 and I3 μm of diameter,represents 4.73% of fibers.About 60% of fibers have a diameter ranging between 4 and 7 μm.The unmyelinated fibers had a diameter ranging between 0.26 and 2.4I μm,with an average of I.03±0.004 μm;area between 0.005 and 0.458 μm^2,with an average of 0.090±0.0007 μm^2;perimeter between I.06 and 7.96 μm,with an average of 3.6I±0.0I5 μm.The histogram showing the relative frequencies of unmyelinated fibers vs their diameter presents a unimodal distribution with a frequency peak between I.0 and I.2 μm of diameter,corresponding to 30% of fibers. From these data we concluded that the distribution of frequency classes vs diameter of myelinated axons is a bimodal one.Such a distribution differs from the unimodal one observed by Schmalbruch (I986) in the myelinated fibers of the same branch of sciatic nerve in the rat.Thus,although the two nerves are macroscopically superimposable,they actually exhibit a different morphometric and,as a consequence,physiological identity.

REFERENCES

Bijlisma W A I983 Muscle & Nerve 6 I04

Girlanda P,Muglia U,Vita G,Dattola R,Santoro M,Toscano A,Venuto C,Roberto M L,Baradello A,Romano M and Messina C Exp Neurol in press

Muglia U,Santoro M,Toscano A,Vita G,Ciriaco E,Bronzetti P,Germanà G and Messina C I985 Proc 39th Symp S.I.S.VET. pp 83-4

Schmalbruch H I986 Anat Rec 2I5 7I

This work was supported by a grant from Sanitary department of Sicilian region - 9/P project.

Motor nerve sprouting after partial denervation in mice: a SEM and TEM study

Kojun TORIGOE

Neuroinformation Research Institute (NIRI), University of Kanazawa School of Medicine, Kanazawa, Ishikawa 920, Japan

ABSTRACT : Motor nerve sprouting in the mouse gastrocnemius muscle was examined by scanning electron microscope (SEM) and transmission electron microscope (TEM) after partial denervation. Terminal sprouts initially grew out from the intact motor endplates with a growth cone and then a Schwann cell migrated into the growing sprout. Being accompanied by a Schwann cell, nodal sprouts branched out from the intact nodes of Ranvier, but they could not be seen having a growth cone in their terminal tip.

MATERIALS & METHODS The medial head of the mouse gastrocnemius muscle is innervated by two muscular nerves. Partial denervation by transecting one of the muscular nerves induces motor nerve sproutings in the intact region innervated by the other. Muscles and nerves of adult male ddY mice (30-45g) were obtained on the 21st postoperative day. The mouse was anaesthetized with 0.5% sodium pentobarbital and perfused through the abdominal aorta with a Karnovsky's fixative. For an observation of terminal sprouts by a SEM, the muscles were longitudinally cut into 50-100 μm thickness with a vibratome. The specimens were post-fixed in 2% osmium tetroxide for 2 hr at 4°C. To remove the extramuscular connective tissue components, the preparations were treated in 8N HCl for 5-7 min at 60 °C. For a SEM study of nodal sprouts, the nerves pre-fixed were quickly removed and immersed in the above post-fixative for 2 hr at 4 °C. The specimens were treated in 5N KOH for 3-5 min at 60°C and immersed in collegenase solution (1mg/ml) for 12 hr at 37°C. The specimens of muscles and nerves were dehydrated through an ethanol series. After being critical - point dried by using CO_2, they were viewed at 15 KV in a HFS-2 SEM. For a TEM observation of nodal sprouts, the nerves fixed were dehydrated and then embedded in Epon 812. Serial ultrasections were examined under JEM-100 B and C microscope after double staining with uranyl and lead citrate.

RESULTS & DISCUSSION **Terminal sprouts** arose from the outer edge of the intact motor endplates and extended onto the muscle fibres. The terminal tip of the sprout was swollen in an amoeba shaped configuration, forming a growth cone (Fig.1). In the intial part of a long terminal sprout the several shorter processes surrounded the sprout, which most likely indicated the processes of a Schwann cell (Fig.2). Such swelling patterns suggested that Schwann cells migrated into the growing sprout.

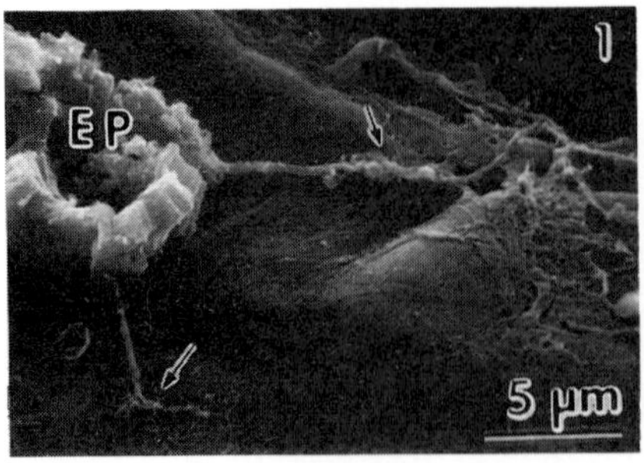

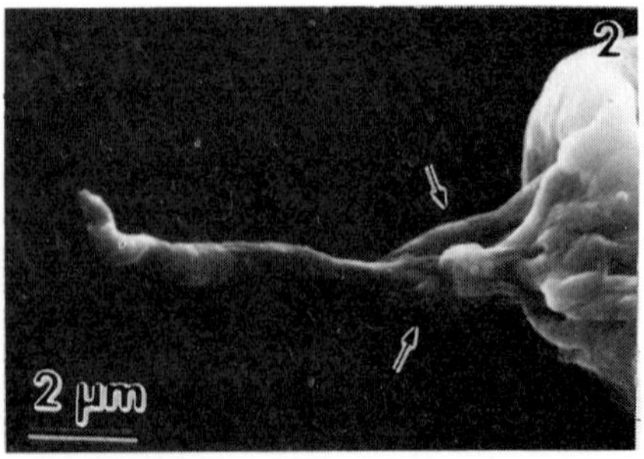

Nodal sprouts appeared to branch out from the stout nerve and run around it (Fig.3). In contrast to the development of terminal sprouts, the nodal ones could not be seen having a growth cone at their tip but being covered over with a cytoplasm of a Schwann cell (Fig.4 A and B). A Schwann cell must play an important role in forming the nodal sprouts, particularly in the initial stage of their development, because the cell encircled a node of Ranvier so as to have a mantis egg capsule like appearance (Fig.5). The SEM observation may be coincident with the TEM one : the node developed to elongate itself with being ensheathed by the high density cytoplasm of a Schwann cell.

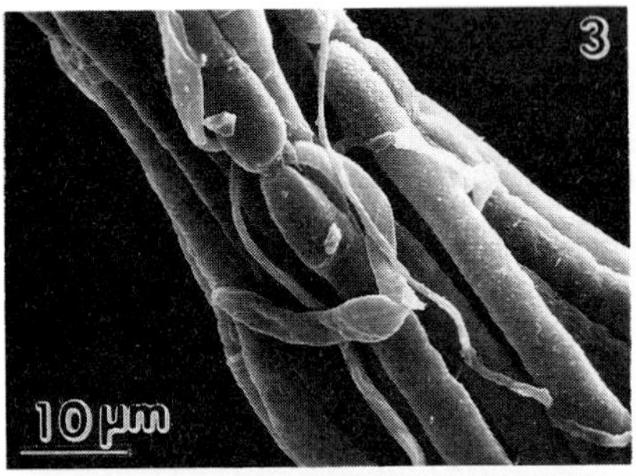

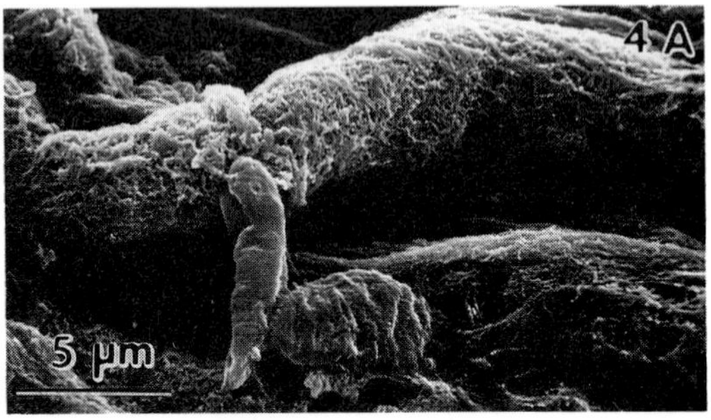

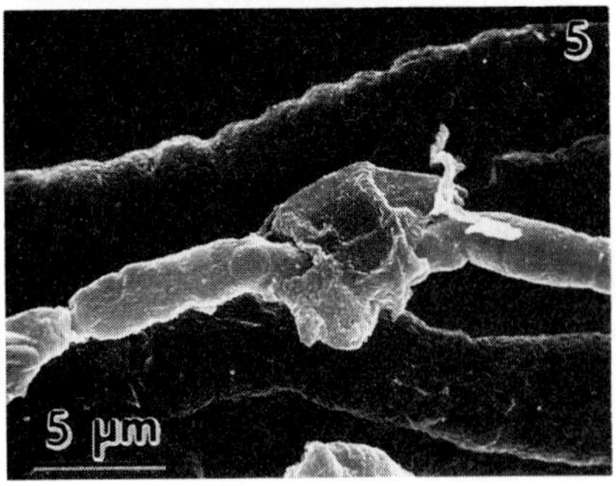

Thus, motor nerve sprouts were not directly guided by any cells, but they may grow out in accordance with some factors which were producd by the denervated muscle region : the factors may be received by a growth cone of terminal sprouts, while they may be retrogradely transported to the nodes of Ranvier so as to induce a nodal sprout.

An immediate effect of noradrenalin transmission blockade on the rat pineal gland

The Institute of Histology and Embryology, Faculty of Medicine, Novi Sad, Yugoslavia

ABSTRACT: An immediate morphodynamic respond of the rat pinealocytes to the injection of an adrenoreceptor antagonist (propranolol 1g/1ml/1oo g of the body weight, intraperitoneally administrated) was studied. The results obtained suggest that noradrenalin might be a pivotal regulative variable which promotes a stress-reactive respond of this organ.

1. INTRODUCTION

The injection act has been found to animate the activity of the pineal gland. Since this multireceptor organ is under a paramount noradrenergic control, we searched into the morphodynamic turnover of the pinealocytes in case of an immediate post-injection stress blockade of noradrenalin transmission.

2. FIGURE PRESENTATION

The postinjection periods observed were 5, 15 and 45 min. 5 min. The absence of cell processes, as well as of the discharge of the content of clear vesicles and structures labeled lipid droplets, which both were the characteristics of the saline-injected rats (1), points out that an initial stress-induced secretory respond of the pineal gland is adrenergically driven. The occurrence of swollen mitochondria (Figure 1) indi-

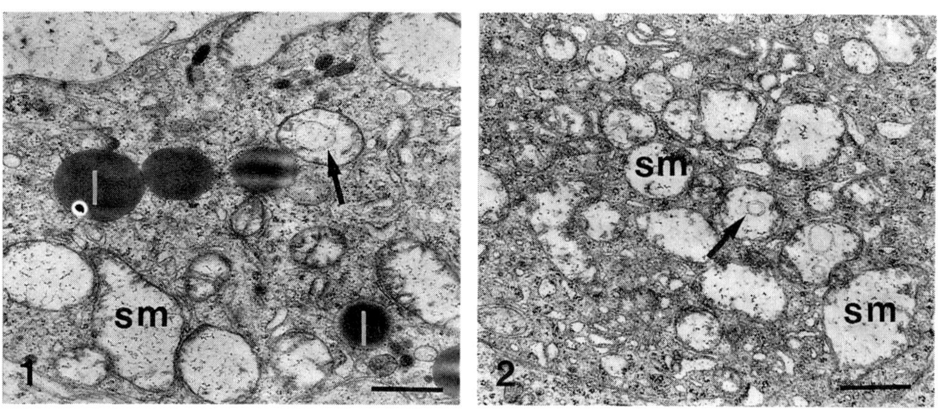

Fig. 1 and 2. Swollen mitochondria (sm), ring-shaped membranous profile in swollen organel (arrow) and structure labeled lipid droplet (l). Scale bar = 1 µm.

cates that the metabolism of these organelles is also under the control of the sympathetic system. 15 min. Scanty ribosomes and granulated endoplasmatic reticulum (GER) show that the adrenoreceptor blockade also restricts the peptidergic pinealocyte activity (Figure 2). 45 min. The augmentation of GER and energized mitochondria (Figure 3) demonstrates that the propranolol effect is relatively short-lasting. The appearence of a ring--shaped membrane in the swollen mitochondria (Figures 1 and 2), as well as of "semiswollen" organelles (Figure 4), indicates that the energized mitochondria might arise through a "self-regenerating" process.

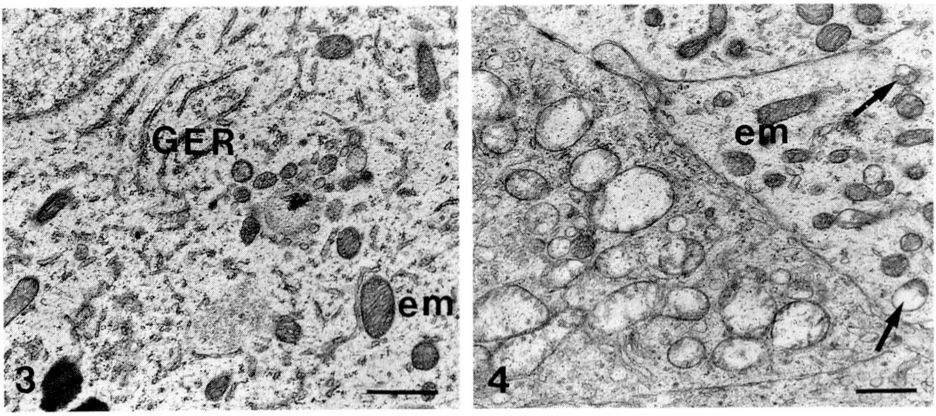

Fig. 3 and 4. Granulated endoplasmatic reticulum (GER), energized mitochondria (em) and those of "semiswollen" shape (arrow). Scale bar = 1 µm.

The results obtained suggest that the immediate stress-reactive respond of the pineal gland to the injection act requires a noradrenergic impelling input to be generated.

Milin J, Martinović J, Demajo M 1984 Arch Anat Microsc Morphol Exp 73 159

Ultrastructural changes in human fetal fibroblasts, chicken fibroblasts, and vero cells after co-cultivation with cells from the kidneys of patients suffering from endemic Balkan nephropathy

P Spasic,* B Keserovic,# N Bojanic,* S Petkovic,§ and K Apostolov ¶

* The Institute of Pathology and Forensic Medicine Military Medical Academy Crnotravska 17, 11002 Belgrade, Yugoslavia
# The Institute of Immunobiology and Viral Biology, Torlak, Belgrade
§ The Department of Urology, University of Belgrade, School of Medicine
¶ Deltanine Research, Brunel University, Uxbridge, Middlesex, England

ABSTRACT: Electron microscopical examination of human fetal fibroblasts, chicken embryo fibroblasts, and a continual line of vero E_6 monkey kidney cells after co-cultivation with cells from kidneys of patients suffering from Endemic Balkan Nephropathy (EBN) showed cytopathogenic effect and pathological changes of nuclei. In all those kinds of cells numerous cytoplasmic vesicles containing virus-like particles corresponded to oncorna viruses were found. It suggests viral etyology of EBN.

The etiology of EBN is still unknown. Certain morphological findings[1] and epidemiological studies[2] suggest a viral etiology.
We analyzed morphological and ultrastructural changes in a culture of cells of human fetal fibroblasts, chicken embryo fibroblasts, and a continual line of Vero and Vero E_6 monkey kidney cells after co-cultivation with cells from kidneys with EBN. By co-cultivating the kidney cells of six individuals suffering from EBN with fibroblast cells of human fetal skin in an in vitro culture, changes in the fibroblasts corresponding to transformation and cytopathogenic effect (CPE) of a certain virus were observed[3]. The transformated cells had a changed morphology and an aberrant karyotype with structural and numerical changes in chromosomes.
The human fetal fibroblasts with the CPE were inoculated with a growth medium in which they grew in a culture of chicken embryo fibroblasts and Vero and Vero E6 cells. The cultures were incubated at 33°C and 37°C. The CPE appeared in all inoculated cultures and at both temperatures.
All of those three cell systems that have appearance of the CPE and the transformed human fibroblasts were analyzed by electron microscope. That analysis showed very large pleomorphic nuclei with extremely irregular profiles and large, marginally situated nucleoli (Figure 1). Numerous vacuoles full of virús-like particles whose shapes and sizes corresponded to oncorna viruses (Figure 2) were observed in the cytoplasms of those kinds of cells. The control cultures were negative. It suggests a viral etiology of EBN.

1 Apostolov K, Spasic P, Bojanic N, 1975 *Lancet* 2 pp 1271-1274
2 Birtasevic et al 1983 *Vojnosanit Pregl* 40:5 pp 319-324
3 Keserovic B, Spasic P, Bojanic N 1986 *Ibid* 43:1 pp 3-7

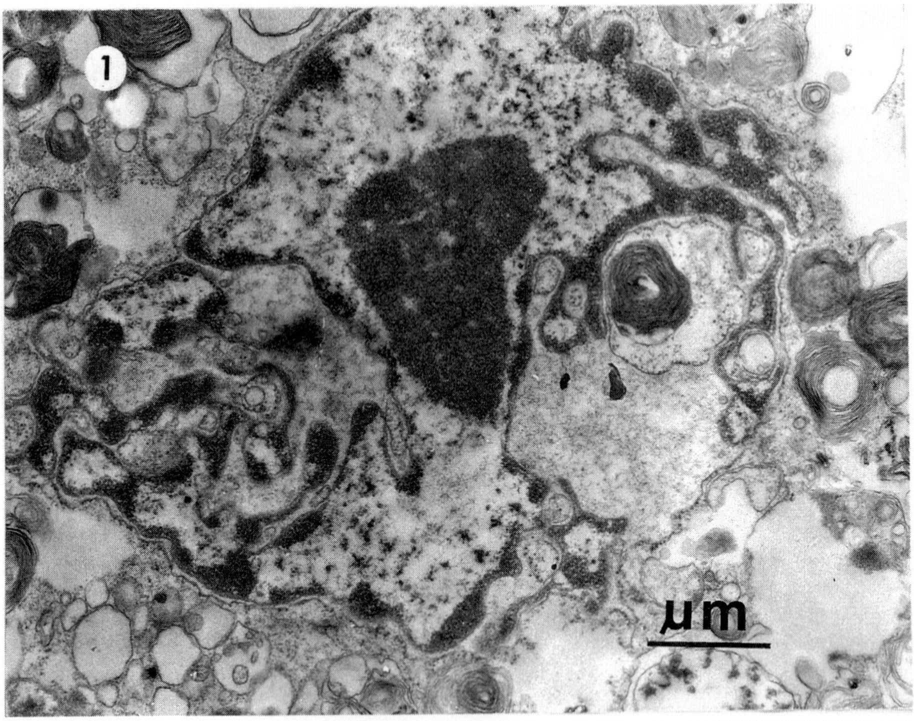
1
µm

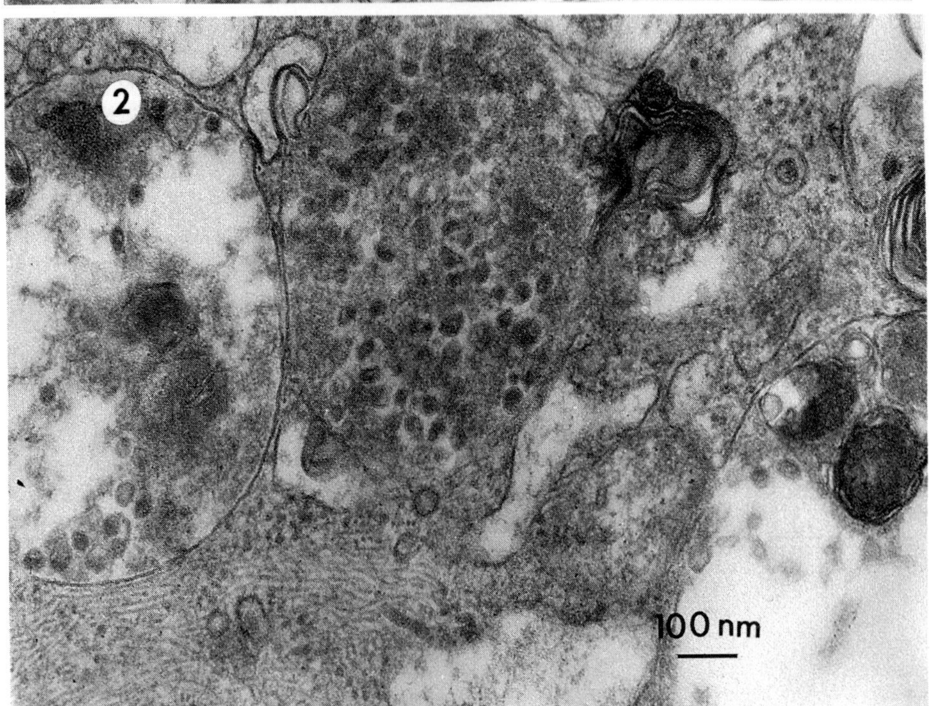
2
100 nm

Effect of smoking on the trophoblastic layers during initial stages of placentation

Ramazan Demir[a], Türkân Erbengi[b], Mehmet Kaya[c], Mine Üner[a]

(a)Dept. of Histol.& Embryol.; Obstet.& Gynecol., Medical Faculty, Akdeniz University, Antalya; (b)Dept. of Histol.& Embryol., Medical Faculty, İstanbul University, İstanbul; (c)Dept. of Histol.& Embryol., Medical Faculty, Çukurova University, Adana.

ABSTRACT: The ultrastructural study of placental villi from smoker showed the following changes in trophoblastic layers: abnormalities of microvilli, increased syncytial lipid droplets and glycogen accumulation, focal infolding of plasma membrane and necrosis, dilatation of RER and SER, degeneration of Langhans cell and thickenning of basal lamina.

1. INTRODUCTION

Many investigations have been made on the clinical aspects cigarette smoking in the various periods of pregnancy. It was pointed out that the smoking causes high rate of perinatal and new born mortality and low birth-weight (Pirani et al.1978; Velde et al.1985). A causative relation has been established between the above mentioned peri and postnatal complications and the placental damage.

In this study, the effects of smoking on the trophoblastic layers in smokers during the initial stages of placentation in pregnancy have been investigated with electron microscope.

2. FIGURE PRESENTATION

Our observations on villus ultrastructure in the human placentas from non-smokers were confirmed to previously described by many authors (Demir, 1979,1980; Kaufmann,1982). Many morphological degenerations were seen on the free surface of the syncytiotrophoblast. Most microvilli were either shorten and swollen or were twisted. In some samples, the syncytial projects were extended to basal lamina between two cytotrophoblastic cells. The syncytial pieces containing the microvilli groups were shattered into pieces and broken off, devastated in the other syncytial elements were poured into IVA. Dilated RER and SER cisternae and vesicles containing slightly dense substances were observed. Increased amount of lipid droplets, accumulated glycogen particles and distrupted cytoplasmic organels were definitly seen. This syncytial devastation was indicating the syncytial necrosis (Fig.1).

It was very interesting to find that a degenerative cytotrophoblastic mass which was passing throught the syncytiotrophoblast and poured into IVA IVA (Fig.2). The damage occured in trophoblastic layers depend on the presence of different substances in tobacco and the hypoxia. A lot of cadmium and polycyclic aromatic hydrocarbons substances presenting in tobacco are important factors for devastation of plasma membrane in trophoblast (Veen et al.1982). It has been suggested that an existence of relationship between the trophoblastic activity and polycyclic aromatic hydrocarbon, the-early-hydrocarbon hydroxylase enzym was induced by tobacco contents (Longo, 1980). In conclusion, the morphological findings of this study suggested that the toxic effects of smoking cause the denaturation.

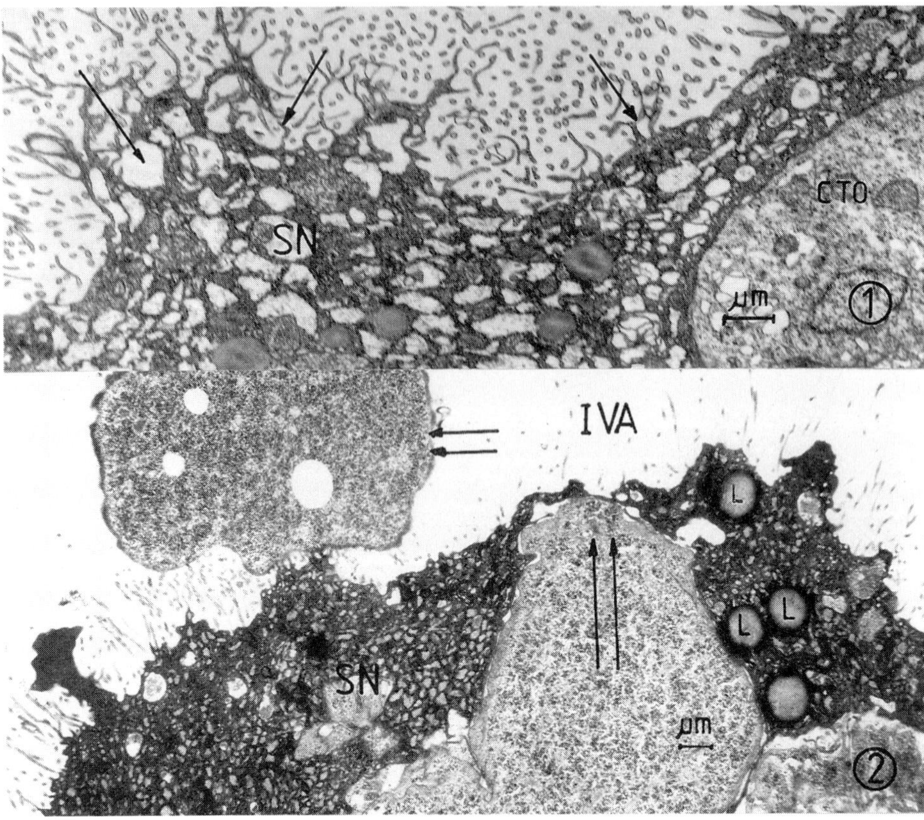

Fig.1 and 2. Trophoblastic damage in placenta from smokers at four week of pregnancy. The syncytial (SN) microvilli, focal infolding of free plasma membrane are seen (arrows). Degenerative trophoblastic mass poured into IVA (double arrows).

We are greatful to TUBİTAK (Scientific and Thecnical Research Council of Turkey) for supporting the project (TAG-527).

Demir, R.(1979). Acta Anat. 105:226-232.
Demir,R.(1980). Acta Anat. 106:18-29.
Kaufmann,P.(1982). Bibliot.Anat. 22:29-39.
Longo,L.D.(1980). Am.J.Obstet.Gynecol. 137:162-173.
Pirani, B.B.K. et al.(1980). Obstet.Gynecol. 52:257-263.
Veen,F. van der Fox,H.(1982). Placenta 3:243-256.
Velde, W.J. et al.(1985). Gynecol.Obstet.Invest. 19:57-63.

Inst. Phys. Conf. Ser. No. 93: Volume 3, Chapter 9
Paper presented at EUREM 88, York, England, 1988

Ultrastructural changes of gastric mucosa in patients with chronic obstructive pulmonary diseases

E Koçyiğit, Y moin, T Erbengi, Y Canberk, S Özdil, K Gazioğlu

Pulmonary Diseases Department and Department of Histology and Embryology, İstanbul Faculty of Medicine İstanbul TURKEY

ABSTRACT: In this research-work the ultrastructural changes of the cells in gastric mucosa were investigated in patients with chronic obstructive pulmonary diseases (COPD). Epithelial cells showed degenerative changes such as obstruction on the apical and dissociation on the lateral aspects of the cells. Intracellular secretory capillaries were narrowed and tubulovesicular structures were increased in parietal cells. But chief cells were at secretory phases. Enterochromaffine-like cells (ECL) and D_1 cells were also showed some degenerative changes in the stomach of these patients.

Spiro (1970) had suggested a prevalance of peptic ulcer in COPD ranging from 4 to 10 % all the way to 90 %. Forward looking studies (Schneider 1963 and Kroeker 1966) make it aparent that about one forth of the patients with serious chronic pulmonary disease have evidence of peptic ulcer at some time during their lives.

Although several studies were carried out, we don't know much about ultrastructural changes. The aim of this research was to investigate the gastric mucosa on ultrastructural level in patients with COPD. For this reason 6 patients were studied. Their diseases were confirmed by physical examination, pulmonary function tests, including vital capasity, forced expiratory volume, flow-volume curves and blood gases. Gastric acidity level and biopsy from the fundus of the stomach were obtained from all patients. They hadn't any other diseases apart from their lung and stomach disorders.

With electron microscopic study, degenerative changes were diagnosed at gastric epithelial cells (Ep). These are destruction on the apical (A), and dissociation on the lateral (L) aspects of the cells (Figure 1).

The secretory system of the parietal cells (PC) was found to be inhibited. Secretory capillaries (Sc) were narrowed and tubulovesicular structures were increased (Figure 2, 3). Degenerative changes such as lysis and swelling of cristae of the mitochondria (Mi) were prominent. The number of myeline figures (➔) and lysosomes (Lys) were found to be increased (Figure 2).

Chief cells (CC) were often found at secretory phase. There were membran lysis at secretory vacuoles (Va) and engolfing (➔) between secretory granules (Figure 4).

ECL cells, known as releasing cells of histamine (Bordi 1983), were full of secretory granules (Sg) and they were at secretory phase (Figure 5).
D_1 cells, known as releasing cells of vasoactive intestinal polypeptide (VIP) (Bordi 1983), also showed degenerative changes including swelling and lysis of the mitochondria and obtained crinophagy (➔) structures (Figure 6)

These changes were found to be related to the duration of lung disease. But no correlation was found between the level of hypoxia, hypercapnia and ultrastructural changes.

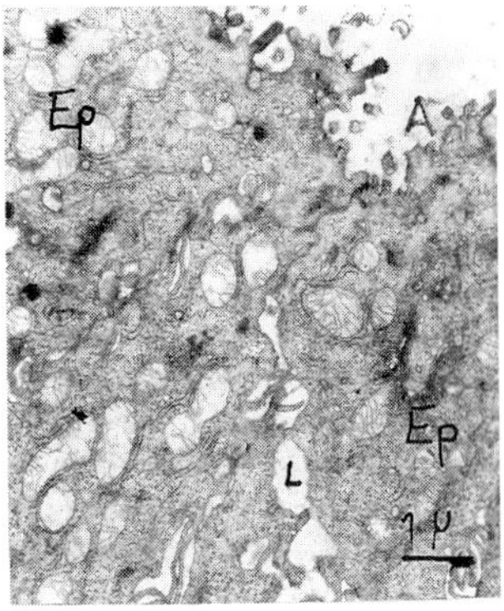

Fig.1.Enlarged intercellular spaces of epithelial cells

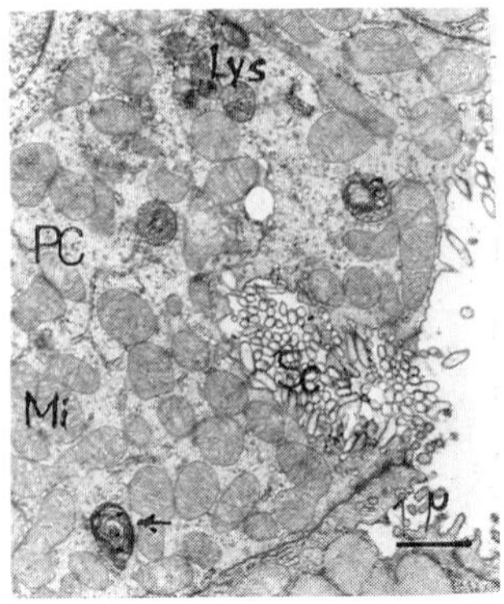

Fig.2.Increased Lysosomes and myeline figure of the parietal cell

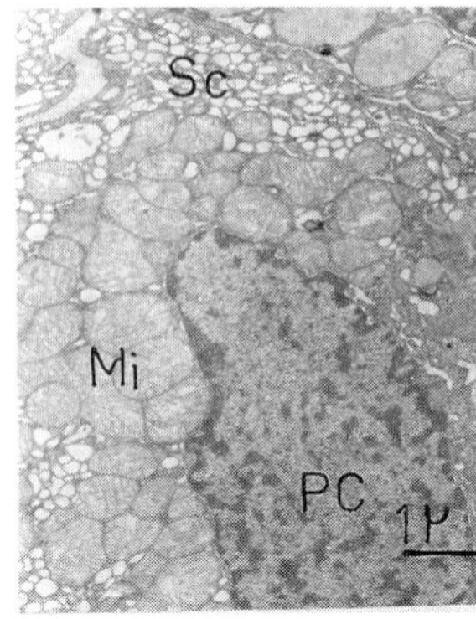

Fig.3.Degenerated mitochondria of parietal cell

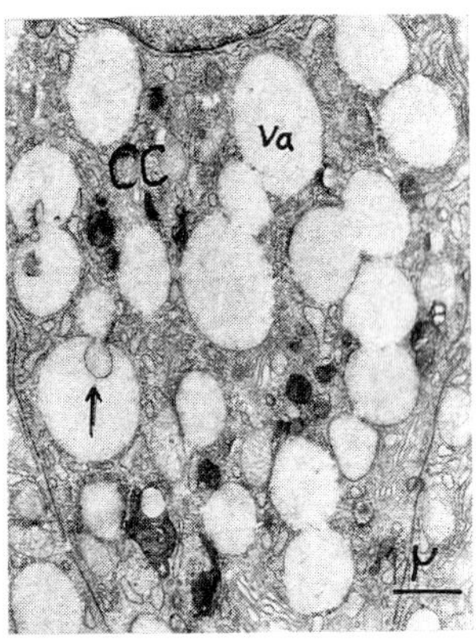

Fig.4.Densely vacuolated chief cell

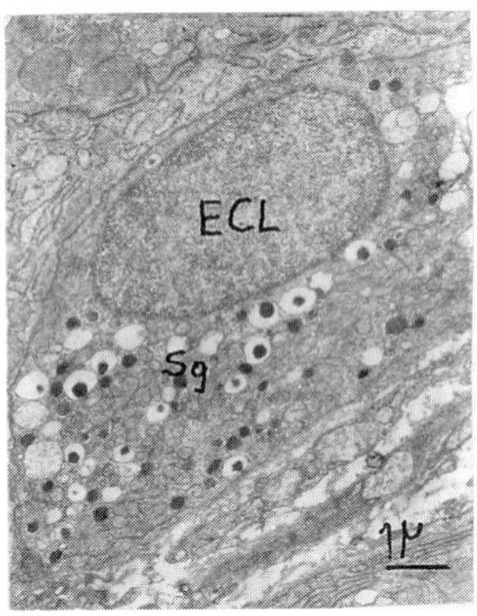

Fig.5.Abundant secretory granules of ECL

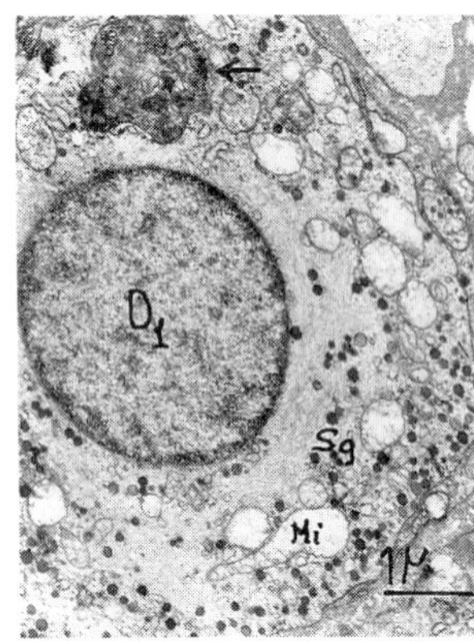

Fig.6.Dense perinuclear fibrillary bundles and small secretory granules of D1 cell

References:

Bordi C, Ravazzola M, Vita O 1983 Ann Pathol *3* 19
Krocker E 1966 Med Clin N Amer *March* 479
Scheneider E, Hyatt R 1963 Jama *186* 1061
Spiro H M 1970 Clinical Gastroenterology *XLII* 296

Tubulization of endothelial glomerular cells

V Vrcelj, P Spasic, N Bojanic

The Institute of Pathology and Forensic Medicine, Military Medical Academy Crnotravska 17, 11002 Belgrade, Yugoslavia

ABSTRACT: Tubulization of glomerular cell is extremly rare. We present a case with such a phenomenon displaid on the surface of the endothelial cells.

Tubulization of glomerular cell is a phenomenon of specialization of the apical surface of the cell membrane in the form of microvilli characteristic for apical part of the proximal tubule cells of the kidney called the brush border.

Tubulization of parietal glomerular cells is noted in various glomerular diseases but it can be found in the urinary pole in normal persons as Ward[4] (1970) and Marcus[3] (1977) described.

In patients with endemic nephropathy, analysed by Apostolov[1] et al (1975), Bojanic[2] et al (1985) described tubulization of visceral epithelial cells (podocytes) and in parietal epithelial cell as well.

The endothelial cytoplasm is highly attenuated and perforated by fenestrations and reacts to stimuli in various ways such as edema, vacuole formation, phagocytosis, "arcades" formation, and separation from the basement mebrane. Whereas tubular metaplasia or tubulization of the parietal Bowman`s capsular epithelium and podocytes have been well documented, there is no reference in the literature to the such a phenomenon of the endothelial cells.

We have noticed this phenomenon during routine electron microscopy of a renal byopsy specimen of 20 years old male with mesangioproliferative glomerulonephritis clinically characterised by haematuria and proteinuria. In this case we have encountered tubulization of parietal epithelium (Figure 1), podocytes (Figure 2) and endothelial cells (Figure 3).

Virus-like particles have been found in cytoplasmic vesicles of mesangial cells, and numerous dense bodies in cytoplasm in the endothelial cells.

We can speculate about the role of viruses which is not excluded in regular microvilli reorganisation of the cell membrane. Such an alteration of the cell membrane can be connected with the etiological agent of the disease.

1 Apostolov K, Spasic P, Bojanic N, 1975 *Lancet* 2 pp 1271-1274
2 Bojanic N et al 1985 *Path. Res. Pract.* 180:3 p 258
3 Marcus PB 1977 *Arch. Pathol. Lab. Med.* 101:12 p 664
4 Ward AM 1970 *J Clin. Pathol.* Vol.23 pp 472-474

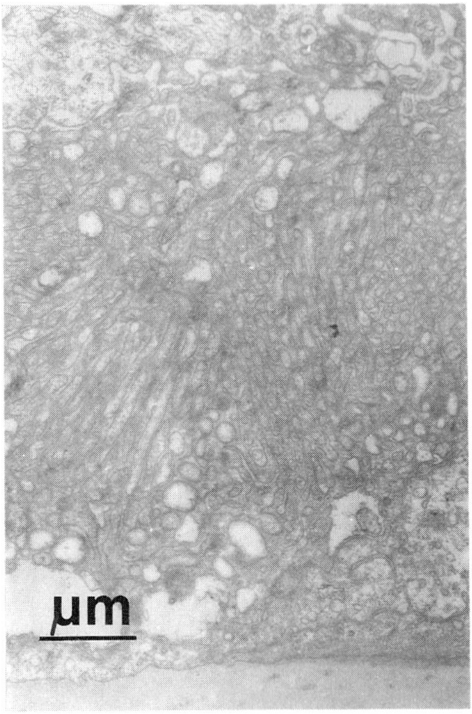

Fig. 1.

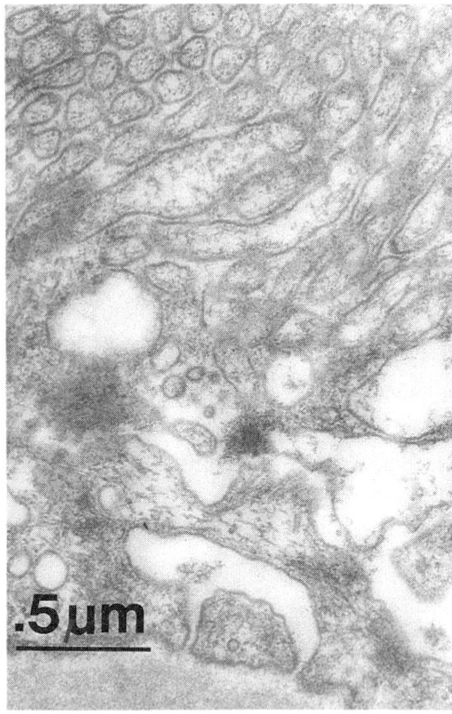

Fig. 2.

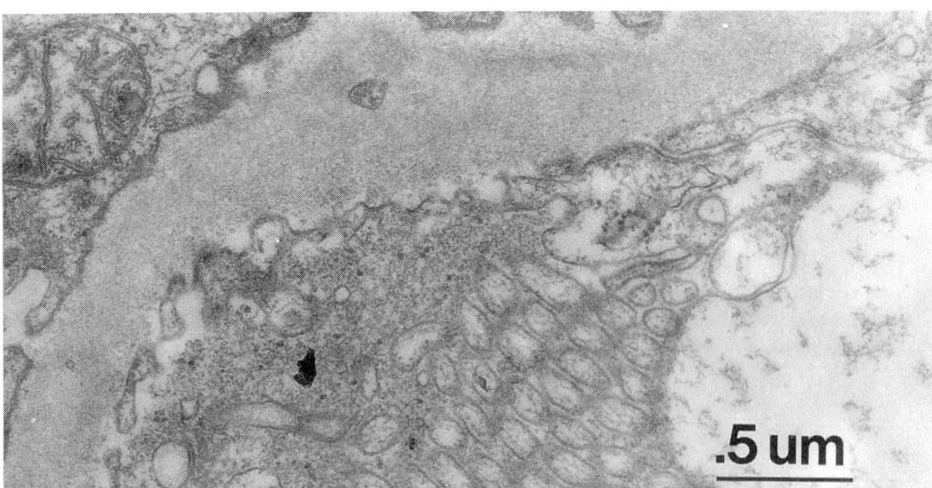

Fig. 3.

Electron microscopical investigation of the endocrine-active pituitary adenomas

T Erbengi, A Erbengi, A Canbolat, E Sencer, S Molvalılar, T Özgen

Departments of Histology and Embryology,Neurosurgery,Internal Medicine,Istanbul Faculty of Medicine and Dept.of Neurosurgery Hacettepe Faculty of Medicine, TURKEY.

ABSTRACT: In this team work,the endocrine active pituitary adenomas were evaluated in patient with medical therapy or without therapy.In sparsely granulated type of Somatotropinomas,large fibrous bodies were observed.In densely granulated growth hormon adenomas, the ultrastructural features of tumour cells were similar to those nontumourous somatotrophs,after treatment.In prolactinoma cases there was a marked reduction in the size of cells after bromocroptine therapy.Our electronmicrographs revealed the presence of intracellular cysts and packages of secretory granules by fibrillary mass in adenoma cells,in the patients with Cushing's disease.

In recent years,there has been many studies on pituitary adenomas reported by neurosurgical centers [Landolt (1978), Rengachary et al (1982)].In addition,ultrastructural observations helped the evaluation of endocrine-active pituitary adenomas in relation to the cell type and specific secretory granules (Asa and Kovacs (1983)).However,the types of adenomas are significantly relevant in the choice of therapy.Bearing this in mind,ourteam evaluated the endocrine-active pituitary adenomas: somatotropinomas (10 cases), prolactinomas (11 cases) and Cushing's disease (6 cases) in patients with medical and radiation therapy or without therapy.In sparsely granulated somatotropinomas (Fig.1),of some cells large spherical bodies (SB) were seen with close approximation to the nucleus.These spherical bodies were composed of microfilaments including mitochondria,endoplasmic reticulum and some secretory granules.In densely granulated growth hormone adenomas,the fine structural features of tumour cells were similar to those nontumourous somatotrophs,after treatment (Fig.2).

Our prolactinoma cases were observed in the sparsely granulated type,containing well developed parallel lamellae of rough endoplasmic reticulum (ER) in the cytoplasm of the tumour cells (Fig.3).In addition,misplaced exocytosis was the most common feature.After bromocriptine therapy,there was a marked reduction in the size of cells and an increase in secretory granules.The nucleus (N) was small with an irregular outline (Fig.4).

We had the opportunity to evaluate the pre-and post-operative endocrine state of six patients with Cushing's disease who were operated by transsphenoidal approach.Our patients seemed to be cured after the operation.Their quite normal responses to dexamethasone suppression tests indicated a successful removal of the pituitary adenoma.Electronmicrographs revealed the presence of the perinuclear fibrillary halo (F) and vesicles,vacuoles in some adenoma cells (Fig.5 a,b),in accordance with the findings of De Cicco et al (1972) and Kovacs et al (1974).Granules were increased in number and localized between the fibrillary masses and the cell membrane.Our two observations in a case of Cushing's disease were not mentioned in the literature, before.One of these findings was the presence of very large vacuoles"Cysts" in the cell cytoplasm.At the periphery of this Cyst(C),there was granule extrusion into the lumen (L).The other interesting finding was the occurence of packages of secretory granules surrounded by accumulated fibrillary material (←) in the cytoplasm of cell poor in organels (Fig.6).

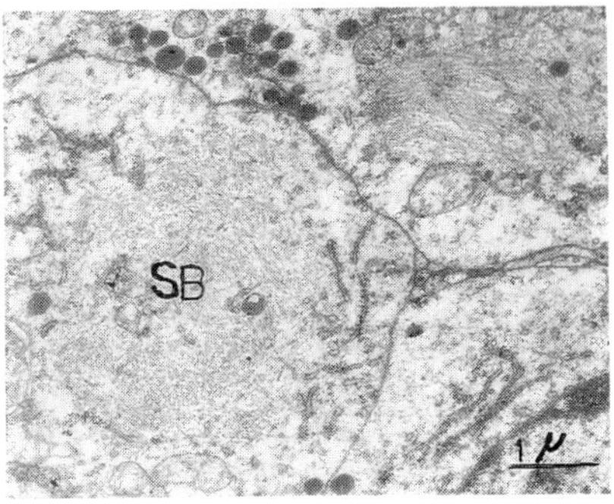

Fig.1.Sparsely granulated somatotropinoma

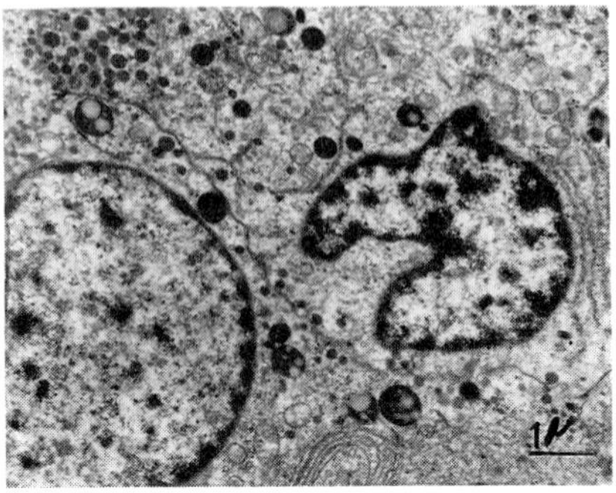

Fig.2.Densely granulated somatotropinoma, after treatment

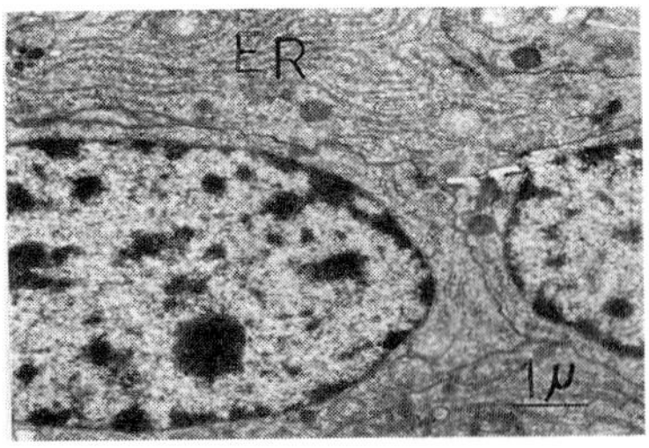

Fig.3.Sparsely granulated prolactinoma

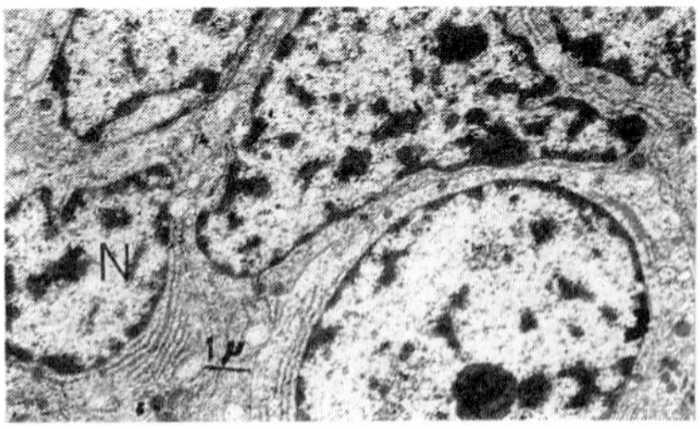

Fig.4. A case of prolactinoma after bromocriptine treatment

Thus,ultrastructural evaluation of endocrine-active adenomas is not only important for diagnose, but also to obtain optimal therapeutic results.

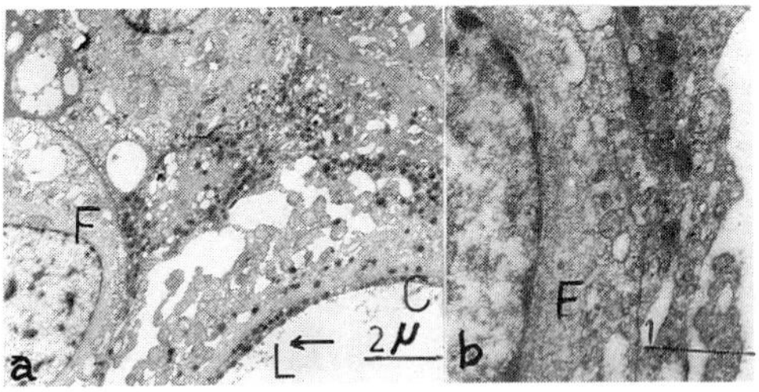

Fig.5. a,b.Cushing's disease. Note Perinuclear fibrillar bundles

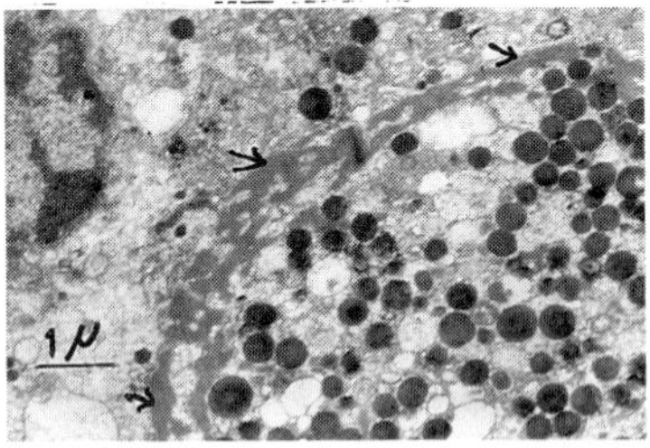

Fig.6.Cushing's disease secretory granules surrounded by fibrillary mass

References:

Asa S L and Kovacs K 1983 Clinics in Endocrinol and Metabol *12* 567
DeCicco F A, Dekker A and Yunis E J 1972 Arch Path *94* 65
Kovacs K, Horvath E, Stratmann I E and Ezrin C 1974 Acta Anat *87* 414
Landolt A M 1978 Advances in Technological Standards of Neurosurgery *5* 3
Rengachary S S, Tomita T, Jefferies B F and Watanabe I 1982 Neurosurgery *10* 242

The case for perfusion fixation of large tissue samples for ultrastructural pathology

MV McCrossan, JC Roberts and HB Jones

Department of Pathology, Smith Kline & French Research Ltd., The Frythe, Welwyn, Hertfordshire, AL6 9AR, U.K.

ABSTRACT: We report the case for the perfusion fixation of large tissue samples of liver and lung from dog, rat and mouse. The technique is simple, quick to perform and yields optimally - preserved specimens for ultrastructural examination. Its major advantage lies in its ability to provide well fixed specimens while avoiding whole organ or whole body perfusion. It has been employed successfully in investigative toxicological pathology and is potentially useful in diagnostic ultrastructual pathology.

1. INTRODUCTION

Tissue samples for ultrastructural investigations are commonly fixed by immersion in an aldehyde fixative or, more rarely, individual animals will be fixed by whole body perfusion. While the latter method yields optimally - fixed tissues for ultrastructural examination they are often unsuitable for routine histological preparation and for studies of pharmacokinetics and drug metabolism. The conflicting requirements for specific fixation regimes or the avoidance of fixation prompted us to address the potential of perfusion fixation of individual lobes of organs (liver and lung) thus permitting the use of the remaining tissue for other purposes. A major benefit of this strategy is that morphological, biochemical and pharmacological data may be obtained from the same animal.

2.MATERIALS AND METHODS

The perfusion apparatus used in this study is the same as that used for whole body perfusion of laboratory animals.
Single lobes of liver (right lateral) or lung (right apical(dog), right anterior (rat and mouse)) were quickly removed at necropsy. A gavage needle (gauge varied according to animal size) was inserted into a major blood vessel (rat and dog) or held against it (mouse) and warm (20-25°C) saline (0.9% w/v NaCl) perfused through the vasculature at a constant pressure of 25 mm Hg. Clearance of blood from the tissue produced local tissue pallor within 10-30 secs. Saline flow was then replaced by fixative at the same pressure which resulted in hardening of the saline-flushed tissue volume within 2-4 minutes. Samples for subsequent processing were taken from this well-fixed tissue. Perfusion pressure and flow rate were important determinants of fixation quality and production of artifactual damage. Rat and mouse liver and lung were perfused at a constant pressure of 25 mm Hg at flow rates of 5 and 9 mls/min respectively. Dog tissues were perfused at the same pressure and at a flow rate of 42 mls/min. Tissue preservation by perfusion fixation was compared with that obtained by immersion of thin (< 1 mm) slivers of tissue taken from the organ lobe immediately prior to perfusion.

3. RESULTS

Superior tissue preservation was obtained by perfusion fixation compared with that obtained by immersion in all specimens examined (Figures 1 and 2). The vasculature was patent, undistended and free of blood. Fixation was uniform throughout. Variable cytoplasmic staining density, seen commonly in immersion - fixed tissue was largely avoided by perfusion - fixation. Occasionally however, a few lucent hepatocytes were seen in otherwise optimally preserved mouse liver. Artifactual distension of the pulmonary microvasculature with endothelial plasmalemmal breaks was noted in early trials which employed elevated perfusion pressures and/or flow rates. The rate of acquisition of well - fixed specimens by perfusion was slower than that achieved by immersion. Nevertheless, 8-12 specimens per hour were prepared by this method.

4.CONCLUSION

Perfusion fixation of isolated lobes of liver and lung is easily performed, is reproducible and provides optimal tissue preservation for ultrastructural investigations.

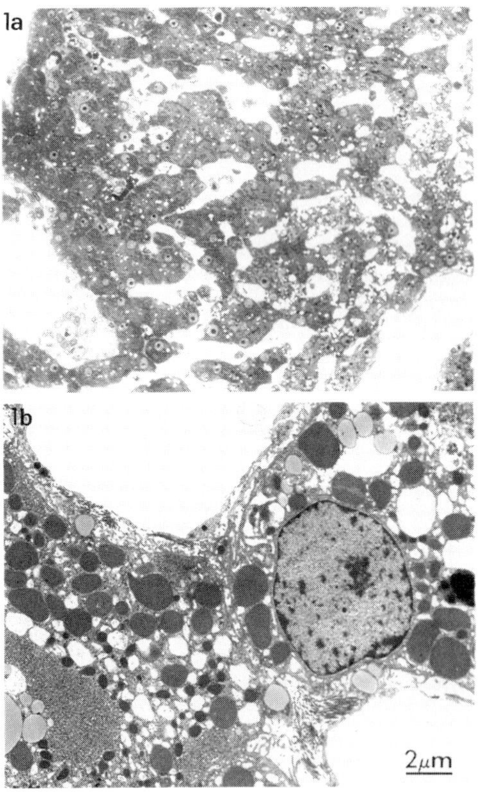

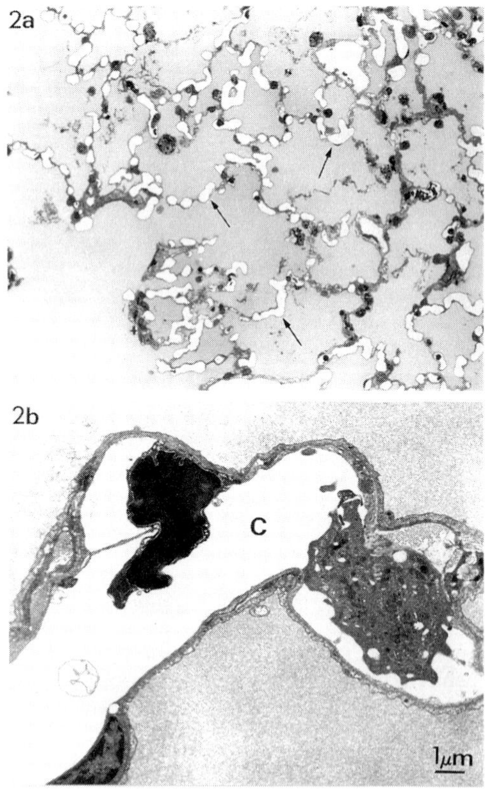

Figure 1. a) Light micrograph of perfusion - fixed dog liver illustrating quality of tissue preservation and centrilobular hepatocellular hydropic degeneration, b) Electron Micrograph - Hydropic degeneration of hepatocytes.

Figure 2. a) Light micrograph of perfusion - fixed dog lung illustrating alveolar oedema and patent, empty, capillaries (arrows). b) Electron Micrograph - perfused capillary (C) showing treatment - related endothelial cell necrosis and alveolar oedema.

Rapid contrasting of extracellular elements in thin sections

KP Dingemans, MA van den Bergh Weerman

Laboratory of Pathology, University of Amsterdam, Academic Medical Centre, Meibergdreef 9, 1105 AZ Amsterdam, The Netherlands

ABSTRACT: Extracellular tissue elements in thin-sectioned material are often poorly and / or capriciously contrasted by conventional uranyl acetate and lead citrate staining. We present a simple, rapid method to overcome these problems.

1. INTRODUCTION

Tannic acid is sometimes used for increasing the ultrastructural contrast of various intra- and extracellular tissue elements. However, the general application of tannic acid contrasting has been hampered by several factors, e.g., the penetration problems that occur when tissue blocks are treated with tannic acid in toto before embedding, and the poor stability of the various mixtures containing tannic acid that are recommended for staining sections.

2. METHOD

We tested tannic acid under different conditions and found the following method for routinely prepared epoxy resin sections. Low-molecular weight tannic acid is dissolved in distilled water (0.05%). This solution is stable for many weeks and only needs to be filtered prior to use. Sections mounted on copper grids (with or without supporting membrane) are treated with the tannic acid solution for 3 min. at 60°C, followed by conventional uranyl acetate and lead citrate staining.

By this simple and rapid method, the contrast of all extracellular elements (collagen - including anchoring fibres -, basement membranes, elastin, proteoglycans, microfilaments, etc.) is consistently and uniformly increased. The contrast of cell components is only slightly increased, except glycogen, which is rendered highly electron-dense, even after block-staining with uranyl acetate.

When longer staining times or higher tannic acid concentrations are applied, the electron density of elastin is greatly increased; under these conditions, tannic acid can be used as a specific elastin stain.

In our laboratory, the method is now routinely used for clinical and experimental material. The illustrations on the next page show some examples of its application.

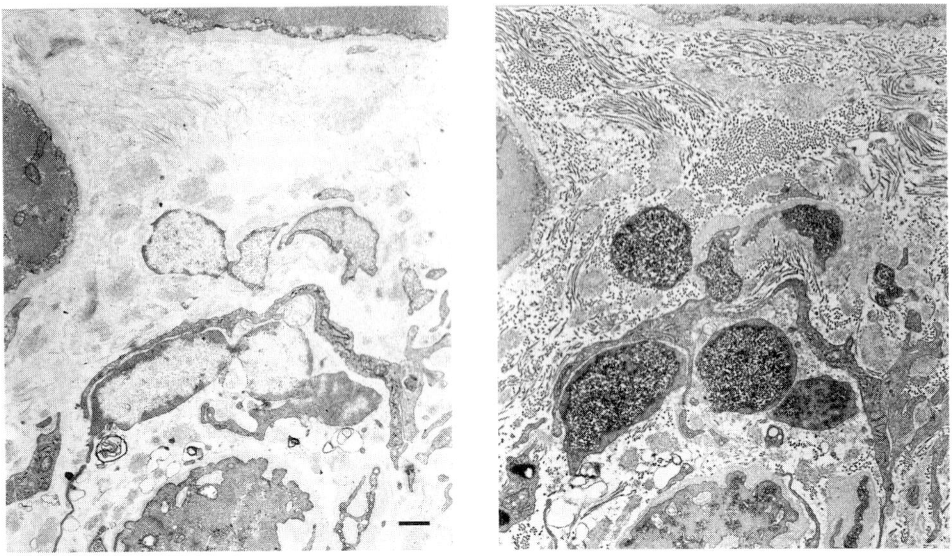

Fig. 1. Myocardial biopsy. In conventionally stained specimen (left), contrast is largely confined to intact cardiomyocytes, parts of which are visible near edges of micrograph. Tannic acid pretreatment of nearby section (right) reveals glycogen contents of extensions of degenerating cardiomyocytes and increased amounts of collagen. Bar represents 1 micron.

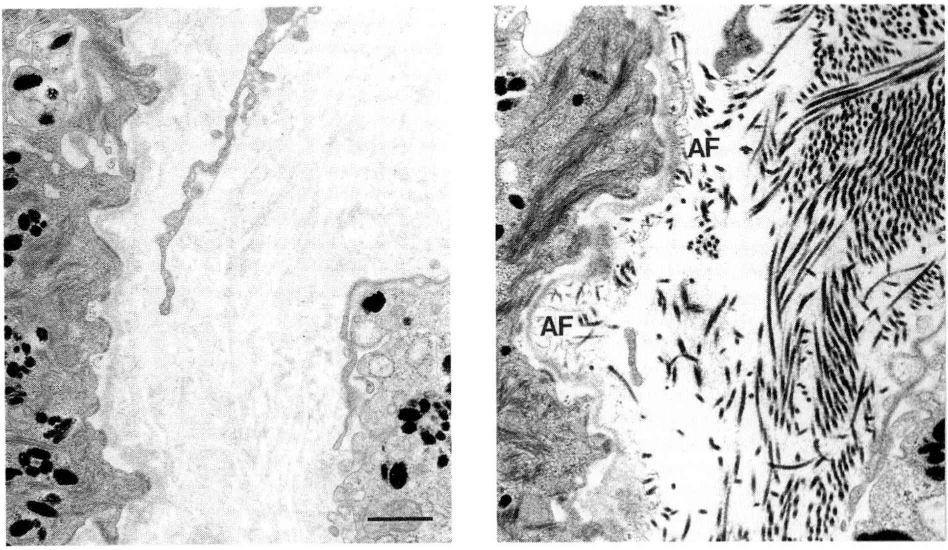

Fig. 2. Skin biopsy. Basal aspect of epidermis is at left; part of dermal melanophage is visible in lower right corner. In this specimen, conventional staining (left) completely failed to contrast collagen fibres (cf. Ghadially, 1982). Tannic acid pretreatment of nearby section (right) has rendered epidermal basement membrane, anchoring fibres (AF), and dermal collagen fibres highly electron-dense. Bar represents 1 micron.

Approximate timetables for embedding media and staining of sections for nematodes: TEM study 17 species for locate carbohydrate moieties on the epicuticle membrane

RYUNTYU, M., Department of Physiology, University of New England, Armidale. N.S.W. 2351. Australia.

The series of experiments shows the results for several resins (Table 1). The duration of OsO_4 fixation was similar to Table as shown in Ryuntyu "Approximate fixative timetable for nematodes". Proc. 9th European Congress on Electron Microscopy, York, U.K.: 1988. The pieces of specimens were transferred from 100% alcohol into various resins. Four grids (10-20 sections each) were prepared for each specimen which stained in the conventional manner (Table 1).

Positive results are given in the form of designations: (a) presence of membrane-like components of the epicuticle; (b) absence of electron-transparent areas with vesicles in the cuticle strata; (c) uniform distribution of electron density over the whole "shape" of the striations and the stratum ribbons (X60,000; X100,000).

The results were considered to be positive: (a) when a strip was obtained of serial sections, i.e. evidence of good penetration by resins, and (b) when sections stained well (7 mins in lead citrate and 12 mins in uranyl acetate). The final series of experiments (Table 2) shows the results for staining of only one type of resin selected in the second series (Table 1).

Table N 1

Key:
* * - for muscles
* I - for selective different cuticular layers.
* °° - for hypodermis and all layers
* ° - for cuticular structures (but not hypodermis).

Table N2

Key:
* I - lead citrate (7 min)]
* II - lead citrate (10 min)]
* III - lead citrate (15 min)
* • - uranyl acetate (7 min)]
* •• - uranyl acetate (15 min)]
* ⁘ - uranyl acetate (20 min)]
* ∞ - uranyl acetate (30 min)]

Results which provided optimal delineation of all cuticular layers

Tables 1-2

Key:
* ++ good result
* +/- unstable result
* \+ not good result

Species:

(1) - A. obtusicaudatus
(2) - X. monohysterum
(3) - P. lobatus
(4) - M. monhystera
(5) - P. redivivus
(6) - P. thornei
(7) - H. avenae
(8) - M. arenaria
(9) - P. nanus
(10) - T. semipenetrans
(11) - S. paratenuicaudata
(12) - A. avenae
(13) - C. papillatus
(14) - R. oxycerca
(15) - C. elegans
(16) - Mermis sp.
(17) - Amphimermis sp.

Table No. 1 GENERAL PROPERTIES OF FREQUENTLY EMPLOYED EMBEDDING MEDIA FOR NEMATODES

Species ♀	Araldite	Epon	Maraglas	Vestopal	Methacrylate
Aporcelaimellus (1)	+I°	++∵°°	+I°	+/-I°	++°
Xiphinema (2)	+∵	+°°	++∵°°	++°	+∵
Paratrichodotus (3)	++∵°	+°°	+/-°°	+/-°°	+∵
Mesorhabditis (4)	+∵	++∵°°	+I*	+°	+∵
Panagrellus (5)	+∵	+°°	+I*	++∵°°	+∵
Pratylenchus (6)	+∵	+°°	++∵°	+/-°°	+∵
Heterodera ♂ (7)	++∵°°	+°	+/-∵	+°	+∵
Meloidogyne ♂ (8)	+∵	+°°	++∵I	+°	+/- ∵
Paratylenchus (9)	+∵	++∵	+I°	+/-°°	+/- ∵
Tylenchus ♂ (10)	+∵	++∵°°	+I	+/-°°	+/- ∵
Seinura (11)	+∵	++°°	+/-°	++∵°	+/- ∵
Aphelenchus (12)	++∵°°	+°°	+/-°	++∵	+∵
Clarcus (13)	+∵	+°°	++∵°°	+/-I	+∵°
Rhabditis (14)	+∵	+°°	++∵°°	+/-I	+∵
Caenorhabditis (15)	+∵	++∵°°	+/-°	+°	+∵
Mermis (16)	+°	++∵°°	+/-°°	+°	+∵
Amphimermis (17)	+I	++I	+I	++∵°°	+∵
Key: ∵ - for moieties on the epicuticle membrane					

Table No. 2 APPROXIMATE STAINING TIME FOR NEMATODES

Species ♀	Hypodermis cell wall	Mito-chondria	Rough endo-plasmic reticulum	Smooth endoplasmic reticulum	Nuclear contents	Muscles	Cutilar layers (++)
Aporcelaimellus (1)	++	++	+	++	+	++	III•
Xiphinema (2)	++	+	+	+	+	++	I•
Paratrichodotus (3)	++	+	+	+	++	++	II ••
Mesorhabditis (4)	+	+	+	++	+	+	I ∞
Panagrellus (5)	+	+	+/-	+/-	+	++	II••
Pratylenchus (6)	+	+	+	+	+	+	II••
Heterodera ♂ (7)	++	+	+	+	+	+	III•
Meloidogyne ♂ (8)	++	+	+	+	++	+	I••
Paratylenchus (9)	++	+	++	+	+	+	I∵
Tylenchus ♂ (10)	+	+	++	+	+	+	III•
Seinura (11)	+	+	+	+	+	++	I••
Aphelenchus (12)	+	++	+	+	++	+	I ∞
Clarcus (13)	++	++	+	+	+	+	II•
Rhabditis (14)	++	++	++	+	+/-	+	I••
Caenorhabditis (15)	++	+	+	+	+	+	I••
Mermis (16)	++	+	++	+	++	++	I••
Amphimermis (17)	++	++	++	+	++	++	II•
Factors upon which evaluation of the results is based	no breaks; dense and essentially layered . No break between cuticle and hypodermis	neither swollen nor shrunk; dense matrix; double membrane and cristae intact	flattened cisternae uniformly arranged in long profiles with attached ribosomes	tubular appearance instead of vesticular; branching tubules with intact membranes not associated with ribosomes	uniformly dense with masses of chromatin scattered adjacent to nuclear membrane	no breaks; no empty spaces; dense long miofibrills Close contact with hypodermis	no breaks; dense and essentially layered cuticle; 3-layer membrane in epicuticle

Cell structure in immobilised systems

Jean-Noël BARBOTIN
Laboratoire de Technologie Enzymatique, Université de Technologie de Compiègne, B.P. 649, 60 206 COMPIEGNE, FRANCE

ABSTRACT : Procedures involving immobilization by whole cell entrapment have been used intensively for biotechnological applications. Various optical and electron microscopy studies have been carried out on the growth and morphology of immobilized prokaryotes (*E. coli, P. nautica*) and eukaryotes (Euglena, hybridomas, hepatocytes, neuroblastomas) cells to investigate the cell distribution within the matrix and to confirm the preservation of their ultrastructure.

1. INTRODUCTION

During the last decade there has been an increased interest in the use of immobilized enzymes, organelles, microorganisms, plant or animal cells (Mosbach 1987). Whole cell immobilization was defined as the physical confinement of cells to a certain region of space with the preservation of some desired activity. Such a process allows a continuous production of a given metabolite, and a high cellular concentration. An understanding of the fundamental properties (physiology, mobility, interactions with the support...) underlying the behavior of immobilized cells is necessary to make them a viable alternative to conventional processes. Generally, the immobilization enhances the functional and the storage longevity (Dainty et al 1986) and it is therefore of some interest to use this technique as a tool for basic studies of the influence of spatial organization and compartmentalization on the cell behavior. The present study outlines the use of microscopy to investigate the distribution of microorganisms growth inside gel beads (Wada et al 1980, Burril et al 1983) and the cell ultrastructure of prokaryote and eukaryote immobilized cells.

2. METHODOLOGY

A multitude of immobilization methods exist : attachment to surfaces, covalent coupling and entrapment within porous matrices. The entrapment within polysaccharides gels is one of the most successfull means of immobilizing cells. This is a simple and gentle process : cell suspension are stirred into solutions of K-carrageenan (a copolymer of D-galactopyranose-4-sulfate and 3-6 anydro-ß-D galactopyranose) or sodium alginate (a copolymer of ß-D-mannuronic acid and α-L- guluronic acid). Then, the respective mixtures are extruded into solutions of potassium or calcium chloride respectively. Gel beads of a mean diameter of 0•5 to 4 mm are obtained in both cases. If cell growth is desired, gel beads are added in a reactor in the presence of nutrients. Cells multiply within the beads giving high cell densities. A qualitative and quantitative evaluation of

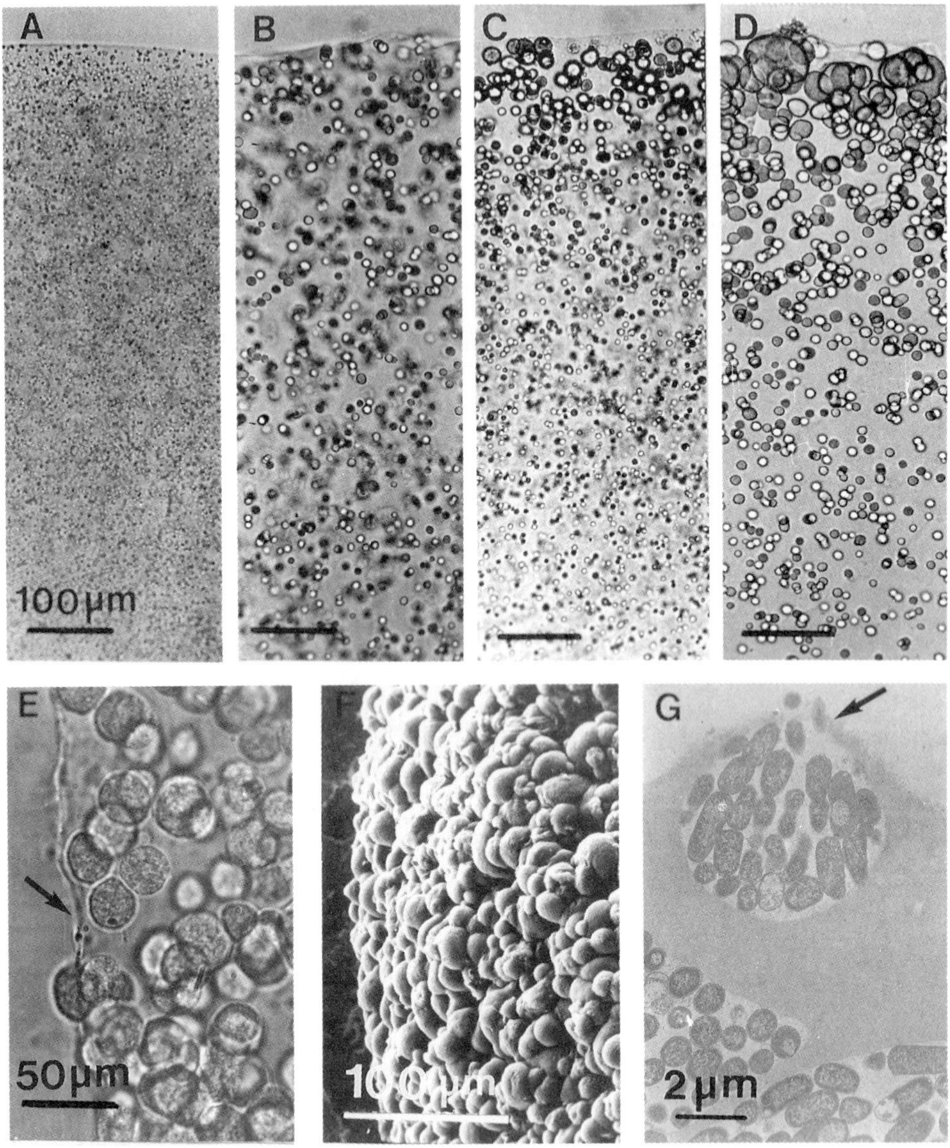

Figure 1 (A, B, C, D and E) : Optical micrographs of sliced K-carrageenan gel beads (glutaraldehyde fixation 0.5 % during 4 h) containing *E. coli* K12/pTG201 microcolonies after different times of incubation : **A** (1•5 h), **B** (3•5 h), **C** (5 h), **D** (10 h) **E** (12 h). In **E** high magnification of the gel surface (arrow) showing the presence of distinct spherical colonies inside the gel.

Figure 1 F : Scanning electron micrograph showing the surface of the gel bead after 24 h of incubation.

Figure 1 G : Electron micrograph of ultrathin section showing the gel rupture on the surface (arrow).

the cell colonization can be approached by using phase contrast light microscope and electron microscopy techniques (gel beads were fixed, dehydrated and embedded in Epon).

3. IMMOBILIZED PROKARYOTES CELLS

3.1. *Escherichia coli* cells : cells of *Escherichia coli* carrying a recombinant plasmid were immobilized in K-carrageenan gel in order to improve the plasmid stability and the expression of the encoded gene (De Taxis du Poët et al 1986, Nasri et al 1987). We have demonstrated that in immobilized continuous culture without selection pressure, plasmid free segregants were not detected even after 240 generations. This appears to be due to the mechanical properties of the gel bead system that allow only a limited number of cell divisions (10-16) to occur in each clone of cells before the clone escapes from the gel bead (Fig.1). The morphological aspects of growing cells have give a basis to explanations dealing with the role played by immobilization on the plasmid stability. The entrapped cells (10^8 per ml) were homogeneously mixed within the gel beads and grew inside distinct cavities (Larreta Garde et al 1981). After 1.5 h and 3.5 h of growth (Fig. 1 A, B), the presence of distinct microcolonies was clearly observed by optical microscopy. After 5 h (Fig. 1 C) and 10 h (Fig. 1 D) the cell cavities were gradually expanded especially near the gel surface. The volume occupied by the cell in the 50 µm thick outer layer was higher than 50 % of the available volume in the bead and a high magnification showed that the cells grew inside distinct cavities (Fig. 1 E). A general view of the external surface of the bead has been observed by scanning electron microscopy (Fig. 1 F). Transmission electron microscopy shows a leakage of bacteria (Fig. 1 G) occuring after 12 h of growth. These observations indicate a limitation of growth in the centre of the gel beads and a continuous renewal of cells near the surface by disrupting the surrounding matrix. The high density of cell (more than 10^{11} cells per ml of occupied volume) has been estimated by electron microscopic examinations (Fig. 2 A, B) and the mechanical constraints inside the gel and the strong physical contacts between immobilized cells have also been observed (Fig. 2 C).

3.2. Denitrifying bacteria : Cells of *Pseudomonas nautica* were immobilized in a carrageenan gel to mimic the behavior of such a denitrifying bacterium in sediment (Bonin et al 1987). Figures 2 D and 2 E show a bacterial colony during exponential growth in a central part of the gel which simulates a microniche in an anaerobic media. Furthermore electron microscopy gives evidence of membrane blebs and vesicle accumulation inside the cavities.

4. IMMOBILIZED EUKARYOTES CELLS

4.1. Algae : Ultrastructural studies of immobilized *Euglena gracilis* cells in calcium alginate stored at 4°C in darkness (Tamponnet et al 1985) indicate that cells were fixed in the same cellular state as when the immobilization occured (Fig. 3 A, B). With green cells we have found that chlorophyll levels of immobilized cells were still 90 % of original levels after three months in storage whereas free cells displayed pheophytinization after only seven days. The involvement of calcium ions in this effect was considered to be of importance and this technique is suggested as being useful for the long-term storage of algal strains in culture collections.

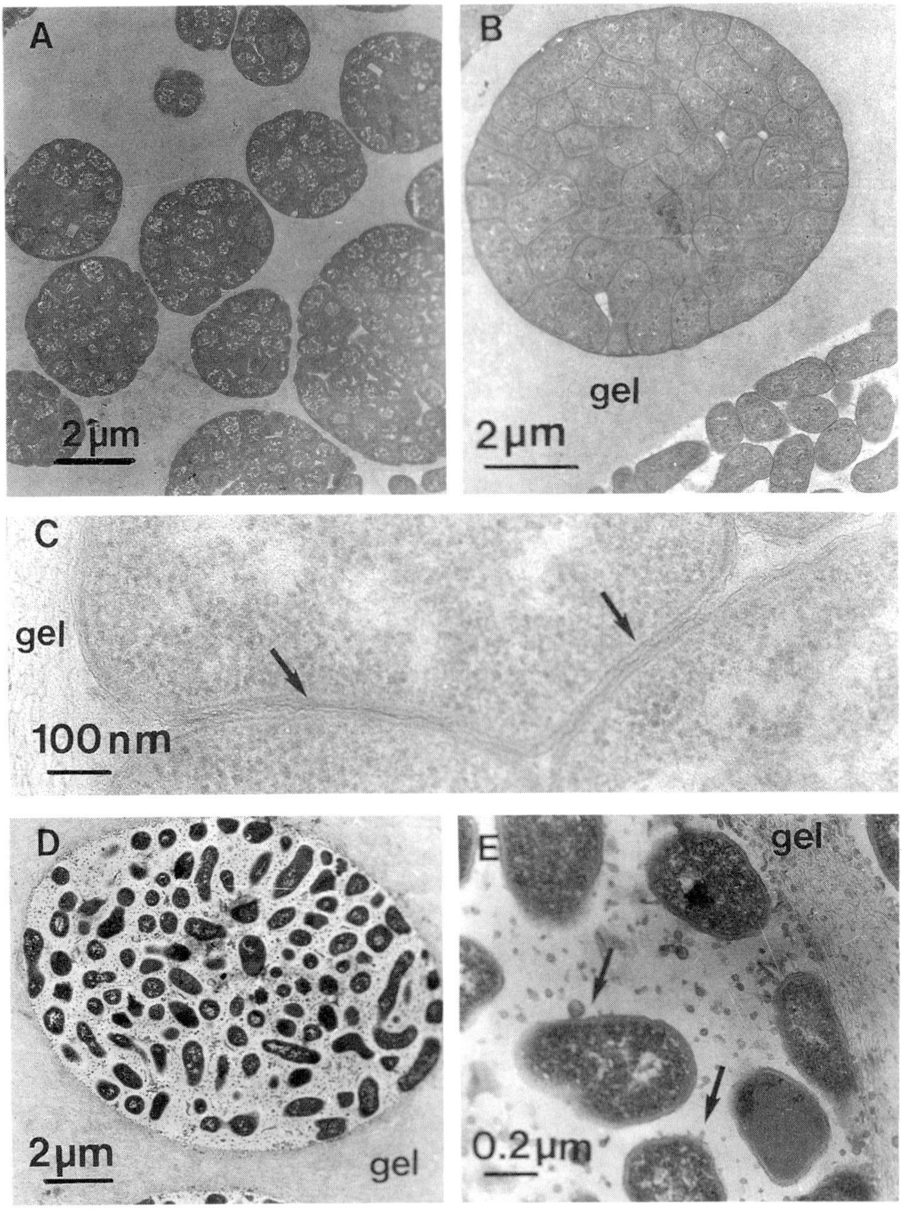

Figure 2 (A, B and C) : Electron micrographs of ultrathin sections through gel beads of immobilized recombinant *E. coli* HB101 after 7 h growth. Note the high density of cells and the spherical aspect of the cavities inside the gel. Figure 2 C shows the strong physical contacts between immobilized cells and the structure of the gel.

Figure 2 D and E : Ultrathin section of immobilized *Pseudomonas nautica* 617 in K-carrageenan gel beads. Note the spherical aspect of cavities. Numerous membrane blebs and vesicles can be observed. Such vesicles are retained inside the gel bead.

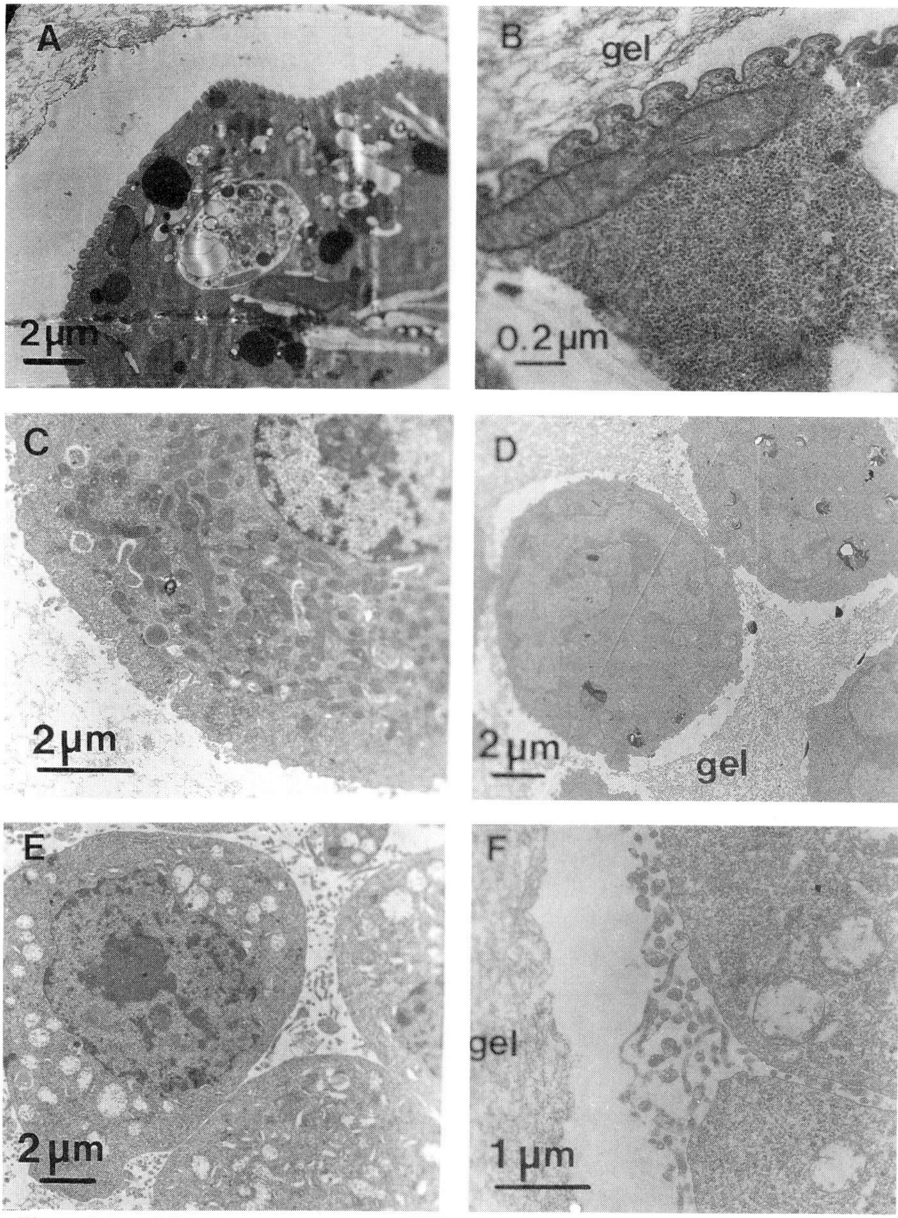

Figure 3 A and B : Ultrastructure of immobilized *Euglena gracilis* cells inside alginate gel bead. Note the presence of a green cell inside a round-shape cavity (**A**) and the integrity of the external pellicle and a mitochondria in an etiolated cell (**B**).

Figure 3 C : Immobilized hepatocyte inside alginate gel bead after 1 month storage.

Figure 3 D : Immobilized hybridomas inside alginate gel bead.

Figure 3 E and F : Immobilized Neuroblastomas cells after 18 h of growing. Note the neuritic growth.

4.2. Rat hepatocytes : isolated rat hepatocytes were entrapped in alginate gel beads (up to 10^5 cells per bead) and have been tested for abilities of gluconeogenesis from alanine and glycogenolysis in response to adrenaline. These physiological activities have been maintained after 15 weeks of storage at 4°C. Electron microscope examination (Fig. 3 C) reveals the ultrastructural integrity of such cells. This system may constitute an interesting approach for *in vitro* detoxification studies.

4.3. Hybridomas : The entrapment of murine hybridomas in calcium alginate gel beads provides a means of obtaining a dense culture (Fig. 3 D) and a concentration of monoclonal antibodies.

4.4. Neuroblastoma cells : Figure 3 E and F illustrated a typical section of alginate beads with mouse neuroblastoma cells after 18 hours of growing. Especially profiles suggestive of neuritic growth can be observed. This approach allows to study neuronal growth and membrane surface properties in a three dimensional lattice which mimic extracellular geometry.

5. AKNOWLEDGMENTS.

The author wishes to gratefully acknowledge the assistance of Mrs M. Velut for electron microscopy studies.

6. REFERENCES

BONIN P, BARBOTIN J.N. , DHULSTER P. and BERTRAND J.C. (1987). Can J. Microbiol. **33**, 276.

BURRIL H.N., BELL L.E. , GREENFIELD P.F. and DO D.D. (1983). Appl. Environ. Microbiol. **46**, 716.

DAINTY A.L , GOULDING K.H. , ROBINSON P.K. , SIMPKINS I. and TREVAN M.D. (1986). Biotechnol. Bioeng. **28**, 210.

DE TAXIS DU POET P. , DHULSTER P. , BARBOTIN J.N. and THOMAS D. (1986). J. Bacteriol. **165**, 871.

LARRETA GARDE V. , THOMASSET B. and BARBOTIN J.N. (1981). Enzyme Microb. Technol. **3**, 216.

MOSBACH K. (1987). Methods in Enzymology New York Academic Press, **135**.

NASRI M. , SAYADI S. , BARBOTIN J.N and THOMAS D. (1987). J. Biotechnol. **6**, 147.

TAMPONNET C. , COSTANTINO F. , BARBOTIN J.N. and CALVAYRAC R. (1985). Physiol. Plant. **63**, 277.

WADA M. , KATO J. and CHIBATA I. (1980). J. Ferment. Technol. **58**, 327.

Detection of proteins expressed in insect cells infected by a genetically engineered virus using immunogold and silver enhancement techniques

J W M van Lent, F A Klinkenberg, K J M Zaal, M Usmany and J M Vlak

Department of Virology, Agricultural University, P.O. Box 8045, 6700 EM Wageningen, The Netherlands

In addition to their use as biological control agents of agriculturally important insects, baculoviruses have become a powerful vector for the high-level expression of heterologous pro- and eukaryotic genes (1,2). Baculoviruses have a number of unique properties that promote their engineering into a valuable expression vector. Two viral genes, polyhedrin and p10, are expressed to very high levels late after infection. In the case of Autographa californica multi-nucleocapsid nuclear polyhedrosis virus (AcMNPV), ca. 25% of the total cell protein is finally polyhedrin and p10. Polyhedrin is the major constituent of the large cuboidal bodies (polyhedra) found in the cell nucleus (Figure 1). These polyhedra contain rod-shaped viral nucleocapsids and are the infectious entities for insect larvae. The other protein, p10, is found associated with large fibrous structures (Figure 1) in the nucleus and cytoplasm in infected cells (3). Other important features of baculoviruses are that the viral DNA genome is expansible, i.e. is able to accommodate additional DNA. In addition, the two major late genes, polyhedrin and p10, are dispensable for virus replication, i.e. infectious virus of the so-called non-occluded form, which is responsible for the cell-to-cell spread of the virus, remains to be produced. These properties allow the manipulation of these viruses by allelic replacement of the polyhedrin and p10 gene by other genes of medical, pharmaceutical or agricultural importance (2).

In this paper we report on the detection of bacterial beta-galactosidase, expressed under the control of the p10 promoter, in AcMNPV recombinant-infected Spodoptera frugiperda cells. The manipulation was carried out in such a way that in the recombinant the N-terminal 53 amino acids of p10 were linked to the 1050 C-terminal amino acids of beta-galactosidase (4). In addition, we document on the consequences of the allelic replacement of the p10 gene by a bacterial beta-galactosidase gene for the fibrous structures and for polyhedron morphogenesis.

In the recombinant virus-infected cells, the fibrous structures had disappeared (Figure 2) confirming the involvement of p10 in these structures. In stead, large granular structures were observed, which were found by immunogold labelling (Figure 3) or immunogold labelling enhanced with silver (Figure 4) to contain beta-galactosidase. The recombinant product was also present in large amounts in the perifery of the cytoplasm, which may explain the secretion of this protein observed. Polyhedra were produced in normal quantity and size. The electron-dense spacers, thought to be precursors of the polyhedron membrane, were absent from cells infected by the recombinant virus; the polyhedra did not contain a membrane.

The immunogold and silver enhancement techniques will appear to be valuable tools in the detection of genetically engineered proteins in recombinant virus-infected cells.

1) Kelly D C 1982 J. gen. Virol. **63** 1
2) Luckow V E and M D Summers 1988 Bio/Technology **6** 47
3) Van der Wilk F, van Lent J W M and Vlak J M 1987 J. gen. Virol. **68** 2615

4) Vlak J M, Klinkenberg F A, Zaal K J M, Usmany M, Klinge-Roode E C, Geervliet J B F, Roosien J and van Lent J W M 1988 J. gen. Virol. **69** in press

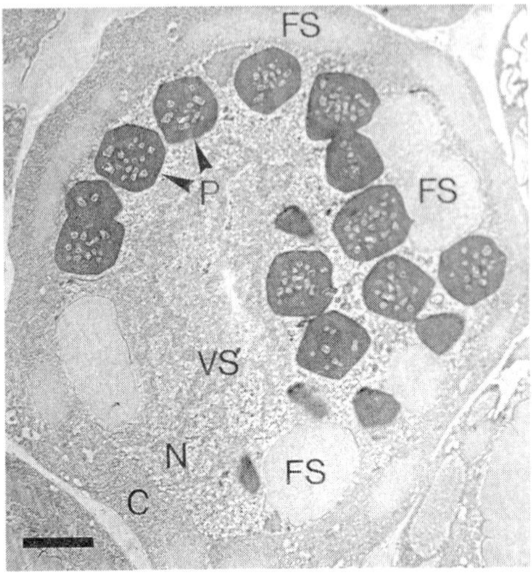

Figure 1. Spodoptera frugiperda cell infected with AcMNPV at 50h after infection. P=polyhedra; FS=fibrous structure; VS=virogenic stroma; N=nucleus; C=cytoplasm. Bar=2µm.

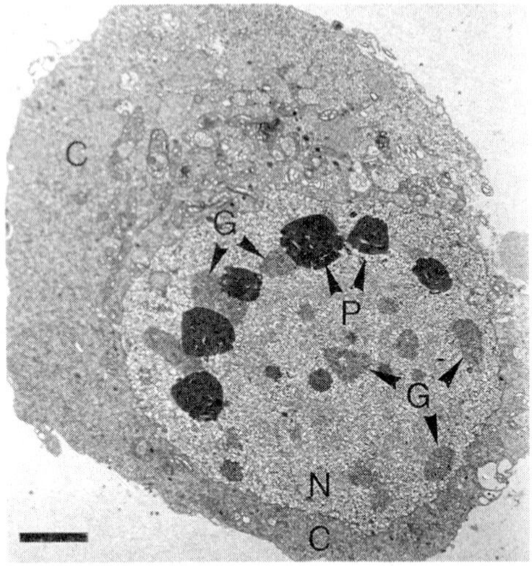

Figure 2. Sp. frugiperda cell infected with a recombinant of AcMNPV at 42h after infection. P=polyhedra; G=granular structure; N=nucleus; C=cytoplasm. Bar=2µm

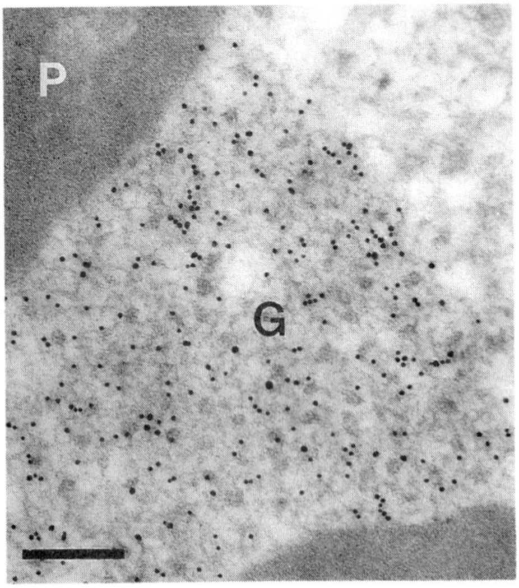

Figure 3. Section with a granular structure (G), treated with anti-beta-galactosidase antiserum and protein A-gold. Bar=0.2µm

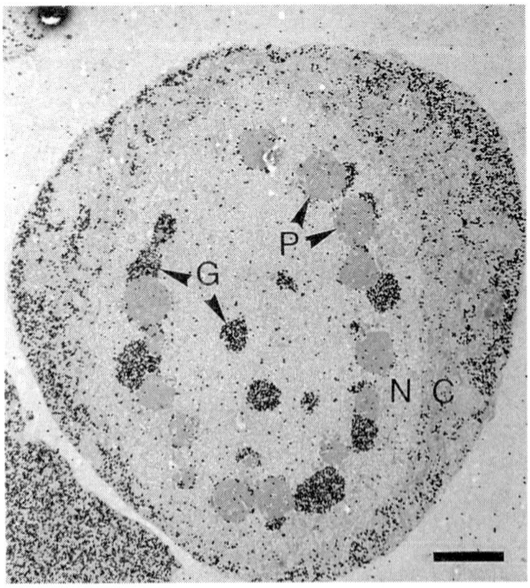

Figure 4. Section of a AcMNPV-recombinant infected cell, treated with anti-beta-galactosidase antiserum and protein A-gold enhanced with silver. Bar=2µm

Candida albicans: ultrastructure and lectin cytochemistry using specimens embedded in acrylic resin

M J Wilkinson[1], A H Everett[2] and P A Hunter[2]

Beecham Pharmaceuticals Research Division, Chemotherapeutic Research Centre[1], Brockham Park, Betchworth, Surrey RH3 7AJ and Biosciences Research Centre[2], Yew Tree Bottom Road, Epsom, Surrey K18 5XQ, UK.

ABSTRACT: A means of preparing cells of C.albicans for transmission electron microscopy is described which overcomes problems and artefacts inherent in previous methods and offers greater sensitivity in the evaluation of chitin distribution when using lecting-gold conjugate.

1. INTRODUCTION

Examining the ultrastructural effects of antifungal agents on C.albicans requires a reliable method of preparing thin sections. Our initial investigations of published methods (Drewe, 1981, Garrison, 1981) showed that preparation with epoxy resins had two main shortcomings: 1) less than satisfactory resin infiltration through the thick cell wall 2) restricted capacity for labelling chitin with lectin-gold conjugate. We therefore examined the potential of acrylic based resin.

2. MATERIALS AND METHODS

C.albicans (ATCC 28366, 73/079 supplied by Dr. F. Odds, Leicester University) was cultured in Sabouraud's broth at 26°C overnight followed by 24 h in Eagle's medium with 10% horse serum at 37°C in microaerophilic conditions. Washed hyphae were fixed with 2.5% glutaraldehyde in 100mM cacodylate buffer pH7.2 for 20 h at 4°C and post-fixed in 1.5% aqueous potassium permanganate for 1.5 h. Following 5 water washes, cells were encapsulated in glutaraldehyde-cross-linked BSA, cut into small ($1mm^3$ or less) blocks and dehydrated in an ethanol series. Infiltration with acrylic resin (L R White, London Resin Co.) was for 1 h followed by 16 h polymerisation in gelatin capsules at 52°C. Infiltration with epoxy resin (Spurr's, Taab, Lemix) was for a minimum of 24 h followed by curing for 16 h at 70°C. The glutaraldehyde-osmium-tannic acid-uranyl acetate method of Persi and Burnham (1981) was also carried out and specimens were embedded in LR White as above. Chitin distribution was assessed by labelling with wheat germ agglutinin conjugated to 10nm colloidal gold (E.Y.Laboratories, San Mateo, USA). Sections on formvar-coated grids were floated on drops of WGA-gold (10µg/ml) for 1 h, washed with PBS and water and stained with uranyl acetate and lead citrate. Controls employing N-acetyl-D-glucosamine (200mM), triacetylchitotriose (8mM) and chitin (50mg/ml) during labelling were carried out.

3. RESULTS

Even the lowest viscosity epoxy resin (Spurr's) failed to produce satisfactory sections and up to 20% of cells exhibited some degree of inadequate resin infiltration. The other epoxies examined produced similar or worse infiltration despite protracted infiltration. Hence, some degree of selectivity was inevitably placed on the ultrastructural assessment of a cell population processed using epoxy resin. The use of LR White acrylic based resin overcame this limitation. Virtually every cell was fully infiltrated. Furthermore, the preservation of fine structure of permanganate post-fixed cells embedded in LR White was good. The plasma membrane, nuclear membrane and organelles such as mitochondria were well preserved (Fig.1). The cell wall exhibited a typical electron dense fibrillar surface layer and a central region of lower electron density. When processed by the G-O-T-O-U method layering of the cell wall was accentuated, as reported by Persi and Burnham (1981). However, the density of the cytoplasm tended to be excessive, possibly because of better preservation of the numerous ribosomes. Artefactual invaginations of the plasma membrane were created, suggesting shrinkage or swelling of the cells occurred during the protracted processing stages involved in the G-O-T-O-U method.

Labelling by WGA-10nm gold of permanganate post-fixed cells embedded in LR White was specific for the cell wall and intense and clearly demonstrated the distribution of chitin components in the cell wall and their concentration within septa (Fig.2). Spurr's resin sections labelled less intensely and background labelling of the resin was apparent. N-acetyl-D-glucosamine, triacetylchitotriose and chitin inhibited specific wall labelling in all cases.

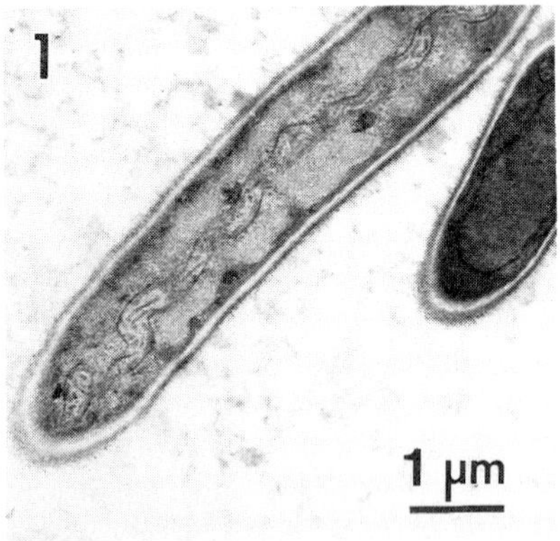

Fig.1 C.albicans hypha in LR White unlabelled, double stained, note filamentous mitochondrion.

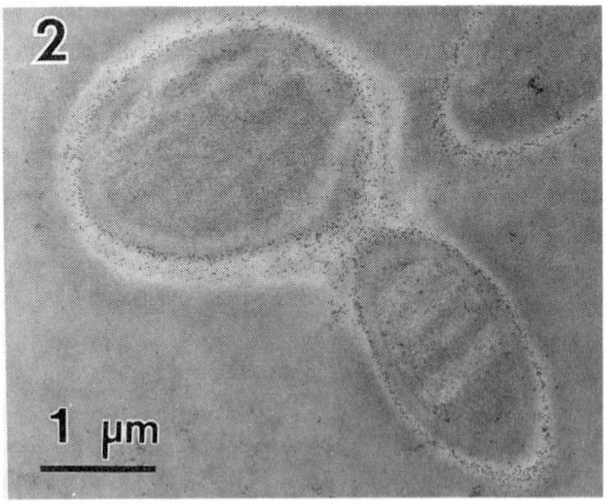

Fig.2 WGA-gold labelled hypha in LR White, unstained section.

REFERENCES

Drewe J A 1981 *Med. Lab. Sci.* **38** 237

Garrison R G 1981 *Yeast Cell Envelopes : Biophysics and Ultrastructure Vol II* ed. W N Arnold (Florida, CRC Press) pp.139-160

Persi M A and Burnham J C 1981 *Sabouraudia* **19** 1

Inst. Phys. Conf. Ser. No. 93: Volume 3, Chapter 10
Paper presented at EUREM 88, York, England, 1988

The fine structure of clean and fouled ultrafiltration membranes

Judith M Sheldon

Department of Plant Sciences, Oxford University, South Parks Road, Oxford OX1 3RA

ABSTRACT: Transmission electron microscopy has been used to examine the fine structure of ultrafiltration membranes. Immuno-chemical staining has been successfully applied to locate protein foulants within the membrane matrix. Freeze-fracture and deep-etching has shown that protein molecules may need to undergo some level of distortion before being able to pass into the membrane.

1. INTRODUCTION

The fouling of ultrafiltration (UF) membranes with protein is a major problem in many separation processes (Fane A G 1986 In: Progress in Filtration and Separation ed R J Wakeman (Amsterdam: Elsevier) pp 101-179). Although there are many data on the effect of such fouling on flux rate and solute rejection, there is less information on the structure of the membrane, the amount of adsorption and the precise location of the protein foulants within them. Here work is reported where a number of different techniques for transmission electron microscopy (TEM) have been modified and used to examine the ultrastructure of both clean and protein fouled UF membranes. Some of these techniques have previously only been used on biological materials.

2. MEMBRANES

Amicon PM10 UF membranes have been studied. These membranes have a molecular weight cut-off of 10kDa, are made from polysulphone and are hydrophobic. Protein fouled membranes were prepared by filtering a BSA solution under cross-flow conditions.

3. METHODS AND RESULTS

Thin section TEM and freeze-fracture combined with deep-etching have been used to examine the ultrastructure of the PM10 membranes. TEM has shown the membranes to be 100μm thick and to have a complex internal structure. The membranes contain a number of chambers which markedly increase in size with distance from the separating (front) surface of the membrane, where they average 500nm in diameter. The membrane matrix also contains many micropores which again increase in size with distance through the membrane, being 5nm in diameter at the separating surface and around 200nm at the back of the membrane. Freeze-fracture and deep-etching reveals that the matrix is made up of a mesh of 15nm thick fibres of polysulphone which, towards the back surface of the membrane, give way to sheets of the membrane material of the same thickness. The size of the micropores within this matrix appears to be governed by the degree of packing of the mesh.

Protein foulants have been located using immuno-chemical staining. The use of both pre-embedding staining and staining ultrathin cryosections have been investigated and the position of the protein within the membrane has been ascertained. Between the separating surface and the first of the chambers, a distance of around 300nm, the BSA is located in the membrane matrix (Figure 1). The bulk of the protein then appears to pass through the chambers (Figure 1) until the back 5-10µm of the membrane where the BSA is again found in the micropores of the matrix. Freeze-fracture and deep-etching has also been used for the examination of protein fouled membranes and has revealed that BSA molecules, which under normal conditions are globular, may only pass through the surface of PM10 membranes after undergoing some degree of distortion (Figure 2).

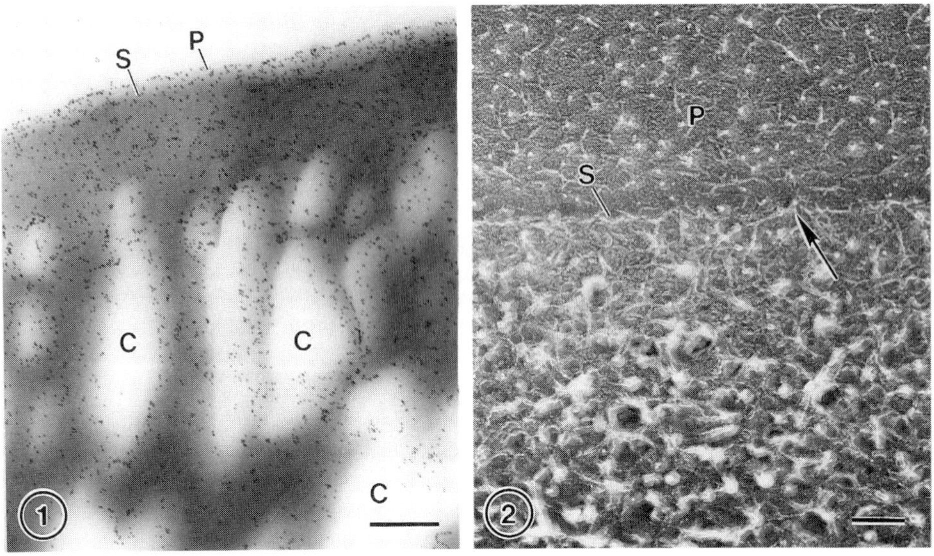

Figure 1. Ultrathin cryosection of a protein fouled PM10 membrane which has been stained with a polyclonal antibody to BSA located with a second species specific antibody conjugated to a 15nm gold particle. A layer of the protein (P) is visible at the separating surface (S) of the membrane and is well stained. Immediately behind the separating surface the staining is in the membrane matrix, but further back it is associating with the walls of the chambers (C). bar = 500nm.
Figure 2. The separating surface (S) of a protein fouled PM10 membrane after freeze-fracture and deep-etching. The protein molecules in the protein layer (P) are filametous in nature and at some points appear to pass through the separating surface (arrowed). The same filamentous molecules are still visible among the fibres of the membrane matrix. bar = 100nm.

This work is supported by the Biotechnological Separations Club (BIOSEP) of the UK Department of Trade and Industry.

Inst. Phys. Conf. Ser. No. 93: Volume 3, Chapter 10
Paper presented at EUREM 88, York, England, 1988

Electron microscopical investigation of blood–surface interaction from different dialysis membranes

A Debrand-Passard, A M Lajous-Petter, H von Baeyer.
Klinikum Rudolf Virchow,Freie Universität Berlin (FRG)

ABSTRACT: Different types of hemo-dialysis membranes were investigated before and after clinical application. The morphology of the wall and the chemical composition of the membranes are responsible for the specific blood-surface interaction. Our results can be correlated with biochemical parameters of biocompatibility.

Transmission (TEM) and scanning (SEM) electron microscopy was used to investigate blood-surface interaction of 7 different types of hollow fiber dialysis membranes (Cuprophan, CU; Hemophan, MC; Cellulose diacetate, CA; Polysulfone, PS; Poly-acrylnitrile, PAN Asahi; Polymethylmethacrylate, PMMA).
After use in routine dialysis the membrane were fixed with 2,5% glutaraldehyde and dried with alcohol . For TEM, the material was embedded in LR White; for SEM the capillaries were cut into halfs with a razor blade and gold sputtered. For TEM a Zeiss EM 10 and for SEM a Cambridge Mark 250 electron microscopes were utilized.

The membranes are different in chemical composition, wall structure and biocompatibility. The latter term indicates the tendency to possible long term complications such as dialysis associated amyloidosis, infections and cachexia.
The membranes allways exhibit depositions of blood components after use however with characteristical distinctions:

1. Asymmetric membranes like PS or PAN show no continuous layer of protein and only very few blood cells.

2. Cellulosic membranes (CU,CA,MC) have a rather uniform behaviour: a continuous layer of protein/fibrin, many leuco-cytes and some thrombocytes.

3. PMMA is characterized by a white mural thrombus predominantly with platelets. Within the first 10 minutes of dialysis the membrane surface shows only platelets.

The morphological correlate of blood-membrane surface interaction is seen as early as 10 min after beginning of dialysis and is completed after 30 min.

These findings explain measurable biochemical parameters of biocompatibility like ß-thromboglobuline as an indicator of thrombocyte adhesion and release of specific proteases from leucocytes, indicating degranulation.

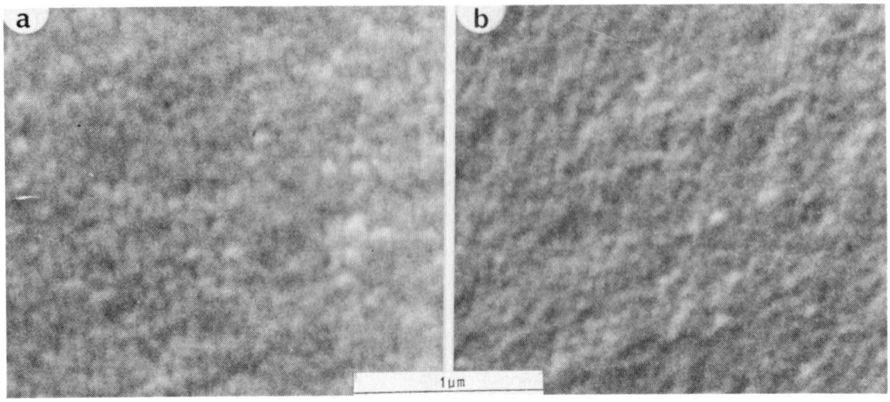

Fig 1: Polysulfone a) before b) after use.
No deposit.

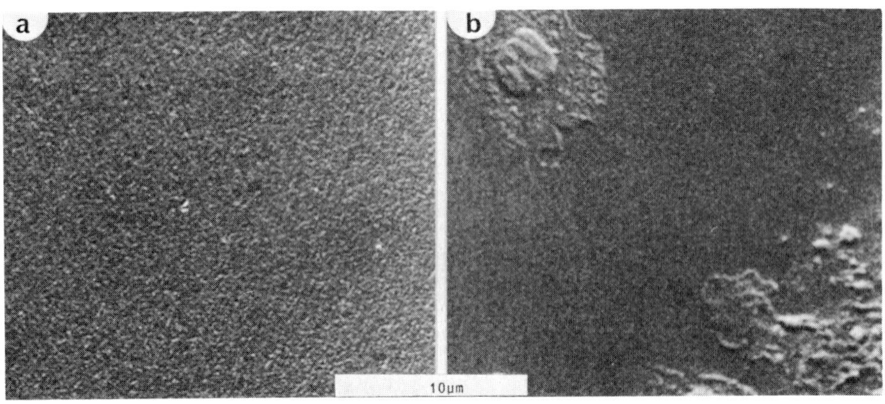

Fig 2: Hemophan a) before b) after use.
b) shows leucocytes and protein/fibrin deposits.

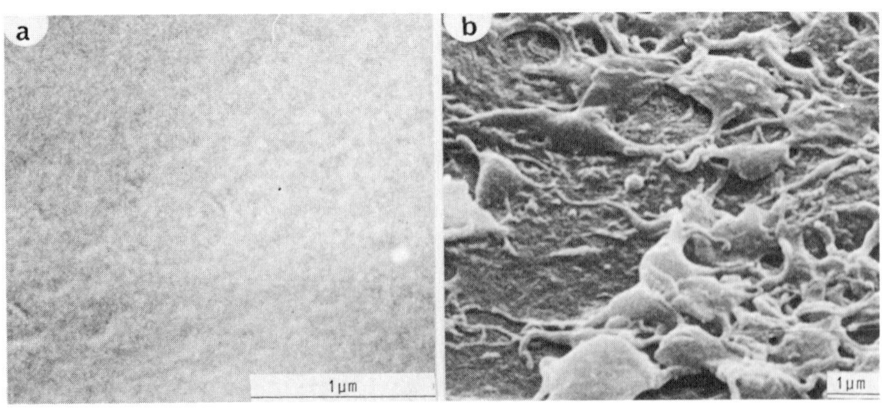

Fig 3: Polymethylmethacrylat a) before b) after use.
b) shows a white mural thrombus particulary platelets.

Inertness of stainless steel structures in rabbit corneas: a SEM-study

W.L. Jongebloed (1), P.L. Cuperus (1), P. van Andel (1), F. Dijk (1) & J.G.F. Worst (2)

(1) Centre for Med. Electron Microscopy / Lab. for Histology, University of Groningen, The Netherlands
(2) Eye Physician and Surgeon, Haren, The Netherlands.

For many years man-made fibres like perlon, supramid and polypropylene have been used as suturing material in ophthalmology, despite the fact that several authors,e.g. Drews (1983), Apple et al. (1984), Jongebloed et al. (1986a),have reported biodegradation of these materials when being used in (human) corneas. Therefore the use of stainless steel (s.s.), 50 um diam., as suture material for I.O.L.- and artificial cornea (K.P.) implantation was introduced (J.G.F. Worst, pers. comm.).
The aim of this study was to investigate behaviour of s.s. sutures in rabbit corneas as fixation suture in K.P.- and I.O.L. implantation, with a possible link to human application.
Hollander rabbit eyes were subjected to extracapsular surgery for either an I.O.L.- or a glass-platinum K.P. implantation. The surgical wounds were closed with s.s. steel sutures. After the implantation, slitlamp inspection followed every other day; no visual signs of infection or repelling reactions were observable. After 1.5,3 and 7 months resp., the corneas were sacrificed, prepared for SEM in the usual way (Jongebloed et al., 1986b), and examined in a JEOL SEM, type 35C, at 15-25 kV.
Results: After 1.5 month the buckle of the s.s. suture was covered with Descemet, containing some fibroblastic- and irregular endothelial cells. After 3 months the covering of the suture appeared to be complete, except for an apparant wound at which the yet denuded Descemet's membrane could be observed, partly covered with filopodia and collagenous material generated by the bordering endothelial cells. These cells were in a high state of activity regarding their numerous microvilli. The 7 months sample revealed a buckle completely covered with Descemet's membrane and newly formed endothelial cells. These new cells had a regular hexagonal shape and ruffled cellborders, like normal endothelial cells.

Legends of photographs:
(1) Covering of buckle (B) of s.s.-suture in rabbit cornea after 1.5 month. Bar = 100 µm. (2) Detail of area B of fig. 1. with Descemet's membrane and some irregular endothelial (ir)- and fibroblastic cells (fb). Bar = 20 µm. (3) Endothelial wound closure at edge of buckle (B) of s.s.-suture after 3 months. Bar = 100 µm. (4) Detail of fig. 3: wound-bordering cells with many microvilli (mi) and denuded Decemet's membrane (*). Bar = 10 µm. (5) Detail of fig. 4 with filopodia (fi), collagenous fibres (cf) and microvilli (mi). Bar = 1 µm. (6) Completely covered buckle (B) of s.s.-suture after 7 months; E = normal endothelial surface. Bar = 10 µm. (7) Detail of fig. 6: newly formed hexagonal endothelial cells (ec) at edge of buckle surface (B); E = normal endothelial surface. Bar = 4 µm.

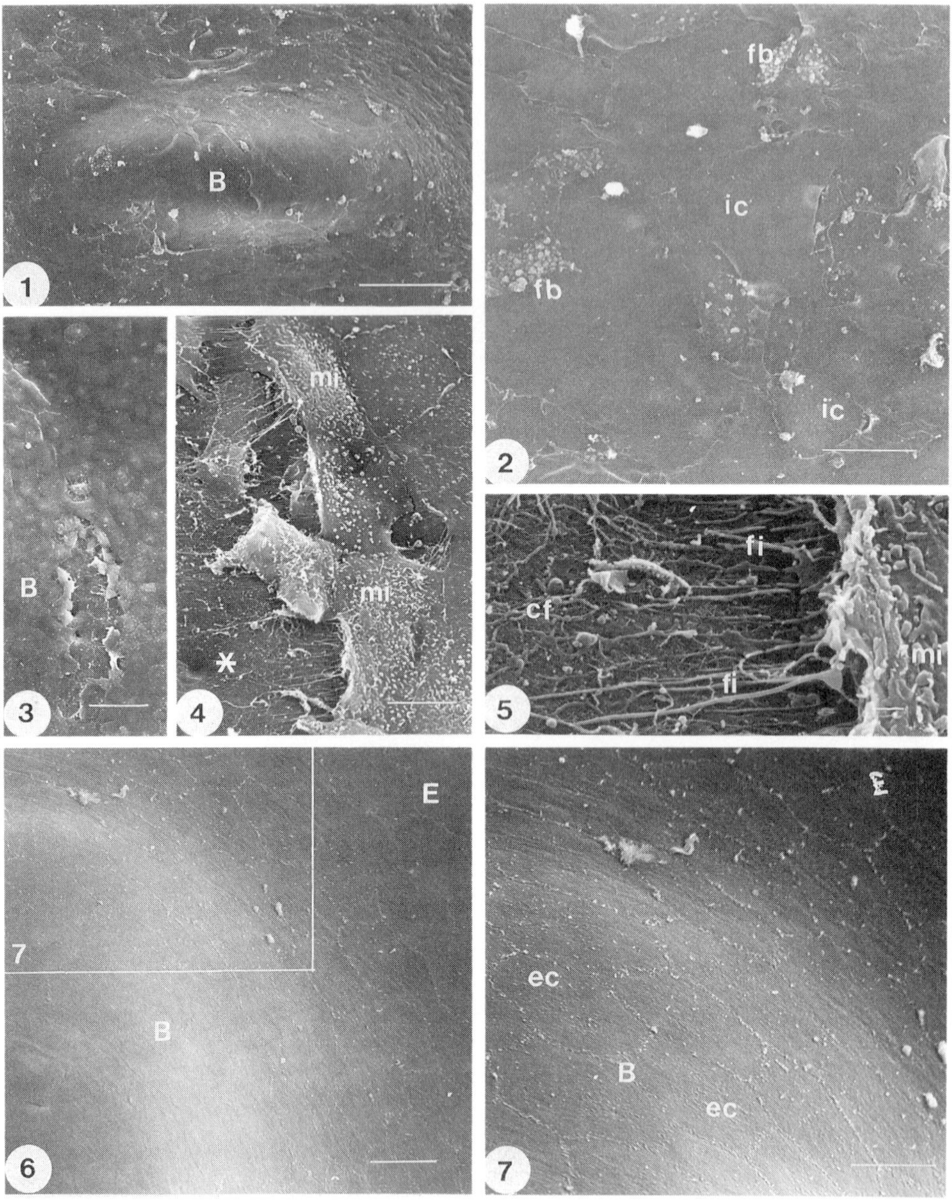

References:

Drews RC (1983). Ophthalmol. 90, 301 pp.

Apple J et al. (1984). Am. Intraocular Impl. Soc. J. 10, 53 pp.

Jongebloed WL et al. (1986a). Doc. Ophthalmol. 61, 303 pp.

Jongebloed WL et al. (1986b). Doc. Ophthalmol. 64, 143 pp.

SEM and LM evaluation of a glass-platinum keratoprosthesis seven months after implantation in rabbits

P.L. Cuperus (1), W.L. Jongebloed (1), P. van Andel (1), M. Kolenbrander (2) & J.G.F. Worst (3).

(1) Lab. for Histology and Centre for Medical Electron Microscopy, University of Groningen; (2) Philips Research Lab, Eindhoven; (3) Jan Worst Research Group, Julianalaan 11, 9751 BM Haren. (THE NETHERLANDS)

A keratoprosthesis (KP) is the last and only surgical resort to gain visus in eyes with a severely damaged cornea when a corneal graft is not available or indicated. Non-transparant corneas, often caused by parasites living in tropical fresh waters, represent an important percentage of the blind in the less developed countries. KP's made of PMMA are difficult to sterilize, however, and vulnerable for scratching. Moreover, these commercially available KP's are too expensive to be of use in countries with a low G.N.P. Therefore, we developed a new KP, made of a glass core melted-in in a platinum cylinder with flange (Fig. 1). Kp's of this type were implanted in eyes of Hollander rabbits by placing them intralamellar in their corneas (Fig. 2 & 3). The KP's were kept in place by four stainless steel traction threads placed around the eyeball (Fig. 3). The purpose of this study was to investigate this type of KP in the rabbit cornea, its acceptance by epi- and endothelium, and its hydro-mechano dynamics in situ. Visual control was carried out by slitlamp inspection every other day; after 1.5, 3 and 7 months resp., a cornea with KP was removed and prepared for (S)EM- and LM in the usual way (Jongebloed et al., Doc. Ophtalmol. 64: 129 pp., 1986).

Results:

Up till now (seven months post-operative), no signs of infection or repelling were observed. The melted-in optics proved to be promising (Fig. 4). No epithelial downgrowth, adverse tissue reaction or extrusion could be detected. LM and SEM revealed that re-endothelialization was found to take place over the newly formed stroma around the trephination the KP fits in (Figs. 5 & 6). At this place, endothelial cells up to only 100 μm from the stromal rim bordering the platinum cylinder of the KP (Figs. 7 & 8), appeared to be anomalous, like in cases of local traumas . From the findings, we conclude that this type of KP (although optically still to be optimized) has been accepted by the rabbit cornea and a clinical trial on human cornea-blind is justified, and currently being undertaken.

CAPTIONS TO FIGURES:

(1) Glass-platinum KP, with melted-in glass core; ⟶ : hole for traction thread; p: platinum; g: glass. Bar = 1 mm. (2) Schematic drawing of KP, anchored intralamellar in cornea of aphakic eye; dotted lines indicate two of the four traction threads. (3) KP in rabbit eye seven months after implantation; ⟶: traction thread. (4) Interferograms of front- (left) and backside of the KP indicating optic regularity. (5) SEM image of KP in rabbit cornea, five months post-operative; p: platinum; g: glass; c: cornea. Bar =1 mm. (6) LM section of cornea of right part of Fig.5. Re-endothelialization has taken place over newly formed stroma; ep: epithelium; en: endothelium; *: original location of flange of KP; d: Descemet's membrane (partly missing due to sectioning). Bar = 1 mm. (7) KP of Fig. 5, viewed from different angle. Re-endothelialized sutures in endothelium can be seen (⟶). Bar = 1 mm. (8) Detail of Fig.7. showing anomalous (a) and normal (n) endothelium at stromal rim bordering platinum cylinder of KP. Bar = 100 μm.

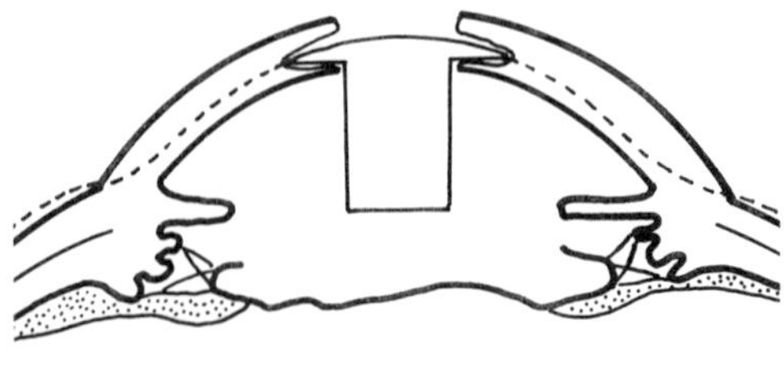

2

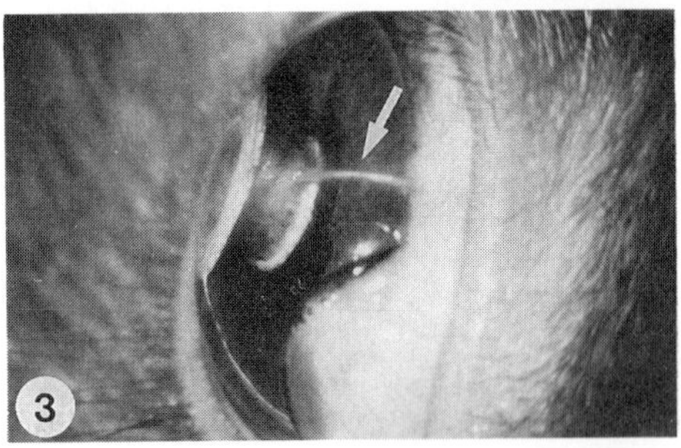

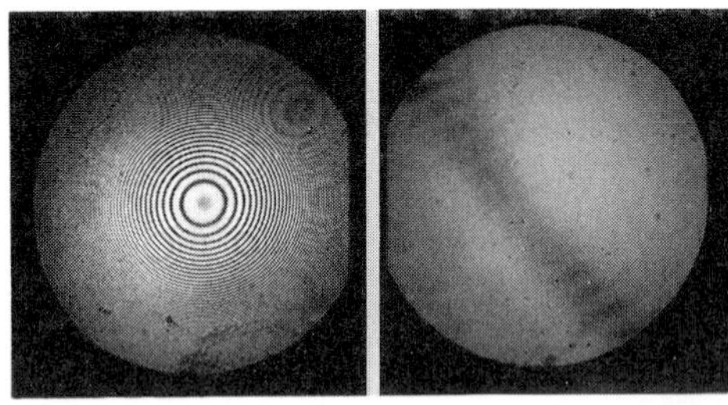

4

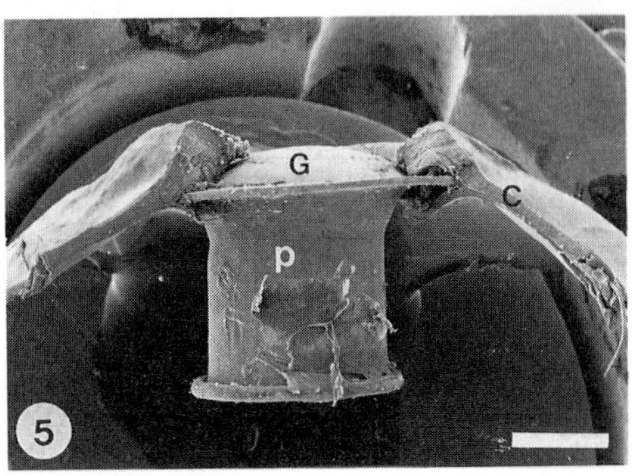

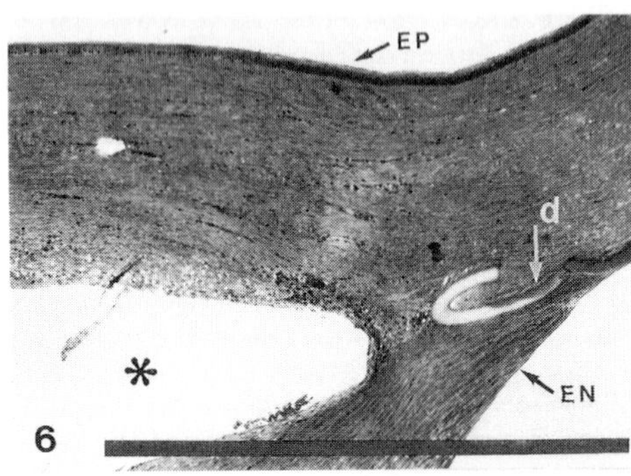

6

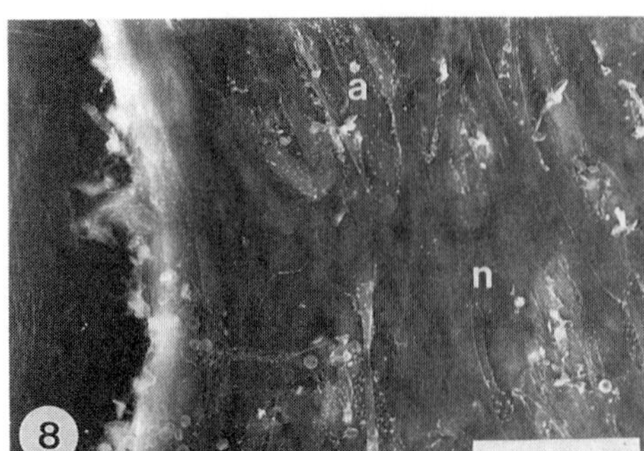

Ultrastructural foundations of liver cells response to cooling

A.S. Kaprelyants

Institute for Problems of Cryobiology and Cryomedicine
310015, Kharkov USSR

Structural and metabolic peculiarities of rat liver cells have been studied by the technique of ultrathin sections, freeze-fracture, freeze-substitution, stereometry and electron-microscopic cytochemistry after local cooling from normothermia down to -180°C. A comparative analysis of the status of membrane structures and cellular organoids has been performed in the frozen state in dynamics and after thawing. A qualitative and quantitative description of the formation of intra-and extracellular ice crystals of various dimensions and differents localization is presented. The cytoplasm and cell nucleus in the frozen state have honeycombed appearance with a considerable deformation of membrane structures and fine elements of the nucleus apparatus followed by shrinkeage due to dehydration. Hepatocytes, Kupffer cells, and sinusoid-circulating blood cells demonstrate differing cryoresistances. After thawing the cells restore their volume and cytoplasm and nucleus structure. However, destructive changes of various degrees which can lead to necrotic processes, are also observed. The dynamics of cell responses and morphogenesis changes occurring in tissues in vivo are described depending on the conditions of cooling and the state of organoids.

Persistent infection with sendai virus is associated with aggregating nucleocapsids

B M Humbel[1], C W Eckerskorn[2], W J Neubert[3]

Max-Planck-Institute for Biochemistry, Dept of Cell Biology[1], Gencenter[2], Dept of Virology[3], 8033 Martinsried, FRG

ABSTRACT: A cell line (Cl-E-8) persistently infected with Sendai virus is compared to an acute infection in the early and the late state. Immunofluorescence and phase contrast data showed in persistently infected Cl-E-8 cells large inclusion bodies which stained for nucleoprotein (NP). With conventional and immuno-electron microscopy these inclusion bodies could be identified as aggregating nucleocapsids.

The persistent infection of Sendai virus (strain 6/94) differs significantly from an acute infection. It shows 1) strongly reduced virus production, 2) noninfectious virus particles (Neubert & Hofschneider 1983), 3) reduced synthesis of viral mRNA, and 4) increased copies of viral genomic RNA. The production of viral proteins differs only slightly from the acute infection. The high amount of viral genomic RNA and the unchanged amount of viral protein is not consistent with the largely reduced production of almost noninfectious viruses. There may be a defect in virus assembly or budding.

Acutely infected cells in a late state show cytopathic effects and loosely packed particles (figure 1 & 2). They have the appearance of nucleocapsids as demonstrated by Bohn *et al* (1986). The persistently infected cells show fewer cytopathic effects. As expected from light microscopical data (not shown), large aggregates could be detected in the electron microscope (figure 3). With immuno-gold labelling on ultrathin cryosections (Tokuyasu 1973) the aggregates could be identified as nucleocapsids, using a NP specific monoclonal antibody (figure 4).

In the beginning of the acute infection of Sendai virus, small clusters of nucleocapsids were detected by immunofluorescence. During the progress of infection the clusters increased in number and sometimes large areas of the cells were stained homogeneously. In the persistently infected cell line Cl-E-8, however, the nucleocapsids are condensed into distinct aggregates. Similar results were reported by Norrby *et al* (1985). They showed in four cases of subacute sclerosing panencephalitis large intranuclear inclusions, detected with antibodies against nucleocapsids of measles virus. The data reported here support the idea that the viral persistence in the cell line Cl-E-8 results from a defect in virus assembly or budding.

REFERENCES

Bohn W, Rutter G, Hohenberg H, Mannweiler K and Nobs P (1986) *Virology* **149** 91
Neubert WJ and Hofschneider PH (1983) *Virology* **125** 445
Norrby E, Kristensson K, Brzosko WJ and Kapsenberg JG (1985) *J. Virol.* **56** 337
Tokuyasu KT (1973) *J. Cell Biol.* **57** 551

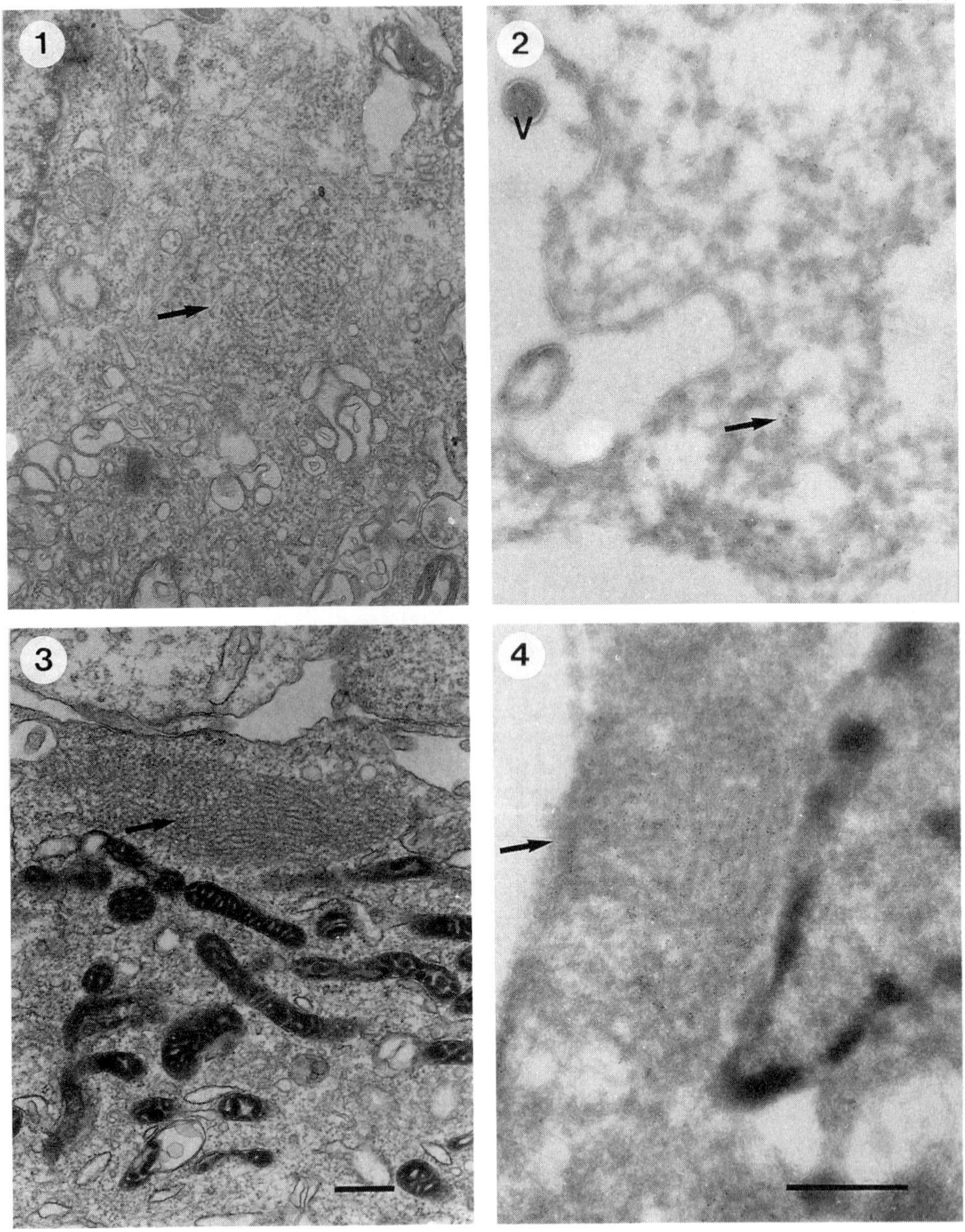

Figure **1** & **2**: Acute infection after 31 h. **1**: Loosely packed particles (→), and **2**: small clusters (→) and virus particles (V) are demonstrated.
Figure **3** & **4**: Cl-E-8 cell line. The aggregating particles (→) were identified as nucleocapsids (**4**). Bar represents 500 nm.

Comparative study of human rotavirus replication in MA104 cells and in the intestinal epithelium of neonatal mice

M Poljšak Prijatelj

Institute of Microbiology, Medical Faculty ,61105, Ljubljana, Yugoslavia

In last years the significant progress has been made in the understanding of the molecular aspects of rotavirus replication in cell culture (Petrie et al 1984). Because the penetration of immunocomplexes has not been satisfactory enough for rotavirus model of replication in pre-embedding systems, we extended our work to post-embedding methods, where the antigens are exposed to immunoreagents when the cells and organelle membranes are cut open during sectioning. We made a comparative study of rotavirus replication in MA104 cells and in the intestinal epithelium of neonatal mice using TEM and SEM.

A pool of rotaviruses was obtained by combining five samples of rotavirus positiv diarrhetic fluid from pediatric patients. Stool specimens containing number of double shelled rotaviruses have been selected for this study. Before inoculation on cell culture the viruses were treated with trypsin. Simultaneously we infected five days old CBA neonatal mice with rotavirus suspension. Mice have been sacrificed after two to ten days p.i. The intestinal tissue has been cut, fixed for ultrastructural or immunocytochemical studies, dehidrated and embedded in ERL (Spurr), or LR Gold (The London Resin Co). The rotavirus antigens in experimental infected cells or mice were detected with immunoperoxidase or protein A-gold methods.

References:

Gouvea V S, Alencar A A, Barth O M, De Castro L, Fialho A M, Aravjo H P, Majerowicz S and Pereira H G 1986 J. gen. Virol. 66 pp 577-581

Petrie B L, Greenberg H B, Graham D Y and Estes M K 1984 Virus Research 1 pp 133-152

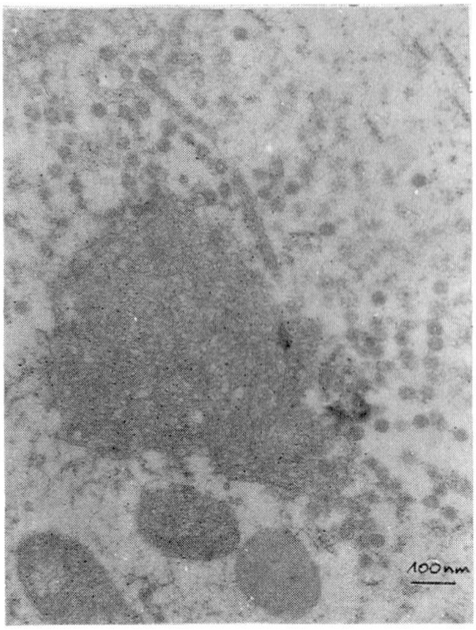

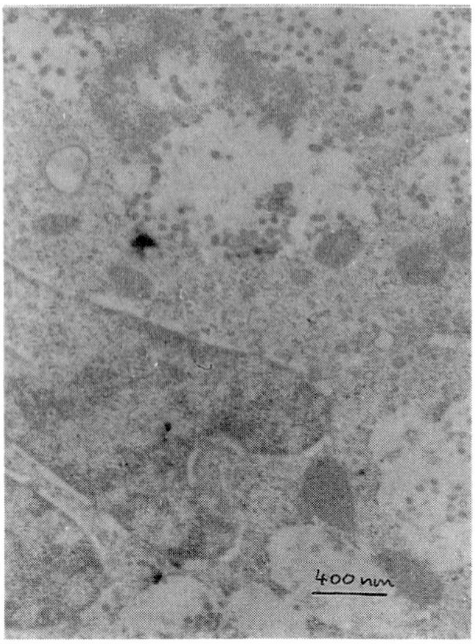

Fig 1 Fig 2

Particles appear to bud from granular viroplasm

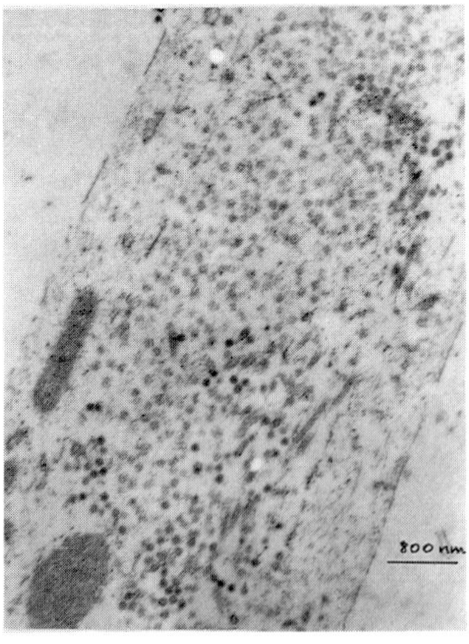

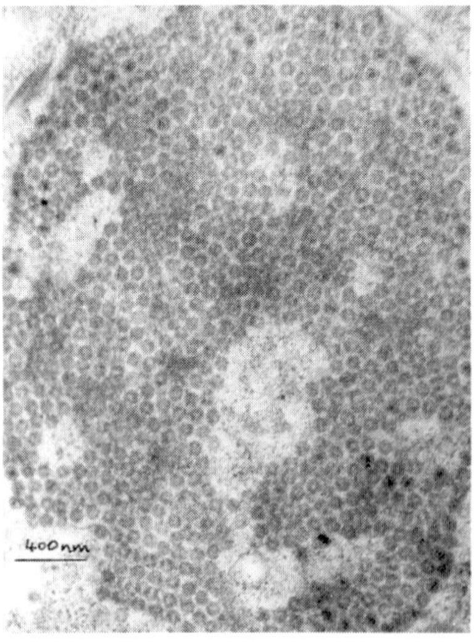

Fig 3 Fig 4

Cell with RER containing viral particles

Immunogold cytochemistry of a plant potyvirus in sections of host tissue: effect of different fixation and embedding methods on labelling with polyclonal and monoclonal antibodies

A Appiano*, P Dell'Orto**, G Viale** and P Braidotti**

*Istituto di Fitovirologia CNR, V. Vigliani 104, 10135 Torino, Italy
**II Cattedra Anatomia e Istologia Patologica Università, Milano, Italy

Potato virus Y° (PVY°) is a filamentous virus, of dimensions ca 730x12 nm. In thin sections, it appears as filaments or dots, depending on the section plane, in the cytoplasm or between the arms of pinwheels, the typical inclusion bodies elicited by this virus in the host cells. Pinwheels consist of a virus-coded protein serologically unrelated to the virus coat protein. Previously (Appiano et al., 1987), with PVY° in tobacco, we failed to obtain immunogold labeling of virus particles with monoclonal antibodies (Mabs) in Epon-Araldite sections, while a polyclonal antiserum (PA) gave strong, specific labeling. ELISA tests made on virus fixed in various concentrations of glutaraldehyde (GA), paraformaldehyde (PFA) or acrolein (Acr) showed that fixation in very low concentrations of GA, but not PFA or Acr, completely blocked reaction of the Mabs. In the present work, infected tissue was fixed in 2% PFA + 0.2% Acr and embedded either in Epon-Araldite or LR White. Both PA and Mabs were tested, using basically the method of Viale et al. (1985).

Epon-Araldite. In spite of the lack of GA and OsO_4, the tissue was acceptably fixed, though most cell membranes had disappeared. Virus particles and pinwheels retained their identity. Compared to GA-fixed samples, the antigenicity to PA was better preserved and the optimum concentration of primary serum could be decreased from 1/100 to 1/500, giving much cleaner labeling on PVY° particles (Fig. 1). With Mabs however, even using undiluted serum and raising the incubation time to 60 h, only weak labeling was obtained (Fig. 2).

LR White. Tissue preservation was rather poor, though virus particles and pinwheels were still clearly recognized. Antigenicity to PA was further improved, and a 1/1000 dilution of primary serum gave highly specific labeling on virus particles, free (Fig. 3) or associated with pinwheels (Fig. 4). With Mabs, specific but weak labeling was obtained, though only by using undiluted serum and a 60 h incubation (Fig. 5).

Our poor results with Mabs suggest that their specific epitopes might be conformation dependent, and then undergo critical modifications when submitted to resin embedding, irrespective of the type of resin used. This hypothesis might be confirmed by the extreme sensitivity of Mab-related

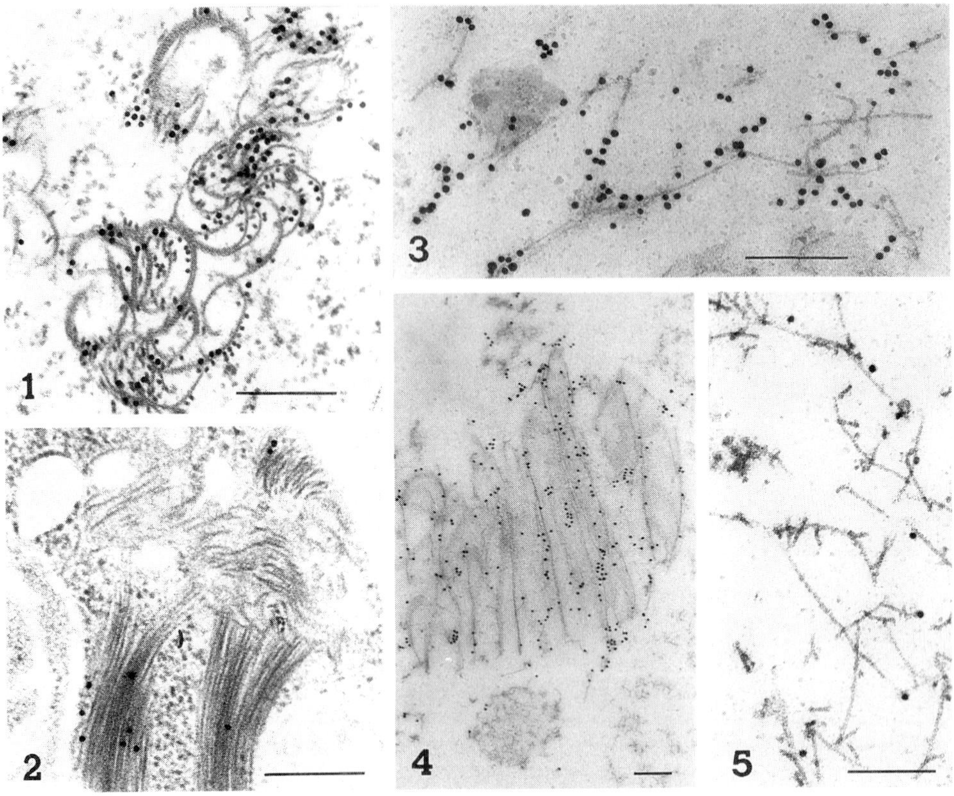

Figs. 1-5. Sections of PVY°-infected tobacco, embedded in Epon-Araldite (1-2) or LR White (3-5), labeled with 15 nm gold. Bar = 250 nm.
Fig. 1. Polyclonal antiserum (PA) labeling on virus particles associated with pinwheels in cross section.
Fig. 2. Monoclonal antibody (Mab) labeling on bundles of PVY° particles.
Fig. 3. PA labeling on isolated virus particles.
Fig. 4. PA labeling on virus associated with pinwheels in oblique section.
Fig. 5. Mab labeling on isolated virus particles.

epitopes to GA and by the fact that Mabs do not react with dissociated virus subunits. Our results may explain why, so far, Mabs have not successfully, to our knowledge, been used in post-embedding labeling studies of plant viruses (for ref. see Appiano *et al*., 1986).

REFERENCES

Appiano A, D'Agostino G, Bassi M, Barbieri N, Viale G and Dell'Orto P 1986 J. Ultrastruct. Molec. Struct. Res. **97** 31

Appiano A, Viale G, Dell'Orto P and Roggero P 1987 Atti XVI Congresso della Società Italiana di Microscopia Elettronica, Bologna, Italy, 77

Viale G, Dell'Orto P, Braidotti P and Coggi G 1985 J. Histochem. Cytochem. **33** 400

Immunoelectron microscopy on tobacco necrosis virus in *Phaseolus vulgaris*: cryoultramicrotomy versus plastic embedding

G. D'Agostino*, P. Dell'Orto**, G. Viale**.

*Istituto di Fitovirologia applicata del CNR.
Via Onorato Vigliani, 104 - 10135 TORINO, Italy
** II Cattedra di Anatomia Patologica, Università di MILANO.
Via A. Di Rudini, 8 - 20142 MILANO, Italy.

The aim of this study was to compare immuno-gold labelling (IGL) on cryosections and on sections of material embedded in LR White resin. The test material consisted of necrotic lesions (NL) induced in P. vulgaris by tobacco necrosis virus (TNV), which has polyhedral particles about 26nm in diameter that are difficult to distinguish from cytoplasmatic ribosomes.

Pieces of leaf each bearing one small necrotic lesion, were fixed in 2.5% glutaraldehyde + 1% acrolein, and either stained in uranyl acetate and embedded in LR White resin or infiltrated with 2.1M sucrose and plunged into liquid nitrogen. For the IGL, we followed the procedure of Griffiths et.al. (1983) with the cryosections, and of Viale et al. (1985) with the LR White sections. The cryo and LR White sections were incubated (using the same dilution of the same antibody) respectively for 30 min and 2 h in the primary antibody, and for 20 min and 1 h in the gold-labelled secondary antibody.

Cryosections generally provide better accessibility of antigens to antibodies, and the efficiency of labelling is reported to be higher than with pre-and post-embedding methods using resin (Tokuyasu, 1986). Nevertheless in our study the results with cryosections (Figures 1,2) were not better than those obtained in LR White resin sections (Figure 3); moreover the virus particles were more clearly seen in the latter. This suggests that, in our case, the plastic embedding procedures did not strongly affect the accessibility of antigens to antibodies, or that the cryotechnique needs further improvement to give significantly better results compared to conventional techniques. IGL with the antibodies employed proved to be a useful method of identifying TNV serologically, but it was not very effective in localizing the virus particles, particularly when these were scattered in the cytoplasm (Figure 4). Pre-embedding RNase treatment of this tissue (Hatta and Francki, 1981) was confirmed as a useful way to distinguish the virus particles from ribo-

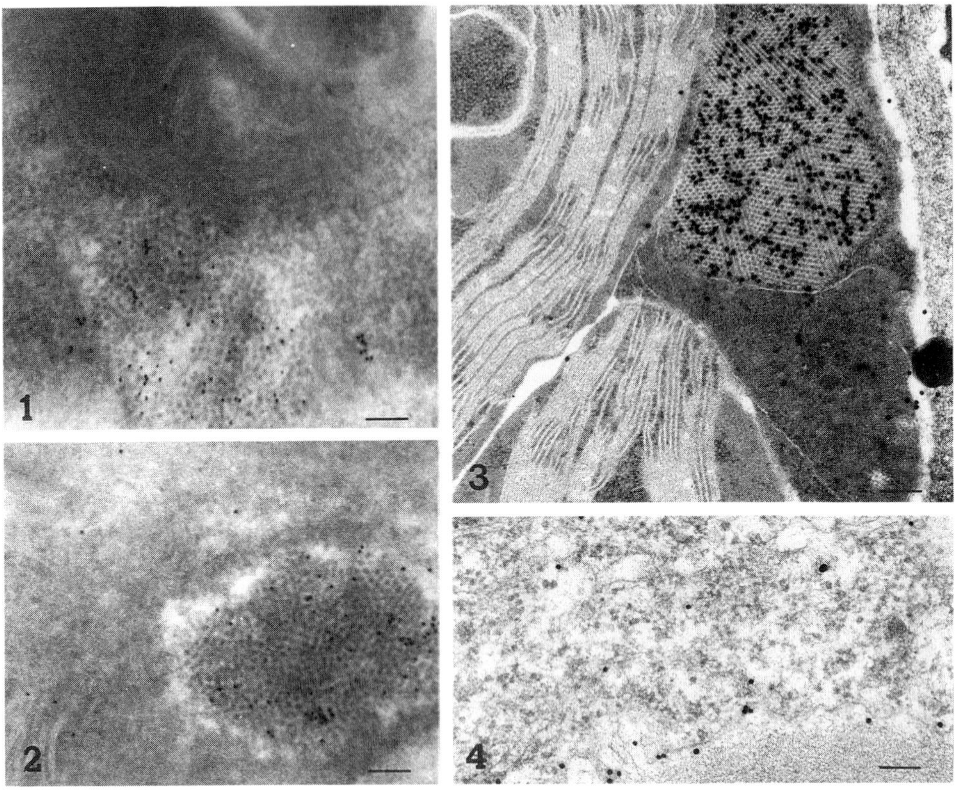

Figs. 1-4. Immuno-gold labelling (IGL) of TNV-induced necrotic lesion in *P.vulgaris* leaves. Bar = 100nm
Figs. 1-2. Cryosections with virus aggregates specifically labelled (gold diameter = 5nm).
Figs. 3-4. LR White resin sections labelled with 10nm gold. Fig. 3. Virus aggregates specifically labelled. Fig. 4. TNV scattered in the cytoplasm, in this case, IGL was not very effective in localizing the TNV particles.

somes (D'Agostino and Pennazio, 1985). We therefore attempted to combine IGL with pre-embedding RNase treatment, but for unknown reasons the viral antigenicity was affected, and so recognition and localization of the virus particles was less successful with these samples.

REFERENCES:

D'Agostino G and Pennazio S 1985 J. Submicrosc. Cytol. 17 229

Hatta T and Francki R B I 1981 J. Ultrastruct. Res. 74 1

Griffiths G, Simons K, Warren G and Tokuyasu K T 1983 Methods Enzymol. 96 466

Tokuyasu K T 1986 J. Microscopy 143 139

Viale G, Dell'Orto P, Braidotti P, and Coggi G 1985 J. Histochem. Cytochem. 33 466

Inst. Phys. Conf. Ser. No. 93: Volume 3, Chapter 11
Paper presented at EUREM 88, York, England, 1988

Purple phototrophic bacteria from laminated microbial communities: a comparison utilizing electron microscopy

I. ESTEVE, J. MIR, & N. GAJU

Department of Genetics and Microbiology and Institute for Fundamental Biology. Autonomous University of Barcelona Bellaterra. Barcelona (Spain)

Phototrophic bacteria from two laminated microbial communities, an evaporite flat (Laguna Figueroa, Baja California, México) and an anaerobic sulfurous Lake (Cisó, Spain) have been characterized by transmission and scanning electron microscopy.

Laguna Figueroa is a lagoonal system comprised of an evaporite flat, a salt marsh and a barrier dune which separates the flat and salt marsh from the ocean. Dessication polygons can be found in abundance at the evaporite flat/salt marsh interface. These polygons are formed by structured communities. These communities appear as distinctly colored layers, one to a few millimeters thick, and are easily seen with the naked eye (1).

Below cyanobacteria, purple bacteria (Chromatium and Thiocapsa) form a red layer. Thiocapsa has been characterized in axenic cultures and by electron microscopy techniques. Thiocapsa is ovoidal and sourrounded by a capsule. The photosynthetic pigments are located in tubular structures which contain bacteriochlorophyll a and carotenoids. Growth of the organisms was accompanied by the oxidation of sulfide, and globules of elemental sulfur were stored within the cells (Figs. 1 and 2).

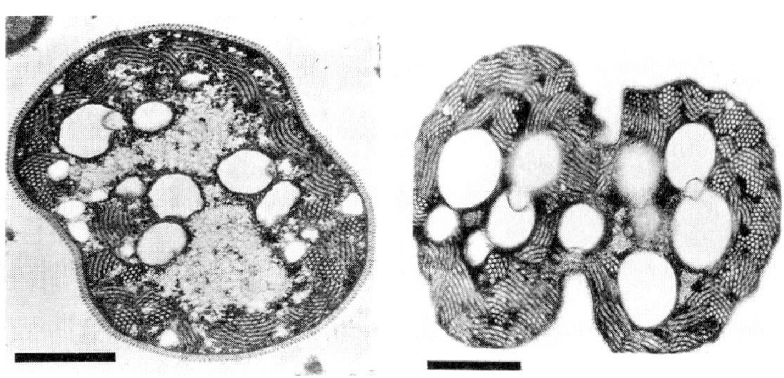

Fig.1 Ultrathin section of Thiocapsa from Laguna Figueroa Bar represents 1 µm.
Fig.2 Cell of Thiocapsa dividing by binary fission. Laguna Figueroa. Bar represents 1 µm. (Courtesy of D. Chase)

In aquatic environments, planktonic phototrophic prokaryotes often form laminated communities, comparable in structure to that of microbial mats (2);Lake Cisó is a good example of these communities. The lake is a small anoxygenic body of sulfurous water, located in the karstic region of Banyoles (Spain). The surface of Lake Cisó shows remarkable color changes, from red to brown, due to massive development of two phototrophic bacteria.

The first one was identified as Chomatium minus, which form a plate with a density of 3 x 10^6 cells/ml at approximately 0.5 m depth. Invariably, the Chromatium bloom is followed by a bust; in less than a week the bright red color can disappear. At least in part, this disappearence is due to predation (Fig. 3) by two newly discovered predatory bacteria (3 and 4).

The second phototrophic bacteria was characterized as Amoebobacter sp., which forms aggregates sourrounded by slime (Fig.4). Ultrathin sections of the cells show gas vacuoles, intracytoplasmic membranes (vesicles) and sulfur globules.

The role of electron microscopy in characterizing microorganisms is well stablished, but nothing is known about relationships among them. The application of electron microscopy techniques to the study of the laminated microbial communities will increase our understanding of early bacterial interactions.

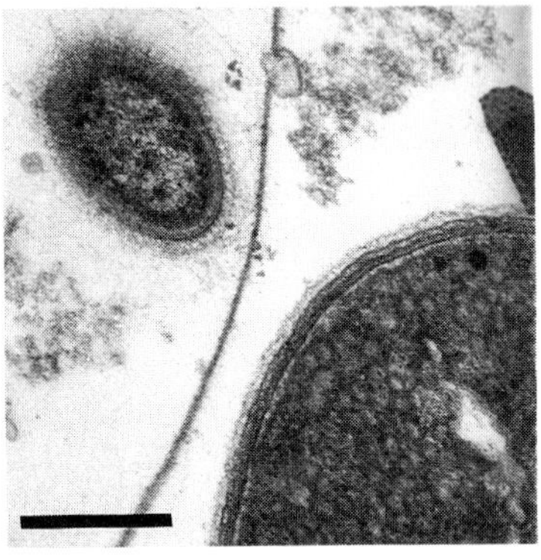

Fig. 3 An ultrathin section of Chromatium minus and one predatory bacterium from Cisó Lake. Bar represents 1 μm.
Fig. 4 Cells of Amoebobacter sp. forming aggregates from Cisó Lake. Bar represents 1 μm.

References

(1) Margulis, L., D. Chase & R. Guerrero (1986). BioScience 36: 160-170.
(2) Guerrero, R. & J. Mas (1988). In Y. Cohen & E. Rosenberg (eds). Microbial Mats. American Society for Microbiology. (in press)
(3) Esteve, I., R. Guerrero, E. Montesinos & C. Abellá (1983). Microbiol. Ecol. 9: 57-64.
(4) Guerrero, R., I. Esteve, C. Pedrós-Alió & N. Gaju (1987). Ann. New York Acad. Sci. 503: 238-250.

Inst. Phys. Conf. Ser. No. 93: Volume 3, Chapter 11
Paper presented at EUREM 88, York, England, 1988

On the diversity of magnetotactic microorganisms found in natural waters from Brazil

M. Farina[1], H.G.P. Lins de Barros[2] and D.M.S. Esquivel[2].

[1]Lab. Microscopia Eletrônica, IBCCF-UFRJ; [2]Centro Brasileiro de Pesquisas Físicas, CBPF, Rio de Janeiro, Brasil.

ABSTRACT: The great variety of magnetotactic microorganisms found until now justifies a better characterization of some parameters as a first step to taxonomy. This work proposes some approaches.

1. INTRODUCTION

Our group has studied for some years many aspects of magnetotactic microorganisms, specially using transmission electron microscopy (TEM) associated to X-ray microanalysis and electron diffraction, to elucidate the nature of the chain of magnectic crystals inside bacteria (Esquivel et al 1983; Farina et al 1983). We have found that most of the samples observed present crystals of magnetite which have been described in litterature as the unique magnetic material produced by these organisms.
Curiously we observed in many samples coming from different regions of brazilian marine coast, the presence of Sulphur and Iron inside crystal regions and some difference in crystallographic characteristics (Farina et al 1986).
At this moment, facing a great variety of magnetotactic in microorganisms, we propose the development of a chart with some data, as a first step to taxonomy, mainly because of the difficulty of cultivation in special media. We choose parameters like: Water characteristics, dimension of the cells, type of wall, number and length of flagella, general ultrastructure, X-ray microanalysis spectrum and dark field image of crystals, volume density, number and dimension of crystals per cell.

2. FIGURE PRESENTATION

The samples observed in our work, until now came from latitudes on the range of 7º to 27º in the South Hemisphere, and a great variety was found. Fig. 1 shows bacteria from 27º latitude concentrated magnetically from a clean preparation in one side of a drop, over a coverslip previously coated with poli-l-lysine. It is easy to see two populations of bacteria. Whole cells from the same sample where observed in TEM operating at 40Kv and contrast was enough to see flagella even without staining.

In Fig. 2 we see two bacteria with two bundles of many flagella on one side of the cell. Operating at 100Kv it was possible to see two chains of crystals oriented in the direction of the bundles (not shown).

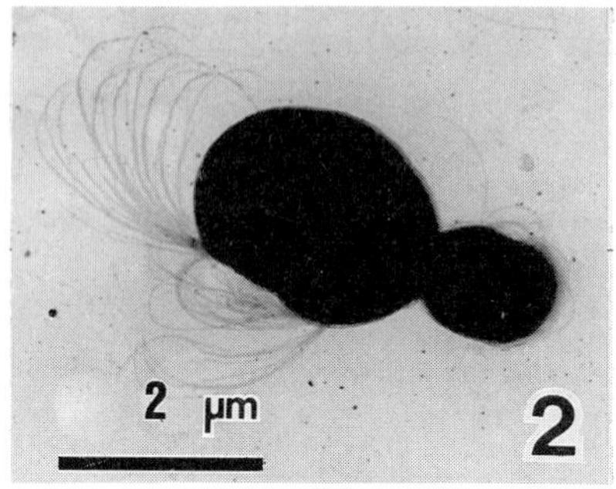

Over all this range of latitudes we could observe another bigger microorganism (Fig. 3) that we believe to be a colony of bacteria. Images with TEM show double membrane bounded aggregates of organisms and X-ray microanalysis shows Sulphur and Iron inside crystaline regions (Farina et al 1986).
After treatment with detergent and magnectic concentration, we could isolate crystals from the microorganisms. Fig. 4 obtained from bacteria near 22º latitude shows crystals with different dimensions coming from different bacteria. These are magnetite crystals (by X-ray microanalysis and electron diffraction), but while the smaller are in the theoretical dimensions of monodomains, the bigger are two domain crystals of magnetite.

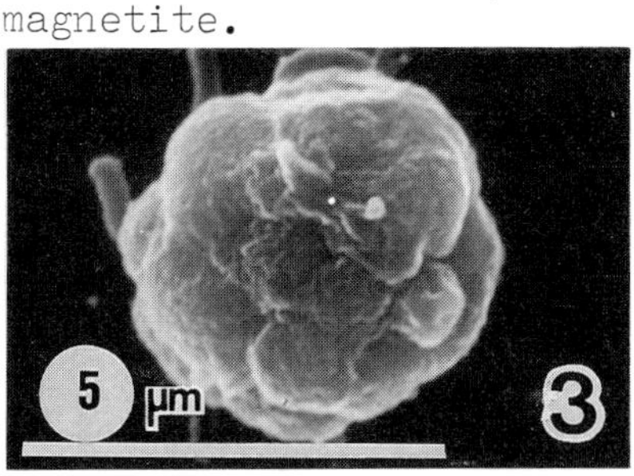

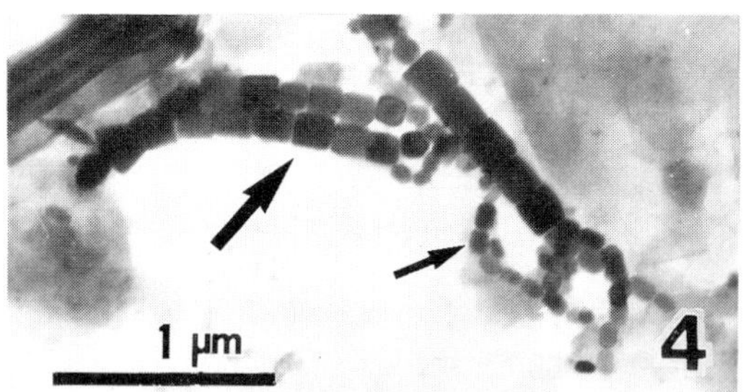

3. CONCLUSIONS

As we can see, some simple techniques directly associated with electron microscopy, can help to penetrate deeper inside the fascinating world of magnetotactic microorganisms.

REFERENCES

Esquivel, D.M.S., Lins de Barros, H.G.P., Farina, M., Aragão, P.H.A. and Danon, J. (1983). Biol. Cell. 47: 227-234.
Farina, M., Lins de Barros, H.G.P., Esquivel, D.M.S. and Danon, J. (1983). Biol. Cell. 48: 85-86.
Farina, M., Sollorzano, G. and Vieira, G.J. Proc. XIth Int. Cong. Electron Microscopy, Kyoto, 1986.

Combined endocytosis and immunoelectron microscopic studies in *Trypanosoma congolense*

U Frevert and E Reinwald

Veterinary Biochemistry, Free University of Berlin, FR Germany

ABSTRACT: The variant surface glycoprotein (VSG) of Trypanosoma congolense was localized intracellularly on vesicles, the tubular membrane system, the golgi apparatus, secondary lysosomes, and the digestive vacuole. To distinguish between VSG import and export, various electron-dense markers were examined for their suitability as endocytosis tracers in ultracryomicrotomy. Protein A gold and peroxidase gold were readily internalized and transported via endocytotic vesicles to secondary lysosomes and accumulated in the digestive vacuole. Anti-VSG-antibodies preincubated with PAG, ferritin, concanavalin A-ferritin and microperoxidase proved less suitable as endocytosis tracers in immunoelectron microscopy.

1. INTRODUCTION

Bloodstream forms of African trypanosomes evade the host's immune response by sequential variation of VSG covering their surface in a 6 nm thick layer of antigenically uniform molecules (Cross 1975). Newly synthesized VSG is transported in a rapid process (Bangs et al 1986) to the flagellar pocket (Steiger and Jenni 1973) where it appears on the cell surface. On the other hand, the flagellar pocket represents the only site of endocytotic uptake of nutrient proteins (Langreth and Balber 1975). To differentiate VSG-tranport to the cell surface from uptake during endocytosis, the trypanosomes were exposed to various electron-dense markers.

2. RESULTS AND DISCUSSION

Intracellular transport of VSG in T. congolense was observed only between flagellar pocket and nucleus. In untreated trypanosomes, anti-VSG antibodies labelled rounded vesicles and flattened membrane compartments near the pocket, secondary lysosomes, the digestive vacuole, golgi cisternae, and the tubular network.

Endocytosis experiments showed that protein A gold (PAG) and peroxidase gold (POG) were internalized by vesicles, were transported to secondary lysosomes, and eventually accumulated in the digestive vacuole.

Combination of endocytosis of PAG and immunogold labelling of VSG indicated that two directions of VSG-transport can be differentiated:
1) Endocytotic vesicles containing PAG budded from the pocket membrane and later fused with the lysosomal complex. These vesicles were only scarcely labelled with anti-VSG.

2) Flattened VSG-labelled vesicles were present in close proximity to the flagellar pocket and to thickenings of the tubular network extending to the lysosomal complex. Since these flattened vesicles were not present during excessive endocytosis, but only in control cells and in trypanosomes incubated with markers binding to the surface coat and therefore causing the formation of filopodia (Frevert and Reinwald 1988), it is assumed that the tubular network and the flattened vesicles belong to the system transporting VSG to the cell surface.

The effectivity of endocytotic uptake of the tracers tested varied considerably. Whereas gold-conjugated protein A and peroxidase were readily internalized, endocytosis of ferritin stopped at the level of endocytotic vesicles, probably due to rapid paralysis of the trypanosomes during incubation. Ferritin-uptake was also low in T. brucei long slender bloodstream forms, whereas short stumpy and culture forms readily ingested this marker (Langreth and Balber 1975). Thus, effectivity of endocytotic uptake seems also to be dependent on the species of trypanosomes and their state of development.

All substances binding to the surface coat were shed rather than taken up: Con A-ferritin was internalized only to the stage of endocytotic vesicles. By formation of filopodia, surface-bound molecules were completely removed from the parasites. Anti-VSG preincubated with PAG also caused the formation of filopodia and was thus removed from the cell surface. In these experiments, gold particles were internalized and accumulated in the digestive vacuole. It remains an open question, however, whether these gold particles represent anti-VSG-PAG-conjugates or free PAG (not bound to anti-VSG), which can readily be endocytosed.

With respect to their contrast on ultracryosections, PAG and POG proved as excellent endocytosis tracers, whereas intracellular ferritin molecules and their conjugates were hardly visible. Detection of intracellular microperoxidase required osmication resulting in a marked reduction of antigenicity, thus excluding this marker from combination with immunoelectron microscopic studies.

Endocytosis seemed to be receptor-mediated, since the clearly accumulated at certain areas of the flagellar pocket and were then taken up into endocytotic vesicles highly concentrated. This finding is in agreement with the results of Fairlamb and Bowman (1980), who proved the ability of T. brucei to internalize tracers in a combined fluid phase and absorptive endocytotic process.

The bigger (10 nm) PAG and the smaller (5 nm) POG were endocytosed in equal amounts. Since the 10 nm PAG allows secondary labelling of ultracryosections with smaller (5 NM) gold particles in a technique similar to double-labelling resulting in high labelling intensity, PAG proved as the best of the tested tracers.

3. REFERENCES

Bangs J D, Andrews N W, Hart G W and Englund P T 1986 J. Cell Biol. 103 255-263
Cross G A M 1975 Parasitology 71 393-417
Fairlamb A H and Bowman I B R 1980 Exp. Parasitol. 49 366-380
Frevert U and Reinwald E 1988 J. Ultrastruct. Molec. Struct. Res., subm.
Langreth S G and Balber A E 1975 J. Protozool. 22 40-53
Steiger R and Jenni L 1973 Trans. Roy. Soc. Trop. Med. Hyg. 67 293-294

Detection of p10 and polyhedrin antigens in cells of *Spodoptera frugiperda* infected with *Autographa californica* nuclear polyhedrosis virus using immunogold and silver staining techniques

F van der Wilk, J M Vlak and J W M van Lent

Department of Virology, Agricultural University, P.O. Box 8045, 6700 EM Wageningen, The Netherlands

Baculoviruses are the most widely studied group of insect viruses to date because of their importance as biocontrol agents of noxious insect pests (1) and their use as vectors for high level expression of pro- and eukaryotic genes (3). The replication cycle of these viruses in insects is rather complex and it is difficult to study the biochemical and structural alterations associated with infection in some detail. Therefore, the viral replication cycle is studied frequently using cultured insect cells (2), the replication of the multiple-nucleocapsid nuclear polyhedrosis virus (MNPV) of Autographa californica in cells of Spodoptera frugiperda (Sf) being the prime example.

The infection cycle is divided into two phases. In the first phase, the rod-shaped nucleocapsids enter the nucleus after invasion of the cell by viropexis of the non-occluded form of the virus. About 10 h after infection condensed areas of virogenic stroma appear which are the sites of viral DNA replication and nucleocapsid assembly. After budding from the cell membrane the new virions are able to infect other cells. Later after infection the budding process is shut off in favour of the occlusion of the rod-shaped virions in the polyhedra. The cells are then characterized by the presence of numerous large proteinaceous crystalline bodies, called polyhedra, in the enlarged cell nucleus (Figure 1). In these polyhedra many virus particles, usually containing multiple nucleocapsids, are occluded. In addition to residual virogenic stroma, large fibrous structures are prominent in the nucleus as well as in the cytoplasm.

Two viral proteins are abundantly synthesized in virus-infected cells, polyhedrin (MW 33 Kd) and a protein of MW 10 Kd or p10. These two proteins account for over 25% of the total cell protein at the end of the infection. Polyhedrin has been shown to be the major constituent of polyhedra, but the function of p10 is unknown. Many other viral proteins are being synthesized during infection, but the function of most of these remain an enigma. The localization of these proteins in infected cells may help to disclose their function. Therefore, we applied immuno-electron microscopy using immunogold labelling and silver enhancement to localize viral antigens in virus infected cells (4, 5).

With polyhedrin antiserum (Figure 2), gold label was found almost exclusively over the paracrystallin matrix of the polyhedra, but not over the fibrous structures associated with polyhedrin morphogenesis and previously thought to consist of precondensed polyhedrin. Silver enhancement greatly facilitated the detection polyhedrin antigens in the cell (Figure 3). In contrast, when p10 antiserum was used (Figure 4), gold label was only found over these fibrous structures in the nucleus and the cytoplasm. This indicates that p10 is a major constituent of the fibrous structures. In the nucleus the fibrous bodies were associated with electron dense "spacers". The "spacers" had a similar structure as polyhedron membranes. The fibrous bodies are therefore thought to be involved in polyhedron morphogenesis as well as maintenance of the cytoskeleton.

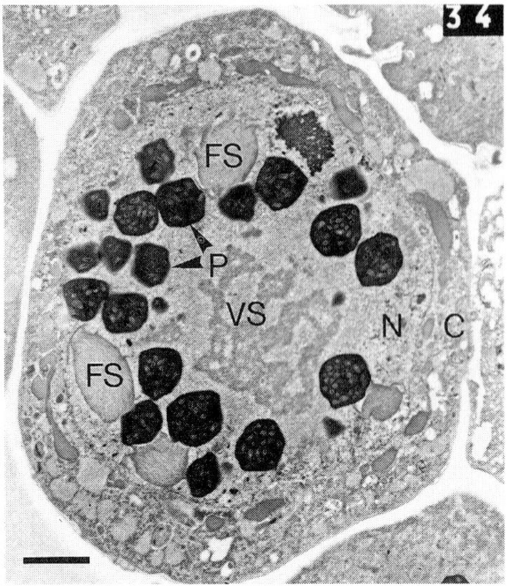

Figure 1. *Spodoptera frugiperda* cell infected with AcMNPV at 50h after infection. P=polyhedra; FS=fibrous structure; VS=virogenic stroma; N=nucleus; C=cytoplasm. Bar= 2μm.

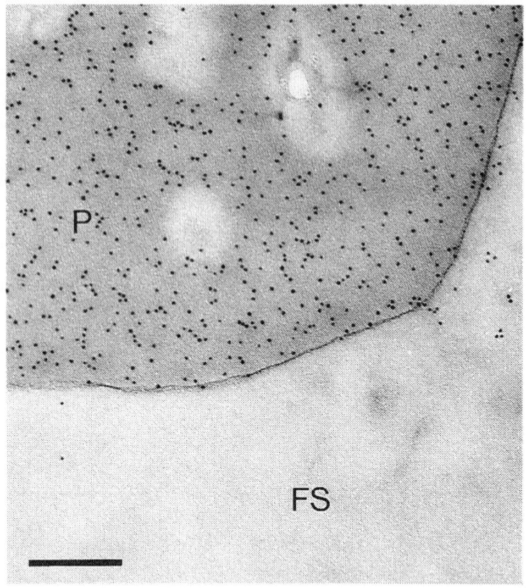

Figure 2. Section with polyhedron (P) and fibrous structure (FS), treated with anti-polyhedron antiserum and protein A-gold. Bar=0.2μm

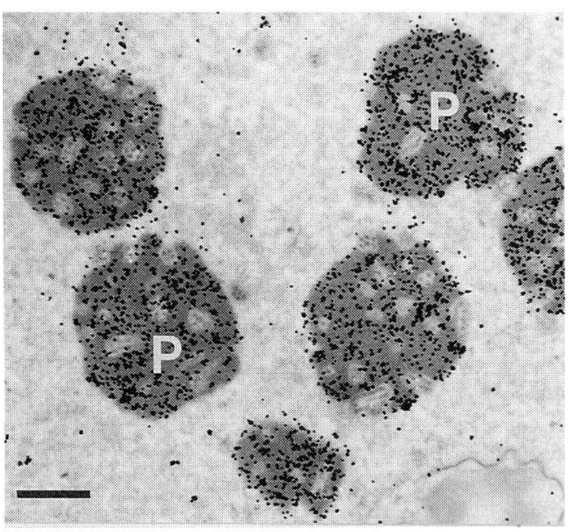

Figure 3. Section with polyhedra (P), treated with anti-polyhedrin antiserum and protein A-gold enhanced with silver. Bar= 0.5μm.

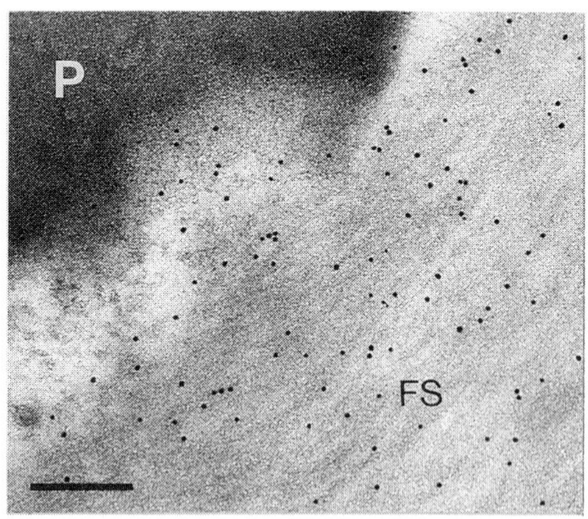

Figure 4. Section with polyhedron (P) and fibrous structure (FS), treated with anti-p10 antiserum and protein A-gold. Bar= 0.2μm.

1) Burges H D (ed) 1981 Microbial control of pests and plant diseases. London: Academic Press.
2) Kelly D C 1982 J gen Virol 63 1
3) Luckow V E, & M D Summers 1987 Bio/Technology 6, 47
4) Van der Wilk F, J W M van Lent & J M Vlak 1987 J gen Virol 68 2615
5) Van Lent J W M, & B J M Verduin 1987 Neth J Pl Path 39 261

Inst. Phys. Conf. Ser. No. 93: Volume 3, Chapter 11
Paper presented at EUREM 88, York, England, 1988

Simultaneous observations of viral antigens on outer and protoplasmic surfaces of HeLa cells

G. Rutter

Heinrich-Pette-Institut f. Exp. Virol. u. Immunol. a.d.Univ. Hamburg, Martinistr. 52, D-2000 Hamburg 20, F.R.G.

ABSTRACT: We present here a method which permits the identification of antigenic determinants on the two surfaces of the dorsal plasma membrane isolated from virus infected HeLa cells. Different labels and labeling procedures were tested for this purpose. Optimal results were obtained with the immunogold labeling method using gold colloid of various sizes. This way, immunologic reagents adsorbed on differently sized gold markers easily discriminate between the presence of an antigen on the external or cytoplasmic surface.

The assembly and release of enveloped viruses as models provide valuable insights into the organization of plasma membrane (PM). For instance, measles virus (MV) particles assemble at the PM and are released from the cells by a budding process (Simon, K. and H. Garoff, 1980). In a previous paper (Rutter et al., 1985) we could correlate the viral structures seen at the protoplasmic surface with antigenic determinants present on the exterior of the cell surface itself. To further characterize the morphological events accompanying the assembly of MV at the PM, we developed a method permitting the concomitant immunocytochemical identification of antigenic structures on both sides of the PM on the same replica (Rutter, G. et al., 1988). Virus glycoproteins appearing at the cell surface were demonstrated by tagging them with rabbit anti-measles antibodies and protein A gold probes (Mannweiler et al., 1982). The cells were stabilized with tannic acid, covered with a cationized coverslip and then split in potassium containing buffer. The membranes adherent to the cationized coverslip were fixed in a formaldehyde-glutaraldehyde mixture and reacted with mouse monoloncal antibodies directed against various structural proteins of measles virus (Bohn et al., 1982). Antibody binding sites present at the protoplasmic surface were visualized either by the antibody bridge method, using normal mouse Ig coupled to gold colloid of different sizes, or by the peroxidase anti-peroxidase procedure. The critical point dried PM were shadowed with platinum/carbon and replicated (Hohenberg, H., 1988). The method permits the isolation of apical PM regions from cultured cells and the identification of structures present at the two faces of plasmalema by immunolabeling. As already shown, it is possible to correlate the distribution of virus antigens labeled with immunogold markers prior to PM preparation with the structures seen on the protoplasmic surface of the plasmalema (Rutter et al., 1985). As the internal structures are also accessible to immunocytochemical identification, we tested and compared here two different methods for tagging the measles virus structures attached to the protoplasmic surface of PM: peroxidase and gold. Both markers proved to be suitable as a second label for our purpose. However, the PAP method seemed to be more sensitive, but using gold markers of different sizes for double labeling gave clearer results.

Since biological membranes are asymmetrically organized, this technique addresses the problems raised by this absolute asymmetry in studying the location of various macromolecules on the two PM surfaces. Making use of double labeling procedures for the antigenic determinants located on the two sides of plasmalemma, one can identify the structures seen at the cytoplasmic surface of PM and correlate them with the presence of antibody binding sites on the exterior of the cell surface itself.

We think that this procedure could be used for some other virus cell systems as well as for morphological studies connected with physiological processes involving the PM.

Bohn, W., Rutter, G., Mannweiler, K.: 1982, Virology 116, 368.
Hohenberg, H.: 1988. In: Verklej, A. ed., Immuno-gold probes in cell biology. CRC Press Inc., Boca Raton, in press.
Mannweiler, K., Hohenberg, H., Bohn, W., Rutter, G.: 1982. J. Microsc., 126, 145.
Rutter, G., Hohenberg, H., Bohn, W., Mannweiler K., 1985. Eur. J. Cell Biol. 39, 443.
Rutter, G., Bohn, W., Hohenberg, H., Mannweiler, K., 1988. J. Histochem. Cytochem. in press.
Simon, K., and Garoff, H. 1980. J. Gen. Virol., 50, 1.

Financially supported in part by Gemeinnützige Hertie-Stiftung, Frankfurt/a.M. The Heinrich-Pette-Institut is financially supported by Freie u.Hansestadt Hamburg, and by Ministerium für Jugend, Familie, Frauen und Gesundheit, Bonn.

Inst. Phys. Conf. Ser. No. 93: Volume 3, Chapter 11
Paper presented at EUREM 88, York, England, 1988

Distribution and alteration of bluetongue virus protein (VP2) within infected cells as determined by immunoelectron microscopy

A.D. Hyatt and B.T. Eaton

Australian Animal Health Laboratories, CSIRO,
P O Bag 24, Geelong, Victoria, 3220, Australia

ABSTRACT: Immunoelectron microscopy has been used to examine some late events in virus morphogenesis in bluetongue virus (BTV) infected cells. Analyses of cells infected with wildtype and variant BTV provide evidence that an outer coat protein VP2 and a nonstructural protein NS1 are added to core-like particles at the periphery of virus inclusion bodies. Further VP2 may be added in the cytosol or whilst the virus is attached to the cytoskeleton. Results suggest that during the final steps in virus morphogenesis VP2 is either removed or undergoes a conformational change.

1. INTRODUCTION

Bluetongue virus (BTV) belongs to the orbivirus genus and the family Reoviridae. The virus contains ten double-stranded RNA segments each of which codes for a single protein. BTV has a diffuse outer coat containing virus proteins VP2 and VP5. The icosahedral core contains major structural proteins VP7 and VP3 and minor structural proteins VP1, VP4 and VP6. Cells infected with BTV also contain non-structural proteins NS1 and NS2.

The cytoplasm of infected cells is characterised by virus inclusion bodies (VIB), virus-specified tubules (VT) and particles. VIB are the presumed site of virus synthesis. Eaton *et al* (1987,1988) and Hyatt and Eaton (1988) have recently provided valuable immunological information on the composition of virus particles and virus-specific structures in BTV-infected cells. In this report we use immunological results obtained from wildtype and variant virus infected cells to provide a preliminary description of late events in virus morphogenesis.

2. METHODS

2.1 Pre and post immunolabelling of infected cells

Cells pre-embedded with lowicryl K4M were labelled with gold conjugated monoclonal antibodies (Mab) or with Mab, biotinylated antimouse antiserum and streptavidin-gold as described by Hyatt and Eaton (1988). Cells destined to be pre-labelled were either (1) extracted with NP40 to reveal the cytoskeleton, washed, incubated and then conventially processed (Eaton *et al*, 1988) or (b) incubated with anti-VP2-gold, washed, fixed in glutaraldehyde, dehydrated and critical point dried from carbon dioxide.

These cells were viewed in a SEM in the backscattered mode.

2.2 Viruses and monoclonal antibodies.

Viruses were examined in the native form (grid-cell-culture-technique, Hyatt et al, 1987), and as purified viruses and virus cores (Eaton et al, 1988). All subsequent immunological reactions were performed as described by Hyatt et al, (1987) and Eaton et al, (1988). Variants were produced as described by Gould et al, (1988). Mabs to VP2, VP7, VP3, NS1 and NS2 were used in the specific incubations (Eaton et al, 1988; Gould et al, 1988; Hyatt and Eaton, 1988).

3. RESULTS AND DISCUSSION

Viruses released from the cell possess an outer fibrillar coat which reacted with Mabs to VP2, NS1 and in some case VP7. When the coat was removed the underlying core particle reacted strongly with Mabs to VP7 and NS1.

Intracellular viruses also possessed VP2, NS1 and VP7 as determined by immunological reactions. VIB reacted with Mabs to VP7, NS2 and to a far lesser extent, VP3. Mabs to VP2 and NS1 failed to react with the internal matrix of VIBs. The results infer that either these two proteins are absent from the matrix or present in a confirmational form unrecognised by the specific Mab. The Mabs do however react at the periphery of VIBs, specifically in locations where viruses are observed emerging from these bodies. Experiments involving BTV variants with an altered VP2 revealed core-like particle aggregates adjacent to VIBs; the particles were surrounded by numerous ribosomes. These particles showed reduced levels of VP2 labelling. Collectively the results indicate that NS1 and VP2 are added at the periphery of VIBs. Viruses in the cytoskeleton distal to VIBs show high levels of VP2 labelling suggesting that additional VP2 is added in the cytosol or following attachment of the viruses to the vimentin containing intermediate filaments (Eaton, et al, 1987). Cytoplasmic viruses (those released from the cell following addition of NP40) contain intermediate levels of VP2.

In contrast to cytoskeleton-associated and cytoplasmic virus particles, viruses released from the cell show reduced binding of anti-VP2 antibodies. This suggests that either VP2 is removed from the developing virus or there is a reorganisation/conformational change in the outer coat. The virus is released from the cell by either budding or extrusion (both modes of release are frequently observed in viable cells). The procedures whereby the virus moves to the edge of the cell for release have yet to be defined.

REFERENCES

Eaton B T, Hyatt A D and White J R 1987 Virology 157, 107
Eaton B T, Hyatt A D and White J R 1988 Virology (In press)
Gould A R, Hyatt A D and Eaton B T 1988 Virology (In press)
Hyatt A D, Eaton B T and Lunt R 1987 J. Micros. 145, 97
Hyatt A D and Eaton B T 1988 J. gen. Virol. (In press)

Immunocytochemical studies on labelling and rearrangement of poliovirus binding sites with Pt/C-Replicas and US

K. Mannweiler, W. Bohn, P. Nobis* and H. Hohenberg

Heinrich-Pette-Institut f.Exp.Virol.u.Immunol.a.d.Univ.Hamburg D-2000 Hamburg 20; *Abt.Molekularbiologie, Univ.Hamburg, FRG.

ABSTRACT: The epitope of the poliovirus receptor demonstrated on Pt/C-replicas after labeling with the MoAb D171 and immunogold probes are patch-like distributed with a similar pattern as adsorbed poliovirus particles. The receptor could not be aggregated with the MoAb D171; a second ligand induced lateral movement in the plan of the PM and strong aggregation of immunocomplexes which were cleared from the cell surfaces either by endocytosis or by extrusion.

A patch-like topographic distribution of poliovirus-binding sites at surfaces of susceptible cells was demonstrated with Pt/C-replicas, ultrathin-sectioning (US) and freeze-fracturing (FF) after labeling with the monoclonal antibody D171 (1) and protein A-gold immunomarkers (2,3) (Fig. 1). Poliovirus particles adsorbed to the cells at 4°C and immunolabeled with rabbit-anti-poliovirus-Ab and pAg show a similar distribution pattern (3,4).

Labeling of native cells with the MoAb D171 at 4°C and subsequently shifting to 37°C for 1h induces no patching of the specifically labeled poliovirus receptor antigen. No morphological signs for an activation of the cell surface mobility like formation of clathrin coated pits or of vesicles could be detected in US. On immunogold-labeled Pt/C replicas the clusters of immunomarker were found to be no longer visible (Fig. 2).

Incubation with a second ligand, an anti-antibody leads to a pronounced aggregation of the receptor D171-complex after shifting to 37°C for 1 h. Large immune complexes located at the cell surface (Fig. 3a) were either endocytosed (Fig. 3b) or extruded from the cells (Fig. 3c).

Financially supported in part by Gemeinnützige Hertie-Stiftung, Frankfurt/a.M. The Heinrich-Pette-Institut is financially supported by Freie u.Hansestadt Hamburg, and by Ministerium für Jugend, Familie, Frauen und Gesundheit, Bonn.

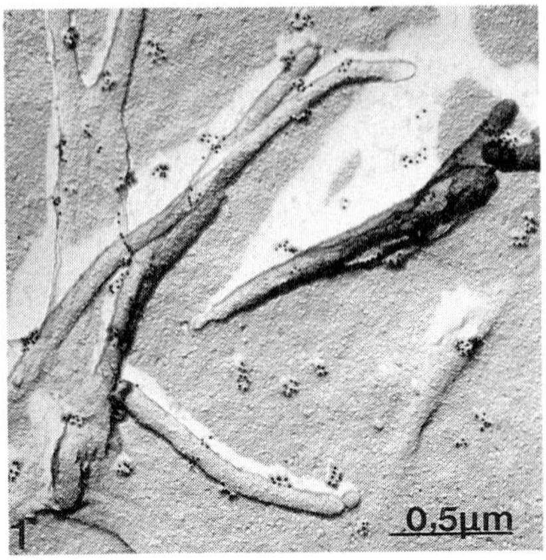

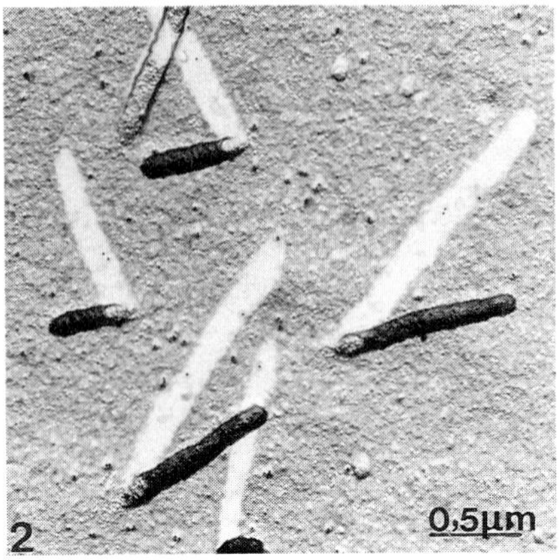

Fig. 1: Patch-like distribution of immunogold-labeled poliovirus binding sites. Paraform-glutaraldehyde-fix. HeLa cells incubated 15 min. with the MoAb D171, labeled with rabbit-anti-mouse antibody and pAg 15 min. each.

Fig. 2: Native HeLa-cells incubated with D171 30 min. at 37°C then aldehyde-fixed and labeled with rabbit-anti-mouse antibodies and pAg. Absence of cluster-like immunogold-labeled receptor domains and of any other kinds of antigenic- and /or structural alterations.

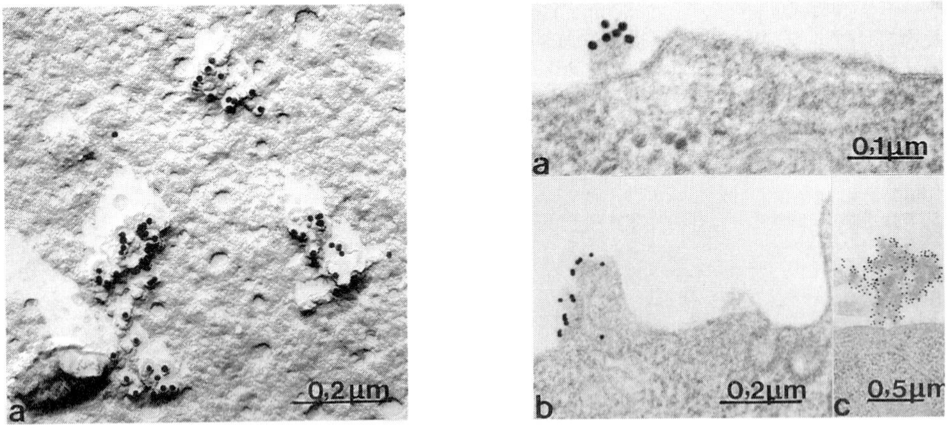

Fig. 3: Rearrangement and aggregation of poliovirus receptor-D171-Ab complex to large immune complexes by a second ligand (anti-antibodies). Native HeLa cells incubated with D171 15 min., and additionally with rabbit-anti-mouse-antibody for 30 min. at 37°C, aldehyde-fixed and labeled with pAg.

1. P. Nobis et al., J. gen. Virol. 1985, 66, 2563.
2. K. Mannweiler et al., J. Microsc. 1982, 126, 145.
3. K. Mannweiler et al., Acta Biol. Hung. 1986, 37, 256.
4. K. Mannweiler et al., Eur.J.Cell Biol. 1987, 44 (Suppl.).

Inst. Phys. Conf. Ser. No. 93: Volume 3, Chapter 12
Paper presented at EUREM 88, York, England, 1988

Low dose electron microscopy for biologists—a practical system

N.G. Wrigley

Division of Virology, National Institute for Medical Research,
Mill Hill, London NW7 1AA, England

ABSTRACT: Specimens for electron microscopy - especially biological specimens - lose fine structural detail on exposure to the electron beam. This 'radiation damage' can be reduced by cutting the total electron dose. However that is no use unless the microscope is also adjusted so that it will image the detail thus preserved. How to achieve those adjustments with sufficient precision is one purpose of this paper. The other is to bring this technology into more widespread use by reducing its seeming complexities, perhaps at the expense of the underlying physics.

1. INTRODUCTION

It has been widely recognised for nearly two decades that specimen damage due to the electron beam is the main limitation to very high resolution in electron microscopy of biological specimens. It is less widely recognised that even at intermediate resolution beam damage can be significant. It is almost unrecognised that, especially with modern electron microscopes, the means to reduce this damage by reducing the electron dose is within the reach of every microscopist. The reason that this low dose technology is still only used by a handful of practitioners is that the underlying physics is undeniably somewhat complex; the purpose of this paper is to make it clearer if possible. It is addressed to the user group who have the most to gain but perhaps the least inclination to bother with the alien physics - the biologists. By eliminating all mathematical equations and other such physical obfuscation I hope to remove the complexity and mystery from the topic.

It is useful to divide the subject into two overlapping parts: that of reducing the electron dose and that of optimising the imaging status of the microscope. The first may be beneficial in nearly all biological applications, even with structures larger than, say, 4 nm; one cannot know without trying. The second becomes important as well if reliable detail below 3 nm is sought, and absolutely essential for anything below 2 nm. The higher the resolution required the more exacting are the demands on the microscopist. 'Resolution' here means real resolution of biologically meaningful fine structure, not merely an electron optically superb picture of - junk. I will begin with three subtopics of underlying importance.

2. BACKGROUND

2.1. The nature of beam damage

Specimen damage, apart from the artefacts of drying, etc, arises from the transfer of energy from the electrons to the specimen. The interaction between beam and specimen is already weak, but with zero interaction there would be no image! We are usually forced to increase the interaction by adding heavy atoms in the form of negative stain in order to have any useful contrast. As a result, precisely because the beam/specimen interaction is increased, the absorption of energy is increased also. That increases beam damage; it also adds to image deterioration through chromatic aberration because the emergent beam contains an increased proportion of loss electrons. Much the same thing occurs - increased interaction plus increased damage - when the electron accelerating voltage is reduced. Good compromises can be achieved with minimal staining (occasionally none) and 100 kV acceleration.

The general form of beam damage is illustrated qualitatively in Figure 1. Each specimen will be different in detail, but it is probably true that structure remaining after high doses is unreliable below about 5 nm. At the other end of the scale it appears from several authors that preservation of 0.5 nm detail requires a subminimal dose of only 4-500 electrons/nm^2 of the specimen. A minimal dose - meaning the dose required to adequately expose the photographic film - is about 1000 electrons/nm^2 for Kodak 4489, and less for Kodak SO-163.

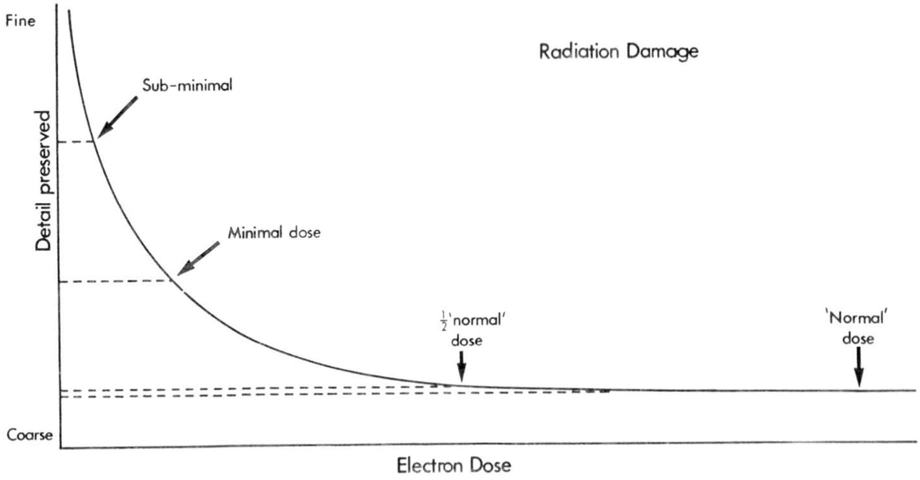

Fig. 1. General form of dose dependent specimen damage.

2.2. The effect of focus.

The contrast referred to above, arising from beam electrons which transfer energy to atomic electrons in the specimen and cause all the trouble, is called scattering contrast. Fortunately, some incident electrons interact with atomic nuclei and do not lose energy but do give rise to phase contrast. This is an interference phenomenon arising from their wave properties and carries all our high resolution information without causing damage. If only all interactions were of this type!

But phase contrast too exacts a price, because it is critically focus-dependent as shown in Figure 2. This is a three dimensional plot of focus vs size of imaged detail vs strength (and sign) of phase contrast. This last is maximum along the curved ridges, but all tail off towards zero (due to imperfect beam coherence) at the front of the diagram, corresponding to the smallest spacings. The valleys correspond to zero phase contrast. What is not shown is that alternate ridges (say the odd-numbered ones) should be drawn as troughs, meaning that contrast is inverted - what should look white in the image appears black or vice versa. Such a gross image distortion, not to mention the valleys where contrast is zero, is quite unacceptable, and is completely unknowable by mere inspection of the micrograph. For example, following the curve corresponding to 1 nm, this detail would have strong inverted contrast at 100 nm overfocus (point A), strong true contrast at 120 nm underfocus (point B) and zero contrast at 210 nm underfocus in the valley at C. (These details depend on the fixed microscope parameters and accelerating voltage, but the general form of Figure 2 remains the same.)

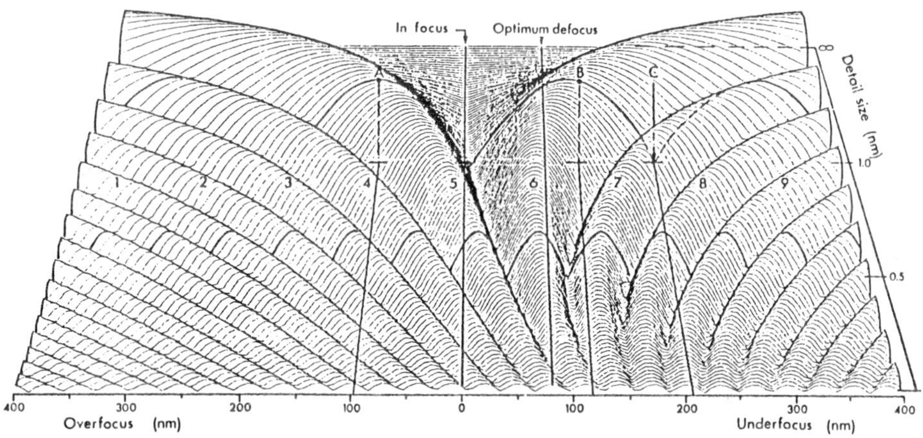

Fig.2. Phase contrast transfer for 100 kV electrons, $C_s = 1.6$ mm $= C_c$.

Fortunately this inconvenient physics gives us one loophole. There is a narrow range of focus, marked on Figure 2 as 'optimum underfocus', where phase contrast does not undergo reversals or zeros over a wide range of detail size. We have to hit this special defocus to avoid the problems above. However, it is a condition which cannot be recognised by eye on the screen. It must be set by introducing a fixed defocus step below 'in focus' (Gaussian focus) which <u>can</u> be recognised by a total loss of contrast in all fine detail. <u>Of course</u>, if there is no meaningful fine detail due to beam damage then none of this matters! But if there is fine detail thanks to successful low-dose operation, we risk scrambling it uninterpretably if we ignore this special focus requirement.

There is one further requirement for optimising the vital phase contrast. To reduce the attenuation of contrast towards the smallest detail (front of Figure 2) beam coherence must be optimised. Apart from special guns - lanthanum hexaboride tips or field emission - this is achieved with a highly excited first condenser, second condenser set well away from crossover, and a thin specimen. With a mean free path of around 50 nm for 100 kV electrons in average biological material, the latter will ensure that electrons scattered coherently by atomic nuclei will not be scattered again, incoherently, by atomic electrons. Here lies the reason why even 'thin' tissue sections produce poor phase contrast and hence not very high resolution - through no fault of the microscopist!

2.3. Low dose and magnification

It seems necessary to scotch once and for all the idea that high magnification is somehow synonymous with high resolution. The very opposite is true. This is because a fixed minimal dose at the image plane is multiplied by the square of the magnification to calculate the dose at the specimen - where the damage is done. Thus every doubling of magnification quadruples the specimen dose. Working at 20,000x magnification represents a 25-fold saving compared to working at 100,000x. Perhaps there is a lack of confidence in photographic emulsions, but they are so efficient that they will comfortably and faithfully record 1 nm detail at a modest 20,000x. This issue of dose versus magnification dominates much of the practical strategy which follows.

3. THE PRACTICAL STRATEGY

As mentioned in the Introduction, all biological applications stand to gain from dose reduction. Only those involving thin specimens - which excludes most thin sections but includes two-dimensional protein crystals, viruses and single macromolecules - stand to benefit from the exacting electron optical adjustments, especially optimum defocus. I will therefore attempt to separate the various sections and summarise the integrated strategy later.

3.1. Dose reduction

Assuming that the dose cannot be reduced below that minimal amount required to expose the film properly, what do we mean by low dose working? We mean elimination of all that wasted dosage incurred, before the picture is taken, in field selection, focussing and astigmatism correction. This can amount to 50-100 times the minimal dose - see Figure 1.

Searching must be done by scanning systematically up and down rows of grid squares, with an illuminated spot smaller than one square to protect adjacent rows. Likely areas can usually be recognised at very low magnification; the dose rate at x1000 is only one thousandth of that at x30,000. Brightness is often much more than adequate and can be reduced by desaturating the gun; reduction by spreading the beam with C2 would put adjacent rows at risk.

Once an area is selected, centre it rapidly, deflect the beam to an adjacent area, and condense the beam to C2 crossover. For this deflection, use the beam shift or tilt system provided on most microscopes, so that the crossover spot is well clear of the chosen area. Go to the photographing magnification (say x30,000) and adjust focus and astigmatism using a supplementary viewing screen fitted at the front edge of the viewing chamber. Alternatively, if image shift is provided as well as beam shift, the off-axis image may be brought back to the screen centre for normal viewing.

When all is ready, blank the beam, either by decentering the condenser aperture to a pre-set stop or by switching a small DC voltage into the gun alignment coils. With the chosen area now protected by blanking, switch off the beam deflection(s), raise the screen, bring in the film and adjust C2 to a pre-set value for correct exposure brightness. Since the built-in shutter is usually below the specimen, it is easiest to set it to a long exposure and time the actual (shorter) exposure manually by unblanking and reblanking the beam. These details will depend on the system provided on the microscope. Much of the mechanics of this scheme was first described by Unwin and Henderson (1975).

3.2. Adding the focus complication

If reliable information on details below 2 nm is sought we must now include the focus requirement described in section 2.2. Focusing (and astigmatism correction) in the casual manner described above will not do, because the chance of getting it accurate enough is negligible. Taking a through-focal series will not do either, because each exposure incurs a further minimal dose which would soon destroy the detail we are trying to preserve. So what is 'accurate enough'? Looking at the focus scale at the front of Figure 2 a departure of only 20 nm either side of the 'optimum underfocus' line takes us well down the sides of the 'optimum' ridge, with a severe loss of phase contrast. The smallest focus step on most microscopes is about 5 nm - a knob that many never touch! - and we must hit optimum within four of these steps. Examining background granularity, the visible 'in focus' condition cannot be seen on the screen with this precision at anything lower than about 300,000x magnification - which is grossly incompatible with the low dose requirement. Although the selected on-axis field of the specimen is protected during focusing, our deflected off-axis field would be right outside the microscope at x300,000. Therefore the image shift facility is now essential, in addition to beam shift; it is standard equipment on the latest microscopes and can be retro-fitted on some older models.

By this means 'in focus' can be found at high magnification with sufficient accuracy. Next, a fixed defocus step must be applied to set 'optimum underfocus' (or any other desired defocus). The size of this step - about 80 nm in Figure 2 - must be determined by laser diffractometry, and once known will remain constant for that microscope at that accelerating voltage. The procedure is given in Chapman (1986) and can be done in another laboratory if a suitable diffractometer is not available. In addition, since the picture will be taken at, say, 30,000x magnification, a correction for the zoom focus error incurred in demagnifying must be determined and added to the defocus step if our focus precision is not to be lost. In practice, both the step and the correction are determined together and applied as one.

Finally, since the whole point of these operations is to maximise the transfer of phase contrast through the microscope, that contrast must itself be maximised by ensuring coherent beam conditions as given in section 2.2. Furthermore a large objective aperture (or none at all) must be fitted; its minimum diameter is calculated from Bragg's law using the electron wavelength (0.0037 nm at 100 keV), the objective focal length (usually around 2 mm) and the smallest detail we wish to resolve. Any misguided use of a smaller aperture will only increase the scattering contrast (which contributes little at high resolution) and cut out the precious phase contrast we have worked so hard to maximise.

4. THE COMPLETE PROCEDURE (Wrigley et al 1983)

It will be clear already that numerous variations of detail are possible in this low dose/accurate defocus procedure, depending on the application and especially onthe microscope to be used. The latest machines have programmed minimum dose systems (MDS) which require operator intervention only for field selection and focus, after the initial settings have been fed in. I will assume below that this is not the case, but that image shift and beam shift coils are fitted below and above the specimen respectively and that some shuttering arrangement exists <u>above</u> the specimen. A graduated knob must also be fitted to the second condenser fine control. Initial settings require only normal alignment, highly excited first condenser (small 'spot size'), large objective aperture and small condenser aperture, the latter chosen such that there is enough illumination at C2 crossover at x300,000. Magnification of x1000, x300,000 and x 30,000 are recommended for the three stages of search, focus and exposure, as listed below.

<u>Search stage</u>. Select x1000 magnification. Set C2 to illuminate less than one grid square. Reduce brightness until just sufficient by desaturating the filament. Defocus the objective grossly for additional contrast (if necessary). Scan systematically up and down rows of grid squares. Centre the selected area. Close the beam shutter immediately. Now, working in the dark, re-saturate the filament, restore nominal objective focus and set C2 to crossover (using graduated knob).

<u>Focus stage</u>. Switch on beam and image shift, pre-set so that cross-over is at the screen centre; the specimen is now protected. Open the beam shutter. Select x300,000 magnification. Viewing the background grain, focus and correct astigmatism iteratively until perfect 'in focus' is obtained. NB: It can take a lot of practice to recognise this condition where grains become so sharp and small that they virtually disappear; perfect adjustment of binoculars and good dark adaptation are needed. Now defocus from this condition by the pre-determined amount for 'optimum underfocus' plus the zoom correction, counting coarse and fine focus steps as necessary. Make sure that the image is not drifting; wait as necessary. Close the beam shutter.

<u>Exposure stage</u>. Select x30,000 magnification. Switch off beam and image shift. Expand the beam with C2 graduated knob to a per-determined level suitable for the exposure. Raise the screen and feed film into position. Open the camera shutter with a long exposure time selected. Time the exposure by hand, by opening and closing the beam shutter.

<u>Footnote</u>. Exposure times should preferably not exceed 2-3 seconds in case of residual drift, and brightness should be low with the beam spread well away from crossover for good beam coherence. That apart, any time/ brightness combination giving the required minimal dose will do. This assumes reciprocity between the two, but recent evidence (Fryer 1987) suggests that this is not so, at least with certain organic crystals.

5. REFERENCES

Chapman S K 1986 Roy Mic Soc Handbook Series Vol 08 pp 60-68 Ox.U.Press

Fryer J R 1987 Ultramicroscopy <u>23</u> 321-328

Unwin N and Henderson R 1975 J mol Biol <u>94</u> 425-440

Wrigley N G Brown E and Chillingworth R K 1983 J Microscopy <u>130</u> 225-232

Electron microscopy of bacteriorhodopsin

R. Henderson

MRC Laboratory of Molecular Biology, Hills Road, Cambridge
CB2 2QH, U.K.

Recent advances in cryo-electron microscopy and computerised image processing are allowing the determination of the high resolution structure of bacteriorhodopsin. In previous work, the structure of bacteriorhodopsin at about 6Å resolution has been determined independently in three different crystal forms (1,2,3). In the present work, bacteriorhodopsin in the naturally occurring purple membrane of Halobacterium halobium has been studied to higher resolution. The 2.8Å projection structure has been determined (4) by recording micrographs at 4°K and carrying out image processing of the entire area of the micrograph.

The collection and processing of high resolution three-dimensional data from tilted specimens is now underway. A persistent problem is that the recorded micrographs are extremely variable in quality in spite of specimens which appear to be nearly identical. The analysis can continue to make progress but the methods would be greatly helped by further improvements in imaging methods. In particular, even the best images so far obtained should be able to be improved on in future.

1. Henderson & Unwin (1975) Nature 257, 28-32.
2. Leifer & Henderson (1983) J. Mol. Biol. 163, 451-466.
3. Tsygannik & Baldwin (1987) Eur. Biophys. J. 14, 263-272.
4. Baldwin, Henderson, Beckmann & Zemlin (1988) J. Mol. Biol., in press.

Alpha helices of TMV visualized by cryo-electron microscopy

T.-W. Jeng[1], R.A. Crowther[2], G. Stubbs[3] and W. Chiu[1]

1. University of Arizona, Dept. of Biochem., Tucson, AZ 85721
2. MRC, Lab. of Molecular Biology, Cambridge, CB2 2QH, U.K.
3. Vanderbilt Univ., Dept. of Mol. Bio., Nashville, TN 37235

Tobacco mosaic virus (TMV) is a helical particle, the structure of which has been determined by X-ray diffraction. We chose it as a test specimen to investigate whether cryo-electron microscopy and computer image processing could reveal internal structure at high resolution in a non-crystalline specimen.

Low dose images of TMV embedded in a thin layer of vitreous ice were recorded at 60,000 magnification in a JEOL100CX electron microscope equipped with a cold stage at -148°C. The best images were selected for processing on the basis of their optical diffractograms. The defocus was determined from the image of the carbon film adjacent to the hole over which TMV particles were suspended. The defocus values of processed images ranged from 5000 to -10000 Å. The initial processing steps included masking out the selected segment of a particle, interpolating the image to make the axis of a selected particle segment parallel to the transform array, straightening the particle segment by a real space correlation and re-interpolation procedure. The computed structure factors from the straightened particle segment were corrected for the effects of defocus by applying a Wiener filter function. The first 6 layer lines were extracted with a tilt and center correction. Subsequently polarity, axial position, azimuthal angle and radial scale factor were determined. We used the X-ray model data to cross check the determination of these parameters as well as the defocus value. Finally, the data set of 12 segments of TMV particles was scaled and the overlapping Bessel functions on each layer line separated by a weighted least squares procedure, which also refined the azimuthal angle and axial origin position for each particle. The refinement of these parameters and the separation of Bessel contributions were repeated alternately until the process converged.

Figure 1 shows the amplitudes and phases of a few layer lines, compared with those from X-ray fibre diffraction. A good match is found in regions where the amplitudes are significantly higher than the background noise. During the processing, the 7th layer line was not used for the parameter determination and the good match between the EM and X-ray data validates our processing scheme. Figure 2a and 2b show sections of a density map of TMV reconstructed from layer lines l=0 to 7 with data extending radially from the meridian to 1/10 $Å^{-1}$. The map shows the 2 pairs of core alpha helices, the C helix and the RNA. These results demonstrate the feasibility of cryo-electron microscopy for determining the secondary structure of helical macromolecular assemblies.

Acknowledgement: This research has been supported by NIH grant RR02250 and a Guggenheim fellowship to W.C. and GM33265(to G.S.).

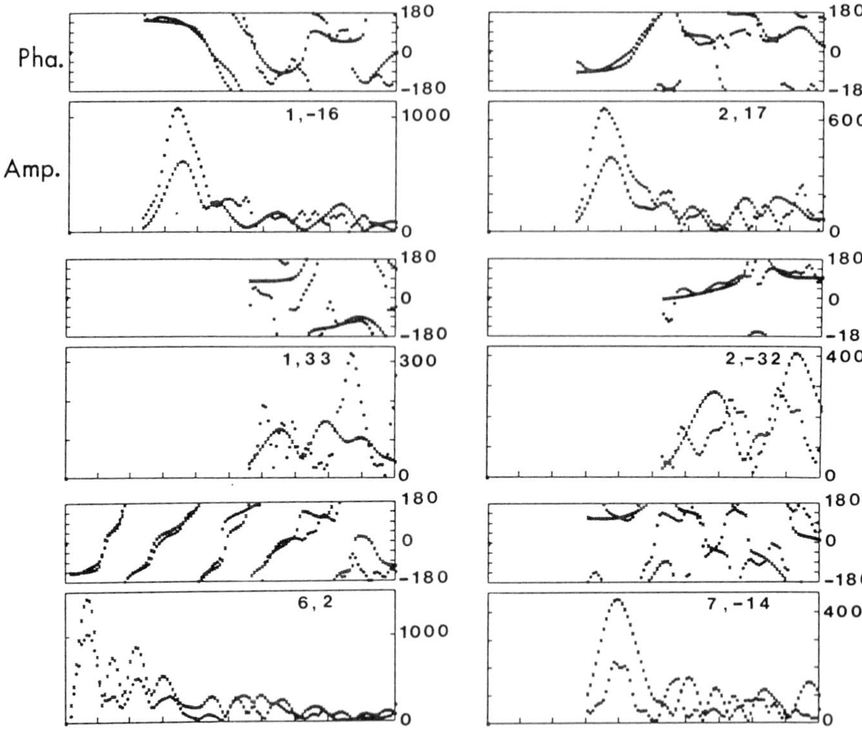

Figure 1. Examples of amplitudes and phases of layer line data of TMV obtained from EM (0) and X ray (x) fiber diffraction. Each tick mark refers to 0.01 $Å^{-1}$ in the horizontal axis. The layer line number and Bessel order are indicated by the pairs of numbers inside the amplitude box.

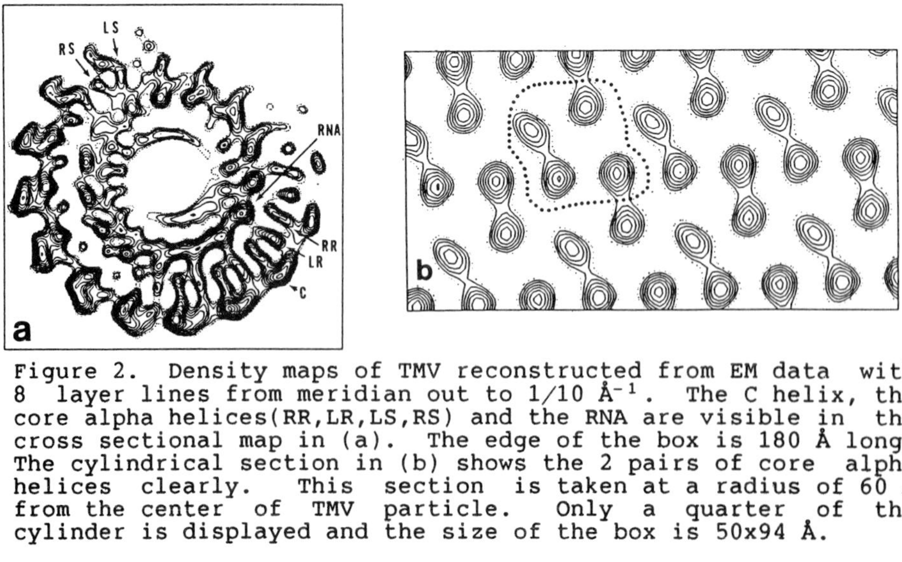

Figure 2. Density maps of TMV reconstructed from EM data with 8 layer lines from meridian out to 1/10 $Å^{-1}$. The C helix, the core alpha helices(RR,LR,LS,RS) and the RNA are visible in the cross sectional map in (a). The edge of the box is 180 Å long. The cylindrical section in (b) shows the 2 pairs of core alpha helices clearly. This section is taken at a radius of 60 Å from the center of TMV particle. Only a quarter of the cylinder is displayed and the size of the box is 50x94 Å.

The probable exit site of the polypeptide in the ribosome: analysis of densities in three-dimensional reconstruction

M Radermacher, J Frank, T Wagenknecht

Wadsworth Center for Laboratories and Research, New York State Department of Health, Albany, N.Y. 12201, USA

ABSTRACT: A reexamination of our reconstruction of the 50S ribosomal subunit from Escherichia coli /5/ reveals a low density channel connecting the probable location of the exit site of the nascent polypeptide chain with the interface canyon in agreement with results of recent immuno electron microscopical studies /1/. In this context, strategies of visualizing 3-D information in electron microscopy need to be reevaluated.

A three-dimensional reconstruction from a tilt series in electron microscopy results in a volume with densities varying continuously in three dimensions. Analysing such a volume poses difficulties. Initially the volume is available as a stack of slices that can be viewed individually or mounted side by side (fig.1). However, to recognize the shape of structures such as cavities or high density areas that continue trough several slices additional means have to be used.

For aiding the analysis of three-dimensional data, surface representation programs have been developed /2,3/. They visualize a surface at a specified density, much as if a solid model had been built from slices using a specific contour level as the definition of the particle's boundary. The proper density threshold for a surface representation can be found from a contoured central slice, as the center level in an area of maximum density gradient.

However such representations have several major disadvantages. The structure is modelled with a precisely defined boundary. First, the information that the model is known only to a limited resolution is lost. Second, cavities whose sizes are on the order of the point resolution value will show densities that are distinct from the surroundings yet not necessarily below the threshold value defined above. In fact, to analyze such variations we found it necessary to vary the threshold in very small increments.

The 50S ribosomal subunit had been reconstructed using a random conical tilt series /4,5/. Images from 489 particles were combined into a single three-dimensional reconstruction. The electron dose that a single particle was exposed to was less than 1000 el/nm^2 (10 $el/Å^2$). The crystallographic resolution of the reconstruction as determined by the 45° phase residual criterion was 3nm, corresponding to a point resolution of 1.8nm.

Using for a threshold the center density in the range of the maximum density gradient, the volume enclosed by the surface (2340 nm^3) /5/ comes out in good agreement with the volume found in solution X-ray scattering studies (2500 nm^3)/6/. In addition, surfaces defined by higher density thresholds were calculated; yet with the viewing angles and threshold increments previously explored a connection with the front part of the particle was recognized only at unrealistically high threshold levels.

Prompted by the results reported in /1/ we reexamined our model using the following strategy: The threshold level was varied in steps of 1.14% of the total density range of the reconstruction. Using the threshold level that rendered the indentation in the back of the particle visible, a series of representations was calculated where parts of the structure were cut away in 0.5nm increments. Thus stepping through the volume we found a connection that starts on the the subunits back, runs parallel to its surface and then leads in a bend towards the interface canyon below the central protuberance and towards the side of the L1 shoulder (fig.2). This pathway agrees with the immunoelectron microscopic findings /1/.

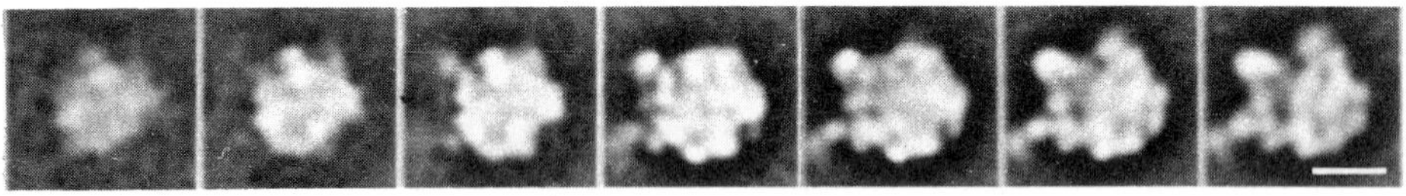

Fig. 1 Series of slices through the three-dimensional reconstruction of the 50S subunit in 1 nm increments. Scale bar 10 nm.

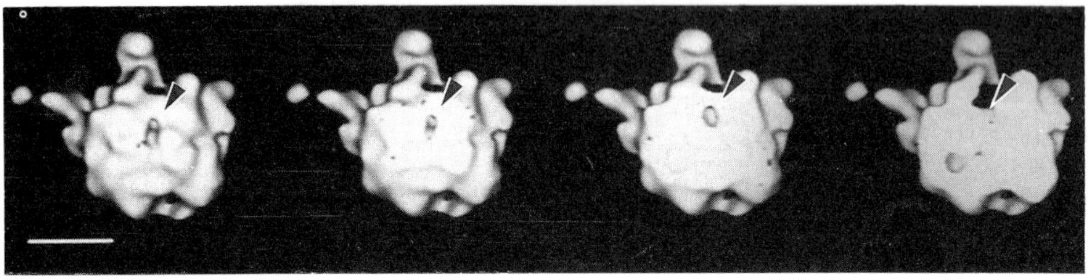

Fig.2 Surface representation of the 50S ribosomal subunit at the density threshold that shows the probable location of the exit site of the nascent peptide. The subunit is cut open in successive planes 1.5 nm apart. Visible is a channel that runs first parallel to the back of the subunit and then curves forward towards the interface canyon. Scale bar 10nm.

1. L.A.Ryabova, O.M.Selivanova, V.I.Baranov, V.D. Vasiliev, A.S.Spirin, FEBS Letters, 226, (1988), 255-260.
2. M.van Heel, Ultramicroscopy 11, (1983), 307-314
3. M.Radermacher, J.Frank, J.Mirosc. 136, (1984), 77-85.
4. M.Radermacher, T.Wagenknecht, A.Verschoor, J.Frank, J.Microsc. 146, (1987), 113-136.
5. M.Radermacher, T.Wagenknecht, A.Verschoor, J.Frank, EMBO J. 6, (1987), 1107-1114.
6. A.Tardieu, P.Vachette, EMBO J. 1, (1982), 35-40
7. Supported by NIH Grant 1R01 GM29169 and NSF grant 8313405

Can electron holography improve EM imaging of biological objects?

Hannes Lichte und Uwe Weierstall
Institut für Angewandte Physik, Universität Tübingen,
D 7400 Tübingen, Federal Republic of Germany

Biological specimen often are very weak pure phase objects. For example, assuming a mean inner potential of 4 eV and a thickness of 5 nm we find a phase modulation of about $2\pi/30$. To achieve a contrast nevertheless, by staining an amplitude contrast is produced bearing the risk of producing artefacts. Using frozen hydrated specimens and large defocusing a well observable phase contrast is given by the phase contrast transfer function (PCTF), however rather imperfect as compared to the ideal Zernike phase contrast: Due to the decrease of the PCTF for spatial frequencies approaching zero, the necessary object dose and hence the danger of radiation damage increase significantly and large area phase contrast is not possible at all. In addition, the oscillations of the PCTF for higher spatial frequencies have to be cut away by an aperture and hence the collection efficiency of the optical system gets very poor; this gives rise to a considerable loss of electrons in the image, increasing quantum noise, and furthermore considerably dampens the phase modulation of the image wave to be analysed.

Electron holography offers a way out of these problems:
By light optical reconstruction using a Zernike phase plate, the phase modulation of unstained bacteriophages can be made visible with high contrast (fig.1). A more quantitative retrieval of amplitude and phase is possible by numerical reconstruction of an electron hologram. The phase modulation by the object can then perfectly be displayed on a monitor showing an image of the object with ideal phase contrast. In particular, large area phase contrast is observed very well (fig.2). For biological specimen the amplitude usually is only of minor interest.

Furthermore, correcting the aberrations of the electron objective lens, the aperture can be opened up, reducing strongly the loss of electrons and the dampening of the signal. A simple estimate shows that the electron dose loaded to the specimen can be lowered by more than one order of magnitude.

Of course, the holographic technique is limited in its overall performance, mainly by lateral resolution, signal resolution and possible specific falsification of the reconstructed image wave with respect to the object. Routinely, the lateral resolution can be made smaller than 0.15 nm hence not limiting the imaging of biological specimen. A major problem, however, is given by the signal to noise ratio in the reconstructed wave limiting the signal resolution; it is determined by the quality of the hologram (H.Lichte, K.-H.Herrmann and F.Lenz 1988), the structural noise of the supporting foil and by some reconstruction noise. As demonstrated in fig.2, the overall noise in the phase of a carbon foil is given by about 2π/50, unfortunately still too bad for uniquely discerning a very weak object lying on the foil. A second problem to be solved is put up by the distortion of the electron microscope and by irregularities of the digitizer (I.Daberkow, W.Dreher, E.Völkl and H.Lichte 1988), because they can pretend phase structures, mainly large area structures, originally not present in the object.

We are indebted to the Koerber-foundation and the Volkswagen - foundation for supporting this work.

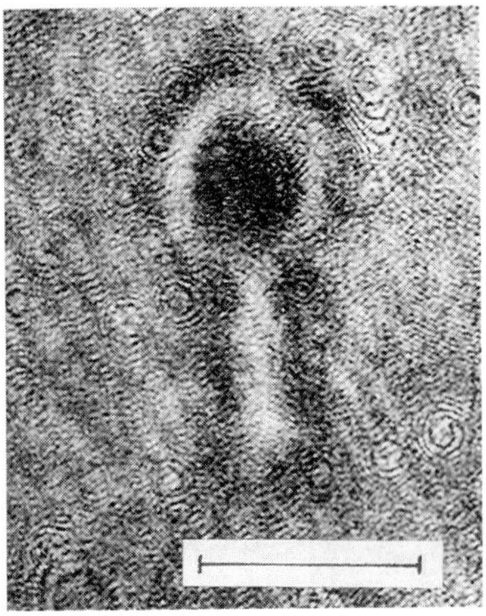

fig.1

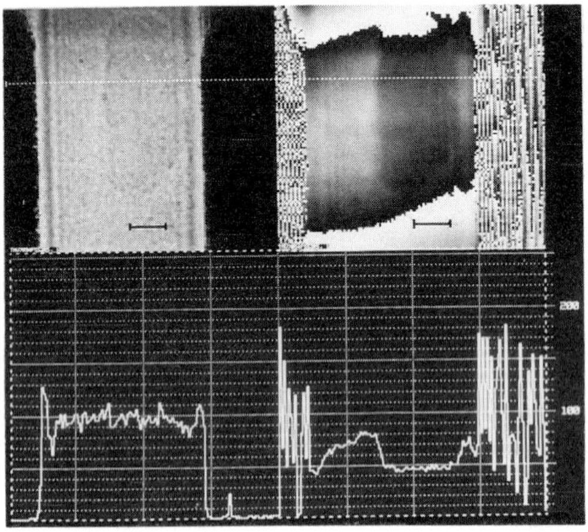

fig.2

fig.1 Light optical reconstruction of an unstained bacteriophage using a Zernike phase plate.
(marker represents 100 nm)

fig.2 Numerical reconstruction of a hole in thin carbon foil covered with another foil.
(marker represents 10 nm)
top lhs: amplitude
top rhs: phase
Unlike in the amplitude, in the phase the hole is descernable with strong large area contrast. The sharp lines of equal phase on top and bottom of the field of view are due to shortcomings of the reconstruction apparatus being improved.
bottom rhs: linescan
The linescan gives the phase distribution quantitatively: the phase shift due to the hole amounts to 2π/12 well above the noise of 2π/50.

Daberkow I, Dreher W, Völkl E and Lichte H 1988 same proceedings

Lichte H, Herrmann K H and Lenz F 1988 same proceedings

Inst. Phys. Conf. Ser. No. 93: Volume 3, Chapter 13
Paper presented at EUREM 88, York, England, 1988

Methods for the study of the three dimensional organisation of the nucleus by scanning electron microscopy

T.D. Allen
Department of Ultrastructure, Paterson Institute for Cancer Research
Christie Hospital, Manchester M20 9BX

ABSTRACT: Nuclei, isolated in conventional biochemical manner using hypotonic buffers and prepared using a cytocentrifuge were 'dissected' by the use of detergent extraction before osmium thiocarbohydrazide infiltration. Internal detail of intact nuclei was revealed using fast atom bombardment or a dry fracture technique.

Conventional transmission and scanning electron microscopy has provided a wealth of detail with respect to cytoplasmic and surface structures, but has been considerably limited with respect to the three dimensional organisation of both chromosomes and interphase nuclei. This lack of information is largely due to the fibrous nature of nuclear components in which the DNA packaging ratios reach 10,000 to 1 in metaphase chromosomes. Chromatin structure has been studied mainly by biochemical techniques which have led to the general acceptance of DNA wrapped around histone protein cores to form nucleosomes, and the nucleosomes themselves arranged in a solenoidal structure which forms the 30 nm 'unit' fibre of both chromosomes and interphase nuclei. Some evidence for chromatin fibre packaging suggests that the chromatin is subsequently organised as loops, anchored in some central matrix within the chromosome, but the overall organisation of chromatin fibres within the interphase nuclei remains largely unexplored.

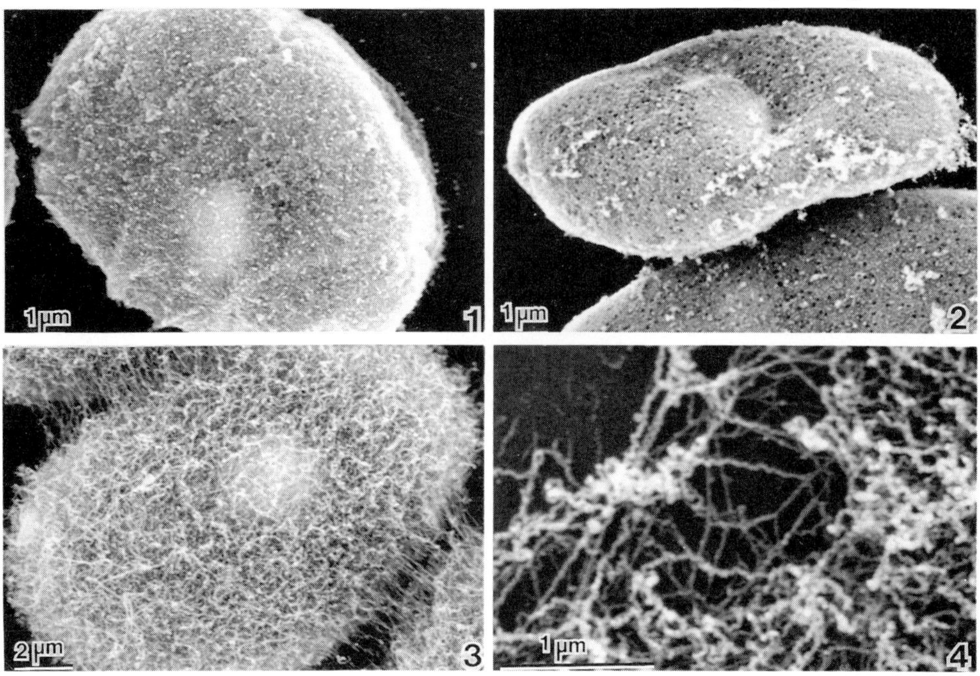

Nuclei were extracted from cultured Indian Muntjac Fibroblasts, after trypsinisation using a hypotonic isolation buffer (Turner and Keohane 1987). The suspension was then spun at 1000 rpm for 5 minutes in a cytocentrifuge (Shandon Scientific Cytospin) onto a 15 mm coverslip without air drying, and the coverslips processed immediately on removal from the centrifuge. Some coverslips were exposed to detergent extraction for different periods of time and subsequently refixed and prepared for SEM using an osmium-thiocarbohydrazide infiltration technique (Allen et al 1985). Two further techniques were used to expose internal nuclear organisation, namely fast atom bombardment (Allen et al 1985) and a dry fracture technique using double sided tape after critical point drying.

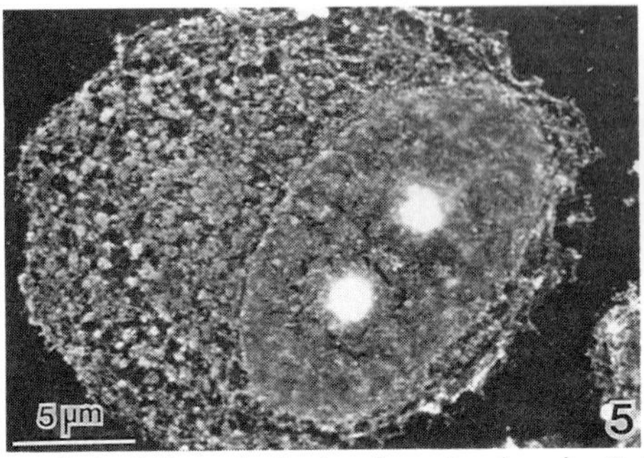

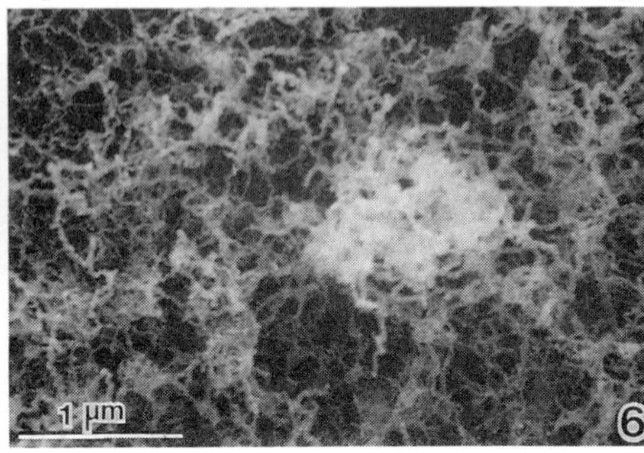

Nuclei liberated in hypotonic isolation buffers show surface structure indicative of retention of cytoplasm (fig 1), including ribosomes and cytoskeletal elements. Increasing time in detergent solutions removes in succession, cytoplasmic contamination (fig 2), the nuclear membrane, the peripheral organisation of the nuclear lamins, and subsequently reveals the organisation of the peripheral chromatin (fig 3). Further extraction results in dispersion of the chromatin, concomitant with the loss of the highest orders of chromatin packaging, and reduction in chromatin supercoiling (fig 4). Alternatively, low angle fast atom bombardment exposes unmodified internal structure, (figs 5 and 6). The use of a dry fracturing technique revealed internal organisation across profiles which encompassed the whole depth of the nucleus (figs 7,8).

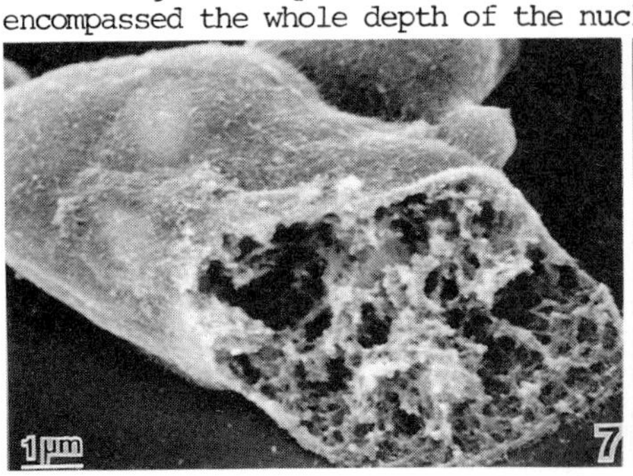

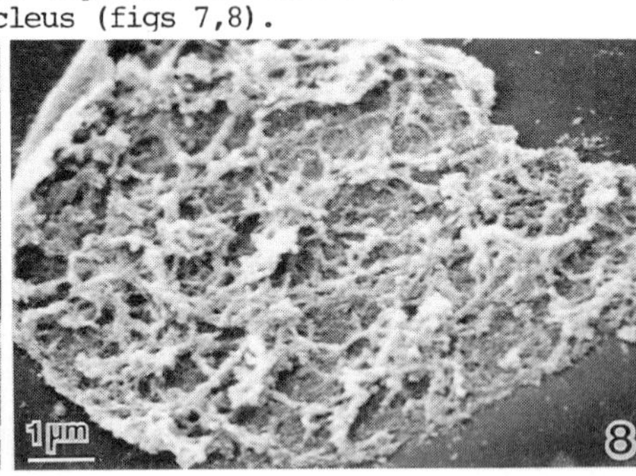

Legends All Muntjac nuclei, extracted in hypotonic isolation buffer.
Fig 1 extracted nucleus fixed in MeAc then Os-TCH. Fig 2 15 mins ddw prior to fixation. Fig 3 15mins 0.4% Sarkosyl extraction, Fig 4 1 hr Sarkosyl extraction. Fig 5, 6, 5 mins Fast atom bombardment (7½° angle). Fig 7, 8, Dry fracture after critical point drying.

References

Allen T.D Jack E.M., Harrison C.J., Claugher D, Scanning Electron Microscopy 1986 1, 301

Turner B.M and Keohane A. 1987 Chromosoma 95 263

Stereological investigations of changes of myocardial cell ultrastructure by diet-induced hypercholesterolaemia and by simultaneous application of two different doses of vitamin A

A.Bózner and J.Dostál

Dept of Pharmacodynamics and Toxicology,Pharmaceutical Faculty, University of J.A. Comenius, Bratislava, ČSSR

ABSTRACT: The study contents results of analysis of myocardial cell ultrastructure changes in japanese quail induced by chronic /103 days/ administration of a cholesterol reach diet / 1 % / and/ or vitamin A.The stereological evaluation of myocardial cell ultrastrcture showed significant changes of mitochondria when cholesterol was combined with vitamin A. The volume of myofibrils was found to be highest after vitamin A only, while was significantly lower in hypercholesterolaemic birds. The results suggest that vitamin A administration during hypercholesterolaemia has favourable effect on the myocardial cell ultrastructure changes.

The changes of myocardial cell ultrastructure by a diet-induced hypercholesterolaemia in japanese quial were investigated stereologically. It is known, that vitamin A has an influence on the membrane stability as well as on the metabolism of the cells / Ott and Lachance, 1984/, however, by application of high doses of vitamin A is known a toxic death in human patients, too / Bürgi et al.,1985/.For this reasons the changes of myocardial cells were studied in controls, in birds treated with cholesterol reach diet /1 % /, in birds treated with vitamin A in doses of 5000 I.U. and of 100.000 I.U. and finally in birds treated both with cholesterol reach diet and vitamine A in both doses.

The results could be summarized as follows:

1/ The chronic administration of a cholesterol reach diet increases the relative volume of mitochondria.The vitamin A treatment has no influence on the mitochondrial volume.The simultaneous administration of vitamin A in hyper - cholesterolaemic birds tends to lower the increased volume of mitochondria / Fig. 1./

2/ Hypercholesterolaemia induces a significant decrease of volume of myofibrils, while vitamin A in higher dose increases this volume.The simultaneous administration of vitamin A and cholesterol inhibits the effect of hypercholestereolaemia on volume of myofibrils / Fig. 2./

3/ The changes in the number of mitochondria in different groups of birds have not unambigous trend if studied in hypercholesterolaemic bird or in birds treated with vitamin A.

Literature:

Bürgi W., H. Kaufmann, H.Fehr, J. Mühlethaler, Chronische letal verlaufende Vitamin-A-Vergiftungen mit haemolytischer Anaemie, Shcweiz.med.Wochenschr.,115,1985,526

Ott D., P.A. Lachance, Vitamin A action on hepatic cholesterol biosynthesis, Nutr.res., 4, 1984, 137-144

Fig. 1

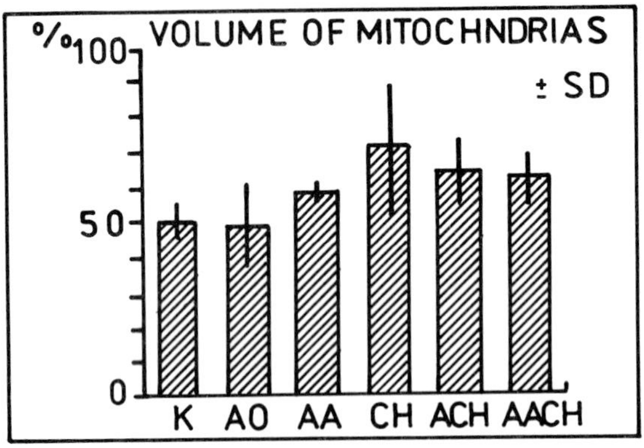

Explanation:
K-control
AO-vitaminA 5000 I.U
AA-vitaminA 100.000 IU
CH-cholesterol 1 %
ACH- vitamin A 5000 I.U + cholesterol
AACH-vitamin A 100.000 IU + cholesterol

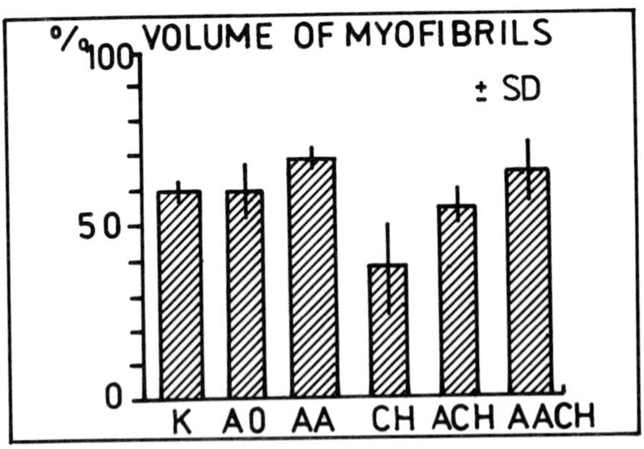

Fig. 2.

Paper presented at EUREM 88, York, England, 1988

Sample preparation for electron microscopy using cryofixation and molecular distillation drying

J.G. Linner, S.A. Livesey, D. Harrison, and T. Schifani

University of Texas Health Science Center at Houston, Cryobiology Research Center, The Woodlands, Texas 77381

ABSTRACT: The combination of cryofixation and molecular distillation drying offers an alternative to wet processing of samples for electron microscopy. This approach is especially suited to application of probing techniques such as colloidal gold immunolabelling and elemental analysis.

1. INTRODUCTION

Preparation of biological specimens which are stable in ultrahigh vacuum and under an electron beam, yet retain an interpretable correlation to the native state is the ongoing aim of electron microscopy.

1.1 CRYOFIXATION

Physical or cryofixation has been described as the optimum tissue preparation method for avoidance of chemical cross-linking and alteration of cellular antigens. Many of today's cryo methodologies, however, include light chemical fixation and one or more aqueous steps that could potentially displace or remove water soluble, noncomplexed proteins or ions.

Cryofixation or ultrarapid cooling of the non-cryoprotected, nonfixed sample achieves almost instantaneous arrest of cellular metabolism without aqueous extraction or crosslinking. Important parameters of metal mirror cryofixation including sample holder design, precooling, sample bounce and thermal sink will be discussed in the context of a new automated cryofixation device.

1.2 MOLECULAR DISTILLATION DRYING

Molecular distillation drying is a process by which cryofixed cellular water is removed from the sample directly without appreciable ice crystal formation or growth. Essential features of the device and process include loading of samples under liquid nitrogen, equilibrium temperature well below the devitrification temperature, ultrahigh vacuum, efficient condensor surfaces and microprocessor controlled sample heating. In this way, the sample is dried without contact with organic solvent. Once dry, the sample can be vapor osmicated and embedded in pure Spurr's resin or held at -30°C and embedded in Lowicryl K_4M in vacuo. Samples therefore, have the inherent benefits of cryopreparation with regard to the retention of water soluble cellular components. In addition, being resin embedded, they are amenable to conventional room temperature sectioning and analysis.

1.3 ANALYSIS

Analysis of the samples for structure, post embedding colloidal gold immunolabelling and electron dispersive spectroscopy will be presented.

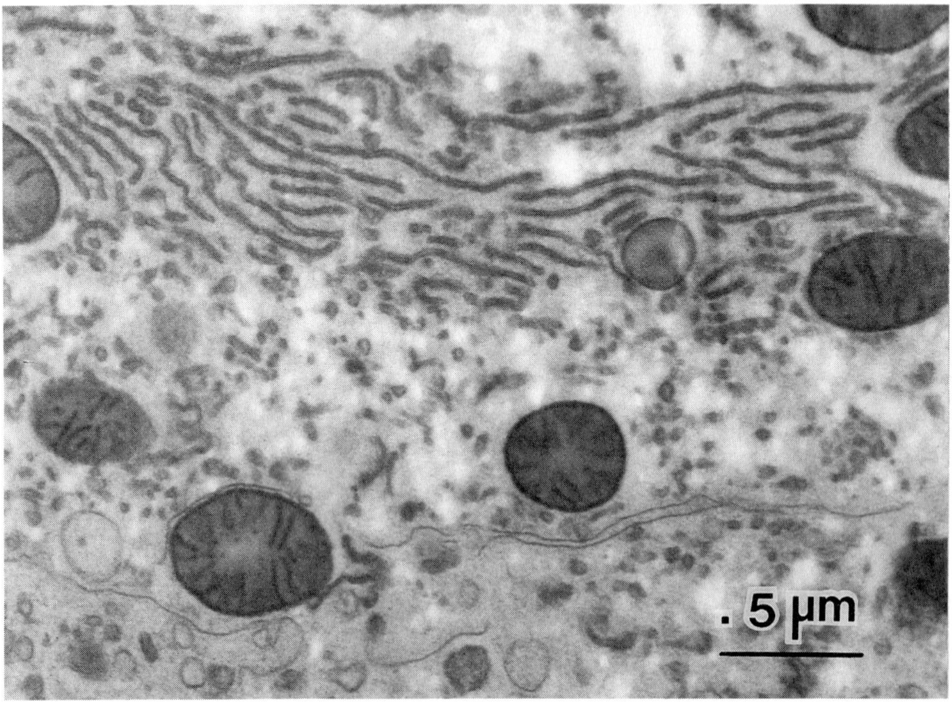

Fig. 1: Normal rat liver, cryofixed, molecular distillation dried, vapor-phase osmicated, and embedded in Spurr's resin.

SEM of wet-etched preparations of bovine mesenteric lymphatic vessels

G McKerr, J M Allen & R Donaghy.

Biomedical Sciences Research Centre, University of Ulster, Newtownabbey BT37 0QB, N.Ireland.

ABSTRACT: Isolated segments of bovine mesenteric lymphatic vessels were cut longitudinally and viewed from their luminal surface by conventional SEM. Preparations subjected to tryptic digestion and subsequent acid hydrolysis (2-20min) revealed detail of both superficial cells and underlying tissues. This technique permits the study both of muscle group distribution and their spatial relationship to the autonomic nerve supply.

1. INTRODUCTION

Attempts to analyse the three dimensional anatomy of functioning bovine mesenteric lymphatic vessels have been fraught with difficulties. The interrelationship of at least three muscle layers plus elements of the autonomic nervous system have confused both manual and computer aided techniques. However, the use of wet-etch techniques (Baulk & Gabella, 1987) applied to open vessel segments has permitted the visualisation of sequentially deeper layers within the vessel wall by scanning electron microscopy (SEM). This technique displays layers in their order of occurrence from the endothelial surface through to the adventitia.

2. METHODS.

Short segments (1cm) of bovine mesenteric lymphatic vessel were removed from recently slaughtered cattle, cleared of adherent fat and pinned in open configuration to a silgard resin base. Each specimen was digested in trypsin (1% w/v; 37°C) for 2min., fixed in glutaraldehyde and post-osmicated. Subsequent tissue etching was carried out (2-20min.) in 8M HCl at 60°C. Etched tissue was then dried by the critical point method and viewed in a scanning electron microscope.

3. RESULTS.

Tissue processed by this technique shows good preservation of ultrastructural detail. Endothelial cells (fig 1), which line the duct lumen, still display the typical surface lamellae and intercellular junctions of control specimens after 2 min. exposure to acid hydrolysis. Etch times in excess of 4-5 min. expose the underlying layers (fig. 2). Immediately below the endothelial covering elongate cells lie in parallel. TEM of these preparations (fig 3) confirm that these spindle shaped structures are indeed smooth muscle fibres. Lateral exposure of the vessel wall at the point of surgical incision displays the complex and branched nature of the autonomic nerve supply (fig 4) and its beaded or varicose nature (inset).

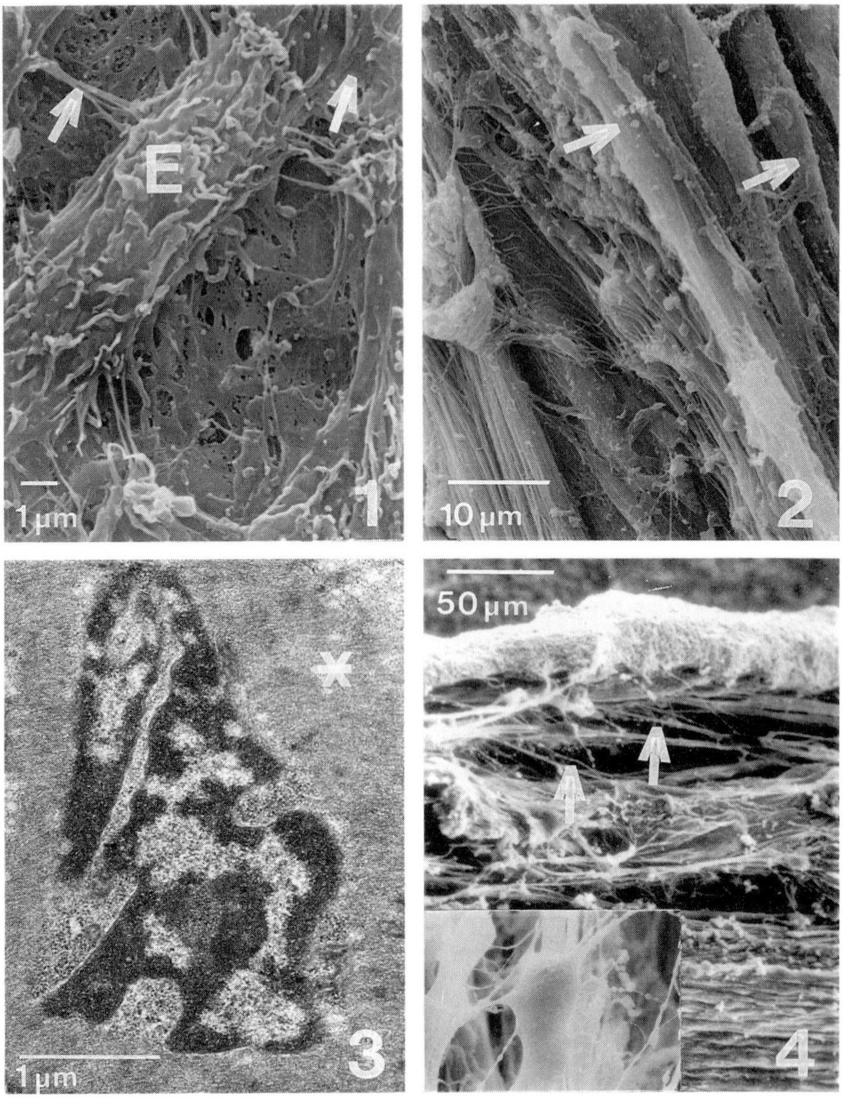

Figure 1. Etch time 2min. Endothelial cells (E) exhibit surface lamellae and intercellular junctions (arrows).
Figure 2. Etch time 2+min. The endothelial surface has now been removed to disclose underlying fibres (arrows).
Figure 3. As for fig 2. TEM of smooth muscle cells. Note myofibrils (*) and irregularly shaped nucleus.
Figure 4. Etched preparation of cut wall showing autonomic nerve fibres (arrows). Inset shows detail of nerve varicosities.

REFERENCE:

Baluk P & Gabella G 1987 *Cell Tissue Res* **250** 551-561

Inst. Phys. Conf. Ser. No. 93: Volume 3, Chapter 13
Paper presented at EUREM 88, York, England, 1988

Neutral red as a fixative additive to enhance contrast of ribosomes and nucleoli in plant cells

June R Lawton

E M Unit, University of Durban Westville, PBag X54001, Durban 4000, South Africa

ABSTRACT : Inclusion of neutral red in glutaraldehyde for fixation of root apices leads to increased contrast of ribosomes and nucleoli. Experiments involving enzymic digestion with RNA-ase before and after fixation suggest that neutral red binds to RNA molecules and renders them highly osmiophilic.

1. INTRODUCTION

Neutral red is a well known vital stain for plant cells and has been used as a fluorescent stain both for lipids (Kirk 1970) and for particles containing RNA (Kurnick 1955). Phenyl neutral red was one of the dyes used by Bünemann and Müller (1978) incorporated into bisacrylamide gel complexes for the separation of DNA molecules on columns, as it intercalates between the G-C linkages of the DNA molecule.

2. MATERIALS AND METHODS

Root tips of *Phaseolus vulgaris* L. var Contender were fixed for 2h in 1% glutaraldehyde to which neutral red (0.05%) was added, post fixed in osmium tetroxide (1h 1%) then dehydrated with ethanol and embedded in low viscosity epoxy resin.

Root tips were treated with RNA-ase (1%) before fixation for 3h at 37° C or after fixation for 8h at 35° C.

3. RESULTS AND DISCUSSION

Apical cells fixed in the presence of neutral red show very increased contrast of ribosomes (Figure 1) and nucleolar granules than those fixed in the absence of neutral red (Figure 2). After incubation of root tip tissue with RNA-ase before fixation cytoplasmic ribosomes are removed and the effect of subsequent fixation in the presence of neutral red is eliminated.

Treatment of root tissue with RNA-ase after fixation in glutaraldehyde containing neutral red does not result in the removal of ribosomes. Similar treatment after fixation in glutaraldehyde alone or containing malachite green removes the cytoplasmic ribosomes. (Malachite green is another of the DNA affinity dyes, intercalating in this case with A-T linkages).

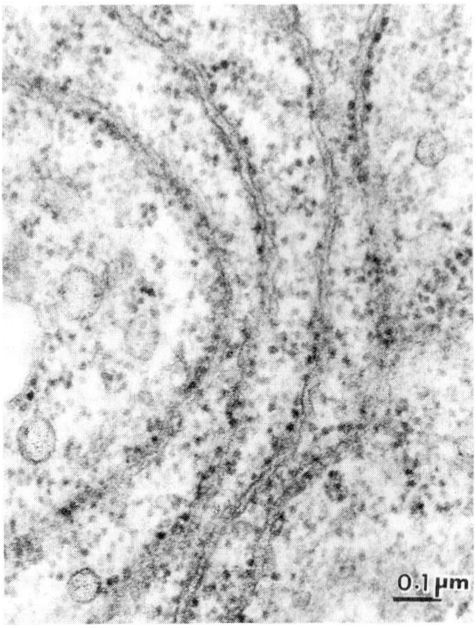

Figure 1. fixed in GANR

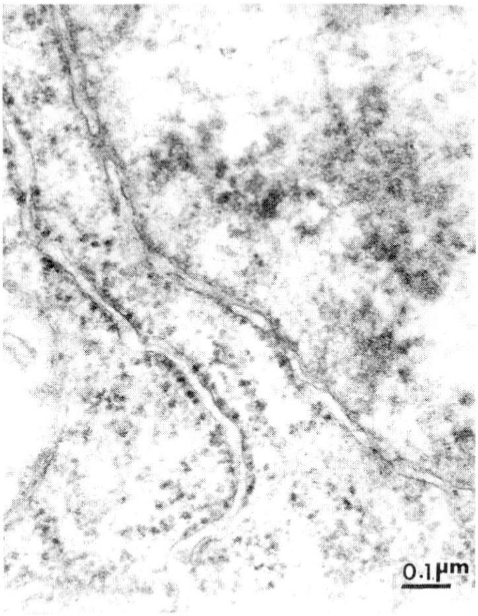

Figure 2. fixed in GA

Figures 1 and 2 - Core cells from root apices of *Phaseolus vulgaris* L.

Similar results have been obtained with roots of *Zea mays* L. and *Lolium multiflorum* Lam.

Neutral red is not an electron dense molecule but does possess several double bonds which can react with osmium tetroxide. It is possible that neutral red intercalates at G-C linkages with the phosphate backbone of the RNA molecule. It is hoped that the use of neutral red during fixation will prove to be a valuable technique for the study of RNA material in the cell.

4. REFERENCES

Bünemann H and Müller W 1978 Nucl.Acids Res. **5** 1059-1074
Kirk P W 1970 Stain Technol. **45** 1-4
Kurnick N.B 1955 Int. Rev.Cytol. **4** 221-268 (p251)

Approximate fixative timetable for nematodes: TEM study 17 species for locate carbohydrate moieties on the epicuticle membrane

RYUNTYU, M., Department of Physiology, University of New England, Armidale. N.S.W. 2351. Australia.

The time factor was considered in 3 series of experiments. The first was carried out to find the best timetable for the secondary fixative (Table). The second was carried out to find resins for the best penetration of the cuticle (Ryuntyu: Approximate timetable for embedding media and staining of sections for nematodes. Proc. 9th European Congress on Electron Microscopy, York, U.K.: 1988). The third was carried out to find the best timetable for stains (Ryuntyu: 1988). Therefore, it is not possible to expect a single OsO_4 fixation of 2 hrs would give simultaneously good results for all cuticular strata. To overcome these difficulties, specimens were fixed in a secondary fixative for up to 2 hrs, as shown in Table. It was found that: 2 hrs OsO_4 fixation was uneven for some strata; 6 hrs was positive for some zones of the basal layer but negative for all other strata. The aims set for embedding concerned the preservation of the epicuticle membranes. Because there was no disintegration of the zone containing cuticular strata, it was considered positive for embedding.

Table: (see next page)

Key:

++	best result
+	good result
+/-	unstable result for some structures
▼	unstable result for all structures
*	hours

Species:

Aporcelaimellus (1)	Tylenchus (10)
Xiphinema (2)	Seinura (11)
Paratrichodotus (3)	Clarcus (12)
Mesorhabditis (4)	Rhabditis (13)
Panagrellus (5)	Caenorhabditis (14)
Pratylenchus (6)	Mermis (15)
Heterodera (7)	Amphimermis (16)
Meloidogyne (8)	Aphelenchus (17)
Paratylenchus (9)	

APPROXIMATE FIXATION TIME FOR SELECTED NEMATODE SPECIES

Species ♀	Time of aldehyde fixation (hours)		Time of osmium fixation (hours)					Hypodermis	Musculs	Cuticular layers			
	2		2	3	4	5	6			Epicuticle	Cortical zone	Median zone	Basal zone
A. obtusicaudatus	*	(1)	▼		*	▼	▼	+	++	+	+/-	++	++
	*			*				++	++	++	++	++	++
X. monohysterum	*	(2)	*	▼	▼	▼	▼	++	++	++	++	++	++
P. lobatus	*	(3)	*	▼	▼	▼		+	+	++	++	+	++
	*						*	++	++	+	+/-	++	++
M. monhystera	*	(4)	*	▼		▼		++	++	++	++	+/-	+
	*						*	++	++	+	+	++	++
	*				*			++	++	++	+/-	+	+
P. redivivus	*	(5)	*	▼	▼	▼		+	+	+	+	++	+
	*						*	++	++	++	++	++	++
P. thornei	*	(6)		*	▼	▼	▼	++	++	++	++	+	+
	*		*					+	++	+	++	++	++
H. avenae ♂	*	(7)		▼	*	▼	▼	+	+	++	+/-	++	++
	*		*					+	+	++	++	++	++
M. arenaria ♂	*	(8)	▼	▼	*	▼		+	+/-	+	++	++	+
	*						*	+	+/-	++	++	+	+
P. nanus	*	(9)	▼	▼		*	▼	++	+	++	+/-	+/-	++
	*				*			++	+	++	++	++	+
T. semipenetrans ♂	*	(10)	▼	▼	▼	*	▼	+	+	++	++	++	++
S. paratenuicaudata	*	(11)		▼	*	▼		+	+	+	++	++	++
	*		*					+	+	+	++	+	+
	*						*	++	++	++	+/-	+/-	+/-
C. papillatus	*	(12)	*		▼	▼		++	++	+	++	+	+
	*			*				+	+	++	+	+	++
	*						*	+/-	+/-	+	++	++	++
R. oxycerca	*	(13)		*	▼	▼	▼	+	+	++	++	+	+
	*		*					++	++	+	+	+	+/-
C. elegans	*	(14)	*		▼	▼	▼	++	+	++	++	+	++
	*			*				++	++	++	+/-	++	+/-
Mermis sp.	*	(15)		*	▼	▼	▼	+	++	++	++	++	++
	*		*					++	+	++	+/-	++/-	++

Microwave enhanced fixation of rat liver

P Wild, M Krähenbühl und EM Schraner

Institute of Veterinary Anatomy, University of Zürich
Winterthurerstr. 260, CH - 8057 Zürich

ABSTRACT Microwave irradiation during fixation for electron microscopy is considered to stabilize cellular structures induced by heat, to enhance cross-linking of proteins by glutaraldehyde, and to lead to improvement of subsequent osmium fixation. The effect is improved by NA/K-phosphate, K^+, Ca^{2+} and Mg^{2+}.

INTRODUCTION

Fixation employing microwave energy is recommended for improved fixation quality in light microscopy (Boon and Kok 1987). Hopwood et al (1984) interpreted cautiously the results obtained by microwave irradiation for electron microscopy whereas Leong et al (1985) or Login and Dvorak (1985) considered microwave energy fixation as a useful method to improve the preservation of cellular structures. The quality of fixation obviously depends on the medium used during irradiation, on postfixation, and on the tissue itself. We will report on the influence of aldehydes, buffers and ions on the ultrastructure of liver cells fixed by microwave enhancement.

MATERIALS AND METHODS

Cubes (1 mm^3) were excised from anaesthetized rats and immediately placed in glass vials containing 10 ml of one of the following mediums:

1. 2.5% glutaraldehyde in a) 0.1M Na-cacodylate, b) 0.1M Na-phosphate, c) 0.1M Na/K-phosphate or d) 0.1M s-collidine
2. 1% glutaraldehyde, 2% formaldehyde in the same buffers as above
3. 0.1M Na-cacodylate or s-collidine containing various concentrations of $CaCl_2$, $MgCl_2$ or KCl
4. 0.1M Na-phosphate or Na/K-phosphate

After a delay of about 30 seconds, the samples were irradiated for approximately 11 seconds in a conventional microwave oven (Miele M696) operated at 700 W, raising the temperature from room temperature to 52-56°C. The samples were left in the warm medium for 5 minutes, quickly washed in the appropriate buffer, postfixed with 1% OsO_4 in the appropriate buffer at room temperature, dehydrated in aethanol and embedded in epon.

RESULTS AND DISCUSSION

Microwave irradiation in the presence of glutaraldehyde in Na/K-phosphate led to excellent preservation of the ultrastructure in liver cells showing distinct membranes of all compartments, dense glycogen and dense cytoplasmic matrix. The results were similar to those obtained by perfusion fixation. The area of good fixation, however, was limited to the periphery (10-15 cell layers) of the tissue block. High quality fixation considering both preservation of membranes and distribution of glycogen was restricted to the outermost layers adjacent to the layers damaged due to cutting forces. The combination of glutaraldehyde and formaldehyde or the use of other buffers led to dilation of ER, loss of membranes (SER), loss of cytoplasmic matrix and of glycogen.

Microwave irradiation in the presence of only a buffer resulted in uniform fixation throughout the whole block: distinct RER and mitochondrial cristae, dense cytoplasmic matrix and glycogen, but hardly recognizable SER. Best preservation was achieved by Na/K-phosphate, worst by s-collidine. Addition of $CaCl_2$, $MgCl_2$ or KCl markedly improved the preservation of intracellular membranes and the retention of the cytoplasmic matrix. $CaCl_2$ and $MgCl_2$ in combination was superior to $CaCl_2$ or $MgCl_2$ alone.

Microwave irradiation leads to stabilization which is sufficient for further processing for light microscopy. It is, however, insufficient for electron microscopy. Thus osmication is essential (Hopwood et al 1984, Marti et al 1987). An important factor in microwave irradiation is the rapid raise in temperature and the rapid conductance of heat. Raise in temperature also leads to enhanced reaction of aldehydes. This certainly is favourable for glutaraldehyde under this condition but not for formaldehyde because of its long reaction time (Fox et al 1985). Stabilization is also improved by K^+, Ca^{2+} and Mg^{2+} possibly in a similar manner as in the course of perfusion fixation (Wild et al 1987).

Boon ME and Kok LP 1987 (Leyden, Coulomb Press) pp 71-84

Fox CH, Johnson FB, Whiting J and Roller PP 1985 J. Histochem. Cytochem. **33** 845-853

Hopwood D, Coghill G, Ramsay J, Milne G and Kerr M 1984 Histochem. J. **16** 1171-1191

Leong AS-Y, Daymon ME and Milios J 1985 J. Pathol. **146** 313-321

Login GR and Dvorak AM 1985 Am. J. Pathol. **120** 230-243

Marti R, Wild P, Schraner EM, Müller M and Moor H 1987 J. Histochem. Cytochem. **35** 1415-1424

Wild P, Bertoni G, Schraner EM and Beglinger R 1987 Micron Microsc. Acta **18** 259-271

Differences in membrane contrast in liver parenchymal cells with varying washing periods in buffer after glutaraldehyde fixation

C E Hulstaert, E H Blaauw, and I Stokroos
Centre for Medical Electron Microscopy, University of Groningen, The Netherlands

When studying the ultrastructure of cells and tissues via ultrathin sections, usually one of the aims of the preparatory procedures is to obtain as much membrane contrast as possible. In this respect a very suitable postfixative is 1% OsO_4 plus 1.5% $K_4Fe(CN)_6$ (de Bruijn 1968; Hulstaert et al 1983).The membrane contrast thus obtained is also dependent on the type of buffer used; cacodylate buffer gives a better membrane contrast than phosphate buffer (Neiss 1984). Nevertheless it occurred in some of our experiments that rat liver tissue treated as described above appeared to have a very poor contrast. Therefore we investigated whether a relation could be found between the membrane contrast and the washing period in buffer (cacodylate and phosphate) after GA fixation. Rat liver was fixed by perfusion with 2% GA in either 0.1M cacodylate or phosphate buffer, pH 7.4 (total fixation period 2 h). Subsequently 100 µm vibratome sections were rinsed for 1, 2, 4, and 24 h at 4° and 20°C in 6.8% saccharose in 0.1M cacodylate or phosphate buffer. Finally the sections were postfixed in 1% OsO_4 plus 1.5% $K_4Fe(CN)_6$ or in 1% OsO_4 alone, both in the same buffer as used before.
The tissue postfixed in OsO_4 plus $K_4Fe(CN)_6$ with cacodylate buffer yielded large variations in membrane contrast with the different washing periods. The highest contrast was obtained 1 h after rinsing (Fig. 1). After 2 h rinsing the cellular membranes and glycogen were without any contrast while the mitochondrial matrix was electron dense (Fig.2). After 24 h rinsing the membranes and glycogen were positively stained while the mitochondrial matrix was less electron dense than after 2 h (Fig. 3). The same tendencies although not as prominent were found when the tissue washings were carried out at 20°C or when a phosphate buffer was used both in combination with a OsO_4 plus $K_4Fe(CN)_6$ postfixation. We have no explanation for the disappearance of the membrane and glycogen contrast after 2 h of rinsing. The electron density of the mitochondrial matrix after 2 h could result from a contraction of the mitochondria; morphometric analysis indicated that these mitochondria are smaller than at 1 and 24 h rinsing.

de Bruijn W C 1968, 4th Eur.Reg.Conf. on Electron Microscopy ed D S Bocciarelli (Roma: Tipogr.Polygl.Vaticana) II pp 65-6
Hulstaert C E, Kalicharan D and Hardonk, M J 1983 Histochemistry 78 71
Neiss W F 1984 Histochemistry 8 231

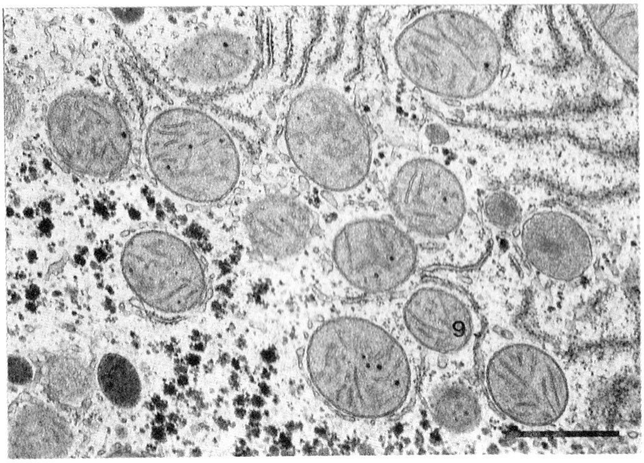

Fig.1. After GA fixation tissue was rinsed for 1 h at 4°C in 6.8 % saccharose in 0.1M cacodylate buffer. Postfixation in 1% OsO_4 plus 1.5% $K_4Fe(CN)_6$. Bar = 1 µm.

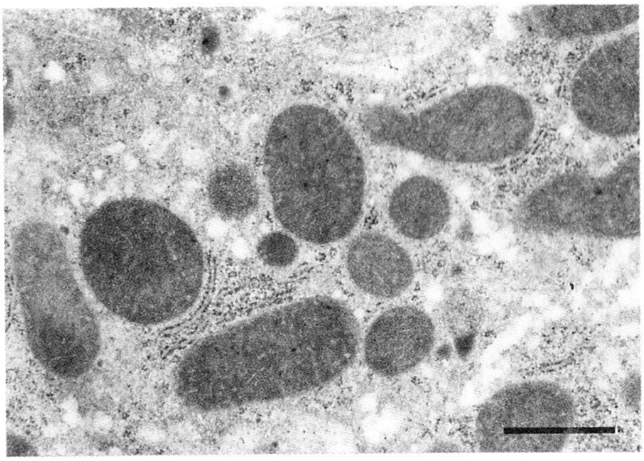

Fig.2. After GA fixation tissue was rinsed for 2 h at 4° C in 8.6% saccharose in 0.1M cacodylate buffer. Postfixation as in Fig.1. Bar = 1 µm.

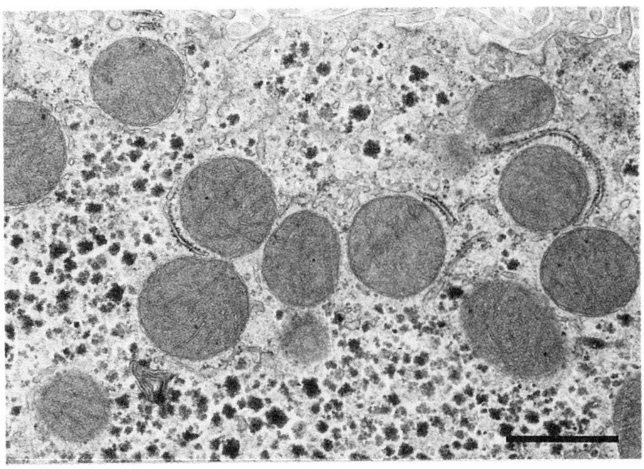

Fig.3. After GA fixation tissue was rinsed for 24 h at 4° C in 6.8% saccharose in 0.1M cacodylate buffer. Postfixation as in Fig.1. Bar = 1 µm.

Methylol-melamine-ether, an electron-transparent substrate for cell culture and for the imaging of individual whole cells consecutively by LM, TEM and SEM

Christel Westphal and Dieter Frösch
(Sektion Elektronenmikroskopie, Universität Ulm, D-7900 Ulm)

ABSTRACT. A simple technique is reported to culture cells on a melamine foil about 80 nm in thickness. Individual cells can be imaged (whole or sectioned) on that foil successively by light, transmission and scanning electron microscopy.

A variety of cells were cultured on a support foil cast from 1% methylol-melamine-ether, dissolved in pure ethanol. When prepared on glass slides or coverslips, this foil can be flamed for sterilization. It is about 80 nm thick, smooth, homogeneous and highly electron-transparent. Yielding excellent whole cell preparations in various kinds of microscopic studies to be performed in one and the same cell, this method is capable of narrowing the gap between light and electron microscopy.
Melamine resins have free amino terminals and are thus able to react with aldehydes in a similar way to proteins. Consequently, cross-links are formed between the cells and the substrate during the process of fixation. Accordingly, the adherence of aldehyde-fixed cells to a melamine foil is probably stronger than to glass or inert plastics. This property may be of some importance for the maintenance of overall cellular topography during dehydration, drying and microscopy. Along with simplicity of handling, this makes the technique introduced here an interesting alternative to other cell culture methods.

Fig. 1. A beating heart muscle cell grown on a melamine foil was identified in the light microscope (a), subsequently, a lettered grid was placed over it (b). After fixation by glutaraldehyde, air-drying and recovery from the glass, the same cell was recorded in the TEM (c) and, after sputter-coating, in the SEM (d).
Fig. 2. Planar view of live fibroblasts (a), and lateral view of the same preparation after fixation and Epon-embedding for perpendicular sectioning (b). Similar preparations processed for TEM are shown in (c) and (d).
Fig. 3. A study of the contacts between cells (3T3) and a melamine foil (arrows) indicates that these occur exclusively in clathrin-coated areas of the lower cell surface.

(For further detail about this technique see Westphal et al., Journal of Microscopy, Oxford, in the press).

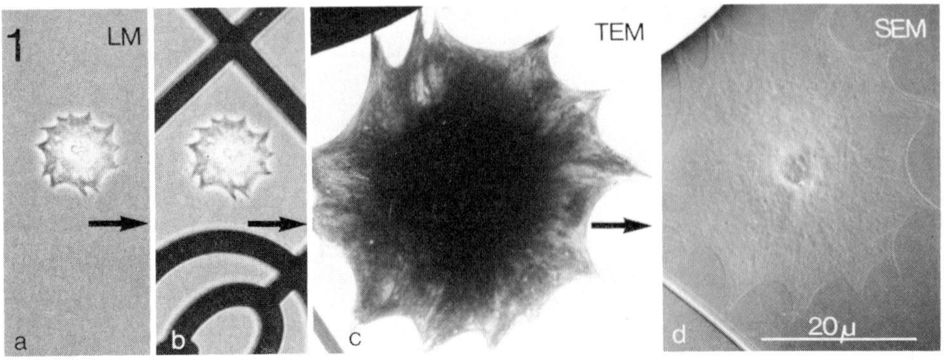
1
LM
TEM
SEM
a
b
c
d
20µ

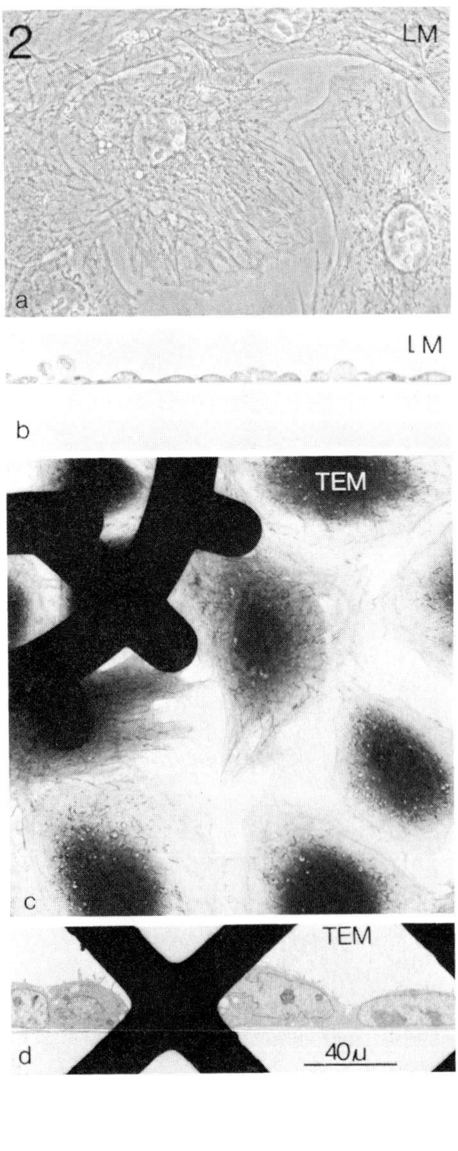
2
LM
a
LM
b
TEM
c
TEM
d
40µ

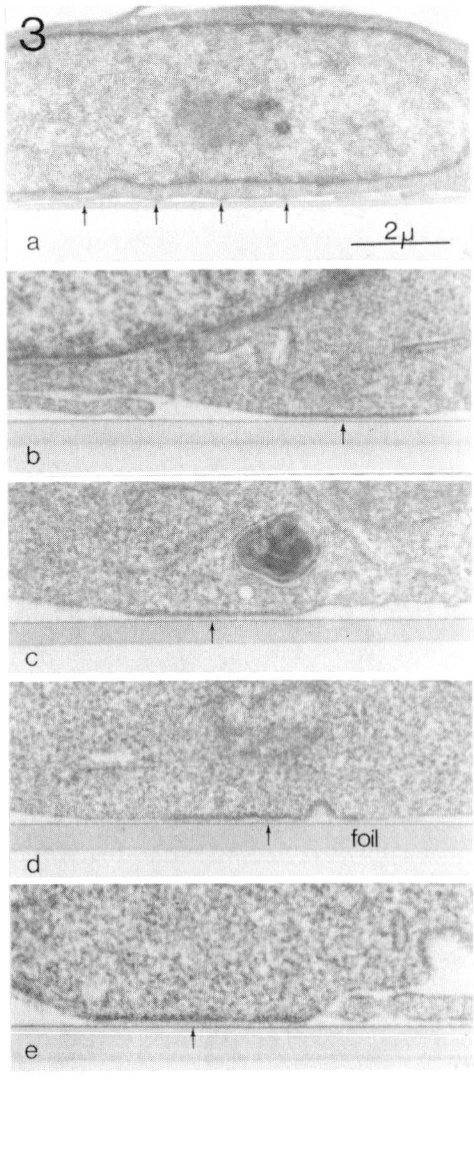
3
a
2µ
b
c
foil
d
e

Paper presented at EUREM 88, York, England, 1988

EDTA-agarose embedding—a new method for ultrastructural examination of tissue culture in situ

Sibylle Widéhn, Gothenburg University, Department of Pathology, Sahlgren Hospital, S-413 45 Göteborg, Sweden.

Ultrastructural studies of tissue cultures have previously been performed on (1) "cell pellets" obtained by centrifugation or on (2) resin blocks containing a monolayer of the in situ culture. The first method results in the loss of the in situ morphology and the second method frequently involves problems of separating the resin block from the culture dish. A new technique for ultrastructural examination of tissue culture in situ, the EDTA-agarose method, is described. By this method the in situ ultrastructure of the tissue culture is well preserved and different areas of the same culture can be used for immunohistochemical analysis.

Inst. Phys. Conf. Ser. No. 93: Volume 3, Chapter 13
Paper presented at EUREM 88, York, England, 1988

Comparison of three corrosion casting resins in some properties of practical relevance

H Ditrich & H Splechtna

Dept. of Comp. Anatomy and Morphology, Institute of Zoology,
Univ. of Vienna, Althanstr. 14, A-1090 Wien, AUSTRIA
Supported by FWF: P6353

ABSTRACT: Mercox*, Mercox diluted with methyl-methacrylate and Tardoplast*, three frequently used microvascular corrosion casting resins are compared in some of their practical properties. While the latter resin is better suited for larger casts, the two others should be prefered in detailed studies on capillary structures.

1. INTRODUCTION

A great variety of casting media and procedures are reported in the literature on the corrosion casting technique (for Refs., see Gannon 1978, Hodde and Nowell 1980, Lametschwandtner et al. 1984, Ditrich and Splechtna 1987). A main question concerning the reliability of this technique relates to the type and properties of the casting medium. Mercox* or Mercox diluted with methyl-methacrylate (Ohtani and Murakami 1978) are widely used in corrosion casting. Tardoplast*, a new casting resin has been introduced recently (Amselgruber et al. 1987). The latter medium is tested in this study and compared with the other two resins mentioned above.

2. MATERIAL AND METHODS

Mercox*<M> (Jap.Vilene Co.- Japan) and Tardoplast*<T> (H.Schumm Chem.Co. - FRG) were prepared according to the manufacturers instructions. Mercox diluted <M+MMA> was prepared by admixture of 25% (vol./vol.) methyl-methacrylate monomer to the Mercox basis resin and addition of 1% to 1.25% (vol./weight) Mercox catalyst/hardener. The resins were either injected as described previously (Ditrich and Splechtna 1987) or poured into silicone forms for TEM embedding. Upon hardening, the resins were tempered in a water bath at 60°C for seven hours, allowed to cool at room temperature and corroded in 10% KOH for three days. Some blocks remained uncorroded as controls (upper row of figures). The blocks were rinsed in tap- (one day) and distilled water (one hour), dried in a desiccator with silicagel blue, mounted, sputtered (6-7 kV, 11-13 mA, 9) and investigated under a JEOL-JSM 35 CF or a ZEISS DSM 950 at 3-15 kV acceleration voltage.

3. RESULTS AND DISCUSSION

All of the three casting resins can replicate vascular systems completely. M and M+MMA may both be used as prefered, being expedient for studies on

* = Reg. Tm.

microvascular morphology of limited areas. Data on some of the technical properties of these two resins are found in the literature (Weiger et al. 1982, 1986). They show a similar structure after corrosion but due to the higher content of softer components (MMA), M+MMA shows stronger surface destruction. This can be disadvantageous in studies on delicate surface sculptures of blood vessels (e.g., endothelial nuclear distribution or vascular cushions, ...), however, the viscosity of the unpolimerized mixture is significantly lower than in the other two resins. Therefore, if highly structured blood vascular systems are to be cast, M+MMA renders best filling as the flow parameters can be kept closer to the physiologic conditions and the risk of extravasates and aneurisms is accordingly lower. The signs of corrosion in T resemble relatively large craters. Such artefacts might be confused with broken branches of cast vessels or endothelial nuclear impressions. However, it seems suitable for larger casts because of its high mechanical stability. Generally, the wide range of applicability and the vivid results should encourage workers in using vascular casting methods.
Several parameters of the three resins are compared in Table 1:
(Values are valid only relative to each other);

	M	M+MMA	T
Injectable time (min.)	7	17	12
Curing time (min.)	0.5	#)	4
Mechanical stability	<	<	++
Shrinkage	<	+	++

Code:
++ = strong
+ = fair
< = low

#) Full polimerization is reached during tempering, the casts can remain glutinous at room temperature.

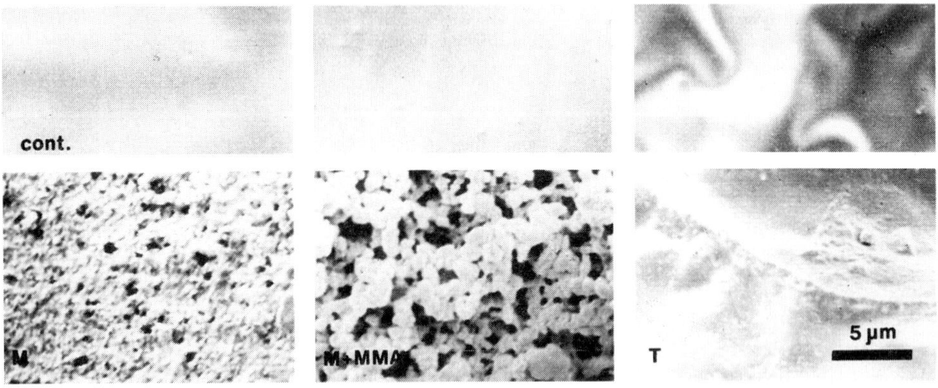

4. REFERENCES

Amselgruber W, Sinowatz F and Bergerson W 1987 Eur. J. Cell Biol. **43** Suppl. **18** 3
Ditrich H and Splechtna H 1987 Scanning Mic. Int. **1**(3) 1339
Gannon B J 1978 Vascular Casting - Principles and Techniques of Scanning Electron Microscopy ed M A Hayat (New York: Van Nostrand) pp 170-193
Hodde K C and Nowell J A 1980 Scanning Elec. Mic. **1980/II** 89
Lametschwandtner A, Lametschwandtner U and Weiger T 1984 Scanning Elec. Mic. **1984/II** 663
Ohtani O and Murakami T 1978 Scanning Elec. Mic. **1978/II** 241
Weiger T, Lametschwandtner A and Adam H 1982 Mikroskopie **39** 187
Weiger T, Lametschwandtner A and Stockmayer P 1986 Scanning Elec. Mic. **1986/I** 243

Preparation of injection replicas for SEM studies of the rat biliary system

JPM Schellens [1)], JG de Haan [2)] and KC Hodde [3)]
Academic Medical Centre, University of Amsterdam, Meibergdreef 15, 1105 AZ Amsterdam, The Netherlands.
1) Laboratory of Histology and Cell Biology, 2) Laboratory of Experimental Medicine, 3) Laboratory of Experimental Surgery.

Nowell (1970) and Murakami (1971) introduced the injection replica method to study the three dimensional organization of small blood vessels in the scanning electron microscope. Recently Murakami (1984) and Yamamoto and Phillips (1984) demonstrated that this method may be applied also in studies of the biliary system. We have tried to evaluate the potential of this method in studying the biliary system of normal rat liver, our ultimate goal being to use the injection replica method in the study of pathological specimens.

Male Wistar rats weighing about 350 g were anaesthetized with sodium pentobarbital. Common bile duct and/or portal vein were cannulated with vein catheters (diameter 0.5 and 1.5 mm respectively). The following methods were tested for filling the biliary system with resin.

1. Straightforward injection via the bile duct, without portal vein cannulation. Test series ranging from 0.15 to 0.75 ml of resin injected were performed.
2. Resin injection preceded by perfusion fixation. In these experiments the animals were heparinized (500 IE per animal) after which the livers were perfused via the portal vein with Hartmann's solution at body temperature, with a pressure of 20 cm H_2O.
   The vena cava was clamped and one carotid artery was opened for drainage. Fixation was performed by perfusion with 1% paraformaldehyde in 0.15 M phosphate buffer pH 7.4 during 2.5 min, keeping temperature and pressure at the same values as during the prerinse. After fixation 0.6 ml resin was injected in about 2 min, during which procedure the liver vasculature was rinsed again with Hartmann's solution via the portal vein.

The resin was prepared by mixing 1.5 part of commercially available resin (Mercox, Dainippon Ink Co, Tokyo) with 1 part methyl methacrylate monomer saturated with catalyst. This final mixture keeps its low viscosity (10-15 cP) for several minutes.

The animals with the liver injected were placed in a warm water bath for 1 h to get the resin fully polymerized. Thereafter the livers were removed and the tissue was macerated in 10% KOH with addition of some detergent at 60°C until no debris was present in the solution. The resulting casts (fig. 1) were thoroughly washed in demineralized water and subsequently dissected under a binocular microscope. Small specimens were cleaned with 4% formic acid and freeze dried. Finally they were sputter coated with gold and observed in an ISI SS 40 scanning electron microscope at an accelerating voltage of 5-10 kV.

fig. 1. Macrophotograph of rat liver cast after injection of 0.6 ml resin in the bile duct. Bile and blood vessel systems both filled to the extent of the whole liver.

fig. 2. Fine meshwork of bile canalicular configuration interspersed between sinusoidal casts. Bar = 20 μm.

fig. 3. Detail of image similar to fig. 2. Bar = 10 μm. S = sinusoidal cast.

From the series without prefixation we learned that at least 0.5 ml of resin has to be used to bring some resin in all liver lobes. However, extensive leakage, mainly into the vascular system, could not be avoided. On the other hand, this leakage turned out to provide the casts with some stability. In some specimens the bile ductular plexus as described by Yamamoto and Phillips (1984) could be observed.

Fixation was introduced to stabilize the tissue before the injection of resin was started. In these prefixed specimens in particular we found the bile canaliculi to be a meshwork of fine tubules with diameters in the order of 2 μm extending from the portal tract and interspersed between sinusoidal casts (figs 2 and 3). As this system was extremely fragile and only supported at places where vascular leakage had occurred, its (inter-)connections were difficult to trace. Inconsistency in filling the biliary system and leakage into the blood vasculature we have not been able to avoid. Nevertheless, the injection replica method may be a very useful method for the study of pathological specimens with aberrant spatial distribution of the biliary system.

REFERENCES

Murakami T 1971 Arch. histol. jap. 32 445

Murakami T, Itoshima T, Hitomi K, Ohtsuka A and Jones AL 1974 Arch. histol. jap. 47 223

Nowell JA, Pangborn J and Tyler WS 1970 SEM/1970/ITT Research Institute, Chicago, Il 60616 pp 249-256

Yamamoto K and Phillips MJ 1984 Hepatology 4 381

2

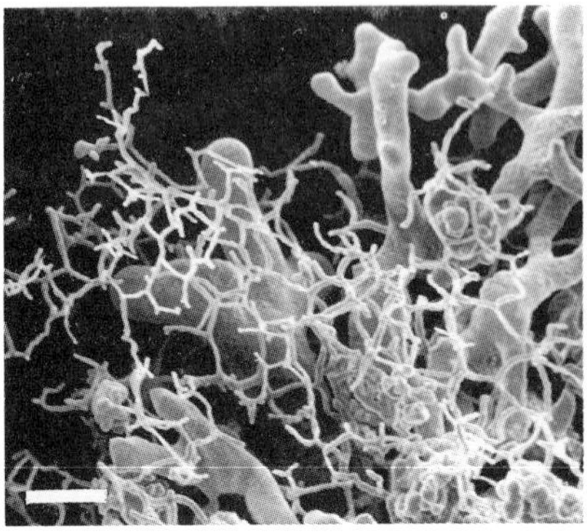

3

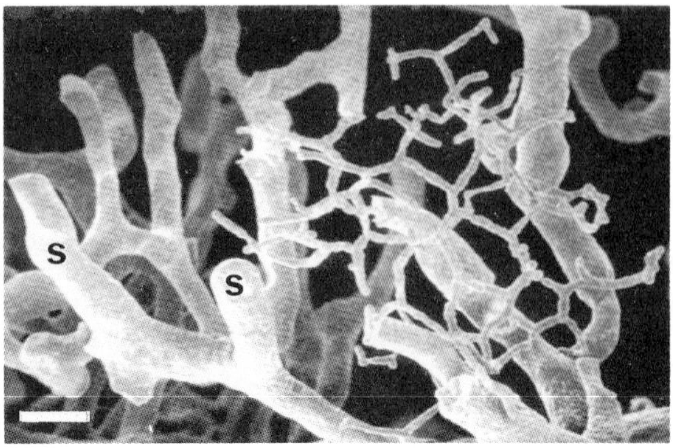

Inst. Phys. Conf. Ser. No. 93: Volume 3, Chapter 13
Paper presented at EUREM 88, York, England, 1988

Electron holographic investigation of the cutting edge of glass knives for ultramicrotomy

R Lauer and K Lickfeld

Physikalisch-Technische Bundesanstalt Braunschweig, and
Universität (GHS) Essen, Med. Mikrobiologie, Fed. Rep. of Germany

ABSTRACT: Small pieces of glass knives glued onto object grids have been investigated by off-axis electron holography. It is found that the profile of the wedges alternates between convex and concave and that the sharpness of the cutting edge approaches atomic range.

Although glass knives have become the main tool in ultramicrotomy, the geometry in the nanometer range near the cutting edge still needs to be investigated. This is due to the difficulties in preparing these bulky, insulating objects for electron microscopy. For the SEM, Lickfeld (1985) prepared profiles of the cutting edge by breaking the wedges of such knives. Profile variations occurring along the wedge could not be detected by this method. With the TEM, on the other hand, the extension of the edge and its notches can be seen, but a thickness profile of the wedge can scarcely be determined. This problem can be solved by means of electron holography.

In off-axis holography, the phase modulated object wave is superposed by a reference wave using a Möllenstedt-biprism. The result is an interference pattern containing fringe bends Δx which are proportional to the object phase $\Delta\Phi$:

(1) $$\Delta x = \frac{1}{2\pi} a \Delta\Phi$$

where a is the fringe distance. According to

(2) $$\Delta\Phi = 2\pi \frac{T}{\lambda} \frac{U_0}{U}$$

the phase shift $\Delta\Phi$ is proportional to the mean inner potential U_0 and the thickness T of the object (λ is the electron wavelength, and U is the accelerating voltage). By interferometric reconstruction as decribed by Hanßen (1982), or by a corresponding digital reconstruction as presented by Ade and Lauer at this conference, the holograms can be transformed into interferograms which have fringes with spacings and directions more convenient for the quantitative phase evaluation than the holograms.

45° glass knives made with an LKB knifemaker were investigated. Small pieces of the cutting wedge (prepared by Lickfeld) were glued onto copper grids. To prevent electric charging in the electron microscope, specimens were coated with a thin conductive layer. Carbon layers proved to be the most suitable.

Preliminary results: As can be seen in fig. 1, even with moderate magnification, the cutting edge is not straight. It shows a series of asymmetric step-like structures somewhat comparable with light microscopical results and the interpretation of these by Persson (1968).

Fig. 2 is a highly magnified micrograph of part of the wedge, and fig. 3 is an electron hologram of the same region. Since the biprism fringes have been directed parallel to the edge in order to cover a large object field, their evaluation is difficult. In the interferometric reconstruction shown in fig. 4, the interference fringes are directed vertically in the region free of matter (bottom) and they are inclined to the right-hand side where the wedge thickness increases (top). It can be seen that the profile is partly flat and partly convex.

The sharpness of the edge can be estimated from the lateral jump of the fringes. Assuming a wedge angle of 60° (Lickfeld 1985), the wedge thickness at the locations where the fringe shift is equal to the fringe distance amounts to 39 nm. The fringe jumps at the edge indicate only a small fraction of this value, and since this value corresponds approximately to the thickness of the conducting carbon layer, it can be concluded that the sharpness of the edge approaches the atomic dimensions.

To determine the wedge angle by holographic means using eq.(2), the mean inner potential U_0 of glass must be known. Measurements of this quantity are under way.

REFERENCES:

Hanßen K-J 1982 Adv. Electronics and Electron Phys. 59, p.1
Lickfeld K 1985 J. Ultrastruct. Res. 93, p.101
Persson A 1968 LKB-Ultramicrotomy course

Fig. 1. Cutting edge of a glass knife with low magnification

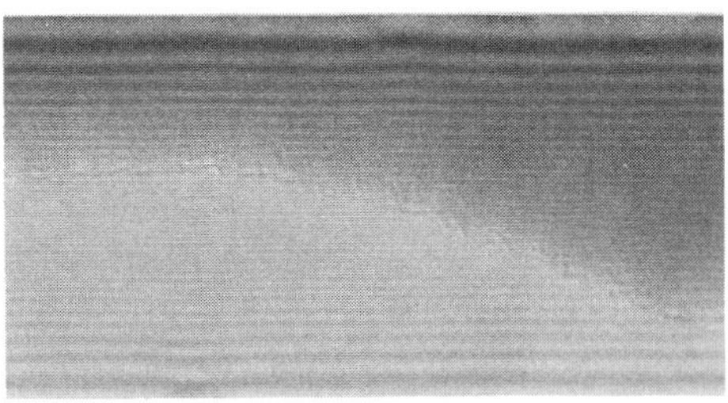

Fig. 3. Electron off-axis hologram of the same detail

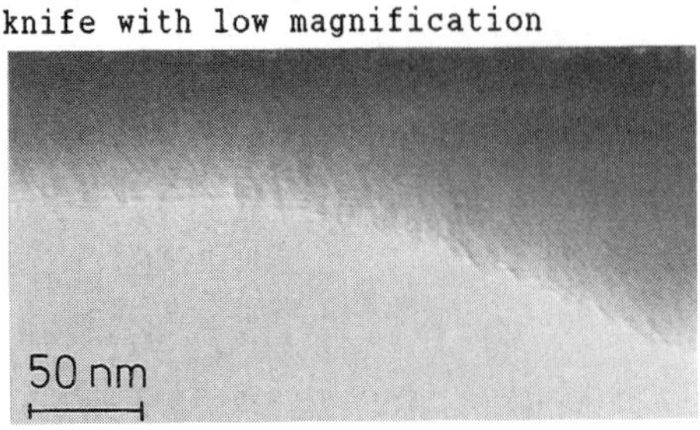

Fig. 2. Magnified detail marked in fig. 1.

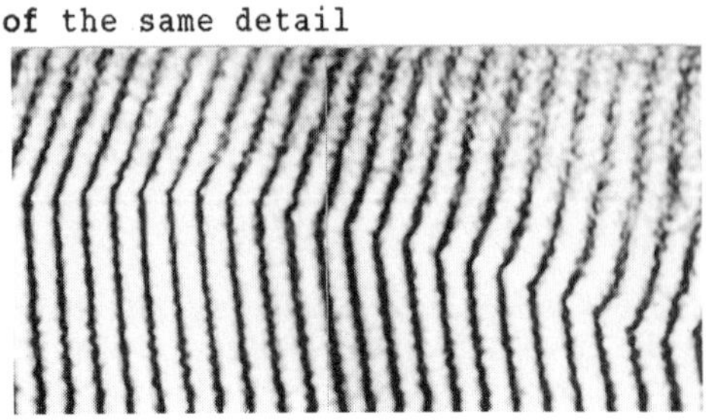

Fig. 4. Interferometric reconstruction of the hologram

The ion beam bombardment apparatus incorporated into scanning electron microscope for observation of biological materials

Y Muranaka[1)], S Ono[2)], N Baba[3)], M Nagase[4)] & K Kanaya[3)]

[1)]Central Lab. for Electron Microscopy of Hamamatsu Univ. School of Medicine, 3600 Handa-cho, Hamamatsu 431-31, Japan, [2)]ELIONIX Inc., [3)]Dept. of Electron Engineering in Kougakuin Univ. and [4)]1st Dept. of Internal Medicine of Teikyo Univ.

1. INTRODUCTION

Ion beam bombardment is a highly effective method to prepare biological specimens for SEM study. It has several advantages; i.e., weak ion beam bombardment enables the uncoating observation (Muranaka et al 1982a) and has the surface cleaning effect on the specimen (Muranaka et al 1982a), relatively strong bombardment facilitates observation of inner structure of the specimen (Muranaka et al 1982b), and tungsten or tantalum ion beam spatter coating is available (Adachi et al 1986). The processes such as SEM observation, the ion beam bombardment and the ion beam spatter coating have been usually performed in the separate chamber of different equipments. In the present study, we devised an apparatus in which the attaching devices such as ion gun and gimbals were incorporated, so that these processes can be consecutively and repeatedly performed without exposing the specimen to air under the same vacuum condition, facilitating integrated and effective observation.

2. MATERIALS AND METHODS

The apparatus is outlined in Fig. 1 and photographed in Fig. 2. To avoid obstacles occurred due to combining the functions, some protecting measures were taken. The ion gun (ELIONIX EIS-1) is attached to the scanning electron microscope (JEOL JSM35) at the side of the specimen chamber with a horizontal bombarding angle of 35 degree, and motor driving gimbals on the specimen stage, target metal for coating, protect board and its remote controlling unit of secondary electron collector and objective lens are incorporated. The ion gun uses argon gas, and is capable of accelerating voltage to the level of 5 KV and of regulating the diameter of beam to desired size. The amount of the beam bombarded to the specimen can be monitored as absorbed electric current.

The organic material used is the glomerulus isolated from the normal rat kidney fixed by perfusion with 2% glutaraldehyde in PBS. The glomeruli were fixed with 1% osmium tetraoxide, block stained with uranium acetate, and then dried at critical point after dehydration with a graded series of alcohols. Using the apparatus devised, the uncoating observation, the surface cleaning, the observation of inner structure of the specimen exposed by ion etching, and the ion beam spatter coating were performed monitoring with SEM.

3. RESULTS

Since monitoring with SEM under the same vacuum condition was possible at all times, corresponding effects to each of ordinary ion beam bombardment methods (Muranaka 1986) were more effectively obtained at most optimal condition. In particular, alternately repeated SEM observation and ion etching on the same specimen enabled to observe the inner structure at various depths from the same original surface region. These results warrant availability of this apparatus in various studies.

4. REFERENCES

Adachi K et al 1986 Proc. 11th Int. Cong. on Electron Microscopy Kyoto **1** pp363-4

Muranaka Y et al 1982a Proc. 10th Int. Cong. on Electron Microscopy Hamburg **3** pp459-60

Muranaka Y et al 1982b Proc. 10th Int. Cong. on Electron Microscopy Hamburg **3** pp449-50

Muranaka Y et al 1986 Proc. 11th Int. Cong. on Electron Microscopy Kyoto **1** pp357-8

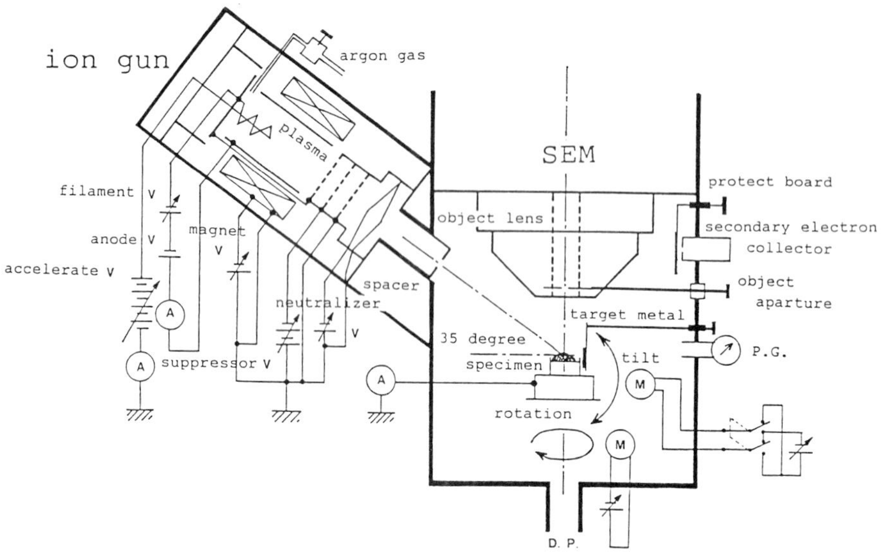

Fig. 1 Diagram of the apparatus

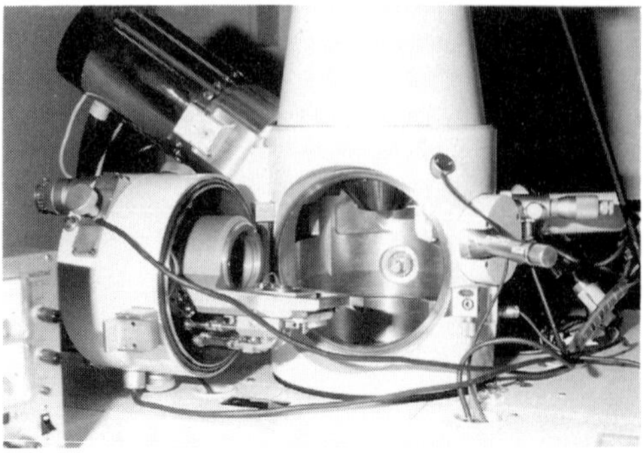

Fig. 2 A: Gross appearance B: Specimen chamber

Inst. Phys. Conf. Ser. No. 93: Volume 3, Chapter 13
Paper presented at EUREM 88, York, England, 1988

Versatile controlling system for preparation techniques in electron microscopy

R Merkl, K Weber and BM Humbel

Max-Planck-Institute for Biochemistry, Dept of Cell Biology, 8033 Martinsried, FRG.

ABSTRACT: An universal software for controlling complex preparation protocols in electron microscopy was developed. Its versatility will be demonstrated on the preparation technique of freeze-substitution. It controls all steps of the preparation protocol from the beginning, i.e., cryofixation, to the end, i.e., polymerisation.

Preparation techniques in electron microscopy are becoming more and more complex. Therefore experiments need to be adapted to automatic control. Low temperature preparation techniques, e.g., freeze-substitution, are good examples of experiments which need elaborate regulation and control of every process parameter. For freeze-substitution there are already two different systems on the market: one according to the protocol of Müller *et al* (1980) and one according to Sitte *et al* (1986). As with every commercial instrument, they have a disadvantage: they are designed for one particular protocol and therefore alteration of the processes is limited. As we are interested in testing different substitution protocols we need a versatile control system. Therefore, a PC based system was developed which not only controls all different steps of the freeze-substitution protocol but controls the whole preparation from cryofixation to final polymerisation. Every single step can be set individually, thus offering the advantage of adapting the protocol to the experimental demands.

Hardware: The system consists of an IBM PC equipped with a Hercules graphic card, a hard disk, an I/O-board, Pt 100 thermoresistors and an interface box to control the magnetic valves and heat cartridges (figure 1).

Software: An *experiment* (e.g. freeze-substitution) is subdivided into different *processes* (e.g. temperature step). Every *process* is active for a certain, presetable time period. It surveys a sensor (Pt 100) and activates the regulating elements (e.g. valves or heat cartridges) according to the sensor signal. A *process* is defined by the following parameters: time period, temperature at the beginning and at the end of the *process*, recording interval and parameters for the regulating elements used in the *process*. Individual *experiments* can be assembled from a subset of pre-defined *processes*. As an example, the *experiment* for freeze-substitution and low temperature embedding according to Humbel and Müller (1984) is depicted in the following table.

Table

Experiment: Freeze-Substitution

process nº:	name	active	temp	function
# 1	step 1	0 - 8 h	-90°C	measures and regulates temperature (#1–#3)
# 2	step 2	8 - 16 h	-60°C	
# 3	step 3	16 - 24 h	-30°C	
# 4	polymer	24 - 72 h	-30°C	UV-lamps: on
# 5	temp	0 - 72 h	-	records temperature

During the *experiment* the corresponding *processes* are interconnected in a loop. Every 50 msec each *process* is checked and the tasks of currently active *processes* are executed. The status of the active *processes* are displayed on the monitor and the sensor readings are stored on the hard disk for further investigations.

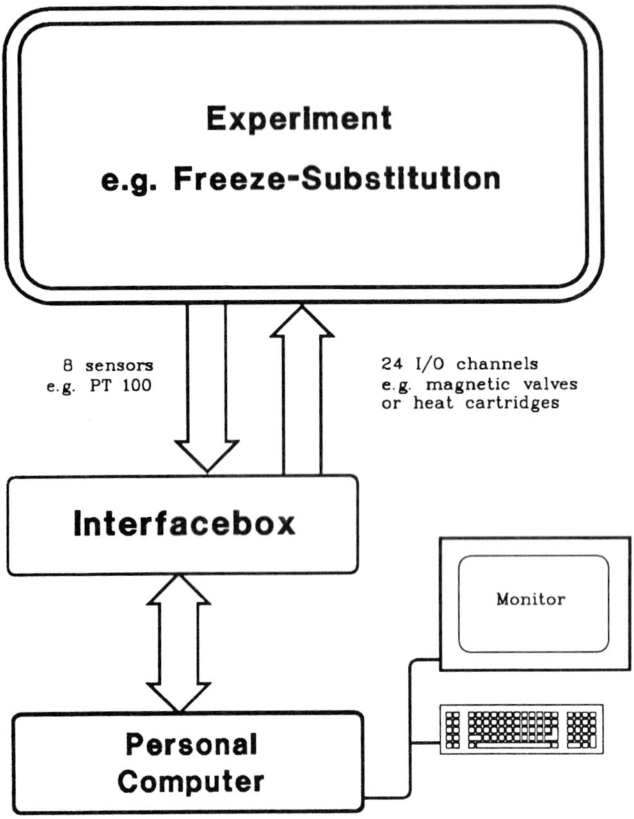

Figure 1: Hardware of the controlling system

REFERENCES

Humbel B and Müller M 1984 *Proc. 8th Europ. Congr. on Electron Microscopy* eds A Csanady, P Röhlich and D Szabo (Budapest) pp 1789 - 1798.

Müller M, Marti T and Kriz S 1980 *Proc. 7th Europ. Congr. Electron Microscopy* eds P Brederoo and W de Priester (The Hague) pp 720 - 721

Sitte H, Neumann K and Edelmann L 1986 *The Science of Biological Specimen Preparation 1985* eds M Müller, R P Becker, A Boyde and J J Wolosewick (AMF O'Hare, USA) pp 103 - 118

Increase of contrast of biological sections by electron spectroscopic imaging

L Reimer, M Ross-Messemer

Physikalisches Institut, Universität Münster, 44 Münster, FRG

Electron Spectroscopic Imaging (ESI) of biological sections can increase the contrast by elastically scattered electrons with zero-loss filtering which also avoids blurring by chromatic aberration of inelastically scattered electrons. Thick sections with a very small fraction of unscattered and elastically scattered electrons can be imaged with ESI at an energy loss of 50-300 eV at the maximum of the broad energy spectrum (Bauer et al. 1987). An elemental mapping can be realized by the digital difference image of ESI beyond and below an ionisation edge (Ottensmeyer and Andrew 1980).
Another possibility of increasing the contrast by non-carbon atoms is the ESI just below the carbon K-edge (Probst and Bauer 1987). ESI of stained and unstained biological sections show the following contrast reversals. Zero- and plasmon-loss filtering results in a positive contrast where areas containing heavy-metal stains but also P and S, for example, appear darker than the embedding. With increasing energy loss ΔE the contrast becomes negative at ΔE = 70-100 eV caused by L and M edges of these elements and the maximum negative contrast (similar to a dark-field image) is observed just below the carbon edge. The contrast again is weakly positive beyond the carbon edge at ΔE = 285 eV and can change the sign again to negative contrast far beyond the carbon K-edge. The energy

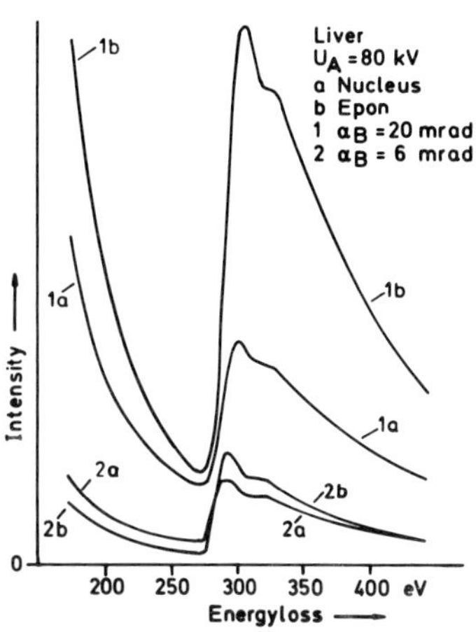

Fig.1a.

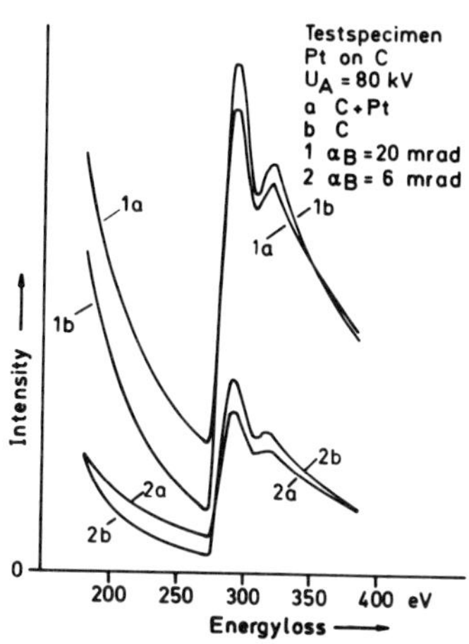

Fig.1b.

Fig.1a. EELS of pure epon and a stained liver nucleus for two objective apertures

Fig.1b. EELS of pure carbon (60 nm) and an additional 3nm Pt-film showing similar contrast reversals losses showing contrast reversal depend on the section thickness, the composition and on the objective aperture α.These contrast reversals can also be recognized in an electron energy loss spectrum (EELS) of a section at positions with pure epon embedding and a nucleus of a liver cell stained by OsO_4 (Fig. 1a). The decrease of EELS intensity of a stained area with increasing ΔE is lower than the decrease of epon or carbon and the ratio is a maximum just below the K-edge. When used a physical simulation of metal-stained sections by platinum-shadowed polystyrene spheres on a carbon substrate, the same contrast reversals in EELS (Fig. 1b) and ESI (Fig. 2) can be observed in and outside the shadow of the sphere. We used this reproducable test specimen for comparing the enhanced negative contrast at ΔE = 250 eV with the positive contrast of an unfiltered image. The contrast is here defined as

$$C = |(I_C - I_{Pt+C}) / I_C|$$

Figure 3 shows measured values of C for platinum films of increasing mass-thickness x_{Pt} on a x_C = 9 μg/cm^2 carbon film in the unfiltered bright-field mode and in the ESI mode with an energy window ΔW= 250 ± 5 eV. In further experiments this increase and reversal of contrast will be measured for different elements to develope a theoretical model for this ESI mode.

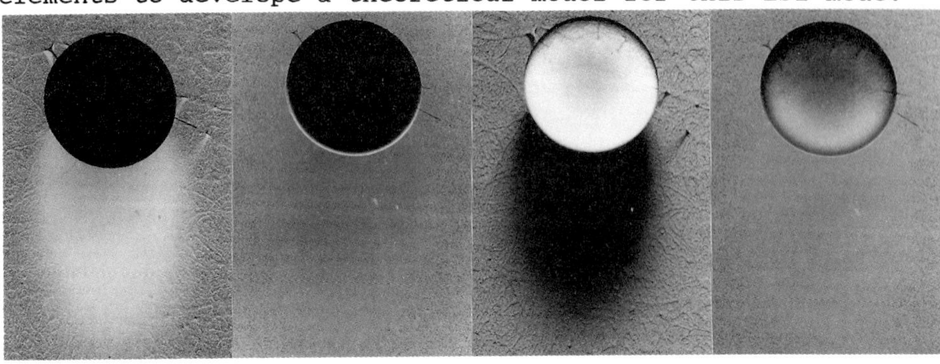

ΔE = 0 eV 50 eV 250 eV 300eV

Fig.2. ESI of 1 μm polystyrene spheres on 60 nm carbon shadowed by 3 nm Pt at increasing energy loss ΔE (α = 4 mrad)

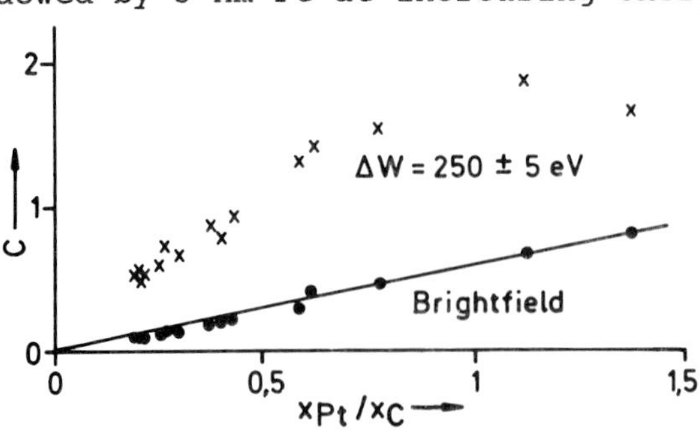

Fig.3. Positive bright-field contrast without energy filtering and negative ESI with ΔW = 250 ± 5 eV of platinum films of mass-thickness x_{Pt} on a x_C = 9 μg/cm^2 carbon film

Bauer R, Hezel U and Kurz D 1987 Optik **77** 171
Ottensmeyer F P and Andrew J W 1980 J.Ultrastr.Res. **72** 336
Probst W and Bauer R 1987 Verh.Deutsche Zool.Ges. **80** 119

Examination of a mineralized organic matrix using the Z-contrast technique

John R. Dunlap

Department of Botany and Program in Microscopy, The University of Tennessee, Knoxville, TN. 37996

ABSTRACT: The Z-contrast imaging technique has been utilized in the present study to further our understanding of fine structure of a mineralized matrix. The technique showed contrast differences between the basic structural components of the matrix thus suggesting differences in atomic composition.

1. INTRODUCTION

The Z-contrast imaging technique utilized in the present study provides a unique opportunity for the examination of a biomineralized material. The Z-contrast technique as reported by Crewe et al (1975) was used to examine heavy atomic number atoms on a low atomic number background. In this study, we are using the technique to examine iron deposition on an organic matrix composed primarily of acidic glycosaminoglycans. Because of the differences in localized atomic number between the organic matrix, which is predominately carbon, and the metal component of the biomineralized material, it should be possible to visually differentiate these components with Z-contrast imaging. This study is focused on the mineralization process in the fresh water algae Trachelomonas which encases itself in a mineralized envelope. Studies on envelope microarchitecture using conventional techniques have been performed (cf. Dunlap et al 1983); however, with these techniques contrast differences are not sufficient for differentiation of the iron deposits from the organic matrix of the envelope (Dunlap and Walne 1985). In order to overcome this obstacle we have utilized Z-contrast imaging.

2. MATERIALS AND METHODS

For this study all images were acquired on an Hitachi H-800 operating at 200 KV. The microscope is equipped with a GATAN model 607 ELS system which is interfaced with an EG&G ORTEC IMAGE MASTER. After digital acquisition, image processing was performed on a Perceptics 9200 Real Time Image Processor.

3. RESULTS AND DISCUSSION

The trachelomonad envelope is an anastomosing complex of mucilagenous strands and associated metal components. Examination of the FE-enriched areas using CTEM (Figure 1) shows that ca. 5nm globular components and ca. 180 nm wide by several micron long fibrillar material are characteristic (Dunlap and

Walne 1985); however, such conventional techniques alone do not make it possible to differentiate the metal component from the mucilagenous component. In the Z-contrast image of the trachelomonad envelope (Figure 2) similar structural features are present and are easily differentiated by contrast. Based on morphological characteristics the fibrillar material is presumed to be the mucilagenous component of the envelope which has been shown to be composed primarily of glycosaminoglycans (Dunlap and Walne 1985). Thus it would have a lower localized atomic number than the metal component of the envelope, in this case Fe. Spectral data generated from areas with a high concentration of the bright spots show that Fe is the predominant element. Based on these results we suggest that the 5 nm bright spots represent localized areas of Fe deposition.

As shown here the Z-contrast technique has proven beneficial towards further elucidation of the fine structural detail of the trachelomonad envelope. Characteristics of the Fe-enriched envelope and possibly the localization of Fe-enriched deposits has been possible. This paper also substantiates the benefits of the Z-contrast imaging technique in biological electron microscopy.

4. REFERENCES

Crewe, A V, Langmore J P and Isaacson M S. 1975. Physical Aspects of Electron Microscopy and Microbeam Analysis. (New York: Wiley) pp 47-62.

Dunlap J R, Walne, P L and Bentley J. 1983. Protoplasma **117**, 97.

Dunlap J R and Walne P L. 1985. J. Protozool. **32**, 437.

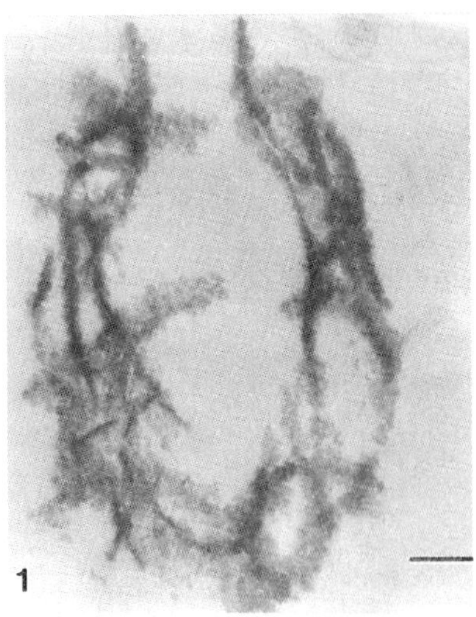

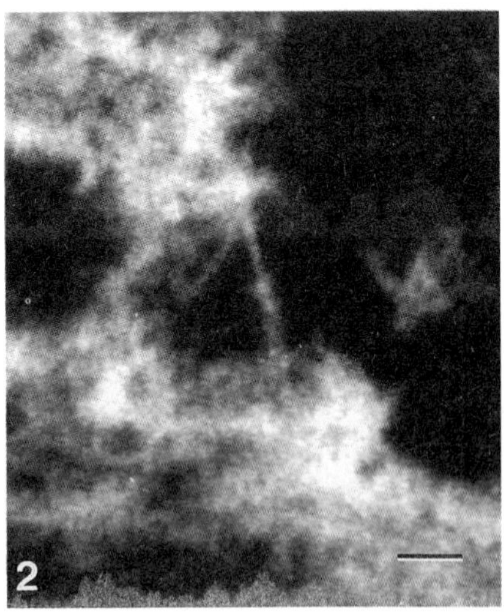

Figure 1. Conventional transmission electron micrograph showing the fibrillar material and areas of differing electron density. Bar= 50nm

Figure 2. Z-Contrast image showing strands of material with associated bright spots. Bar= 50nm

Pure Z-contrast: can it be achieved with a double deflection electron spectrometer?

M. Haider

European Molecular Biology Laboratory, Postfach 102209, D-6900 Heidelberg

ABSTRACT: An approach toward obtaining pure Z-contrast images by means of a Double Deflection Electron Spectrometer is proposed.

1. INTRODUCTION

A thickness independent imaging mode for studies of the morphology and concentration of thin unstained biological sections has certain advantages, because the thickness variation can cause a stronger contrast than the varying composition of the object. A technique with which one should achieve this, is Z-contrast, proposed first by Crewe et al (1975) and later applied to biological objects by Carlemalm and Kellenberger (1982). They used the annular dark-field detector signal divided by the signal obtained from the inelastically scattered electrons. Unfortunately, the dark-field detector is placed in front of the spectrometer in a STEM, and therefore it records not only elastically scattered electrons (Haider 1987). Thus, the signal obtained with these two detectors is still dependent on the thickness of the object, as has been shown by Reichelt and Engel (1984).

Egerton (1982) proposed two different ways of mixing three signals (dark-field, inelastic and bright-field) in order to obtain a thickness independent signal. One of these methods has already been applied by Jeanguillaume and Tence (1987). We propose to use only two signals which still lead to pure Z-contrast .

2. METHOD

In order to optimize the detector parameters of the two signals (the acceptance angles and the energy window of the inelastic detector) we calculated the partial scattering cross-sections of the most important elements of biological objects. Taking into account the finite illumination angle of the primary beam and the multiple scattering processes, we had to carry out following convolutions

$$B_0(\theta)\ [\otimes I(\theta)]^m\ [\otimes E(\theta)]^n \ .$$

With $B_0(\theta)$ the electron distribution within the primary beam and $I(\theta)$ and $E(\theta)$ the cross-sections, described by Eusemann et al (1982), for the m-times inelastic and the n-times elastic scattered electrons. This convolution could be carried out by analytical Fourier-transformation, multiplying the various components and doing the back-transformation numerically. The energy distribution, described by Wall et al (1974), of the m-times inelastically scattered electrons has been achieved by an m-fold numeric convolution.

These calculation suggest that combining the filtered dark-field image with the inelastic image a new image can be obtained which is nearly independent of thickness. This is illustrated in Fig.1b, where the somewhat changed instrumental parameters are also given. The same result can be obtained by using the three signal technique, proposed by Egerton (1982) .

Approaching the original idea of dividing the "elastic" signal, obtained by the annular dark-field detector in front of the spectrometer, by the "inelastic" signal, we designed and built a new spectrometer which allows us to obtain a ***filtered*** dark-field image in a STEM. This pure elastic image divided by the inelastic, obtained with a reduced energy window (ΔE_W = 5 - 40 eV) produces a thickness independent signal which is only dependent on the averaged atomic number composition.

The spectrometer has been especially constructed (Haider 1987) and mounted onto our Cryo-STEM (Jones et al 1985) for two purposes. First, for analytical modes, like spectra acquisition and elemental maps using a parallel recordings system, and second to explore new imaging modes e.g. filtered dark-field alone or in combination with the inelastic signal to give pure Z-contrast. In combination with the detection system of the spectrometer we are now able to acquire three different images simultaneously (the data acquisition system would allow the parallel acquisition of four images).

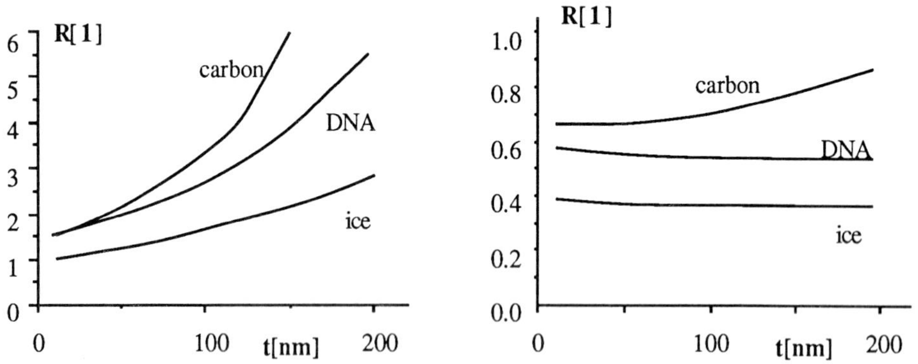

a) β_0 = 10 mrad β_A = 7.5 mrad
β_1 = 13 mrad β_2 = 130 mrad
ΔE_w = 5 - 70 eV

b) β_0 = 10 mrad β_A = 130 mrad
β_1 = 13 mrad β_2 = 130 mrad
ΔE_w = 5 - 40 eV

Fig.1: Dependence of the signal on the thickness of a carbon-layer. a) with the conventional ratio-contrast or b) the filtered dark-field divided by the inelastic signal. The parameters shown describe the illumination-angle β_0, the inner and outer acceptance-angle β_1 and β_2 of the annular detector, the acceptance-angle β_A of the spectrometer and the energy window ΔE_W.

REFERENCES:

Carlemalm E. and Kellenberger E. 1982 *EMBO Journ.* **1** 63
Crewe A.V., Langmore J.P. and Isaacson M. 1975 *Phys. Aspects of EM and Microbeam Analys. (Wiley, N.Y.)* 47
Egerton R.F. 1982 *Ultramicroscopy* **10** 297
Eusemann R, Rose H. and Dubochet J. 1982 *J. of Microscopy* **128** 239
Haider M. 1987 *PhD Thesis, Techn Hochschule Darmstadt*
Jeanguillaume Ch. and Tence M. 1987 *Ultramicroscopy* **23** 67-76
Jones A.V., Homo J.C., Unitt B.M. and Webster N.1985 *Journ. Micros. Spectros. Electroniques* **10** 361.
Reichelt R. and Engel A. 1984 *Ultramicroscopy* **13** 279
Wall J., Isaacson M. and Langmore J.P. 1974 *Optik* **39** 359

Digital processing of scanning electron microscope images

E. Wisse and R.B. De Zanger

Laborarory for Cell Biology and Histology (VUB), Laarbeeklaan 103, 1090 Brussels, Belgium

The secondary electron signal of a scanning electron microscope is digitized, in order to optimize observations during a microscopical session. We have combined hard- and software components for the acquisition and processing of TV (512 x 512) and high resolution images (1000 x 1000 or 2000 x 2000), enabling us to directly study the result of online digital image processing during microscope operation. After the study of the enhanced image, the information is stored digitally, or is transferred to the high resolution photomonitor of the scanning electron microscope for conventional photographic recording.

The hardware consists of a 68020 CPU based Masscomp 5520S computer system with 4Mb RAM memory, a data acquisition processor, a graphics processor with two frame buffers for images with 8 bit / 256 grey values and a high resolution display monitor (910 x 1150). Storage and data exchange facilities are provided by a 71 Mb hard disk, a 60 Mb tape drive and a 650 kb floppy disk. The system is equipped with Imaging Technology video digitizing boards: analog I/O, an ALU, and two memory mapped frame buffers for TV images of the IP 512 series. At present, our Philips SEM 505 scanning electron microscope is connected to the system. The open architecture of the system allows connections to other scanning or TV-equipped microscopes, including light microscopes. Our local facility is embedded in a UNIX-supported network, enabling us to communicate with other image-processing computers in the building or the campus by an Ethernet connection.

The data acquisition processor contains 12 bit AD/DA convertors with a maximum sampling rate of 333 kHz and a set of 5 programmable counter timers for generating the sampling frequency patterns. The data acquisition processor frees the CPU from routine tasks during acquisition. The timers are synchronized by the line and frame blanking pulses of the SEM. By measuring the increment in the vertical drive voltage and the time between line blanking pulses, the computer can determine the settings of the scan generator of the microscope. In this way, the interfacing between the different hardware components is kept relatively simple and the control panel of the microscope remains completely functional.

In our setup, real time image processing is supported by software running under a real time version of the UNIX operating system. The system can be efficiently programmed in C or Fortran. Special programs have been written to make an optimal use of the available RAM. At present, we have several image processing programs available, such as integration of 256 TV frames (see Figs 1 and 2), running averaging (2 - 256 frames), low pass and high pass filters, a median filter, unsharp and crispening filters, together with programs for scaling and clipping. Although the machine is not a dedicated "image processor", the integration of 256 consecutive images takes only 16 sec, filtering programs take about 5 - 8 sec, whereas scaling and clipping take only 1 sec.

An independent graphics processor facilitates the almost simultaneous display of the images, while acquisition or processing is in progress. A "still" image can be build up during a single slow scan. The results of image processing can be stored digitally, in compressed form, on hard disk or tape streamer. Through a DA converter, via the video boards, the results of processing can also be recorded on a Umatic video recorder, or the high resolution photomonitor of the SEM. Images can be ideally adapted for printing on a monochrome video graphic printer. Our present storage facilities have a capacity of about 400 to 450 compressed 512 x 512 TV images on a 60 Mb tape cartridge, costing less than 4 BF for storage-tape per image.

The proven advantages of the described instrumentation concern the possibility to have good TV images at magnifications up to 40,000 x, and the availability of large (25 x 38 cm), high resolution images with remarkable contrast during microscopic observation. These conditions allow a faster and better judgement of the details and the quality of the image and the specimen, than the use of a photographic routine print. The procedure allows the detailed online study of the preparation, by avoiding time-consuming darkroom procedures, and offers the possibility to store images digitally on relatively cheap tape cartridges or other storage media, such as optical disks.

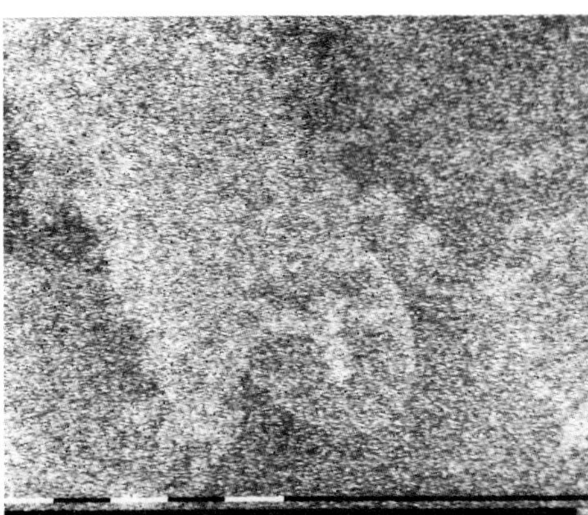

Fig. 1 Single 512 x 512 frame, grabbed from a sequence of TV images, showing a rat liver sinusoid containing a Kupffer cell and a red blood cell.
Initial magnification 10,000 x.

Fig.2 Same preparation, but the image is the result of an averaging procedure by integrating digitally 256 single TV frames of the same quality as the one shown in Fig. 1.

Paper presented at EUREM 88, York, England, 1988

A fast algorithm for the computer processing of periodical structures

J.P. Secilla and J.L. Carrascosa (*)

IBM Scientific Centre - P. Castellana, 4 - 28046 Madrid, Spain
(*) Centro de Biología Molecular - Univ. Autónoma - 28049 Madrid, Spain

ABSTRACT: A new algorithm for the computer processing of periodical structures is presented. Its main characteristics are described, as well as a comparison with a traditional algorithm. This new method has proven to be both fast and reliable, and thus very appropriate for environments with low processing power, such as Personal Computers.

1. INTRODUCTION

Images of periodical structures can be processed in two main ways: either in the Fourier domain or in the spatial domain. In the first case, the Fourier transform of the image is calculated, and both the module and the phase of the reflections of the crystal are extracted. An inverse Fourier series of these data will yield a filtered image.

It is also possible to use averaging to process the image in the spatial domain: an specimen is selected and the cross correlation function (CCF) between this new image and the whole image is calculated. This CCF will show strong peaks in the locations of individual specimens of the crystal, thus enabling to average them and greatly reduce the overall SNR.

The processing time for these two methods is, however, high. In the case of Fourier domain processing, two Fourier transforms have to be computed. In the case of spatial averaging, the CCF is usually implemented through three Fourier transforms. In both methods, additional processing is required to find the reciprocal lattice information in the first case and to find the peaks of the CCF in the second. This implies that, unless a powerful computer is available, the processing of the image will be rather slow.

2. THE NEW ALGORITHM

However, taking into account the fact that the specimens are periodically arranged, the processing times can be diminished drastically without degrading the results obtained, since only a small portion of the pixels of the periodical image must be examined to locate the individual specimens. Thus, it would be possible for the user to mark the rough positions of the specimens of the crystal and then have the computer refine around those positions, looking for the displacement that yields a higher similarity between that specimen in particular and the one that was selected as a template to look for. Repeating the process over all the image would yield a set of specimen positions that could be used to average all of them.

Nevertheless, the above process would be rather cumbersome, since it would require the work of a human operator. So, we have automated the process so that the user only has to indicate initial data to the computer. The algorithm would be as follows:

1. Select an specimen of the crystal as a template.
2. Mark the position of two specimens which are adjacent to the original one and in the direction of the lattice vectors. The position of these two specimens give an initial estimate of the lattice vectors, **a** and **b**. Add this specimen to a list of processed specimens and keep its position in a list of positions.
3. While there are specimens left do the following:
    a. Using the current estimate of the lattice vector **a** and **b**, look for all the specimens that are adjacent to the already processed ones. Use the expected position for each specimen, i**a** + j**b**, as the starting point for an steepest descent method that

minimises the sum of the euclidean distances between the pixels of the template and the image. Once a minimum is found, add its position to the list of positions and add this specimen to the list of processed specimens.

b. Using the list of positions of processed specimens, use a least squares error (LSE) method to find a new estimate for the vectors **a** and **b**.

4. Using the list of positions, average all the specimens. If a new periodical image is wanted, repeat this averaged specimen according to the last estimate of the lattice vectors.

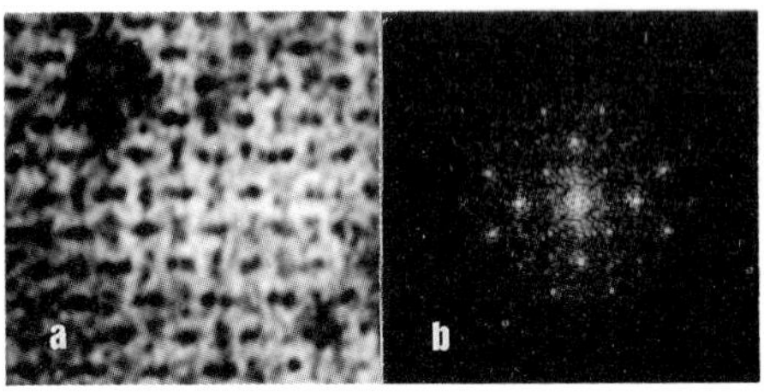

Fig. 1: a) Original image of a tetragonal crystal of connectors of phage Φ29 b) Its Fourier transform.

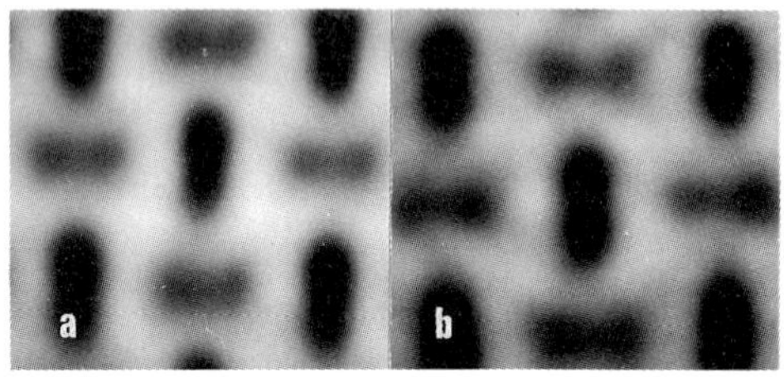

Fig. 2: a) Image filtered in the Fourier domain b) Image filtered with the new algorithm.

3. PERFORMANCE AND RESULTS

The performance of the algorithm has been tested using a tetragonal crystal of connectors of phage Φ29, which is shown in Figure 1a (the size of this image is 256 x 256 pixels). The Fourier transform of this image is shown in Figure 1b. Figure 2a represents an image which has been filtered in the Fourier domain. Figure 2b represents an image which has been filtered using the new algorithm. The results, as can be seen, are similar for both methods, although the image filtered with the new algorithm shows a bit more detail, since small displacements of individual specimens from their theoretical lattice positions have been corrected.

The processing times, however, are notably inferior in the case of the new algorithm. All the tests have been carried out on an IBM PC/AT microcomputer, which is capable of performing a 256 x 256 FFT in about three minutes. Processing the image in the Fourier domain took about six minutes. Using a traditional averaging method would have taken about nine minutes. However, our algorithm processed the image in just forty seconds, with results which are better than those obtained using conventional Fourier space processing.

4. CONCLUSIONS

We have presented a fast algorithm for the processing of periodical images which is especially suited for low power computers. We have implemented it in an image processing workstation based on an IBM Personal Computer AT.

The results obtained with the use of this algorithm are comparable, and even better, than those obtained using traditional methods, whose execution times are one order of magnitude bigger.

5. REFERENCES

Amos, L.A., Henderson, R. and Unwin, P.N.T., 1982, Prog. Biophys. Mol. Bio. 39 , 183-231.

Carazo, J.M., Santisteban, A. and Carrascosa, J.L., 1985, J. Mol. Bio. 183 , 79-88.

Powell, M.J.D., in J. Walsh (Ed.), 1966, *Numerical Analysis: an introduction*, 143-157.

Saxton, W.O. and Baumeister, W., 1982, J. Microscopy 127 , 127-138.

Secilla, J.P., Garcia, N. and Carrascosa, J.L., Signal Processing, 1988, in press.

Inst. Phys. Conf. Ser. No. 93: Volume 3, Chapter 14
Paper presented at EUREM 88, York, England, 1988

Effects on image classification by multivariate statistical methods of different normalization procedures and factorial representations

B L Trus[1], M Unser[2] & A C Steven[3]

[1]Computer Systems Lab, DCRT; [2]Biomedical Engineering Branch, DRS;
[3]Lab of Physical Biology, NIAMS; NIH, Bethesda, MD 20892, USA

ABSTRACT: Using a set of images of negatively stained proteins known to fall into two subtly different classes, we have tested how the ability to discriminate between them was affected by the use of different normalization procedures (constant min/max (CMM) or constant mean/variance (CMV)) and factorial representations (correspondence analysis (CA) or principal components (PC)). With CMM and PC, the leading factor was, in fact, spurious. With these data, the best discrimination was achieved with PC and CMV, although the improvement over CA and CMV was only slight.

1. INTRODUCTION.

The benefits of image averaging depend crucially on all images averaged being intrinsically alike. Multivariate statistical analysis (MSA)[1] provides a means of identifying sets of images that are formally homogeneous. With MSA, the images are represented as coordinates in an abstract (factorial) space, and assigned to (homogeneous) sets by some criterion of local clustering in this space. The dimensionality of the space used is the number of factors (N_f) deemed to convey significant differences i.e. not just random noise. Genuine but subtle distinctions may be difficult to discern in the presence of high noise levels, even for the powerful techniques of MSA. With this consideration in mind, we have used a model system of experimental data to evaluate how discriminatory power may be affected by two aspects of the overall analysis - how the digital images are normalized, and the choice of factorial space in which they are represented.

2. RESULTS AND DISCUSSION

The images used were of "distal half-fiber" protein of bacteriophage T7[2] (Fig 1). The half-fiber consists of four aligned globules of similar sizes (Fig 1b), and the pseudo-twofold axis at its center makes it difficult to determine a half-fiber's orientation ("up" or "down") from the image itself, although this decision may readily be made on the basis of its position relative to the rest of the complex (Fig 1a). A set of "up" and "down" half-fibers, aligned by correlation techniques, provides an ideal system to test the efficacy of MSA classification techniques. These data were analyzed according to both normalizations and both factorial representations. For each case, the percentages of dynamic power (λ) in the three first factors are listed in Table 1. These numbers are generally used in assessing whether factors are significant. As an index of how well the two subsets were distinguished in a given factor, we used the measure β which represents the ratio between the inter-subset variance (ie the Euclidean distance between their means) and the intra-subset variance. Values of β close to zero imply no distinction between the sets, whereas higher values correspond to progressively better discrimination.

Table 1

Normalization	PC or CA	λ1, β1	λ2, β2	λ3, β3
CMM	PC	7.43%, 0.10	5.16%, 1.44	4.30%, 0.09
	CA	5.34%, 1.62	4.50%, 0.05	4.43%, 0.03
CMV	PC	5.26%, 2.19	4.26%, 0.10	4.23%, 0.00
	CA	5.13%, 1.99	4.26%, 0.07	4.17%, 0.03

With CMM and PC, we obtained the surprising result that the primary factor did not distinguish between the sets (β=0.10, cf. Fig 2a), and we conclude that this factor was spurious. Thus the numerical scaling of aligned particle images can result in a factor that is more pronounced than a genuine, albeit slight, difference among them. CMM normalization is sensitive to the optical densities of single points (the extrema), whereas CMV normalization depends on the entire distribution. CMV is therefore more stable and, we consider, a better procedure for analyses of this kind.

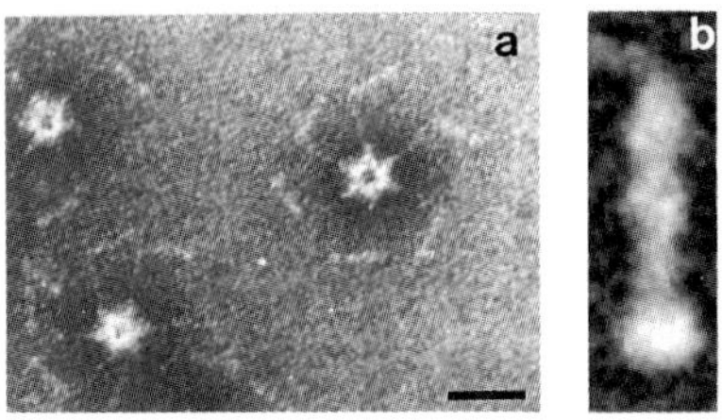

Fig. 1 : (a) Negatively stained "tail" complexes of phage T7, bar=20nm; (b) "distal half-fiber" component after correlation averaging, bar=5nm.

Comparing the β-values for PC and CA with CMV normalization (Table 1), both representations indicate that only one factor is significant, and slightly better discrimination between the subsets is achieved with PC. The two-factor plot for the (PC & CMV) analysis (Fig 2a) shows clear segregation of the data in the first dimension but random mixing in the second. Interestingly, CA did not produce a spurious first factor when CMM normalization was used, presumably because CA implicitly effects an operation similar to imposing a constant mean. However, the increased randomization introduced by CMM normalization must be expressed at some level, and presumably accounts for the lower value obtained for the discrimination index β (1.62 vs 1.99).

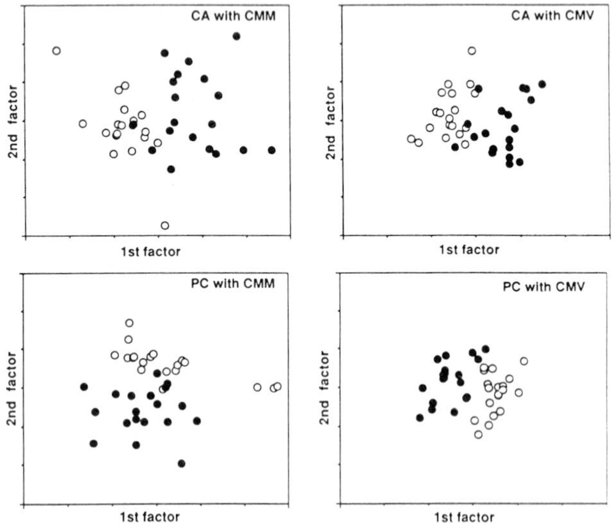

Fig. 2 : (a)-(d). Two-factor plots for sets of distal half-fiber images in the "up" (open circles) and "down" (filled circles) configurations, analyzed according to two normalizations and two factorial representations.

1. Van Heel, M. and Frank, 1981 Ultramicroscopy **6** pp 187-194
2. Steven, A.C., et al, 1988 J.molec. Biol. - in the press.
3. Hotelling, H., 1933 J. Educ. Psychol. **24** pp 417-441 498-520.

Digital structural analysis of helical biological specimens obtained by electron microscopy

K. Kanaya, N. Baba, M. Ogasawara, R. Fukatsu*, T. Obara** and H. Mori***.

Kogakuin Univ. Shinjuku-ku, Tokyo, *Sapporo Med. Coll., Sapporo, **Dept. Psychiat. and Neurol., Hokkaido Univ., ***Dept. Clin. Pathol., Tokyo Metrop. Inst. Gerontol., Tokyo; Japan.

The digital structural analysis method with the aid of a scanning densitometer with the computer system was successfully applied to helical neurofibrillary tangles in the brain in cases of senile dementia of the Alzheimer and Gum Parkinson type.

According to the Fourier method (Kanaya et al., 1985 & 87), the computer-generated image can then be obtained as $\rho(r)=\sum \rho_j (r-r_j)$;

$\rho_{hkl}(\mathbf{r})=f(hkl)\cos\{2\pi(h\cdot u+k\cdot v+l\cdot w)+\alpha_{hkl}\}$.

where $f(hkl)=|\mathbf{F}(\mathbf{q}_{hkl})|/F(0,0)$ with $\alpha_{hkl}=\tan^{-1} Fi/Fr$, in which α_{hkl} is 0 (positive contrast) or π (negative contrast).

Fig.1 shows a digital processing result for electron micrographs of Alzheimer's neurofibrillary tangle,, which are taken at 75 kV for the positively stained specimen. Fig.2 shows the result of Parkinson dementia specimen by the improved preparation of negative-staining method. Fig.4 A is a paired helical model consisting of double stranded ropes for a Alzheimer type specimen and B,a double paired helical model for the Guam Parkinson type. Fig.3 shows the result of the electron micrographs shadowed by tungsten on the same unstained Alzheimer specimen(Fig.1). The unidirectionally shadowed specimen shows that the filaments are clearly left-handed. Fig.5 shows the synthetic images for the Guam Parkinson based on the double paired helices assuming the 15 subunits per one rope in the unit cell. The contrast variations due to the folded helices and the multiple stripe appearance at diffrent rotation angle Ψ_0 satisfactorilly could explain the proposal model.

From the electron micrographs and the subsequent digital processing results, it was confirmed that tangles from a Alzheimer specimen are constituted of a pair of ropes with 78~90 nm pitch and 11 nm width in the tetragonal unit cell a = b = 11 nm, and similar tangles from a Parkinson specimen are stranded by double folded ropes with 4.5nm width in which one pair is 58~80 nm pitch and another pair of insight ropes a half of that in the unit cell $2r_0$ = 15nm and $2r_0'$ = 7nm. Futhermore, stereographic micrographs and their processed lattice images obtained by ion beam sputter deposition technique show the helical filament to be wound in a left-handed manner.

Kanaya,K et al.,1985,Micron microsc.Acta,16:17-32;1987,18:131-146.
Obara,T et al.,1986. J.Neurol.Sci.,27:173-181.

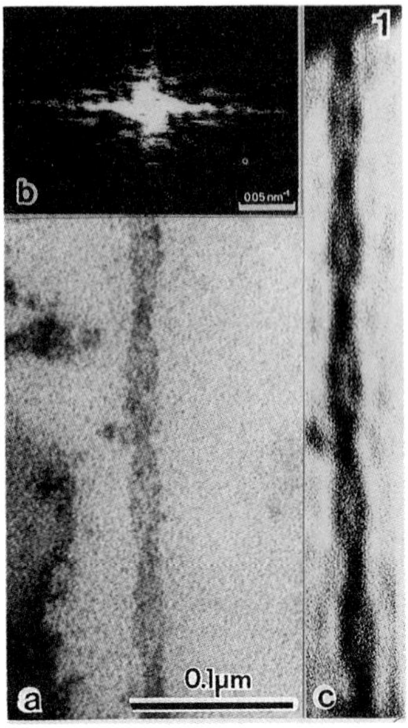

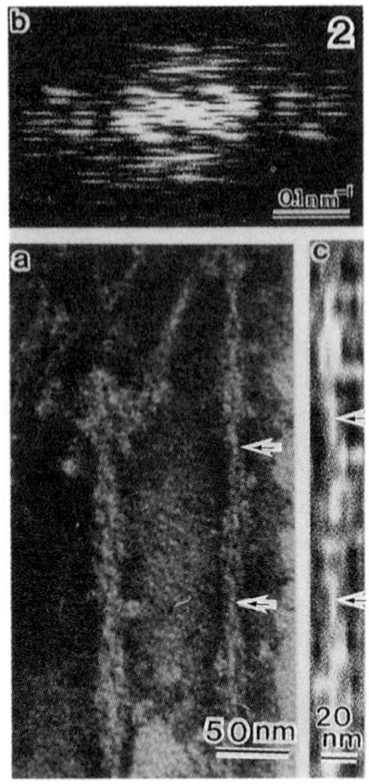

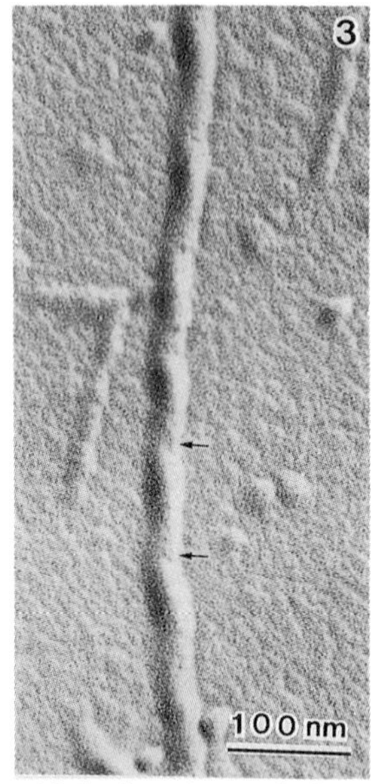

Fig. 1 and 2 Digital processing results for a Alzheimer's and Guam Parkinson type neurofibrillary tangles, respectively, (a) is the original, (b) its computer-generated diffractogram and (c) the computer-generated lattice image. Fig. 3 The Alzheimer's neurofibrillary tangle shadowed by tungsten showing the paired helical filaments.

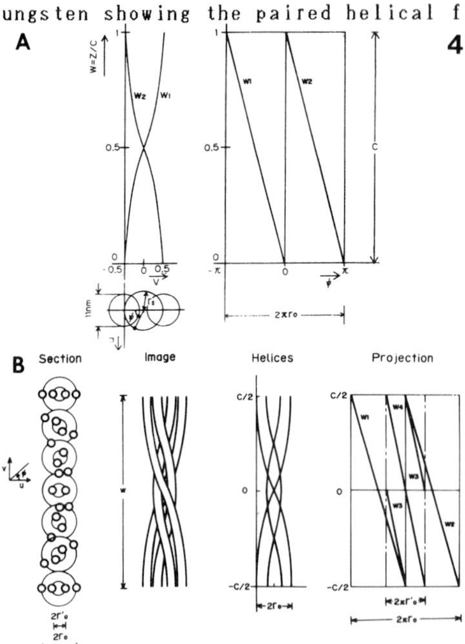

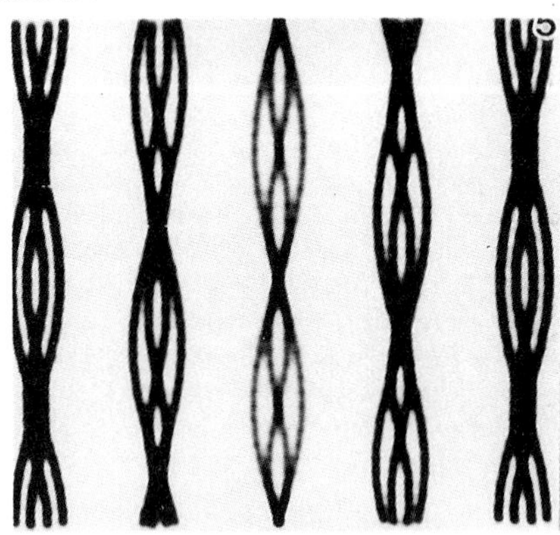

$\phi_0 = 0 \quad \pi/4 \quad \pi/2 \quad 3\pi/4 \quad \pi$

Fig. 4 Two kinds of paired helical filament models for the neuro-fibrillary tangles incubated with the Alzheimer A & Guam Parkinson type diseases B, respectively. Fig. 5 The synthetic images of neuro-fibrillary tangles for a Guam Parkinson type specimen.

Inst. Phys. Conf. Ser. No. 93: Volume 3, Chapter 14
Paper presented at EUREM 88, York, England, 1988

Improvements of a serial section reconstruction system to be applicable to more complicated structures

N Baba, M Baba*, M Osumu*, S Nakamura** and K Kanaya

Dept.Electr.Eng.,Kogakuin Univ.,Shinjuku-ku,Tokyo, *Dept.Biol.,Japan Women's Univ.,Bunkyo-ku,Tokyo,**Dept.Pathol.,Kouchi Medical School,Japan

ABSTRACT: The serial section reconstruction system has been improved to be applicable to more genuine structures. A marked improvement is the smoothing of surfaces of reconstructed objects by averaging coordinates of contour points, which reduces artificial roughness of the surface. Furthermore, the system has been improved so as to automatically reproduce plural structural objects with branchs if only contour line data are provided.

Recording photographs of serial thin sections of a specimen with a microscope is one method to study the three-dimensional structure of an object. Computer graphics is suitable for observing a three-dimensional object reconstructed from serial section images (Nakamae et al. 1985; Menhardt et al.1986; Yaegashi et al.1987; Baba et al.1986a,b), and the computer-assisted method is advantageous for quantifying three-dimensional structure. The present enhanced version is capable of smoothing the surface of reconstructed object by averaging coordinates of contour points, and cutting or sectioning a reconstructed object which generates the cut end together with the whole parts remaining by cutting. 'Smoothing of the surface' does not refer to 'tiling' of the 3-D surface defined by serial contour lines with polygons, triangles, nor the smoothing processing for calculating brightness distribution on the 3-D surface to be displayed after the 'smooth tiling'. Instead, 'smoothing' proposed here smooths the reconstructed object itself by correcting or reducing roughness of the surface due to artifacts arising during specimen preparation or photography with a microscope, or due to a difficulty of completely matching and alignment of serial contours. Also, the system has a capability of displaying a reconstructed result including modeled objects in simple forms such as three-dimensionally distributed particles in spheres, micro tubules in thin curved rod or line and so on. Furthermore, the system has been improved so as to automatically reproduce plural structural objects with branchs if only contour line data are provided. Using binary image processing, the relations between every contour line in a section image and that in the upper or lower section image are examined for all pairs of contour lines between the upper and lower sections, and proper pairs of contour lines are chosen and then the contour lines are connected.

1. COMPUTER SYSTEM

The system consists of a LUZEX 5000 image processor (NIRECO corporation) and a TV image digitizer. It also includes a quality colour display device (CRT), that is useful for distinguishing various structural parts of a shaded stereo image, and for displaying a transparent model.

2. APPLICATION

The present system has been applied to study an intracellular structure of a yeast target cell (Saccharomyces cerevisiae cell) induced by mating pheromone, α factor(Baba and Osumi 1984). Various organelles (nucleus,N; vacuole,V; mitochondria,M; lipid,L) could be clearly displayed through the outer cell membrane and vesicles(Ves) distributed in three-dimension observed. Photos show 3 stages of projection formation from Fig.1 to 3 . The smoothing of the surface of the reconstructed cell membrane by computer processing mentioned above made it more reliable to observe the intracellular structural objects through the outer cell membrane structure (by transparent display). In the previous stage, the roughness of the outer surface that became conspicuous by the shading disturbed the observation of the intracellular structure.

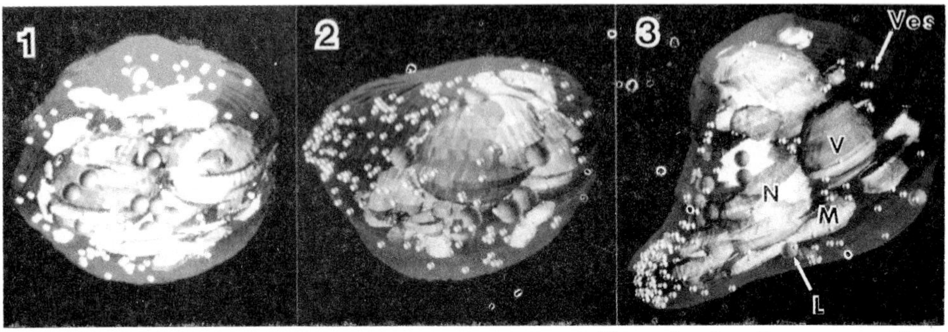

Fig.1 Reconstructed result of yeast target cells showing 3 stages of projection formation; Various organelles (nucleus,N; vacuole,V; mitochondria,M; lipid,L) could be clearly displayed through the outer cell membrane and the distribution of vesicles(Ves) also seen.

Baba M and Osumi M 1987 J.Electron Microsc.Technique 5 249-261

Baba N,Nakamura S,Kino I and Kanaya K 1986a J.Electron Micriosc.Technique 3 401-406

Baba N,Baba M,Osumi M,Nakamura S,Imamura M,Koga M and Kanaya K 1986b Sci. Form 2 49-58

Menhardt W,Lockhausen J,Dallas W J and Kristen U 1986 Micron and Microsc. Acta **17** 349-357

Nakamae E,Harada K,Kaneda K,Yasuda M and Sato A 1985 Trans.Inform.Proc. Society of Jpn **26** 181-188

Yaegashi H,Takahashi T and Kawasaki M 1987 J.Microscopy **146** 55-65

Inst. Phys. Conf. Ser. No. 93: Volume 3, Chapter 14
Paper presented at EUREM 88, York, England, 1988

Inverse transform of an image autocorrelation function

N Baba and K Kanaya

Department of Electrical Engineering, Kogakuin University, Nishishinjuku 1-24-2, Shinjuku-ku, Tokyo, 163-91 Japan

ABSTRACT: A method of the inverse transform of an image autocorrelation function has been established. The serious problem in the inverse transform has been overcome by the method of least squares, which is a restriction in which the sign of Fourier coefficients obtained by Fourier transform of the modified autocorrelation function must be plus. An arbitrary modification of the autocorrelation can be made, which generates an image satisfying the modified autocorrelation function.

A method for the inverse transform from the autocorrelation function to the image was already devised to remove noise from a single image (Baba et al. 1985). The reversible transform between an image and image autocorrelation function can be realized by performing, in reverse, the computing procedure of the autocorrelation function performing two Fourier transforms, which is based on the convolution theorem. However, when the autocorrelation function is modified, there are some restrictions. The most serious restriction is that the modified function must be made by a summation of only even function of the cosine and the sign of the cosine must be plus. Therefore, in general, coefficients of the cosine (Fourier coefficients obtained by Fourier transform of the modified function) can not be obtained simply by Fourier transform of the modified function because some of coefficients may be minus.

In this study the above problem has been solved by the method of least squares and the iterative computer processing. In the following, to simplify the discussion one-dimensional case is considered, but the conclusion is easily applicable to the two-dimensional case (image).

If the modified autocorrelation function is denoted by $R_j(x_j)$ and $R_j(x_j)=\sum_i a_i\cos(q_i x_j)$ the coefficients a_i can be determined by the method of least squares $\sum_j\{R_j-[\sum_i a_i\cos(q_i x_j)]\}^2 \Rightarrow \min.$

If one coefficient a_l is noticed and other coefficients are left, a_l is solved as: $a_l=A_2/(2A_1)$; $A_1=4\sum_j\cos^2 q_l x_j$;

$$A_2=4\sum_j\{R_j-[a_0+\sum_{i=1}^{l-1}2a_i\cos(q_i x_j)+\sum_{l+1}^{n}2a_i\cos(q_i x_j)]\}\cos(q_l x_j).$$

When the sign of a_l is minus, 0 is set for the value. Then, all coefficients a_i can be obtained by iterating the above calculation through all coefficients. Therefore, arbitrary design of the autocorrelation function can be realized under some restrictions. Figures are examples for the application of the noise reduction in computer simulation. Two samples of signal were employed as shown in Fig.a and d. Noisy signals were generated by random numbers as shown in Fig.b and e. The noise signal autocorrelation was modified according to the scheme described in previous paper(Baba et al.1985). An example of the modified function is shown in Fig.b. As shown in Figs.c and f, the noise was considerably reduced.

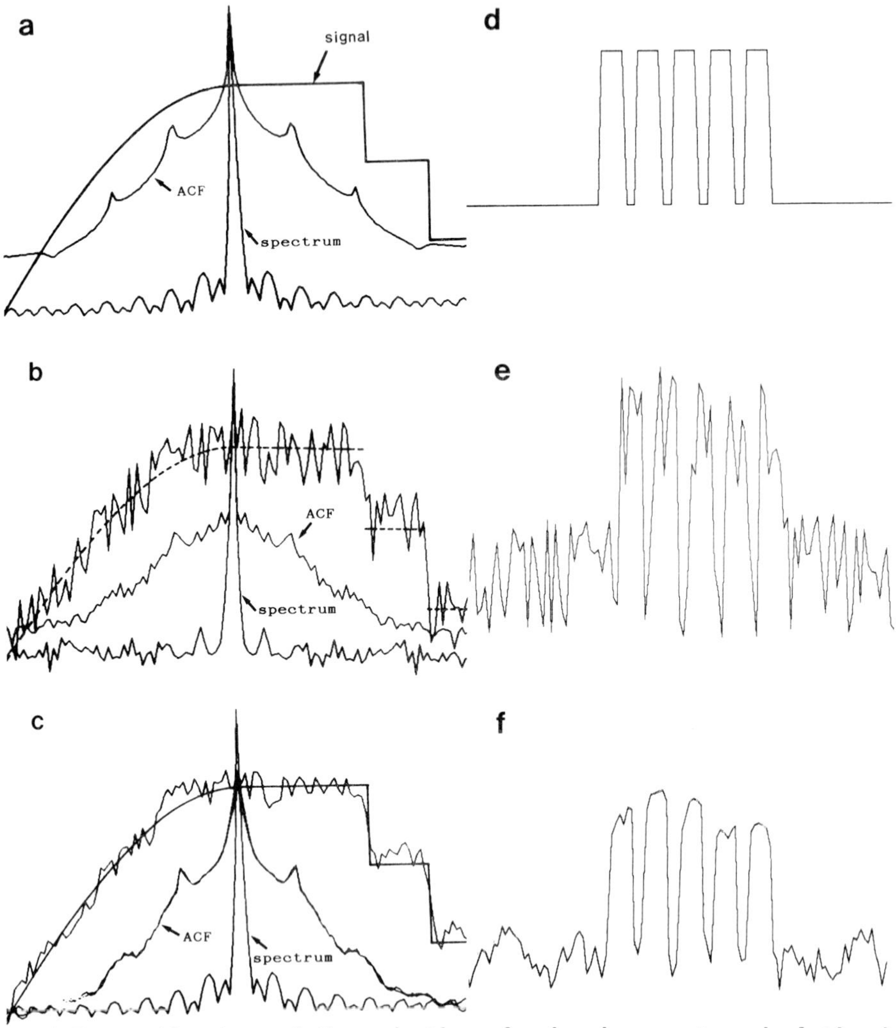

Fig.1 Two applications of the reduction of noise in computer simulation by the inverse transform of modified autocorrelation functions.

Baba N,Oho E and Kanaya K 1985 J.Electron Microsc.Technique 2 431-438

Structural study of actin bundle by cryo-electron microscopy

M.F. Schmid[1], T.-W. Jeng[1], P. Matsudaira[2], J. Bordas[3] and W. Chiu[1]

1. University of Arizona, Dept. of Biochem., Tucson, AZ 85721
2. MIT, Whitehead Institute, Cambridge, MA 02142
3. MRC/SERC Biology Support Group, Daresbury, Cheshire, U.K.

F-actin plays a central role in many facets of cell motility in both muscle and non-muscle cells. Actomyosin generates continuous and variable movement by an arrangement of adjustable crossbridges, but the acrosomal bundle of some marine invertebrate sperm cells uses actin differently. In these systems, a fixed set of configurations is achieved by a change in the twist of the actin filaments in the bundle. This change propagates down the length of the bundle to produce a transition from a compact coil under the sperm head to a 60 μ process protruding from the head. Several proteins in the bundle serve to crosslink the actin and to generate or communicate the change in twist between the actin filaments in the bundle.

To understand how actin can show such variable behavior, it will be necessary to study the structure of actin itself and the interactions of these proteins in situ. Unfortunately, the actin filament exhibits variability in its helical parameters. This limits the resolution to which the actin molecule can be reconstructed. This variability extends to observations of the actin filament in vitreous ice, and may represent a dynamic situation which is present in solution. In the acrosomal bundle, the variability of actin pitch is much reduced, and the structure of actin and its crosslinking proteins can be determined to a higher resolution.

A suspension of true discharge bundles was placed on a microscope grid, blotted and quickly dropped into liquid ethane near its freezing point. The grids were examined in a JEOL 100CX microscope equipped with a cold stage. Figure 1 shows a scanned image of the bundle in vitreous ice. Protein is dark. The images were subjected to a lattice distortion correction procedure based on cross correlation. Figure 2 is a representation of the IQ values in the computed Fourier Transform of this image after correction (Henderson et al. 1986). Data out to 1/17 $Å^{-1}$ are included. To examine the phase relationships between the reflections, six images in the same orientation were evaluated for the presence of symmetry along the lattice directions in the plane of the images. Table 1 shows the RMS minima in degrees found in the phase origin searches for these possible conditions. It indicates that a good two-fold screw axis exists along the helix axis, but the crystallographic quality of this axis will not be certain until it is examined in three dimensions. We are currently assessing the orientations of tilted images with respect to the view shown in Figure 1 to build a three dimensional data set.

Acknowledgement: This research has been supported by NIH grant no. RR02250.

Henderson,R.H., Baldwin,J.M., Downing,K.H., Lepault,J. & Zemlin, F. (1986) Ultramicroscopy 19, 147.

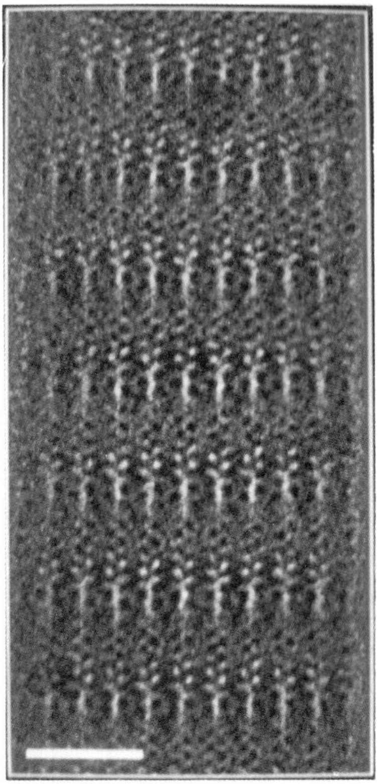

Figure 1. Scanned image of an acrosomal bundle embedded in vitreous ice. The scan raster is 8.3 Å/pixel. Scale bar represents 500 Å. Bundle axis is vertical.

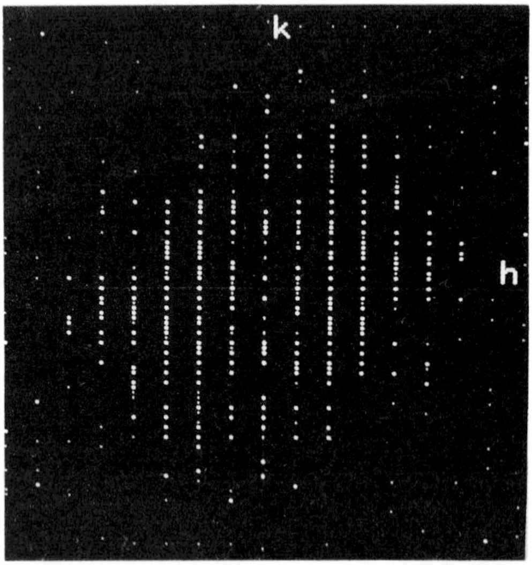

Figure 2. Reciprocal lattice plot of the transform of the image shown in Fig 1. Points indicate where the background-subtracted, sinc-fitted vector sum of the transform near the predicted lattice position is greater than 2X the value of the local background (IQ=4). Brighter points are at least three times the background (IQ<4). $h=1/145Å^{-1}$ and $k=1/891Å^{-1}$

Table 1 RMS phase residuals in degrees after phase origin search for various possible symmetry axis conditions. All pairs of reflections both of whose members had IQ less than or equal to 4 were compared at equal weight at a maximum resolution of 30 Å. The bundle axis is along k.

Condition	Image 1	2	3	4	5	6
2-fold along k	47	47	44	42	46	43
2_1 along k	16	8	13	10	8	6
2-fold along h	42	38	36	40	38	38
2_1 along h	46	42	44	40	38	42

Structural studies of the flagellar filaments of some archaebacteria

D. Typke, M. Nitsch, A. Möhrle, R.Hegerl, M. Alam, D. Grogan and J. Trent
Max-Planck-Institut für Biochemie, D-8033 Martinsried, West-Germany

Like eubacteria, many archaebacteria possess flagella (1-5). Up to now, relatively little is known about their flagellar apparatus. It is established from protein-chemical work for *Halobacterium halobium* that its flagellar filaments are composed of 3 glycoproteins (3,6). Recently also the primary structure of these proteins has been determined (7). We have started to analyse by electron microscopy and image analysis the structures of the flagellar filaments of 3 species from different branches of the archaebacterial kingdom: *Halobacterium halobium*, *Pyrococcus woesei*, and *Sulfolobus* sp. (strain B12). The goal is the determination of the 3D structures and possibly an answer to the question how the corkscrew shape of the flagellar filament is related to its molecular structure.

Electron microscopy: In the first experiments, whole cells in their growth medium (the flagella partly removed from the cells by mechanical shearing forces) were used. They were prepared for electron microscopy using Valentine's method and negatively stained with several different stains (uranyl acetate, ammonium molybdate, phosphotungstate). Later cesium chloride gradient purified flagellar suspensions were used for EM preparation. The archaebacterial flagellar filaments exhibit a helical structure. However, the corkscrew shape implies deviations from helical symmetry about the filament axis (8,9). We assume that this deviation is smaller than the resolution that can be attained with negative stain. For the flagella under investigation no straight forms are known, or conditions under which they become straightened experimentally. However, the EM images often contain parts of the flagellar filaments with small and slowly varying curvature. This kind of filament, from images with 200 to 500 nm underfocus, was selected. *Image analysis* was done using the SEMPER-V program system with minor changes. The first step was straightening of the fibres. SEMPER provides a procedure (SST) for this, which traces a polynomial through a series of points defining the fibre axis and then interpolates the image data along straight lines perpendicular to this curve within a strip of selectable size. The definition of axial points by eye or by the centre-of-mass condition (even with averaging over a specific length) turned out to be insufficiently accurate to produce straight filaments. In a reiterated run of SST, we therefore used the peaks of the cross-correlation function of a cutout with the whole fibre to improve the straightening. The pictures of fibres were evaluated by calculating the power spectra, Fourier-filtered images and correlation functions to obtain information about the helical parameters necessary for 3D-reconstruction.

Preliminary results: A common feature of the flagellar filaments of the 3 species is that the subunits are arranged in 3 intercalated helices. The filaments of *H. halobium* and *P. woesei* are very similar in size (ca. 10 nm diameter) and helical pitch (ca. 17 nm). Their structure seems to be rather compact; the helical arrangement is nearly invisible in the original images. The Fourier-filtered images are nearly identical for both species, showing clearly thehelical arrangement. The fine structure seems to be rather complicated, a long-range helical period could not be determined, neither by cross-correlation nor from the power spectra. The *Sulfolobus* B12 filaments clearly exhibit helical organization already visible in the original images (Fig. 1a). The size (ca.13nm diameter) and helical pitch (ca.15 nm) are different from the other two species. Figs. 1b and c show a series of Fourier-filtered images and the power spectrum of one filament. In Fig. 1d all Fourier coefficients on the layer lines containing the strongest peaks were used to obtain the filtered image. This image suggests that the 3 intercalated helices are quasi-continuous with no obvious subunit structure.

References:

(1) Zillig,W.et al., Zbl.Bakt.Hyg., I.Abt.Orig. **C3**, 304-317 (1982)
(2) Zillig,W.et al., System.Appl.Microbiol. **4**, 88-94 (1983)
(3) Alam,M., Oesterhelt,D., J.Mol.Biol. **176**, 459 (1984)
(4) Zillig,W.et al., System.Appl.Microbiol. **9**, 62-70 (1987)
(5) Grogan,D., unpublished results
(6) Wieland, F. et al., J. Biol. Chem. **260**, 15180 (1985)
(7) Sumper,M., Biochem.Biophys.Acta **906**, 69 (1987)
(8) Klug,A. in: Formation and Fate of Cell Organelles (K.B.Warren ed.) pp.1-18, New York and Lond. Acad.Press
(9) Calladine,C.R., J.Theor.Biol. **57**, 469-489 (1976)

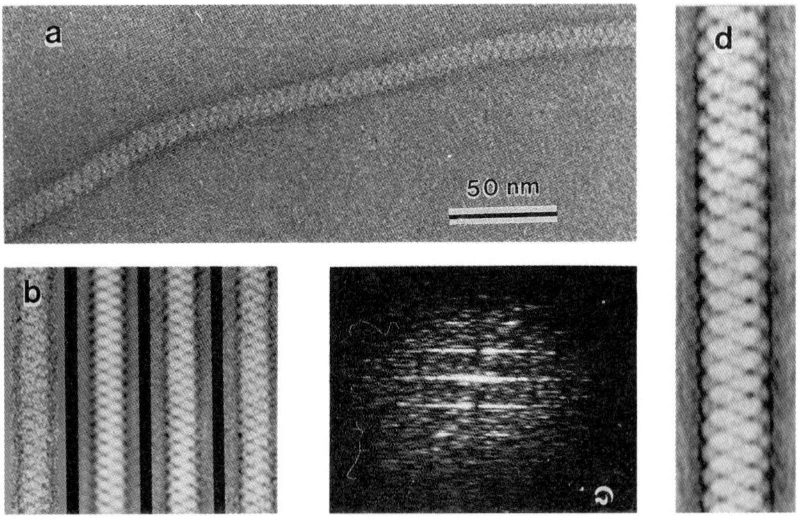

Fig.1: *Sulfolobus* B12 flagellar filament: (a) cutout of original image; (b) series of Fourier-filtered images, utilizing an increasing number of Fourier coefficients (left: original); (c) power spectrum of a straightened fibre; (d) Fourier-filtered image, utilizing the Fourier coefficients on the layer-lines containing the strongest peaks.

Inst. Phys. Conf. Ser. No. 93: Volume 3, Chapter 14
Paper presented at EUREM 88, York, England, 1988

Three-dimensional reconstruction of the HP1-layer of *Deinococcus radiodurans* embedded in Cd-thioglycerol

U. Jakubowski, R. Hegerl, H. Formanek*, S. Volker, U. Santarius, and W.Baumeister.
Max-Planck-Institut für Biochemie, D-8033 Martinsried, FRG
*Botanisches Institut der Universität München, D-8000 München, FRG

Negative staining with conventional embedding media such as uranylacetate or phosphotungstate is - in spite of being an extremely successful technique - subject to severe limitations. With most negatively stained specimens resolution is restricted to 1.2 - 1.8 nm and even if a higher resolution is attained problems of interpretation arise from concurrent positive staining or due to stain level effects (1). Frozen hydrated preparations combine an excellent preservation of the structure (as attested by electron diffraction) with an (in principle) straightforward interpretation of image densities. Practical problems, however, such as mechanical instabilities of the cryoholders or experimental difficulties in controlling the thickness of the ice film, usually with the consequence of excessive inelastic scattering, prevent us from taking full advantage of this technique. Glucose embedding is a very convenient technique which has been applied with particular success to membrane crystals (2). With proteins not embedded in a lipid matrix aurothioglucose is the preferred alternative (3).

We have made some efforts to find other embedding media capable of avoiding specimen dehydration, even upon exposure to vacuum. Thioglycerol proved to be a good choice in that respect. The use of Cd-thioglycerol - or any mixtures of thioglycerol with Cd-thioglycerol - increases contrast.

We have used the HPI layer, a two-dimensional protein crystal from the cell envelope of *Deinococcus radiodurans* (4) as a test specimen for evaluating the properties of this embedding medium. The outer surface of this layer is extremely prone to interface denaturation (5); if, in a negatively stained preparation, this surface comes to rest on the carbon support, it shows up completely flattened in a three-dimensional reconstruction. Hence, a comparison of three-dimensional reconstructions of different oriented layers serves as a sensitive test for the structure preserving properties of an embedding medium.

Images of Cd-thioglycerol embedded HPI-layer sheets were drawn from several tilt 'series', each of them comprising only three or four projections in order to keep the electron dose within tolerable limits. These tilt series were sorted into two data sets according to the two possible orientations of the layer on the grid as judged by the handedness. Other criteria were resolution as indicated by diffraction spots and a 'safe' contrast transfer function, i.e. no intruding zeros to a resolution of approx. 1 nm. From each projection, an area of 2048 by 2048 pixels was digitized and subjected to correlation averaging. The obvious problem is, of course, the correct combination of the projections in the course of the three-dimensional reconstruction. First, we have determined effective tilt angles and the effective tilt axis azimuth for each tilt series, taking into account specimen inclinations. In the second step, one lattice base vector from each projection was defined to be parallel to the x-axis while the other one is rotated anti-clock-wise. After relating the tilt axis azimuth to the new x-axis and taking the effective tilt angles as found before, all data were related to a common coordinate system. The distribution of projection directions does not correspond to single axis tilting and is virtually random. Then the three-dimensional reconstruction was performed as usual using the hybrid Fourier space /real space approach (6). Any projections which did not correlate satisfactorily in the course of the alignment and normalisation procedure were rejected. The present 3-D model (Fig.1) is therefore based on 15 projections only; we are currently in the process of including a much larger number of projections into the reconstruction which will further improve the structure.

(1) Engel,A., Baumeister,W., and Saxton,W.O., Proc. Nat. Acad. Sci. USA 79 (1982), 4050-4054.
(2) Unwin,P.N.T. and Henderson, R., J. Mol. Biol. 94 (1975), 425-440.
(3) Rachel,R., Jakubowski,U., Tietz,R., Hegerl,R., and Baumeister,W., Ultramicroscopy 20 (1986), 305-310.
(4) Baumeister,W., Karrenberg,F., Rachel,R., Engel, A., Ten Heggeler,B., and Saxton, W.O., Eur. J. Biochem. 125 (1982), 535-544.
(5) Baumeister,W., Barth,M., Hegerl,R., Guckenberger,R., Hahn,M., and Saxton,W.O., J. Mol. Biol. 187 (1986), 241-253.
(6) Saxton,W.O., Baumeister,W., and Hahn,M., Ultramicroscopy 13 (1984), 57-70.

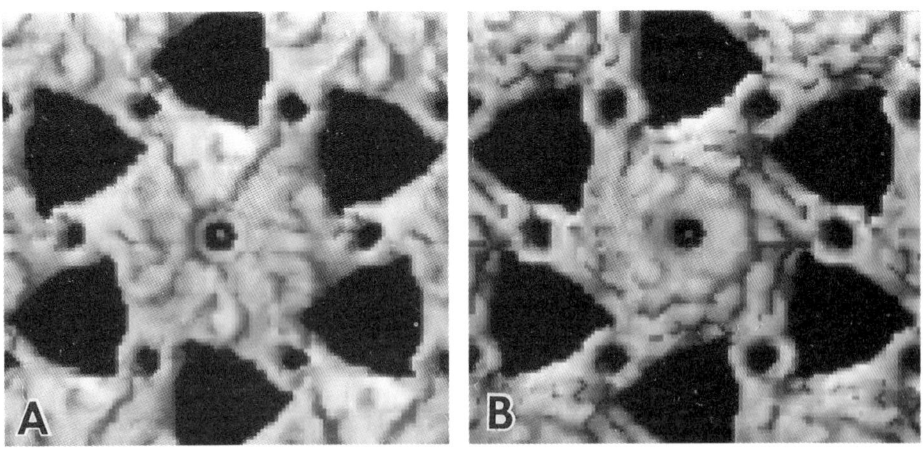

Fig.1: Preliminary 3-D model showing (A) the outer and (B) the inner surface of the HPI layer. The reconstruction very clearly depicts the central pore and shows no indication of serious flattening.

Cobalt-tetrammine-ATP induced crystalline arrays of Na,K-ATPase

E. Skriver[1], A. B. Maunsbach[1], H. Hebert[2], G. Scheiner-Bobis[3] and W. Schoner[3]

Department of Cell Biology[1] at the Institute of Anatomy, University of Aarhus, DK-8000 Aarhus C, Denmark, Department of Medical Biophysics[2], Karolinska Institutet, S-10401 Stockholm, Sweden, and Institute of Biochemistry and Endocrinology[3], Veterinary Medical School, Justus-Liebig-University Giessen, D-6300 Giessen, FRG

ABSTRACT: Two new types of membrane crystals were induced in purified Na,K-ATPase membranes with cobalt-tetrammine-ATP, which is a stable MgATP complex analogue. The building blocks of the crystalline arrays correspond to $(\alpha\beta)_2$ dimers of the enzyme protein indicating that α-α interaction may be important in the pumping process.

1. INTRODUCTION

Membrane-bound Na,K-ATPase purified from kidney outer medulla forms two-dimensional crystals during incubation with vanadate and magnesium or other ligands which favour the E_2 conformation of the enzyme (Skriver et al. 1981, 1985). Image analysis has previously revealed two types of membrane crystals: p1 crystals where the unit cell contains one $\alpha\beta$ protomer of the enzyme and p21 crystals where the a unit cell contains an $(\alpha\beta)_2$ dimer (Hebert et al. 1982). In this work we report two new types of crystalline arrays of Na,K-ATPase which have dimers as building blocks and which assemble during incubation with cobalt-tetrammine-ATP which is a stabile MgATP complex analogue. This ligand binds to the low affinity ATP binding site of Na,K-ATPase forming a stable ADP-$Co(NH_3)_4$-phosphoenzyme complex in the E_2 conformations of the enzyme (Scheiner-Bobis et al. 1987)

2. MATERIALS AND METHODS

Purified membrane-bound Na,K-ATPase from rabbit renal medulla was incubated in 0.1 mM $Co(NH_3)_4ATP$ at 4 °C or first at 37 °C for one hour and then at 4 °C. Specimens were negatively stained with 1% uranyl acetate at different intervals. Selected areas from ordered arrays were digitized and processed with correlation averaging methods using the "EM" program system (Hegerl and Altbauer, 1982).

3. RESULTS AND DISCUSSIONS

Following incubation with $Co(NH_3)_4ATP$ two new types of membrane crystals were observed (Figs. 1 and 2). Image analysis demonstrates that the repeating building block in both types is a dimer. In projection maps each unit in the dimers has the same triangular shape as seen before in vanadate crystals. However, instead of close end-to-end contact as in p21 crystals induced with vanadate the units in the dimers are in close contact along their long sides. In Fig. 3 the dimers form a crystalline array showing projection symmetry p2 and unit cell dimensions a=119 Å, b=86 Å, γ=107°. In Fig. 4 the dimers are arranged with projection symmetry p4 with four pairs of enzyme units arranged in a ring and with unit cell dimensions a=b=136 Å, γ=90°. Since the relationship between the protomers in the dimers in vanadate-induced crystals is different from that in crystals formed in presence of $Co(NH_3)_4ATP$, the present observations indicate that the E_2 conformations may also be different. The extensive dimer formation in $Co(NH_3)_4ATP$-induced crystals is consistent with the possibility that two α-subunits of Na,K-ATPase cooperate during the pumping process.

REFERENCES

Hebert H, Jørgensen P L, Skriver E and Maunsbach A B 1982 Biochim.Biophys. Acta **689** 571

Hegerl R and Altbauer A 1982 Ultramicroscopy **9** 109

Scheiner-Bobis G, Fahlbusch K and Schoner W 1987 Eur. J. Biochem. **168** 123

Skriver E, Hebert H and Maunsbach A B 1985 In: The Sodium Pump (eds: I. Glynn and C. Ellory), The Company of Biologists Ltd. Cambridge, pp.37-4

Skriver E, Maunsbach A B and Jørgensen P L 1981 FEBS Lett **131** 219

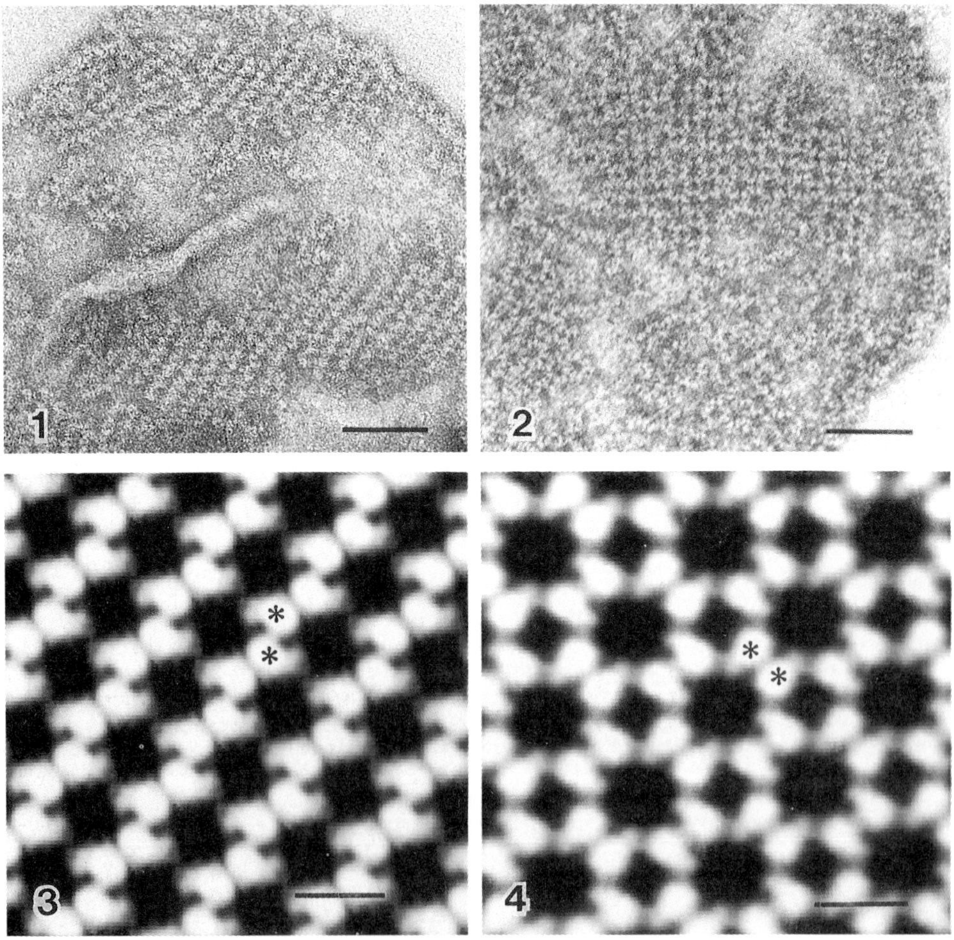

Figs. 1. and 2. Electron micrographs of the two types of two-dimensional crystalline arrays of Na,K-ATPase induced by cobalt-tetrammine-ATP. Bar: 500 Å.

Figs. 3. and 4. Averaged structures from Figs. 1 and 2. Asterisks (*) indicate the paired units in a dimer. Bar: 100 Å.

Three-dimensional structure analysis of renal Na,K-ATPase

H Hebert[1], E Skriver[2] and A B Maunsbach[2]

[1]Department of Medical Biophysics, Karolinska Institutet, Box 60400, S-104 01 Stockholm, Sweden and [2]Department of Cell Biology at the Institute of Anatomy, University of Aarhus, DK-8000 Aarhus C, Denmark

ABSTRACT: Low resolution 3-D models of Na,K-ATPase obtained from two different types of two-dimensional membrane crystals show rod-like protein regions extending 100 Å perpendicularly to the membrane. In p21 crystals these are symmetry-related and form pairs corresponding to $(\alpha\beta)_2$-dimers. A peripheral domain is interpreted as the β-subunit while the rest of the models corresponds to the α-subunit.

1. INTRODUCTION

Membrane-bound Na,K-ATPase, the Na,K-ion pump, can form two-dimensional membrane crystals after incubation with different ligands under specific conditions (Skriver et al. 1981; Skriver et al. 1985). Two crystal forms with two-sided plane group symmetry p21 and p1 have been observed (Hebert et al. 1982). The crystals are suitable for structure analysis with electron microscopy and image processing. In this investigation we compare the results from an analysis of the p1 crystals with our previously reconstructed 3-D model for the p21 crystal (Hebert et al. 1985).

2. MATERIALS AND METHODS

Na,K-ATPase was isolated from the outer medulla of rabbit kidney. The enzyme was incubated at 4°C with 3 mM NH_4VO_3 and 3 mM $MgCl_2$ in 10 mM imidazole buffer, pH 7.0. Uranyl acetate (1%) was used for negative staining and tilt series of the specimens were recorded in a JEOL 100 CX electron microscope at angles between -60° and +60°. The digitized electron micrographs were processed with the "EM" system (Hegerl and Altbauer 1982). Each projection was either Fourier filtered or subjected to correlation averaging. The calculated 3-D volumes were presented as sections on a plotter and shaded surface models on a graphics display system.

3. RESULTS AND DISCUSSION

Electron micrographs of negatively stained two-dimensional crystals of membrane-bound Na,K-ATPase (Fig. 1) typically have a resolution limit of 20 - 25 Å. The 3-D models reconstructed from the tilt series (Fig. 2) show rod-like protein regions with a size expected for the αβ-protomer of the enzyme which has an estimated molecular weight for protein of 146 kD. The rods are slightly flattened and inclined relative to the normal of the membrane plane. For the p21 crystal (Fig. 2a) adjacent rods are related by two-fold symmetry. Protein units are paired together to form $(\alpha\beta)_2$-

dimers by stain deficient regions on one side of the lipid bilayer. A dip with an extension of about 40 Å observed in the contrast variation curve of the model is interpreted as the position of the lipid bilayer. Thus, the protein is asymmetrically distributed in the bilayer in accordance with conclusions from a sequence analysis of the α-subunit (Shull et al. 1985). The best rotational alignment between the models is obtained when the directions of the unit cell axes coincides. Apart from the central elliptic domain, which is elongated along the a-axis of the p1 crystal, the αβ-protomer contains two peripheral parts laterally shifted along the b-axis. The larger of these has its equivalence in the 3-D model obtained from the p21 crystal. This domain is interpreted as the β-subunit while the rest of the protein constitutes the α-subunit.

4. REFERENCES

Hebert H, Jørgensen P L, Skriver E and Maunsbach A B 1982 *Biochim. Biophys. Acta* 689 571
Hebert H, Skriver E and Maunsbach A B 1985 *FEBS Lett.* 187 182
Hegerl R and Altbauer A 1982 *Ultramicroscopy* 9 109
Shull G E, Schwartz A and Lingrel J B 1985 *Nature* 316 691
Skriver E, Hebert H and Maunsbach A B 1985 *The Sodium Pump* eds I Glynn and C Ellory (Cambridge: The Company of Biologists Ltd.) pp 37-44
Skriver E, Maunsbach A B and Jørgensen P L 1981 *FEBS Lett.* 131 219

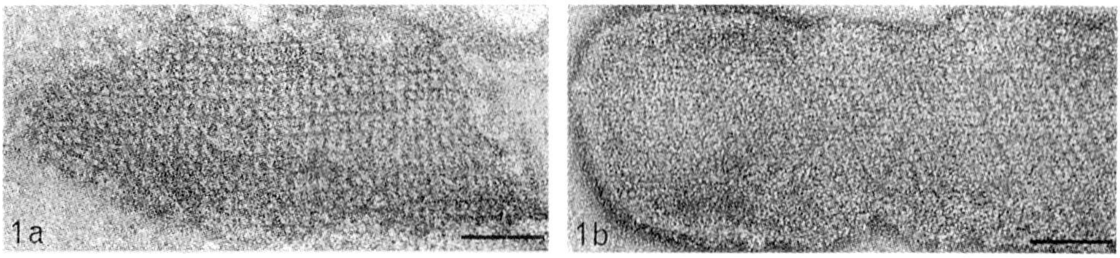

Fig. 1 Electron micrographs of negatively stained two-dimensional membrane crystals of Na,K-ATPase of the a) p21 and b) p1 type. Bar 50 nm.

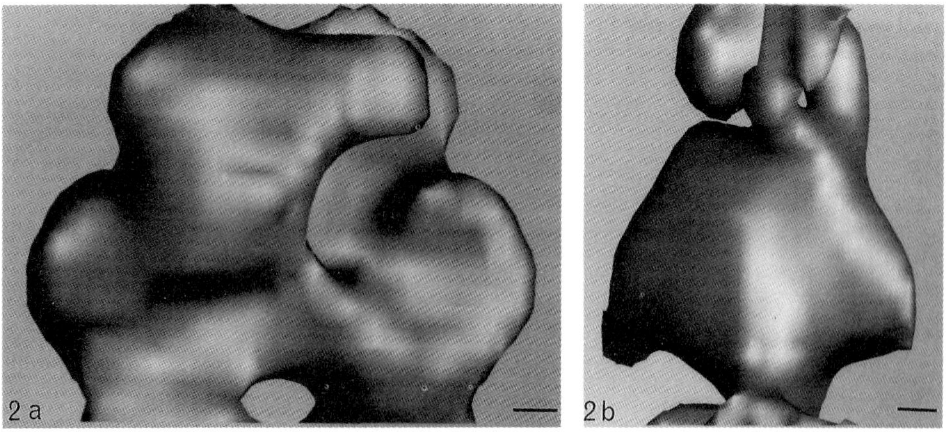

Fig. 2 Three-dimensional models of Na,K-ATPase obtained from crystals of the a) p21 and b) p1 type. The model is seen along the membrane plane. Bar 1 nm.

Inst. Phys. Conf. Ser. No. 93: Volume 3, Chapter 14
Paper presented at EUREM 88, York, England, 1988

Comparative electron microscopic studies of crystallized ribosomal 50S subunits after ERL-embedding and freeze-fracturing

H.Renz, L.Bohrer and B.Tesche

Fritz-Haber-Institut der Max-Planck-Gesellschaft, Faradayweg 4-6, D-1000 Berlin 33

In 1980 A.E.Yonath et al. were successful in the in vitro crystallization of ribosomal 50S subunits of Bacillus stearothermophilus*. The large needle-like 3-D crystals (approx. 0.6x0.15x 0.1 mm) are too thick to be studied directly by electron microscopy. Therefore we used ultrathin uranyl-stained sections and replicas to determine the subunits topography and their orientation in the crystal.

To prepare ultrathin sections, crystals were fixed in 0.4% glutaraldehyde, dehydrated in ethanol and embedded in Spurr's resin. Carbon replicas were obtained by rapid-freezing the crystals in liquid propane without any pretreatment, fracturing at -130°C and shadowing with Pt/C.

With either method we identified at least two different projections of the crystal: a) views revealing a hexagonal-like arrangement (Figs. 1a, 2a) and b) views showing double strings of alternating units with distances of 25 and 50 nm (Figs. 1b, 2b). The unit-distances of the sections and the replicas were in good agreement. With both methods zigzag patterns were recognized (Figs. 1a, 2a). Optical diffractions of micrographs (inserted in figs.) showed a resolution of 3.5-4 nm in resin-embedded crystals and up to 4 nm in replicas of cryofixed crystals. Salts of the precipitation medium, preserving the crystal structure, were probably responsible for the lower resolution of the replicas.

Digital image processing of replica tilting series and sections of embedded crystals will probably allow us to reconstruct a 3-dimensional model of the crystal.

The labelling of specific subunit proteins with protein A-gold-coupled antibodies on sections should confirm our 3-D model.

*Crystallized subunits were kindly provided by Prof.Wittmann, MPI für Molekulare Genetik, Berlin

Yonath,A.E., Müssig,J., Tesche,B., Lorenz,S., Erdmann,V.A. and Wittmann,H.G., 1980,Biochem. International I: 428-435

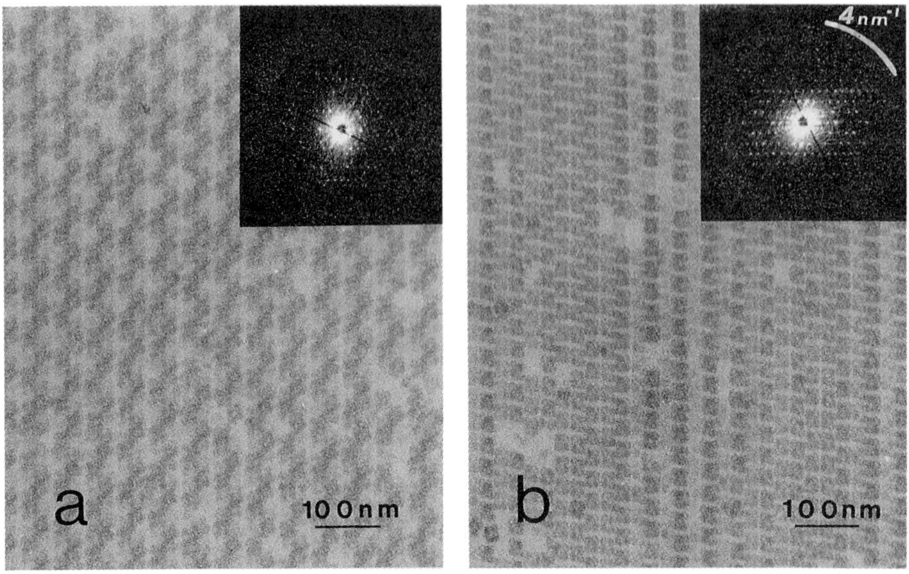

Fig.1: Positively stained sections and optical diffraction patterns of embedded crystallized ribosomal 50S subunits

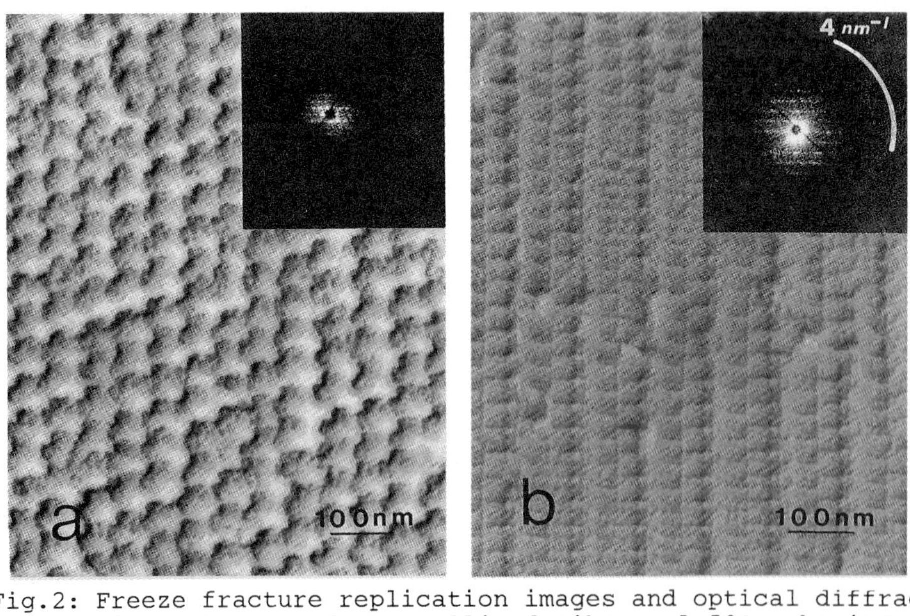

Fig.2: Freeze fracture replication images and optical diffraction patterns of crystallized ribosomal 50S subunits

Three dimensional electron crystallographic analysis of the T4 DNA helix-destablizing protein gp32*I

R. A. Grant[1], M. F. Schmid, D. A. Rankert, and W. Chiu Department of Biochemisty and Department of Molecular and Cellular Biology, University of Arizona, Tucson, AZ 85721

1. Current address: Department of Molecular Biology, Research Institute of Scripps Clinic, La Jolla, CA 92037

Gp32*I is a major proteolytic fragment of the T4 DNA-helix destabilizing protein. It retains the intact molecule's ability to bind cooperatively to single-stranded DNA. In low salt, gp32*I forms thin multilayered crystals that diffract to high resolution (Chiu and Hosoda, 1978). A previous electron microscope study of these crystals determined the crystals' space group to be $P2_12_12$ and produced a 3-D reconstruction of negatively stained crystals based on that determination (Cohen et. al., 1983). We report here the 3-D reconstruction of crystalline areas with $P2_1$ symmetry. The $P2_1$ cell represents 1/2 of a $P2_12_12_1$ cell that grows in 1/2 unit cell steps in the thin dimension of the gp32*I crystal.

The hypothesis that the crystals' space group was $P2_12_12_1$ was the result of the observation that the projection symmetry of an untilted, negatively stained crystal could change across a step in the crystal, as demonstrated in Figure 1. The Figure 1a shows the 2-D reconstruction obtained from the thinner area of such a crystal, Figure 1b shows the reconstruction of the thicker area. The thinner area has pg projection symmetry along the long (63 Å) axis in projection, quantitated by a phase residual of 6.3° for the phase relationships between appropriate reflections. The corresponding residual for p2 symmetry, which should be present if the area processed was either a full or partial $P2_12_12$ cell, is 24°. The thicker area has pgg symmetry, quantitated by a p2 phase residual of 8.3° and pg phase residuals along the long and short axes of 6.2 degrees and 6.3° respectively when weak reflections forbidden by pgg are omitted from the data. (The observation of the forbidden reflections in data with otherwise excellent symmetry is probably due to uneven staining.) The pg reconstruction represents the projection of a 1/2 unit cell step of a $P2_12_12_1$ unit cell. The complete unit cell, with pgg projection symmetry, can be generated from the 1/2 cell layer by applying the screw axis along the c axis of the crystal. Figure 1a was assumed to be the projection of 1/2 of a $P2_12_12_1$ cell, and the projection of the full cell was simulated by rotating Figure 1a around the appropriate axis and superimposing the rotated and unrotated projections. The result of the simulation, shown in Figure 1c, is very similar to experimental data with pgg symmetry (Figure 1b).

Crystalline areas with pg projection symmetry and the characteristic appearance of Figure 1a were reconstructed in 3-D. Four tilt series from 3 negatively stained crystals were used. The final 3-D data set contained contributions from 111 images, and was merged in $P2_1$ with an average phase residual of 11.5°. The resulting reconstruction is shown in Figure 2.

Chiu, W., and Hosoda J. (1978) J. Mol. Biol. 122: 103-107.
Cohen, H. A., Chiu, W., and Hosoda. J. (1983) J. Mol. Biol. 169: 235-248

(a) (b) (c)

FIGURE 1

The symmetry of a gp32*I crystal can change across a step. The thinner area (a) has pg symmetry. The thicker area (b) has pgg symmetry. If the pg projection is assumed to represent a 1/2 unit cell step in a $P2_12_12_1$ cell, the projection of a full cell can be simulated by applying the 2_1 screw axis along the screw axis parrallel to the projection axis. The result of such a simulation is shown in (c). The average phase residual between (b) and (c) is 6.9 degrees.

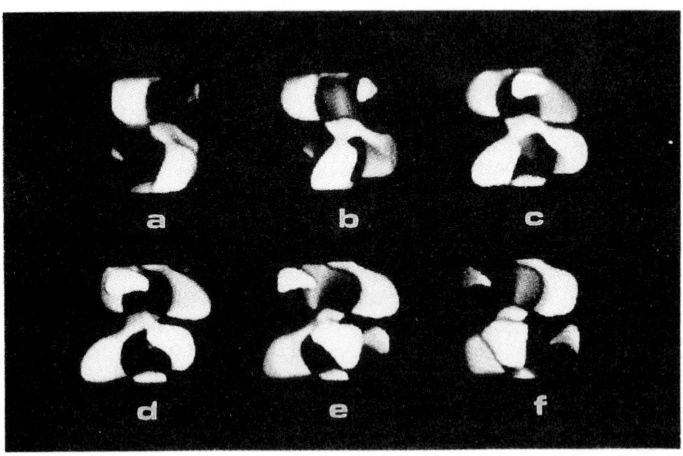

FIGURE 2

Computer surface representation of the 3-D reconstruction of the $P2_1$ cell. This cell represents 1/2 of the full $P2_12_12_1$ cell, and contains 2 gp32*I molecules related by the 21 screw axis. Frames a-f are views of the unit cell space 30 degrees apart as the cell is rotated about the 2_1 screw axis. The molecule is V-shaped, with two large "domains" connected by a smaller "domain" at the base of the V. This relationship between "domains" is most obvious in frame d.

Inst. Phys. Conf. Ser. No. 93: Volume 3, Chapter 14
Paper presented at EUREM 88, York, England, 1988

A common three-dimensional structure for the connector of unrelated viruses

JL Carrascosa, LE Donate, L Herranz, JP Secilla[1] and H Fujisawa[2].

Centro de Biología Molecular (CSIC-UAM), Universidad Autónoma, 28049 Madrid. [1]IBM Madrid Scientific Centre.- Pº de la Castellana 4, 28046 Madrid. [2]Department of Botany, Faculty of Science, Kyoto University, Kyoto 606.

ABSTRACT: The three-dimensional structure of bacteriophage T3 connector has been obtained up to 2.3 nm resolution by electron microscopy and image reconstruction. The data obtained have been compared to those obtained for bacteriophage ϕ29 connector. The similarities between both assemblies reveal the possible presence of common domains to carry out the functions of these viral components.

1. INTRODUCTION

The morphogenesis of viral particles has been studied in detail using complex DNA bacteriophages as adequate models. There is a structural component, called the connector, that plays an important role during viral head assembly and DNA encapsidation,(reviewed by Bazinet and King,1985; Carrascosa, 1986). Until now, three-dimensional data for the connector was only available for bacteriophage ϕ29 (Carazo et al.,1985, 1986a) a virus infecting Bacillus subtilis, with a prolate head and a complex tail comprising 12 appendages. To search for common domains among the different connectors, we have studied the 3D-structure of the connector of bacteriophage T3. This virus infects Escherichia coli and its head is icosahedral with a tail structure quite different from that of bacteriophage ϕ29.

2. RESULTS

The connector of bacteriophage T3 is build up by protein gp8 dodecamers. In purified concentrated solutions, ordered two-dimensional aggregates of gp8 dodecamers can be obtained (Nakasu et al.,1985; Carazo et al.,1986b). The crystals were found to be tetragonal with a=b=15.9±0.2nm. Diffraction patterns from such crystals revealed spots extending up to (1/2.3)nm^{-1}. To calculate the three-dimensional reconstruction of the gp8 connector tilted views were taken in the electron microscope between ±60 degrees at an electron dosage around 2-3x10^3 e nm^{-2}. The calculated density map (Figure 1) revealed that there were two connectors building up the unit cell that were oriented in opposite directions. The basic features of each reconstructed connector were identical: the height was found to be 10.9nm with a wide domain (14.4nm in diameter) showing twelve morphological units, and a narrower domain (9.7 nm in diameter). A channel with a variable diameter was located along the longitudinal axis of the particle.

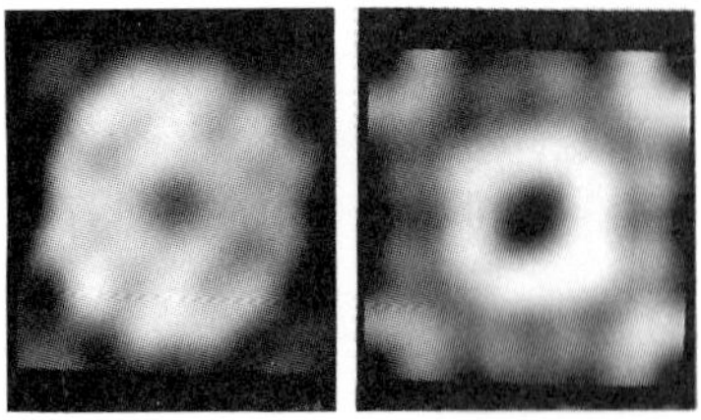

Figure 1. Sections trough the wider domain (1.4nm) and the narrower one (9.7nm) of the reconstructed T3 connector.

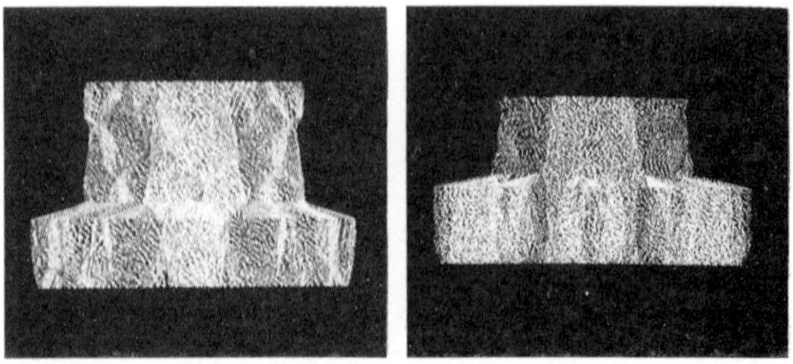

Figure 2. Traslucent view of the connectors of phage T3 (left) and Ø29 (right).

3. DISCUSSION

The comparison of the three-dimensional reconstruction of phage T3 connector with phage ϕ29 connector shows a series of common characteristics (Figure 2). Both structures present two domains: One has twelve units in the periphery (where it attaches to the 5-folded head vertex), that are paired towards the center. The other domain is narrower and present 6-fold symmetry. The main difference between both connectors is related to the morphology of the central channel: In the ϕ29 connector it is fully open, while the T3 connector presents a strangling region below the connection between the two structural domains.

The structure of the channel can be correlated with the connector function: In the case of phage ϕ29, the connector was in an active state (i.e., it was ready for DNA traslocation), and it presents the channel fully open along the particle. For the phage T3, the connector was extracted from particles with the DNA securely packaged inside the head. In this case, the channel presents a different conformation with a region that could prevent the DNA movement along the channel. In fact, when the connector of phage ϕ29 is extracted from viral particles, the channel is also closed (Carazo et al., 1985), thus suggesting that the state of this channel plays an important role for DNA translocation.

4. REFERENCES

Bazinet, Ch. and King, J. 1985. Ann. Rev. Microbiol. 39, 109-129.

Carazo, J.M., Santisteban, A. and Carrascosa, J.L. 1985. J. Mol. Biol. 183, 79-88.

Carazo, J.M., Donate, L.E., Herranz, L., Secilla, J.P. and Carrascosa, J.L. 1986a. J. Mol. Biol. 192, 853-867.

Carazo, J.M., Fujisawa, H., Nakasu, S. and Carrascosa, J.L. 1986b. J. Ultrastruc. Res. 94, 105-113.

Carrascosa, J.L. 1986. Electron Microscopy of Proteins Vol. 5 (Harris and Home, eds.). pp. 37-70.

Nakasu, S., Fujisawa, H. and Minagawa, T. 1985. Virology 143, 422-434.

Oblique S-layer lattices from *Bacillaceae*

D Pum, M Sára, P Messner, UB Sleytr

Zentrum für Ultrastrukturforschung, Universität für Bodenkultur,
A-1180 Wien, Austria

ABSTRACT: The oblique surface layers (S-layers) of *Bacillus stearothermophilus* strain NRS 2004/3a and *Bacillus coagulans*, strain Hammer, E38-66 have been characterized by electron microscopy supplemented by optical and computer image analysis.

Many eubacteria and archaebacteria possess as the outermost cell envelope component a crystalline array of macromolecules (Sleytr and Messner, 1983). These surface layers (S-layers) cover the entire cell and may have hexagonal, square or oblique symmetry. The morphological units are composed of hexamers, tetramers or dimers respectively. So far studies on the structure, chemistry and assembly of S-layers have predominantly been performed on p6 and p4 lattices. Here we describe the structure and self assembly process of oblique (p2) S-layer lattices of two *Bacilli*.

The S-layer of *Bacillus stearothermophilus* strain NRS 2004/3a (Messner *et al*, 1986) and *Bacillus coagulans*, strain Hammer, E38-66 have been characterized by electron microscopy supplemented by optical and computer image analysis. The S-layers composed of glycoprotein subunits have oblique (p2) symmetry. The size of the oblique unit cell of *B. stearothermophilus* is 11.6 nm by 9.4 nm with a base angle of 78° and of *B. coagulans* 9.4 nm by 7.4 nm with 80°. Both flat sheets and cylindrical mono- and double layered self assembly products were found with *B. stearothermophilus*. All double layers revealed a back-to-back orientation. It was further possible to divide the observed Moirê patterns in five superposition types with respect to their angular and lateral displacement. In vitro self assembly of isolated subunits of S-layers of *B. coagulans* only lead to the formation of double layer assembly products. These were oriented in a back-to-back orientation and could be divided into two classes. For both strains three dimensional models of the mass distribution were reconstructed from tilt series of negatively stained self assembly products using Fourier domain techniques. Specific labelling experiments with antibodies and electron dense markers allowed to localize the carbohydrate moities of the S-layer glycoprotein subunits.

Sleytr UB and Messner P 1983 *Annu. Rev. Microbiol.* **37** pp 311-339
Messner P, Pum D and Sleytr UB 1986 *J. Ultrastruct. Mol.Struct. Res.* **97** pp 73-88

High resolution electron microscopy of influenza virus internal structures

L.J. Calder and R.W.H. Ruigrok

Division of Virology, National Institute for Medical Research, Mill Hill, London NW7 1AA, U.K.

Influenza A virus was negatively stained with sodium silicotungstate, and penetration of the stain into particles shows a variety of internal structures. One of these, which is thought to be the viral M protein (Wrigley et al., 1986) is just beneath the membrane and looks like a finger-print (Figure 1). Another structure looks like a coil (Figure 2) and was long thought to be the viral ribonucleoprotein, e.g. Murti et al., 1980. However, we have used low dose imaging combined with accurate defocussing methods (Wrigley et al., 1983) to show that the two structures are composed of very similar units. The finger-prints (Figure 1) consist of an array of parallel strands with a spacing of 4 nm. Each strand is composed of units also 4 nm apart. The coils (Figure 2) have similar strands but in pairs. The strands within a pair are 4 nm apart and are composed of units 4 nm apart. The spacing of the strand pairs is 10.5 nm. Figure 3 shows viral particles with half finger-print, half coil structures demonstrating that coils and finger-prints are manifestations of the same internal structures.

We have tried to identify the units by immunogold labelling of the virus using monoclonal antibodies against M protein and nucleoprotein, the two major internal proteins of influenza virus. However, the viral particles could only be labelled after treatment with detergent which also destroyed the internal structures, so rendering the labelling experiments inconclusive. Finally, we isolated, purified and incorporated M protein into liposomes. The resulting arrays (Figure 4a) strongly resemble the internal structures of viral particles and could only be labelled with anti M monoclones (Figure 4b), not anti NP antibodies.

Murti K.G., Bean Jr., W.J. and Webster R.G. 1980. Virology 104 224-229.

Wrigley N.G., Brown E.B. and Chillingworth R.K. 1983. J. Microsc. 130 225-232. See also abstract N.G. Wrigley in these proceedings.

Wrigley N.G., Brown E.B. and Skehel J.J.1986. Electron microscopy of influenza vius. In: Electron microscopy of proteins (eds. J.R. Harris and R.W. Horne) Academic Press, London.

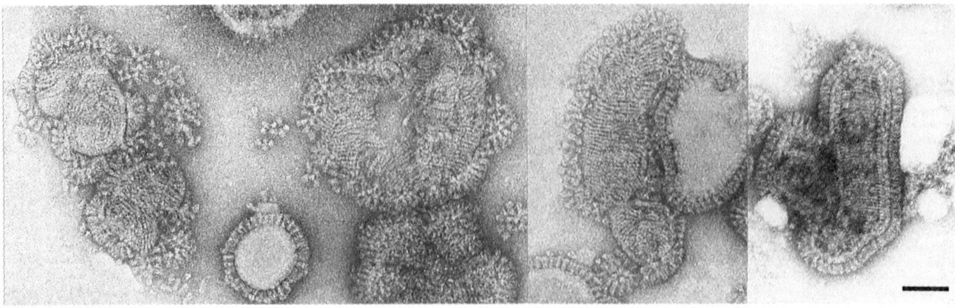

Figure 1.　　Bar represents 50 nm

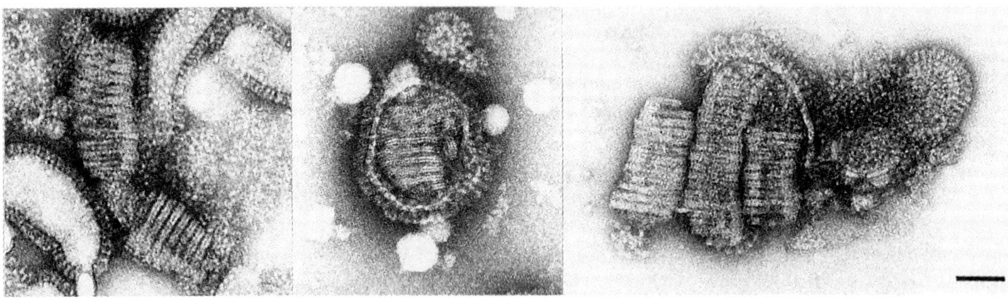

Figure 2.　　Bar represents 50 nm

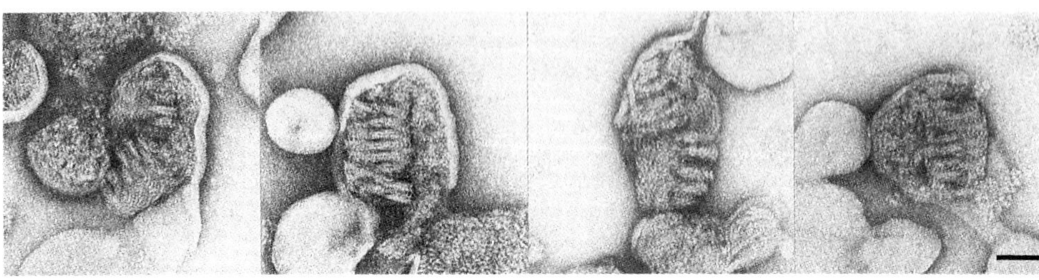

Figure 3.　　Bar represents 50 nm

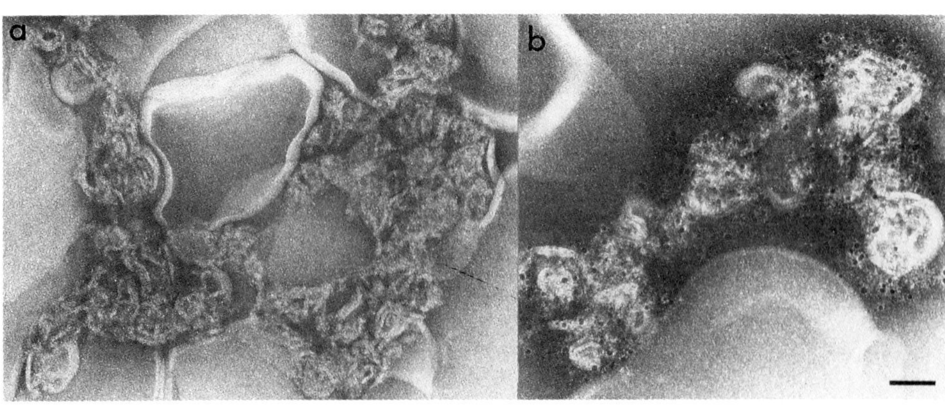

Figure 4.　　Bar represents 50 nm

The structure of oligomeric α-toxin from *Staphylococcus aureus* in human platelet membranes

A. Olofsson, U. Kavēus and H. Hebert

Department of Medical Biophysics, Karolinska Institutet, Box 60400
S-104 01 Stockholm, Sweden

ABSTRACT: The cytolytic protein α-toxin from *Staphylococcus aureus* is considered to be an important virulence factor. An annular oligomeric form of the toxin observed after interaction with human platelet membranes has been analyzed by electron microscopy and image processing introducing a new method for rotational alignment of individual particles. The oligomer is shown to consist of six protein units arranged around a central protein deficient domain.

1. INTRODUCTION

α-toxin from *Staphylococcus aureus* consists of a single polypeptide chain with known amino acid sequence having an M_r of 33 000 (Gray and Kehoe 1984). The toxin can also form oligomers especially after interaction with eukaryotic cell membranes or with liposomes. Molecular weight determinations and earlier electron microscopy observations has indicated that an oligomer may consist of six identical protein units (Freer et al. 1968; Füssle et al. 1981). The aim of the present work is to obtain structural information about α-toxin which is important for the understanding of the formation of ion-permeable pores and the molecular mechanism behind cell lysis.

2. MATERIALS AND METHODS

Staphylococcus aureus strain Wood 46 was used to obtain α-toxin which was purified by isoelectric focussing and gel filtration (Thelestam 1983). The protein (670 μg/ml) in trisbuffered Saline, pH 7.0, was incubated with human platelets for about 30 min at 22°C. The specimens were negatively stained with phosphotungstic acid, pH 7.2, and observed in a Philips EM301 at 75 000 X magnification. Selected electron micrographs were scanned and subjected to image processing using the "EM" program system (Hegerl and Altbauer 1982) running on a microVAX II. Single particle averaging was performed introducing a new method for rotational alignment. The rotational search was converted to a translational one by calculating the convolution between the modulus square of the Fourier transforms written on polar form. The odd men out algorithm of Unser et al. (1986) was used to select the most significant molecules of the set.

3. RESULTS AND DISCUSSION

Electron micrographs of membrane fragments showed ring shaped particles with a stainfilled core. Selected particles were subjected to the alignment procedure and a preliminary average was calculated. After a refinement step including the 18 most significant particles the structure in Fig. 2 was obtained. Six protein peaks were clearly visible demonstrating the arrangement of the α-toxin molecules in the oligomer. The resolution was estimated at 20 Å by comparing two independent averages. For a pH of 7.2 of the phosphotungstic acid the diameter of the ring is 75 Å and that of the stainfilled core is about 25 Å. These figures became larger when the pH was lowered. This may be related to an increased transmembrane flux of molecules. An increased sensitivity of human erythrocytes towards hemolysis by α-toxin at low pH was in fact observed by Bhakdi et al. (1984).

4. REFERENCES

Bhakdi S, Muhly M and Füssle R 1984 Infect. Immun. 46 318.
Freer J H, Arbuthnott J P and Bernheimer A W 1968 J. Bacteriol. 95 1153
Füssle R, Bhakdi S, Sziegoleit A, Tranum-Jensen J, Kraug T and Wellensiek H 1981 J. Cell Biol. 91 83
Gray G S and Kehoe M 1984 Infect. Immun. 46 615
Hegerl R and Altbauer A 1982 Ultramicroscopy 9 109
Thelestam M 1983 Biochim. Biophys. Acta 762 481
Unser M, Steven A C and Trus B L 1986 Ultramicroscopy 19 337

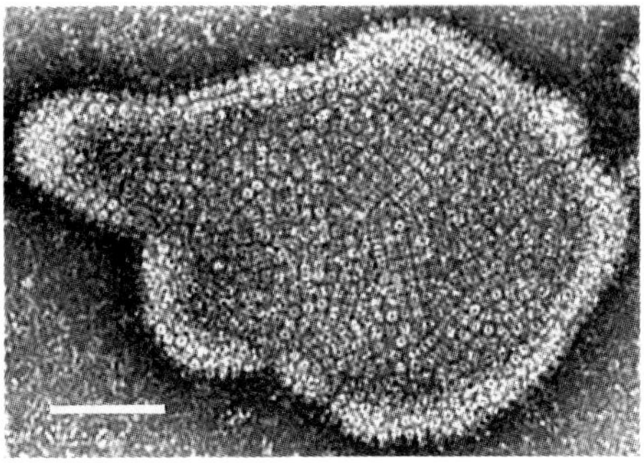

Fig. 1 Electron micrograph showing a membrane fragment from a human platelet containing ring shaped oligomers of α-toxin. Scale bar 50 nm.

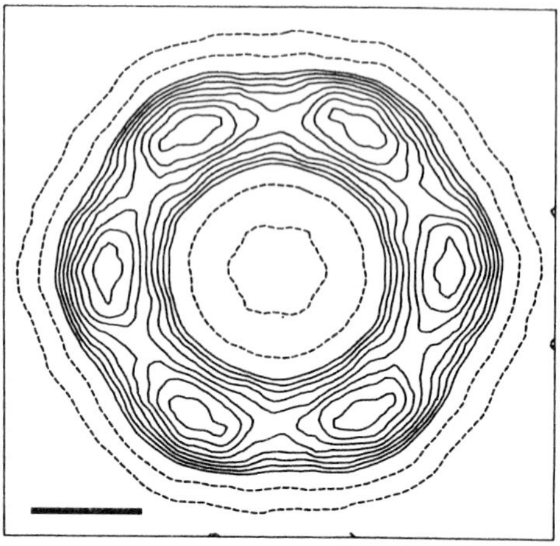

Fig. 2 Refined average. Contour levels have been drawn for the protein rich region. The increment between the dashed lines is 5 times that of the solid lines. Scale bar 20 Å.

Electron microscopic investigation of DNA structure

P J Highton, R J Myers and Y Chang

Department of Molecular Biology, University of Edinburgh, King's Buildings
Mayfield Road, Edinburgh EH9 3JR

ABSTRACT: By making heteroduplexes between the DNA molecules of lambdoid bacteriophages we have shown that their genomes are composed of 33 functional segments, which have been exchanged between phages during their evolution. The number of alleles of the segments may be correlated with the likely evolutionary pressures. By partial denaturation we have divided the phage lambda genome into 105 homostable regions. The correspondence between homostable regions, segments and genes suggests that variations in stability have biological significance.

1. INTRODUCTION

The structure of DNA is commonly considered to have been determined by Watson and Crick (1953) by X-ray diffraction. What more then can electron microscopy tell us? Well what Watson and Crick did in fact was to build a molecular model, the double-helix, the parameters of which were consistent with those that could be deduced from the diffraction patterns from DNA fibres. This did not tell us the positions of any of the atoms in the structure, or anything about the nucleotide sequence of DNA, but it did offer a ready explanation for the way in which information could be encoded, and passed on from generation to generation, by the separation and replication of two complementary polynucleotide chains. This concept alone was sufficient to give rise to the science of molecular genetics, and subsequently to the techniques of gene manipulation.

One of the first successes of electron microscopy in this field was the demonstration that DNA molecules were in fact long strands, with about the width predicted by the Watson and Crick model (Beer 1961). Later on, with the general adoption of the protein monolayer technique developed by Kleinschmidt (1959) it became easily possible to measure the lengths of molecules, and to determine whether they were linear, circular or super-coiled circular molecules. These were also the first direct images of single molecules of any sort. However, fascinating as these pictures were, and still are, they only show us the gross anatomical features of the molecules, which simply reflect the structure and life cycle of the organisms from which they come, or to put it another way, the particular packages which have evolved to protect and propagate the DNA molecules.

2. PROBING THE NUCLEOTIDE SEQUENCE

The raison d'être of a DNA molecule is that it contains information, which permits its own propagation, and that of the package/cell/organism which surrounds it. To get at this information we need to probe 'inside' the molecule. To do this Beer and Moudrianakis (1962) proposed that we should first split it down the middle along its length, by breaking the hydrogen bonds which hold together each pair of nucleotides, to produce two covalently bonded single polynucleotide chains. The information, ie the sequence of the nucleotides, in these chains could then be determined by specifically labelling each nucleotide with a stain containing heavy atoms. Unfortunately this proposal has never become a practical reality. As far as I am aware the best results obtained are a demonstration of specific staining by Whiting and Ottensmeyer (1972), using an osmium based stain that Beer and I developed for thymidylic acid (Highton et al 1968). However, the idea has again been raised in recent discussions of the chemical sequencing of the human genome.

Those wishing to probe inside DNA molecules by electron microscopy have had to be satisfied with a lesser prize than the nucleotide sequence. However, the development of the techniques of partial denaturation mapping, and heteroduplex formation, have made it possible to ask questions about the overall distribution of the nucleotide pairs along a molecule, and about the sequence similarities between molecules. In recent years we have been using these techniques to study the structure and evolution of the genomes of the lambdoid bacteriophages which infect *Escherichia coli* and live within our intestines.

Partial denaturation mapping (Inman 1966) depends on the fact that although a DNA molecule can be split down the middle by breaking the hydrogen bonds between the complementary nucleotides, the strength of the bonding is found to vary along the length of the molecule. Thus by raising a solution of DNA molecules to high pH, it is possible to separate the two polynucleotide chains in a number of discrete steps.

Heteroduplex formation (Westmoreland et al 1969) depends on the fact that the completely separated polynucleotide chains of two different types of molecule will bond together if they contain regions of similar nucleotide sequence. Thus if a solution containing a mixture of two such types of molecule is raised to high pH and then returned to neutrality, chance collisions between chains will result in the formation of both homoduplex and heteroduplex molecules. In neutral aqueous solutions separated chains are partly folded up by the formation of short imperfect intrachain duplex regions resulting from chance internal sequence complementarities. This folding interferes with interchain duplex formation, but it may be prevented by the inclusion of a concentration of formamide sufficient to destabilise the intrachain duplexes, but not sufficient to prevent the formation of longer more perfect duplexes between chains.

3. THE SEGMENTED FORM OF THE LAMBDOID PHAGE GENOMES

In our study of the structure and evolution of the genomes of the lambdoid bacteriophages we began by making heteroduplexes between the DNA molecules of a group of six phages. (Heteroduplexes between some of these DNA molecules were previously measured by Simon et al (1971).) From these we concluded that the positions of the regions of nucleotide sequence homology between the six DNA molecules divide them into 33 segments, most of which have a number of alternative forms (alleles), which must be functionally homologous in general, but which differ in nucleotide sequence and length (Figure 1). Comparison of the positions of the boundaries between segments in phage lambda, with the positions of the genes (Sanger et al 1982), showed that each segment is probably a gene or a group of genes, and that each phage genome is a different combination of the alleles of the segments (Figure 2). These results suggest that the phages evolved from a common ancestor.

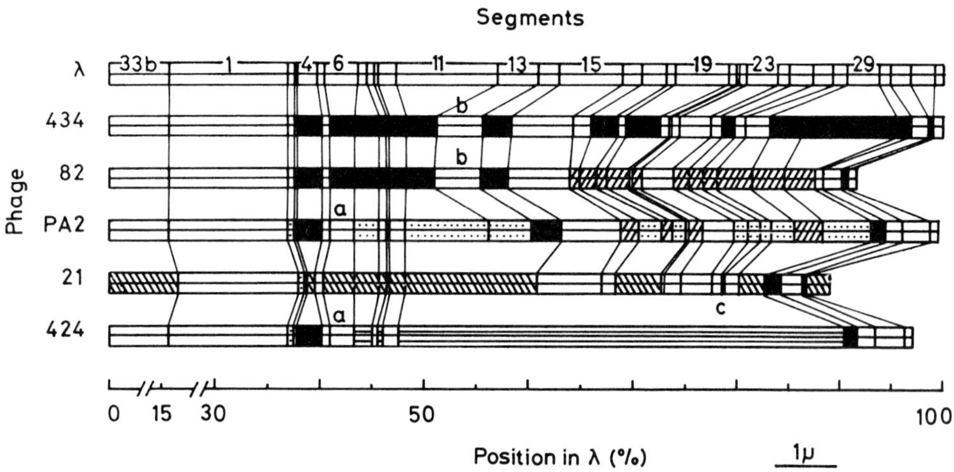

Fig. 1. The homologies between the six phage DNA molecules. Corresponding segments with the same marking form a double-helix in a heteroduplex. Segments marked a, b or c have small deletions or insertions.

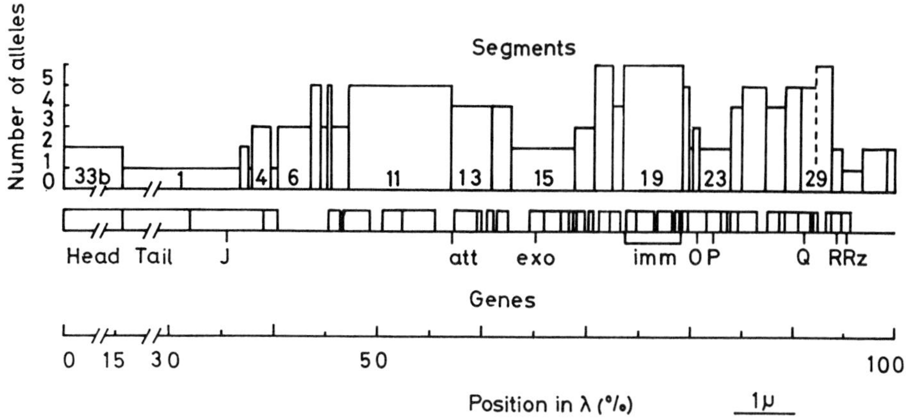

Fig. 2. The correspondence between segments (upper half) and genes (lower half) in the lambda genome, showing the number of alleles of each segment present in the six phages. The genetic map is based on that of Sanger et al (1982). A more detailed map is shown in Figure 4.

The phage DNA molecules can exist either within free phage particles, or as part of the bacterial host genome, and by considering this dual nature we have been able to draw a correlation between the number of alleles of a segment and the strength of the selective pressures likely to have produced diversity. We believe that diversity has evolved because it permits superinfection of hosts carrying an integrated phage genome.

In reaching these conclusions it was assumed that in a double-helical region of a heteroduplex the two nucleotide sequences have high homology *ie* the two genomes have the same allele, and that in an unbonded region the two sequences have low homology *ie* the two genomes have different alleles. This interpretation is supported by comparison of the sequences that have been determined for a few segments in phages 434 (Grosschedl and Schwarz 1979, R.Yocum personal communication, Campbell *et al* 1986), phage 82 (Moore *et al* 1981), phage 21 (Franklin 1984) and phage PA2 (Blasland *et al* 1986), with the sequence of the corresponding segments of phage lambda (Sanger *et al* 1982). However, we need to know that the assumption is likely to be generally true, and formamide provides a sensitive and quantitative probe of sequence homology. By raising or lowering the formamide concentration after forming the heteroduplex, but before spreading it, we can determine at what concentration a double-helical region denatures, and at what concentration an unbonded region forms a double-helix. Thus by varying the formamide concentration by 20%, we concluded that it is very likely that the sequences of all six phages have at least 9% more homology in a segment with the same allele, than in a segment with a different allele. Generally the difference in homology may be much greater, but this level is sufficient to indicate that exchanges of segments must have occurred during the evolution of these phages from a common ancestor.

To demonstrate this we require groups of four phages which exhibit all four possible combinations of two alleles of two separate segments. To find such groups we isolated fourteen new lambdoid phages, and studied six of them in detail. The groups found cannot have evolved from a common ancestor, without exchange of segments, unless extensive sets of nucleotide sequence changes occurred twice, which would be very unlikely.

4. THE STABLITY OF THE GENOME SEGMENTS

The determination of the nucleotide sequence of the phage lambda DNA molecule (Sanger *et al* 1982) allows us to consider another aspect of evolution, namely the GC content of the segments *ie* the percentage of GC nucleotide pairs in a segment. Electron microscopy indicates that GC content is conserved within alleles (Highton and Whitfield 1975), and the sequences of the *cI* and *cro* genes of phage 434 (Grosschedl and Schwarz 1979, R.Yocum personal communication) show only 3% difference in GC content from those of phage lambda, although the sequences have very low homology.

The strength of the bonding between the two polynucleotide chains of a DNA molecule increases with the GC content, and is an important physical property of the molecule. The fact that the two chains may be readily split apart is the basis of the process of information propagation, and its use for the production of proteins. Variations in GC content may represent a fine level of control in these processes and are likely to have been selected in evolution. It is of interest to ask whether electron microscopy does in fact show variations in bonding strength along the DNA molecule which are related to the

GC content, and whether these variations are correlated with the segments and genes.

By partially denaturing the lambda DNA molecule, in a series of solutions of increasing denaturation strength, we have been able to divide it into 105 homostable regions, grouped into 23 stability levels (Figure 3). The relative stabilities show a good correlation with the GC content. The stabilities can in fact be expressed in terms of GC content, by comparison with the differential melting profile obtained for phage lambda DNA molecules by Vizard and Ansevin (1976). To obtain this the temperature of a solution of molecules was raised slowly and the change in absorbance at 270 nm, which results from the separation of the two polynucleotide chains, was measured.

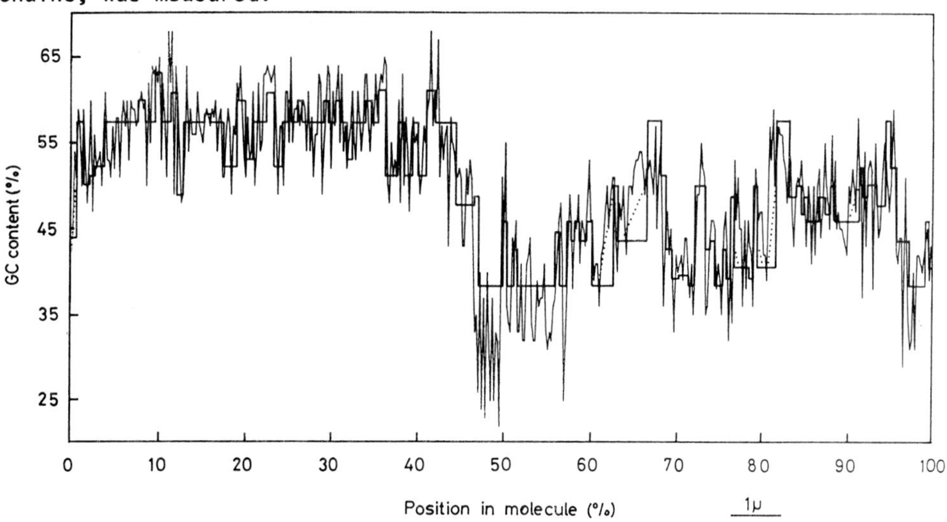

Fig. 3. The correspondence between stability (thick line) and GC content (thin line) in the DNA molecule. The GC content is a 100 nucleotide block average deduced from the sequence of Sanger et al (1982). Stability is expressed as GC content by using the differential melting profile obtained by Vizard and Ansevin (1976). The lowest level of stability corresponds to 38% GC or less. Dotted lines indicate that a gradation in stability was detected.

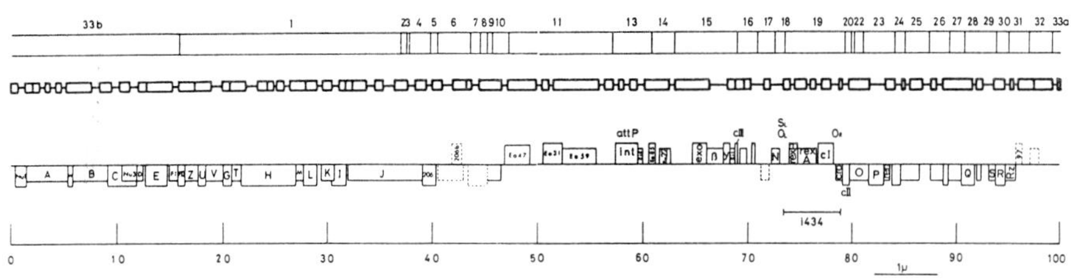

Fig.4. The correlation between the homostable regions (middle line) genetic segments (upperline) and genes (lower line). The genetic map is based on that of Sanger et al 1982. (To help to relate the stability map to Figure 3, regions flanked by two others of lower stability are drawn as single horizontal lines.)

Figure 4 shows the correlation between the homostable regions and the genetic segments and genes. The level of correspondence between the positions of the changes in stability and the ends of segments or genes, suggests that the variations in stability along the molecule may have a biological significance.

REFERENCES

Beer M 1961 J.Mol.Biol. 3 263
Beer M and Moudrianakis E N 1962 Proc.Nat.Acad.Sci. 48 409
Blasland A J, Marcotte W R and Schnaitman C A 1986 J.Biol.Chem. 261 12723
Campbell A, Ma D P, Benedik M and Limberger R 1986 Banbury Conference Reports (New York: Cold Spring Harbor Laboratories)
Franklin N C 1984 J.Mol.Biol. 181 75
Grosschedl R and Schwarz E 1979 Nucleic Acids Res. 6 867
Highton P J, Murr L B, Shafa F and Beer M 1968 Biochemistry 7 825
Highton P J and Whitfield M 1975 Virology 63 438
Inman R B 1966 J.Mol.Biol. 18 464
Kleinschmidt A K and Zahn R K 1969 Z.Naturforschung 14b 730
Moore D D, Denniston K J and Blattner F R 1981 Gene 14 91
Sanger F, Coulson A R, Hong G F, Hill D F and Petersen G B 1982 J.Mol.Biol. 162 729
Simon M N, Davis R W and Davidson N 1971 The Bacteriophage Lambda ed A D Hershey (New York: Cold Spring Harbor Laboratory) pp 313-328
Vizard D L and Ansevin A T 1976 Biochemistry 15 741
Watson J D and Crick F H C 1953 Nature 171 737
Westmoreland B, Szybalski W and Ris H 1969 Science 163 1343
Whiting R F and Ottensmeyer F P 1972 J.Mol.Biol. 67 173

The electron histochemistry of proteoglycans in connective tissues

J.E. Scott
Chemical Morphology, Cell and Structural Biology, Chemistry Building, Manchester University, Manchester M13 9PL.

ABSTRACT: The domains of high mol. wt. anionic polysaccharides, nucleic acids, proteoglycans etc. collapse on staining with cationic reagents, or on embedding with plastic. Reagents (Cupromeronic blue, etc.) were designed to control and limit this phenomenon. Used in critical electrolyte concentration (CEC) methods, in conjunction with specific enzymes, binding sites of proteoglycans on collagen fibrils are identifiable, and a map of such binding sites has been drawn up.

Proteoglycans (PGs) are large, water soluble, highly negatively charged biopolymers that are found in connective tissues of the most primitive as well as the most developed animals (Poole, 1986). Their physical properties are similar to those of many other high molecular weight polyanions,- e.g. alginates, pectates, carrageenans, nucleic acids, mucins,etc., and consequently they share many special problems involved in localising them by electron microscopy. They are characteristically highly swollen in aqueous environments, occupying domains which are largely water, even at the highest concentrations found in most tissues. Their numerous negative charges are balanced by e.g. Na^+, K^+, Ca^{++} & Mg^{++}, and in this form they are totally insoluble in non-aqueous solvents, such as those used to dehydrate tissues prior to plastic embedding. They are then precipitated more or less randomly, during which process the domain they occupy collapses, producing an artefact which has unclear relevance to the situation which existed *in vivo*. They lose their molecular morphology and are simultaneously translocated. They neither look like their functional forms, nor are they necessarily placed at their functional locations. In these respects they differ fundamentally from e.g. collagen fibrils which are coherent, permanent structures, visible as such without staining.

There are thus three practical problems associated with the electron histochemistry of tissue PGs (Scott, 1985).

1) It is necessary to render them visible (i.e.stain them), preferably specifically.

2) Their molecular shape should be retained, as much as possible.

3) They must be fixed vis-a-vis other tissue elements. Of the three problems, this is currently least important. Standard fixatives, e.g. glutaraldehyde, give acceptable results.

1) Staining of polyanions It is a completely general property of all polyanions that they bind, and are precipitated by, cationic reagents such as Alcian blue, cetyl pyridinium, Ruthenium red, etc. All polyanions thus can form stoicheiometric complexes containing colour (and/or electron density, in the form of chelated heavy metals). They are then easily visible by light or electron microscopy. Unfortunately, there are two drawbacks, viz., it is a completely unspecific phenomenon (all polyanions participate), and the polyanion domain collapses on precipitation in much the same way as in non-aqueous solvents, i.e.the molecular morphology becomes grossly distorted. However, by suitable choice and design of the cationic reagent, the latter defect may be at least partly remedied, and by delivering the reagent under the right conditions, the specificity for given polyanions can be much improved. Although there is still far to go before the polyanion can be visualised in its *in vivo*, fully expanded three dimensional state, recently developed procedures (Scott, 1985) have produced interesting and probably biologically important findings. The intrinsic specificity of the reagents is uncovered and exploited by the critical electrolyte concentration (CEC) approach. This involves competing with the dye for anionic sites using e.g. Mg^{++}

ions. Quaternary ammonium dyes such as Cupromeronic blue are displaced with greater difficulty from polyanions with sulphate ester groups (e.g. chondroitin sulphate, heparin) than from polycarboxylates (e.g. hyaluronan, sialo-proteins). This pattern is decided by the Mg^{++} ion, the dye is a more or less inert indicator of negative charge. The effect is sharp, all-or-none, at a well defined electrolyte concentration, hence the term 'critical' electrolyte concentration (e.g. Scott, 1974). Using Cupromeronic or Alcian blue, sulphated polyanions are the last to retain their stain as the Mg^{++} concentration increases, i.e. they are selected from among the general class of polyanions (Scott, 1974). With the help of specific enzymes such as keratanase and chondroitinase AC, used in conjunction with CEC methods, most glycosaminoglycans in connective tissues can be identified and localised. These aspects have been reviewed (Scott, 1985).

2. Application of staining methods to connective tissues.

It was often speculated that PGs were specifically associated with collagen fibrils, but techniques were inadequate to establish the point, and knowledge of the chemistry of PGs was insufficient to illuminate even the barest outlines. Advances in both fields coincided, and the first identification of a specific binding site on the type 1 collagen fibril for a characterised PG (the 'small' dermatan sulphate proteoglycan) was made in 1981 (Scott and Orford, 1981),

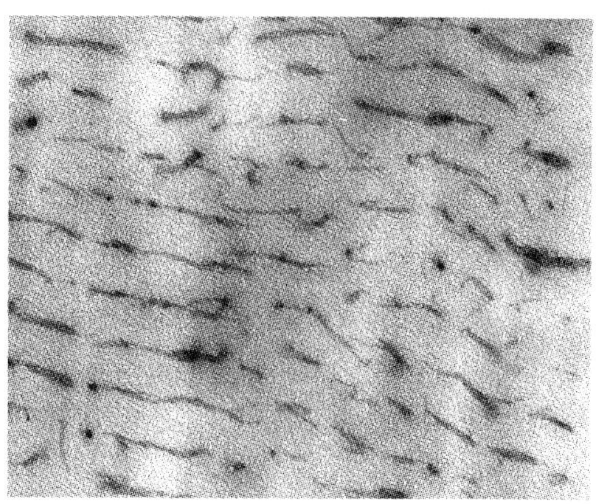

Fig. 1

Orthogonally arrayed (right→ left) dermatan sulphate PG filaments against lighter collagen fibrils (running top→ bottom), in calf tendon. Stained with Cupromeronic blue in 0.3M $MgCl_2$ (x 120,000).

(Fig. 1). This binding site (the d and e bands) in the 'gap zone' of the fibrils was found to be similarly occupied in several other tissues containing mainly type 1 collagen (tendon, sclera, skin, cornea),but not, apparently, in bone. It was suggested that the function of this PG:collagen association was to inhibit mineralisation of soft connective tissues, such as skin and tendon (Scott & Haigh 1985a).

As well as the d & e bands, the a & c bands can also associate with a PG which contains keratan sulphate instead of dermatan sulphate, but so far this has been seen only in the corneal stroma (Scott & Haigh 1985b)). Mice corneas do not contain keratan sulphate, and the a & c bands therein are not occupied. Based on CEC results and enzyme digestions, the four binding sites on corneal collagen fibrils have been assigned to four distinct and characterised PGs, giving rise to the first 'map' of PG binding sites on collagen fibrils (Scott & Haigh, 1985b (Fig 2). This implies

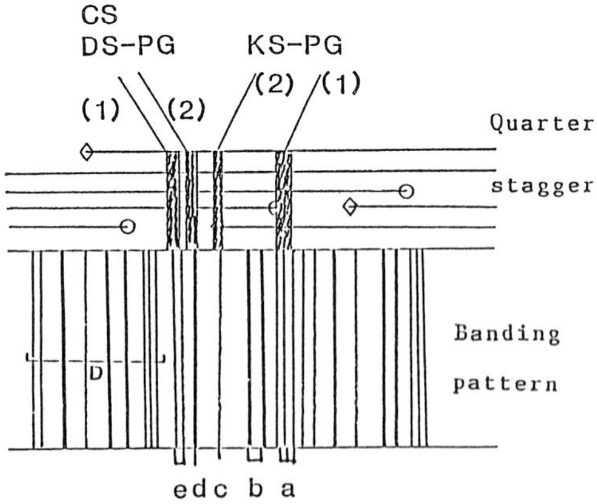

Fig.2. Map of binding sites of PGs along the collagen fibril, displayed against the quarter-stagger arrangement of collagen molecules, to correlate with the a-e banding pattern. PGs were stained with Cupromeronic blue, a-e bands with UO_2^{++}. CS = chondroitin sulphate, DS = dermatan sulphate, KS = keratan sulphate. See Scott, 1988 for a review.

two important corollaries, viz a), each PG has its specific binding site, and in the absence of that PG, the site will be unoccupied (Scott 1986).

This map has been found to hold good in several mammalian species, and something similar is observed in a very simple animal, sea cucumber (holothurian), even though the structures of the glycosaminoglycuronans (and probably the collagens) differ considerably from those of more advanced creatures (Scott 1988). PG:collagen interactions, as uncovered by electron histochemistry thus appear to follow a persistent biological pattern, throughout much of evolution.

References

Poole R A 1986 Biochem. J. 236 1
Scott J E 1985 Collagen Res. Rel. 5 541
Scott J E Trans. Biochem. Soc. 1 787
Scott J E and Orford C R 1981 Biochem. J. 197 213
Scott J E and Haigh M 1985a Biosc. Repts. 5 71
Scott J E and Haigh M 1985b 5 765
Scott J E 1986 Ciba Foundation Symposium No. 124 Functions of Proteoglycans (Chichester, J.Wiley) pp104-124
Scott J E 1988 Biochem. J. in press

The molecular organization of amylopectin in starch

G T Oostergetel and E F J van Bruggen

Biochemisch Laboratorium, Rijksuniversiteit Groningen, The Netherlands

Starch granules are the major form of packaging storage carbohydrates in green plants. They are semi-crystalline and composed of the linear poly-α-(1→4)-glucan amylose and the branched poly-α-(1→4),α-(1→6)-glucan amylopectin. The molecular organization of amylopectin in starch granules from different botanical sources was studied by (cryo) transmission electron microscopy and cryo electron diffraction. Granule fragments were obtained by applying a limited acid hydrolysis followed by wet mashing (Yamaguchi et al 1979). In electron micrographs of negatively stained or frozen hydrated starch fragments crystalline lamellae are visible that comprise the linear segments of the amylopectin molecules (Figure 1). Using optical diffraction (Figure 2A) a lamellar spacing ranging from 9,3nm in potato starch to 10.4nm in barley starch could be deduced from the electron micrographs. These values correlate well with the position of a Bragg peak at ~0.1nm^{-1} in small angle X-ray diffractograms of starch suspensions (Figure 2B). The crystalline lamellae which have a thickness ranging from 2.5 to 3.5nm are oriented in two main directions with a relative angle of ~45°. Electron diffractograms of frozen hydrated starch fragments (Figure 3) were recorded according to the procedure described by Chanzy et al (1977). These diffractograms, showing regular order to 0.2nm and indicating a strongly preferred orientation of the helices containing the linked glucose residues, are very similar to X-ray fibre diffractograms of B-amylose fibres (Wu and Sarko 1978). This suggests that the linear segments of the amylopectin molecules in the lamellae are all oriented parallel to the fibre axis and therefore not normal to the plane of the crystalline lamellae. Observation of stereo micrographs revealed that bends in the lamellae cause the appearence of different lamellar orientations. This type of organization is remarkably similar to that found in certain synthetic polymers like polyethylene (Bassett 1984). A schematic model for the organization of amylopectin in starch that accounts for our observations is shown in Figure 4.

REFERENCES

Basset D C 1984 Crit. Rev. Solid St. Mat. Sci. **12** 97
Chanzy H, Guizard C, Vuong R 1977 J. Microscopy **111** 143
Wu H -C H, Sarko A 1978 Carbohydr. Res. **61** 7
Yamaguchi M, Kainuma K, French D 1979 J. Ultrastr. Res. **69** 249

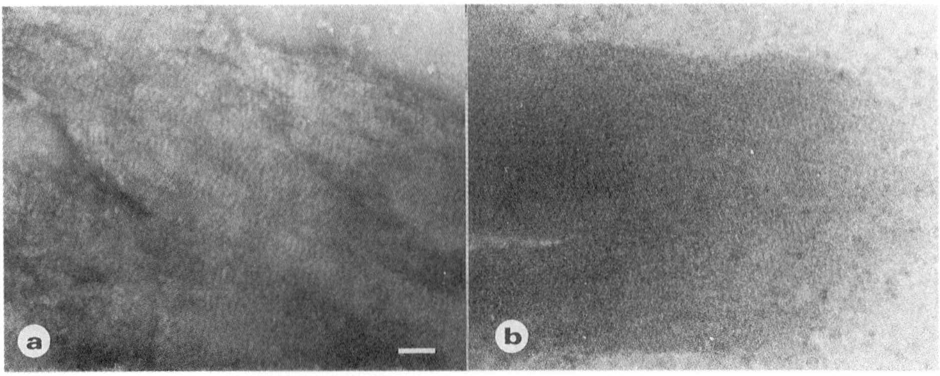

Figure 1. Potato starch granule fragment negatively stained with uranyl acetate (A) and frozen hydrated (B); Bar 50nm.

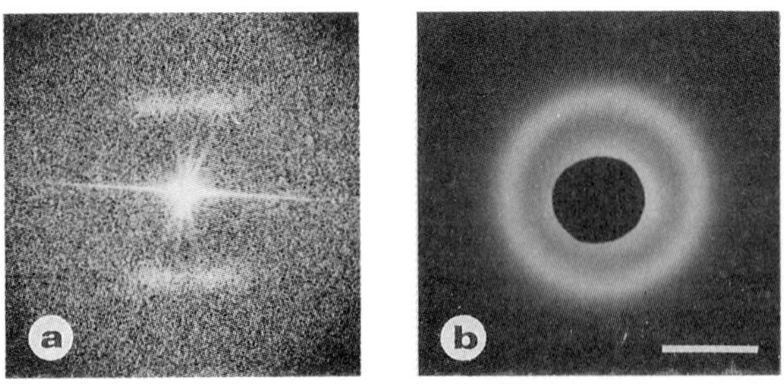

Figure 2. Optical diffractogram of an electron micrograph as in Figure 1A (A) and small-angle X-ray diffractogram of potato starch (B). Bar $0.1nm^{-1}$.

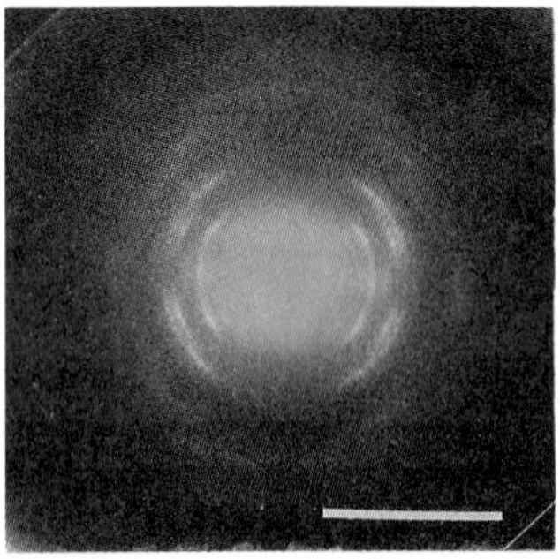

Figure 3. Electron diffraction pattern of a frozen hydrated potato starch fragment. Bar $4nm^{-1}$.

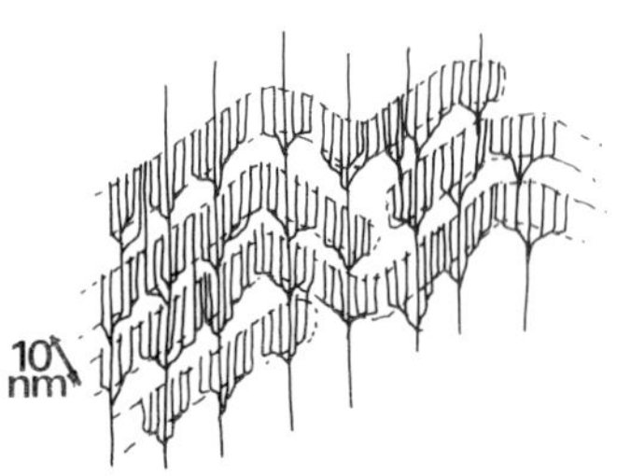

Figure 4. Model for the lamellar organization of amylopectin in starch.

The fine structure differentiation of the plasma membrane ATPase from *Sulfolobus acidocaldarius* with electron spectroscopic imaging

H. Lünsdorf, M. Lübben# and G. Schäfer#

GBF-Gesellschaft für Biotechnologische Forschung, Mascheroder Weg 1, D-3300 Braunschweig, #Institut für Biochemie, Medizinische Universität Lübeck, Ratzeburger Allee 160, D-2400 Lübeck 1

The plasmamembrane-associated ATPase from the thermoacidophilic archaebacterium Sulfolobus acidocaldarius has been isolated and purified to homogeneity (1,2). The enzyme has been investigated with the Zeiss EM 902 using negatively stained samples (3% uranylacetate, pH 4.5). Because of the instrumental possibilities to image single molecules (a) in the elastic bright field ($\Delta e = 0$ eV) and (b) with element-specific inelastically scattered electrons of uranium atoms (Electron Spectroscopic Imaging mode = ESI; $\Delta e_U = 114$ eV) new possibilities arise to describe the quaternary structure of the oligomeric molecule. Thus differentiation of structural details is possible because of the two imaging modi. Opposite to the situation of taking pictures in the bright field at slight underfocus in order to get a higher structural contrast ESI images are taken at the Gaussian focus. This will make it possible to balance phase contrast artifacts from the bright field with structural details of high resolution ESI-images. Thus the arrangement of single polypeptides within the enzyme molecule is much more impressive. Further, topographical relations of the polypeptides can be distinguished, i.e. the relative spatial arrangements of subunits within the enzyme molecule is distinctly presented. The subunit arrangements of S. acidocaldarius ATPase are shown in Fig.1. The overall projection contours of the molecules are of hexagonal to trigonal appearance (Fig. 1,c-g). With ESI the six peripheral protein masses as well as the smaller central ones can be contoured more clearly compared to those in the elastic bright field (Fig.1,c-i) thus revealing the quaternary structure in a more plastic way. Side views can be recognized as two layers of protein masses (Fig.1,h,i) especially with the ESI-mode. Thus the enzyme reveals the same architecture as it is known from many procaryotic and eucaryotic F_1-type ATPases (3-5) and presumably the same stoichiometry of α- and ß-subunits (6,7) is given.

(1) M. Lübben and G. Schäfer (1987) Eur. J. Biochem. **164**, 533-540
(2) M. Lübben, H. Lünsdorf and G. Schäfer (1987) Eur. J. Biochem. **167**, 211-219
(3) H. Tiedge, G. Schäfer and F. Mayer (1983) Eur. J. Biochem. **132**, 37-45
(4) F. Mayer, D.M. Ivey and L.G. Ljundahl (1986) J. Bacteriol. **166**, 1128-1130
(5) T. Wakabayashi, M. Kubota, M. Yoshida and Y. Kagawa (1977) J. Mol. Biol. **117**, 515-519
(6) H. Lünsdorf, K. Ehrig, P. Friedl and H.U. Schairer (1984) J. Mol. Biol. **173**, 131-136
(7) H. Tiedge, H. Lünsdorf, G. Schäfer and H.U. Schairer (1985) Proc. Nat. Acad. Sci. USA **82**, 7874-7878

H. Lünsdorf, M. Lübben# and G. Schäfer#

GBF-Gesellschaft für Biotechnologische Forschung, Mascheroder Weg 1, D-3300 Braunschweig, #Institut für Biochemie, Medizinische Universität Lübeck, Ratzeburger Allee 160, D-2400 Lübeck 1

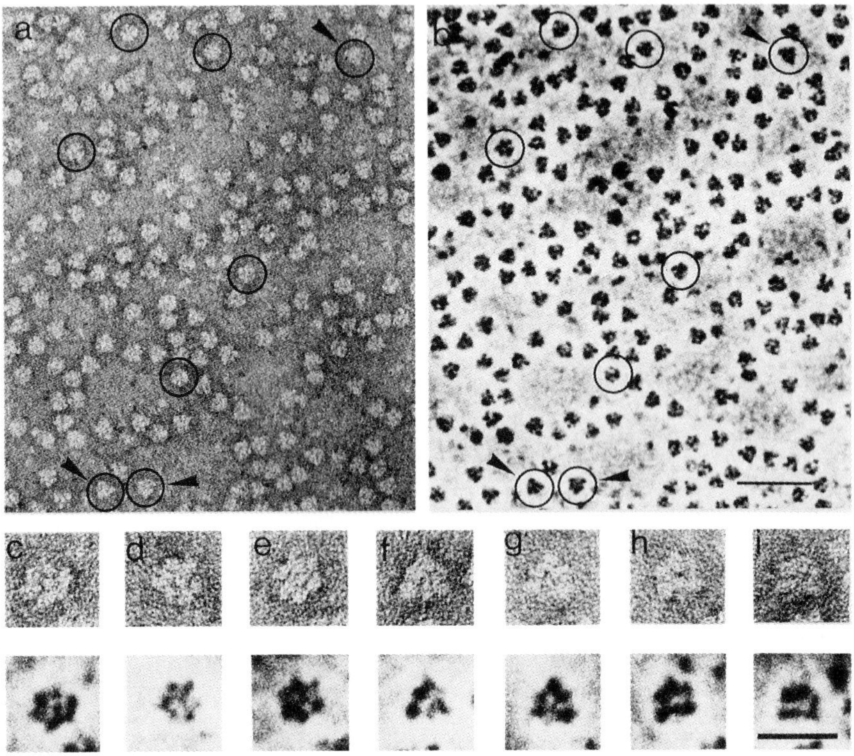

Fig.1: (a) General view of S. acidocaldarius-ATPase molecules imaged in the elastic bright field (Δe=0 eV), (b) Electron Spectroscopic Imaging of the ATPase, Uranium-specific inelastically scattered electrons (Δe = 114 eV). Encircled molecules = hexagonal top views; arrowhead marked circles = trigonal top views. Bar: 50 nm. (c-e) Gallery of different projections of the ATPase-molecules, showing the structural difference in more detail. (c-e) hexagonal top views, (f,g) trigonal top views, (h,i) side views. Upper row of the gallery: elastic bright field; lower row: electron spectroscopic images. Bar: 20 nm.

Inst. Phys. Conf. Ser. No. 93: Volume 3, Chapter 15
Paper presented at EUREM 88, York, England, 1988

Structural studies on *Prosthecochloris aestuarii* chlorosomes

W Wullink and E F J van Bruggen

Biochemisch laboratorium, Rijksuniversiteit Groningen, The Netherlands

We are interested in the structure of the complexes involved in the photosynthesis of the green sulfur bacterium Prosthecochloris aestuarii. These complexes are localized in specific cell organelles called chlorosomes (Fig. 1). In 1980 a structural model for the chlorosomes in Chlorobium limicola based on freeze-fracturing studies was proposed (Staehelin et al 1980). In this model a crystalline base-plate (consisting of Bchla-protein) is thought to be localized between the chlorosomal core (containing rod-shaped Bchlc-protein) and the cytoplasmic innermembrane (probably containing the reaction centres). A Bchla-protein corresponding with the base-plate protein in C. limicola was isolated from P. aestuarii. This protein was crystallized and its structure was determined at 0.19 nm resolution by X-ray diffraction (Fenna et al 1974, Tronrud et al 1986).
We found that P. aestuarii chlorosomes contain a crystalline base-plate quite similar to that in C. limicola. Freeze-fractured samples show base-plates with a periodicity of about 6.5 nm as determined by optical diffraction. Rod-shaped elements, about 11 nm in diameter, and arrays of circular particles (heart to heart distance about 14 nm) were also found: probably reaction centre complexes and associated proteins. One of the goals of this study is to make a 3-dimensional model for the individual complexes and their arrangement in the chlorosomes. Since periodic objects offer the best possibilities for high resolution computer-reconstruction we aim for obtaining the reaction centre complexes in regular arrays. Considering the possibility of isolating the membrane embedded reaction centre complexes and associated proteins in regular arrays, when isolated in connection with the crystalline baseplate, we started to isolate whole chlorosomes including small patches of the cytoplasmic membrane to which they are attached. We applied a mild fixation treatment (2% formaldehyde) or a treatment with a reversible crosslinker (dithiobis[succinimidyl propionate]) to the whole P. aestuarii cells before disrupting them. Fractions obtained after (partial) purification were analysed by electron microscopy using negative staining techniques and polyacrylamide gelelectrophoresis (PAGE). The isolated chlorosomes turned out to be attached to membrane patches, in which membrane embedded complexes (possibly reaction centre complexes) with a diameter up to 18 nm could be visualized (Fig. 2). Unfortunately the protein complexes did not show up in regular arrays. Upon partial destruction of the fixed chlorosomes rod-shaped elements with a 11 nm diameter and a 3 nm striation perpendicular to their main axis were found (Fig. 3). A relatively pure fraction of protein particles

(diameter about 8.5 nm) was isolated from decrosslinked chlorosomes (probably consisting of baseplate proteins or parts of reaction centre complexes) (Fig. 4). SDS-PAGE(8%) analysis of the isolated chlorosome fraction gave 3 major bands (40, 42 and 45 kDa), the purified 8.5 nm-protein fraction showed only one band (42 kDa). Some functionally characterized fractions of P. aestuarii (PP and Complex I), isolated by the group of Dr. J. Amesz (Biofysica, RU Leiden) showed to contain the same but also many more bands.

We gratefully acknowledge the gift of P. aestuarii cells and some cell fractions by the group of Dr. J. Amesz, the assistance of J. Haker, K. Gilissen and F. Dijk and the support by the Netherlands organisation for chemical research (SON) with financial aid from the Netherlands organisation for scientific research (NWO).

REFERENCES

Fenna R E, Matthews B W, Olson J M and Shaw E K 1974 J. Mol. Biol. **84** 231
Staehelin L A, Golecki J R and Drews G 1980 Biochim. Biophys. Acta **589** 30
Tronrud D E, Schmid M F and Matthews B W 1986 J. Mol. Biol. **188** 443

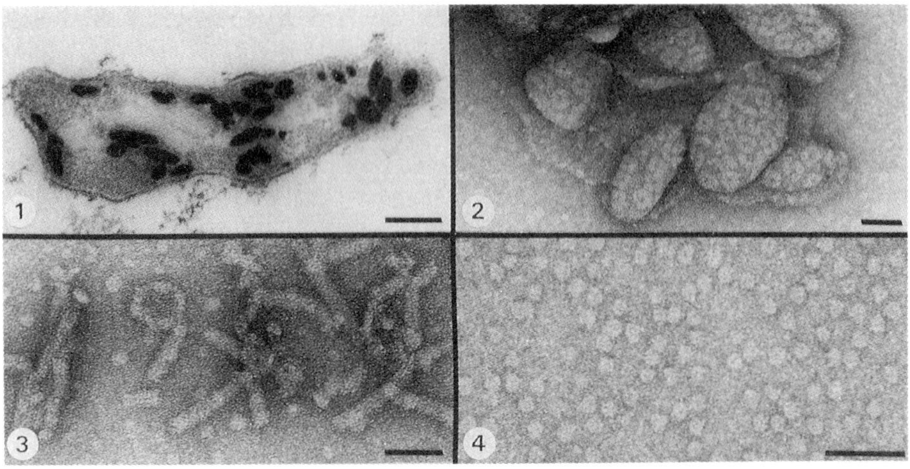

Fig. 1. Thin section of a P. aestuarii cell fixed with $KMnO_4$. The darkly stained organelles are the chlorosomes (bar = 200 nm);
Fig. 2. Purified formaldehyde fixed chlorosomes: membrane embedded complexes are visible;
Fig. 3. Rod-shaped elements released from fixed chlorosomes;
Fig. 4. 8.5 nm particles isolated from decrosslinked chlorosomes.
(Figs. 2, 3 & 4: Negative staining with uranyl acetate, bar = 40 nm).

Paper presented at EUREM 88, York, England, 1988

Isolation of several forms of nucleic acid molecules from nuclear polyhedrosis virus of *Porthetria dispar* L

P G Ploaie

The Research Institute of Plant Protection, 71592 Bucharest-Băneasa, Romania

ABSTRACT: The nucleic acid of the nuclear polyhedrosis virus of Pd was found to be formed from three forms of molecules: linear and circular supercoiled molecules and rare molecules suggesting a mRNA.

1. INTRODUCTION

The nuclear polyhedrosis virus of Pd is a baculovirus with rod-shaped particles of 400-428 nm in length and having a diameter of 60 nm (Ploaie and Hauser 1972). Although this virus and other baculoviruses have been studied for a long time little is known about their molecular biology. Double stranded DNA molecule in circular or linear forms have been detected for several baculoviruses by electron microscopy or by sedimentation analyses (Burgess 1977;Vlak and Odink 1979). This paper represents an attempt to reveal the morphological aspect of DNA molecule of Pd baculovirus.

2. MATERIALS AND METHODS

The nucleic acid was purified from virus particles by SDS and heating treatments followed by low speed centrifugation. For visualization of DNA molecules a simple method was used (Ploaie and Hauser 1972).

3. RESULTS

The results illustrated in Figure 1 show the presence in the virus particles of three forms of molecules:linear forms (a) circular supercoiled forms (b) and rare small molecules that suggest a mRNA with biomolecules (ribosomes) binded on it. The most frequent molecules were linear. Measurement on 30 well separated linear molecules gave range of length from 6.5 to 15 µm and an approximate MW of 12.7 to 29.4 x 10^6 daltons.

4. REFERENCES

Burgess S 1977 J. gen. Virol. 37 501-10
Ploaie G P and Hauser R E 1972 Rev. Roum. Virol. 9 151-5
Vlak J M and Odink K G 1979 J. gen. Virol. 44 333-47

Structural basis for the vesicoelasticity of the human erythrocyte membrane: folded spectrin hypothesis

Betty W. Shen

Argonne National Laboratory, Argonne, Illinois 60439, USA

ABSTRACT: We have examined the structure of spectrin, which is a major component of the filamentous network underlying the human erythrocyte membrane, either as solublized dimers and tetramers or as a constituent of intact, as well as fragmented, skeletons. Electron micrographs of negatively stained specimens showed that spectrin molecules freshly liberated from the membrane bilayer assume a condensed structure. These condensed spectrin molecules then undergo gradual extension over a several-minute to hours period into the long, thin, flexible filaments commonly observed in the low-angle platinum-shadowed preparations of spectrin that were isolated by low-ionic-strength extraction. Our data support the hypothesis that the spectrin molecules assume a condensed rather than extended structure at the surface of the membrane in the erythrocyte at rest and suggest that the reversible interconversion of spectrin from a condensed to an extended conformation could be the molecular basis for the remarkable deformability and elasticity long observed in the membranes of intact red cells under mechanical shear stress.

The membrane of the human erythrocyte is reinforced at its cytoplasmic surface by a network (or skeleton) of peripheral proteins consisting of spectrin, actin, and band 4.1. This network is linked to the membrane bilayer through specific protein-protein interactions (1). It is well documented that this skeleton is responsible for the mechanical stability and deformability of the erythrocyte. However, the molecular basis for its elasticity is unclear. We have shown previously that the membrane skeleton is organized into short actin filaments interconnected by multiple strands of spectrin tetramers and that the spectrin molecules in the least perturbed skeletons assume a condensed structure that extend into long, thin, flexible filaments accompanying the expansion of the skeletal network (2,3). We now present direct evidence that spectrin molecules are in a condensed form when they are freshly released from the membrane bilayer irrespect of their preparation procedures.

Spectrin dimers and tetramers were extracted from the isolated membrane (i.e., ghosts) by low-ionic-strength buffer or by exposing membrane ghosts to excessive mechanical pressure. Immediately upon their separation from the bilayer and at fixed intervals afterward, the specimens were visualized by negative staining with uranyl acetate on fenestrated carbon films. When extracted with low-ionic-strength buffer on ice for a short interval, the spectrin-actin network dissociates from the membrane bilayer in large pieces, leaving band 4.1 and ankyrin, the linkage proteins, largely attached to the surface of the inverted vesicles. These large two-dimensional aggregates of spectrin-actin complexes display granular features of uniform size over the entire surface of the array. These aggregates dissociate gradually in low-ionic-strength media into smaller particles with a similar granular nature. These granules eventually unfold into long, thin, flexible filaments of approximately 2000Å long with a broad spectrum of intermediates displaying both granular and filamentous features. The size of the smallest particles that totally retain the granular features corresponds to to the size of spectrin tetramers with a calculated thickness of 30Å.

The soluble proteins released from membrane ghosts after exposing to excessive mechanical pressure at 0°C in phosphate buffered saline (PBS) are mainly spectrin tetramers. The spectrin released after a short period of pressure treatment appears to be compact and closer to symmetrical than elongated, whereas the spectrin solubilized after longer pressure treatment exhibits various degrees of unfolding. The mechanism by which spectrin unfolds in PBS under excessive mechanical pressure is currently under investigation.

REFERENCES

1. B. W. Shen. In Red Blood Cell Membranes: Structure, Functions and Clinical Implications. P. Agre and J. Parker, eds Marcell Dekker Inc. (In Press).
2. B.W. Shen et al., J. Cell Biol. (1984) 99, 810.
3. B.W. Shen et al., J. Cell Biol. (1986) 102, 997.
4. This research was supported by PHS Grants HL 33254 (BWS) and HL 30121 and by the U. S. Department of Energy under contract No. W-31-109-ENG-38. Valuable editorial assistance by Susan H. Barr is acknowledged.

Gold immunolabelling of skelemin in cryosections of human skeletal muscle

J Walsh, J J Fulthorpe, M J Cullen, M G Price* and J B Harris

Muscular Dystrophy Group Research Laboratories, Newcastle General Hospital, Newcastle upon Tyne, NE4 6BE, UK and *Research Institute of Scripps Clinic, La Jolla, CA 92037, USA

ABSTRACT: Gold conjugated secondary antibody labelling has been used to localise skelemin in healthy human skeletal muscle. The label was concentrated in a band, 0.1 - 0.15 μm wide, at the level of the M-line.

1. INTRODUCTION

Skelemins are high molecular weight polypeptides (220,000 and 200,000 mol. wt.) found in mammalian striated and smooth muscle (Price 1984). They have recently been shown to be immunologically distinct from numerous proteins of similar molecular weight, and not to be oligomers of other muscle proteins (Price 1987). Immunofluorescence studies have shown that skelemins are localised at the M-line in a 0.4 μm wide band. In the present study we have used colloidal gold immunolabelling to identify further the location of the skelemins in human skeletal muscle.

2. METHODS

Skelemins were partially purified from bovine ventricular myocardium (Price 1984) and the antibodies raised in New Zealand White rabbits (Price 1987). The antibodies were lyophilised in La Jolla and reconstituted in Newcastle.

The human biopsy muscle (quadriceps) was fixed in 2% paraformaldehyde + 0.1% glutaraldehyde in PBS at 4°C for 2h. Small blocks were infiltrated with 2.3M sucrose overnight, placed on stubs and frozen by plunging into LN_2. Sections were cut on a Reichert FC4D cryoultramicrotome set for a specimen temperature of -90°C and a knife temperature of -110°C. The grids with the cryosections were inverted onto drops of 2% gelatine on parafilm over ice for at least 10 min, then transferred to drops of PBS/FCS containing 0.15% glycine (blocker). After three PBS/FCS rinses the sections were incubated with anti-skelemin (0.2 mg/ml) for 60 min at room temperature. Other sections were incubated with pre-immune serum.

After six more PBS/FCS rinses the sections were incubated with the 10 nm gold conjugated secondary antibody (BioClin goat anti-rabbit IgG) in Tris buffered saline (pH 8.2) for 60 min. The sections were further rinsed in PBS, followed by distilled water, and then 'embedded' and adsorption stained with a methyl cellulose/uranyl acetate mixture (0.9 cm^3 2% methyl cellulose + 0.1 cm^3 4% aqueous UA) on ice. The grids were then blotted dry and stored before examination in the electron microscope (JEOL 1200 EX).

3. RESULTS AND DISCUSSION

Gold particles were mostly concentrated in a band situated at the level of the M-line and in the area of the A-band bare of cross-bridges immediately either side of it (Figs 1, 2). The width of this band was approximately 1.0 - 1.5 µm. The label was very sparse in the rest of the A-band, I-band and Z-line. In the control, using pre-immune serum, gold particles were very sparsely and randomly distributed.

EM immunolabelling confirms unambiguously that skelemins are localised at the M-line in human skeletal muscle. Its precise position and structural role at the M-line has yet to be determined.

Fig. 1. Gold immunolabellin of skelemin. The label is a the level of the M-line (M). Scale bar = 0.5 µm.

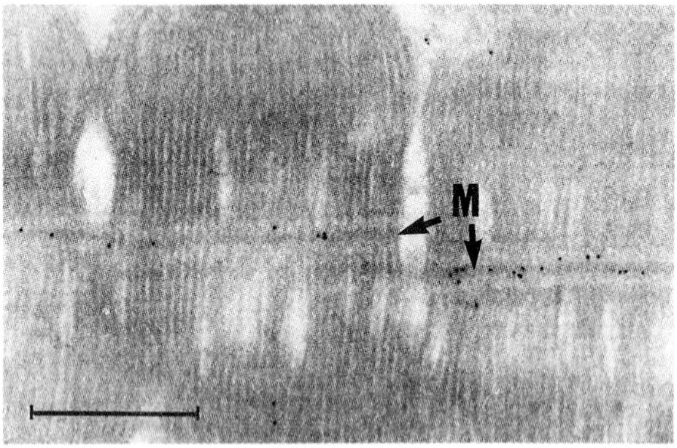

Fig. 2. Centre of a sarcomere at higher magnification showing the gold particles close to the M-line (M). Scale bar = 0.2 µm.

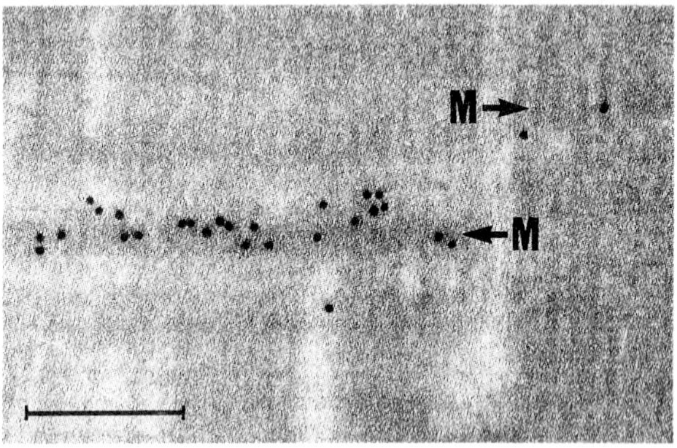

4. REFERENCES

Price M G 1984 Am. J. Physiol. 246 H566
Price M G 1987 J. Cell. Biol. 104 1325

Model building of arthropod hemocyanins

T. Wichertjes, J.F.L. van Breemen, M.M.C. Bijlholt, W. Keegstra and E.F.J. van Bruggen

Biochemisch laboratorium, Rijksuniversiteit Groningen, The Netherlands

We are interested in the relation between the structure and the function of the various hemocyanin molecules occurring in certain Arthropoda. Their function is to transport oxygen. Depending on the species these hemocyanin molecules are made of 1, 2, 4, 6 or 8 hexameric units in a specific mutual orientation. The atomic structure of the single-hexameric hemocyanin from the spiny lobster Panilurus interruptus has been determined by X-ray diffraction in combination with amino acid sequence studies (Gaykema et al. 1984). Others as well as we ourselves analysed the structure of several multi-hexameric hemocyanin molecules by electron microscopy and single particle averaging of the molecular profiles. In all cases the profiles of the hexameric units appeared to be very similar to that of the single hexameric P. interruptus hemocyanin (Fig. 1). Based on the X-ray data we made very precise polyurethane models of the P. interruptus hemocyanin monomers. Hexamers having 32 pointgroup symmetry and a staggered arrangement were assembled from these monomers (Fig. 3). Amino acid sequence studies indicate a strong homology for crustacean (crabs, lobsters) hemocyanins. Thus the P. interruptus hemocyanin models can be used to make models of crustacean multi-hexameric hemocyanin molecules by fitting them to the averaged molecular profiles (Fig. 2 & 4). For the Chelicerata (spiders, scorpions, horseshoe crabs) the monomeric model had to be modified because of a missing polypeptide loop. Here too models were made by fitting them to the averaged profiles obtained from electron microscopy. Arthropod hemocyanins are microheterogeneous. They are made of up to eight different monomeric and dimeric subunits, each having a specific role in the architecture and function of the multihexameric molecules (Linzen et al. 1985). From immuno-electron microscopy the position of the subunits is known a.o. for the four-hexameric Eurypelma californicum (tarantula) hemocyanin (Markl et al. 1981). Based on the amino acid sequence homology we assume a similar "hemocyanin"-fold for the different subunits. Therefore our next step will be the implementation of the known X-ray structure for a hexamer in the by electron microscopy best studied multi-hexameric structures using computer graphics. This will enable us to study the interhexameric contacts and other structural features at the level of the involved amino acids.

We gratefully acknowledge the assistance of J. Haker, K. Gilissen, J. Spoelstra & H.J. van der Velde and the support by the Netherlands organization for chemical research (SON) with financial aid from the Netherlands organization for scientific research (NWO).

REFERENCES

Gaykema W P J, Hol W G J, Vereijken J M, Soeter N, Bak H J & Beintema J J, 1984 Nature, Lond. **309** 23

Linzen B et al. 1985 Science **229** 519

Markl J, Kempter B, Linzen B, Bijlholt M M C & Van Bruggen E F J 1981 H.-S. Z. Physiol. Chem. **362** 1631

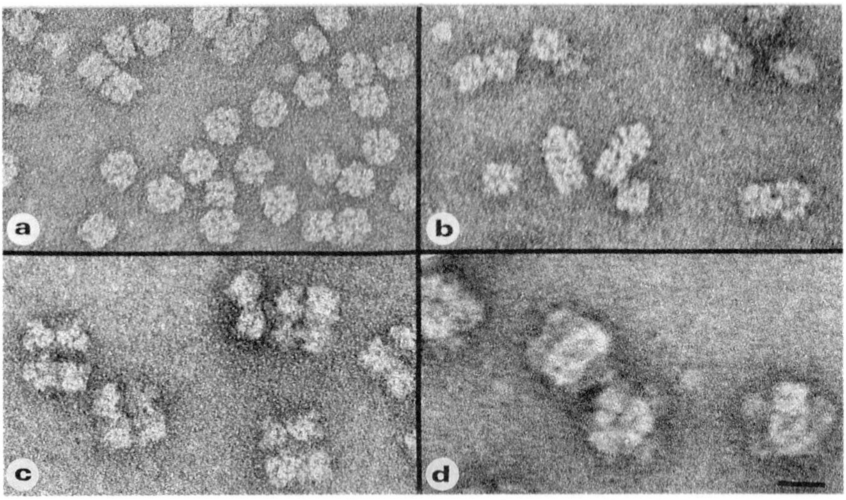

Fig. 1 a: single hexamers of P.interruptus hemocyanin, b: two-hexamers of C.pagurus hemocyanin, c: four-hexamers of E.californicum hemocyanin, d: eight-hexamers of L.polyphemus hemocyanin, negative staining with uranyl acetate, bar = 50 nm.

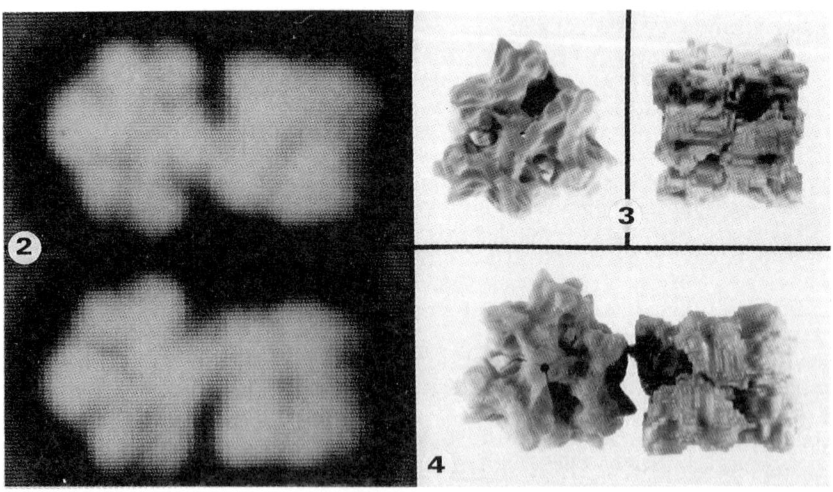

Fig. 2: Two views as a result of single particle averaging of the two-hexamer hemocyanin molecular profiles of C.pagurus.
Fig. 3: Hexagonal and rectangular view of the polyurethaan model of P.interruptus hemocyanin based on the X-ray data.
Fig. 4: A polyurethaan model of the two-hexamer hemocyanin of C.pagurus.

Electron microscopy of alcohol oxidase from *Hansenula polymorpha*

J Vonck and E F J van Bruggen

Biochemisch laboratorium, Rijksuniversiteit Groningen, The Netherlands

Microbodies containing catalase and a flavin oxidase are called peroxisomes (Tolbert 1981). Very large peroxisomes are found in certain yeast cells when they are grown in media containing methanol as the only carbon and energy source. These organelles contain alcohol oxidase and catalase, both involved in the first oxidation step of the substrate methanol. Inside the peroxisomal membrane the enzymes occur (partly ?) in a crystalline state (Veenhuis et al. 1981).

Several research groups at our university collaborate in a research programme concerning the mechanisms involved in the development and functioning of these peroxisomes for mainly biotechnological applications. The determination of the structure of alcohol oxidase from the yeast Hansenula polymorpha by a combination of X-ray diffraction and electron microscopy is one of the side-projects of this programme.

Alcohol oxidase (AOX) was a gift from the group of professor Harder. The molecules are known to be octameric and have a molecular weight of about 670 kDa (Kato et al. 1976). Upon negative staining with uranyl acetate the single molecules are observed as roughly square profiles of 13 nm x 13 nm with indications for a 2- and 4-fold substructure (figure 1). We selected 1200 profiles, which are now subjected to single particle averaging procedures in combination with correspondence analysis.

Aiming for a higher resolution we are trying to make 2-dimensional crystals, which turned out to be very difficult. Contrarily 3-dimensional crystals are easily formed in the presence of polyethylene glycol (concentration > 0.2% w/v). These crystals resemble the in vivo crystals (figure 2). Electron diffraction of such crystals shows a resolution of 2.4 nm (figure 3). We are now continuing the studies of thin 3-dimensional crystals using cryo-electron microscopy and cryo-electron diffraction. The idea is to use these results as start data for the refinement of high resolution data from X-ray diffraction. Unfortunately larger 3-dimensional crystals did not diffract X-rays. We are investigating the reason for this.

Still some 2-dimensional crystals have been obtained in a hardly reproducible way. They also need polyethylene glycol (0.05% w/v) and in addition the presence of folds in the supporting film (figure 4). They are, however, poorly ordered which makes them less suitable for further work.

In addition the crystallization process can lead to the formation of tubular structures (figure 5).

REFERENCES

Kato N, Omori Y, Tani Y & Ogata K 1976 Eur. J. Bioch. **64** 341
Tolbert N E 1981 Ann. Rev. Biochem. **50** 133
Veenhuis M, Harder W, Van Dijken J P & Mayer F 1981 Mol. Cell. Biol. **1** 949

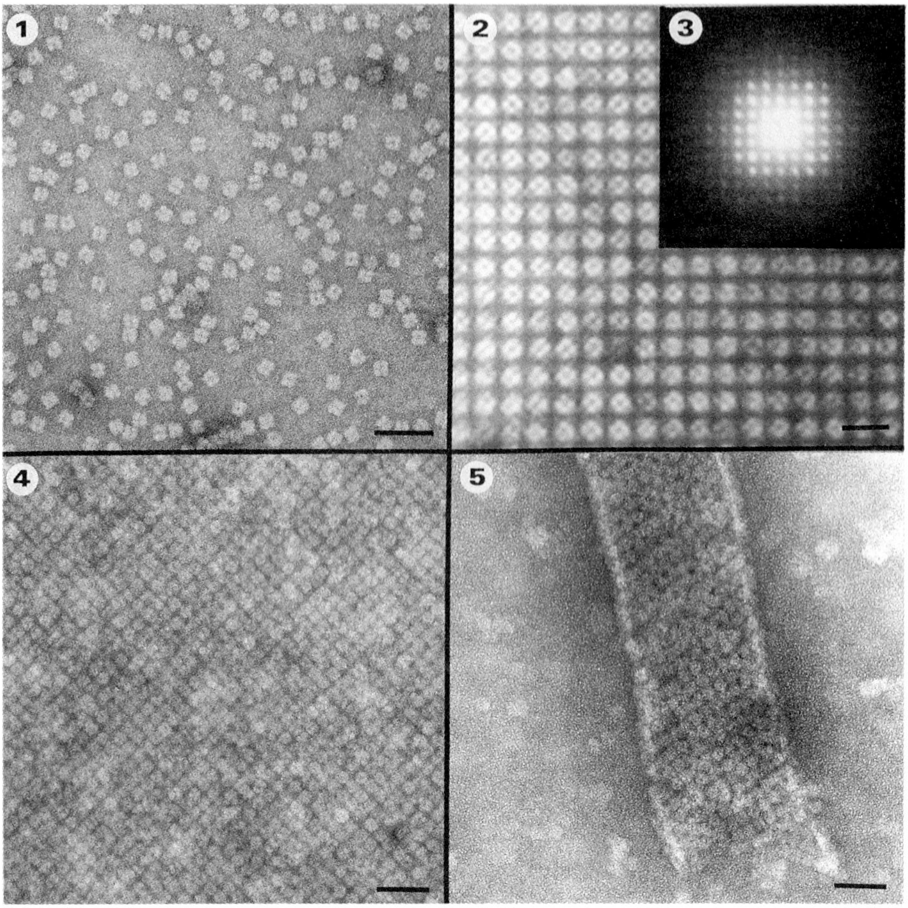

Electron microscopy/diffraction of alcohol oxidase (AOX) from the yeast Hansenula polymorpha:
Figure 1: Single AOX molecules negatively stained with uranyl acetate. Bar = 50 nm.
Figure 2: In vitro made 3-dimensional AOX crystal negatively stained with uranyl acetate. Bar = 30 nm;
Figure 3: Electron diffraction pattern of a similar negatively stained crystal. The hardly visible 7th order of diffraction corresponds with a resolution of 2.4 nm.
Figure 4: 2-dimensional crystal of AOX negatively stained with uranyl acetate. Bar = 50 nm.
Figure 5: Tubular aggregate of AOX negatively stained with uranyl acetate. Bar = 50 nm.

Specific decoration sites on frozen-hydrated crystals of heavy riboflavin synthase

S. Weinkauf[1], A. Bacher[2], K. Schott[2], I. Wildhaber[3], W. Baumeister[3], L. Bachmann[1]

Institut für Technische Chemie[1] und Institut für Organische Chemie und Biochemie[2], TU München, D-8046 Garching, FRG
Max-Planck-Institut für Biochemie[3], D-8033 Martinsried, FRG

Heavy metal shadowing reveals the surface topography whereas decoration reflects the spatial distribution of surface regions with different physical-chemical properties, i.e. the nucleation probabilities of the condensing material. Nucleation phanomena play an important part in replica formation under freeze-etching conditions, but the correlation of surface relief and decoration pattern on hydrated biomolecules is not yet possible /1/.
In the present work, the decoration behaviour of different metals on heavy riboflavin synthase (HRS) was investigated. The molecule is essentially a spherical capsid of 60 identical ß subunits with icosahedral symmetry /2/; it has narrow channels running along the 5-fold symmetry axes /3/. The crystals have a hexagonal layer packing /4/. The orientation of the icosahedra within the orthogonal ac-plane alternates along the c-axis (Fig.1c). The freeze-etched crystals of HRS were either shadowed with Pt/C or decorated with Au, Pt or Ag. The diffraction pattern reflects the alternating orientation of the particles (Fig.1a). Decoration replicas with Au (Fig.2a), when processed by correlation averaging /5/, portray the icosahedral symmetry of the particles and their orientation in the crystal lattice (Fig.2b). Decoration with Pt leads to a similar metal distribution, whereas Ag (Fig.3) clusters accumulate at sites where 3- or 5-fold symmetry axes penetrate the particle surface. Au occupies areas located about halfway between these symmetry centres (Fig.3c).
These results indicate that the decoration is a sensitive method for the investigation of frozen-hydrated protein surfaces with high resolution. It may be a useful technique for studying the molecular symmetry and the crystal packing of proteins which form crystals too small for X-ray crystallographic analysis. On the other hand HRS is an ideal test specimen for investigating decoration phanomena, since an atomic model will be available shortly. This will allow us to correlate the observed decoration phanomena with certain physico-chemical properties of the protein surface.

/1/ Bachmann L, Becker R, Leupold G, Barth M, Guckenberger R and Baumeister W, Ultramicroscopy 16 (1985) 305.
/2/ Bacher A, Ludwig H C, Schnepple H and Ben-Shaul Y, J.Mol.Biol. 187 (1986) 75.
/3/ Ladenstein R, Bacher A and Huber R, J.Mol.Biol. 195 (1987) 751.
/4/ Ladenstein R, Meyer B, Huber R, Labischinski H, Bartels K, Bartunik H D, Bachmann L, Ludwig H C and Bacher A, J.Mol.Biol. 187 (1986) 87.
/5/ Saxton W O and Baumeister W, J.Microscopy 127 (1982) 127.

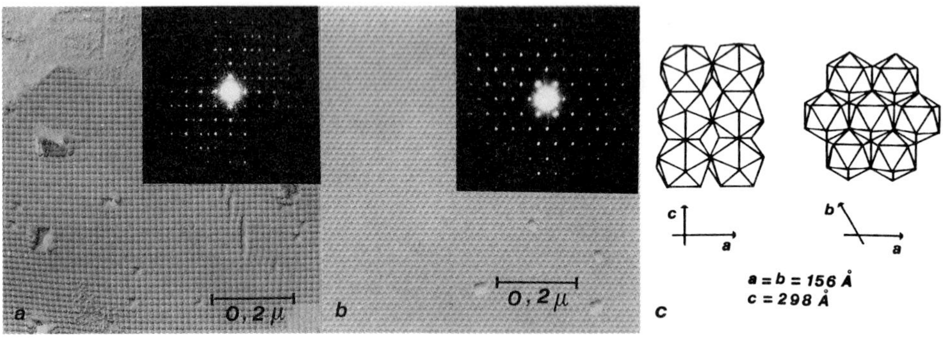

Fig. 1 Freeze-etched and Pt/C shadowed crystal planes of HRS: (a) orthogonal ac-plane; (b) hexagonal ab-plane; (c) packing model of HRS crystal.

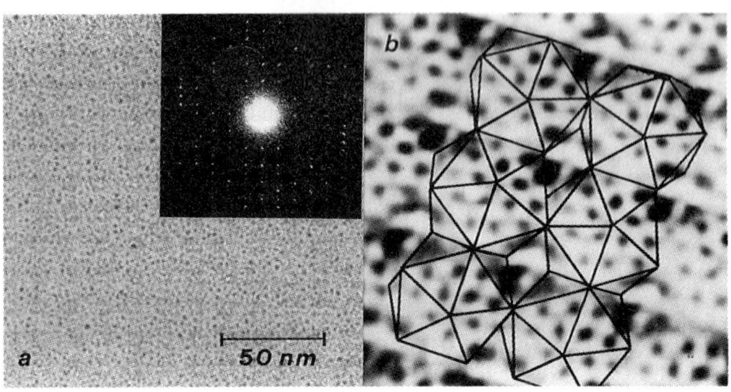

Fig. 2 (a) ac-crystal plane decorated with 1,3 Å Au; (b) averaged image with an overlay of the packing model.

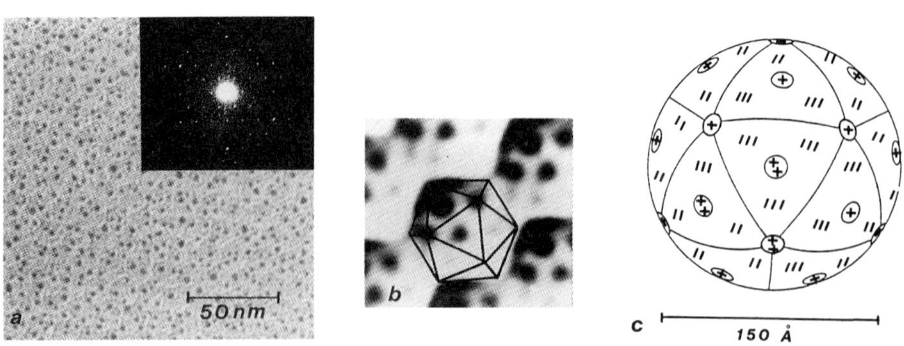

Fig. 3 (a) ab-crystal plane decorated with 4 Å Ag; (b) averaged image; (c) model of the enzyme complex with Au (///)- and Ag (+++)-decoration sites.

Inst. Phys. Conf. Ser. No. 93: Volume 3, Chapter 15
Paper presented at EUREM 88, York, England, 1988

Structure of the photosystem I complex from a thermophilic cyanobacterium

Egbert J. Boekema

Fritz-Haber-Institut, Faradayweg 4-6, D-1000 Berlin 33 (West)

The study of isolated membrane proteins as single particles is rather difficult because these molecules easily aggregate with their hydrophobic parts in a random orientation. Crystallization into two-dimensional sheets is the best way to overcome most of the difficulties and since 1970 some 20 different membrane proteins have been crystallized after purification. However, for many important membrane proteins, no good crystals are (yet) available. As an alternative, the preparation of single particles was optimised and we found that the negative staining technique gives good specimens if the membrane proteins are prepared in the presence of detergents in a concentration above about two times the cmc (critical micellar concentration).

Photosystem I (PS I), which catalyses the light-induced transfer of electrons from special chlorophyll molecules in photosynthetic membranes, was isolated and purified from the cyanobacterium Synechococcus sp. (Boekema et al. 1987). It was prepared for electron microscopy in the presence of either 1.2% octylglucoside or 0.03% dodecylmaltoside (Figure 1). PS I molecules have the shape of a flat cylinder with a height of 6.5 nm. The diameter depends on the size of the detergent molecules which are attached in a boundary layer; it is about 19 nm for octyl glucosid and about 21 nm for dodecylmaltosid.

Computer image analysis of about 2000 top view projections showed that the PS I is a trimer of three very similar if not identical units. Moreover, the top views could be separated into two groups mainly differing in handedness and which represent the two different ways the top views can be attached to the carbon support (Figure 2). In combination with biochemical data it is concluded that each monomer of approx. 300 kDa contains two large subunits of about 80 kDa, which are largely located on the outer ends of the monomers. The trimeric complex likely also reflects the organisation of PS I in the photosynthetic membrane.

ACKNOWLEDGEMENTS
The isolation and biochemical characterization has been carried out by Drs. J.P. Dekker, M. Rögner and I. Witt at the Max-Volmer-Institute of Prof. H.T. Witt, Technical University Berlin. The author thanks his colleagues for their generous supply of the material.

REFERENCES

Boekema E J, Dekker J P, van Heel M G, Rögner M, Saenger W, Witt I and Witt H T 1987 FEBS Lett 217 283

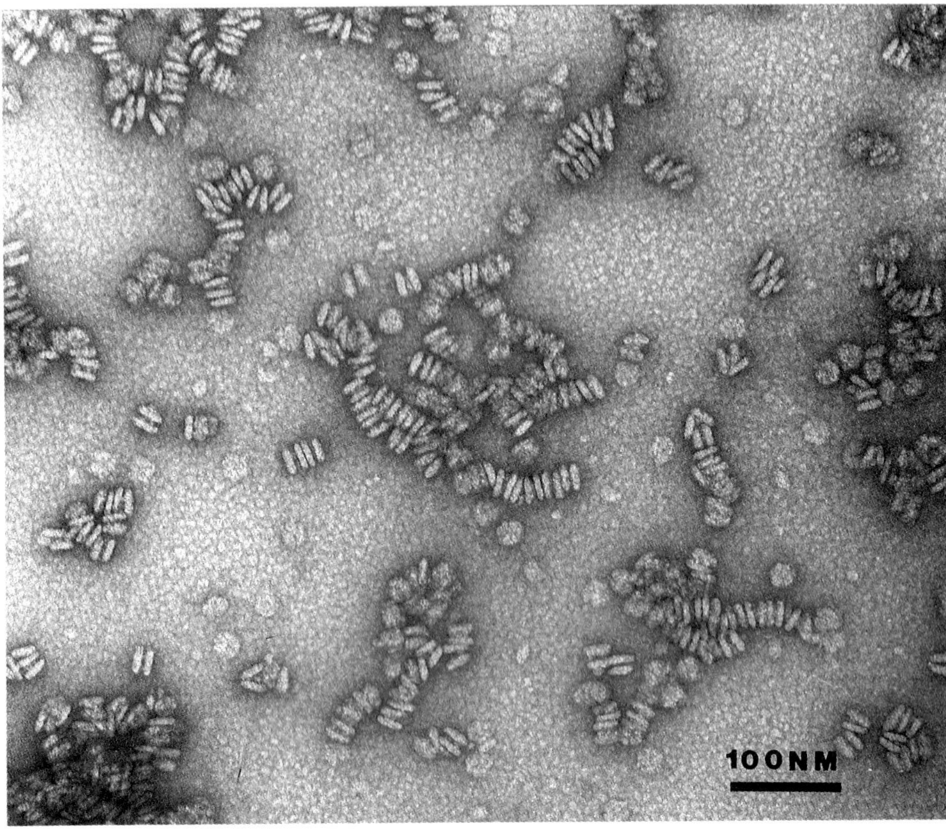

Figure 1. Electron micrograph of negatively stained PS I molecules. Trimeric molecules frequently aggregate when lying on their side.

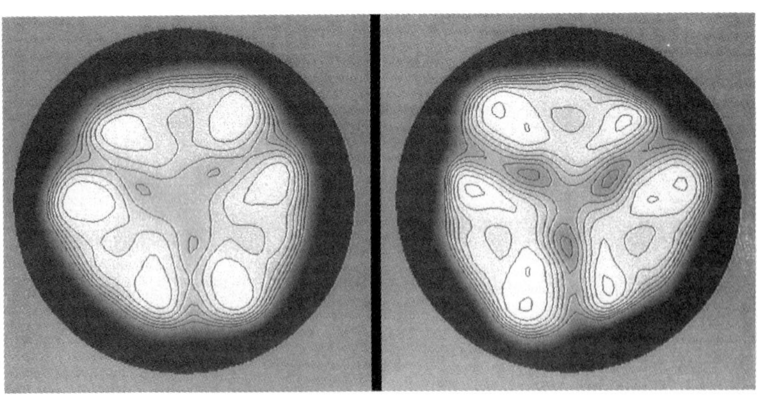

Figure 2. Image analysis of trimeric PS I particles. After image analysis (repeated alignments in combination with multivariate statistical analysis) 2000 top view projections could be separated into two groups that differ in handedness ("face-up" and "face-down").

Function-dependent configurational changes of bacterial ribulose-1,5-bisphosphate carboxylase/oxygenase

A Holzenburg and F Mayer*

Maurice E.-Müller-Institute for High Resolution Electron Microscopy at the Biozentrum of the University of Basel, Klingelbergstrasse 70,CH-4056 Basel, Switzerland
*Institute f.Microbiology, University of Göttingen,Grisebachstr.8,D-3400Göttingen, F.R.G.

ABSTRACT: The key carboxylating enzyme of the reductive pentose phosphate cycle D-Ribulose-1,5-bisphosphate carboxylase/oxygenase [*RuBisCO*] isolated from the chemolithoautotrophic, H_2-oxidizing bacterium *Alcaligenes eutrophus* H16 exists in three stable functional states: As the active enzyme [**Ea**] (in the presence of Mg^{2+} and HCO_3^-); as inactivated enzyme [**Eia**] (in the absence of Mg^{2+} and HCO_3^-) and as the enzyme locked in an *in vitro* transition state [**CABP-E**] (**Ea** fully saturated with the transition state analogue 2-carboxy-D-arabinitol-1,5-bisphosphate [CABP]).Three different functional state-correlated configurations of the enzyme were revealed by electron microscopy of 2D crystals and single molecules, and by X-ray crystallography of 3D crystals of the **CABP-E**.

The structure of the L8S8-RuBisCO (8 large and 8 small subunits; Mw 536,000 Da; overall dimensions of the **Ea**: a x b x c = 12.5 x 12.5 x 10.5 nm) has been the subject of several investigations (see Holzenburg et al., 1987). Recently each of the three well defined functional states of RuBisCO was crystallized into a monolayer (2D) or thin multilayer (3D) directly on the support film used for electron microscopy (Holzenburg, 1988). Electron micrographs of selected 2D crystals were subjected to further analysis by optical diffraction and digital image processing (Van Heel and Keegstra, 1981). The stoichiometry of RuBisCO under crystallization conditions was shown not to be altered. A summary of the electron microscopical crystal data is given in Table 1. The **Ea**-molecules within the correlation averaged (Saxton and Baumeister, 1982) image sums depicted in Figs.1a and 1b show a distinct central depression surrounded by protein masses that differentiate to eight "arms"(4 LSUs from the upper L4S4 half and 4 from the lower L4S4 half) towards the periphery. In the fourfold rotational averaged sum (Fig.1a) a handedness is observed within the protein masses surrounding the central pit due to preferential staining of a L8S4 unit. The handedness is even stronger pronounced in the p4 sum of the rotationally shadowed 2D **Ea**-crystal (Fig.1c). The structures that are arranged in a handed manner are interpreted as SSUs as they are located on top of the molecule and exhibit a diameter of around 3 nm which corresponds exactly to the value reported by Bowien et al. (1976). The diameter of the "S4-ring" (about 9 nm) also matches the previous data. Interestingly the upper layers (data not shown) of the electron density map published by Holzenburg et al. (1987) also suggest this type of arrangement of the SSUs though the data were obtained with the **CABP-E**. The lack of information about the LSU and the periphery of the molecule is due to self-shadowing phenomena. The dimensions of the central depression (diameter around 2 nm) are similar to those obtained after negative staining. Without symmetrization, the image sum of the **CABP-E** still shows a central depression but instead of the eight arms, only faintly expressed protrusions are observed (Fig.1e). The molecules are aligned in two antiparallel rows, row "A" with the tip of the trigonal symmetry elements pointing upwards, and row "B" pointing downwards. The tetragonal symmetry present in the **Ea**-molecules has therefore been substituted by trigonal and/or pentagonal symmetry elements in case of the **CABP-E**. These symmetry elements are enhanced after imposing the mirror symmetry (Fig.1f) and coincide with those of the ideal transmission images obtained from the X-ray data of the **CABP-E** (Holzenburg et al., 1987) where a 3.6 nm shift between the lower and upper L4S4 halves of the molecule was proposed. Similar features also hold true for the **Eia**-sum (Fig.1d).The "A" and "B" rows, as well as the eight protrusions are also present; the stain pit is even more pronounced and elongated. The reproducible resolutions of the unsymmetrized image sums are: 1.7 nm for the **Ea** (negatively stained), 3.2 nm for the **Ea** (rotationally shadowed), 1.9 nm for the **CABP-E**, and 2.4 nm in case of the **Eia**. From electron micrographs of freeze-dried and with W/Ta unidirectionally shadowed single molecules the average diameter of the **Ea** and the **CABP-E** were determined to be 15nm , and 18 nm respectively.

In conclusion RuBisCO appears to undergo a drastic configurational change upon transition from the **Ea** to the **CABP-E**. The upper and lower L4S4 halves of the molecule are laterally shifted against each other by as much as 3.6 nm while the location of the SSUs on top of the LSUs is predicted to remain the same. For the **Eia** it was concluded from 2D crystals and single molecules that the molecule has the dimensions a x b x c = 15.0 x 12.5 x 10.5 ± 0.5 nm. The changes in dimensions are most probably a result of a sliding- and rotating-layer configurational change between the L4S4 halves of the molecule, with the shift being in the range of 2 - 2.5 nm. A movement of the SSUs relative to the LSUs cannot be excluded.

This work was supported by a grant for Förderung der wissenschaftlichen Forschung in Niedersachsen (A.H.). Additional funds were provided by the Swiss National Science Foundation (grant # 3.524-0.86) and by the M. E. Müller Foundation of Switzerland (A.H.).

REFERENCES:

Bowien B, Mayer F, Codd GA and Schlegel HG 1976 Arch. Microbiol. **110** 157

Holzenburg A, Mayer F, Harauz G, Van Heel M, Tokuoka R, Ishida T, Harata K, Pal GP and Saenger W 1987 Nature **325** 730

Holzenburg A 1988 Methods in Microbiology **20** ed F Mayer (London: Academic Press) in press

Saxton WO and Baumeister W 1982 J. Microsc.**127** 127

Van Heel M and Keegstra W 1981 Ultramicroscopy **7** 113

Table 1: Crystal data.

	staining method	unit cell dimensions in nm ± 0.5 a	b	projection symmetry	α	maximum crystal-inherent resolution in nm
2D Ea	negative staining	11.7		p4mm	90°	2.3
	rotatory shadowing	11.7		p4	90°	5.8
3D Ea	negative staining	11.7		p4mm	90°	2.3
2D CABP-E	negative staining	25.0	12.0	p2	84°	3.1
2D Eia	negative staining	24.5	12.25	p2*	84°	3.1

* presence of distinct twinned maxima

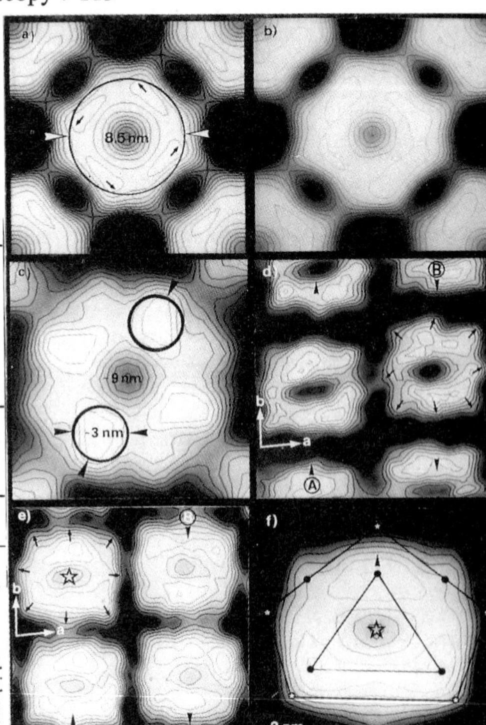

Electron microscopic studies of antibody-gold complexes as prepared by a solid-phase technique

G. Tischendorf and W. Kapfer

Department of Ultrastructure Research and Electron Microscopy, Div. of Biol., Freie Universität Berlin, Königin-Luise-Str. 12-16, 1000 Berlin 33

Gold particles in the "3-5" nm category (Au_3, Au_5) are generally considered to be ideal markers in the electron microscopic immunocytochemistry. Their diameters are considerable smaller than that of IgG-molecules or Fab- or Fc-arms. Therefore, steric hinderances during immunolabelling are not expected when gold of this size is used. On the other hand the gold particles are suggested to impair biological activities of the antibody if their binding sites interfere with functional sites such as those required for antigen binding or interactions with Fc-receptors. The binding sites of gold particles on antibodies are unknown. Therefore, we prepared gold labeled antibodies on a "solid-phase" basis for comparative analysis of Au_3 and Au_5 attachment points on negatively contrasted rabbit IgG and human IgM.

Negatively charged gold particles were prepared according to Slot and Geuze (1985). Labelling of immunoglobulins was performed by adsorption of gold particles to IgG and IgM attached to negatively charged carbon films.

An average of 40% of the IgG´s contained attached gold particles (Fig.1). Au_3 bound in Fab- or Fc-arms and also in the "center" of the structure (Fig. 2). About 50% of the immunoglobulins showed double attachment. The rarely finding of three particles, each on a separate arm of the IgG-molecule, suggests interference of 3 nm gold particle attachment points with functional sites of both Fab and Fc, probably impairing biological activities of the immunoglobulin. Au_5 bound as single particles, and as far as was observed, mostly in the "center" of the IgG-molecule, except for a few gold particles which bound at the periphery (Fig.3). Therefore, 5 nm gold might be a label much more neutral to the function of the antibody.

About 70% of the IgM-molecules bound gold (Fig.4). IgM showed attachment of from one to eleven 3 nm particles which were distributed over the Fab- and Fc-arms of the pentamer (Fig.5). The attachment of two to four gold particles was most common. However, IgM binds from three to four 5 nm gold particles, rarely one, two or five (Fig.6). The 5 nm gold particles appeared to be attached predominantly in the "center" of each of the five IgG-like substructures.

3 nm and 5 nm gold particles each appeared to bind to areas of the IgM molecule which were comparable to IgG, suggesting, therefore, a general principle of interaction between immunoglobulins and a single class of gold particle.

Slot, I.W. and Geuze, H.I. (1985) Eur. J. Cell Biol. 38, 87-93

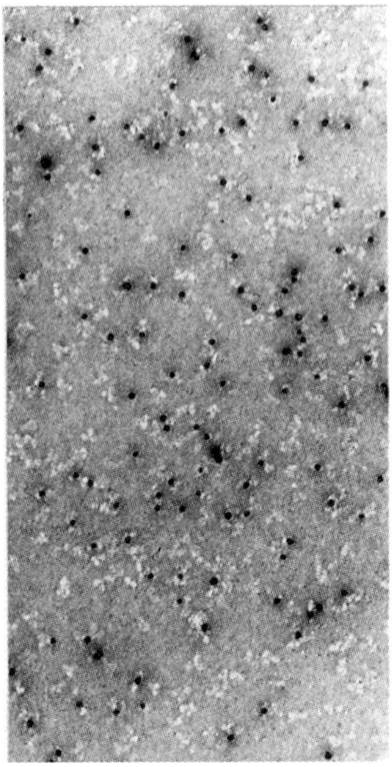

Fig.1 IgG and IgG·Au_5 — 50 nm

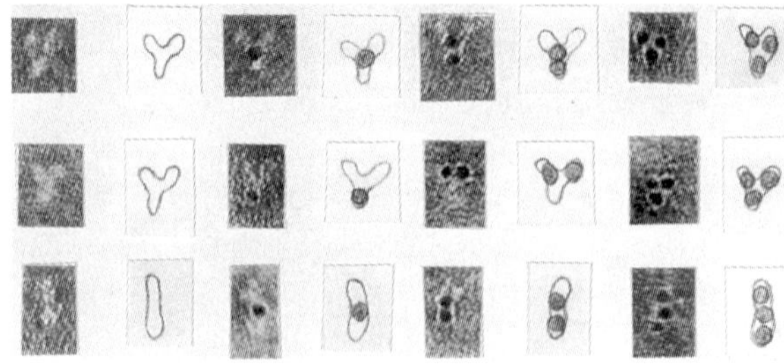

Fig.2 IgG and IgG·Au_3

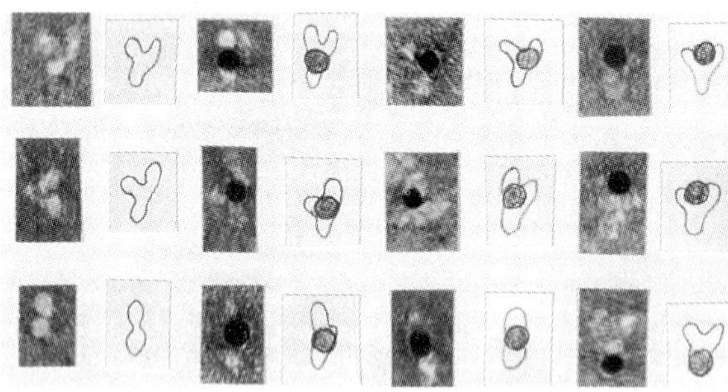

Fig.3 IgG and IgG·Au_5 — 10 nm

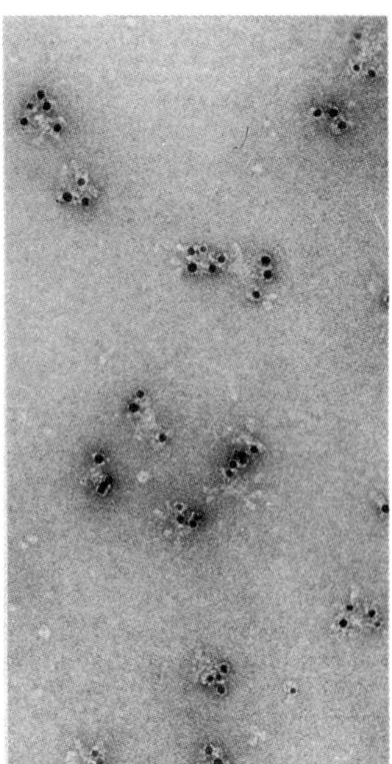

Fig.4 IgM·Au_5 — 50 nm

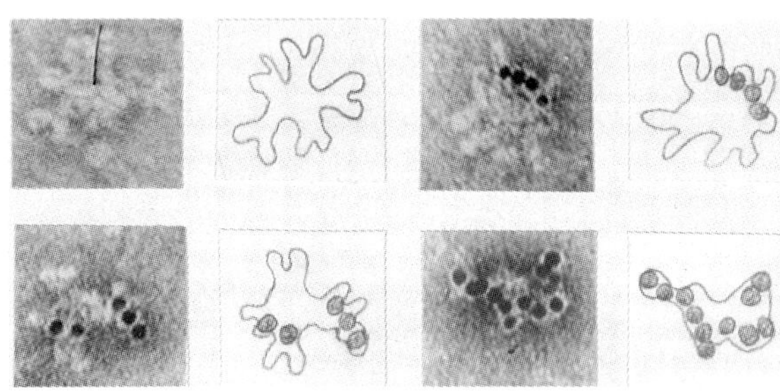

Fig.5 IgM and IgM·Au_3

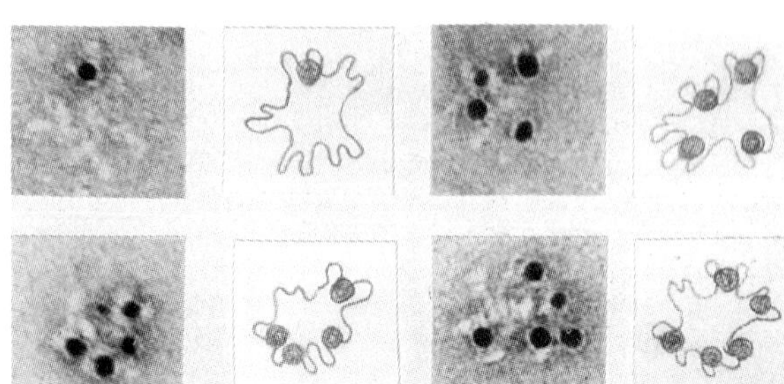

Fig.6 IgM·Au_5 — 10 nm

Inst. Phys. Conf. Ser. No. 93: Volume 3, Chapter 16
Paper presented at EUREM 88, York, England, 1988

Analysis of membrane structure and function

A.J. Verkleij

Department of Molecular Cell Biology, University of Utrecht, Padualaan 8, 3584 CH Utrecht, The Netherlands

ABSTRACT: In this presentation I will describe and discuss the contributions and value of electron microscopy in the understanding of membrane structure and dynamics, in particular membrane domains, lateral phase separation as a result of fluid-solid transitions, lipid polymorphism and their possible relevance for biological processes.

1. ELECTRON MICROSCOPY

Electron microscopy is potentially a very powerful technique to unravel membrane architecture and dynamics. In the early days of electron microscopy some expectations were not fulfilled because of preparation procedures including chemical fixation, dehydration and embedding methods, which induced many artefacts and placed severe limitations upon sample choice and interpretation of results. Lipids are extracted by traditional dehydration techniques and relocation of lipids and also proteins can occur. Freeze fracture, however, has been proven to be a powerful method to study membrane architecture, since the samples under investigation are rapidly frozen. Physical fixation by rapid freezing is now seen as the most reliable method with which to trap membrane molecules in their native state. Because of the ice crystallization during freezing particularly in tissue samples, a compromise had to be reached, i.e. freezing which was proceeded by a slight chemical fixation in glutaraldehyde coupled with cryoprotection with glycerol. However, nowadays the technology is available to freeze the specimen rapidly which excludes large ice formation and thus makes the use of chemical fixation and cryoprotectants unnecessary. Rapid freezing, which literally 'traps' the living state within less than a msec, i.e. 5.5 x 10^{-4}•Csec^{-1} at 40 μm to 10^{-6}•Csec^{-1} at 10 μm from the contact with the freezing source (Kopstad and Elgsaeter 1982). This implies that dynamic events occurring in the msec range and longer can in principle be visualized by electron microscopy. Such freezing rates have been found to be essential to prevent the lateral phase separation of lipids caused by fluid-solid transitions and hexagonal H_{II}-lamellar transitions and permit the visualization of fusion intermediates (Knoll et al. 1987).

Rapid freezing is now also used in combination with freeze substitution at low temperature, which gives us an alternative insight in membrane events. In addition to the gain in structural preservation by rapid freezing both freeze fracturing and freeze substitution can be combined with immuno-gold labeling and autoradiography, greatly enhancing the extent of techniques available to the electron microscopist.

2. MEMBRANE STRUCTURE AND DYNAMICS

The generally accepted consensus regarding the structure of biological membranes has been the result of a variety of approaches including monolayer studies, electron microscopy, biochemical analysis, X-ray diffraction, differential scanning calorimetry, nuclear magnetic resonance and fluorescence microphotolysis, etc.

In their 'fluid mosaic model' biomembranes are built up of (glyco)proteins and (glyco)lipids (Singer and Nicholson 1972). The lipids are organized in a fluid bilayer in which the transmembrane proteins are freely mobile in the plane of the bilayer. Membrane lipids diffuse very rapidly in the plane of the bilayer. Information concerning this property has been obtained by fluorescence photobleaching recovery. In general, lipid diffusion is 3-10 times slower in biomembranes than in bilayer vesicles, i.e. diffusion of lipid about 1.0 μm^2/sec and 10.0 μm^3/sec at 37°C in the model system and the biomembrane, respectively. Protein diffusion is generally more restricted (see review Edidin 1987).

Lateral diffusion of membrane protein has also been studied by electron microscopy on the mitochondrial inner membrane using the freeze fracture technique (Höchli and Hackenbrock 1976). The mitochondria were first cooled down below the phase transition of the lipid. Upon warming up to physiological temperatures, this lateral phase separation is reversed and the former random distribution recovers. In order to determine the lateral diffusion coefficient of IMPs in the mitochondrial inner membrane, Sowers and Hackenbrock (1981) combined electrophoresis and freeze-fracture electron microscopy. An electric current was passed through a suspension of purified inner membrane vesicles, which caused an electrophoretic migration of intramembrane particles in the membrane plane into a single, crowded patch facing the positive electrode. The membrane vesicles were quenched at specific times after removing the electrophoretic force while the particles were diffusing back into a random distribution. A diffusion coefficient of the IMPs at 8.3 x $10^{-2}/\mu m^2$/sec at a temperature of 20°C. This diffusion coefficient for IMPs is consistent with reported values for a variety of proteins for various cellular membranes. In conclusion, lipids of protein (transmembranes) can diffuse rather fast, the former about 100-1000 times faster than the latter. This would predict a homogenous distribution of membrane molecules in the plane of the bilayer, consistent with the fluid mosaic model.

However, there are many examples of proteins and lipids being restricted in lateral motion. A prime example is the tight junction in epithelial cells, which acts as a barrier preventing proteins moving from the apical to the basal-lateral membranes of the cell. It has also been shown that a proportion of the lipids (the glycolipids) are not free to move through this barrier, whereas other lipids classes are (Van Meer et al. 1986). This has been attributed to the fact that the glycolipids are present in the outer leaflet (lipid asymmetry). So it is thought that the outer leaflet lipids are more restricted than the lipids in the inner leaflet. The reason for this is unknown but has to be sought in the structure of the tight junction itself. Alternatively, protein molecules can be non-covalently linked to the cyto- or membrane skeleton, such as is the case for the erythrocyte membrane. In that membrane the linkage between membrane proteins is known in some detail (Branton et al. 1981). Part of the anion pump protein band 3 is anchored to membrane skeleton via ankyrin. Binding of ligands (hormones, growth factors) and antibodies with membrane receptor proteins can lead to non-covalent linkage to clathrin network, a coated pit, which will form coated vesicles by endocytosis.

Clustering of receptors may also occur independent of coated pits. Using label fracture it has been found that EGF receptors form micro-aggregates upon stimulation with EGF. This EGF receptor clustering is thought to be associated with the interaction with other cytoskeleton elements, most likely actin filaments (Boonstra et al. 1988). Clustering of membrane receptors has also been found during capping of lymphocytes. Domain formation can be found in areas where two membranes make close contact. The most prominent example is the gap junction where IMPs are aggregated at the contact area. Another example are the stacked membranes in the chloroplast, where for instance photosystem II is exclusively located in these membranes. Lastly, a protein domain has been found in the membrane of Halobacterium halobium, where bacteriorhodopsin is organized in a paracrystalline organization.

3. LATERAL PHASE SEPARATION

Lipids in biological membranes are fluid, which allow fast lateral diffusion. In several prokaryotic membranes the lipids start to solidify below the growth temperature. This can be detected by DSC (Steim et al. 1969) and X-ray diffraction (Engelman 1970). With freeze fracturing one can visualize aggregation of IMPs upon lowering the temperature (Verkleij and Ververgaert 1975). The membrane proteins and fluid lipids are squeezed out of the domains of solid lipids. Eventually if all the lipids are solid the IMPs are seen to be aggregated in large clusters. It has to be said that fast freezing is required to detect a homogenous distribution of IMPs at 37°C. With conventional freezing in a mixture of liquid and solid nitrogen IMP aggregation is obtained also if one freezes from 37°C (Knoll et al. 1987), caused by the slower rates of freezing. Alternatively the homogeneous distribution can be obtained at 37°C if the cells are fixed with glutaraldehyde before conventional freezing.

That fluidity is essential for biological functioning and is not only clear from the fact that the cold shock alters membrane permeability but also by the fact that if one grows bacteria at lower temperature than 37°C the lipid composition is corrected to maintain the correct fluidity. The membrane becomes enriched in unsaturated fatty acids (Melchior 1982). IMP aggregation has been observed also in eukaryotic membranes but its starts at lower temperatures 5-10°C (see Verkleij and Ververgaert 1975). This is consistent with the phase transition temperature of the lipids in these membranes. So it is found that below 10°C the IMPs of mitochondrial, alveolar, nuclear membranes cluster.

Particle aggregation has also been found under pathological conditions in the sarcolemma membrane of cardiac muscle cells. This phenomenon is here induced isothermally by Ca^{2+} overload in the cytoplasm to levels in the blood (1.3 mM) after a period of ischemia followed by Ca^{2+} rich perfusion (Verkleij and Post, 1987). This can be explained as follows. The membrane phospholipids are also asymmetrically distributed with the negatively charged phospholipids in the cytoplasmic leaflet (Post et al. 1988). Upon Ca^{2+} entry, the negatively charged phospholipids start to solidify. Thus, solidification of phospholipids in one leaflet can induce lateral phase separation and IMP aggregation.

A similar phenomenon has been found in the erythrocyte membrane, however, only when actin and spectrin have been removed. So, it is found that in inside-out vesicles of actin-spectrin stripped ghost, Ca^{2+} and low pH can induce IMP aggregation. This phenomenon was initially explained as a result of the precipitation of the residual actin-spectrin (Elgsaeter et al. 1976). However, this is more likely due to a lipid effect, i.e. lipid solidification since this IMP aggregation does also occur after complete removal of actin and spectrin. Moreover, IMP aggregation in these membranes can also be induced by lowering the temperature (Gerritsen et al. 1976). It is of interest to note that the cholesterol/phospholipid ratio is rather high, i.e. 0.9. This suggests that cholesterol is not able to obliterate phase transitions and thermally induced IMP patching. Thus, phase changes are also possible in the erythrocyte, however, they are obscured and no visible IMP aggregation is seen because of the interaction of IMPs (band 3) with the cytoskeleton elements spectrin-ankyrin-actin. In addition, spectrin phospholipid may also be involved in this restriction (Mombers et al. 1979).

The fact that phospholipids in biomembranes are fluid at physiological temperatures indicates that this physico-chemical state must be of relevance for membrane functioning. Biomembranes where lipids become solidified and exhibit lateral phase segregation become leaky for ions, indicating that fluidity is of importance for the semi-permeable properties of membranes. In eukaryotic cells endocytosis is prevented at low temperatures indicating that fluidity is essential for fusion processes. The activities of many membrane enzymes have been shown to be strongly affected by the transition from the fluid to solid state. This all indirectly shows that fluidity is essential for membrane functioning. At present there is no proof that a fluid-solid transition of lipids is of importance for membrane functions. It is of interest to note the alveolar membranes are very rich in dipalmitoylcholine which has a transition at 41°C. It is conceivable that in these membranes the fluid-solid transition of this lipid plays a regulatory role in the compressibility of membrane surfaces. Moreover, it cannot be excluded that locally solid-fluid transitions may have a regulatory function in various membrane processes. In this way it is reasonable to assume that Ca^{2+} at the cytoplasmic surface of cell membranes where the negatively charged phospholipids are located, i.e. PS en PI can solidify these lipids and may in turn be regulatory in membrane enzyme activities.

In conclusion, fluidity is essential for membrane functioning. The fluid-solid transition has not yet been shown to be of biological direct relevance except under pathological conditions. However, it cannot be excluded that in some membranes locally fluid-solid transitions are important, for instance for the regulation of permeability properties and membrane enzyme activity.

4. LIPIDIC PARTICLES AND FUSION

The hypothesis that lipids preferring the hexagonal H_{II} phase and/or inverted micelle configuration play a crucial role in membrane fusion has been strongly supported by experiments with model membrane systems. This was first demonstrated with vesicles composed of an equimolar mixture of cardiolipin and egg-PC (see Verkleij 1984). These vesicles in fact fuse in a non-leaky fashion (Wilschut et al. 1982). More relevant for biological fusion events are mixtures of PE and negatively charged phospholipids. Ca^{2+}, which is well-known to promote biological membrane fusion, triggers bilayer to non-bilayer (H_{II} and/or inverted micelle) transitions isothermally in these mixtures (Cullis et al. 1982). An increase in temperature also promotes such a transition. It is therefore logical to suppose that the presence of Ca2+ or an increase in temperature allows the non-bilayer tendency of endogenous lipids to be expressed, thus promoting the fusion event. These predictions hold for a variety of mixture when Ca^{2+} is added (see for reviews Cullis et al. 1982; Verkleij 1984), such as PE/PS, PE/PI, PE/PA, PE/cardiolipin, PE/PC/cholesterol/PS and when the temperature is raised in PE/PC/cholesterol. In all these cases the incubations result in the formation of larger structures showing that net fusion has occurred.

In all these mixtures undergoing fusion this event is associated with the appearance of lipidic particles sometimes localized in regions corresponding in the fusion interface. In a review (Verkleij 1984), it has been demonstrated that the different particle types, which confused the issue in the early days of the discovery of these lipidic particles, can be interpreted as a reflection of different stages during the fusion event.

5. BIOMEMBRANE FUSION

Initially it has been generally assumed that membrane proteins, both the extrinsic or membrane skeleton proteins as well as the intrinsic membrane spanning proteins, have to be cleared from the fusion site before actual fusion can start. This has been deduced from freeze-fracture studies, where IMP-clearing and a so-called diaphragm at the site of fusion was seen. However, these studies have been carried out using chemical fixation and cryoprotectants. Such diaphragm formation is not seen using fast freezing without these pretreatments. Thus, only a small area about 10 nm in diameter is necessary for fusion, so-called local point fusion (Chandler and Hauser 1980).

The visualization of lipidic particles in artificial membranes as possible fusion intermediates has led to further research for such structures during biological membrane fusion. In biological systems it is in principle difficult to trap fusion intermediates, because; (1) of the short life time, (2) the problem to synchronize the fusion event, and (3) the amount of fusion events per surface. Even with exocytosis only a few hundred vesicles at most per cell involved in fusion. However, Schmidt et al. (1983) have been able to catch exocytosis in chromaffin cells using fast freezing cleavage and statistical analysis. After stimulation of the exocytotic activity in a controlled way by adding carbachol, the plasma membrane displayed structural features which strongly resemble or better which are identical to the intermediates seen in model systems, i.e. lipidic particles and vulcanoe-like protrusions. Such structures have also been visualized using fast freezing during fusion of membranes in blood platelets upon stimulation (Hols et al. 1985). Budding of viruses has been shown to be associated with well-defined particles being fusion intermediates at the nexus of the plasma membrane and the membrane of the budding influenza virions. Membrane intercalated particles as fusion intermediates have also been found on liposomes and the apical membrane of MDCK cells during fusion of liposomes with these cells (Knoll, in preparation). In summary, well-defined particles have been seen in some systems during fusion.

6. CONCLUSIONS

Electron microscopy is of great value in the analysis of membrane structure, dynamics and function. This is due especially to cryotechniques, freeze fracturing (etching) and freeze substitution, in combination with rapid freezing. Cryofixation reliably preserves membrane structure (space resolution) and can follow dynamic events (time resolution) such as lateral movement and fusion. Many concepts such as lipid bilayer organization, membrane heterogeneity or membrane domains, lateral phase separation, lipid polymorphism, local point fusion, lipidic particles etc. have been derived and/or supported by this morphological approach. Freeze fracturing, cryosectioning, freeze substitution in combination with immunogold labelling and autoradiography give us additional powerful tools for the analysis of membrane architecture and function, since we are now in principle able to identify and to follow membrane components during their lifetime.

References

Branton D, Cohen CM and Tyler J 1981 Cell **24** 24

Boonstra J, Van Belzen N, Van Bergen en Henegouwen PMP, Hage WJ, Van Maurik P, Wiegant FAC and Verkleij AJ 1988 CRC in press

Chandler DE and Hauser JF 1980 J Cell Biol **86** 666

Cullis PR, Hope MJ, de Kruijff B, Verkleij AJ and Tilcock CPS 1985 Phospholipids and Cellular Regulation (New York: CRC Press) Vol 1 p

Edidin M 1987 Curr Top Membr Transp **29** 91

Elgsaeter A, Shotton D and Branton D 1976 Biochim Biophys Acta **426** 101

Engelman DM 1970 J Mol Biol **47** 115

Gerritsen WJ, Verkleij AJ and van Deenen LLM 1976 Biochim Biophys Acta **555** 26

Höchli M and Hackenbrock CR 1976 Proc Natl Acad Sci USA **73** 1636

Hols H, Sixma JJ, Leunissen-Bijvelt J and Verkleij AJ 1985 Tromb Haemost **54** 574

Knoll G, Verkleij AJ and Plattner, H 1987 Cryotechniques in biological electron microscopy (New York: Springer Verlag) pp 258

Kopstad G and Elgsaeter A 1982 Biophys J **40** 163

Melchior DL 1982 Curr Top Membr Transp **17** 263

Mombers C, Verkleij AJ, de Gier J and van Deenen LLM 1979 Biochim Biophys Acta **551** 271

Post JA, Langer GA, Op den Kamp JAF, Verkleij AJ 1988 submitted

Schmidt A, Patzak A, Lingg G, Winkler H and Plattner H 1983 J Cell Biol **32** 31

Singer SJ and Nicolson GL 1972 Science **75** 720

Sowers AE and Hackenbrock C 1981 Proc Natl Acad Sci USA **78** 6246

Steim JM, Tourtelotte ME, Reinert JC, McElhaney RN and Rader RL 1969 Proc Natl Acad Sci USA **63** 104

Van Meer G, Gumbiner B and Simons K 1986 Nature **322** 639

Verkleij AJ 1984 Biochim Biophys Acta **779** 43

Verkleij AJ and Post JA 1987 Lipid metabolism in normoxic and ischaemic heart (Darmstadt: Steinkopff Verlag) pp 85

Verkleij AJ and Ververgaert PHJTh 1975 Ann Rev Phys Chem **26** 101

Wilschut J, Holsappel M and Jansen R 1982 Biochim Biophys Acta **690** 297

The cytoskeleton

Frans C.S. Ramaekers

Department of Pathology, University of Nijmegen,
Geert Grooteplein Zuid 24, 6525 GA Nijmegen, The Netherlands

Within the cytoskeleton of the cell three different types of filaments can be distinguished on basis of their ultrastructural appearance and biochemical composition.

1. Microfilaments (5 nm), containing actin and which assemble into bundles frequently called stress fibers. These latter structures may be associated with myosin, α-actinin and tropomyosin, and various other proteins (Fig. 1a).
2. Microtubules (25 nm) which consist of α- and β-tubulin (Fig. 1b).
3. Next to microfilaments and microtubules, filaments measuring 8-11 nm in diameter are commonly seen in mammalian cells. These so-called intermediate-sized filaments (Wang et al 1985), which often constitute a considerable part of the intracellular matrix (Fig. 1c), are extremely insoluble and show a protein composition which is completely different from that of microfilaments and microtubules.

Possible functions of these cytoskeletal and contractile structure in eukaryotic nonmuscle cells have been the object of many studies. It is now generally accepted that microfilaments are involved in the realization of cell shape and in cell dynamics. Processes like cell contraction and elongation, exocytosis, membrane ruffling and movement on a substrate find their molecular basis in the interaction of continuously aggregating microfilament bundles with the plasma membrane.

Although microfilament bundles (stress fibers) appear to be highly organized structures (Fig. 2), they can disaggregate rapidly into an amorphous meshwork, mainly observed in rounded cells that do not adhere strongly to a substratum. This meshwork is commonly located just below the cell membrane and is also seen in areas of membrane ruffling in motile cells. The two states of microfilament aggregates should be considered as being in equilibrium and interconvertable. Bundles of microfilaments give the cell structural support, while the microfilaments building up the meshwork are responsible, at least partly, for cellular motility processes.

As a rule microtubules exist as singlets in the cytoplasm and play a role both in cell division and in the determination of cell shape. They are involved in motility processes that are either based on microtubule assembly or tubular sliding mechanisms. It has been proposed that microtubules are involved in cell elongation, while microfilaments are responsible for cytoplasmic contractions. The growth of axons, for example, depends on the elongation of microtubules, since the disassembly of these filaments by mitotic drugs blocks axonal extension and promotes axonal retraction in vitro.

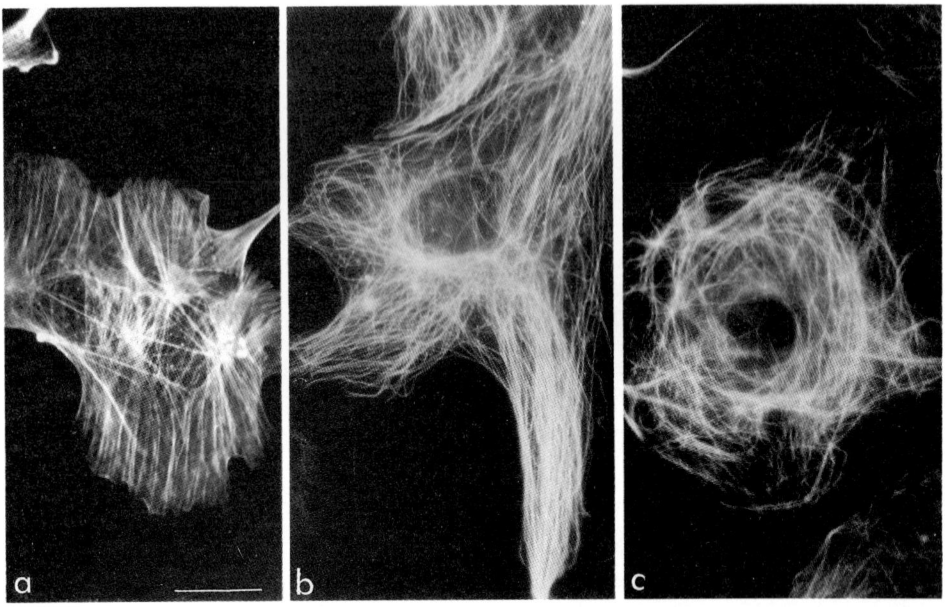

Fig. 1. Different cytoskeletal structures visualized in cultured lens cells stained for actin (a), tubulin (b) and vimentin (c) using the indirect immunofluorescence technique. Bar = 20 micron.

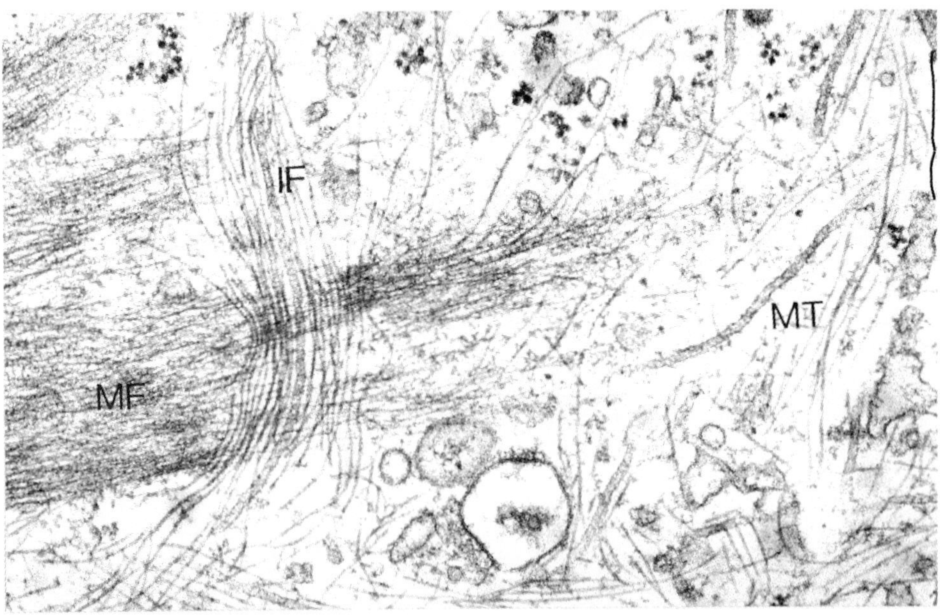

Fig. 2. Electron micrograph of a section of a cultured lens cell, showing bundles of microfilaments (MF), intermediate-sized filaments (IF) and microtubules (MT). Bar = 0.5 micron.

The function of intermediate filaments is still a matter of debate. Biochemical and immuno(cyto)chemical studies have demonstrated that five different types of intermediate filament proteins (IFP) can be distinguished in mammalian cells. These include the cytokeratins, vimentin, desmin, glial fibrillary acidic protein (GFAP) and the neurofilament triplet.

These different types of IFP are distributed in a tissue-specific manner, and from Table 1 it may be obvious that the subdivision made between tissues on basis of their IFP content corresponds strikingly well with the major tissue classification based on histologic principles.

Monoclonal and polyclonal antibodies to the different types of IFP could be shown to react in a tissue specific manner, meaning that in general antibodies to cytokeratin react only with epithelial tissues, antibodies to desmin only with muscle tissues, GFAP antibodies react only with glial cells and neurofilament antibodies react with neural cells. Antibodies to vimentin normally react only with cells and tissues of mesenchymal origin, but for this type of IFP some exceptions are known (see below). The different types of normal tissues denoted in Table 1 can thus be distinguished by immunocytochemical methods, using specific antibodies to IFP. Using such antibodies also the different types of intermediate filaments can be distinguished at the ultrastructural level using the immunogold-labeling technique (Fig. 3).

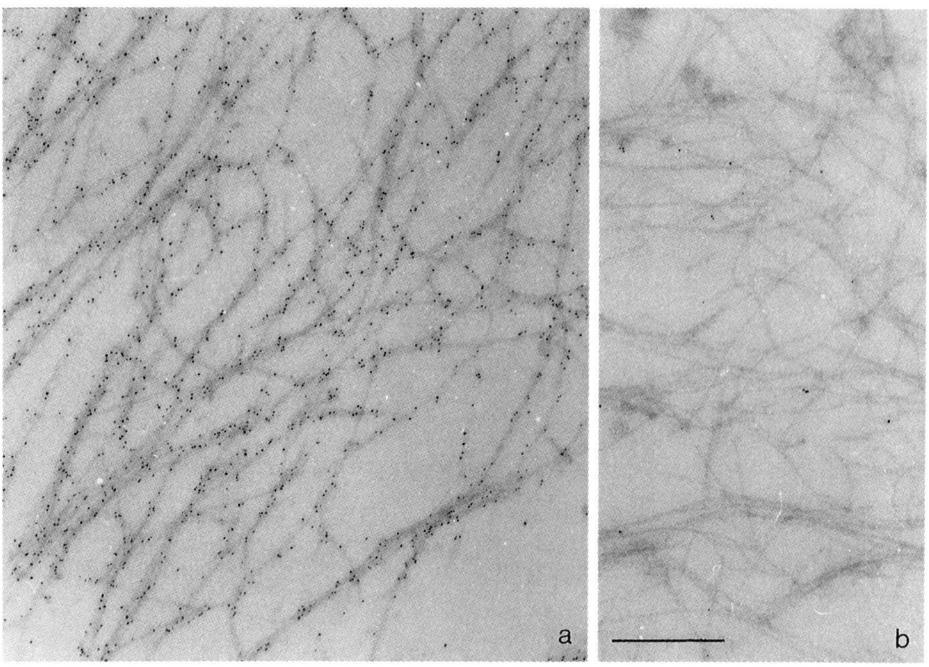

Fig. 3. Immuno-electron microscopy of a BHK cell culture, showing immunogold-labeled desmin filaments (a). The cells were incubated with a polyclonal rabbit antiserum to desmin and thereafter with a second goat-anti-rabbit Ig antibody labeled with 5 nm gold particles. (b) Control incubation in which only the second antibody was applied. Bar = 2 micron.

Exceptions to the rule as outlined in Table 1 have also been described. These include some epithelioid tissues, such as the seminiferous epithelium, lens epithelium, pigment epithelium in the retina, ciliary epithelium and epithelial cells covering the iris which do not contain cytokeratins but posses vimentin IFP. Mesothelial cells have been suggested to co-express cytokeratins and vimentin, a phenomenon also seen in some other types of epithelial cells.

Table 1. Tissue- and tumour specificity of intermediate filament proteins.

Type of IF protein	Tissue type	Tumour type
(cyto)keratins	epithelial tissues	carcinomas
vimentin	mesenchymal tissues	lymphomas, sarcomas
desmin	muscle tissues	myosarcomas
GFAP	astroglial cells	astrocytomas
neurofilament proteins	nerve tissues	some neural tumours

Some types of neurons have shown to be completely devoid of intermediate filaments, while some types of vascular smooth muscle cells contain either vimentin alone or vimentin next to desmin.
Furthermore, it should be kept in mind that certain tissue types in developing embryos may contain IFP different from those present in their adult counterparts. For example myoblasts contain vimentin instead of desmin and neonatal glial cells may contain vimentin in addition to GFAP. Some types of kidney epithelial cells transiently express vimentin during embryogenesis. Finally, cells brought into tissue culture commonly express vimentin next to their cell type specific IFP.
It has been shown that generally tumour cells retain their original IFP, with the exception of a few types of neoplasms. Therefore, antibodies to IFP are widely applied in tumour histopathology and cytopathology (Osborn & Weber 1983; Ramaekers et al 1983).

Cytokeratins are a family of IFP. In contrast to other types of intermediate filaments, cytokeratin filaments are characterized by a remarkable biochemical diversity, represented in human epithelial tissues by at least 19 different cytokeratin polypeptides ranging in molecular weight between 40 and 68kDa, designated 1 to 19, cytokeratin 1 being the polypeptide with the highest molecular weight and highest isoelectric pH and cytokeratin 19 being the polypeptide with the lowest molecular weight and lowest isoelectric pH (Moll et al 1982). These polypeptides are not expressed randomly throughout epithelia but occur in cell-type specific combinations. Also the diverse types of carcinomas differ in their cytokeratin polypeptide content. The polypeptide patterns of epithelial tumours are either identical with the cytokeratin pattern present in the cell of origin or at least closely related to it. The cytokeratin polypeptide patterns (as obtained by two-dimensional gel electrophoresis) can be used for the identification or at least a subclassification of epithelial tissues and carcinomas.
Several monoclonal antibodies to different cytokeratins have been prepared and different reactivities of such antibodies with different subtypes of epithelial tissues and epithelial tumours have been noted

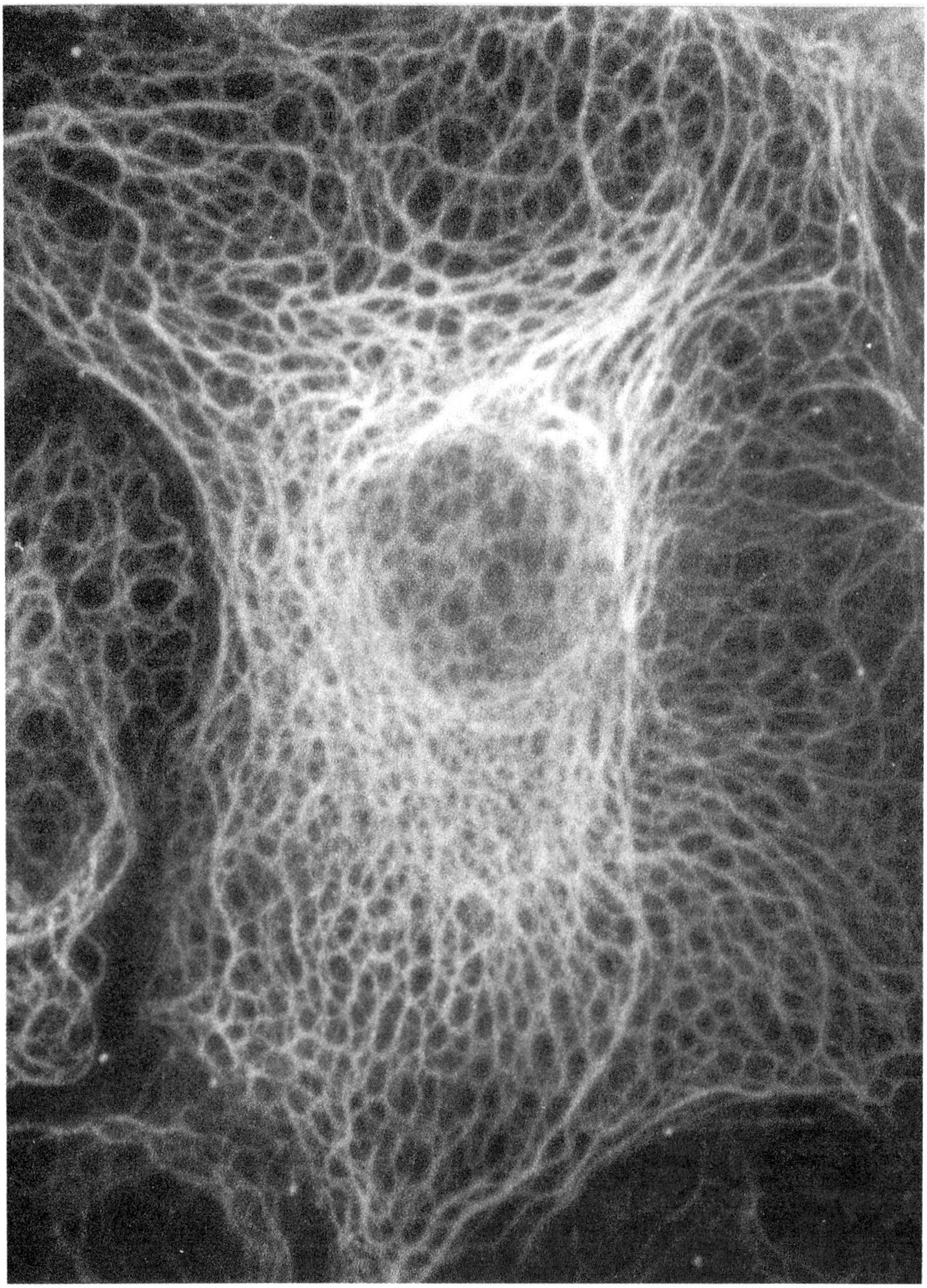

Fig. 4. Cytokeratin filaments in a cultured human hepatoma cell as detected by a monoclonal antibody to cytokeratin 18.

(Cooper et al 1985). Chain-specific cytokeratin antibodies have also been shown to be able to detect minor subpopulations of cells present in complex epithelial tissues or in heterogeneous carcinomas.

REFERENCES

Cooper D, Schermer A, Sun T-T 1985 Lab. Invest. **52** 243
Moll R, Franke WW, Schiller DL, Geiger B, Krepler R 1982 Cell **31** 11
Osborn M, Weber K 1983 Lab. Invest. **48** 372
Ramaekers FCS, Puts JJG, Moesker O, Kant a, Huysmans A, Haag D, Jap PHK, Herman CJ, Vooijs GP 1983 Histochem. J. **15** 691
Wang E, Fischman D, Liem RKH, Sun T-T (Eds) 1985 Intermediate filaments. Ann. N. Y. Acad. Sci., Volume 455.

Inst. Phys. Conf. Ser. No. 93: Volume 3, Chapter 16
Paper presented at EUREM 88, York, England, 1988

Structural differentiation of membranes in a water-shunting epithelial complex of an insect

D. Thomas, J.F. Hubert, A. Cavalier and J. Gouranton

Laboratoire de Biologie Cellulaire, U.A. 256 CNRS
Université de Rennes I, Campus de Beaulieu
35042 Rennes-Cédex, France.

A large body of evidence indicates that the increase in water permeability induced by antidiuretic hormone (ADH) on target epithelial barriers is the result of the insertion in plasma membranes of particle aggregates which probably represent water channels. In amphibian urinary bladders, these structures are observed in freeze-fractured preparations but they represent only 1 to 2% of the apical surface of ADH-sensitive epithelial cells so that their biochemical nature is still undetermined (Chevalier et al 1985).

The digestive tract of some homopteran insects presents a highly specialized structure called filter chamber. The purpose of this structure is thought to be the shunting of the excess dietary water from the initial part of the midgut to the terminal part of the midgut and the proximal regions of the Malpighian tubules.

Previous electron microscopical studies of the filter chamber of Cicadella viridis revealed very thin epithelia with numerous tubular or lamellar basal infoldings and apical microvilli (Gouranton 1968).

Freeze-fracture experiments (Figure 1) carried out on these filter chambers reveal an arrangement of intramembrane particles on the whole membrane surface of these microvilli and basal infoldings. The **E** face exhibits a regular network with a constant repeat unit of 8 nm. The **P** face shows numerous aggregated particles. These structures are quite similar to ADH-induced particle aggregates described in amphibian urinary bladder (Chevalier et al 1979).

The particle containing membranes were isolated and purified. A negatively stained membrane preparation contains mostly vesicles and rods (Figure 2). Membranes appear spotted with numerous stain excluding regions which correspond to a fine and irregular pattern of surface projections. These projections have a 10 nm diameter and extend radially about 5-6 nm from the membrane surface demarcated by the stain.

Freeze-dried, shadowed membrane preparations also display vesicles and rods (Figure 3). Two kinds of membrane surface are observed : a mostly rough surface covered with irregularly clustered particles or an essentially smooth surface with crystalline subunits (periodicity = 9 nm) appearing as undulations. Investigations of folded membranes produced by broken vesicles revealed that these two types of surface correspond respectively to the outer and to the inner membrane surfaces.

These results are in agreement with the freeze-fracture observations. The numerous particles seen on the **P** fracture face correspond to those which are observed after shadowing on the inner surface of isolated membranes and which protrude into the cytoplasm. The regular network seen on the **E** fracture face is related to the undulations observed after shadowing on the outer surface of isolated membranes. Thus the membrane structure is assymetric.

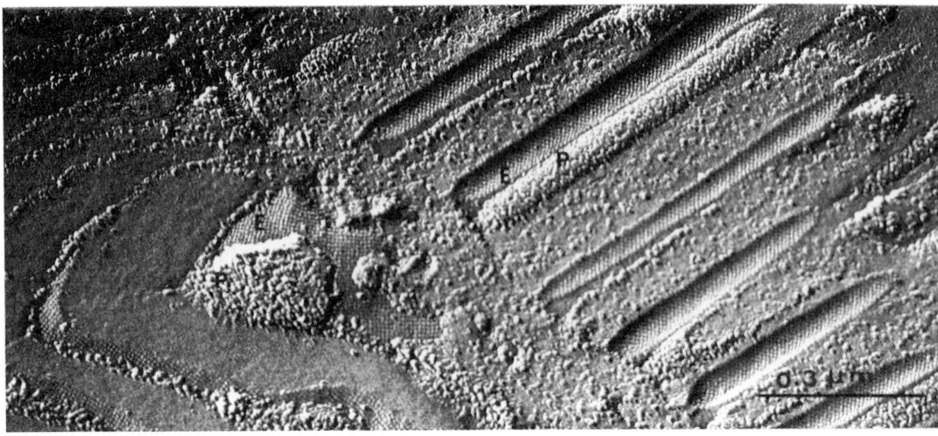

Figure 1 - Freeze-fracture of a Malpighian tubule cell. On the right : microvilli. On the left : basal infolding membranes.

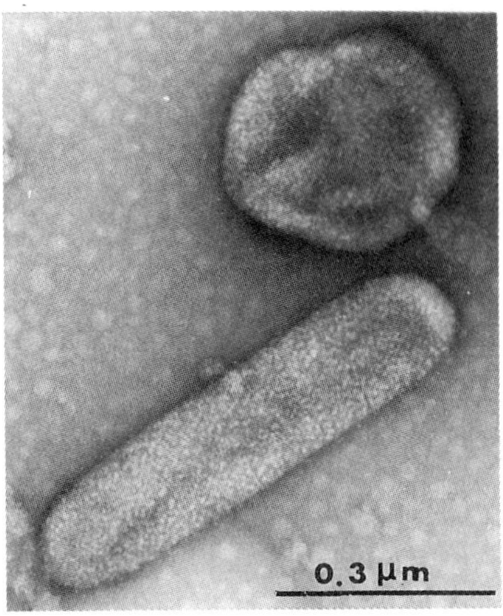

Figure 2 - Negatively stained membranes.

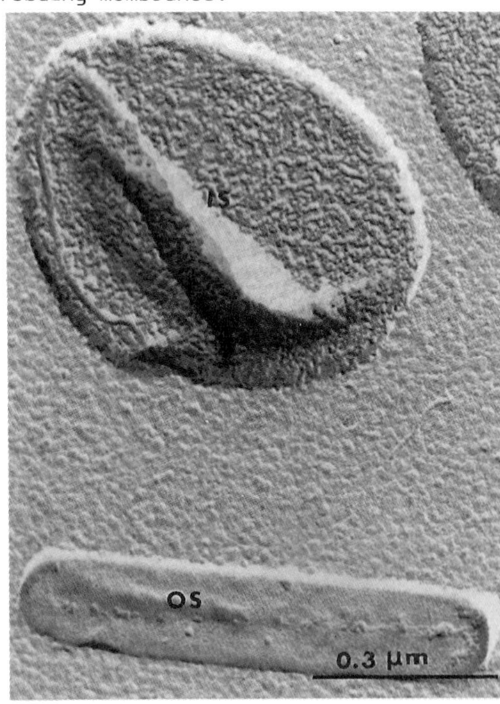

Figure 3 - Freeze-dried shadowed membranes. IS : inner surface; OS : outer surface.

The membrane differentiation described here is found only in the filter chamber of the digestive tract and is probably involved in water transit, a function attributed to this complex. The present material may provide a good model for the study of the possible relationship between intramembrane particles and water permeability.

Chevalier J, Parisi M and Bourguet J 1979 Bio. cell. **35** 207

Chevalier J, Pinto Da Silva P, Ripoche P, Gobin P, Wang X, Grossetete J and Bourguet J 1985 Bio. cell. **55** 181

Gouranton J 1968 J. Microscopie **7** 559

Paper presented at EUREM 88, York, England, 1988

Localization of cytoskeletal proteins in *Dictyostelium discoideum*

Marijke Brink

Max-Planck-Institut für Biochemie, D-8033 Martinsried, Federal Republic of Germany

ABSTRACT: Cells of *Dictyostelium discoideum* were prepared for immunofluorescence- and immunogold-labelling of actin, α-actinin, gelation factor, myosin, capping protein and severin. Labelling of cytoskeletons and of non-extracted cells were studied.

Cytoskeleton preparations of *Dictyostelium discoideum* show the organization of the microtubule system, a network of filaments between them, a dense texture of filaments at the periphery of the cells and microfilament bundles in spike-like extensions (Claviez *et al* 1986). Two questions remained unsolved until now. 1. What is the nature of the filaments that we see: do they all consist of actin or does *Dictyostelium discoideum* possess an intermediate filament system as well? 2. Where are actin-binding proteins located? Their localization might give insight into their in-vivo functions in rearranging actin and regulating motility.

Cytoskeletons (Figures 1-4) were prepared by extracting cells adhering to pioloform films on gold grids with 1% Triton X-100 and 0.05% glutaraldehyde in a cytoskeleton stabilizing buffer, followed by postfixation with 0.5% glutaraldehyde and staining with 2% ammonium phosphotungstate. Parallel preparations for fluorescence-microscopy, stained with TRITC-labelled phalloidin, showed that microspikes (Figure 1) contain filamentous actin. In addition, fluorescence was seen at the periphery and at the broader moving front of the cells, from which we conclude that the dense meshwork seen in the electron-microscope (Figure 2) consists of filamentous actin as well. The character of the other filaments that are seen, e.g. in the area of the microtubules is not yet known (Figure 3). The applicability of immunolabelling to cytoskeleton preparations has been shown for tubulin (Figure 4), and should be possible for actin and actin-binding proteins. Work on the labelling of cytoskeletons with antibodies against actin and actin-binding proteins is in progress. Monoclonal antibodies against α-actinin, a 120kD gelation-factor, severin, capping protein and myosin are available. Binding sites of anti-α–actinin, anti-gelation-factor and anti-myosin antibodies were determined by rotary-shadowing of antibody-antigen complexes. Binding of the antibodies to the functional site of the antigen would make them inadequate for the labelling of cytoskeletons.

Whereas immunolabelling of cytoskeletal preparations would give information about the location of proteins that are bound to the cytoskeleton, staining of non-extracted cells (cryosections, methanol-treated whole cells) detects the unbound, Triton-extractable proteins as well. Up until now actin-binding proteins have only been studied on cryosections and methanol-treated whole cells. At the light-microscopical level diffuse cytoplasmic staining was found for α-actinin, gelation-factor, severin, and monomeric myosin. Labelling with an antibody that recognizes filamentous as well as monomeric myosin resulted in staining of the cell periphery. Finally the capping protein is clearly enriched in the moving front of the cells.

REFERENCES:

Claviez M, Brink M, Gerisch G 1986 *J. Cell Sci.* **86** 69

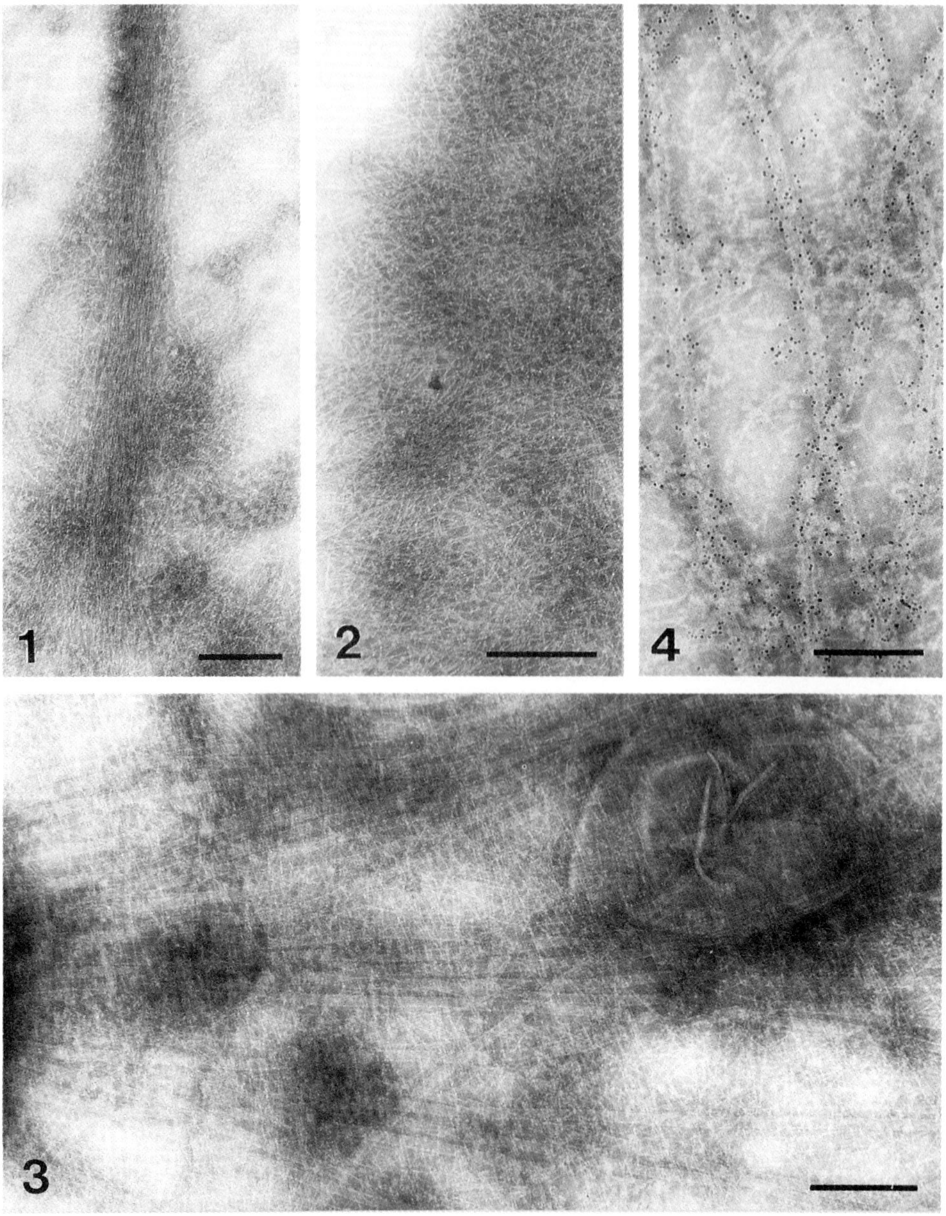

Figure 1-4: Negatively stained cytoskeleton preparations of *Dictyostelium discoideum* showing microfilament bundles in a microspike (1), the filament network in the cortex (2), filaments that embed microtubules (3) and the labelling of microtubules with a monoclonal antibody coupled to 5nm gold (4)(Bars 0.25μm).

Inst. Phys. Conf. Ser. No. 93: Volume 3, Chapter 16
Paper presented at EUREM 88, York, England, 1988

Fracture-flip: where freeze-fracture is turned upside down and comes from behind the looking glass to reveal the cell surface

Pedro Pinto da Silva, Catarina Anderson Forsman and Kazushi Fujimoto

Membrane Biology Section, Laboratory of Mathematical Biology, National Cancer Institute, Frederick Cancer Research Facility 538/104, Frederick, MD 21701 USA

It is now twenty years since I presented the experimental proof of membrane splitting to the International Botanical Congress at Seattle. Over these years we looked at membrane faces as signals of an enigmatic reality that appeared as visually close (its beauty, its crispness) as it was intellectually remote. The characterization of the freeze-fracture image, first attempted by freeze-etch cytochemistry, continued through indirect biochemical routes (membrane recombinants) and, in recent years, through the development of fracture-label (Pinto da Silva et al., 1981; reviewed in Pinto da Silva, 1987a,b) and label-fracture (Pinto da Silva and Kan, 1984). The paradox of the freeze-fracture image--images almost tactile, yet unrelatable to any common experience--was not resolved. Could we pass this virtual reality, this image beyond a mirror? What transformation (transformation in a mathematical sense) was necessary?

We did it and we report here. The transformation is an inversion, almost the inversion of a matrix. It is a corollary of label-fracture. With label-fracture we showed that the split membrane halves remain attached to the replica. Thus, exoplasmic halves of a membrane can be labeled (pre- or post-fracture) and superimposed, coincident images of the label and of the Pt/C replica of the E face are observed. We reasoned: if the outer half of the membrane remains attached to the replica it should be possible to stabilize the fracture face with (electron translucent) carbon, thaw, wash and invert the replica (FLIP) to expose the actual surface of the membrane. Then, we could obtain its high resolution cast by Pt/c evaporation.

The sequence of steps is illustrated below. Fracture-flip is now giving us in a constant, rapid stream images of a new, intimate world. To interpret the replicas some adaptation is required. For instance: cells remain attached to the replica by their P faces or by cross fractured interfaces; transmembrane proteins that, on fracture, partition with the P half are not be seen. Stereo pairs provide striking images of this new reality in a way that never happened with freeze-fracture. The cell surface is the surface of an object, intrinsically familiar as any; in contrast, the fracture face is but the signal of a process, a shadow flickering in Plato's cave.

The routine resolution of Fracture-Flip is about 2-3 nm, much higher than that of SEM of biological specimens. Fracture-flip allows also the observation of the surfaces of intracellular membranes (after isolation of organelles) and is easily combined with colloidal gold cytochemistry. We present our initial results to this meeting (Anderson Forsman & Pinto da Silva; Fujimoto & Pinto da Silva).

References:
Pinto da Silva P, Parkison C and Dwyer N 1981 Proc. Natl. Acad. Sci. U.S.A. **78** 343
Pinto da Silva P and Kan FW 1984 J. Cell Biol. **99** 1156
Pinto da Silva P 1987a Electr. Microsc. Membrane Proteins (Harris JR and Horn RW, eds) Academic Press: New York **6** 2-38
Pinto da Silva P 1987b Adv. Cell Biol. (Miller K, ed) J.A.I. Press: Greenwich, Conn. **1** 157-190

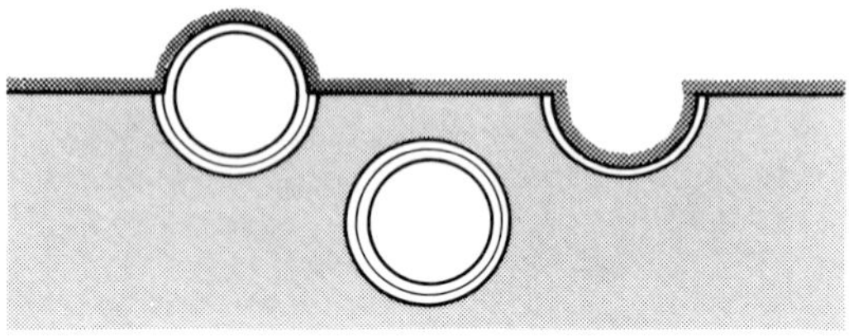

Cells are fractured and carbon cast

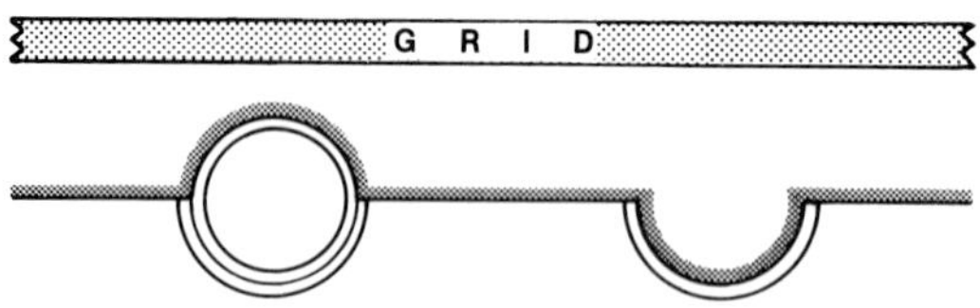

Specimens are thawed, washed and picked on grids

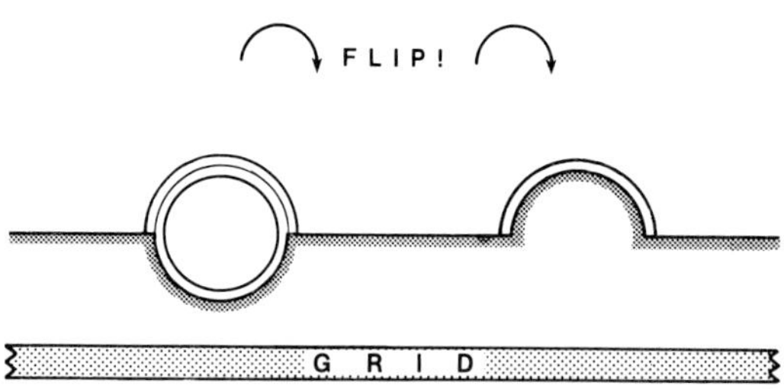

Membrane surfaces are now on top

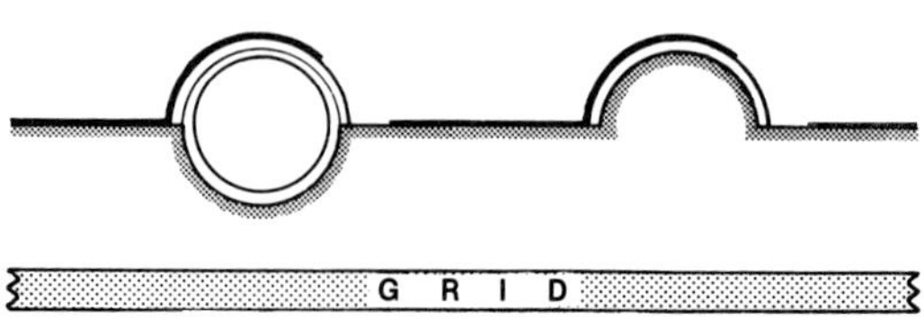

Pt/C images cell surfaces; Look!

Progress in the reconstruction of *E coli* B Omp F porin with lipids into regular single-layered membranes

A Holzenburg, A Engel and U Aebi

Maurice E.Müller-Institute for High Resolution Electron Microscopy at the Biozentrum of the University of Basel, Klingelbergstrasse 70, CH-4056 Basel, Switzerland

ABSTRACT: A novel flow dialysis device is presented which has successfully been employed for the reconstitution of *E. coli* B Omp F porin with lipids into regular single-layered porin sheets. Since good crystalline order is achieved even without lipolysis, this method seems to be a powerful tool for future investigations of structure-function relationships of porin and membrane proteins in general.

Reconstitution of solubilized membrane proteins into lipid vesicles allows to perform functional studies in an environment that mimics the situation *in vivo*. In addition, the formation of ordered protein-lipid arrays is a prerequisite for structural studies. Highly ordered 2D arrays within single-layered membranes that are sufficiently large for patch clamping experiments facilitate both, structural as well as functional studies.

A first step towards these goals is a simple and reproducible purification procedure that warrants full biological activity of the protein of interest. Outer membrane porin (Omp F) from *Escherichia coli* B was purified in a single step after solubilization according to Garavito and Rosenbusch (1986) with octyl (polydisperse)-oligooxyethylene [octyl-POE] employing preparative free-flow electrophoresis (Holzenburg et al., in preparation). The integrity of porin and bound lipopolysaccharide-moieties [LPS] which are known to be essential for the biological activity (c.f. Lugtenberg and Van Alphen, 1983) was maintained, thus providing the basis for structural studies with functional implications.

A second step is the reproducible formation of well-ordered 2D arrays. Besides the starting conditions the rate of detergent removal and the temperature critically affect the formation, ultimate size and order of protein-lipid arrays. To precisely control these two parameters we built a flow dialysis [FD] device (Fig.1) which allows generation of highly reproducible temperature profiles as a function of time. This is achieved with a microprocessor (Z-80) controlling a Peltier element. The dialysis buffer is brought to a preselected temperature during its travel through a meandering path before entering the dialysis chamber. The flow rate is adjusted via a peristaltic pump that is controlled by the Z-80 as well. Detergent concentration can be varied with time using a gradient mixer. 250 µl homogeneous porin-LPS (2 mg protein/ml HEPES-buffer containing 0.5 % v/v octyl-POE) were mixed (Dorset et al., 1983, modified) with 250 µl of L-α-dimyristoylphosphatidylcholine [DMPC] (2mg/ml HEPES-buffer containing 1.5 % v/v octyl-POE) and dialyzed for 17 hrs against HEPES-buffer (20mM HEPES, 10mM $MgCl_2$, 0.1M NaCl, 0.2mM EDTA, 3mM NaN_3, pH 7.0) without octyl-POE at 37 °C (flow rate: 0.9ml/min), for 24 hrs against distilled water (37 °C, 0.9ml/min), followed by a period of 6 hrs against distilled water (flow rate: 0.4 ml/min) during which the temperature was linearily decreased to 21 °C. Samples were prepared for electron microscopy as descibed earlier (Dorset et al., 1983), however, uranyl formate (0.75 % w/v, pH 4.25) was used in this case.

By means of this novel FD it was possible to yield ordered arrays within single lipid bilayers (Fig 2). The sharp first order reflections of optical diffractograms recorded from typically 5-µm-diameter areas (Fig.2, insert) indicate excellent long range order. Diffraction pattern from higher magnification (50k x) micrographs extend to $(2.2\ nm)^{-1}$ (Fig.4a). A correlation-averaged projection normal to the membrane plane is shown in Fig. 4b. For comparison a typical preparation employing the same starting conditions but conventional equilibrium dialysis (Dorset et al., 1983) is displayed in Fig. 3. As indicated by arrowheads single-layered membranes are apparent in the peripheral vesicle areas only.

This work was supported by the Swiss National Science Foundation (grant # 3.524-0.86) and by the M.E.Müller Foundation of Switzerland.

REFERENCES:

Dorset DL, Engel A, Häner M, Massalski A and Rosenbusch JP 1983 J.Mol.Biol.**165** 701

Garavito RM and Rosenbusch JP 1986 Methods in Enzymology **125** eds S Fleischer and B Fleischer (Orlando: Academic Press) pp 309-328

Lugtenberg B and Van Alphen L 1983 Biochem.Biophys.Acta **737** 51

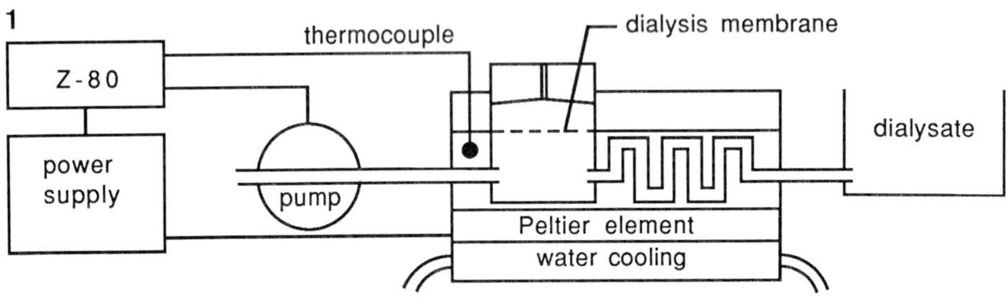

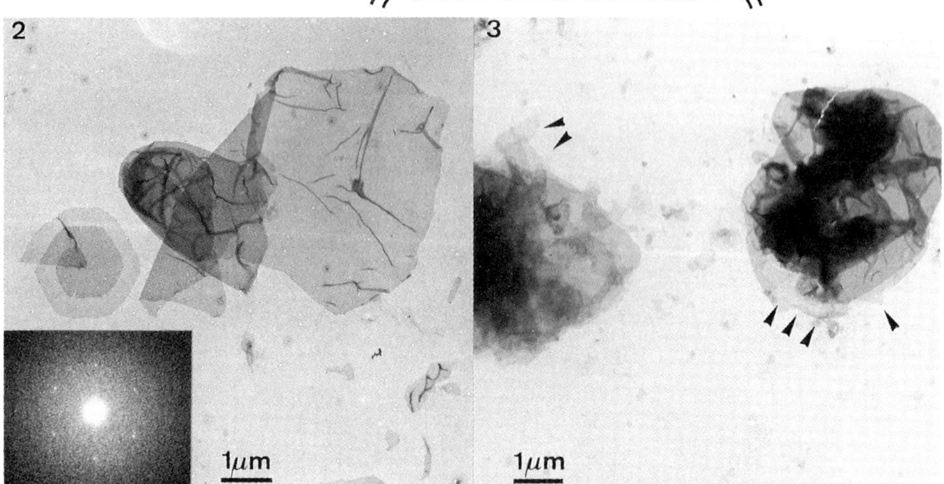

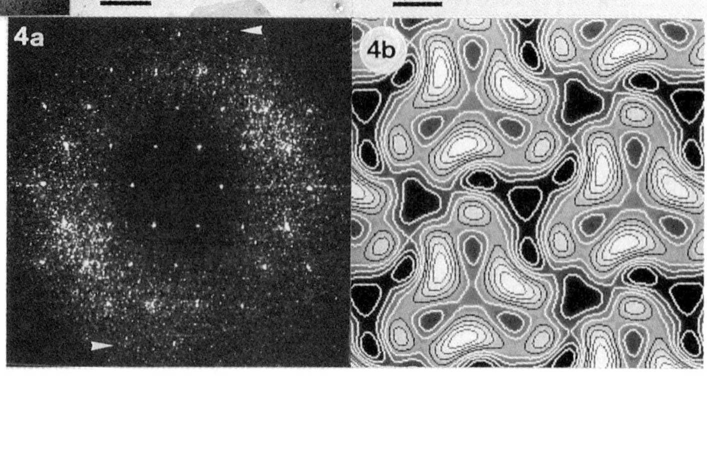

Differential effects of adriamycin and daunomycin on the cell shape and plasma membrane ultrastructure of human erythrocytes

G Arancia, A Molinari, P Crateri, A Calcabrini, L Silvestri°, G Isacchi°

Department of Ultrastructures, Istituto Superiore di Sanità, Viale Regina Elena 299, 00161 Rome and °Department of Human Biopathology, Section of Hematology, University of Rome "La Sapienza", Italy

ABSTRACT: The effect of adriamycin and daunomycin on the cell shape and plasma membrane ultrastructure of human erythrocytes were investigated by scanning electron microscopy and freeze-fracturing. The two anthracycline congeners induced different morphological and ultrastructural changes suggesting a different interaction with the plasma membrane.

1. INTRODUCTION

Adriamycin (ADR) and daunomycin (DAU), two congeners of the anthracycline class widely used against a variety of solid tumors and leukemias, have a very similar chemical structure, differing by only a hydroxyl, but display quite a different pharmacological activity. Their cytotoxic effect is mainly due to the capability of intercalating into the double-stranded DNA thus inhibiting nucleic acid synthesis. However, a large body of evidence has demonstrated that the interaction with the plasma membrane plays a crucial role in the mechanism of cytotoxic activity exerted by both ADR and DAU. Therefore, the differential pharmacological activity of the two drugs might be ascribed to different modalities of interaction with the plasma membrane. In order to verify this hypothesis, we carried out a comparative freeze-fracture (FF) and scanning electron microscopy (SEM) study on the effects of ADR and DAU on human erythrocytes.

2. RESULTS AND DISCUSSION

Fresh human erythrocytes were treated in vitro for 2 h at 37°C with 50 and 100 µM ADR and DAU before processing for SEM and FF. SEM observations revealed a remarkable dose-dependent effect induced by both drugs on the cell shape. However, ADR produced a discocyte to stomatocyte transition at 50 µM and a mottled appearance at 100 µM (Fig. 1), whereas DAU transformed discocytic cells to cupped forms only at the higher concentration (Fig. 2). Moreover, DAU induced an intense vesiculation in the concave area of the cell surface detectable either by SEM (Fig. 3) or by FF (Fig. 4). This vesiculation process was never observed in ADR treated cells. FF revealed a very different effect of the two substances on the plasma membrane ultrastructure. In fact, on both exoplasmic and protoplasmic fracture faces of cells treated with 100 µM DAU, the density and distribution of the intramembrane particles (IMP) appeared to be very similar when compared to control cells, whereas 100 µM ADR produced a very

particular IMP distribution characterized by the presence of numerous smooth tracks (Fig. 5) on both fracture faces.
Our ultrastructural observations, together with biochemical data, suggest that DAU binds to red blood cell membrane producing alterations which mainly consist in a membrane vesiculation inside the cell whereas ADR is incorporated in the lipid bilayer inducing remarkable changes in cell shape and plasma membrane molecular organization. In addition to confirm a significant interaction of ADR and DAU with the cell membrane, these findings might account for their different cytotoxic activity.

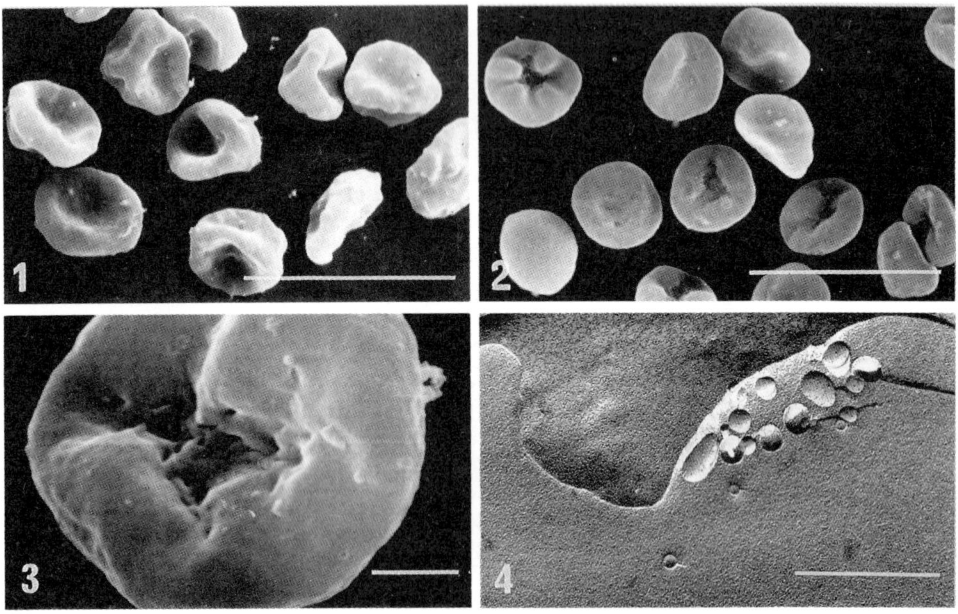

Fig. 1. Cells treated with 100 μM ADR exhibited a mottled appearance.
Fig. 2. The treatment with 100 μM DAU induced the formation of cupped cells. Fig. 3. A vesiculation process was detectable in the concave area of cells treated with 100 μM DAU. Fig. 4. In cross-fractured erythrocytes treated with 100 μM DAU numerous vesicles were observed inside the cell. Figs. 1, 2: Bars= 10 μm. Figs. 3, 4: Bars= 1 μm.

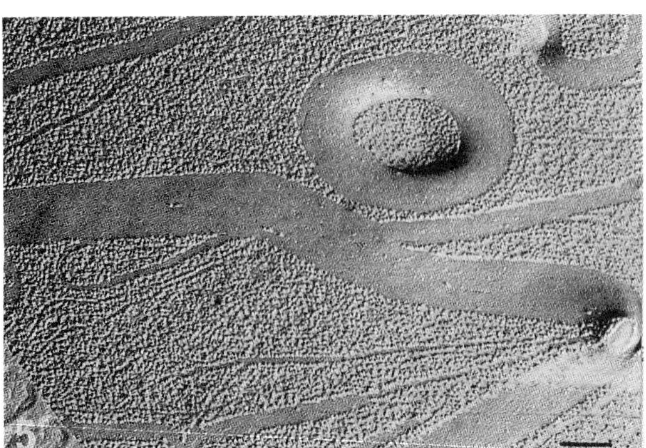

Fig. 5. Particular of a protoplasmic fracture face of the plasma membrane of erythrocyte treated with 100 uM ADR. Numerous smooth tracks with a regular geometry are clearly visible. Bar= 0.1 μm.

Native membrane crystals of *Bordetella pertussis*' anion-selective porin

A C Steven[1], B L Trus[2], M J Brennan[3], M E Bisher[1] & M Kessel[1]

[1]Lab of Physical Biology, NIAMS & [2]Computer Systems Lab, DCRT, NIH;
[3]Div of Bacterial Products, CBER/FDA, Bethesda, MD 20892, USA

ABSTRACT: Cells of two non-pathogenic strains of Bordetella pertussis are covered with a crystalline surface lattice. Combining structural analysis with biochemistry, we have established that this lattice is a natural membrane crystal of a 40kDa protein recently found by others to be an anion-selective porin. Its rectangular 8.3 x 14.3nm unit cell contains two differently oriented porin trimers arranged according to p2 symmetry: in projection, they present the typical porin motif of stain-penetrable triplets. Four pathogenic strains contain the same protein in similar amounts but without crystalline ordering.

1. INTRODUCTION

The principal route by which small soluble molecules of up to ~650Da traverse the hydrophobic barrier posed by the outer membrane of Gram-negative bacteria is by diffusion through aqueous pores formed by "porin" proteins (rev. Nakae, 1986). Porins share certain characteristic properties: for instance, they are very abundant at up to ~10^5 copies per cell, have M_r's of 30-40 kDa, tend to form trimers, and their secondary structure is predominantly β-sheet. Structure-function relationships in this extensive family of prokaryotic channel-forming proteins have attracted much interest. Hitherto, however, porosity studies have been conducted mainly with 'black' lipid films exposed to purified porins, while structural analysis has focused on synthetic crystals of detergent-solubilized porins, and there has been a lack of evidence directly linking the two lines of investigation.

2. RESULTS AND DISCUSSION

When contrasted by low-angle rotary shadowing, B. pertussis cells of strains 10109 or Tohama III (either whole mounts or intact envelope preparations) are seen to be covered with patches of a regular surface lattice (Fig 1a). After negative staining, the periodic structure is barely discernible (Fig 1b), but it is clearly revealed by optical diffraction (Fig 2). The patterns contain two reciprocal lattices deriving respectively from the upper and lower layers of the flattened envelopes: they correspond to a rectangular unit cell of 14.3 x 8.3nm. Correlation averaging (Fig 3) reveals that the lattice observes p2 symmetry, with the unit cell containing two differently oriented copies of the "stain-penetrable triplet" motif (Steven et al, 1977) that is characteristic of porin trimers, and which Engel et al (1985) have shown to represent the projection of a transmembrane channel.

SDS-PAGE analysis of envelopes differentially extracted with Zwittergent-3,14 shows that the periodic lattice is composed of a 40kDa protein with the same electrophoretic properties as the protein that Armstrong et al (1986) have purified from another strain

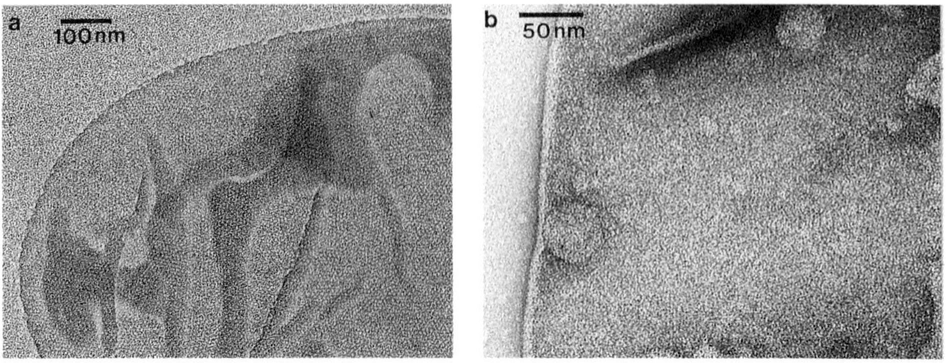

Fig. 1 : Cell envelopes of B. pertussis 10109 prepared by (a) low-angle rotary shadowing with platinum; (b) negative staining with uranyl acetate.

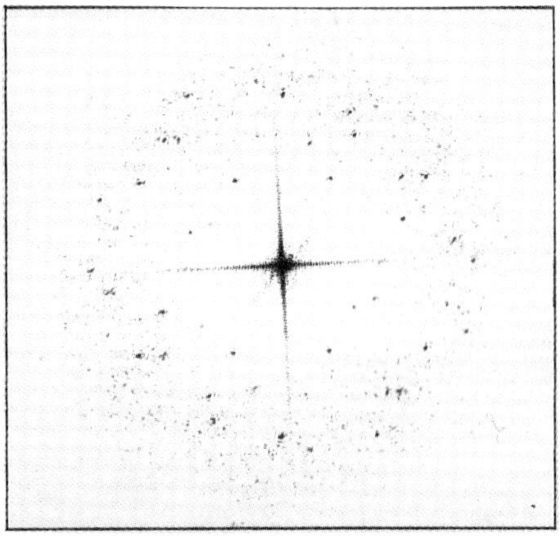

Fig. 2 : Optical diffraction pattern of negatively stained cell envelope: periodic reflections extend to ~0.5nm^{-1}.

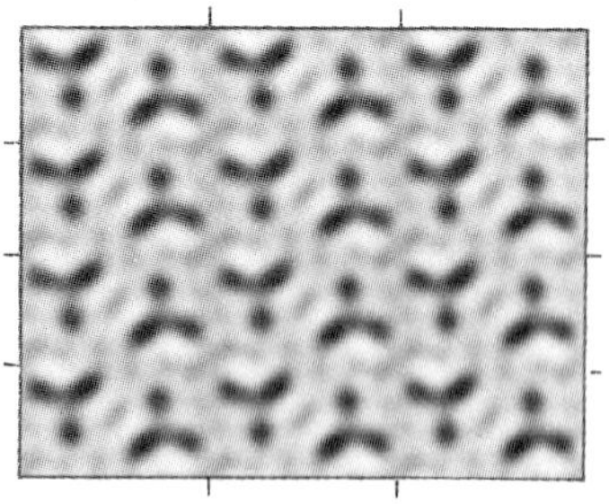

Fig. 3 : Projection of negatively stained surface lattice (p2 symmetry), showing two porin trimers per unit cell.

of B.pertussis, and shown to be an anion-selective porin. We have also found the same protein in similar amounts in four other strains that do not exhibit crystalline ordering. Because the bacteria grow at the same rate whether the 40kDa porin is crystalline or not, we infer that its functionality is not affected by crystallinity.

This observation of an evidently functional porin that is naturally crystalline and whose structure resembles those of other porins in synthetic crystals strengthens the proposition that the latter do indeed represent functional states. This resemblance further implies that the conferring of charge-selectivity (most porins studied to date are non-selective) entails minor tuning and not a radical change of the archetypal porin structure.

Armstrong S K, Parr T R, Parker C D and Hancock R E W 1986 J Bact **166** 212
Engel A, Massalski A, Schindler H, Dorset D L and Rosenbusch J P 1985 Nature(London) **317** 643
Nakae T 1986 Crit Rev Microbiol **13** 1
Steven A C, Ten Heggeler B, Mueller R, Kistler J and Rosenbusch J P 1977 J Cell Biol **72** 292

Structure of ordered acetylcholine receptors in membranes

M.Giersig, *W.Kunath, *H.Sack-Kongehl and F.Hucho

Institut für Biochemie, Freie Universität Berlin,Thielallee 63, D-1000 Berlin 33
*Fritz-Haber-Institut der Max-Planck-Gesellschaft,Faradayweg 4-6, D-1000 Berlin 33

The structure of the native acetylcholine receptor (AChR) as obtained from micrographs of receptor-rich membranes exhibits five stain-excluding regions in a ringlike arrangement [1]. The variation in the stain distribution around the individual molecules did not allow the identification of the different subunits which constitute the pentameric molecule. To obtain the structure of the AChR with higher resolution we tried to prepare two-dimensionally ordered molecules within the membrane.

The ordering of AChR was performed by extracting the peripheral membrane proteins [2] and subsequent treatment with phospholipase A_2 [3]. Depending on the concentration of phospholipase we succeeded in growing tetragonal and hexagonal ordered crystals with a crystallographic resolution of about 2.5 nm (fig.1.).

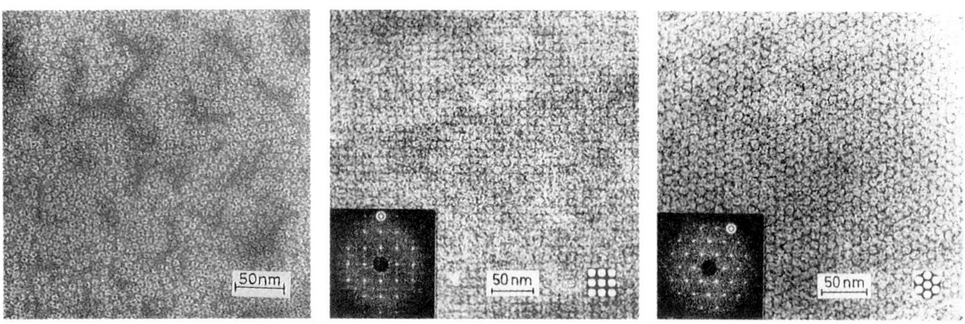

Fig.1. AChR-molecules in membranes (a) as single molecules, (b) in tetragonal order, (c) in hexagonal order. The power spectra (insert) exhibit diffraction spots out to $1/3.1\ nm^{-1}$ and $1/2.7\ nm^{-1}$ respectively.

For image processing we applied two methods. First we used correlation averaging [4] to obtain the mean unit cell structure. But the averages reflected the two crystal symmetries rather than any intrinsic structure. Supposing that the crystals may be composed by the molecules in different orientations we treated the unit cells independently like randomly oriented single molecules. Using the same method as for the single molecules in [1], namely aligning the unit cell images with the aid of the rotational correlation matrix (to be independent of any reference) and averaging the circular harmonic components of the images in the Fourier space, we obtained averages shown in fig.2. All averages have been taken from 400 individual images. They agree in the pentameric structure, but due to the fact that the molecules are arranged to paracrystals with a random orientation (fig.3), the resolution in the averages could not yet be enhanced.

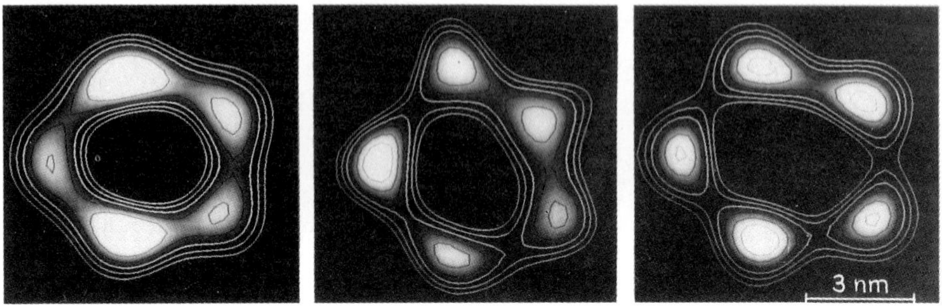

Fig.2. Averages, each obtained from 400 molecules and unit cells resp. as in fig.1. (a) Single-molecule average, (b) and (c) Unit-cell averages from the tetragonal and hexagonal paracrystals.

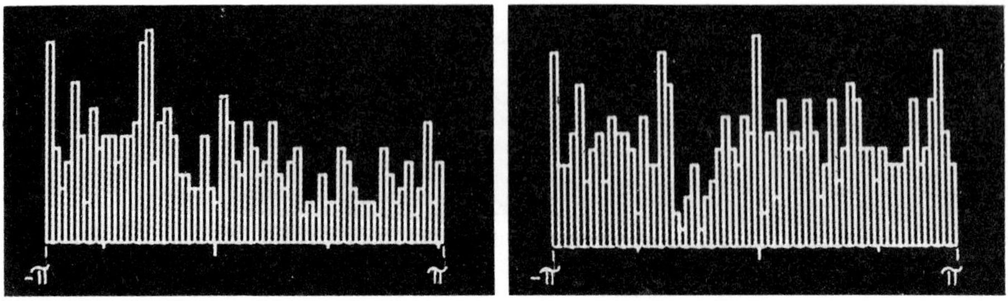

Fig.3. Histogram of the orientation of 400 unit cells within the tetragonal and hexagonal paracrystal respectively.

References

[1] Giersig M. Kunath W. Sack-Kongehl H. and Hucho F. 1986 in: "Nicotinic Acetylcholine Receptor" ed.A. Maelicke 7-17, Springer-Verlag, Berlin/Heidelberg
[2] Porter S. Froehner S.C. 1983 J.Biol.Chem. 258, 10034-10040.
[3] Mannella C. 1986 Meth. Enzymol. 125, 595-613.
[4] Saxton W.O. and Baumeister W. 1982 J.Microsc. 127, 127-138

Inst. Phys. Conf. Ser. No. 93: Volume 3, Chapter 17
Paper presented at EUREM 88, York, England, 1988

Fracture-labelling of terminal glycoconjugates in atrial specific granule membranes

N J Severs

Department of Cardiac Medicine, Cardiothoracic Institute, London SW3 6HP

ABSTRACT: Fracture-labelling shows atrial specific granule membranes to be rich in terminal glycoconjugates. These secretory granules are a product of the Golgi, and resemble in composition the plasma membrane with which they fuse during exocytosis.

Atrial specific granules (ASGs) are peptide hormone-containing secretory vesicles found in muscle cells of the atrium. The hormone secreted by these cells, atrial natriuretic peptide (ANP), has powerful diuretic and hypotensive effects. The mechanisms of processing, packaging and release of ANP have not been investigated in detail but presumably follow the same membrane pathways (i.e. endoplasmic reticulum → Golgi apparatus → secretory granule → plasma membrane) as documented for other peptide hormone-secreting cells. Here the structure and composition of the ASG membrane is investigated by freeze-fracture and by fracture-labelling using the lectin, wheat germ agglutinin (WGA) as a probe for terminal glycoconjugates.

Perfusion-fixed atrial samples from rat and rabbit hearts were glycerinated and frozen in propane. Standard freeze-fracture replicas were prepared in a Balzers BAF 400T machine. For fracture-labelling (Torrisi and Pinto da Silva, 1984) the frozen samples were crushed under liquid nitrogen, thawed in glutaraldehyde/glycerol and rinsed in polyethylene glycol and ammonium chloride solutions. Cytochemical detection of WGA binding sites was done by treating the samples first with WGA (0.25 mg cm^{-3}, 1h) and then with ovomucoid-gold (ovo-Au) complexes (1h). Controls received i) lectin treatment in the presence of excess N-acetylglucosamine followed by ovo-Au, or ii) ovo-Au alone. The samples were processed for thin sectioning and examined in a Philips EM 301.

In standard freeze-fracture replicas, ASGs are seen to be abundant in the vicinity of the Golgi, apparently elaborated from dilated cisternae at the trans face (Fig. 1). In rabbit, the mean true diameter of the ASG is 344 nm, the P-face intramembrane particle density is ~1000 μm^{-2} and the E-face 400 μm^{-2}. An additional vesicle type of similar dimensions (accounting for < 15% of the granule population) has smooth P and E-faces. Freeze-fractured samples treated with WGA and ovo-Au show extensive labelling of ASG 'E-faces'; other intracellular membranes and cross-fractured cytoplasm are, by contrast, poorly labelled (Fig. 3). As in other cell types (Torrisi and Pinto da Silva, 1984), the 'E-faces' of plasma membranes are as heavily labelled as those of secretory granules (Fig. 4). Controls treated with excess N-acetylglucosamine show a >85% reduction in gold particle density (Fig. 5). In the absence of prior WGA, there is virtually no binding of ovo-Au.

A function unique to the Golgi apparatus is the attachment of terminal sugars (N-acetylglucosamine and sialic acid) to glycoproteins. Terminal glycosylation of glycoproteins destined for the ASG membrane renders this membrane similar to the plasma membrane with which it eventually fuses. How incorporated membrane is recycled from the plasma membrane, and the role, if any, of the smooth-faced vesicles in this process remain to be determined.

REFERENCE

Torrisi MR and Pinto da Silva P 1984 J. Cell Biol. **98** 29

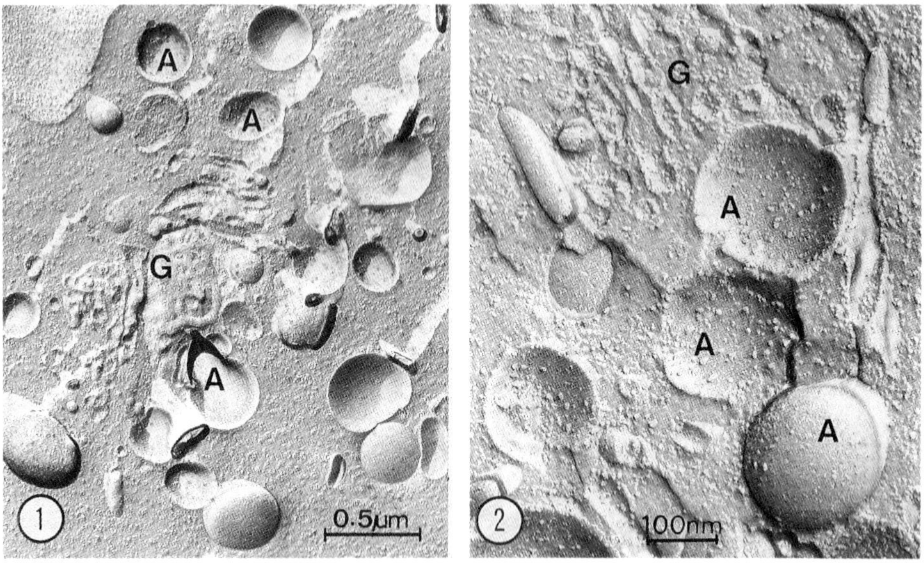

Figs 1 and 2. Survey and high power freeze-fracture replicas illustrating structural features of ASG membranes (A) and Golgi apparatus (G), in rabbit atrial muscle cells.

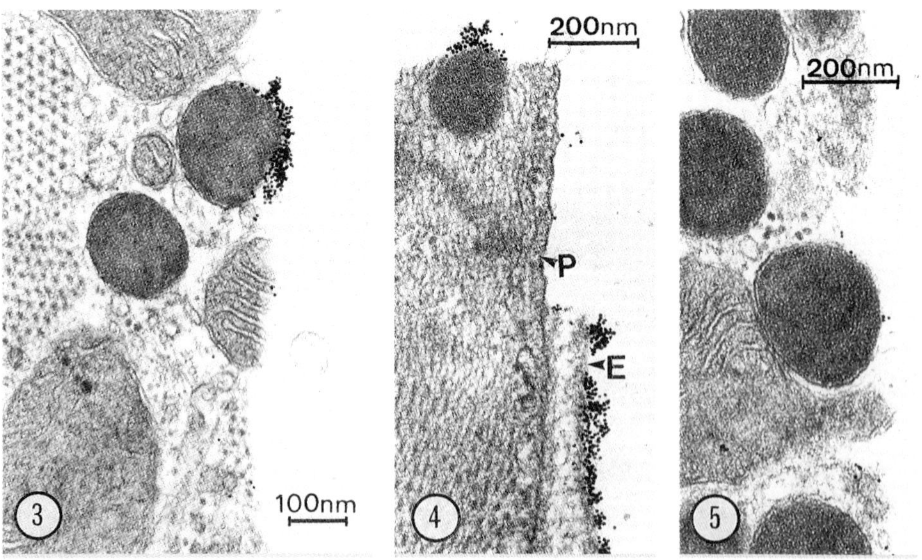

Figs 3-5. Colloidal gold-labelled WGA binding sites. Note extensive labelling of ASG membrane 'E-faces' in Figs 3 and 4. In Fig. 4, the 'E-face' (E) of the plasma membrane is heavily labelled whereas the 'P-face' (P) shows sparse labelling. Fig. 5 is a control (excess N-acetylglucosamine) (Rat atrial muscle cells).

Ultrastructural demonstration of MAP 1 in tracheal ciliated cells

L Stockinger, W Sellner, Ch Oberkanins* and G Wiche*

Institute of Micromorphology and Electronmicroscopy, *Institute of Biochemistry, University of Vienna, A-1090 Vienna, Austria

ABSTRACT: The localization of MAP 1 in rat and porcine tracheal ciliated cells is shown by immunoelectronmicroscopy and immunoblotting.

1. INTRODUCTION:

The predominant microtubule associated proteins in normal tissue are the HMW proteins MAP 1 and MAP 2. It was of interest to look if proteins related to the MAP's are structural components of the ciliary microtubules as well. Here we studied MAP 1.

2. METHODS:

Tracheas were fixed in 4% glutaraldehyde and 0,5% paraformaldehyde and embedded in LR-White. Thin sections were incubated in 1) polyclonal rabbit antibodies to brain MAP 1; 2) affinity purified rabbit antibodies to brain MAP 1 (Wiche et al 1983) and rabbit preimmunserum for 15 hours at +4°C. Detection of the antibodies was done by biotin-streptavidin-gold system (5 nm).
We modified the methods of Hastie et al (1986) and Anderson (1977) to isolate cilia from cells of the porcine trachea. This system is based on a permeabilization of the plasma membrane wich allows exposition of ciliary shafts to relatively high concentrations of calcium ions. Little mechanical stress leads to isolation of cilia from their cells. The resulting preparation was purified by a 1,5/2.0 M discontinuous sucrose gradient. At the 1,5/2,0 M interface we found the majority of cilia with only little cytoplasmic contaminants (Figure 2). The proteins of this fraction were separated by SDS-PAGE. For immunoblotting the proteins transferred to nitrocellulose sheets were incubated in the same sera we discribed above.

3. RESULTS:

Ciliated cells of rat and porcine tracheas showed labelling of the ciliary shaft and the basal bodies of the cilia. The antigen was located closely to the microtubules. The basal feet and ciliary rootlets showed no reaction. Only few gold particles were found over the cell body (Figure 3).
In immunoblotting experiments antibodies 1) and 2) were reactive with ciliary proteins of molecular weight corresponding to brain MAP 1 (Figure 1).

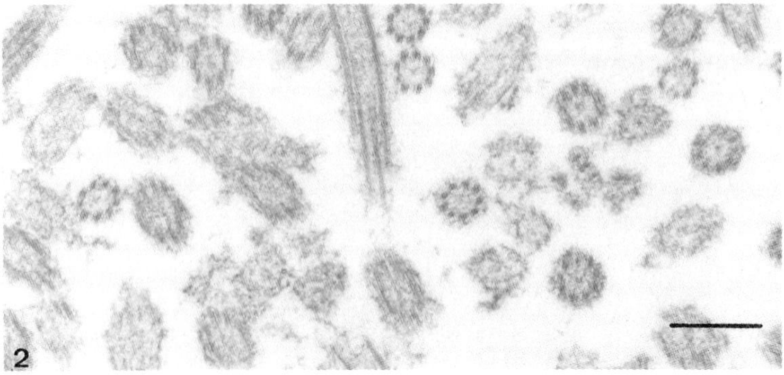

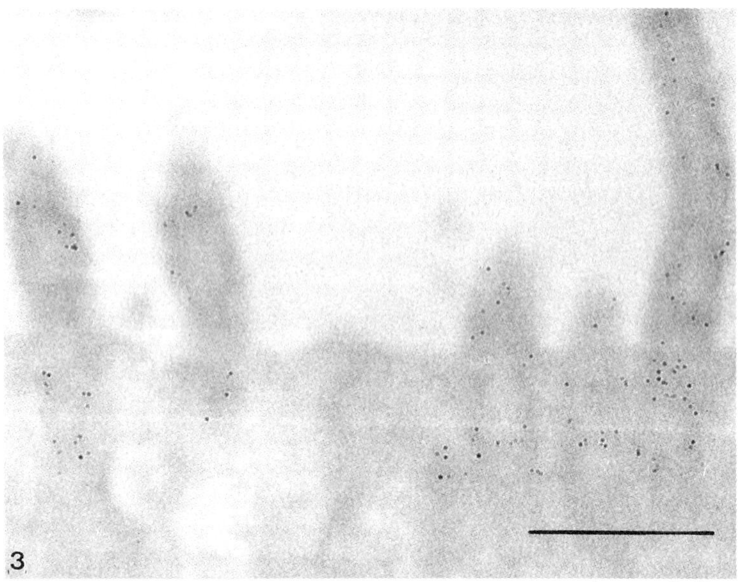

bars 0.5µm

4. FIGURES:

1)lane a:Coomassie-blue staining of ciliary proteins
  lane b:immunoblotting with antibody 2)
  lane c:immunoblotting with rabbit preimmunserum
2)electron micrograph of the purified fraction of cilia
3)immunolocalization of MAP 1 in the upper part of a porcine tracheal ciliated cell

5. REFERENCES:

Anderson R G W, JCB 74,1977 pp 547-560
Hastie A T et al, Cell Motility 6, 1986 pp 35-34
Wiche G et al, EMBO-J 2, 1983 pp 1915-1921

This work was supported by "Fonds zur Förderung der wissenschaftlichen Forschung, Projekt 5751 "

The phase transition of N-Palmitoyl-D-Sphingomyelin studied by negative staining

J.R. Harris, North East Thames Transfusion Centre, Crescent Drive, Brentwood, Essex CM15 8DP.

ABSTRACT: It is shown by negative staining with uranyl acetate at controlled temperatures that the dynamic phase transition of N-palmitoyl-D-sphingomyelin can be readily revealed by transmission electron microscopy. The undulatory (ca 25mm periodicity) $P\beta'$ pre-transition phase is readily distinguished from the smooth crystalline ($L\beta'$) and liquid-crystalline ($L\alpha$) phases. Spontaneous vesicularization of the phospholipid occurs at high temperatures.

1. INTRODUCTION

Natural sphingomyelin (SM) from bovine brain and other tissues is not chemically pure due to the presence of fatty acid of varying chain length and saturation. The ultrastructure of the $P\beta'$ pre-transition phase of bovine brain SM in negatively stained specimens has been reported by Harris (1986). This study has now been extended to chemically pure N-palmitoyl-D-sphingomyelin. The crystalline to liquid-crystalline phase transition temperature for N-palmitoyl-D-SM is 41°C (Calhoun and Shipley, 1979). Thus, it was hoped that the smooth crystalline $L\beta'$ phase might be detected at low temperatures by negative staining, with the undulatory $P\beta'$ pre-transition phase at intermediate temperatures and the smooth $L\alpha$ liquid-crystalline phase at higher temperatures.

2. RESULTS AND DISCUSSION

Specimens negatively stained with 2% aqueous uranyl acetate have been prepared from aqueous suspensions of N-palmitoyl-D-SM, following incubation of both stain and lipid at temperatures ranging from 4°C to 100°C. At 4°C, (Figure 1) most of the multi-lamellar bodies are smooth surfaced ($L\beta'$) although occasional particles do show the undulatory $P\beta'$ conformation. At 20°C and 40°C (Figures 2 and 3) all the phospholipid multilamellar bodies show the undulatory $P\beta'$ pre-transition phase. At 60°C, (Figure 4) the $P\beta'$ phase is less evident and is totally absent at 80°C (Figure 5). Some indication of SM vesicularization is present at 80°C, and is even more pronounced at 100°C (Figure 6). Spontaneous vesicularization of bovine brain SM has been observed to occur at 60°C (Harris 1986). The results presented show that negative staining of phospholipids can be performed at widely ranging temperatures, and that the presence of the undulatory $P\beta'$ pre-transition phase of sphingomyelin can be easily determined and distinguished from the smooth $L\beta'$ and $L\alpha$ phases.

3. REFERENCES

Calhoun W I and Shipley G G 1979 Biochemistry 18 1717-1722
Harris J R 1986 Micron Microsc. Acta 17 175-200.

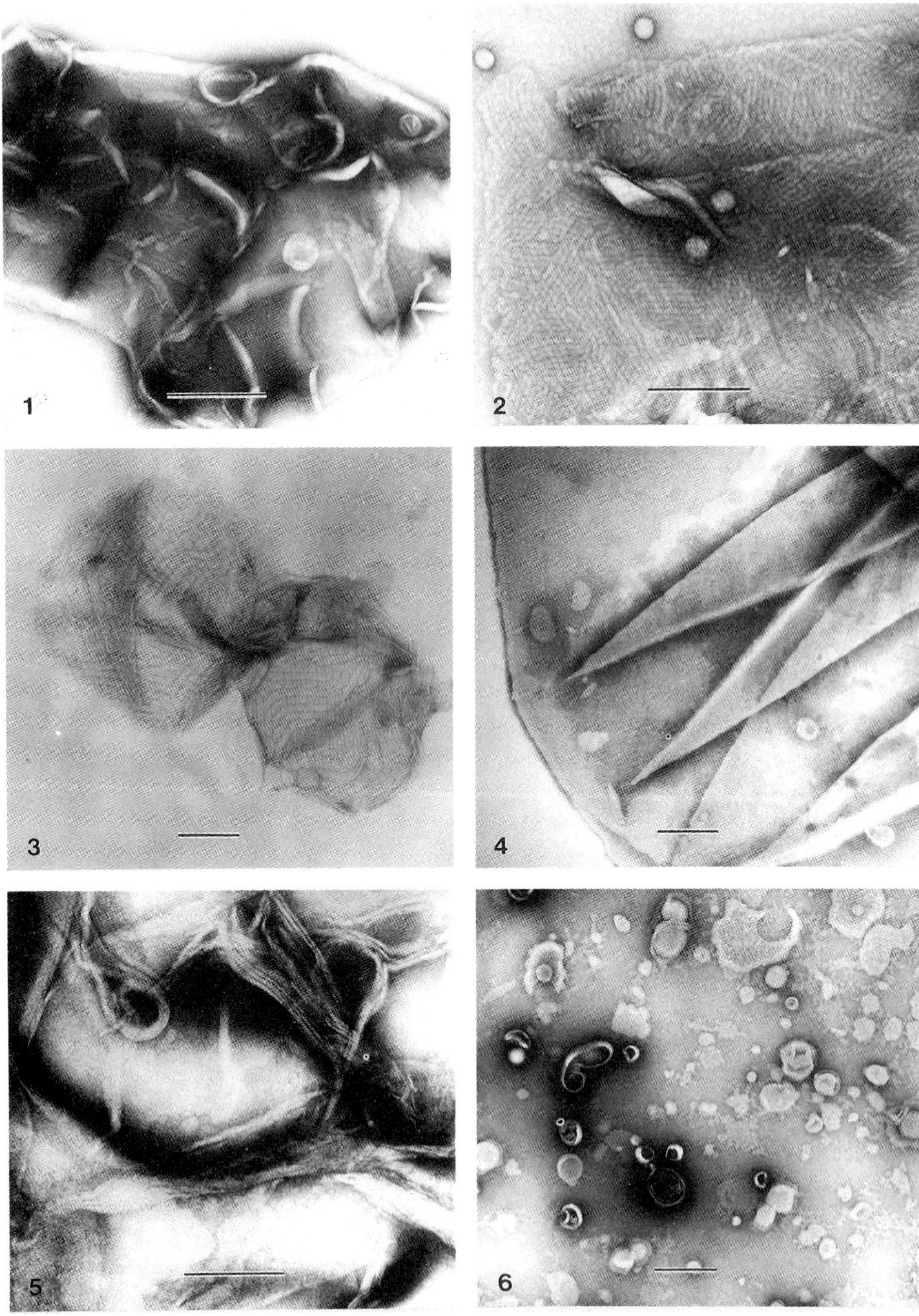

Figures 1 to 6 show representative examples of N-palmitoyl-D-SM negatively stained with uranyl acetate after equilibration at 4°C, 20°C, 40°C, 60°C, 80°C and 100°C, respectively. The scale markers indicate 200nm.

Fracture-flip views the surface of sperm cells at macromolecular resolution

Catarina Andersson Forsman and Pedro Pinto da Silva

Membrane Biol. Sect/Lab. Mathemat. Biol., NIH/NCI-FCRF, Frederick, MD 21701 USA

The boar sperm plasma membrane is morphologically and functionally differentiated into specific regions. We use Fracture-Flip (Andersson Forsman and Pinto da Silva, 1988; Pinto da Silva et al., these proceedings) to examine at macromolecular resolution the surface of boar spermatozoa. New aspects of the surface of sperm cells are revealed by fracture-flip. Our routine resolution is about one order of magnitude better than that provided by high resolution scanning electron microscopy of biological specimens.

Semen from mature boars, Sus Scrofa, was washed in PBS, fixed in 1.5% glutaraldehyde in PBS, rinsed, glycerinated and fracture-flipped (Andersson Forsman and Pinto da Silva, 1988).

We summarize our observation as follows:
1) over the acrosomal area the surface of the plasma membrane is covered by a high density of 15-20 nm particles. The rim of this anterior portion has few or no particles (Fig. 1).
2) a high density of similar particles exists over the post-acrosomal area, except for a particle-free rim (approx. 100 nm wide) above the posterior ring and the cords (Fig. 2).
3) at the outer surface the cords are represented by arrays of particles (20-25 nm diam.).
4) the neck displays irregular patterns of heterogenous particles (8-12 nm diam.).
5) the surface of the tail contains numerous particles (12-15 nm) aligned into rows often running helical-parallel down the length of the tail (Fig. 3).
6) small (50x200 nm) rectangular arrays of 9-12 nm particles are found over both the neck and tail.

The particles that we observe over the surface of the sperm cell are likely to represent surface (glyco)proteins. They can be visualized due to the high resolution of fracture-flip images. We propose fracture-flip as an alternative to high resolution scanning microscopy. SEM, even when high resolution instruments are used, is hindered by preparatory procedures of coating. Combination of colloidal gold cytochemistry and fracture flip will lead to molecular resolution maps of the surface topochemistry of sperm cells.

References
Andersson Forsman C and Pinto da Silva P 1988 J Cell Sci, Submitted
Pinto da Silva P, Andersson Forsman C and Fujimoto K, these proceedings

Freeze-fracture analysis of gap-junction alterations induced by hypoxia and hyperkalemia in isolated perfused rat hearts

P A P M Aertgeerts, D W Scheuermann, A M G L De Mazière

The Institute of Histology and Microscopic Anatomy, University of Antwerp, Groenenborgerlaan 171, B-2020 Antwerp, Belgium

ABSTRACT: The configuration of gap-junctional particles has been studied quantitatively after increasing periods of hypoxic, hyperkalemic perfusion of isolated rat hearts. Hyperkalemia alone, or combined with hypoxia for up to 20 min, produces no visible gap-junctional alterations. After 30 min hypoxia, the connexons are significantly, but reversibly tighter packed. Following 40 min hypoxia, further compaction of the gap-junctional particles often results in hexagonal arrays. Since reoxygenation turns more gap junctions into these crystalline arrays, this configuration may represent a final stage in the particle rearrangement accompanying electrical uncoupling.

1. INTRODUCTION

The observed enhancement of the electrical resistance between myocardial cells during progressive periods of hypoxia (Wojtczak 1979), is supposed to be due to a gradual reduction of the number of open cell-to-cell channels located in each of the aggregated particles of the gap junction. To determine whether a closure of the channels, concomitant with a rearrangement of gap-junctional particles is involved, the present quantitative freeze-fracture study of the particle configuration in myocardial gap junctions during increasing periods of hypoxia was undertaken. In order to prevent irregular bursts of contractions, the hearts were subjected to K^+ arrest, a condition provoking membrane depolarization without uncoupling the cells (De Mello and van Loon 1987), and which was studied separately.

2. MATERIALS ANS METHODS

Rat hearts were perfused on a Langendorff cannula during an equilibration period of 10 min with a Krebs-Henseleit solution (KH), bubbled with 95% O_2 : 5% CO_2, and containing (in mM) 118.0 NaCl, 4.7 KCl, 2.5 $CaCL_2$, 1.2 $MgSO_4$, 1.2 KH_2PO_4, 25.0 $NaHCO_3$ and 11.0 glucose, at 37°C and pH 7.4. The perfusion was then continued either with KH for 10 min, with KH containing 16 mM K^+, also for 10 min, or with 16 mM K^+-containing, glucose-free KH gassed with 95% N_2 : 5% CO_2, for 10, 20, 30 or 40 min with or without subsequent reoxygenation for 20 min. After perfusion-fixation, myocardial tissue from the left ventricle was processed for freeze-fracture as described by De Mazière et al (1985). Data from measurements of centre-to-centre distances between neighbouring particles and packing density were pooled per group and evaluated using Student's t-test.

3. RESULTS AND DISCUSSION

In hearts perfused with normal KH, the gap-junctional particles are regularly arrayed with a centre-to-centre distance (mean $\pm$ SD) of 9.17 $\pm$ 1.52 nm (n=2,702) into small, randomly oriented domains. These domains are separated by smooth aisles (Fig. 1), resulting in a mean density of 8,490 $\pm$ 600 particles/μm^2 (n=85). After hyperkalemic perfusion, a similar configuration is observed with equal particle density (8,420 $\pm$ 620/μm^2; n=87) and spacing (9.15 $\pm$ 1.51 nm; n=2,644). This consorts with the apparent independence of the permeability of cardiac gap junctions on the membrane potential (Weingart 1986, De Mello and van Loon 1987). In hearts perfused with hypoxic KH for only 10 or 20 min, the normal appearance of the gap junctions persists. After 30 min hypoxia, some gap junctions with compacted particle domains (Fig. 2) appear among the normal ones, which is reflected in an increased overall density of 9,050 $\pm$ 760 particles/μm^2 (n=36), while the centre-to-centre distance is reduced to 8.93 $\pm$ 1.44 nm (n=1,118). Upon reoxygenation, all gap junctions are reverted to the control type. After 40 min hypoxia, the particle packing is further condensed to 10,390 $\pm$ 790/μm^2 (n=29) by the disappearance of all aisles. Some of these junctions consist of an uninterrupted crystalline array (Fig. 3). Reoxygenation after 40 min hypoxia augments the fraction of gap junctions in the crystalline configuration, without further decreasing their particle spacing.

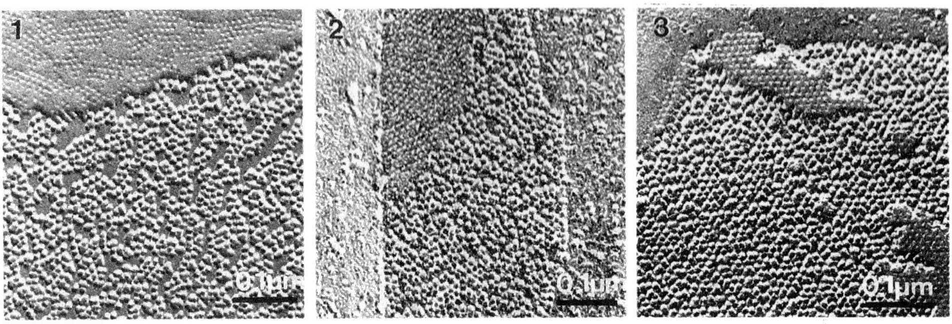

Figs. 1-3. Cardiac gap junctions after control perfusion (Fig. 1), after 30 min hypoxia (Fig. 2) and after 40 min hypoxia (Fig. 3)

The observed sequence of changes during progressive uncoupling by hypoxia suggests essentially a compaction of the particles, which initially tends to disturb the existing short-range order, but ends up by fitting all particles, possibly with minimal spacing, in a hexagonal lattice.

4. REFERENCES

De Mazière A M G L, Scheuermann D W and Aertgeerts P A P M 1985 J. Microsc. 137 185

De Mello W C and van Loon P 1987 Cell Biol. Int. Rep. 11 1

Weingart R 1986 J. Physiol. (London) 370 267

Wojtczak J 1979 Circ. Res. 44 88

Inst. Phys. Conf. Ser. No. 93: Volume 3, Chapter 17
Paper presented at EUREM 88, York, England, 1988

Post mortem changes in surface-appearance and cross-links of stereocilia in guinea-pig organ of corti examined by high resolution scanning electron microscopy

M P Osborne, S D Comis and J O Pickles

Department of Physiology, The Medical School, The University of Birmingham, Birmingham B15 2TJ UK

ABSTRACT: Post mortem changes in stereocilia were studied at time periods from 15 min to 4 h after circulation arrest. Within 15 min the surfaces of stereocilia from both inner and outer hair cells in the basal turn of the cochlea were more granular and links between stereocilia were thicker and fewer than normal. With increasing time, surface irregularities of stereocilia became more pronounced, more stereocilia became fused, links became more globular and fewer in number and damage spread progressively to more apical regions of the cochlea.

1. INTRODUCTION

Acoustic trauma (Pickles et al 1987a), certain antibiotics and anticancer agents (Pickles et al 1987b; Comis et al 1986) initially produce subtle changes in surface texture and cross-links of stereocilia; later complete destruction of hair bundles can occur. To determine if these changes are due to progressive necrosis we have investigated post mortem (pm) changes of stereocilia up to 4 h following circulation arrest.

2. MATERIALS AND METHODS

Guinea pigs were killed by an overdose of pentobarbitone sodium. The time of cardiac arrest was noted and cochleas were fixed in glutaraldehyde 15,30,60,120 and 240 min thereafter. Cochleas were opened, dehydrated in acetone, critical point dried with liquid CO_2, and sputter coated with platinum. Specimens were examined in a JEOL 120CXII TEMSCAN with a secondary electron detector and accelerating voltage of 40 kV.

3. RESULTS AND CONCLUSIONS

By 15 min pm there were marked changes in stereocilia of hair cells in the basal turn; few changes were present in more apical regions. These changes became more obvious and spread more apically along the cochlea with time, but even by 4 h pm hair cells at the apex appeared virtually normal. Initial changes included an increase in granularity of the surface of stereocilia and swelling of lateral crosslinks accompanied by their fracture (Figures 1,2). Upward pointing tip links were also swollen, elongated, but less prone to fracture. By 1-2 h changes in stereocilia were more pronounced. These changes included a further increase in surface granularity and a marked swelling of bases of stereocilia at their points of insertion into the cuticular plate.

Lateral links were almost completely destroyed; tip links were fewer in number, particularly in the basal regions of the cochlea.

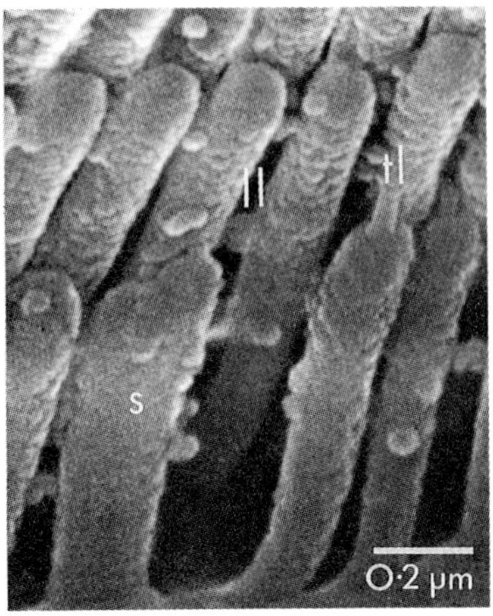

Fig. 1. Stereocilia 15 min pm. Note granular surface, lateral links (ll), tip links (tl), fused stereocilia (s).

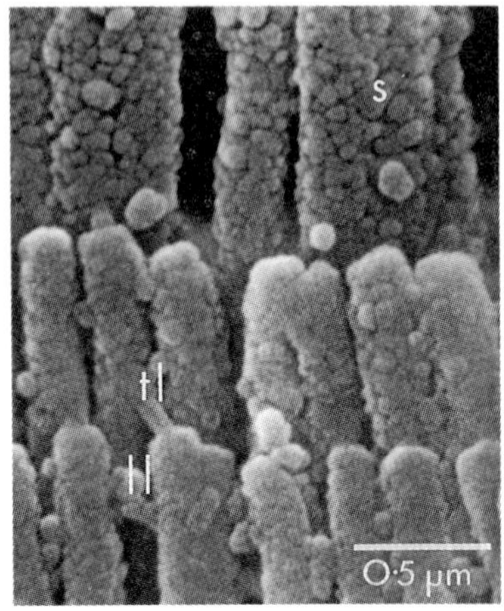

Fig. 2. Stereocilia 2 h pm. Surface is more granular; lateral links (ll) are more globular.

Specimens 4 h pm showed further exaggeration of the above changes accompanied in many cases by collapse of hair bundles and a greater tendency for stereocilia to fuse. In these extensively damaged hair bundles lateral and upwards pointing links were largely destroyed.

These studies show that some previously described changes induced by acoustic stimulation or drugs are similar to pm changes described here. The changes in surface texture and crosslinks probably reflect alterations in the physiocochemical characteristics of the glycocalyx coat of stereocilia accompanying death.

4. ACKNOWLEDGEMENTS

We thank the MRC, the Royal National Institute for the Deaf and the Hearing and Speech Trust for financial support.

5. REFERENCES

Comis S D, Rhys-Evans P H, Osborne M P, Pickles J O, Jeffries D J R and Pearse H A C 1986 J. Laryngol. Otol. 100 1375

Pickles J O, Osborne M P and Comis S D 1987a Hear. Res. 25 173

Pickles J O, Comis S D and Osborne M P 1987b Hear. Res. 29 237

Specialization of endoplasmic reticulum in the syncytiotrophoblast of human placental villi

Türkân Erbengi[a] and Ramazan Demir[b]

[a]Department of Histol. & Embryol., Medical Faculty, Istanbul University;
[b]Depart.of Histol.& Ebriyol.,Medical Faculty, Akdeniz University, Antalya

1. INTRODUCTION

The surface of human placenta has been extensively studied with electron microscope. Kaufmann and Stegner (1972) explained a decrease in the rough endoplasmic reticulum (ER) from the early pregnancy to the begining of third trimester. It was pointed out that vesicular from of rouge ER was found mainly in the syncytium before the tenth week of pregnancy (Kaufmann and Stegner, 1972), from the end of the first trimester to term, narrow channels and dilatation of the ER were observed (Schuhmann and Wynn, 1980). However it has no yet clarified the functional significance of the differentiation of syncytial ER. In this study we have examined the specialization of ER depending on the gestational age.

2. FIGURES PRESENTATION

The ER of syncytiotrophoblastic cells shows some different morphological features according to the gestational age. Syncytial ER is essantially rough surfaced; irregularly arranged as labyrinth-like channels (Fig.1), dilatation of adjacent parallel tubules (Fig.2) or very dilated saccules (Fig.3) may fill up most of the cytoplasm, but there are also lamellar structures of more or less flattened cisternae. In addition, vesicle-like appearances of smooth ER are present in different parts of syncytiotrophoblast. In very early stages of pregnancy, especially the invasive syncytiotrophoblast has most labyrinth-like channels form of ER (Fig.1). From week 6 to the end of the second trimester of pregnancy, syncytiotrophoblast was characterized by fairly uniformly distributed the ER, narrow labyrinth-like channels absent, but adjacent parallel tubules of its own predominate. At the term, placental villi show a high sacculation for ER, especially overlying the cytotrophoblast cells and vesells, in syncytiotrophoblast (Fig.3).

Regional specialization in the syncytiotrophoblast has been explained by Dempsey and Luse (1971). According to our results, the appearance along the length of the labyrinth-like channels, the parallel tubules and the saccules suggest specialized ER for different functions depending on the gestational age and fetal requirements. These functions can be able to summarize as the absorption, the secretion, the transportation and a change from protein metabolism to energy production (Schweikhart and Kaufmann, 1977; Demir, 1980; Schuhmann and Wynn, 1980).

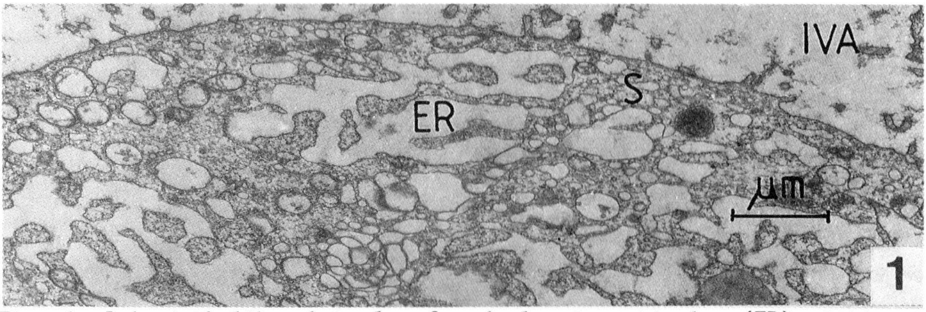

Fig. 1. Labyrinth-like channels of endoplasmic reticulum (ER) within the syncytiotrophoblast (S) can be seen. Placenta from 4 week of pregnancy.

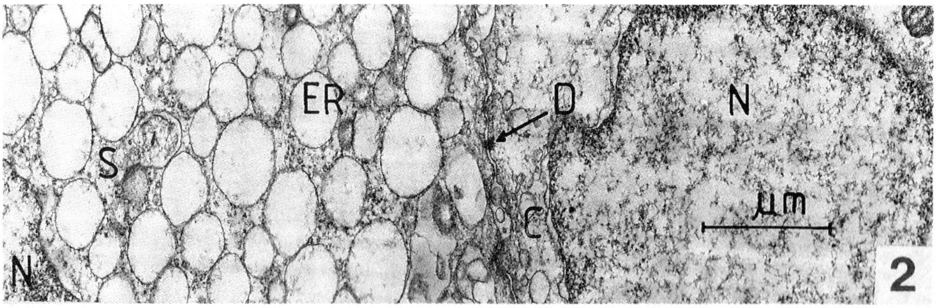

Fig. 2. Two trophoblastic layers in a placental villus from 6th week of gestation. Numerous cross-sections of tubular form of endoplasmic reticulum (ER) in syncytium (S) are clearly seen.

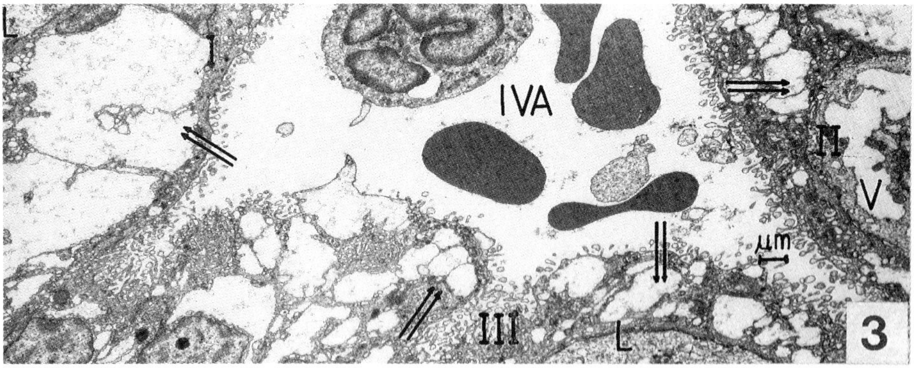

Fig. 3. Fragment of the walls of three placental villi (I,II,III) from the 36th week of gestation. Very dilated endoplasmic reticulum forming saccules (arrows) in syncytium overlying Langhans cells (L) and vessels.

Demir, R. (1980) Acta Anat. 106:18-29
Dempsey, E.W. and Luse, S.H. (1971) J.Anat. 108:545-561
Kaufmann, P. and Stegner, H.E. (1971) Z.Zellforsch. 135:361-382
Schuhmann, R.A. and Wynn, R.M. (1980) Placenta 1:345-353
Schweikhart, G. and Kaufmann P (1977) Ar.Gynaekol. 222:213-230

Inst. Phys. Conf. Ser. No. 93: Volume 3, Chapter 17
Paper presented at EUREM 88, York, England, 1988

BOSS—A new membrane substructure in insect thermoreceptors (goniometer analysis of cryosubstituted specimens)

R A Steinbrecht

Max-Planck-Institut für Verhaltensphysiologie, D-8130 Seewiesen, FRG

ABSTRACT: The sensory ending of a thermoreceptor cell in the silkmoth *Bombyx mori* shows extensive lamellation. The membranes of these lamellae are studded with electron-dense knobs (bosses) on their external face in an orthogonal pattern. The structure is named BOSS (**B**ossed **O**rthogonal **S**urface **S**ubstructure) and is supposed to play a role in the primary receptor processes of this thermoreceptor.

Insect antennae are multimodal sense organs equipped with mechano-, chemo-, hygro-, and thermoreceptor cells. The sensilla styloconica of *Bombyx mori* contain 2 hygro- and 1 thermoreceptor (Fig. 1) (Steinbrecht and Kittmann, 1986). The sensory endings of the latter show an unusual membrane substructure which in this paper for the first time is described at high resolution.

Antennae of *Bombyx mori* were cryofixed by immersion into liquid propane at 90 K, cryosubstituted in acetone (+2% OsO_4) at 194 K, and embedded in Epon at room temperature. Thin sections were stained with U/Pb and studied in a Zeiss EM 10A equipped with a high resolution goniometer.

The outer leaflet of the plasma membrane of the thermoreceptor lamellae bears electron-dense knobs in orthogonal array (Fig. 2, 5). This structure is named BOSS (see above). The diameter of the knobs is ~8 nm, the period width of the array in the two axes is ~15 nm and ~10 nm, respectively. The BOSS-arrays on two facing membranes of neighbouring lamellae are always oriented in parallel, those of the two membranes of the same lamella often are not. The axes of the array may be slightly curved, so that a given orientation is not maintained over a large area. Goniometer-tilting of cross-sectioned lamellae shows that the described configuration is typical for the whole lamellar region and that other "structures", such as septa between lamellae or continuous dense coats on the plasma membranes are due to superposition of obliquely oriented rows of "BOSSes" (Fig. 3, 4). Earlier papers on insect thermoreceptors (reviewed by Steinbrecht, 1984) mention "septate structures", which are, however, poorly preserved after chemical fixation. Possibly BOSS-membranes prevail in insect thermoreceptor endings and are of importance for the primary receptor processes.

REFERENCES

Steinbrecht R A 1984 In *Biology of the Integument* ed J Bereiter-Hahn, A G Matoltsy, K S Richards (Heidelberg: Springer) Vol **1** pp 523-53

Steinbrecht R A and Kittmann R 1986 Verh. Dtsch. Zool. Ges. **79**, 111

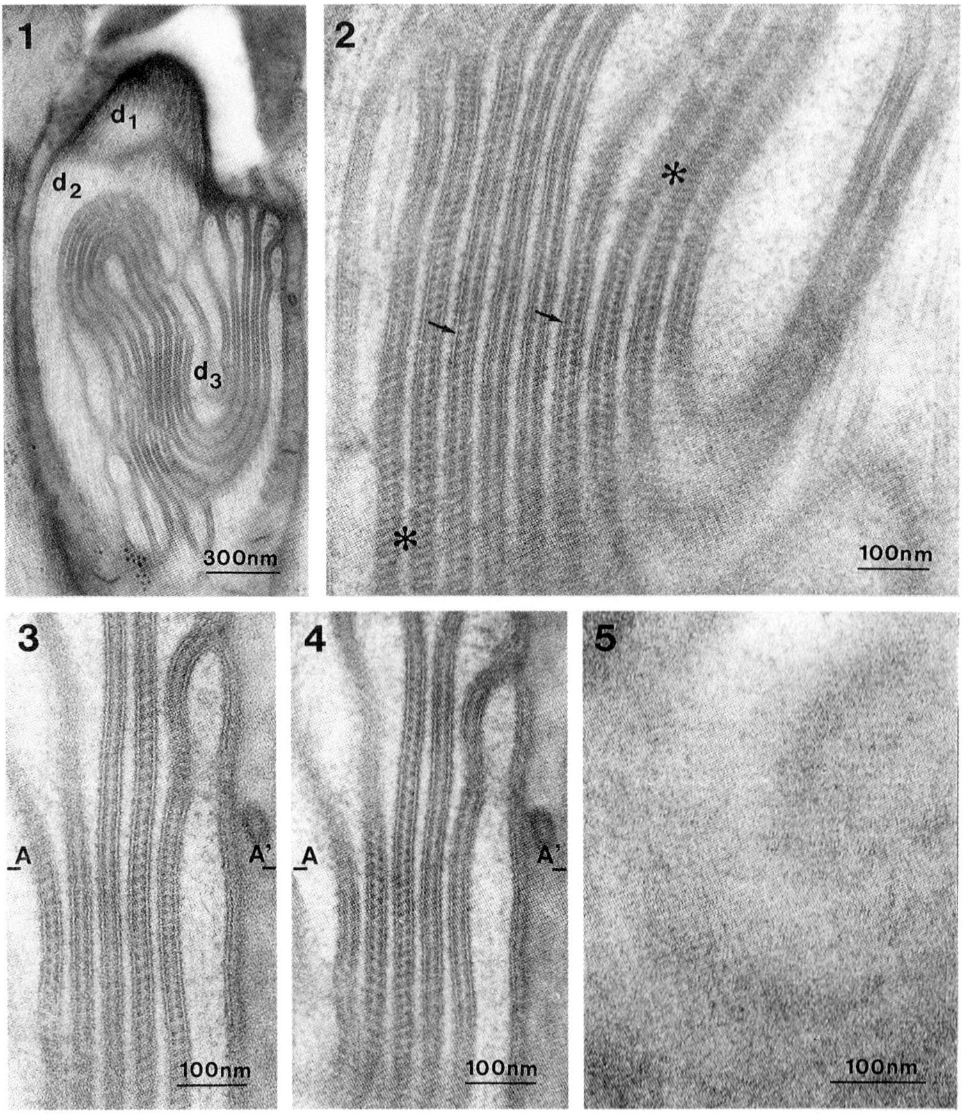

Fig. 1. Bombyx mori, s. styloconicum. Longitudinal section through sensory endings. Two hygroreceptor dendrites (d_1, d_2) are seen and a thermoreceptor dendrite (d_3), which displays extensive lamellation.

Fig. 2. Lamellae sectioned at various angles at high magnification. BOSSes (→) are resolved only under favourite viewing angles. In tilted lamellae, superimposed BOSSes may appear as oblique septa (★).

Figs. 3, 4. Detail of Fig. 1 at two tilt angles around axis A-A'. BOSSes which stand out clearly in Fig. 3 (0°) are superimposed to a dense coat in Fig. 4 (-30°) and vice versa.

Fig. 5. Tangential section reveals orthogonal BOSS-array.

Inst. Phys. Conf. Ser. No. 93: Volume 3, Chapter 17
Paper presented at EUREM 88, York, England, 1988

The surface of human blood cells as revealed by fracture-flip

Catarina Andersson Forsman and Pedro Pinto da Silva

Membrane Biology Section/Laboratory of Mathematical Biology, NIH/NCI-FCRF, Frederick, MD, 21701, USA

We use Fracture-Flip (Andersson Forsman and Pinto da Silva, 1988; Pinto da Silva et al., these proceedings) to examine at macromolecular resolution the cell surfaces of the human erythrocytes, lymphocytes and thrombocytes.

Human lymphocytes and thrombocytes were isolated using Ficoll-Paque (Pharmacia), fixed in 1.5% glutaraldehyde in PBS, rinsed in the same buffer and glycerinated. The specimens were fracture-flipped (Andersson Forsman and Pinto da Silva, 1988; Pinto da Silva et al., these proceedings).

At low magnification the surface of erythrocytes appears smooth. Closer examination of the surface of erythrocytes reveals a high density of very small (3-5 nm) granularities (Fig. 1) which contrasts the smoother adjoining crossfractured glycerol.

By Fracture-Flip the lymphocytes appeared as dome shaped structures, with many cross fractured cell accidents (e.g. microvilli) (Fig. 2). These villi dry, contract, shrivel and the plasma membrane detaches and reveals dried cytoplasm. The surface of human lymphocytes is covered by 10-17 nm particles distributed over the cell body and the microvilli at a density of approx. 1500/μm^2.

The surface of thrombocytes displays a flocculent texture, with loose aggregates (about 25-30 nm diameter) of smaller structures, at a density of approx. 2500/μm^2 (Fig. 3). Fracture-Flip easily shows the indentations and openings of the open canalicular system (OCS).

We show that carbon casting of freeze-fractured cells prevents deformation of cell shape. Pt/C replication (unidirectional or rotary) of the cell surface easily reveals new, high resolution, images of the cell surface. In Figure 4 we show the distribution of IgG Fc-receptors as labeled by immunocytochemistry colloidal gold. The combination of Fracture-Flip with colloidal gold cytochemistry permit the localization of cytochemically defined structures to morphologically defined structures.

References

Andersson Forsman C and Pinto da Silva P 1988 J. Cell. Sci., Submitted
Pinto da Silva P, Andersson Forsman C and Fujimoto K 1988, these proceedings

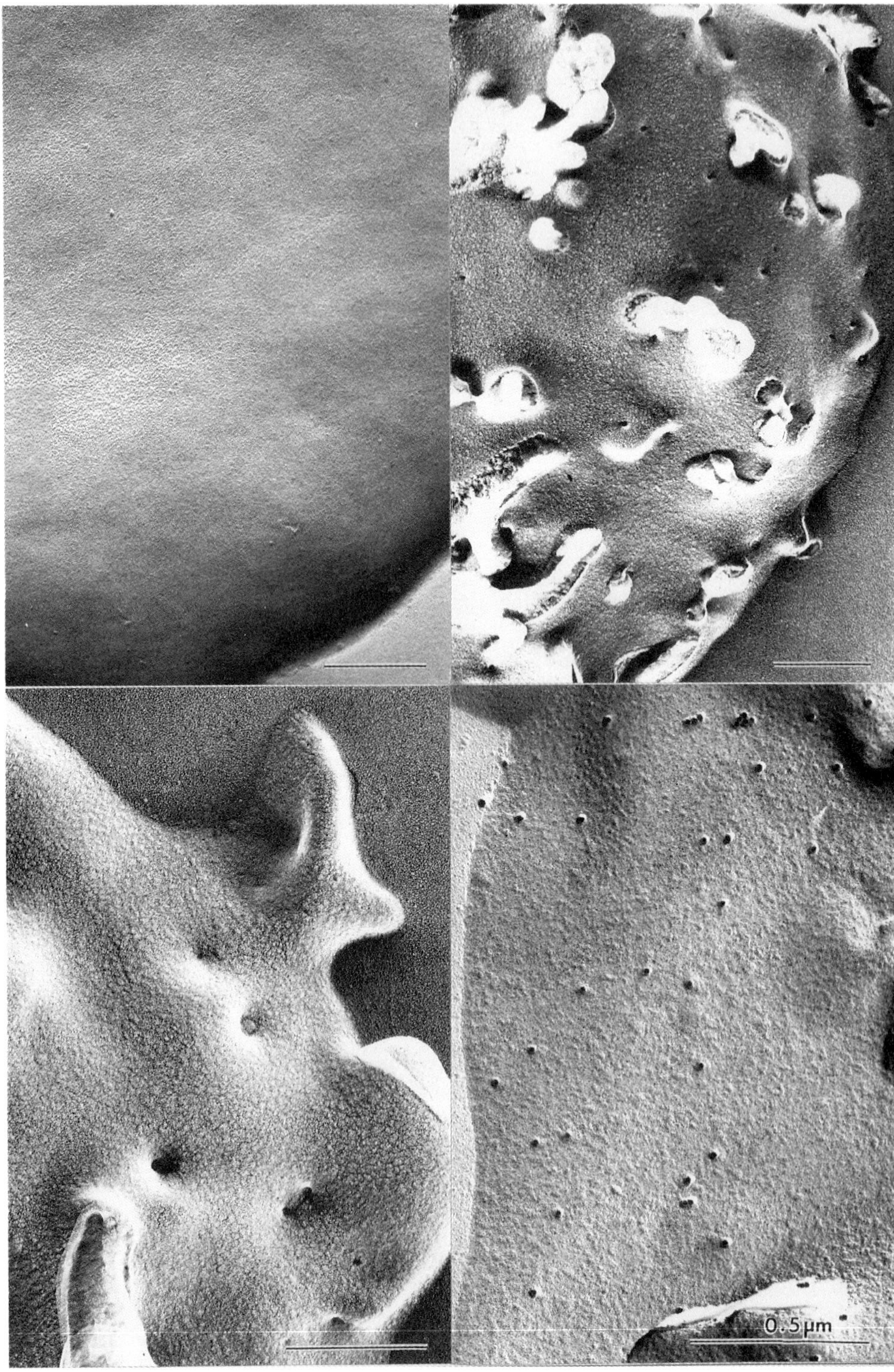
0.5 µm

Macromolecular dynamics of the cell surface during coated pit formation revealed by fracture-flip

Kazushi Fujimoto and Pedro Pinto da Silva

Section of Membrane Biology, Laboratory of Mathematical Biology, National Cancer Institute, Frederick, MD 21701, USA

Fracture-Flip (Andersson Forsman and Pinto da Silva 1988; Pinto da Silva, Andersson Forsman and Fujimoto, these proceedings) is used to provide macromolecular resolution views of the cell surface of rat macrophages during spreading. This process involves the assembly and endocytosis of numerous clathrin coated vesicles (Araki and Ogawa 1986).

Pulmonary alveolar macrophages were plated into culture dishes containing gold double fracture device specimen carriers, allowed to attach on carriers for 30 min at 4 °C, and immediately incubated for 15-60 min at 37 °C to spread. After incubation, the macrophages were fixed with 2 % glutaraldehyde for 60 min. Fracture-Flip was performed (Pinto da Silva, Andersson Forsman and Fujimoto, these proceedings).

We will illustrate the following sequence of events:

I. After 30 min at 4 °C macrophages remain rounded; the entire surface of the membrane is covered by numerous particles (10-25 nm; a density of > 800/μm^2).

II. Further incubation for 15 min at 37 °C leads to cell spreading into flattened hemispheres. Most particles are now restricted to the central region of the adherent surface. They are aggregated into irregularly shaped patches (area 0.5-4 μm^2).

III. After 30 min at 37 °C, particle aggregation becomes more pronouced with the particles restricted to circular (diameter 100-200 nm), often confluent, areas (Fig. 1). These circular patches exhibit a continuum of concavity and illustrate aspects of the cell surface during coated pit formation (Fig. 2).

IV. After 60 min at 37 °C we see steeper, narrower pits in addition to circular concavities. Most particles are now seen in the intervening areas of the membrane surface as well as uniformly distributed over the peripheral areas of the adherent membrane and the free surface of the macrophage.

We believe that the particles revealed by fracture-flip represent, at least in part, the recycling membrane proteins including several different receptors (e.g. IgG, LDL, transferrin receptors). This can now be elucidated by the combination of colloidal gold cytochemistry, fracture-flip, label-fracture and fracture-label.

Andersson Forsman C and Pinto da Silva P 1988 J Cell Science submitted.
Araki N and Ogawa K 1986 Acta histochem cytochem 19, 31

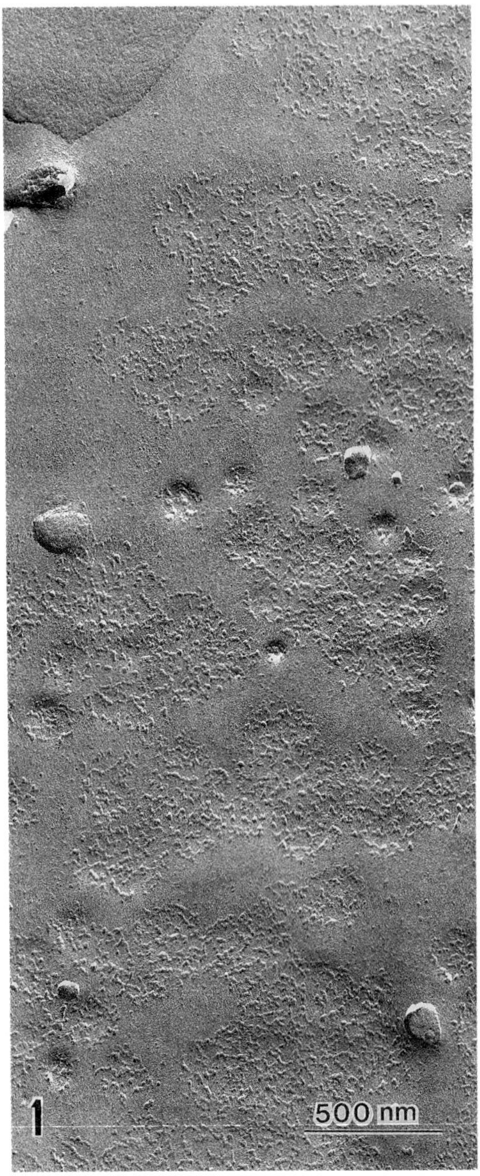

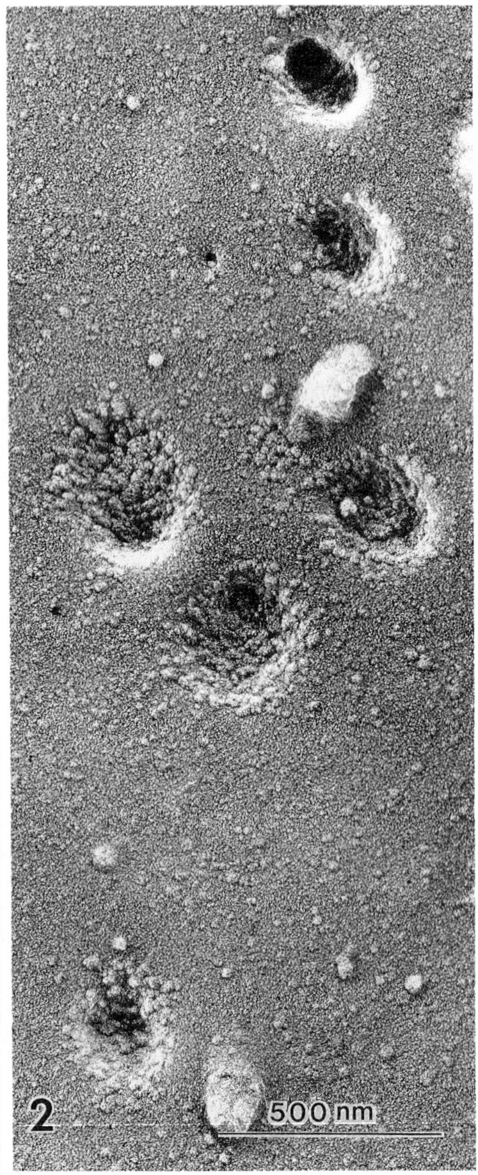

Endocytosis of cationized ferritin in Paramecium cells

Elżbieta Wyroba
Nencki Institute of Experimental Biology, Warsaw, Poland

ABSTRACT: Cell surface binding and endocytosis of cationized ferritin (CF) was observed in axenic ciliates Paramecium aurelia. After incubation with CF for 1.5-3 min, tracer was accumulated in coated profiles (pits at the parasomal sacs) located close to the membrane and next delivered to the cisterns and vacuoles of different diameter. The process of CF internalization was affected by pretreatment of the cells with beta-receptor blocker dichloroisoproterenol shown previously to be the inhibitor of phagocytosis in Paramecium, however labelling of the cell surface and parasomal sacs was still observed.

1. INTRODUCTION

Cationized ferritin as an exogenous tracer of negatively charged groups and a marker of adsorptive pinocytosis may give some information on the endocytic pathway of the small (about 12 nm) polycationic molecules.

2. MATERIALS and METHODS

Starved Paramecium aurelia cells (299s - axenic strain) have been incubated with CF (0.05 mg per ml) for 1.5-15 min. After incubation cells were fixed for 1 h in 3.6% glutaraldehyde in 0.15 M cacodylate buffer, pH 7.4. Following fixation, the cells were rinsed twice with buffer and postfixed with 1% OsO_4 for 45 min. The samples were then dehydrated in ethanol and Epon embedded. The uncontrasted sections were examined in Jem 100 B electron microscope.

3. RESULTS and DISCUSSION

After incubation for 1.5-3 min CF was bound evenly to the entire cell surface including cilia (Fig. 1). Such a location of cationized ferritin indicates that anionic sites are evenly distributed all over the entire cell surface of Paramecium contrary to the results obtained in mammalian tissues (van Deurs and Nilausen 1982, Geuskens 1986) and Amoeba (King and Preston 1977). Since a uniform pattern of CF labelling was also observed in Tetrahymena (Nilsson and van Deurs 1983) it may be postulated that a homogenous distribution of negatively charged groups on the cell surface is a characteristic feature of ciliates surface membrane. Cationized ferritin was found to be internalized via coated pits (Fig. 2) at parasomal sacs. Parasomal sacs are identations of plasma membrane associated with ciliary basal bodies shown previously to be the sites of endocytosis of horse-radish peroxidase in Paramecium (Allen and Fok 1980) and cationized ferritin in Tetrahymena (Nilsson and van Deurs 1983). Similarly, polycationic dye ruthenium red was found to bind to the membrane of parasomal sacs (Wyroba, unpublished), thus indicating on the role of these structure in the cation-selective uptake. After incubation with CF for 3-9 min the marker was found in cisterns and vacuoles of different diameter. Some of them contained myelin-like figures (Fig. 3). In the vacuoles ranging from 0.3 μm to 1 μm CF was membrane associated. It has been previously shown that process of endocytosis in Paramecium may be inhibited by the action of beta-adrenergic antagonists (Wyroba 1986, Wyroba 1987). To check whether beta-blockers influence endocytic pathway of cationized ferritin, the cells pre-exposed to 70 μM l-dichloroisoproterenol for 20 min were subsequently incubated with CF for various time intervals. It has been found that labelling of cell surface and coated pits with cationized ferritin was not altered (Fig. 4), however the delivery of the tracer to the intracytoplasmic structures was inhibited.

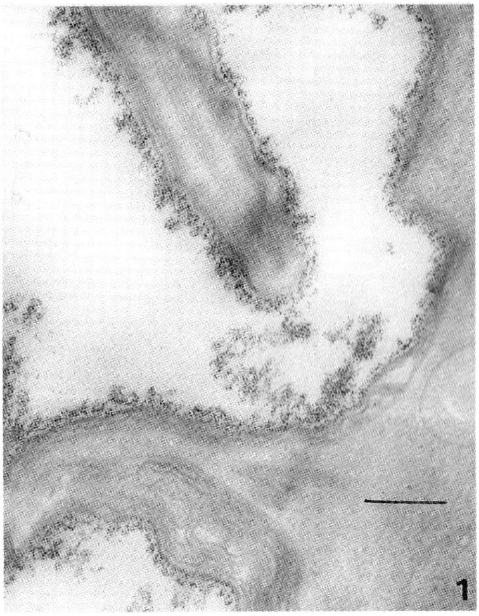

Fig.1.Labelling of Paramecium cell surface after a 1.5 min incubation with CF.Scale bar represents 0.25μm

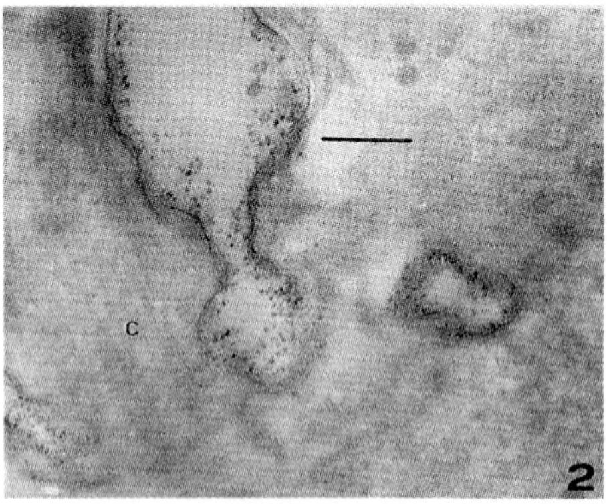

Fig.2.Section through cilium (c) and parasomal sac.A 3-min incubation with CF.Bar equals 0.1μm

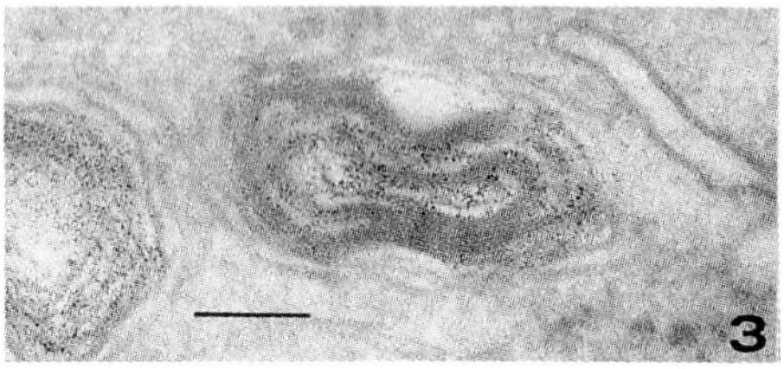

Fig.3.CF in cistern after 9 min incubation.Scale bar equals 0.1μm

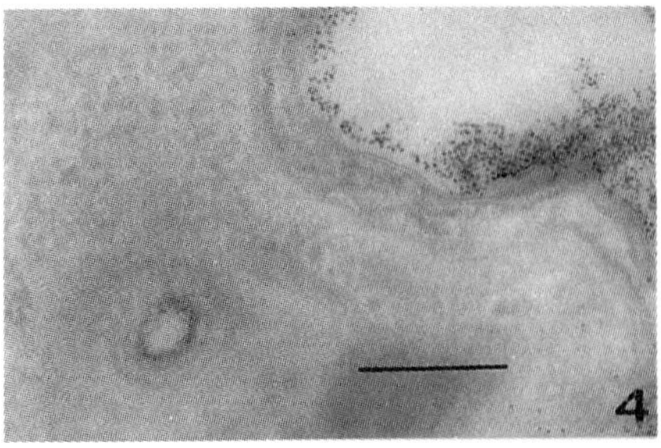

Fig.4.Cell pretreated with dichloroisoproterenol and incubated with CF (3 min).Scale bar represents 0.25μm

REFERENCES

Allen R D, Fok A K 1980 J.Cell Sci.45 131
Geuskens M J 1986 Submicr.Cytol.18 225
Nilsson Y R, van Deurs B 1983 J.Cell Sci. 63 209
King C A, Preston T M 1977 J.Cell Sci. 28 133
Van Deurs B, Nilausen K 1982 J.Cell Biol. 94 279
Wyroba E 1986 Acta Protozool. 25 167
Wyroba E 1987 Cell Biol.Int.Rep. 11 657

Route of intracellular maturation and sorting of two herpes simplex virus type I glycoproteins as found by simultaneous immunogold staining of ultrathin cryosections

M.H.Nielsen*, L.Bastholm*, S.Chatterjee***, B.Norrild**.
*Institute of Pathological Anatomy and ** Medical Microbiology, University of Copenhagen, 11 Frederik D. V Vej DK-2100 Copenhagen, Denmark
*** Dept. of Pediatrics University of Alabama, Birmingham.

ABSTRACT: Simultaneous immunocytochemical triple-staining of ultrathin cryosections of Herpes Simplex Virus I infected cells indicate, that virus glycoprotein C and D are formed in Golgi vesicles and are transported to the cell surface along the cytoplasmic microtubules. A varying number of virus particles located in the cytoplasm and outside the cell were stained with anti glycoprotein C and D but none of the intranuclear virus particles .

1. INTRODUCTION

The mechanism of intracellular maturation and sorting of Herpes Simplex Virus type I (HSV) glycoproteins is not known in details. Light microscopic studies indicate that some glycoproteins may be present in the perinuclear area and that they usually are found in the Golgi region as well as on the cell surface where they may be evenly scattered or confined just to zones of intercellular contact (Norrild et al 1986). By immunocytochemical methods and electron microscopy have we tried to identify the intracellular route for virus glycoproteins specially emphasizing their relations to cytoplasmic microtubules.

2. MATERIAL and METHODS

Human embryonic lung cells (MR C5) were infected with HSV and 12 hours later fixed in buffered 2 % glutaraldehyde/ 2 % paraformaldehyde for 10 minutes, scraped of the culture dish, embedded in 10% buffered gelatine, cryoprotected in 2,3 M buffered sucrose, and frozen in liquid nitrogen prior to ultramicrotomy (Sorvall model 5000). Ultrathin cryosections collected on formvar coated grids were simultaneously triple-immunogold stained with monoclonal antibodies specific to HSV glycoprotein C (gC)(Koga et al. 1986), glycoprotein D (gD)(Koga et al. 1986) and tubulin (Amersham) as described (Bastholm et al. 1987, 1988).

3. RESULTS and COMMENTS

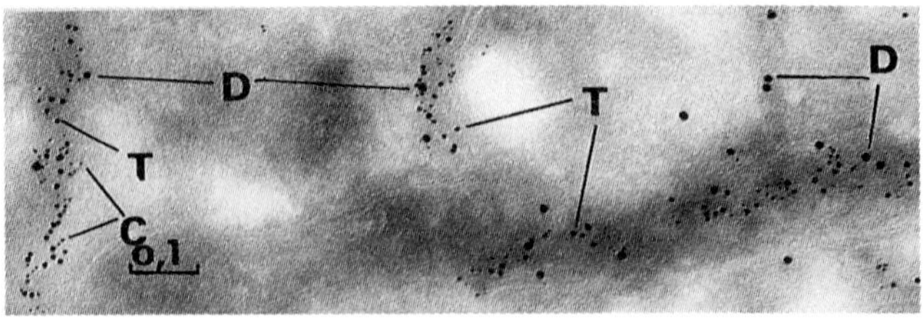

Fig.1. Microtubules simultaneously immunostained for tubulin(T)(10nm gold), glycoproteins C(C)(5 nm gold) and D(D)(15nm gold).

Application of specific monoclonal antibodies to gC and gD resulted always in immunological staining of virus particles located outside the cell, and of some virus particles in the cytoplasm, but never of the incomplete virus visible in the nucleus. Positive staining was also obtained for Golgi vesicles, for the cytoplasm surrounding microtubules and for the external coat of the cell membrane. Application of antibodies to α and ß tubulin gave immunological marking of the moderately dense cytoplasmic microtubules, near the Golgi region and the cell membrane, but never in labelling of Golgi vesicles or virus particles inside or outside the cell. Our findings indicate that the virus gC and gD are sorted in the Golgi vesicles and are transported to the cell surface along the cytoplasmic microtubules. Virus particles follow in some cases a route via the Golgi region but it is uncertain if they normally obtain gC and gD coating here or at the cell surface.

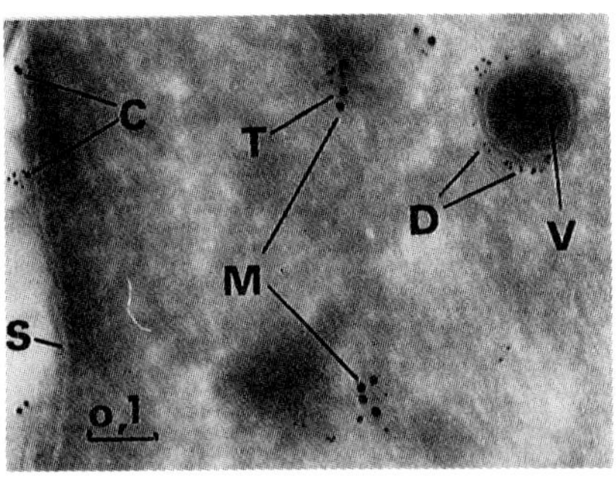

Fig.2. Simultaneous immunostain for tubulin(T) (15nm gold) glycoproteins C (C)(10nm gold) and D(D) (5nm gold). V marks virus, M crosscut microtubules, S the surface membrane.

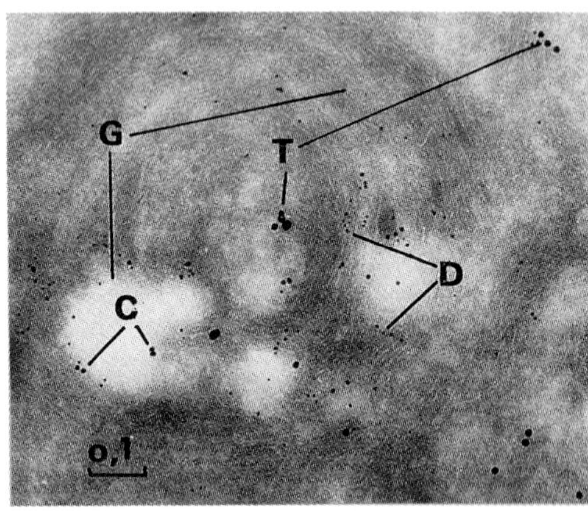

Fig.3. Simultaneous immunostain for tubulin(T)(15nm gold), glycoprotein C(C)(10nm gold) and D(D)(5nm gold). G marks Golgi vesicles.

REFERENCES

Bastholm L, Nielsen M H, Larsson L-I 1987 Histochem. 87 229
Bastholm L, Nielsen M H, Chatterjee S, Norrild B 1988 EUREM
Koga J, Chatterjee S, Whitley R J 1986 Virology 151 85
Norrild B, Letho V-P, Virtanen I 1986 J.Gen.Virol. 67 97

Effects of monensin on the Golgi complex and corticosteroidogenesis

M.M.Magalhães and Maria C. Magalhães

Institute of Histology and Embryology of Faculty of Medicine and Center of Experimental Morphology of the University of Oporto (INIC), Portugal

The role of the adrenal Golgi complex on the steroidogenesis in still unknown, although the involvement of this organelle in steroid secretion has been suggested (Fawcett et al., 1969) and it presents marked alterations under experimental conditions that alter the normal steroidogenesis (Magalhães and Magalhães, 1969; 1972). Monensin is a carboxylic ionophore which profoundly alters the Golgi complex fine structure of different tissues and blocks some cell functions, such as terminal glycosylation and sulfation, membrane regeneration, and vesicular transport (Tartakoff, 1980).Therefore,the study of the effects of monensin on the adrenal Golgi complex and on adrenal secretion could clarify the relationship between Golgi complex and corticosteroidogenesis.

Cells from rat adrenal in primary culture, adrenal slices, and adrenals from intact rats were used in this study. Monensin was added to the cell culture medium (1 μM) during 60 minutes and to the incubation medium of the slices (5 μM) during 15 minutes. For the in vivo experiments, adult rats were injected intravenously with 0.65 mg/Kg/body weight of monensin and sacrificed 10 minutes later. The corticosterone concentrations in culture media, with and without monensin, were determined by a fluorometric method (Mejer and Blanchard, 1973).

The Golgi complex of the zona fasciculata cells from untreated material presented stacks of flattened cisternae and small coated vesicles in close proximity to them (Figures 1 and 3). The zona fasciculata cells in all experimental models treated with monensin presented marked alterations of the Golgi complex (Figures 2 and 4). This appeared disorganized, with the cisternae irregularly dilated or replaced by smooth-surfaced vesicles, sometimes with vacuolar appearance. The mitochondrial matrix was more electron dense in the monensin-treated cells. The biochemical data showed a marked decrease in average corticosterone concentrations in the incubation media after monensin treatment (Table I).

Table I

Effects of monensin on corticosterone concentration in culture medium (μg/ml)*

Control	1.77±0.3
Monensin	0.73±0.05

*Adrenal cells in primary culture were maintained during 6 days. Monensin was then added to the medium (1 μM) during 60 minutes

These data show that monensin disrupted the adrenal Golgi complex and induced a marked decrease in adrenal steroidogenic activity which needs further investigation.

Fawcett DW, Long JA, Jones AL 1969 *Recent Prog. Horm. Res.* 25 315
Mejer LE, Blanchard RF 1973 *Clin. Chem.* 19 710
Magalhães MC, Magalhães MM 1969 *Lab. Invest.* 21 491
Magalhães MC, Magalhães MM 1972 *Endocrinology* 90 444
Tartakoff AM 1980 *Int. Rev. Exp. Pathol.* 22 227

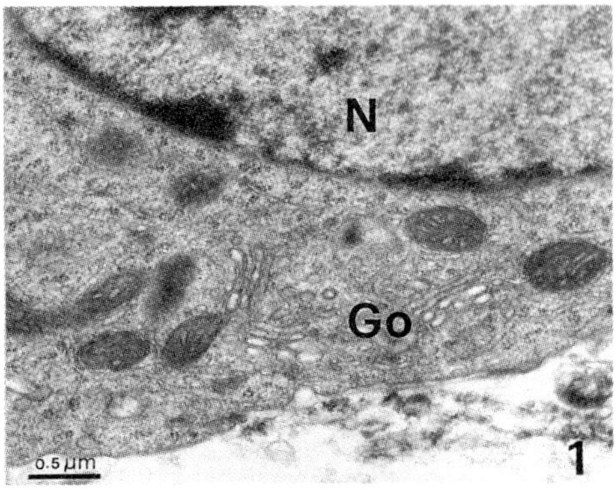

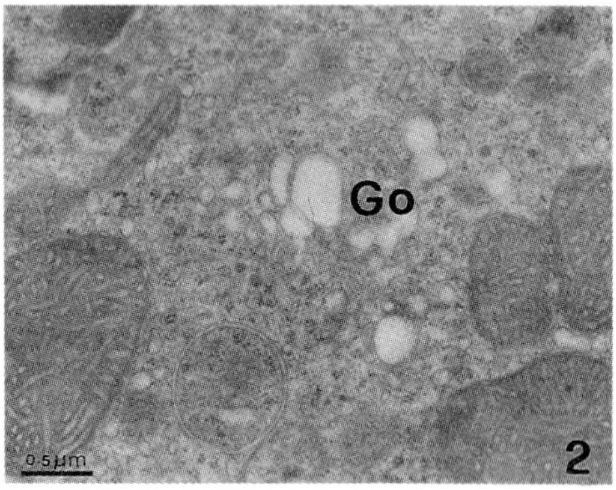

Figure 1 - Cortical cells in tissue culture during 6 days. The Golgi complex (Go) is well preserved. N - nucleus. Fixation: glutaraldehyde plus osmium tetroxide. Staining: uranyl acetate and lead citrate.

Figure 2 - Cortical cells in tissue culture treated with monensin during 60 minutes on day 6. The Golgi complex (Go) appears disrupted with large vesicles. Fixation and staining as for Figure 1.

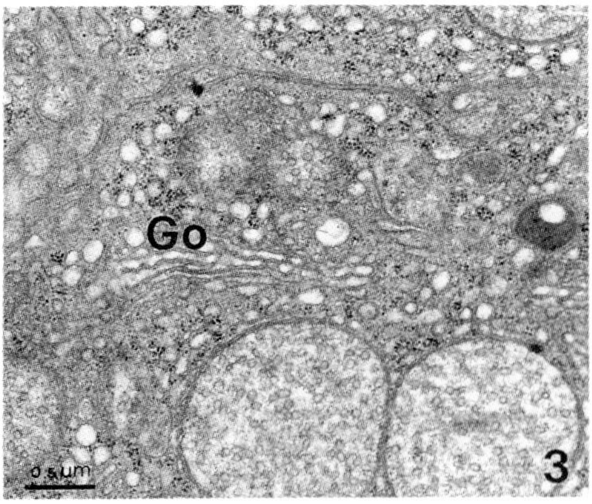

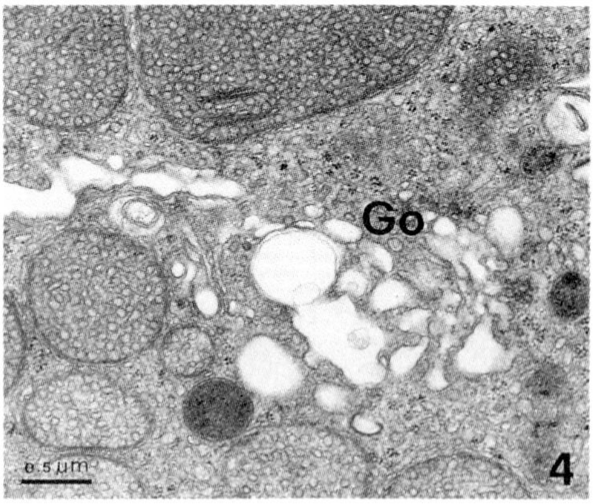

Figure 3 - Rat control. The Golgi complex (Go) presents stacks of cisternae and coated vesicles. Fixation and staining as for Figure 1.

Figure 4 - Rat injected with monensin and sacrificed 10 minutes later. The Golgi complex (Go) appears disrupted with large vesicles. Fixation and staining as for Figure 1.

Approximate osmolarities of vehicles for nematodes: TEM study 17 species for locate carbohydrate moieties on the epicuticle membrane

RYUNTYU, M., Department of Physiology, University of New England, Armidale. N.S.W. 2351. Australia.

A series of experiments was designed for TEM to preserve the cuticle in the best possible condition by careful monitoring and stabilisation of the level of osmolarities between solutions.

Plant parasites are treated as a distinct group; total osmolarities were determined as for specimens inhabiting "meristematic cellular fluid", 350-400 mOsM. Accordingly, all stages of TEM treatment were on the basis of total osmolarities of 350-400 mOsM (Table). The osmolarity value was not allowed to exceed ± 30 mOsM, as variations greater than ± 50 mOsM cause artefacts in membrane-like structures (Glauert, A.M. 1965. "The Fixation and Embedding of Biological Specimens." In: "Techniques for EM", (Ed. D.H. Kay), Oxford: 166-212; (1970) "Practical Methods in EM", Vol. I, North-Holland Publ. Co., Amsterdam) similar to those on the cuticle (Ryuntyu, 1978. "General Methods of the Histochemical Treatment for Zooparasitic and Plant Parasitic Nematodes". Published by The Central Office of the All-Union Institute of Scientific and Technical Information, Monogr. Suppl. Series, Academy of Sciences Ukrainian SSR, Kiev, USSR: 2-162). Calculations were based on Bone's Tables supplemented by Hayat's Tables (Hayat, M.A. 1981. "Principles and Techniques of EM: Biological Applications." Edward Arnold Ltd., London).

Free-living species were determined as a group from "fresh water" or "running water" environment. In the absence of reports on their osmolarity values, it was resolved to view Hayat's (1981) "delicate specimens" and their total osmolarities of 300-350 mOsM.

With such aims in mind, it would have been reasonable to expect uniform quality of EM chemical treatment for all parts of the hydrostatic skeleton. The reaction of hypodermal syncytium to fixatives remains unknown. Probably they are a very special epithelium tissue which, through their functional features, are active between the liquid crystal cuticle and the coelomic liquid pressure. All nematodes were kept in solutions with constant level of osmolarity: 5 mins (first washing buffer before primary fixation); 2 hrs (primary fixative); 5-10 mins (2nd washing buffer after primary fixation); 5 mins (3rd washing buffer before secondary fixation); 2-6 hrs (second fixative). When an identical buffer was used for all fixative, specimens were not treated in the third washing buffer.

TABLE: APPROXIMATE OSMOLARITIES OF VEHICLES FOR NEMATODES (see next page).

TABLE Species ♀	Wash in buffer (first)	Total Osmolarity (mOsM)	Vehicle Osmolarity (mOsM)	Aldehyde fixation buffer	Total Osmolarity (mOsM)	Vehicle Osmolarity (mOsM)	Wash in buffer (second)	Total Osmolarity (mOsM)
Aporcelaimellus obtusicaudatus	0.1M CACODYLATE BUFFER + 0.05% $CaCl_2$ + 0.17M SUCROSE pH 7.2	340	+10	1.8% GLUTARALDEHYDE (GA) + 0.1M CACODYLATE BUFFER + 0.05% $CaCl_2$ pH 7.2	350	-10	0.1M CACODYLATE BUFFER + 0.05% $CaCl_2$ + 0.17M SUCROSE pH 7.2	340
Xiphinema monohysterum	0.1M CACODYLATE +0.05% $CaCl_2$ pH 7.2 +0.23M SCUROSE	400	0	2% GA + 0.1M CACODYLATE pH 7.2	400	-10	0.1M CACODYLATE +0.05% $CaCl_2$ pH 7.2 0.23M SUCROSE	390
Paratrichodorus lobatus	0.1M CACODYLATE + 0.15M SUCROSE pH 6.85	362	-22	1% GA + 0.1M CACODYLATE pH 6.85	340	+10	0.1M CACODYLATE +0.05% $CaCl_2$ + 0.18M SUCROSE pH 6.85	350
Mesorhabditis monhystera	0.1M CACODYLATE + 0.05% $CaCl_2$ + 0.17 M SUCROSE pH 7.2	340	+10	1.8% GA + 0.1M CACODYLATE + 0.05% $CaCl_2$ pH 7.2	350	-10	0.1M CACODYLATE +0.05% $CaCl_2$ + 0.17M SUCROSE pH 7.2	340
Panagrellus redivivus	0.1M CACODYLATE + 0.05% $CaCl_2$ + 0.13 M SUCROSE pH 7.4	330	+20	1.8% GA + 0.1M CACODYLATE + 0.05% $CaCl_2$ pH 7.4	350	-10	0.1M CACODYLATE + 0.05% $CaCl_2$ + 0.13M SUCROSE pH 7.4	340
Pratylenchus thornei	0.1M CACODYLATE +0.05% $CaCl_2$+0.23M SUCROSE pH 7.2	400	0	2% GA + 0.1M CACODYLATE pH 7.2	400	-10	0.1M CACODYLATE + 0.05% $CaCl_2$ + 0.2 3M SUCROSE pH 7.2	390
Heterodera avenae ♀♂	0.1M CACODYLATE + 0.15M SUCROSE pH 6.85	362	-22	1% GA + 0.1M CACODYLATE pH 6.85	340	+10	0.1M CACODYLATE + 0.05% $CaCl_2$ + 0.18M SUCROSE pH 6.85	350
Meloidogyne arenaria ♀♂	0.1M CACODYLATE + 0.15M SUCROSE pH 6.85	362	-22	1% GA + 0.1M CACODYLATE pH 6.85	340	+10	0.1M CACODYLATE + 0.05% $CaCl_2$ + 0.18M SUCROSE pH 6.85	350
Paratylenchus nanus	0.1M COLLIDINE + BUFFER + 0.13M SUCROSE pH 7.4	370	+30	3% GA + 0.1M COLLADINE + 0.05% $CaCl_2$ pH 7.4	400	- 20	0.1M COLLIDINE BUFFER + 0.13M SUCROSE pH 7.4	380
Tylenchus semipenetrans ♀♂	0.1M COLLIDINE + 0.13M SUCROSE pH 7.4	370	+30	3% GA + 0.1M COLLIDINE + 0.05% $CaCl_2$ pH 7.4	400	-20	0.1M COLLIDINE + 0.13M SUCROSE pH 7.4	380
Seinura paratenuicaudata	0.1M PHOSPHATE BUFFER (SÖRENSEN) + 0.09M SUCROSE pH 7.4	330	+20	1.8% GA + 0.1M SÖRENSEN pH 7.4	350	-10	0.1M CACODYLATE + 0.05% $CaCl_2$ 0.13M SUCROSE pH 7.4	340

Nuclear matrix structural elements determining stability of cenome of hybrid somatic cells

A.Ju.Kerkis, N.S.Zhdanova, N.B.Khristolubova

Institute of Cytology and Genetics of the Academy of Sciences of the USSR, Siberian Branch, Novosibirsk 630090, USSR

ABSTRACT: Deep invagination of inner membrane of nucleic envelopment not specified up to now were found in cells of hybrid clones obtained by fusion of fibroblasts from Chinese hamsters and mouse hepatoma. The structures connected with nuclear matrix of hybrid cells formed additional inner surface of nucleus necessary for well spatioally tegulated localisation of "superfluous chromosomes" in the nucleus. This was demonstrated by the analysis of serial sections, autoradiographical and hystochemical investigations.

As a rule,in cultures of somatic cell hybrids segregation of the chromosomes of one of the parents is observed. Nuclear matrix seems to be a cellular structure considerably affecting the hybrid genotype stability.Its elements act as a nuclear skeleton and so probably ensure the functionally regulated localisation of chromosomes in the hybrid cells.

We studied ultrastructures of nucleus and internal matrix of hybrid clones obtained by the fusion of long cultured fibroblasts of Chinese hamster and mouse hepatoma cells ($BWTG_3$) using different methods, namely, electron microscopic analysis,serial sections, morphometry,electron autoradiography and histochemistry.

Nuclei in the cells of the most of the studied clones were found to contain numerous structures not specified up to now,having a form of round caverns seen in cross sections as surrounded by single-layer membranes with an inner diameter of 40 nm (Fig 1). In some clones the propotion of such nuclei was as high as 70 per cent.An analysis of serial sections as well as results obtained by means of H^3-tymidine incorporation using method invented in our laboratory (Khristolubova and Kerkis, 1972) proved these structures to be long interior invaginations of inner membranes of nucleic envelopment surrounded by close-fitting weakly condensed replicating chromatin (Fig 1). Specific reaction for transcriptionally active chromatin regions (Frenster, 1971), confirmed the adjoining chromatin to be transcriptionally active.Treatment of cells first with DNAse, then with Triton X-100,and then with the two together showed the formation described above to be in fact abnormally developed elements of the nuclear matrix (Fig 2).

Thus, the structures descovered form additional inner surfaces of nuclei, which seem to be indispensable for attachment of "superfluous chromosomes" in the hybrid nucleus. Although our data show a considerable increase of the inner surface of the nuclear membrane, the volume and the area of the external surface of the nuclear membrane remain the same.

These structures seem to determine the stability of genome of hybrid cells and regulate the rate of chromosome segregation in these cells.

REFERENCES.

1. Frenster J.H. 1971. Cancer Res. 31 pp 1129-1133.
2. Khristolubova N.B., Kerkis A.Ju. 1972. Proc. Fifth Europ. Cong. on Electron Microscopy. pp 228-229.

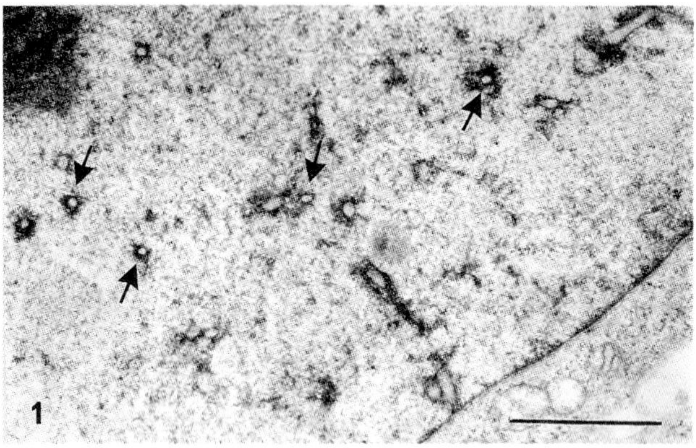

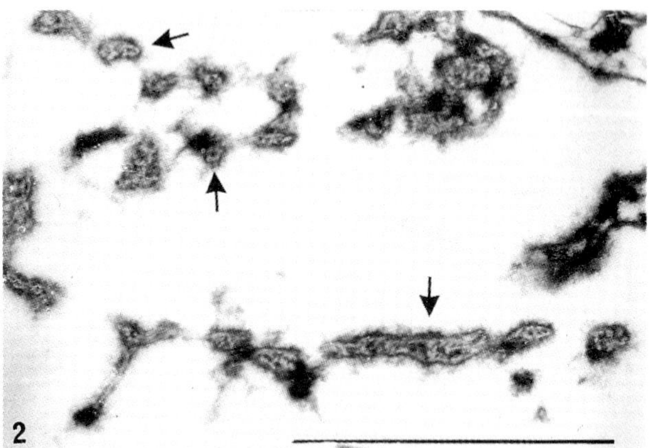

Fig.1. Internuclear structures (↑) cross-section.
Fig.2. The same structures with nuclei treated by Triton X-100 and DNAse Bar 1 μm.

Three-dimensional localisation of muscle spectrin in normal and Duchenne human muscle fibres

A J North, P M Lackie and D M Shotton

Department of Zoology, Oxford University, South Parks Road, Oxford OX1 3PS

ABSTRACT: We are determining the three-dimensional localisation of muscle spectrin in normal and Duchenne human muscle fibres using immunochemical labelling with anti-human spectrin monoclonal antibodies. The overall distribution of spectrin was demonstrated at the light microscope level by indirect immunofluorescence. To provide a more detailed picture this study is currently being extended to the level of the electron microscope using immunogold labelling of ultrathin cryosections. Work is at present being performed on healthy human muscle fibres, and will then be extended to diseased biopsy specimens.

1. INTRODUCTION

While significant information has now been obtained about many of the proteins responsible for integrating the cellular architecture of healthy muscle fibres, by holding the contractile myofilament bundles in register and attaching them to the sarcolemma, little work has yet been done on spectrin. We wish to determine the detailed ultrastructural localisation of muscle spectrin molecules in normal human muscle fibres, in order to elucidate the functional role of this highly specialised membrane associated actin-binding protein, and then to investigate changes in spectrin abundance and distribution in various human diseases, particularly Duchenne muscular dystrophy.

2. METHODOLOGY

We have approached these questions by combining a variety of immunocytochemical techniques, beginning with indirect immunofluorescent staining of fixed or unfixed cryostat sections of normal human muscle. Using the supernatants of several monoclonal antibodies raised in our laboratory against the cross-reacting beta chain of human erythrocyte spectrin as the primary antibody and a fluorescein isothiocyanate (FITC)-conjugated secondary antibody we have revealed the distribution of spectrin in the region of the sarcolemma. A useful positive control in this and all of the following experiments was provided by the monoclonal antibody NOQ7.5.4D which is directed against the heavy chain of the slow myosin isoform of human skeletal muscle and gave a highly specific pattern of labelling demonstrated at the level of the electron microscope (Semper *et al* 1988).

We are currently attempting to extend our studies to the electron microscope level using immunogold labelling of ultrathin cryosections. However, the labelling of spectrin was significantly reduced by treatment with mild fixatives, such as periodate-lysine-paraformaldehyde or 2% formaldehyde plus 0.1% glutaraldehyde (after a pretreatment with ethylacetimidate to reduce the degree of cross-linking). In order to determine the best possible fixation and labelling protocol and to link our light microscopy results with electron microscopy we studied immunogold-labelled ultrathin cryosections by electron microscopy and semi-thin cryosections from the same tissue block labelled under the same conditions and visualised by silver enhancement using epi-polarisation microscopy.

Parallel studies on gold-labelled human erythrocyte ghosts were also undertaken, as this system has been extensively studied and provides a much greater concentration of spectrin within each section. This allows a better evaluation of the degree and specificity of antibody labelling.

3. CONCLUSION

By the methods we have discussed we hope to build up a detailed picture of the three-dimensional localisation of muscle spectrin in healthy human muscle fibres. We then wish to reveal any changes in this distribution which are associated with Duchenne muscular dystrophy and to study the interactions of spectrin with other skeletal muscle proteins. Parallel studies on normal and dystrophic chicken muscle fibres will be undertaken.

4. REFERENCES

Semper A E, Fitzsimons R B and Shotton D M 1988 *J. Neurol. Sci.* **83** 93

An unusual microtubule complex in murine megakaryocytes

SL Green and JM Radley

Peter MacCallum Cancer Institute, 481 Little Lonsdale Street, Melbourne, Victoria 3000, Australia

ABSTRACT: Unusual microtubular complexes occurring in mature megakaryocytes have been investigated. The complexes consisted of staggered rows of microtubules in an electron-dense hexagonal matrix. Fine electron dense strands connected individual microtubules with the surrounding dense matrix. Dense particles sometimes adorned the occasional microtubule bundle.

1. INTRODUCTION

Microtubules are involved in the formation of processes by megakaryocytes, a morphological change which takes place prior to platelet liberation. In general, microtubules follow the long axis of the processes, either singly or in small clusters. In a recent study (Radley et al 1987) it was noted that microtubules occasionally formed bundles in which tubules were separated by a dense band. An examination of this complex is reported here.

2. MATERIALS AND METHODS

Bone marrow was obtained from the femurs of 3-6 month-old Balb/c mice and incubated at 37°C for several hours. Mature megakaryocytes which developed processes were collected onto poly-L-lysine covered slides, fixed in 2% buffered glutaraldehyde then processed for routine TEM. In addition, some mice were perfusion fixed via the dorsal aorta with a solution containing 2% paraformaldehyde and 2% glutaraldehyde. Femurs were removed, decalcified and processed for TEM as above. Mice were often injected 7-8 days prior to experimentation with 5-fluorouracil (150 mg/kg) to enhance the yield of megakaryocytes.

3. RESULTS AND DISCUSSION

Microtubular complexes were seen in vitro and in vivo but were only present in mature megakaryocytes. They were quite rare, usually limited to several per cell. In longitudinal section, complexes appeared as microtubules arranged in parallel rows with dense bands separating neighbouring microtubules (fig 1). Cross sections revealed 10 nm filaments hexagonally arranged around each microtubule. Filaments were usually joined together by an electron-dense material, giving an overall appearance of a honeycomb-like network (fig 2). Microtubules lying at the periphery of the bundles were only half enmeshed by the intermicrotubular network. Fine strands linked microtubules to the surrounding matrix. Although this is the first report of microtubular complexes in a haemopoietic cell, similar microtubule arrangements have been described in several other types of cell

(Bird 1984; Katsuda et al 1987; Nishikawa and Kitamura 1982; Reaven et al (1983). A comparatively unusual feature of the megakaryocytic microtubule complex was the association of electron-dense particles with approximately 60% of the bundles (fig 3). Only Katsuda et al (1987) reported a similar finding and claimed that the particles were ribosomes. They suggested that newly synthesized tubulin from the ribosomes may participate in the formation of the complex. This may also be the case for the megakaryocytic microtubule complex. Although various hypotheses have been put forward (Nishikawa and Kitamura 1984, 1985; Reaven et al 1983), the exact physiological function of these complexes is yet to be resolved.

4. REFERENCES

Bird MM 1984 J Ultrastruct Res, 89 123.
Katsuda S, Okada Y and Nakanishi I 1987 Cell Biol Int Rep. 11 103.
Nishikawa S and Kitamura H 1982 Anat Rec, 202 305.
Nishikawa S and Kitamura H 1984 J Electron Microsc, 33 34.
Nishikawa S and Kitamura H 1985 Archs Oral Biol, 30 1.
Radley JM, Hartshorn MA and Green SL 1980 Thromb Haemost, 58 732.
Reaven E, Jensen CG, Spicher M and Azhar S 1983 J Ultrastruct Res, 83 284.

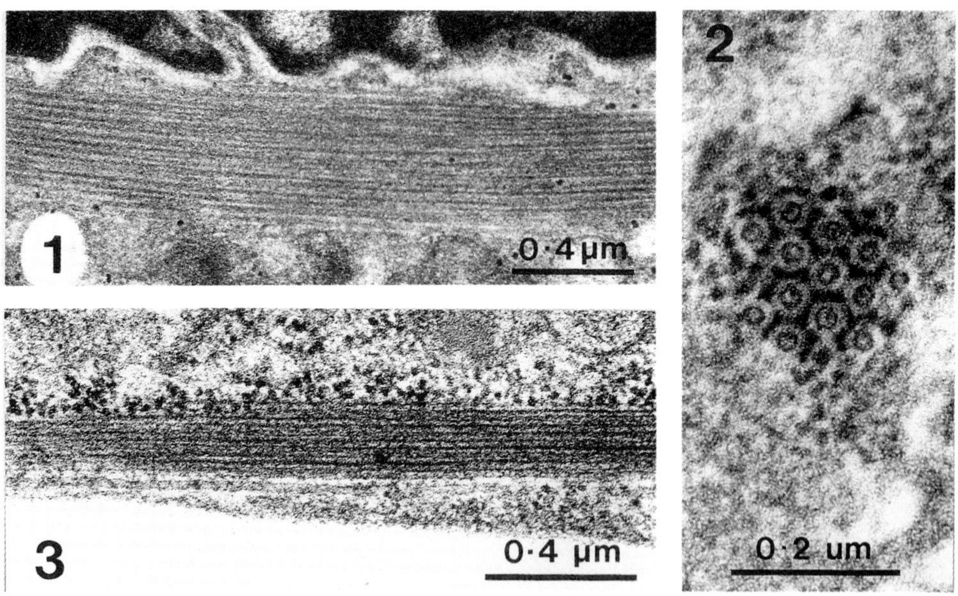

Fig. 1. Longitudinal section of a microtubular complex illustrating the parallel array of microtubules and dense bands. In vivo preparation.

Fig. 2. Transverse section of a microtubular complex. Individual microtubules are surrounded by a hexagonal dense matrix. In vitro preparation.

Fig. 3. Longitudinal section of a microtubular complex revealing its association with dense particles. In vivo preparation.

Figs. 1-3. Uranyl acetate and lead citrate stained.

Inst. Phys. Conf. Ser. No. 93: Volume 3, Chapter 18
Paper presented at EUREM 88, York, England, 1988

SEM studies of the ventilatory system of a highly active insect muscle

K Kral and H Bradacs

Institute of Zoology, University of Graz, A-8010 Graz, Austria

ABSTRACT: The large air sacs surrounding the indirect flight muscles in the propodeum of the hornet contain numerous inverse tracheas, ingrowths from the air sac wall. Anastomoses of these structures form an internal framework to hold up the air sacs and prevent them from collapsing.

1. INTRODUCTION

In the tracheal system of insects, there are air sacs in certain sections that are expansions of the larger tracheas. In insects that are strong fliers, these air sacs are very well developed. In the pterothorax of the hornet, as in other Hymenoptera, large air sacs are found close to the indirect flight muscles. The original function of the air sacs was surely concerned with gas transport; specialized functions have come to include acting as a resonance body, heat insulator, and filler of body cavities. The intima of the tracheas is provided with taenidias, thick spiral structures that prevent the tracheas from collapsing. The air sacs have thin walls and large volume and so the question arises as to how they keep from collapsing. It has been suggested that a negative ambient pressure keeps the air sacs expanded (Dreher 1936), as is the case with the hemolymph in the cockroach (Davey and Treherne 1964). Observations made in the course of our work offer a plausible explanation of how the air sacs could be kept expanded.

2. MATERIALS and METHODS

The experimental animals were adult hornets, Vespa crabro, caught during foraging flights. The cuticula was removed from the propodeum and the pterothorax bisected. The specimens were treated as usual for scanning electron microscopy (SEM) and viewed with a Zeiss DSM 950.

3. RESULTS and DISCUSSION

Fig. 1 shows a view into an air sac in the dorsolateral area of the propodeum. On the right side of the picture, the air sac lies on the left dorsal longitudinal muscle. Numerous inverse tracheas pass through the air sac; these are ingrowths of the air sac wall. Here, the intima is turned to the outside and the epithelium to the inside. The inverse tracheas grow in all directions and anastomose; their diameter is 10-30 µm. Rarely, short, narrow commissures of approximately 5 µm are seen. Occasionally, a branch is seen to dead end. There are often openings in the inverse tracheas, or a short stump, open at the end, branches off. Figure 2 shows higher magnification of the inside of an air sac wall and the inverse tracheas growing out of it. The intima has well developed taenidias that are somewhat irregular only in certain places. The surface of the intima shows glomerular structure throughout.

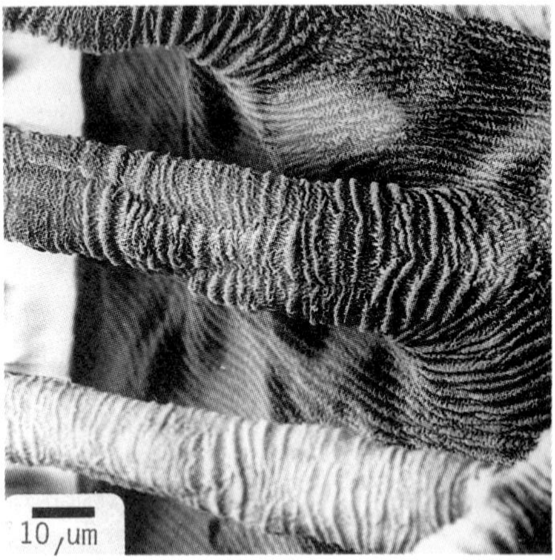

Fig. 1. A view into an air sac of the propodeum. Inverse trachea pass through air sac. W - air sac wall

Fig. 2. Higher magnification of inside of air sac wall and inverse trachea showing taenidia

The arrangement of the inverse tracheas gives them the appearance of a supporting framework. It is assumed that the thoracic air sacs of Hymenoptera are not subjected to any massive pumping movements (Weis-Fogh 1964); ventilation is via the large abdominal air sacs that are connected to the thoracic air sacs and fill them with air. The thoracic air sacs could function as air reservoirs proximal to the muscles, guaranteeing constant oxygen supply to the flight muscle with its high respiratory activity, and prevented from collapsing by the framework of inverse trachea. Weber (1933) suggests the name air chamber for all such air sacs that are so structured that they cannot collapse, as for example those connected to the tympanal organ.

4. REFERENCES

Davey K G and Treherne J E 1964 J.Exp.Biol. 41 513
Dreher K 1936 Z. Morphol.Ökol. Tiere 31 608
Weber H 1933 Lehrbuch der Entomologie (Jena: G. Fischer)
Weis-Fogh T 1964 J.Exp.Biol. 41 207

Inst. Phys. Conf. Ser. No. 93: Volume 3, Chapter 18
Paper presented at EUREM 88, York, England, 1988

Mast cell supplasmalemmal cytoskeleton in exoytosis. Effect of incubation in low and high calcium medium

E Holm Nielsen, K Braun and T Johansen

Institute of Anatomy and Cytology and Institute of Pharmacology, Odense University, Campusvej 55, 5230 Odense M, Denmark

ABSTRACT: The effect of depletion and loading of cell calcium seemed to indicate a relationship between the calcium ion concentration and the rearrangement of the subplasmalemmal cytoskeleton, which occurs during compound 48/80 stimulated exocytosis.

1. INTRODUCTION

In mast cells the subplasmalemmal network consists of a thick mat of crossing and dividing actin filaments forming a barrier which apparently has to be depolymerized or rearranged before a secretion can take place (Allison 1973, Stossel 1981, Bennett 1984, Hansson et al. 1984, Nielsen and Jahn 1984). A rise in intracellular calcium ion concentration is known to be essential for the breakdown of the rigidity of actin (Bennett 1984), and also histamine release is dependent on intra- or extracellular calcium (Douglas 1974, Chakravarty and Yu 1984). Therefore there might be a connection between calcium ion concentration, actin state and histamine release. By scanning electron microscopy (SEM) we have studied the subplasmalemmal network in secreting and non-secreting mast cells incubated in normal and high calcium concentrations and without calcium. The histamine release was measured according to the method of Shore et al. (1959).

2. RESULTS

Incubated with 2.5 mM calcium the network forms small openings through which the cell membrane can just be seen. In secreting mast cells these openings are wider and exocytotic openings are surrounded by areas of uncovered cell membrane. After depletion of calcium by incubation with 5 mM EGTA the network is very dense and individual filaments are hardly recognizable. Compound 48/80 releases only little histamine and the openings between the filaments are small (Fig. 1). Incubation with 5 mM calcium does not effect the network in non-secreting cells, but in secreting cells lysis loosens the attachment of the filaments to the membrane and large areas of bare membrane are revealed, and also the histamine release is very high (Fig. 2).

3. CONCLUSION

The results support the view that in rat mast cells a stimulus for secretion is followed by rearrangement of the subplasmalemmal filaments so that the network no longer prevents fusion between the exocytotic membranes. It also seems that calcium has to be present for these changes to occur. This is supported by the experiments with calciumloaded cells, in which the binding of actin to the cell membrane loosens during exocytosis. This effect of calcium may not be an effect on the microfilaments themselves, but perhaps on calcium-dependent actin-crosslinking proteins.

4. REFERENCES

Allison A C 1973 Virchows Arch. 45 313
Bennett J-P 1984 Biochem. Soc. Transact. 12 963
Burgoyne R D and Cheek T R 1987 Nature 326 448
Chakravarty N and Yu W-J 1984 Agents and Actions 14 386
Douglas WW 1974 Secretory Mechanisms of Exocrine Glands eds N A Thorn, O H Petersen (New York: Academic Press) pp 119-29
Hansson H A, Lonnroth I, Rozell B and Tinberg H 1984 Acta path. microbiol. immunol. scand. sect. B 92-93
Nielsen E H and Jahn H 1984 Virchows Arch 45 313
Shore P A, Burkhalter A and Cohn V H J Pharmacol. exp. ther. 127 182
Stossel T P 1981 Methods in Cell Biol. 23 215

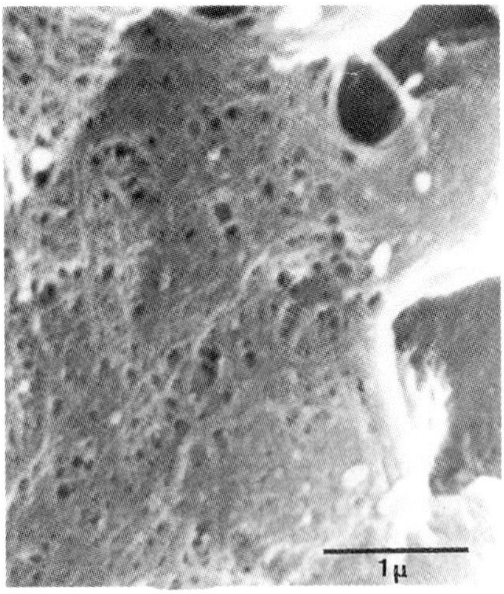

Fig. 1. Subplasmalemmal cytoskeleton from lysed mast cell incubated for 3 h with 5 mM EGTA before compound 48/80 stimulated exocytosis.

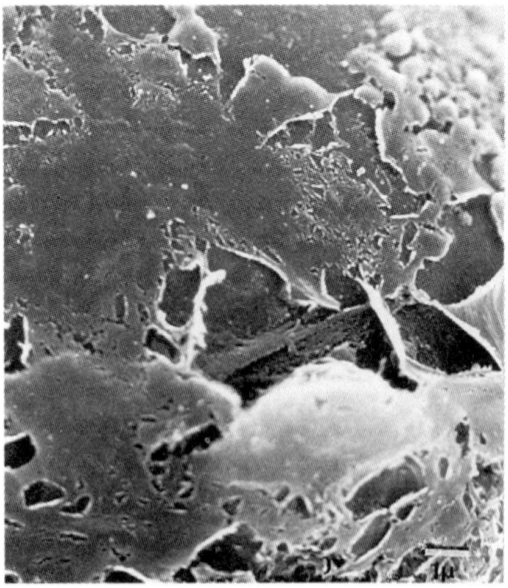

Fig. 2. Large bare areas of cell membrane from lysed rat mast cell incubated for 3 h with 5 mM calcium before compound 48/80 stimulated exocytosis.

Paper presented at EUREM 88, York, England, 1988

Effect of influenza virus on the organization of cytoskeleton associated with cell membranes

F Čiampor, O Križanová and E Závodská

Institute of Virology, Slovak Academy of Sciences, Bratislava, Czechoslovakia

ABSTRACT: The cytoskeleton prepared from influenza virus-treated membranes revealed an increase in amount of actin detected by DNase inhibition test and immuno-gold electron microscopy. Actin filaments often appear to interact and terminate at sites of cell surface receptors for influenza virus. Catalytic subunit of protein kinase causes structural changes in arrangement of haemagglutinin spikes on influenza particles and virus similarly as Ca/CaM-dependent protein kinase increases ATPase activity of cytoskeleton.

1. INTRODUCTION

The association of virus penetration, assembly and replication with cytoplasmic architecture may represent the use of elements in which highly specialized mechanisms for normal cell function have been coopted for virus growth. The influenza A virus polypeptides M, NP and HA_0 are associated with the host cell cytoskeleton during vegetative infection /Laevitt et al 1981/. Superprecipitation of the cytoskeleton prepared from chick embryo cells is stimulated by addition of the influenza virus /Križanová et al 1986/.

2. RESULTS AND DISCUSSION

The cytoskeleton /CS/ isolated from virus-treated plasma membranes by the method of Wolosin et al /1983/ contained more proteins and more actin than the CS isolated from uninfected membranes. Virus particles were found to be associated with CS. Autophosphorylation of CS, especially of the polypeptides 24K, 45K, 65K and 105-110K, was increased in CS prepared from virus-treated membranes. Superprecipitation of CS prepared from chick embryo cells was stimulated by addition of the virus.
Electron microscopy of the Triton X-100 insoluble residue from virus-treated plasma membranes showed more structures of microfilament bundles than Triton X-100 insoluble residue from control membranes. Actin filaments, detected by anti-actin rabbit antibodies and protein A-colloidal gold /10-12 nm/ in electron microscope, appear to interact and terminate at the sites of cell surface receptors for influenza virus. Actin polymerization and reorganization induced by interaction of the virus with corresponding cell receptors causes redistribution of receptors /Fig. 1 and 2/.

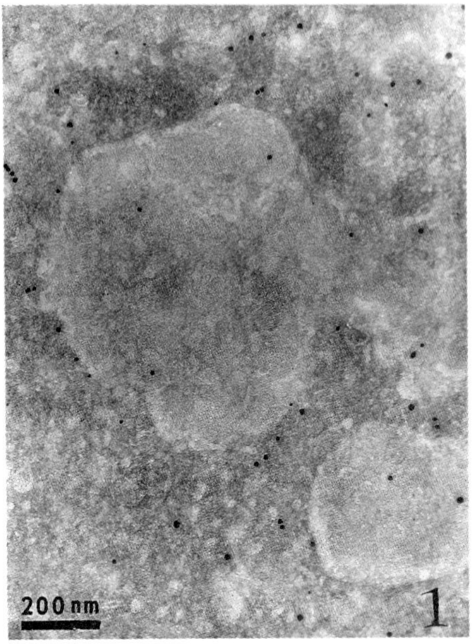

Fig.1. CS isolated from control membranes

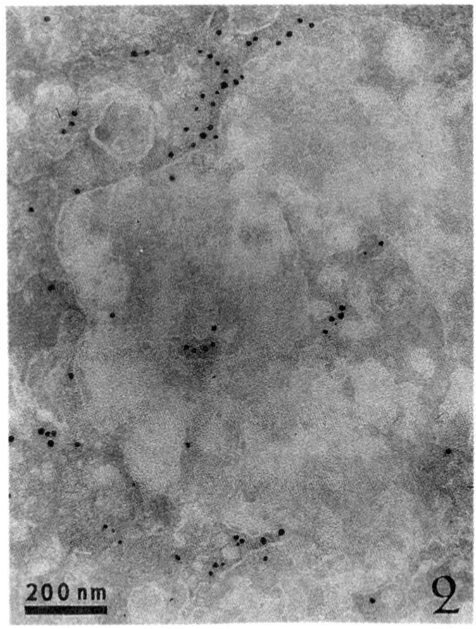

Fig.2. CS isolated from virus treated membranes

Anti-actin rabbit Ab/Protein A-colloidal gold 2 % PTA, Philips EM 300, 80 kV

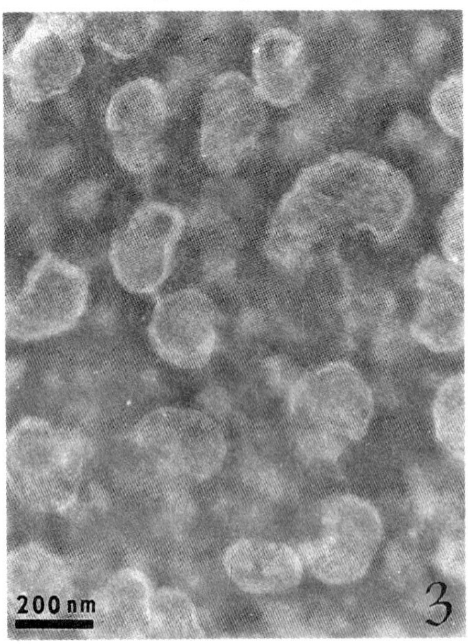

Fig.3. Influenza virus A/WSN treated with catalytic subunit of protein kinase, 2 % PTA

Catalytic subunit of protein kinase causes structural changes in arrangement of HA spikes of influenza particles /Fig.3/. The influenza virus similarly as Ca/CaM-dependent protein kinase increase ATPase activity of CS.

The relationship between phosphorylation of CS elements and virus induced contraction remains to be elucidated.

Križanová O, Čiampor F, Závodská E, Matis J and Stanček D 1986 Acta virol 30 273-280

Laevitt J, Bushar G, Mohanty N, Mayner R, Kakunaga T and Ennis F A 1981 The Replication of Negative Strand Viruses pp 353-362

Wolosin J M, Okamoto C, Forte T M and Forte J G 1983 Biochim biophys Acta 761 171-182

Inst. Phys. Conf. Ser. No. 93: Volume 3, Chapter 18
Paper presented at EUREM 88, York, England, 1988

Phragmoplast microtubules—their role in the formation of the cell plate

M Wierzbicka

Institute of Botany, University of Warsaw. 00-927 Warsaw, Poland

ABSTRACT: Cytokinesis after lead treatment ($PbCl_2$ and $Pb(NO_3)_2$) was studied in Allium cepa L. root tip cells. It was found that phragmoplast microtubules are not necessary for exocytotic vesicles to gather in the central part of the cell (which until now has been commonly accepted), but that they are necessary for the cell plate to assume its proper shape and position.

1. INTRODUCTION

Due to environmental pollution, studies were undertaken to trace the path of Pb^{2+} in the root (Wierzbicka 1987a b) and the effect of lead on mitosis (Wierzbicka 1988). The formation of binuclear cells with a vesicle body between daughter cells was a particular case of disturbance of cytokinesis. Detailed analysis of this type of cells was the aim of this study.

2. MATERIAL AND METHODS

Allium cepa L. adventitious roots were used in the experiment. Lead was administered in the form of $PbCl_2$ and $Pb(NO_3)_2$ at Pb^{2+} concentrations of 1.0, 2.5 and 3.0 mg dm^{-3} for 1, 3, 6, 12 or 24 hours. Squashed slides stained with acetoorcein or PAS + Erlich's hematoxylin were analysed. After fixing the root tips in glutaraldehyde and osmium tetroxide, ultrastructural studies using routine electron microscopy techniques were conducted.

3. RESULTS AND DISCUSSION

Binuclear cells arose after treatment with 2.5 and 3.0 mg dm^{-3} Pb^{2+}. They were most numerous after 24 hours of incubation. A vesicle body between the daughter nuclei was clearly visible in the light microscope, in part of these cells. A positive PAS reaction suggested that this body contained polysaccharides. Electron microscopy showed that the vesicle body was composed of two types of vesicles (Fig. 1). One type, the larger ones with a light profile, was derived from the ER, the other, smaller, with a dense content, was from the dictyosomes. Identical vesicles were seen in the phragmoplasts of control cells. The difference lay in the lack of microtubules. Microtubules were well visible in control cells whereas they did not occur either inside or near the vesicle body.

The lack of microtubules in a cell which had accumulated the entire necessary material for forming the cell plate in the central part of the cell seems to be the direct reason that these vesicles could not be arranged in the equatorial plane of the cell, and thus, a normal cell plate

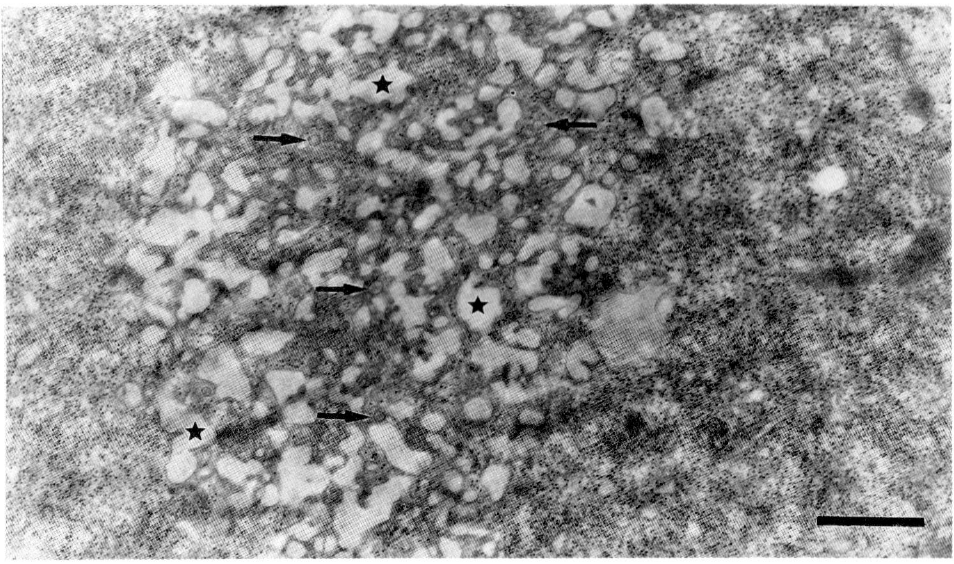

Fig. 1. A vesicle body between the nuclei of a binuclear cell. Large vesicles - asterisks; small vesicles - arrows. Incubation for 6 hrs in 2.5 mg dm^{-3} $PbCl_2$. Bar = 1 µm

could not be formed.

It is very interesting that the exocytotic vesicles accumulated between the daughter nuclei without the aid of microtubules. It has been commonly thought until now that the phragmoplast microtubules direct the movements of the exocytotic vesicles during the formation of the cell plate. The observations presented in this study indicate that microtubules are not needed to accumulate the vesicles which will form the future cell plate. However, microtubules are necessary to form the appropriate shape and position of the cell plate during cytokinesis.

4. REFERENCES

Wierzbicka M 1987a Can. J. Bot. 65 1851
Wierzbicka M 1987b Plant Cell Environ 10 17
Wierzbicka M 1988 Caryologica (in press)

Recent developments in ultrastructural cytochemistry of phosphatases and oxidases using the cerium technique

H.D. Fahimi, S. Angermüller, E. Baumgart, A. Völkl

Department of Anatomy, II. Division, University of Heidelberg, Im Neuenheimer Feld 307, D-6900 Heidelberg, Fed.Rep.Germany.

ABSTRACT
Selective staining of cell organelles for specific enzyme functions facilitates not only the qualitative morphological analysis of that intracellular compartment but may also be useful for quantitative morphometric investigation. Indeed, if sufficient contrast in sections is available, the morphometry can be automated by the use of television based image analysis systems as described recently for peroxisomes (Beier & Fahimi, 1986, 1987). In this review some of the recent developments in electron microscopic cytochemistry of phosphatases and peroxisomal oxidases are presented with particular emphasis on the cerium technique. The selective staining of cell organelles is considered in each case.

1. INTRODUCTION

The criteria for the validation of enzyme histochemical techniques were carefully reviewed by Stoward (1980). They included: precision, reproducibility, specificity and validity. The advantage of methods which meet those criteria is that they can be used not only for qualitative but also for quantitative analysis by microspectrophotometry. There are however, many histochemical techniques which permit qualitative evaluation of the distribution of staining patterns both at light and electron microscopic level. By adapting such enzyme-histochemical methods for selective staining of cell organelles, the quantitative analysis of that particular intracellular compartment by morphometry will be markedly facilitated (Fahimi 1980). Indeed if sufficient contrast is available in sections, the morphometry can be performed by television-based automatic image analysis, as described recently for peroxisomes (Beier & Fahimi 1986, 1987). In this short review some of the recent developments in electron microscopic cytochemistry of phosphatases and oxidases are presented with particular emphasis on the cerium technique, as used for the investigation of peroxisomal enzymes

2. CERIUM TECHNIQUES

Cerium was introduced originally for the ultrastructural cytochemical detection of sites of formation of hydrogen peroxide in leukocytes (Briggs et al. 1975). It forms a fine homogenous electron dense precipitate which is ideal for electron microscopic cytochemistry, being comparable to the reaction product of oxidized 3,3'-diaminobenzidine (DAB). Veenhuis et al. (1980) described the feasibility of cerium also for the localization

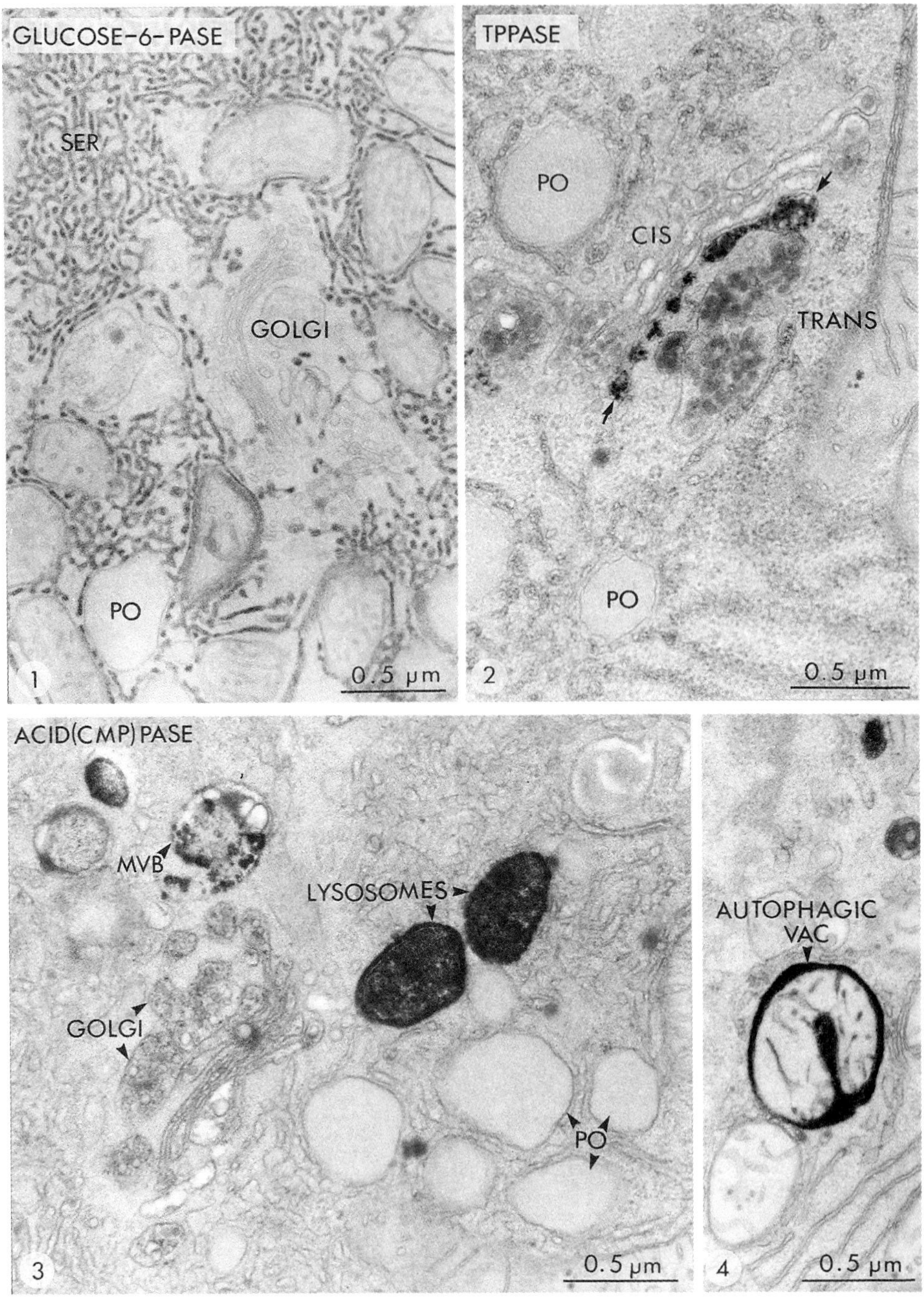
GLUCOSE-6-PASE
SER
GOLGI
PO
1
0.5 µm
TPPASE
PO
CIS
TRANS
PO
2
0.5 µm
ACID(CMP)PASE
MVB
LYSOSOMES
GOLGI
PO
3
0.5 µm
AUTOPHAGIC VAC
4
0.5 µm

of phosphatases. Since then cerium has become the medium of choice for the localization of phosphatases (Robinson & Karnovsky 1983) and it has been shown to be superior to the classical lead-technique for this purpose (Angermüller & Fahimi 1984).

2.1. Glucose-6-phosphatase

Fig. 1 shows the localization of glucose-6-phosphatase by the cerium-method with distinct staining of the rough and smooth endoplasmic reticulum in a rat liver parenchymal cell. The enzyme is excluded from the cisternae and the vesicles of the Golgi complex which are negative.

2.2. Thyamine pyrophosphatase

Incubation of sections in a medium for thyamine pyrophosphatase shows the prominent staining of the first trans-most Golgi cisterna without any reaction in the remaining portions of the Golgi complex (Fig. 2). The peroxisomes and mitochondria are also unstained although a very weak reaction is noted in the endoplasmic reticulum.

2.3. Acid phosphatase

The prominent staining of lysosomes incubated in the cerium chloride medium for acid phosphatase with cytidine monophosphate as substrate is shown in Fig. 3 which in addition, reveals marked heterogeneity in the intensity of reaction in multivesicular bodies and lysosomal vesicles at the trans-aspect of the Golgi complex. A few electron dense granules are also seen in the dialated cisternae at the trans-side of the Golgi. An autophagic vacuole containing a partly disintigrated mitochondrion and a heavy acid phosphatase reaction is shown in Fig. 4. The major advantage of the cerium technique over the classical lead-method for phosphatases lies in the scarcity or the absence of nonspecific deposits and precipitates in sections (Figs. 3 and 4).

2.4. Uricase

The cerium technique has been used also for the localization of peroxisomal oxidases. For this purpose the tissue should be preferably fixed for a very short time (5 min.) with a low concentration of glutaraldehyde (0.25%) (Angermüller & Fahimi 1986). Fig. 5 shows the localization of urate oxidase in peroxisomes of rat liver. The enzyme is confined to the core region in the matrix of peroxisomes. The Tris-maleate buffer used originally in the cerium method has an inhibitory effect upon the activity of uricase and indeed the Pipes buffer has proven much superior for this purpose. The fine structural details of localization of uricase in crystalline cores of rat liver peroxisomes are shown in Fig. 6. The enzyme reaction product is localized in the lumen of small 5 nm tubules sparing the larger (20 nm) space between them.

2.5. Xanthine oxidase

Another enzyme of purine catabolism xanthine oxidase was also discovered in the cores of rat liver peroxisomes (Angermüller et al. 1987), showing a similar distribution as uricase with weaker intensity (Figs. 7 and 8).

2.6. D-amino acid oxidase

The localization of this enzyme was studied recently in peroxisomes of rat liver and kidney (Angermüller & Fahimi 1988a). By converting the reaction product of cerium with DAB to a coloured compound the distribution of the enzyme can be visualized also by light microscopy (Angermüller & Fahimi 1988b). This procedure revealed that D-amino acid oxidase is

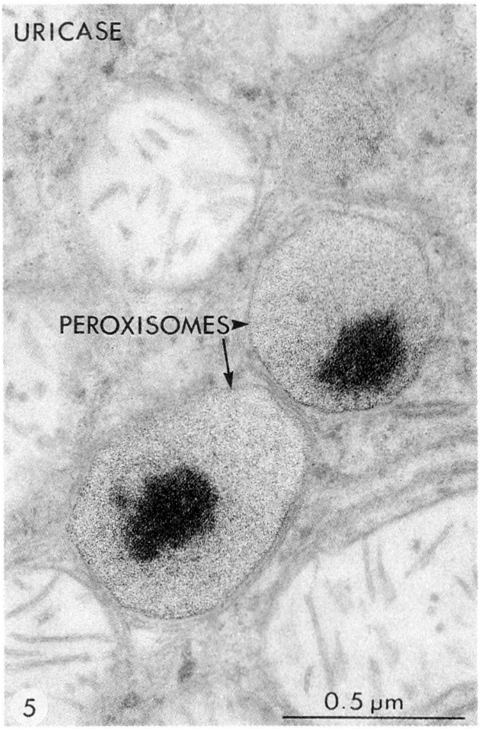
URICASE
PEROXISOMES
5
0.5 μm

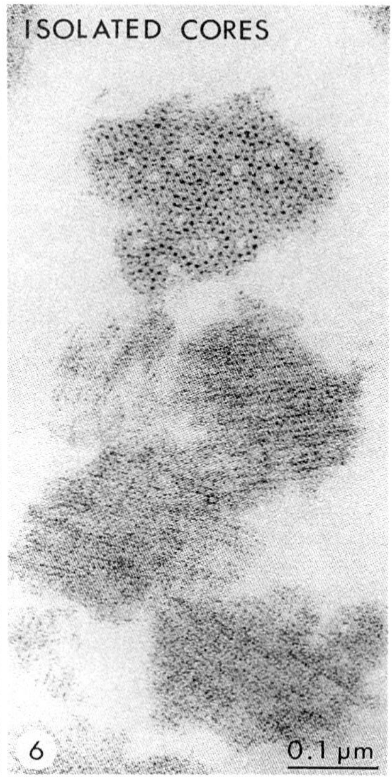
ISOLATED CORES
6
0.1 μm

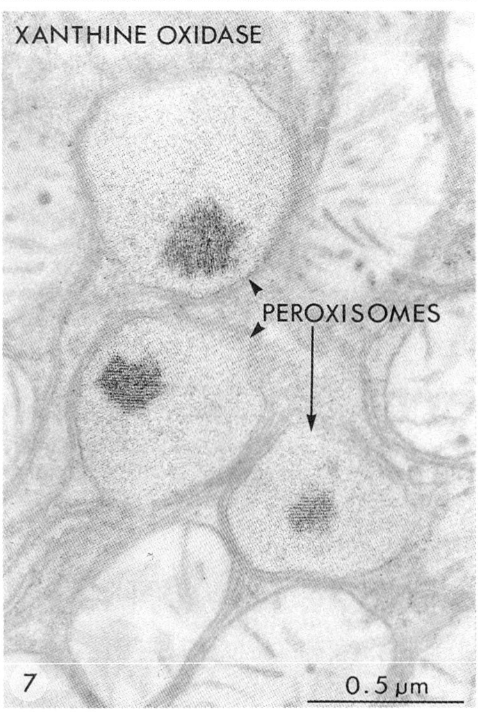
XANTHINE OXIDASE
PEROXISOMES
7
0.5 μm

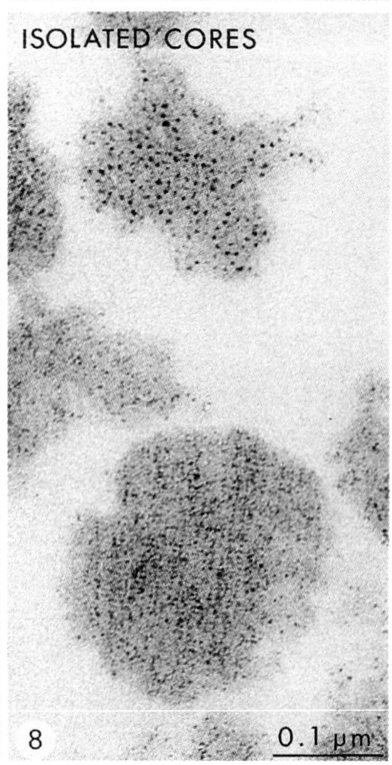
ISOLATED CORES
8
0.1 μm

heterogenously distributed in liver lobule with strongly and lightly stained cells side-by-side (Fig. 9). Moreover, within the same cell the peroxisomes show a marked difference in the intensity of staining for this enzyme (Fig. 12).

These observations demonstrate the advantages and the versatility of the cerium technique in the study of phosphatases and oxidases. This method has proven particularly valuable for the investigation of structure and function of peroxisomes.

Supported by a grant from the Deutsche Forschungsgemeinschaft Bonn - Bad Godesberg (Vo 317/3-1).

3. REFERENCES

Angermüller S and Fahimi H D 1984 Histochem. **80** 107

Angermüller S and Fahimi H D 1986 J. Histochem. Cytochem. **34** 159

Angermüller S and Fahimi H D 1988a Histochem. **88** 277

Angermüller S and Fahimi H D 1988b J. Histochem. Cytochem. **36** 23

Angermüller S, Bruder G, Völkl A, Wesch H and Fahimi H D 1987 Europ. J. Cell Biol. **45** 137

Beier K and Fahimi H D 1986 Cell Tissue Res. **246** 635

Beier K and Fahimi H D 1987 Cell Tissue Res. **247** 179

Briggs R T, Drath D B, Karnovsky M L and Karnovsky M J 1975 J. Cell Biol. **67** 566

Fahimi H D 1980 Trends in enzyme histochemistry and cytochemistry. Ciba Foundation Symposium 73, pp 33

Robinson J M and Karnovsky M J 1983 J. Histochem. Cytochem. **31** 1197

Stoward P J 1980 Trends in enzyme histochemistry and cytochemistry. Ciba Foundation Symposium 73 pp 11

Veenhuis M, Van Dijken J P and Harder W 1980 FEMS Microbiol. Lett. **9** 285

4. FIGURE LEGENDS

Fig. 1: Section of rat liver incubated in the cerium chloride medium for localization of glucose-6-phosphatase. Note the prominent staining of rough and smooth endoplasmic reticulum (SER) in this micrograph. The Golgi complex is negative. PO: peroxisome.

Fig. 2: Localization of thyamine pyrophosphatase in rat liver. Note the staining in the trans-most cisterna of the Golgi complex with a weak reaction in endoplasmic reticulum. PO: peroxisome.

Figures 3 and 4: Localization of acid phosphatase with cytidine monophosphate (CMP) as substrate. Two lysosomes are uniformly positive, while a multivesicular body (MVB) shows a speckled appearance. Note the heavy reaction for acid phosphatase in an autophagic vacuole in Fig. 4.

Figures 5 and 6: Localization of urate oxidase in rat liver. Note the strong reaction in the crystalline cores of peroxisomes. In Fig. 6 isolated cores were incubated in the cerium medium for uricase and the reaction product is seen in small 5 nm tubules sparing the larger 20 nm space between them.

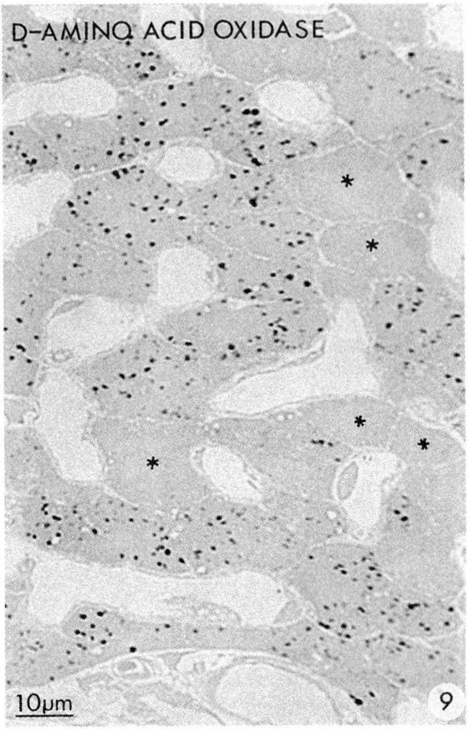

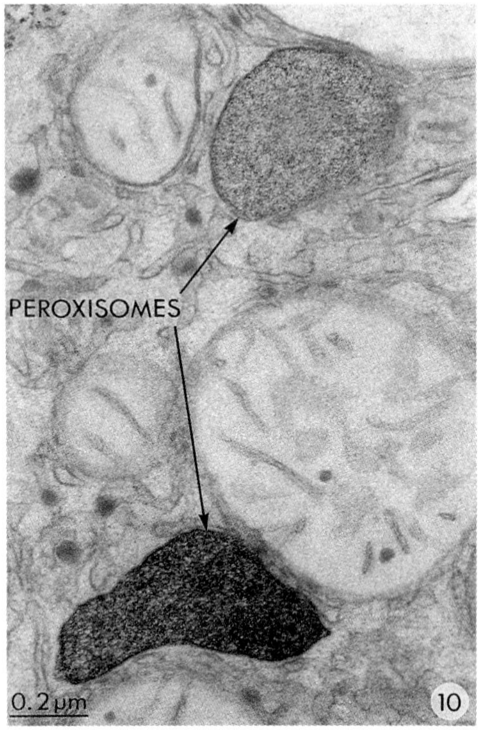

Figure legends continued:

Figures 7 and 8: Localization of xanthine oxidase in rat liver. The positive reaction is observed in the cores of rat liver peroxisomes showing a similar distribution as uricase. In Fig. 8, the localization of xanthine oxidase in isolated cores is demonstrated.

Figures 9 and 10: Localization of D-amino acid oxidase with D-proline as substrate in rat liver. Fig. 9 shows the light microscopic appearance with marked heterogeneity in the localization of the enzyme in different hepatocytes. Cells with light and heavy staining and few negative cells (*) are seen side-by-side. Figure 10 shows that at the electron microscopic level also heterogenously stained peroxisomes are present within the same cell.

Quantitative analysis of microscopical images

HK Koerten[1], HJ Tanke[2], MJ van Noord[1] and LA Ginsel[1]

[1] Center for Analytical Electron Microscopy and [2] Department of Cytochemistry and Cytometry, University of Leiden, Rijnsburgerweg 10, 2333 AA Leiden (The Netherlands)

1. INTRODUCTION

The laboratory contribution to clinical diagnosis is generally based on two types of investigation: morphological studies concerning organs, tissues, and cells, and chemical analysis of body fluids, e.g. blood, lymph, and urine. The advances made in the field of clinico-analytical procedures during the last two decades have been remarkable. The knowledge that body fluids reflect the function of many organs, led to the development of accurate biochemical and serological methods for the determination of electrolyte, protein, enzyme, and hormone level. Progress in the area of diagnostic quantitative analysis of individual cells from biopsied, aspirated, or exfoliated material or blood has been relatively slow compared with whole-tissue analysis. Visual inspection of microscopical preparations of cells and tissues is of course often sufficient for the detection of changes in cells and cellular constituents. For this purpose, two kinds of microscopical method are in principle available: light microscopy (LM), which permits observation of individual cells and to a certain extent of cell organelles, and electron microscopy (EM), which allows observation of details escaping the spatial resolution of the light microscope.
If the direct interpretation of LM and EM images by the observer is too limited, specialized equipment and analytical techniques can be used to extract reliable quantitative information from the preparations. This paper gives a review of automated image analysis in the field of light microscopy, as well as discussion of recent developments in automated on-line image analysis and X-ray microanalysis in electron microscopy.

2. IMAGE ANALYSIS IN LIGHT MICROSCOPY

In light microscopy, quantitative analysis of clinical or biological preparations has been facilitated by the development of techniques for the cytochemical staining of macromolecules based on specific physical or chemical interactions with absorbing, fluorescent, or radioactive

reagents. These methods made it possible to mark small cellular components in a way giving rise to signals with a very low intensity. Changes in signal intensities due to experimental or clinical variations are often too small to be detected by the naked human eye, this problem has been solved by recently developed microscope equipment used in combination with micro electronics and computer technology.
The distinction between small experimental or clinical alterations and normal biological variations must be made on the basis of statistical methods, which means that large numbers of measurements must usually be performed. Because these measurements are generally time consuming and therefore expensive, the use of automated or semi-automated selection procedures is often the only acceptable way to obtain sufficient data. Moreover, since automated selection of particles is more objective, interpretation of morphologic features is improved by the use of these techniques.
The development of reproducible staining techniques on the one hand and of automated selection of structures of interest combined with the use of statistical methods on the other, has opened a completely new field of research in the last two decades, i.e., automated image analysis in microscopy. The application of image analysis in combination with a specific quantitative staining method for the demonstration of macromolecules in intact cells was described by Ploem et al. (1986). These authors carried out a fully automated image analysis with the Leyden Television Analysis System (LEYTAS) on 1,500 cervical smears quantitatively stained for DNA by the acriflavine-Feulgen-Sits procedure.
The LEYTAS system classifies such cytological preparations according to the frequency of cells with an abnormal DNA content. During this procedure artifacts such as dirt and overlapping cell nuclei are efficiently eliminated by the use of software algorithms (Meyer 1979), (Smeulders 1983). The measurements made with this fully automated analysis system led to less than 2% false-negative and approximately 12% false-positive classifications, which means that inclusion of this kind of system among the routine diagnostic procedures for automated recognition and measurement of cells has come within reach.

3. COMBINATION OF LIGHT AND ELECTRON MICROSCOPY

The use of light-microscopical staining techniques in electron microscopy, especially transmission electron microscopy, is limited by the fact that only a few of the LM staining procedures give precipitates with sufficient electron density and/or selectivity for transmission electron microscopy (TEM). Under certain conditions, however, there are some exceptions to this rule, for example provided by the use of a specimen table permitting combined and simultaneous use of light and scanning electron microscopy (LM/SEM) (Ploem and Thaer 1981). With this specimen table, internal structures can be observed through the cell surface by using the light-microscopical part, which provides information about macro molecules present in the cells that can be used for cell identification, and at

the same time the cell surface can be studied with the SEM part.
Quantitative staining with Feulgen-acrivlavine-SO_2 according to Tanke (1979) made it possible to use DNA cytofluorometry to study chicken erythrocytes, mouse liver cells, and human lymphocytes (Fig. 1) with a semi-automated microfluorometer attached to the combined instrument (Wouters 1982).

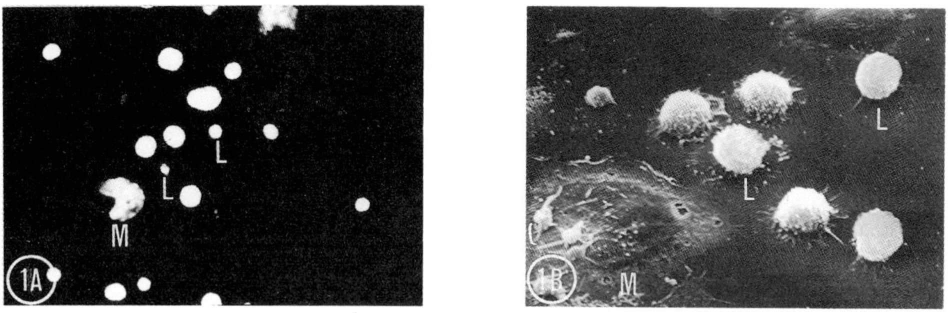

Fig. 1. Combined LM/SEM micrographs of human lymphocytes (L) and monocytes (M) stained with acriflavine-Feulgen and prepared for SEM. 1a. LM fluorescence image (660x) 1b. SEM image of the same group of cells (2200x) (courtesy by dr. Wouters).

This instrumental configuration revealed the presence of cells with diploid, tetraploid, or octaploid DNA, and another investigation showed that normal cervical cells can be distinguished from benignly and malignant cervical cells on the basis of the light-microscopical characteristics. The surface morphology of the normal cells, as observed with the SEM part of the combined microscope, was found to differ from that of abnormal and malignant cervical cells (Wouters 1986).

4. IMAGE ANALYSIS IN TRANSMISSION ELECTRON MICROSCOPY

Although image analysis in light microscopy is in principle based on the presence of chemical dyes in certain cell organelles, the spatial resolution of the light microscope does not provide data on organelles of individual cells. At present quantitative data on individual cell organelles can only be obtained by analysis of TEM images. For these investigations the same criteria must be satisfied as for light microscopy i.e., the grey values and contrast must be at a level permitting selection of the relevant structures by the computer.
However, TEM images generally show many overlapping grey values, which are complicated enough not constant within any given population of cell organelles. The degree of contrast is influenced by the quality of the sections and the chosen staining procedure. When the selection is done in electron micrographs, an additional variable is introduced by the photographic technique. The natural contrast associated with the presence of chemical elements in organelles is generally insufficient to select the latter from the surrounding cytoplasm. To obtain or enhance selective contrast in a given type

of cell organelle, cytochemical staining procedures are usually indispensable.

In an earlier study, our group assessed the value of LEYTAS with that of a semi-automated method using a digitizing tablet, for automated image analysis in electron microscopy in terms of changes in the lysosome population of differentiating monocytes (Gravekamp et al. 1982). This study was done in transmission electron micrographs of mouse blood monocytes and peritoneal exudate macrophages cytochemically stained for the demonstration of peroxidase activity (Fig. 2). Identification of the organelles was based on intensity thresholding, size, and the shape factor.

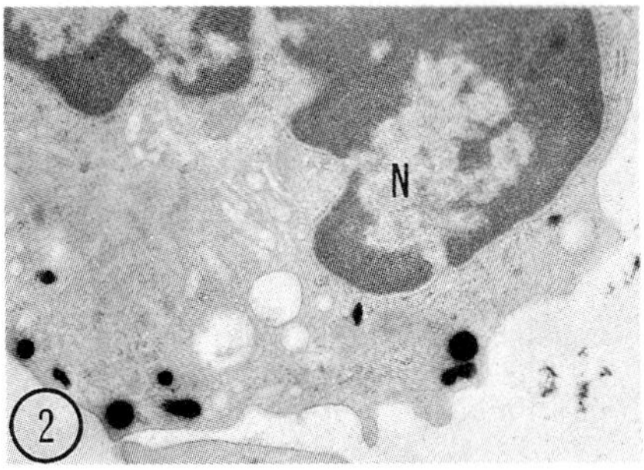

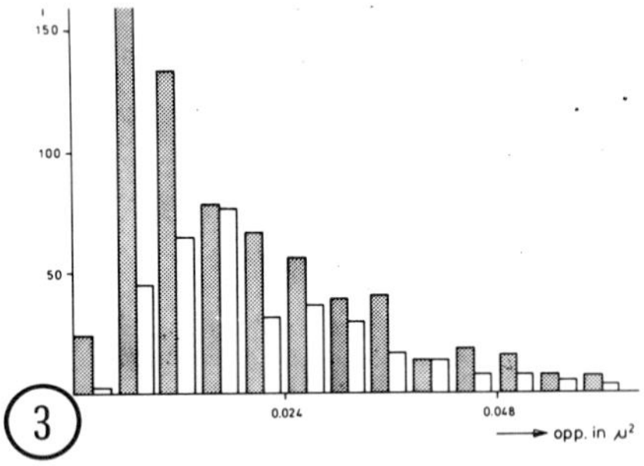

Fig. 2. Part of a monocyte with Peroxidase-positive lysosomes. N = Nucleus (28000x).

Fig. 3. Frequency distribution of the area of the lysosome population of monocytes and monocyte derived macrophages

The results show that cytochemically stained cell organelles can be distinguished in electron-microscopical images and can provide morphometrical data. The results of the study showed that monocyte-derived macrophages have fewer lysosomes than blood monocytes do. Evaluation of the morphometric data e.g. the area of the lysosomes, showed that this difference was due mainly to the smaller lysosome population of the former (Fig. 3). Comparison of the results with those obtained with the digitizing tablet showed close similarity, which means that preference may safely be given to the automated image analysis of electron micrographs which has the great advantage of being much less time consuming.

5. IMAGE ANALYSIS IN COMBINATION WITH X-RAY MICROANALYSIS

So far, we have discussed the use of quantitative cytometry in light microscopic images generated by cytochemically stained biological structures, and have shown that automated morphometrical analysis can be performed in electron micrographs of cytochemical stained cells. This brings us to on-line processing of electron-microscopical specimens. Various applications of on-line methods have been reported by our group. The recognition and classification of particles in metals, coal and urban dust (van Noord 1984), and an on-line procedure for quantitative X-ray microanalysis, on biological

material, using a semi automated selection procedure (de Bruijn 1987).
Another application was developed for the automated recognition of asbestos fibres (van Noord 1987), which are determined on-line on the basis of their strong natural contrast and specific relative ratio between length and diameter. The computer program controlling the selection procedure was designed such that crossing fibres are recognized as individual fibres and processed accordingly. Here too the coordinates of detected structures are stored for later X-ray micro-analysis. For the latter, the electron beam is guided to the position of the fibre and the röntgen spectrum is obtained while the beam is moving along the skeleton of the selected fibre (Fig. 4). The various types of asbestos recognized are automatically grouped. The results of this study are of great importance for environmental studies, since it is known that certain types of asbestos fibre are responsible for the development of malignant diseases.

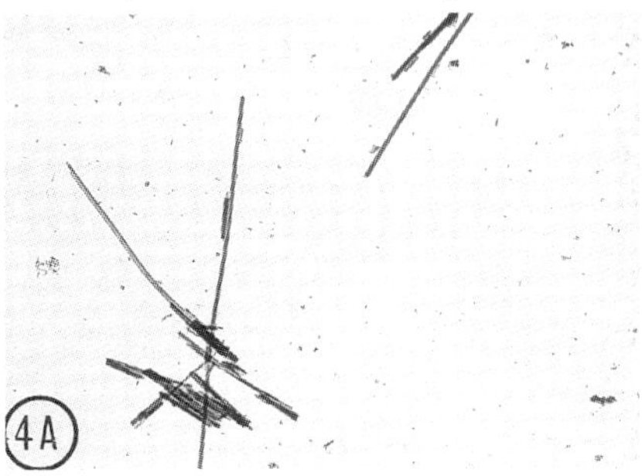

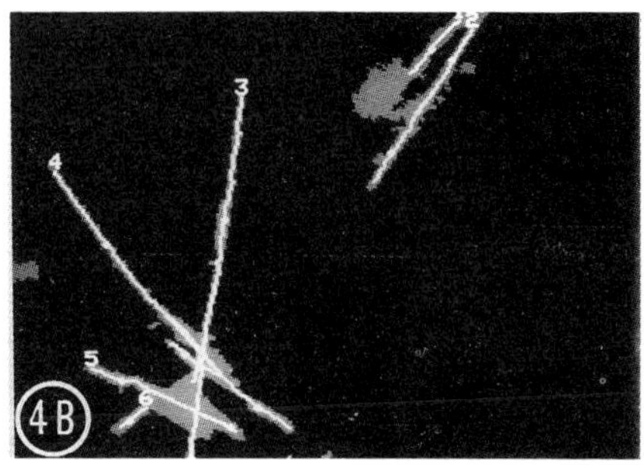

Fig. 4a. Transmission electron micrograph of chrysotile asbestos fibres. (3600x). 4b. The same group of fibres visualized on the computer monitor. The fibres are recognized and numbered, and the skeleton marking the position for X-ray micro-analysis is projected onto each fibre.

Recently, we developed particle recognition programs which select, measure, and chemically identify sub-microscopical features or sub-cellular structures automatically and on-line in the electron microscope. In the first run, the particles are detected on basis of morphological criteria. For each image field the coordinates of the objects are stored by the computer and when the choice of a field by a morphologically based selection procedure has been completed, the electron beam is automatically focussed and automatically guided to the position of the objects under study. Chemical information is then obtained by X-ray microanalysis, and on the basis of this second step the features are regrouped and a final identification is made. The reliability of the final selection is almost 100%.
In a preliminary investigation this method was applied to ultrathin unstained sections of macrophages containing iron-rich inclusion bodies, and the values obtained were analyzed according to statistical programs performed by the same computer.

6. CONCLUDING REMARKS

The data presented here show that quantitative cytometry performed with the combined light and electron microscopes is an important technique for diagnostic pathology and cytology. The combination of quantitative cytochemistry and automated image analysis in light microscopy yields an abundance of objective information about cells in a relatively short time. An important advantage of light-microscopical image analysis is that large fields of cells can be processed in one run, allowing the unbiased detection of extremely rare (one in a million) abnormal cells.

The application of light-microscopical image analysis techniques in electron microscopy has made it possible to measure the DNA contents of cells by combining LM and SEM, and analyze these values in relation to the characteristics of the fine structure of the cell surface. But as we have shown, use of the LEYTAS system also makes it possible to obtain morphometric data from transmission electron micrographs of cytochemically stained cells.

Furthermore, the combination of image analysis with X-ray microanalysis opens wide perspectives for on-line quantitative morphometrical and chemical analysis of electron-microscopical preparations. This technique enables the investigator to measure cellular components and to decide with certainty whether a specific particle belongs to the population of interest. The automation of these analytical systems can be expected to lead ultimately to quantitative analysis of many cellular structures in one specimen without human interference.

7. REFERENCES

Bruijn de WC, Koerten HK, Cleton-Soeteman MI, Blok-van Hoek CJG 1987 Scanning Microscopy Vol1 No4 pp 1651-1667

Gravekamp C, Koerten HK, Verwoerd NP, de Bruijn WC, Daems WTh 1982 Cell Biol. Int. Rep. **6** 656

Meyer F 1979 J Hist. Cytochem. **27** 128

Noord van MJ, Blok-van Hoek CJG, Lampers R 1984 Proc. eight. Europ. cong. Electr. Micr. ed A Csanady (Budapest) 317

Noord van MJ, Vriens CWF, Drenth L 1987 Ultramicr. **21** 207

Ploem JS, van Driel-Kulker AMJ, Goyards-Veldstra L, Ploem-Zaayer J, Verwoerd NP, van der Zwan M 1986 Histochem **84** 549

Ploem JS, Thaer A 1981 Proc. Royal. Microsc. Soc. **16** 253

Smeulders AWM 1983 Thesis University of Leiden

Tanke HJ, van Ingen EM, Ploem JS 1979 J Hist. Cytochem. **27** 84

Wouters CH, Koerten HK 1982 Cell Biol. Int. Rep. **6** 955

Wouters CH, Hesseling SC, Daems WTh, Ploem JS 1986 J Histochem **84** 445

Differential localization of glycoconjugates recognized by Lens culinaris lectin in various cellular compartments of absorptive enterocytes

M Pavelka and A Ellinger

Institute of Micromorphology and Electron Microscopy, University of Vienna, Schwarzspanierstrasse 17, A - 1090 Vienna, Austria

ABSTRACT: Glycoconjugates binding at diverse affinities with the mannose-/glucose-/N-acetyl-glucosamine-specific lectin of Lens culinaris were localized in compartments of rat absorptive enterocytes.

1. INTRODUCTION and METHODS

The mannose-, glucose- and N-acetyl-glucosamine (man,glc,glcNAc)-specific Lens culinaris lectin (LCA) has been characterized as particularly recognizing the tri-mannosidic core region of N-glycosidically linked glycans and binding with high affinity only in the cases a fucose is attached to the asparagine-linked glcNAc (Debray et al 1981, Kornfeld et al 1981). We localized binding sites for the lentil lectin in rat absorptive enterocytes by means of a preembedding affinitycytochemical technique using 10µm thick cryosections of pre-fixed tissue (4% paraformaldehyde/0.5% glutaraldehyde). The cryosections were incubated in LCA-peroxidase conjugates (50µl/ml phosphate-buffered saline) for 4h at 20°C; subsequently, the peroxidase activity was visualized by means of the diaminobenzidine reaction (0.5mg DAB/ml Tris HCl buffer, 20µl 1% H_2O_2/ml). After post-fixation in OsO_4, the specimens were dehydrated and embedded in Epon. The LCA binding reactions were further analyzed with the aid of graded series of competitive sugars (0.001-0.6M) added to the incubation media.

2. RESULTS and DISCUSSION

In the mature absorptive enterocytes, LCA binding reactions were localized in segments of the endoplasmic reticulum (ER), in Golgi apparatus cisternae (G), various vesicles, lysosomes, and apical and basolateral plasma membrane portions (Fig.1). The reactions were slight and were easily eliminated by addition to the incubation media of 0.2M solutions of the competitive sugars man, glc or glcNAc; higher concentrations of competitive sugars were required for inhibitions of the plasma membrane reactions and those in lysosomes and various vesicles; the highest sugar concentrations (0.5M solutions of man, glc or glcNAc) were necessary for the abolition of the intense reactions in the cis and medial Golgi cisternae. In the presence of 0.3-0.4M solutions of man, glc or glcNAc, reactions were apparent only in cis and medial Golgi apparatus cisternae and some Golgi-associated vesicles (Fig. 2).
It is likely that in absorptive enterocytes, just as in other cells, a range of glycoconjugates exist which all are recognized by the lentil lectin but bind with the lectin at diverse affinities. The differentia-

ted patterns obtained in our inhibition studies indicate that, in the absorptive enterocytes, glycoconjugates, which bind with LCA at high affinity, possibly corresonding to core-fucosylated N-linked glycans, are concentrated in cis and medial Golgi apparatus cisternae and in some of the Golgi associated vesicles.

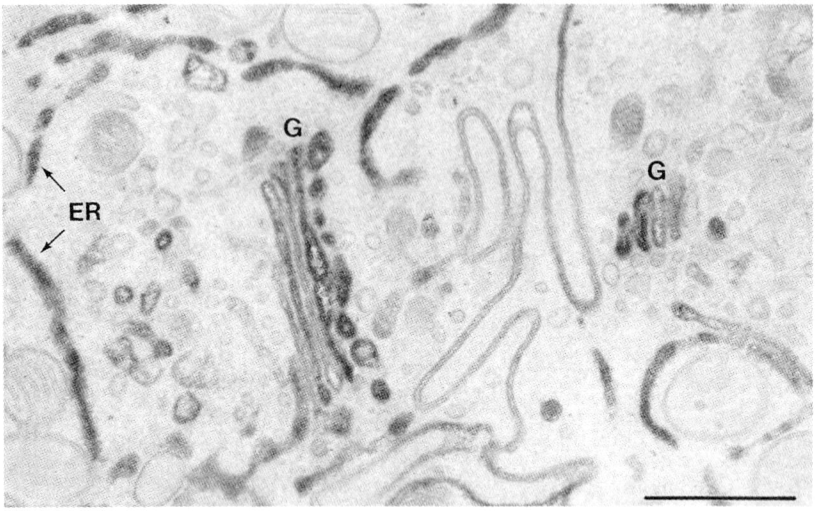

Fig. 1
LCA

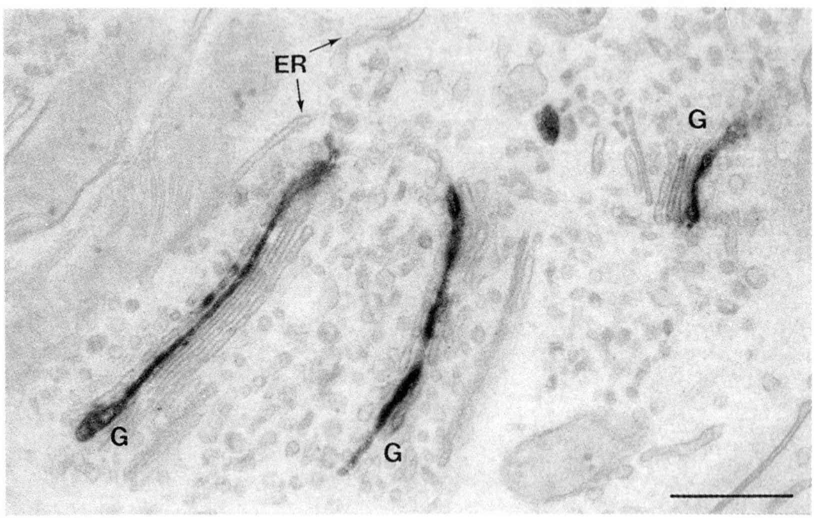

Fig. 2
LCA +
0.3M
GlcNAc

bars 0.5μm

REFERENCES:

Debray H, Decout D, Strecker G, Spik G and Montreuil J 1981 Eur J Biochem **117** 41

Kornfeld K, Reitman ML and Kornfeld R 1981 J Biol Chem **256** 6633

This work was supported by the "Hochschuljubiläumsstiftung der Stadt Wien".

Inst. Phys. Conf. Ser. No. 93: Volume 3, Chapter 19
Paper presented at EUREM 88, York, England, 1988

Immunoelectron microscope study of degranulation in human neutrophils

J.E. Beesley, E.M. Cramer*, D.Y. Mason**,
The Wellcome Research Laboratories, Beckenham, Kent, U.K.,*Hopital Bichat, Paris, France,**Haematology Dept, John Radcliffe Hospital, Oxford, U.K.

ABSTRACT: Colloidal gold immunoelectron microscope techniques have confirmed the localisation of elastase in the primary granules of resting neutrophils. Degranulation has been studied by double labelling. Primary and secondary granules secrete into phagosomes very early in the phagocytosis of latex microbeads. Secretion by the secondary granules is altered by pretreatment of the microbeads.

Colloidal gold is a distinctive, particulate marker for electron immunocytochemistry. It is particularly suited for double labelling experiments and quantification enabling the fluctuations of two proteins in a cell to be examined.

The enzyme elastase is important in both the normal and pathological functions of the neutrophil. Previous attempts to localise this protein have relied upon subcellular fractionation, which may cause organelle disruption and therefore secondary redistribution of their contents, and enzymatic cytochemical techniques which possess poor specificity. We have used colloidal gold techniques in conjunction with a monoclonal antibody NP57 to localise elastase in human resting neutrophils and confirmed the localisation in double labelling experiments with a polyclonal antibody against myeloperoxidase, a specific marker of primary granules. Furthermore, we have carried out double labelling in combination with rabbit antibody against lactoferrin, a marker for secondary granules, to compare the quantitative distribution of these two proteins during phagocytosis of latex microbeads.

Human blood was harvested on Heparin and neutrophils were isolated by the dextran-Radioselection sedimentation technique. Some aliquots were fixed directly with freshly prepared 4% formaldehyde in sodium cacodylate buffer, others were permitted to phagocytose latex microbeads, o.6 um diameter, approximately 100 particles per neutrophil before fixation. For some experiments the latex microbeads were pre-incubated with 20% normal human serum, in others the latex microbeads were used without pretreatment.

The cells were embedded in 10% gelatin and cryoprotected with 2.3M sucrose before freezing in liquid nitrogen slush and preparing ultrathin frozen sections. The sections were incubated with specific antibody and the gold probe before contrasting in uranyl acetate and embedding in methyl cellulose. For double labelling experiments the sections were incubated with a mixture of mouse and rabbit antibodies followed by a mixture of two different-sized gold probes, one coated with antimouse IgG, the other with antirabbit IgG.

Our results show that elastase is localised in the large granules of the neutrophil. Double labelling showed that this colocalised with myeloperoxidase. Further double labelling experiments showed that the labelling for elastase and lactoferrin was separate and was confined to different types of granules, the lactoferrin occurring in the small granules (Fig. 1).

As early as 30 seconds after initiation of phagocytosis, latex beads had been drawn into phagosomes. Some phagosomes contained mainly lactoferrin, others mainly elastase but the majority contained both elastase and lactoferrin (Fig. 2). Quantitative examination showed that there was no difference in the elastase content of phagosomes containing either serum pretreated or untreated latex microbeads. There was, however, twice as much immunolabelling for lactoferrin in the phagosomes of untreated beads compared with those of treated beads.

In conclusion, we have confirmed by using the high resolution colloidal gold immunolabelling technique the localisation of elastase in the primary granules of neutrophils. We have further shown, that very early in phagocytosis, both primary and secondary granules secrete their contents into the phagosome. Finally, it appears that the ratio of the secretions from the two granule populations depends upon the treatment of the latex particles.

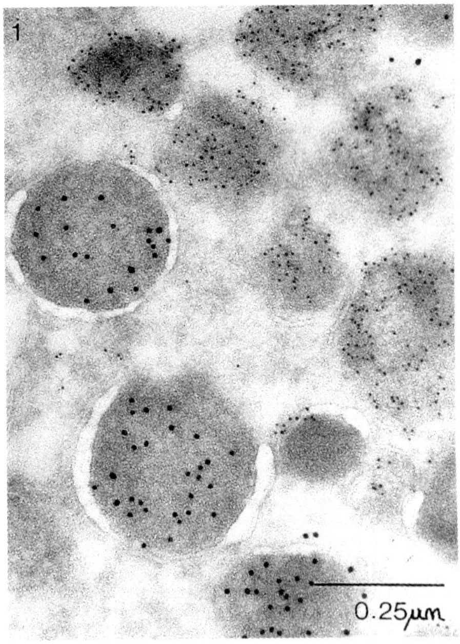

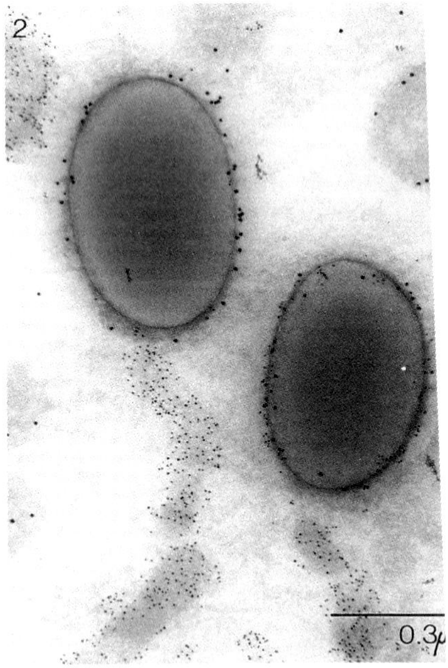

Fig. 1 Double labelling experiments showing that elastase (15nm probe) occurs in the primary granules whereas lactoferrin (5nm probe) is restricted to the secondary granules.

Fig. 2 Thirty seconds after initiation of phagocytosis the majority of phagosome contain both elastase (15nm probes) and lactoferrin (5nm probes).

Inst. Phys. Conf. Ser. No. 93: Volume 3, Chapter 19
Paper presented at EUREM 88, York, England, 1988

Immunoelectronmicroscopy on cryosections section permeability to specific antibodies, protein A-gold complexes and ferritin conjugated IgGs

Y-D Stierhof[*] and H Schwarz[o]

[*]Max-Planck-Institut für Entwicklungsbiologie; [*]Hygiene-Institut, Abteilung für medizinische Virologie und Epidemiologie der Viruskrankheiten, [o]Max-Planck-Institut für Biologie, D-7400 Tübingen, Fed Rep Germany

ABSTRACT: Unconjugated antibodies seem to penetrate well-preserved cryosections to a low extent, ferritin conjugated IgG to a even lower extent, while protein A- 4 nm-gold is restricted to the section surface.

1. INTRODUCTION

Using transverse sectioning of plastic embedded ultrathin and semithick immunolabelled cryosections prepared according to Tokuyasu we could show that penetration of protein A-gold (pAg) complexes is obviously less than is generally believed (Freudl et al 1986, Stierhof et al 1986): pAg complexes of different size are not able to enter the IgG labelled section significantly but follow surface irregularities. However gold particles can be detected within insufficiently fixed and/or visibly damaged sections. As an example hemoglobin in chicken erythrocytes fixed in different ways was detected by a specific rabbit antiserum and pAg (Figure 1-3).

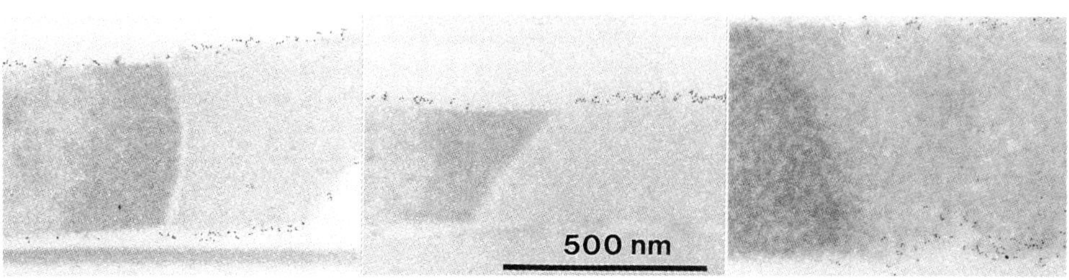

Fig 1: Fixation
0.05% GA + 2% FA
4 nm-gold

Fig 2: Fixation
8% FA
4 nm-gold

Fig 3: Fixation
PLP (McLean and Nakane 1977)
4 nm-gold

RESULTS & DISCUSSION

In contrast to pAg complexes (Figure 1) ferritin conjugated secondary antibodies are able to penetrate identically GA-FA fixed and labelled sections as shown in Figure 4 (elastic brightfield image, $\Delta E = 0$ eV). At $\Delta E = 115$ eV uranyl acetate staining of labelled sections facilitates identification of both structural details (e. g. membranes) and ferritin molecules (Figure 5).

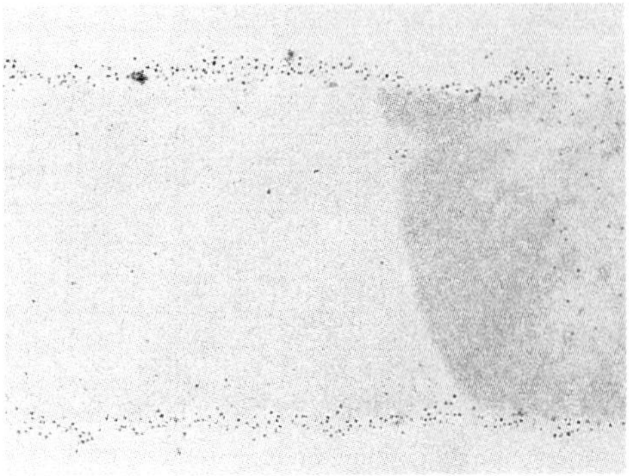

Fig. 4: Fixation 0.05% GA + 2% FA
Ferritin conjugated goat anti-rabbit IgG
ΔE = 0 eV

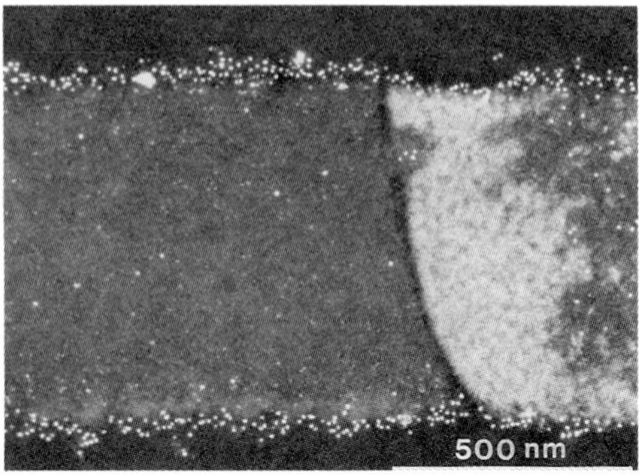

Fig. 5: ΔE = 115 eV

To find out the extent to which primary antibodies are able to penetrate melted cryosections we use the following approach: Semithick cryosections were cut, placed on a fixed gelatin layer and labelled with the specific antibodies. Such sections were again infiltrated with sucrose, frozen and cut perpendicular to the original section plane. After blocking sections were incubated with pAg and embedded in methyl cellulose. In GA-FA fixed sections gold particles bind nearly exclusively to the original section surface (Figure 6). Obviously antibody penetration is negligible in this example. FA fixed sections show a slightly increased intracellular labelling (Figure 7). Using fluorescein labelled secondary antibodies as a small and sensitive probe the same results were obtained.

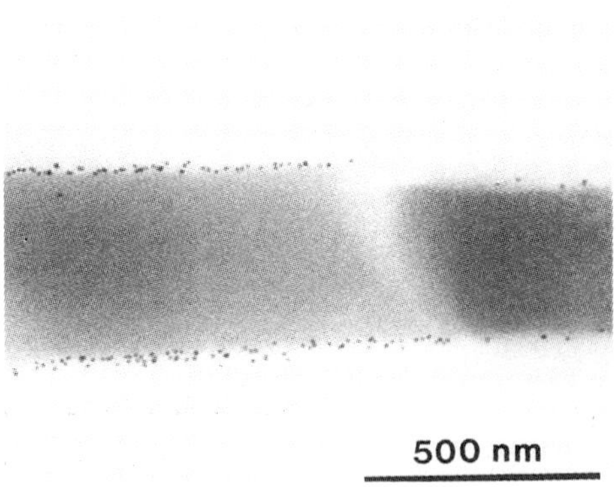

Fig. 6: Fixation 0.05% GA + 2% FA

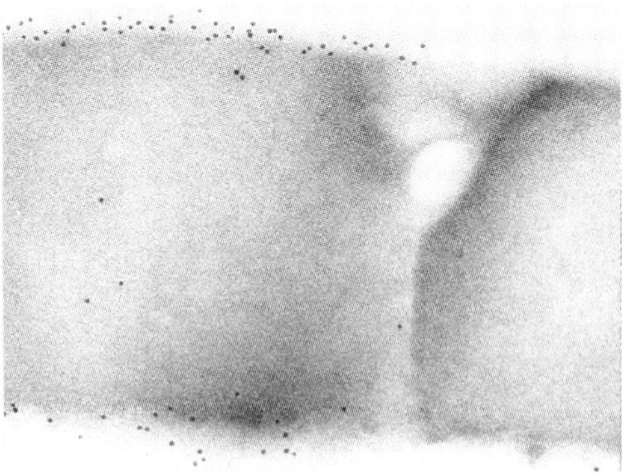

Fig. 7: Fixation 8% FA, 8 nm-gold

REFERENCES
Freudl R, Schwarz H, Stierhof Y-D, Gamon K, Hindenach I and Henning U 1986 J. Biol. Chem. 261, 11355
Mc Lean I W and Nakane P K 1977 J. Histochem. Cytochem. 22, 1077
Stierhof Y-D, Schwarz H and Frank H 1986 J. Ultrastr. Molec. Str. Res. 97, 187

Distribution of DNA stained with osmium amine complex in nuclei of three leucemic cell-lines during interphase and mitosis

D Ploton, M Menager, P Jeannesson* and JJ Adnet.

Laboratoire d'Histologie, Unité INSERM U314, UFR de Médecine, Reims, France and *Laboratoire de Biochimie, UFR de Pharmacie, Reims.

ABSTRACT: Specific staining of DNA with osmium amine complex was performed on sections of three leucemic human and murine cell-lines (K562, HL60 and L1210). We focused our study on the distribution of DNA during interphase mainly relatively to the nucleolus. The structure of chromosomes was also compared during all the steps of mitosis from one cell-line to the others.

1. INTRODUCTION

Interphasic nucleolus is characterized by the presence of well defined components i.e : chromatin (peri and intranucleolar), fibrillar centres, dense fibrillar component and granular component. From our earlier studies it appeared that the distribution of these components seems typical of a given cell-line. Moreover, during mitosis, nucleolus disaggregates when chromosomes condense. When we studied this phenomenon by using a silver-staining specific for fibrillar components we showed typical behaviour of nucleolar remnants (Ploton et al 1987). We also observed that chromosome morphology was, in some cases, very complex mainly during metaphase, anaphase and telophase. In order to get complementary data we used a specific staining for DNA which allowed us to describe, with a high resolution, the structure and conformation of chromosomes during all the steps of mitosis.

2. MATERIALS AND METHODS

Thin sections of Epon embedded glutaraldehyde fixed cells were specifically stained for DNA as previously described (Derenzini et al 1982).

3. RESULTS AND DISCUSSION

3.1 Interphase nucleoli (Figures 1 and 2)

Each cell-line is characterized with a typical repartition of DNA within nucleoli. In HL60 (Fig. 1) very few clumps of DNA are evidenced. Some of them are localized in contact with faintly stained roundish zones probably representing DNA in fibrillar centres. In K562 cells intranucleolar DNA constitutes an irregular network. In L1210 cells (Fig. 2) perinucleolar DNA shows very regular invaginations which give this nucleolus a very typical shape. The median part of the nucleolus is devoid of DNA excepted in some roundish zones where loosened DNA is seen.

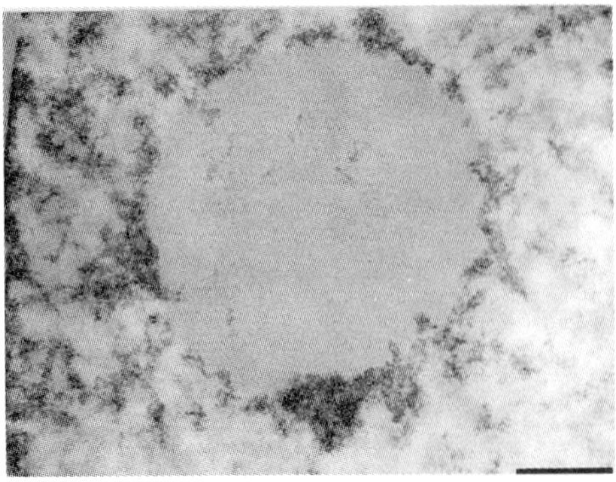

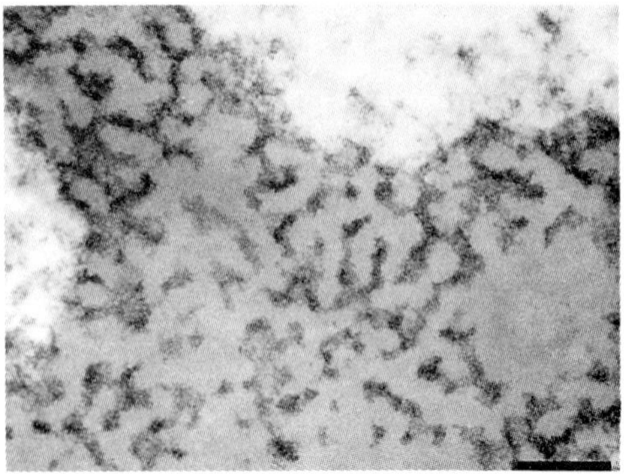

Figure 1 : Nucleolus in a interphase HL60. Bar = 0.5 µm

Figure 2 : Nucleolus in a interphase L1210-cell. Bar = 0.5 µm

3.2 Mitosis (Figures 3 and 4)

In K562 and HL60 cells chromosomes are frequently regular with smooth contour (Fig. 3). At the opposite, chromosomes in L1210 cells show crenated contours (Fig. 4) in which RNP are localized. In all metaphasic chromosomes, nucleosomes are well evidenced. During late telophase numerous roundish RNP structures are seen in the three cell-lines among the decondensing chromosomes.

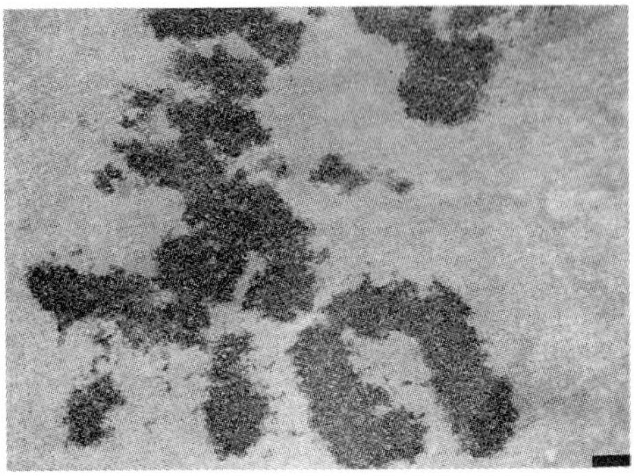

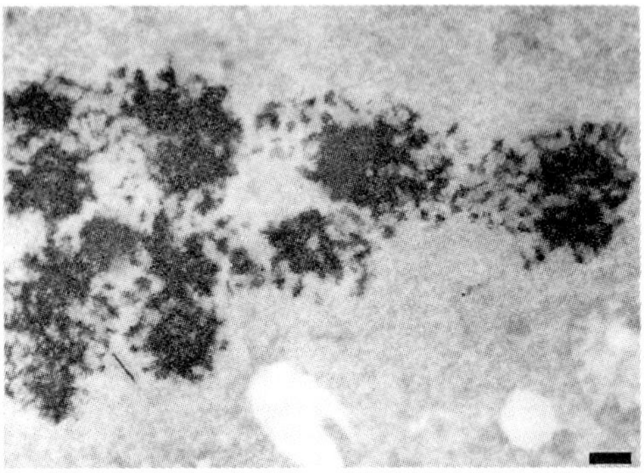

Figure 3 : Metaphase HL-60 cell. Bar = 0.5 µm

Figure 4 : Metaphase L1210 cell. Bar = 0.5 µm

In conclusion, these studies give us very informative data concerning the structure of interphasic nucleoli and on the typical shape and structure of mitotic chromosomes in these three leucemic cell-lines.

ACKNOWLEDGEMENTS

We acknowledge Pr. M Derenzini for the staining with osmium amine complex.

BIBLIOGRAPHY

Derenzini M et al. 1982 J. Ultr. Res. 80 133-147

Ploton D et al. 1987 Chromosoma 95 95-107

Paper presented at EUREM 88, York, England, 1988

An "in vitro" study of RNA synthesis in adrenal cortex. Effects of ACTH and actinomycin D

Maria C.Magalhães, A.Bonito-Vitor and M.M.Magalhães

Institute of Histology and Embryology of Faculty of Medicine of Oporto and Center of Experimental Morphology of the University of Oporto (INIC), Portugal

In previous radioautographic studies, the effects of ACTH on RNA synthesis and migration were studied by using slices of young rat adrenal cortex incubated with [^{3}H]-uridine. We noticed that the adrenocorticotrophic hormone decreased significantly the uptake of uridine by adrenocortical cells after 15, 30 and 45 min of chase time incubation. In order to clarify the ACTH effects referred to above, we have repeated those experiments in the presence of the nucleolar transcriptional inhibitor, actinomycin D.

Nine young rats were sacrificed by decapitation, and the adrenals were removed and sliced. Slices were incubated in Krebs-Ringer bicarbonate solution with glucose, at 37^{0}C, gassed with a 95% O_2 and 5% CO_2 mixture and containing 5,6-[^{3}H]-uridine (100 µCi/ml). After 30 min (pulse time) some slices were removed from the medium and fixed, while the remainder were transferred either to incubation medium without tritiated uridine but containing nonradioactive uridine (100 µg/ml) and actinomycin D (25 µg/ml), or to an incubation medium identical to the latter to which ACTH (100 mU/ml) was added. Slices from both groups were removed from the media and fixed, after 30 and 60 min chase time, in 2.5% glutaraldehyde and post-fixed in 1% osmium tetroxide and Epon embedded.

Sections from both experimental groups were processed for light and electron microscope radioautography with exposures of 20 and 98 days, respectively.

A qualitative appraisal of radioautographs revealed that the presence of actinomycin D in the medium produced a marked decrease in the number of silver grains occurring in controls at pulse time and that the presence of ACTH enhanced that decrease (Figures 1 and 3); however in the case of 60 min chase incubations ACTH provoked an increase of radioautographic reactions over that occurring in the presence of actinomycin D only (Figures 2 and 4). Quantitative radioautographic analysis at LM level confirmed the qualitative findings. Actinomycin D alone produced a decrease of nuclear labelling after 30 min of chase (6.78 vs. 13.68 silver grains/1000 µm^2 at pulse time) or 60 min (6.5 vs. 13.68). The addition of ACTH led to a more marked fall of silver grains at 30 min (3.56 vs. 6.78) but after 60 min the labelling was significantly higher (63.32 vs. 6.5).

These results confirm the precocious inhibitory effects of ACTH on RNA synthesis in adrenal cells and further show that this is followed by a late stimulation which is not prevented by actinomycin D.

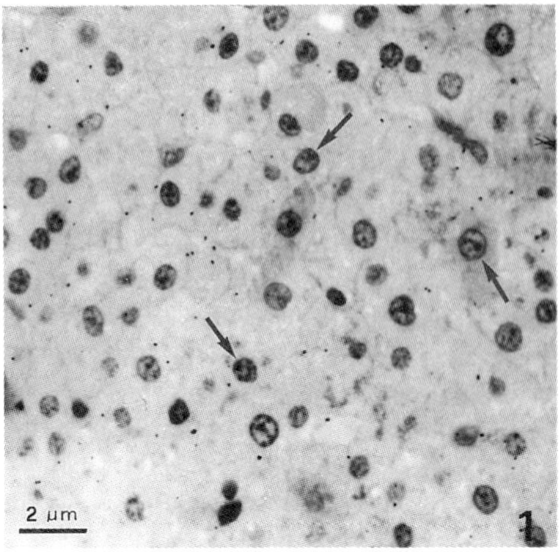

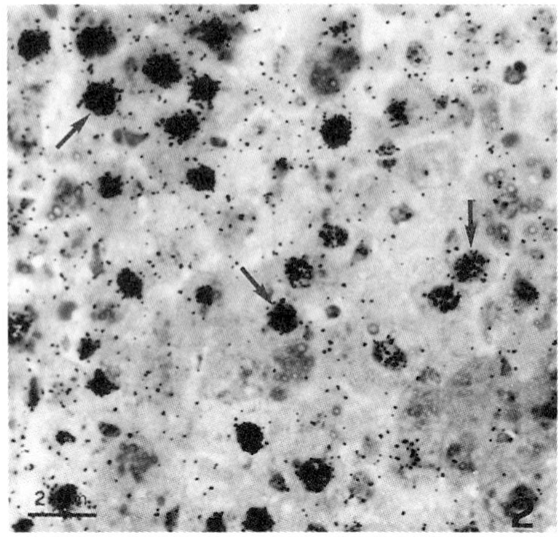

Figure 1 - Light (LM) radioautograph of zona fasciculata cells pulse-labelled with [^{3}H]-uridine for 30 min and chased for 30 min with actinomycin D and ACTH. Nuclei are weakly labelled (arrows). Exposure 20 days.

Figure 2 - LM radioautograph of zona fasciculata cells pulse-labelled with [^{3}H]-uridine for 30 min and chased for 60 min with actinomycin D and ACTH. Nuclei are intensely labelled (arrows). Exposure 20 days.

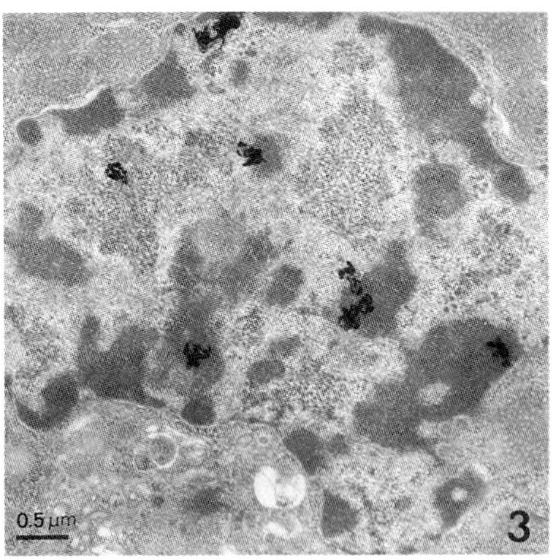

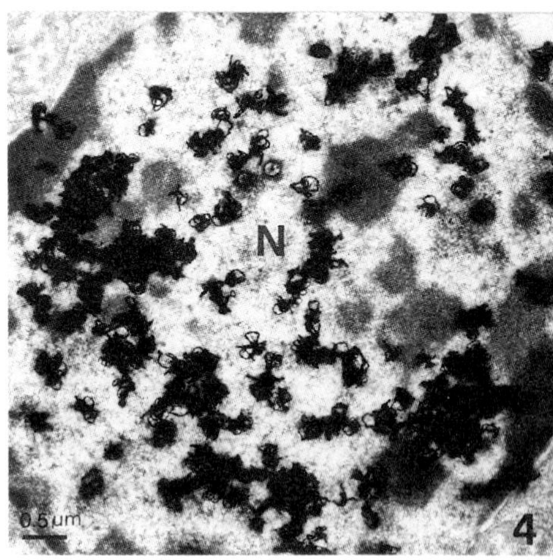

Figure 3 - Electron microscope (EM) radioautography of the zona fasciculata cells pulse-labelled with [^{3}H]-uridine for 30 min and chased for 30 min with actinomycin D and ACTH. Silver grains are scarces. Development with Kodak D_{19b}. Exposure 98 days.

Figure 4 - The same as for Figure 2. Intense labelling is observed over nuclear components. N - nucleus. Development with $Kodak_{19b}$. Exposure 98 days.

Nuclear RNA is removed by phospholipase C: cytochemical quantitative evaluation

N. Zini, G. Mazzotti, A. Galanzi, R. Rizzoli, A.M. Martelli, L. Neri, L. Manzoli and N. M. Maraldi

Inst. of Citomorfologia C.N.R. c/o Inst. Codivilla Putti, Bologna and Inst. of Anatomia Umana, University of Bologna, Italy

ABSTRACT: A significative reduction of the RNA labelled with RNase-colloidal gold has been observed in nuclei and nuclear matrices treated with phospholipase C as well as after phospholipase C digestion on thin sections of specimens embedded in London White (LW), suggesting that phospholipids should be involved in hydrophobic interactions between nucleic acids and matrix proteins.

1. INTRODUCTION

Phospholipids are present in membrane deprived isolated nuclei, and in purified nuclear matrices (Manzoli et al. 1979). The matrix bound phospholipids can be selectively removed by digestion with phospholipase C. This digestion causes also the release of a great amount of the nuclear matrix linked DNA and RNA (Cocco et al. 1980). A cytochemical ultrastructural investigation utilizing RNase conjugated with colloidal gold has been performed in order to determine whether a selective removal of RNA occurs after digestion with phospholipase C.

2. MATERIALS AND METHODS

Nuclei were isolated from rat liver in the presence of Triton X 100. Nuclar matrices obtained essentially as previously described (Cocco et al. 1980) were digested with phospholipase C. Glutaraldehyde fixed samples were embedded in Epon and LW. Thin sections were incubated with RNase-colloidal gold complex (Bendayan 1981). Thin sections of LW embedded nuclei were digested with phospholipase C before the labelling . Quantitative evaluation of the gold particle distribution was done with a Cambridge Quantimet 970 image analyzer.

3. RESULTS AND DISCUSSION

The labelling obtained with RNase-colloidal gold appears very reduced in phospholipase C digested nuclear matrix in comparison with control (Figs. 1,2). Quantitative analysis indicates that while the label at the level of the inner matrix (IM) shows a reduction of about 60%, a reduction of about 90% occurs in the nucleolar remnant (NR). Phospholipase C digestion results effective also on thin sections of nuclei embedded in LW. The label decreases to about 85% both on the nucleolus (N) and on the inter-chromatin ribonucleoproteins (RNP) (Figs. 3,4). These data confirm that phospholipids should be involved in hydrophobic interactions between nucleic acids and nuclear matrix, allowing the maturation and transport of the transcripts.

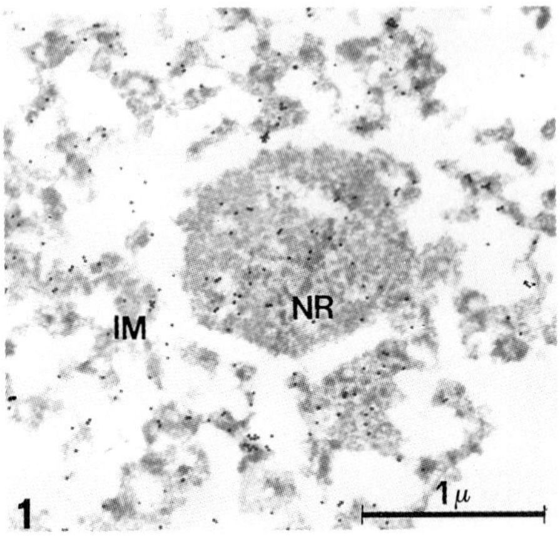

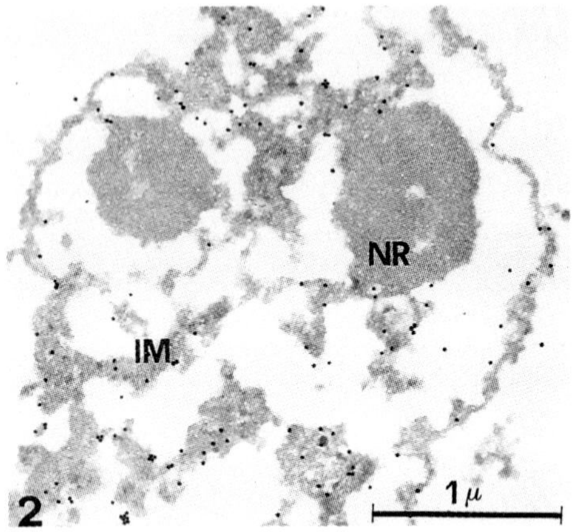

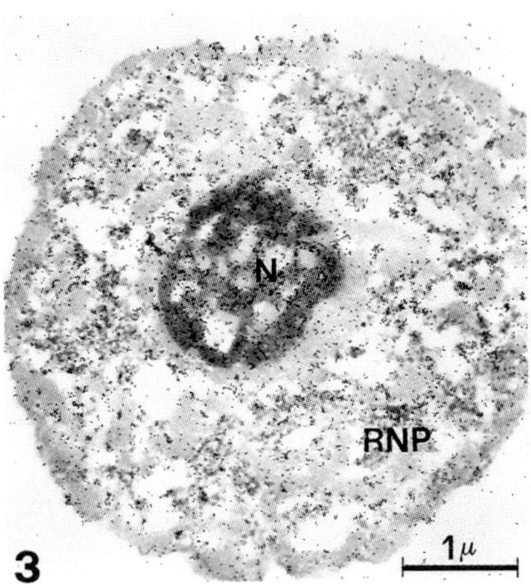

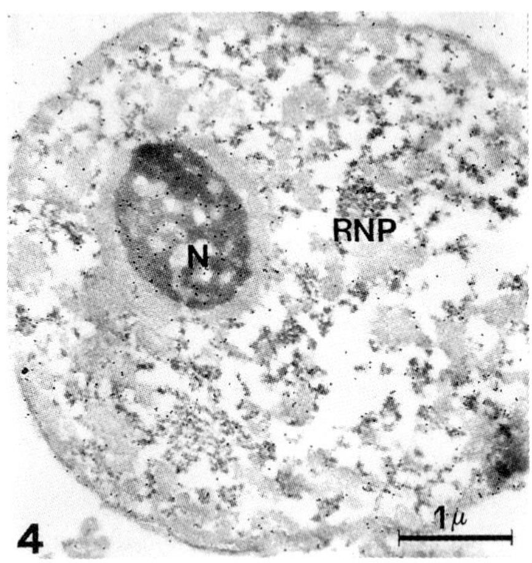

Figs. 1,2 Nuclear matrix embedded in Epon; RNase-colloidal gold (Fig.1, control; Fig. 2, phospholipase C digestion in suspension before fixation).

Figs. 3,4 Rat liver nuclei embedded in LW; RNase-colloidal gold (Fig.3, control; Fig. 4, phospholipase C digestion on thin sections)

4. REFERENCES

Bendayan M 1981 *J. Histochem.Cytochem.* 29 531.

Cocco L, Maraldi N M, Manzoli F A, Gilmour R and Lang A 1980 *Biochem.Biophys.Res. Commun.* 96 890

Manzoli F A, Capitani S, Maraldi N M, Cocco L and Barnabei O 1979 *Adv. Enzyme Regul.* 17 175

Cytochemical, X-rays microanalytical and biochemical studies on rat oocyte nucleolus at the antral follicle stage

N Antoine*, C Quintana** and A Vigneron*

*Laboratoire de Biologie cellulaire et tissulaire, Université de Liège, 4000-Liège, Belgique - **Centre de Biologie cellulaire CNRS, 94205-Ivry, France

Rat oocyte nucleolus originally has a reticulated structure essentially composed of the classical nucleolar components: strands of dense fibrillar component, small fibrillar centres and aggregates of granules (Fig. 1).

During follicular growth and oocyte maturation, the nucleolus undergoes various morphological changes accompanied by significant modifications in nucleolar transcriptional activity (Antoine et al ,1987).

In fact, at the end of oocyte maturation, the oocyte nucleolus becomes made up of a homogeneous compact mass which exhibits vacuoles (Fig. 2). No relation between the presence of vacuoles and the process of follicular atresia has been found.

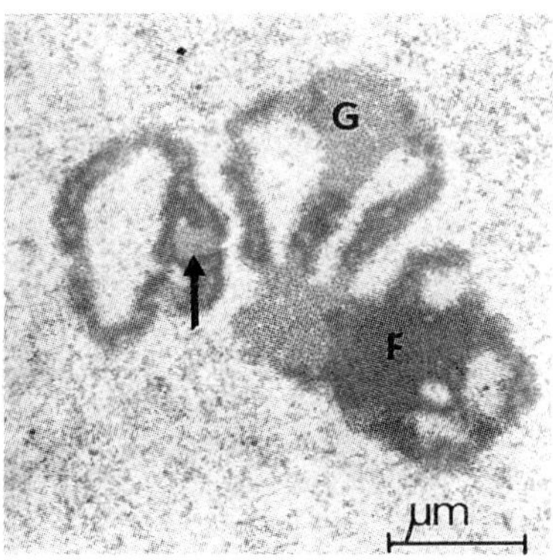

Fig. 1. The oocyte nucleolus at the primordial follicle stage is composed of dense fibrillar component (F), small fibrillar centre (arrow) and granular aggregates (G).

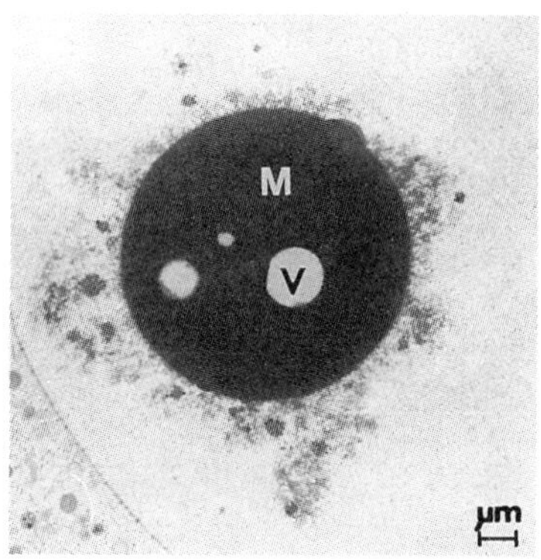

Fig. 2. The oocyte nucleolus at the antral follicle stage is essentially composed of a homogeneous proteinic compact mass (M) and exhibits vacuoles (V).

Cytochemical studies performed at the photonic and ultrastructural level suggest that the nucleolar mass essentially consists of acidic and no argyrophilic proteins. However these cytochemical methods do not allow us to precise the exact chemical nature of these proteins.

Ultrastructural X-rays microanalytical study provides new informations concerning the composition of the homogeneous mass (Fig. 3). This technique reveals with precision the occurence of metals and principally an important quantity of S in the nucleolar mass. The proteins are thus rich in thiol groups and disulfure links which could be responsible for the high compaction of the nucleolar mass at this stage of oocyte maturation (Quintana et al, 1987).

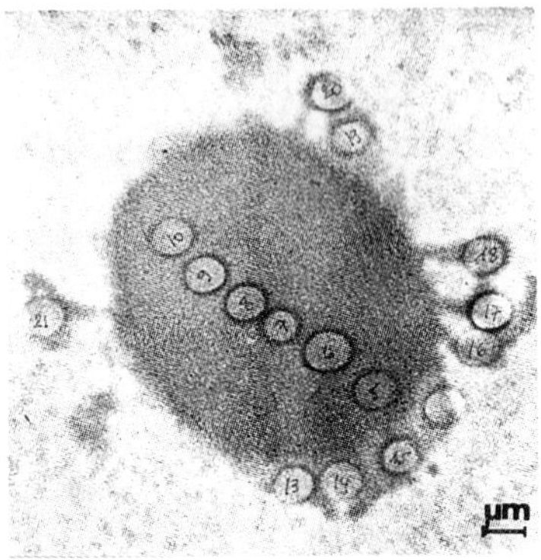

Fig. 3. Oocyte nucleolus at the antral follicle stage microanalytically studied with 400 nm spotes.

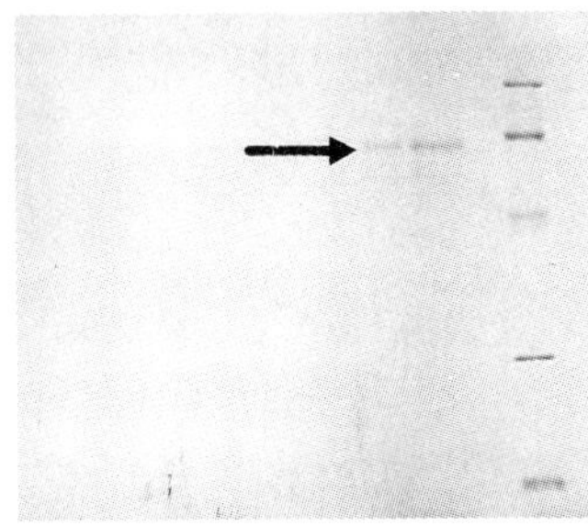

Fig. 4. SDS-polyacrylamide gel electrophoresis reveals a major nuclear polypeptide of MW 65,000 (arrow).

Finally, oocyte nuclei and nuclei of follicular cells were isolated and comparative SDS-polyacrylamide gel electrophoresis of total nuclear proteins was performed. The gel patterns reveal that these two kinds of nuclei essentially differ by a major polypeptide of MW 65,000 in oocyte nuclei (Fig. 4).

In a next future, this polypeptide will be isolated in order to produce antibodies which can be used to localize this polypeptide in situ by immunocytochemical method performed on ultrathin sections of rat oocytes.

Antoine N, Lepoint A, Baeckeland E and Goessens G 1987 Biol. Cell 59 107
Quintana C, Olmedilla A, Antoine N and Ollacarizqueta A 1987 Biol. Cell 61 115

N. Antoine is fellow of the IRSIA and is grateful to J. Greffe for the efficient technical assistance and to the "Fonds de la Recherche Scientifique Médicale" for the financial support (grant n° 3.4512.86).

Incorporation of lipid precursors in friend cells. Subcellular localization of ^{3}H-glycerol and ^{3}H-myo-inositol

M. Mazzoni[1], V. Bertagnolo[1], A. Galanzi[2], D. Ricci[1], A.M. Billi[2], L. Cocco[3], N.M. Maraldi[4], S. Capitani[1]

Istituti di Anatomia Umana Normale, Università di Ferrara[1], Bologna[2] e Chieti[3], Istituto di Citomorfologia del C.N.R., c/o Istituti Codivilla-Putti, Bologna[4], Italy.

ABSTRACT: Tritiated glycerol and myo-inositol incorporated by Friend cells are autoradiographycally revealed in all cell compartments. The relative nuclear labelling is higher for myo-inositol than for glycerol, and increases after short-time erythroid differentiation.

1. INTRODUCTION

Phospholipids are involved in the control of key metabolic events, like the modulation of replication and transcription in vitro (Manzoli et al., 1985) and the transduction of agonist-dependent intracellular signals (Berridge and Irvine, 1984). In particular inositol lipids have been demonstrated to generate second messengers in response to a number stimuli, and recent data have indicated that nuclei isolated from Friend Erythroleukemia Cells (FELC) can phosphorylate phosphatidylinositol, with a variable pattern related to the cell differentiation state (Cocco et al., 1987). To study the metabolic significance of nuclear lipids, and to gain additional information of the role of inositol lipids in targeting differentiation-linked cellular events, we have analyzed the subcellular distribution of labelled glycerol and myo-inositol in control and hexamethylenbisacetamide (HMBA)-induced FELC.

2. MATERIALS AND METHODS

FELC were cultured in RPMI 1640 + 10% FCS. Logarythmically growing cells, at a density of 2×10^{5}/ml, were labelled for 24 hrs either with 3 μCi/ml of ^{3}H-glycerol (2.5 Ci/mmol) or ^{3}H-myo-inositol (85 Ci/mmol).
Erythroid differentiation was triggered with 4 mM HMBA for 2 hrs, and cell samples were fixed with 2.5% glutaraldehyde in phosphate buffer, pH 7.2 and embedded in Araldite. Semi-thin and thin sections were coated with monolayers of Ilford-4 emulsion and exposed for 15 and 90 days, respectively.

3. RESULTS AND DISCUSSION

The grain density is higher with ^{3}H-glycerol, according to the fact that inositol lipids are a few percent of total lipids. The subcellular distribution of the grains indicates that glycerol and inositol-derived molecules are present in both nucleus and cytoplasm (Fig. 1 and 2). Computer assisted image analysis demonstrated that the nuclear grain density, compared to that of the cytoplasm, is higher for inositol than for glycerol (1.20 and 0.516 N/C ratio, respectively). In addition, the inositol nuclear labelling increases to 2.03 N/C ratio after short-time induction with HMBA. These results indicate that the cell nucleus is involved in the intracellular processing of these precursors and suggest that inositol lipids might be implicated in the early nuclear events of erythroid commitment.

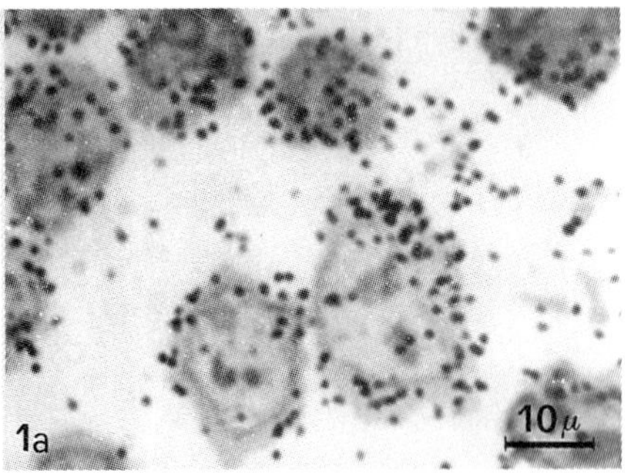

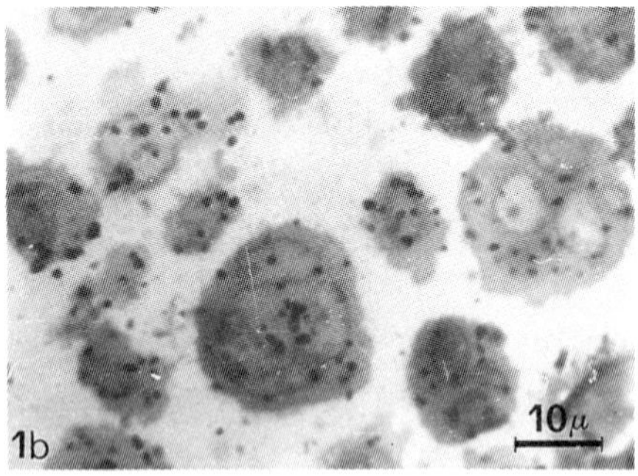

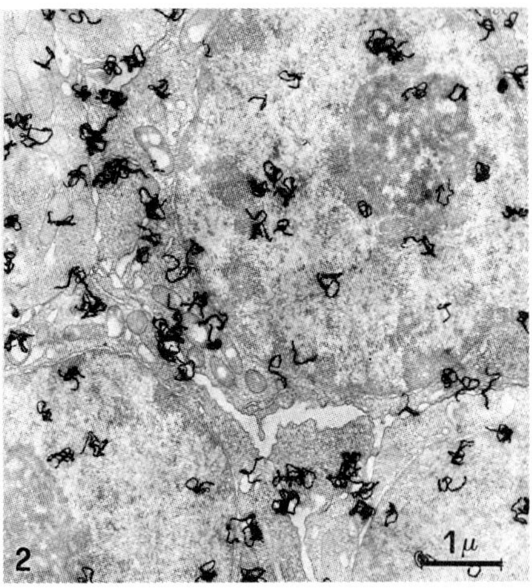

Fig.1 Autoradiography on semithin sections of FELC incubated with ^{3}H-glycerol (a) and ^{3}H-myo-inositol (b).
Fig.2 Electron microscope autoradiography of ^{3}H-glycerol labelled FELC, showing the subcellular distribution of the isotope. Similar results were obtained with ^{3}H-myo-inositol (not shown).

4. REFERENCES

Berridge M J and Irvine R F 1984 Nature 312 315.

Cocco L, Gilmour R S, Ognibene A, Letcher A J, Manzoli F A and Irvine R F 1987 Biochem. J. 248 765.

Manzoli F A, Maraldi N M and Capitani S 1985 in "The Pharmacological Effect of Lipids II" Ed JJ Kabara (Champaign: AOCS Publ) pp 133-56

Cytochemical localization of CA++—ATPase activity in the rat adenohypophysis

G. El-Sherif and E. Bácsy
Institute of Experimental Medicine, Hungarian Academy of Scinces, H-1450 Budapest P.O.B. 67, Hungary

ABSTRACT: Ca^{++}-ATPase was localized electron microscopically in the rat anterior pituitary with a lead precipitation method. Specific reaction deposit was found at the outer surface of both the parenchymal (PC) and the folliculo-stellate cells (FSC). Some of the cells contained traces of reaction product in the endoplasmic reticulum or Golgi elements as well. Omission of Ca^{++} or ATP, or substitution of -glycerophosphate (GP) for ATP prevented lead deposition almost completely. It is suggested that the Ca^{++}-pump of glandular cell membranes could partly be visualized.

1. INTRODUCTION

Although the fundamental role of intracellular Ca^{++} concentration in stimulus secretion coupling is well known, the histochemical localization of Ca^{++}-ATPase activity in the pituitary gland is still an unsolved problem. We used recent modification of the lead citrate technique with the aim of obtaining information on the localization of the Ca^{++}-ATPAse activity in the PCs.

2. MATERIALS and METHODS

Rats were perfused with a cold fixative of 1% formaldehyde and 1% glutaraldehyde in 0.1 M cacodylate buffer pH 7.3 containing 7.5% sucrose and 67 mM $CaCl_2$. 50 um chopper sections of the adenohypophysis were preincubated without substrate for 30 min, then incubated for Ca^{++}-ATPase activity according to Takano et al., 1986, for 45 min at 37 C. As controls, Ca^{++}-free, substrate-free media and medium with GP instead of ATP were used. Slices were osmicated, embedded in Araldite 6005 and examined in JEM-100C electron microscope.

3. RESULTS

Lead phosphate deposit was found between the cell membranes of most of the contacting cells, irrespective wheather one of them was a FSC or both were PCs. Lead precipitate was usually found on cell surfaces set free occasionally by the perfusion pressure. In some cells, a faint reaction deposit was also observed in Golgi saccules or ER cisternae. There were short discontinuities of the reaction product at the cell membranes. Some tangential sections of cell borders suggest a special pattern of distribution of the enzyme within the membrane. Heavy lead phosphate deposit occured on the endothelial cells and in the perivascular space of the sinuses. Elimination of Ca^{++} from the tissue (with EDTA) and from the substrate medium resulted in drastic decrease in the amount of reaction product; some precipitate was still found in the membrane of endothelial cells and in the perivascular space. Incubation in a substrate-free medium, substitution of ATP with GP as substrate prevented lead deposition on the cell membranes.

4. DISCUSSION and CONCLUSION

We suggest that the lead precipitate found at the surfaces of both the PCs and FSCs reflect the localization of a Ca-dependent ATPase. Our method seems to be at the sensitivity limit that would enable one to localize the intracellular Ca^{++}-ATPases. We think that the lead deposit in the perivascular spaces due to less specific nucleotidases as proposed in the kidney by Hardonk and Bakker, 1987.

The activity of the FSC membranes agrees with the findings of Bambauer et al., 1985, who found cytochemical Ca^{++}-ATPase activity only in this localization in guinea pigs. On the other hand, we can firmly state the presence of this histochemical activity also at the parenchymal cell membranes: it appeared when their surfaces were set free of contact with other cells or when they contacted each other without the intervention of a FSC process and in some instances, fine extensions of the FSC revealed different degrees of activity at their surfaces bordering at 2 different parenchymal cells.

REFERENCES

H.J. Bambauer, S. Ueno, H. Umar, and M. Ueck (1985): Histochemistry, 83, 195-200.

M.J. Hardonk and W.W. Bakker (1987): Histochem J, 19: 616-617.

T. Takano, H. Ozawa, and M.A. Crenshaw (1986): Cell Tissue Res, 243, 91-99.

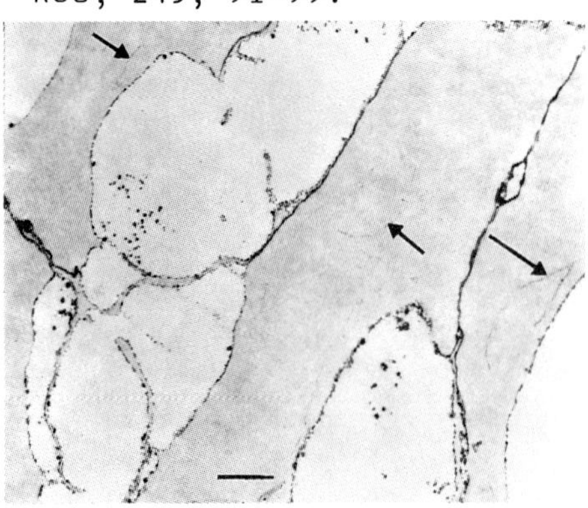

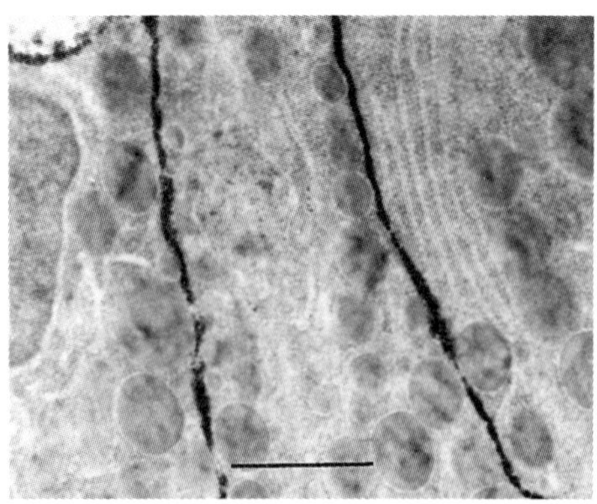

Fig.(1): Ca -ATPase activity between the cell membranes of FSCs, PCs and faint reaction deposit in Golgi and ER elements (arrows). Uncontrasted.

Fig.(2): Enzyme activity between the PCs.contrasted. Bars =1 um

Inst. Phys. Conf. Ser. No. 93: Volume 3, Chapter 20
Paper presented at EUREM 88, York, England, 1988

Interactions of lectins with specific granule types in granular leukocytes

T Daimon, K Kawai and K Uchida

Department of Anatomy, School of Medicine, Teikyo University, Kaga, 2-11-1, Itabashi-ku, Tokyo, 173, Japan

ABSTRACT: Characteristic binding of lectins to granules in granulocytes of several laboratory animals was studied by a post-embedding method at an ultrastructural level. Azurophilic granules of neutrophils showed a strong affinity for wheat germ lectin(WGL) and concanavalin A(Con A). Specific granules of neutrophils were virtually unlabeled with WGL, but showed a weak affinity for Con A. Eosinophilic granules were stained with both WGL and Con A. Basophilic granules were not stained with WGL but were weakly labeled with Con A. These studies indicate that WGL is reliable markers for lysosomes.

Morphological heterogeneity of the granules of granulocytes has been studied by means of electron microscope and enzyme cytochemical techniques. Recently, lectins has been used for hematological studies at a light microscopic level and Lee et al.(1987) found that some lectins selectively react with a subset of granulocytes. In the present study we have made use of recent advances in electron microscope cytochemical procedures to detect lectin binding sites in the granules of granulocytes.

Buffy coats of the centrifuged blood of rabbits, guinea pigs, rats and chickens were fixed with 4% formaldehyde and 0.5% glutaraldehyde and were embedded in Lowicryl K4M or LR White. The labeling of thin sections from resin-embedded materials was performed as a direct technique(Roth 1983) using WGL-gold complex and Con A-ferritin complex. The concentration of lectins varied from 10μg/ml to 50μg/ml. Following incubation thin sections were stained with uranyl acetate and lead citrate, and examined with a JEM-100C transmission electron microscope.

Neutrophils(Heterophils, polymorphonuclear leukocytes) of rabbits possess two types of granules. Azurophilic granules are large and dense. Specific granules are smaller and less dense than azurophilic granules. The labeling pattern obtained with the WGL-gold staining sequence on thin sections from rabbit neutrophils is illustrated in Figure 1. Azurophilic

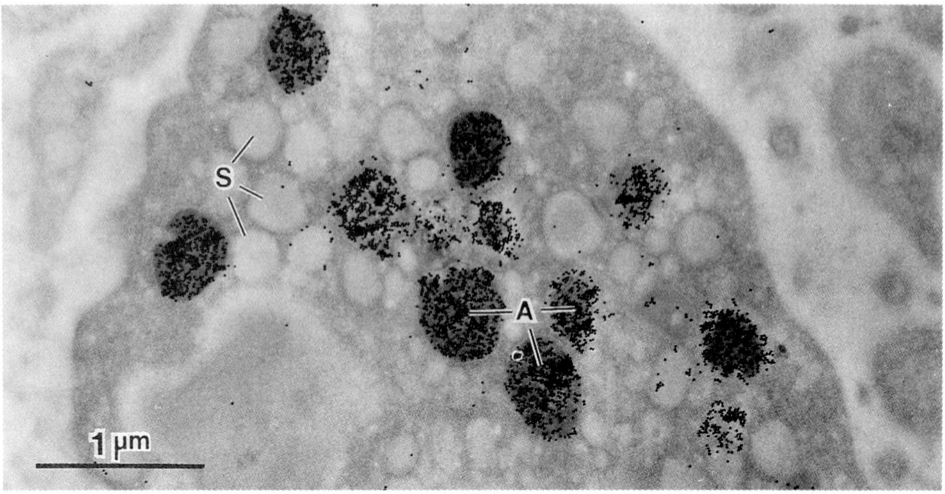

Fig. 1. Detail from a rabbit heterophil stained with WGL-gold. Azurophilic granules(A) are positive. Specific granules(S) appear unstained.

granules were stained heavily. In contrast specific granules were not stained. These staining patterns were essentially the same as those of guinea pig, rat and chicken neutrophils. In rats and guinea pigs, the granules of eosinophils reveal a central core. Binding sites for WGL were present over the core, whereas the surrounding matrix was not stained. Chicken eosinophils have only homogeneous granules which lack a central core. Gold particles were present uniformly over the matrix of granules. The granules of basophils were not stained with WGL-gold. When ultrathin sections were stained with Con A-ferritin, Con A was located over the azurophilic granules of neutrophils in rabbits, guinea pigs, rats and chickens. The eosinophilic granules of all animals examined in the present study were stained intensely with Con A-ferritin conjugates. The ferritin particles were also localized over the granules of basophils in rats and chickens.

By applying a post embedding lectin staining technique we localized binding sites for WGL and Con A in the granules of granular leukocytes. The results reveal major differences in the WGL binding characteristics of azurophilic granules and specific granules in neutrophils. As the azurophilic granules contain acid phosphatase and other acid hydrolases, WGL can be used as markers for lysosome.

References

Lee M C, Turcinov D and Damjanov I 1987 *Histochemistry* **86** 269
Roth J 1983 *J Histochem. Cytochem.* **31** 987

Inst. Phys. Conf. Ser. No. 93: Volume 3, Chapter 20
Paper presented at EUREM 88, York, England, 1988

Histochemical study on the epiphyseal cartilage during pre- and postnatal periods of development in rat

İsmail Üstünel and Ramazan Demir

Depart. of Histol.Embryol., Medical Faculty, Akdeniz Universty, Antalya

ABSTRACT: Acide and alkaline phosphatases, ATPase, 5'-nucleotidase enzyme activities were significantly increased in epiphyseal cartilage of proliferative, hypertrophic and ossification zones according to the developmental age. The synovial membrane showed positive reaction for acide phosphatase, ATPase, 5'-nucleotidease enzymes but not alkaline phosphatase.

1. INTRODUCTION

Many investigations have been made on the biochemical parameters regulating bone and cartilage growth (Bona et al.1965; David et al.1983; Delbrück, 1970; Fischer,1980). The role of the phosphatases has been particularly discussed by different invertigators (Gedikoğlu,Ö. et al.1987). It was pointed out that developing bone and actived or hypertropich cartilages had high phosphatase activity. These enzymes catalyzed the mineral accumulation in ossification and other processes (Kuhlman,1965; Wergedal, 1969; Woltgens et al.1970). The role of enzymes is well known during bone growth. However, enzyme activity changes and distribution in cartilage growth with pre-and postnatal development ages have not been investigated in detail. The present study aims to describe the activity and distribution of ATP'ase, 5'-nucleotidase, alkaline and acide phosphatases in epiphyseal portions of tibiae and femora, and the synovial membrane during pre-and postnatal periods of development in rats.

2. MATERIALS AND METHODS

In this study, eight male fetus between 17-20 days of prenatal age obtained from four pregnant rats and twenty male offspring animals between 0-19, 20-39 and 40-60 days of postnatal age were used. The epiphyseal cartilage tissue speciments were sectioned in coronal plane at 10 micron in a microtome-cryostat at -20°C. Sections were also decalcified through the use of 15% EDTA at 4°C for one week. The histological methods were aplied for alkaline and acide phosphatase (Gomori,1952), ATPase (Padykula and Herman,1955), 5'-nucleotidase (Wachstein and Meisel,1957) activities.

3. RESULTS PRESENTATION

The methods used for the demonstration of ATPase, 5'-nucleotidase, alkaline and acide phosphatase activity revealed the cellular enzymatic reactions observed in different level of the tissue. The results are summarized in Table-I on a 0 to +++ scale.

Table-I. Distribution and relative activity of ATP'ase, 5'-nucleotidase and phosphatases in the epiphyseal cartilage during pre-and postnatal periods of development in rats.

		Alkaline Phosphatase				Acide Phosphatase				ATP'ase				5'-nucleotidase			
Epiphyseal Zones		17-20 days Fetus	0-19 days Postnatal	20-30 days	40-60 days	17-20 days	0-19 days	20-39 days	40-60 days	17-20 days	0-19 days	20-39 days	40-60 days	17-20 days	0-19 days	20-39 days	40-60 days
Articular Cartilage	SM	0	0	0	0	+	++	+++	++	0	+	+	+	0	+	+	+
	TnZ	0	0	0	0	+	++	++	+	+	++	+++	+++	+	++	+++	+++
	RZ	0	0	0	0	+	++	++	+++	0	+	+	++	+	+	++	++
Epiphyseal Plate	EZ	-	++	+++	++	-	++	++	+++	-	+	++	++	-	++	++	++
	SOC	-	+	++	+	-	++	++	+++	-	+	+++	++	-	+++	+++	++
	TrZ	0	0	0	0	0	0	+	+	0	+	0	0	0	+	++	++
	PZ	+	+	++	++	+	+	++	+++	+	+	++	+++	+	++	+++	+++
	HZ	+	++	+++	++	+	++	++	+++	+	++	++	+++	+	++	+++	+++
	MZ	0	0	0	0	0	0	0	0	0	0	0	0	0	0	0	0
	POZ	+	+	++	+	+	++	++	+++	+	++	++	++	+	++	++	++

SM=Sinovial membrane, TnZ=Tangenital zone, RZ=Radiar zone, EZ=Epiphyseal zone, SOC=Seconder ossification center, TrZ=transforming zone, PZ=Proliferative zone, HZ=Hypertrophic zone, MZ=Mineralizing zone, POZ=Primer ossification zone.
O=negative, (+)=slightly positive, (++)=positive, (+++)=strongly positive.

Alkaline phosphatase activity was observed as cytoplasmic reaction products which observed particularly at the periphery of the cells. Acide phosphatase activity varied fairly within the different zones of epiphyseal cartilage . ATPase activity products apeared at intra-and extracellulary in the most zones but not in the transforming and mineralizing zones in both the pre and postnatal periods. 5'-nucleotidase activity was observed in all zones of cartilage except the mineralizing zone (Table-I).

REFERENCES

Bone,C. et al. (1965). Acta Histochem. 21:98-119.
David,J.S. et al. (1983). Mineral Elect.Metab. 9:28-37.
Delbrück,A. (1970). Enzymol.Biol.Clin. 11:130-153.
Fischer,G. (1980). Acta Anat. 106:150-157.
Gedikoğlu,Ö. et al. (1987). T.J.Res.Med.Sci. 5(4):539-542.
Gomori,G. (1952). Microscopic Histochemistry.Univ. Chicago press, Chicago.
Kuhlman,R.E. (1965). J.Bone Jt.Surg. 47:545-550.
Padykula,H. and Herman,E. (1955). J.Histochem.Cytochem. 3:161-169.
Wachstein,M. and Meisel,E. (1957). Am.J.Clin.Path. 27:13-23.
Wergedal,J.E. (1970). Calc.Tiss.Res. 3:55-73.
Woltgens,J.H.M. et al. (1970). Calc.Tiss.Res. 4.Suppl. 39.

Morphometric evaluation of myelinated and unmyelinated fibres in the frog habenular commissure

M. Kemali and V. Guglielmotti
Istituto di Cibernetica del CNR, 80072 Arco Felice (Naples), Italy

ABSTRACT: The area of a semithin section of the frog habenular commissure was calculated. Myelinated and unmyelinated fibres were also counted. The myelinated fibres, which belong to the olfactory system, are of enormous size but few in number with respect to the unmyelinated ones.

INTRODUCTION

In a previous paper (1) we demonstrated, by the injection of Horse - radish Peroxidase, the crossing fibres of the olfactory system of the frog which traverse the habenular commissure, and by electron microscope and conventional histology, we demonstrated that these fibres are myelinated and disposed at the periphery of the commissure (2). The remaining fibres are commissural (3).

The purpose of this work was to study, by morphometry, the number of myelinated and unmyelinated fibres of the frog habenular commissure.

MATERIAL AND METHODS

Frogs of the species Rana esculenta were used. They were anaesthetized with MS 222 (Sandoz) 1:3,000 and perfused transcardially with 1% paraformaldehyde plus 1% glutaraldehyde in phosphate buffer at pH 7.4 and postfixed in 2% osmium tetroxide. The habenular commissure was dissected out and the piece underwent uranyl acetate block staining. Semithin sections were cut transversely, 1 µm thick, and the myelinated axons were counted directly (Fig. 1). By superimposing a grid formed by small squares of known size on a picture of a transversely cut semithin section of the commissure, the area of that particular section was calculated. The result was approximately the same when applying two further methods: weighting a piece of paper, of known specific weight, on which the commissure had been outlined

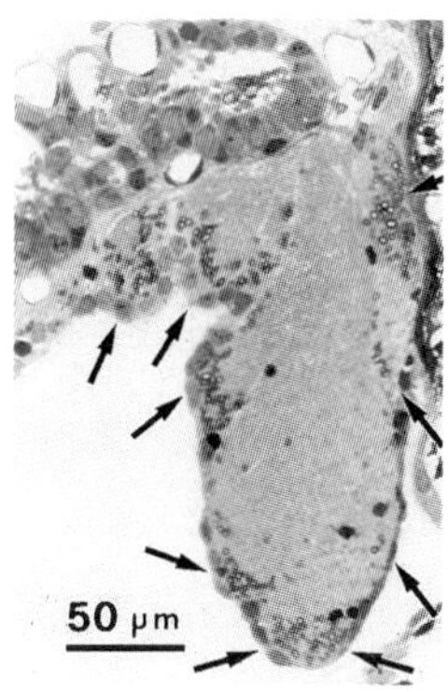

Fig. 1 - Semithin section of the frog habenular commissure cut transversely 1 µm thick. The myelinated fibres are visible at this enlargement (arrows).

by "camera lucida" and, secondly, the estimation of the same area using a planimeter. Ultrathin sections counterstained with lead citrate were observed on a Philips 400 Electron Microscope. The number of the unmyelinated fibres was counted on specimens from electron microscope photographs and referred to the area of the semithin section habenular commissure. The average percentage space occupied by glia and border epithelium was subtracted in order to obtain the real number of myelinated fibres (Fig. 2). The electron microscope gave the approximate average size of the unmyelinated axons with respect to the myelinated ones (Fig. 3).

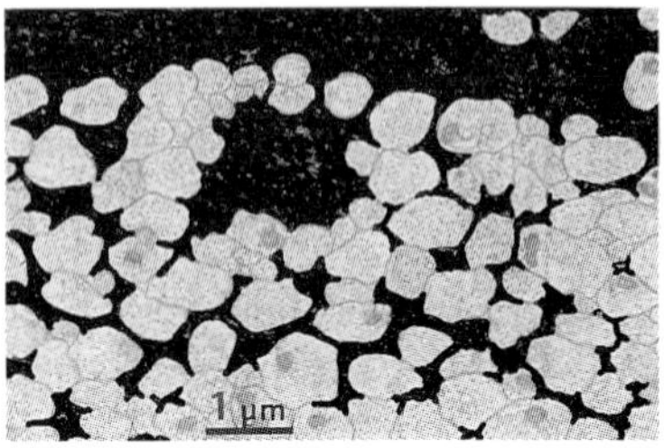

Fig. 2 - Ultrathin section showing at the electron microscope, unmyelinated fibres of the habenular commissure. In pen are outlined glial profiles.

RESULTS AND DISCUSSION

The results of our calculations gave an area of the semithin section of the habenular commissure which was approximately 20,000 μm^2. The number of unmyelinated fibres was 9,000 and that of myelinated fibres was 1,620. From these figures we had already subtracted the area occupied by glial and by epithelial cells at the border, estimated to be 50 % of the semithin section area. By calculating the average size of the unmyelinated and myelinated fibres we see that the latter are of enormous size (5 times greater) but are few in number. We propose then that the olfactory crossing fibres, which become myelinated once they cross the commissure (1), have such a high conduction velocity that it compensate for their low number.

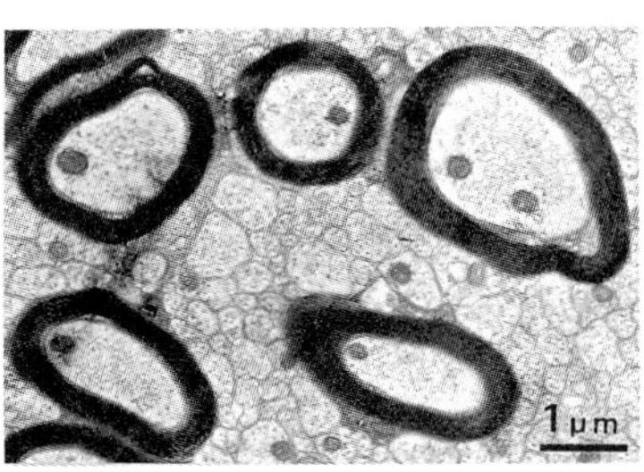

Fig. 3 - Ultrathin section of the habenular commissure seen at the electron microscope. Compare the size of myelinated and unmyelinated axons.

ACKNOWLEDGEMENT

We gratefully ackowledge the technical assistance of Mr. G. Dafnis, of the Stazione Zoologica of Naples, at the electron microscope.

Literature

(1) J. Comp. Neurol. 263, 400, 1987.
(2) In: Structure and function of cerebral commissure 15, 1979.
(3) Neuroscience, 522S, 1987.

Surface areas and intramembrane particles in renal basolateral membranes

Flemming Ø. Mathiasen[1], Elisabeth Skriver[1], Hans Jørgen G. Gundersen[2] and Arvid B. Maunsbach[1]

Department of Cell Biology[1] at the Institute of Anatomy, and Stereological Research Laboratory[2] at the Institute of Clinical Experimental Research, University of Aarhus, DK-8000 Aarhus C, Denmark.

ABSTRACT: The basolateral membrane of the medullary thick ascending limb in the rabbit kidney has a surface area of 1.31 ± 0.31 10^6 μm^2/mm tubule length and exhibits 5.16 ± 0.77 10^3 intramembrane particles/μm^2. It is suggested that each intramembrane particle corresponds to 2 αβ-units of Na,K-ATPase.

1. INTRODUCTION

The basolateral membrane (BLM) of cells in many transporting epithelia has an amplified surface area and shows a high concentration of Na,K-ATPase. In isolated and purified BLMs from the medullary thick ascending limb of the renal tubule this enzyme can be visualized as intramembrane particles (IMPs) by freeze-fracture methods (Maunsbach et al. 1984). The present investigation was carried out to determine the the surface area and the number of intramembrane particles (IMP) in the basolateral cell membranes in the medullary thick ascending limb in the rabbit kidney and to relate these parameters to the known Na,K-ATPase content of the BLM.

2. MATERIALS AND METHODS

The area of the basolateral cell membranes was determined by ultrastructural stereology using "vertical" sections and a cycloid test system (Baddeley et al. 1986). The frequency of intramembrane particles was determined by freeze-fracture of corresponding tubule segments from the same kidney.

3. RESULTS AND DISCUSSION

The BLM area was 1.31 ± 0.31 (SD) 10^6 μm^2/mm tubule length (mean of 6 animals). The frequency of IMPs was 5.16 ± 0.77 10^3 particles/μm^2 corresponding to $6.76 \cdot 10^9$ particles/mm tubule length. Mernissi and Doucet (1984) have determined that $2.13 \cdot 10^{-14}$ moles of ^{3}H-ouabain are bound to the basolateral membrane/mm tubule length. This corresponds to 1.9 ouabain binding sites per intramembrane particle. Since the great majority of IMPs in the basolateral membranes of the thick ascending limb are assumed to represent Na,K-ATPase units and since several studies have demonstrated that one ouabain molecule is bound per αβ-unit of the enzyme, the present study suggests that one intramembrane particle in the BLM corresponds to 2 αβ-units of Na,K-ATPase.

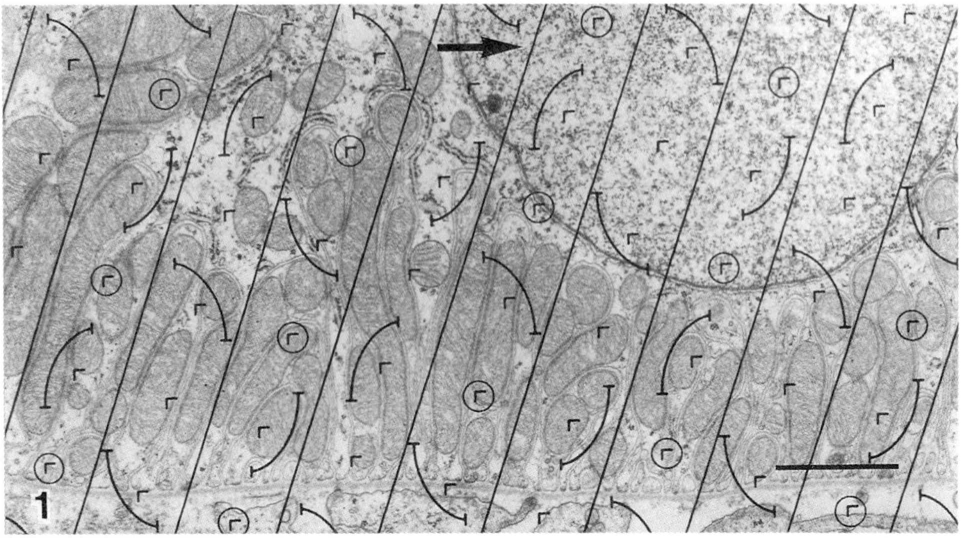

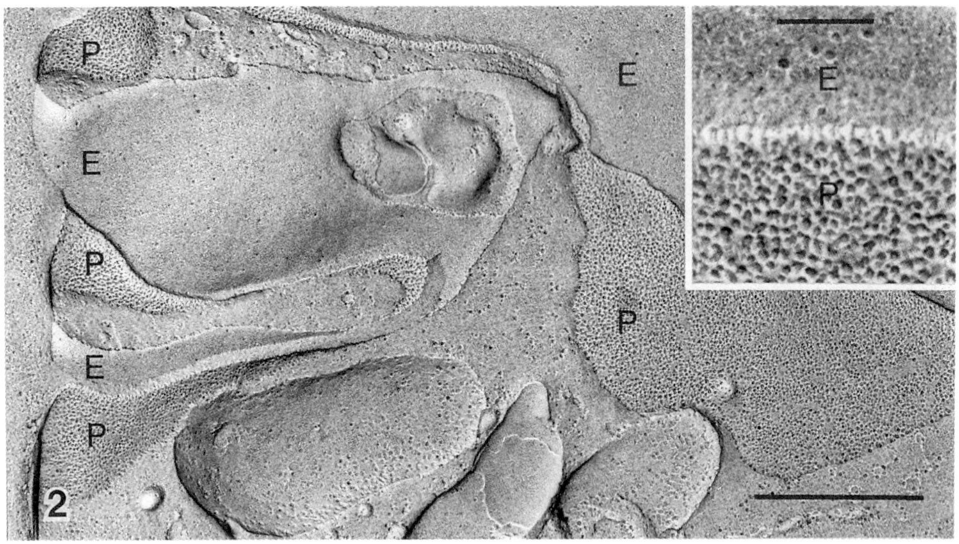

Fig. 1. Part of "vertical" section through basolateral membranes in medullary thick ascending limb with a superimposed cycloid test system. The oblique lines are not part of the test-system. The arrow indicates the "vertical" direction. Bar: 1 μm.

Fig. 2. Freeze-fracture replica of basolateral membranes after rotary shadowing. P- and E-faces are indicated. Bar: 0.5 μm. Inset: P- and E-faces. Bar: 0.1 μm

4. REFERENCES

Baddeley A J, Gundersen H J G and Cruz-Orive L M 1986 J. Microscopy **142** 259

Maunsbach A B, Skriver E, Hebert H and Jørgensen P L 1984 Verh. Anat. Ges. **78** 55

Mernissi G El, Doucet A 1984 Am. J. Physiol. **247** (Renal Fluid Electrolyte Physiol. **16**) F158

Immunogold triple labelling of cell surface-associated antigens on human suspended cells

GC Manara, C Ferrari*, C Torresani and G De Panfilis

Departments of Dermatology and *Histology, Parma University, Parma, Italy

ABSTRACT: A method is described for transmission immunoelectronmicro=scopy (IEM) which allows the simultaneous detection of three distinct cell surface-associated antigens on the same cell. The method here de=scribed is simple and performed by means of purified, biotinylated and gold granules (GG)-coupled monoclonal antibodies (MAbs). As GG are ele=ctron dense and uniform in size, they are ideally suited for multiple marking of cell surface antigens.

1. INTRODUCTION.

In last years, GG have been largely utilized as markers, in transmission IEM, in order to detect cell surface-associated antigens by means of pre--embedding approaches. Investigations have been performed both on suspen=ded cells (Schmitt et al, 1983) and on tissue sections (De Panfilis et al, 1986). Moreover, both single (Manara et al, 1986) and double (Dezutter--Dambuyant et al, 1985) labelings have been utilized. In the present study we obtained the contemporaneous detection of three distinct cell surface antigens on the same cell, by using three different sized GG.

2. MATERIALS AND METHODS

Specimen cells were obtained by venous blood of healthy donors and by nor=mal human skin suspensions. Cells were prefixed in 0.1% glutaraldehyde and incubated in PBS containing 1% BSA and 10% heat-inactivated human AB se=rum. Subsequently sample cells were subjected to the following six incuba=tions: 1) purified first MAb; 2) goat anti-mouse IgG coupled to 40 nm GG; 3) mouse unreactive IgG; 4) biotinylated second MAb; 5) 20 nm GG-labeled streptavidin; 6) 5 nm GG-labeled third MAb. Cells were then routinely pro=cessed for transmission electron microscopy.

3. RESULTS AND DISCUSSION

At low magnification (inset Fig 1), same GG were identifiable, but the in=dividual markers could not be differentiated. However, at higher magnifi=cation they could be distinguished (Figs 1 and 2): 40 nm, 20 nm and 5 nm GG labeling the purified first MAb, the biotinylated MAb and the GG-linked MAb respectively. In conclusion, a triple marking of cell surface antigens has been achived for the first time using MAbs and GG of different size. The commercial availability of purified MAbs, biotinylated MAbs and GG-coupled MAbs, and the use of a blocking incubation with mouse unreactive IgG provided specificity of the labeling.

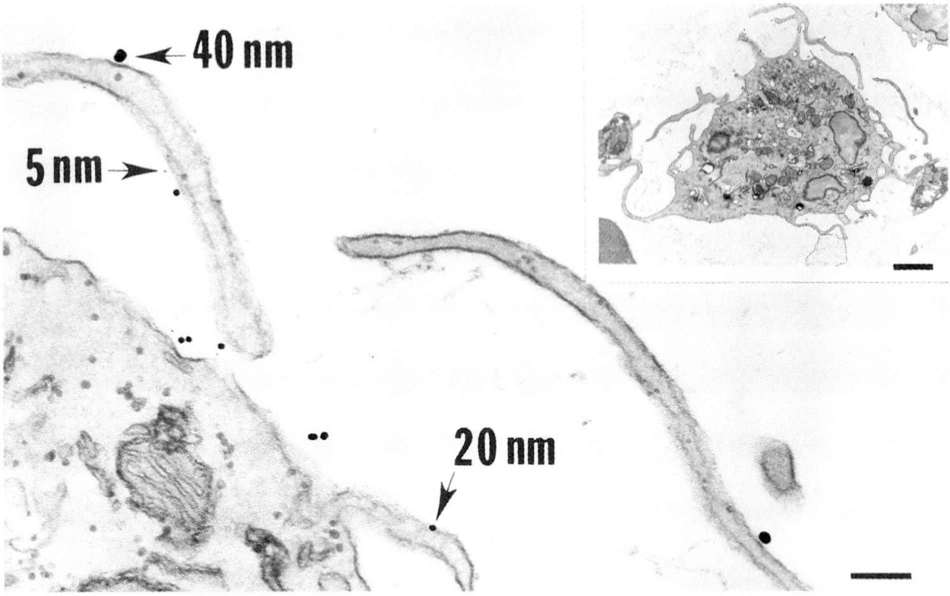

Fig 1. Natural killer cell coexpressing CD16 (40 nm GG), HNK-1 (20 nm GG) and CD8 (5 nm GG) antigens. Bar = 0.25 µm. Inset bar = 1 µm.

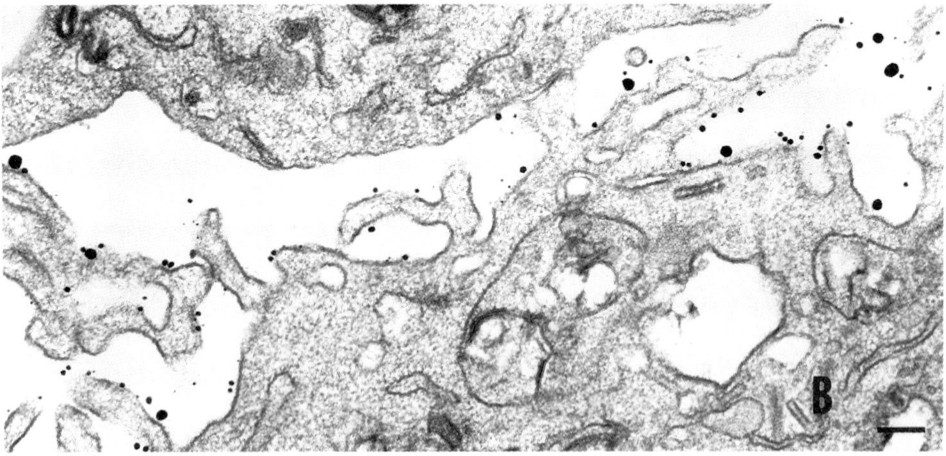

Fig 2. Langerhans cell coexpressing CD1a (40 nm GG), HLA-DR (20 nm GG) and CD4 (5 nm GG) antigens. B = Birbeck granule. Bar = 0.25 µm.

4. REFERENCES

De Panfilis G et al. 1986 J. Invest. Dermatol. 87 510
Dezutter-Dambuyant et al. 1985 J. Invest. Dermatol. 84 465
Manara GC et al. 1986 Scand. J. Immunol. 23 225
Schmitt D et al. 1983 J. Invest. Dermatol. 80 344

Freeze-substitution and immunolabelling

H. Schwarz and B.M. Humbel

Max-Planck-Institut für Biologie, D 7400 Tübingen; Max-Planck-Institut für Biochemie, D 8033 Martinsried, FRG.

ABSTRACT: Freeze-substitution as a useful preparation method for immuno labelling is described. The article focuses on the influence of chemical fixatives added to the substitution solvent. The less fixative added the higher the labelling density. However, relocation of membrane bound antigen may occur if fixation is omitted.

Freeze-substitution in combination with low temperature embedding (Müller et al 1980, Humbel & Müller 1984, Hunziker et al 1984) is an alternative to cryoultramicrotomy or low temperature embedding of prefixed material. The advantages of freeze-substitution are: i) Cryofixation allows arrest of biological samples in a precise physiological state. ii) Dehydration process starts at low temperature by dissolving the ice which is different to dehydration at room temperature. iii) During substitution the chemical fixatives penetrate the specimen prior to crosslinking which is initiated by raising the temperature. The fixative then reacts over the whole sample at the same time. iv) As already demonstrated completely unfixed material can be substituted and embedded at low temperature (Humbel et al 1983, Engfeldt et al 1986). With freeze-substitution and immunolabelling it is possible to overcome at least some of the difficulties in detection of antigens which are sparse or sensitive to chemical prefixation (Carlemalm et al 1986, Engfeldt et al 1986, Humbel & Schwarz 1988).

E.coli bacteria fixed during freeze-substitution with formaldehyde/-glutaraldehyde (Figure 2) or uranylacetate alone (Figure 3) show a higher labelling density of the outer membrane protein OmpA then cells prefixed with formaldehyde/glutaraldehyde (Figure 1). The labelling efficiency is even higher after freeze-substitution without any chemical fixatives (Figure 4) but it is more difficult to maintain an optimal ultra-structure. Moreover, antigen relocation may occur during the conventional sectioning process; OmpA leaks out and forms a concentration gradient on the support film in the case of Lowicryl K4M or directly on the section with the more hydrophobic Lowicryl HM20 (Figure 5).

REFERENCES

Carlemalm E, Villiger W, Acetarin JD and Kellenberger E 1986 Science of biological specimen preparation eds M Müller, RP Becker, A Boyde and JJ Wolosewick. SEM Inc. pp 147-154
Engfeldt B, Hultenby K and Müller M 1986 Acta Path.Microbiol.Immunol. Scand.Sect.A 94 313

Hunziker EB, Herrmann W, Schenk RK, Müller M and Moor H 1984 J.Cell Biol.98 267
Humbel B, Marti T and Müller M 1983 Beitr.elektronenmikroskop.Direktabb. Oberfl. 16 ed G Pfefferkorn pp 585-594
Humbel B and Müller M 1984 Proc. 8th Europ.Congr. on Electron Microscopy eds A Csanady, P Röhlich and D Szabo (Budapest) pp 1789-1798
Humbel BM and Schwarz H 1988 Immuno-gold labeling methods eds AJ Verkleij and JLM Leunissen (Boca Raton: CRC Press) in press.
Müller M, Marti T and Kriz S 1980 Proc. 7th Europ.Congr. Electron Microscopy eds P Brederoo and W de Priester (The Hague) pp 720-721

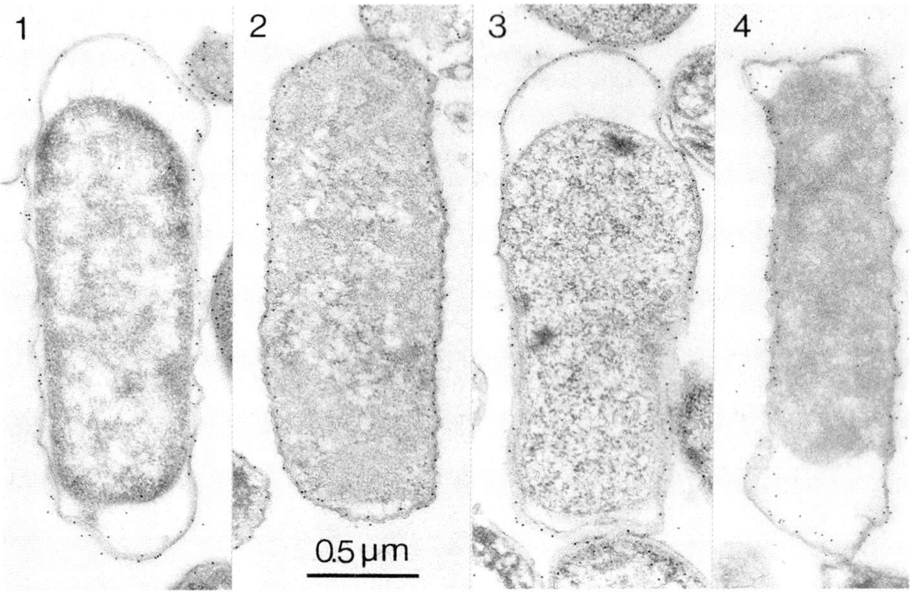

Figures 1-4: Detection of OmpA in E.coli wild type cells after freeze-substitution and Lowicryl HM20 embedding using rabbit antiserum against OmpA and protein A-8 nm gold.

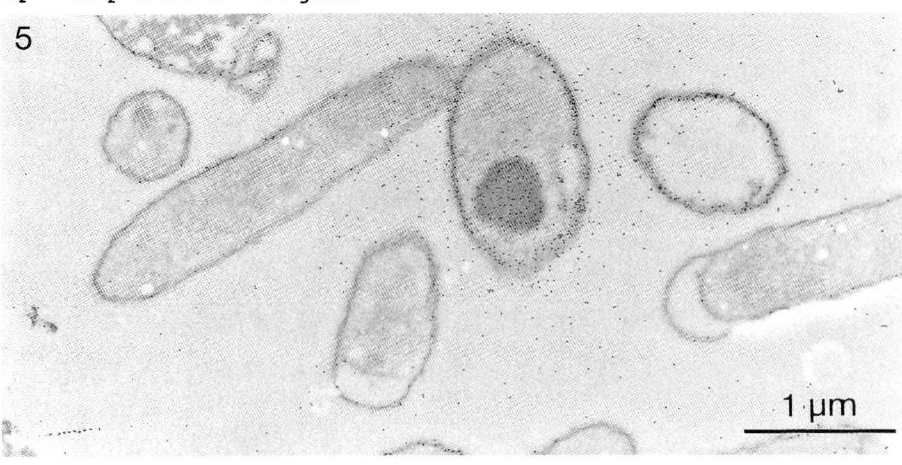

Figure 5: Unfixed OmpA overproducing cells prepared as in Figure 4.

Inst. Phys. Conf. Ser. No. 93: Volume 3, Chapter 21
Paper presented at EUREM 88, York, England, 1988

Testing and loss of gold particles in replica immuno-gold cytochemistry

H.Hohenberg and G.Rutter

Heinrich-Pette-Institut f. Exp. Virol. u. Immunol. a.d.Univ. Hamburg, Martinistr.52, D-2000 Hamburg 20, F.R.G.

ABSTRACT:An indispensable prerequisite for the application of the immuno-gold replica technique is the stability of the gold particles in the course of the replica preparation. If this stability is not given, reproducible studies of antigen distribution and quantification is not possible. In our testing system various agents used for the replica cleaning and detachement can be checked and we could show that under special conditions there is no loss of gold particles from the replica.

Many important cellular functions are initiated and regulated via interactions taking place at the cell surface, therefore studies on surface antigens and their distribution have become a field of increasing applications. Prerequisites for these studies are routine preparation techniques which preserve the ultrastructure and the topography of the biological surfaces and of the immunomarker. We could show that the combination of different cell preparation methods with the replica and immuno-gold technique (1) is suitable for the localization and distribution analysis of cell surface antigens at high resolution in the transmission electron microscope as well as for making a direct correlation between these labeled surface antigens and both intramembranous (2) and submembranous (3) structures and the cytoskeleton(4).
An essential prerequisite for the application of the immuno-gold replica technique as an exact and routine method is the stability of the gold in the replica; this is an significant factor, because neither the detachement nor the cleaning procedures, required in the course of replica preparation,should result in a loss of gold particles, i.e. neither the location nore the size of the gold markers should change during these procedures.
Loss of gold particles can take place during the detachement of the labeled and the shadowed specimen from the support to which it is affixed, e.g. a plasma membrane labeled at the surface which is attached to a coverslip and shadowed on the protoplasmic surface (3). This specimen must be detached from the coverslip with hydrofluoric acid (HF); the unshadowed exoplasmic surface is exposed briefly to the HF (30-60 seconds).
The following comparative investigations have been regarded as indications for probable or possible loss of gold in the hydrofluoric acid: the membrane described above was allowed to

remain in the HF for a period exceeding that required for the detachement process, and pieces of the replica were mounted on an EM-grid after certain fixed time intervals. It was then possible to ascertain by direct comparison whether or not the labeling density had altered in the time t from t_1 to t_x. The results showed that the membrane preparation could be left in the HF for up to 30 minutes without resulting in a loss of gold particles (see figs.1 and 2).

Platinum/Carbon replica of the protoplasmic surface of an isolated membrane labeled on the exoplasmic surface after 1min (fig.1) and after 30min (fig.2) on HF. N=nucleocapsid,x27.000.

In order to test the stability of the gold particles in the cleaning step of the replica preparation, glass coverslips were coated with rabbit IgG and protein-A gold, shadowed with Pt/C and floated off onto HF and exposed to the cleaning agents normally used for biological material, i.e. chromic acid (30%) or sodium hypochlorite (6%). The results are shown in the figs. 3-6: the results archived with chromic acid show that there is no direct loss of gold particles, they stay anchored in the replica, but after 2,5h the particel diameters have changed. After 5 hours the gold is almost entirely dissolved. Sodium hypochlorite did not have a similar dissolution effect. Fig.6 shows the gold probe after 20 h on hypochlorite.

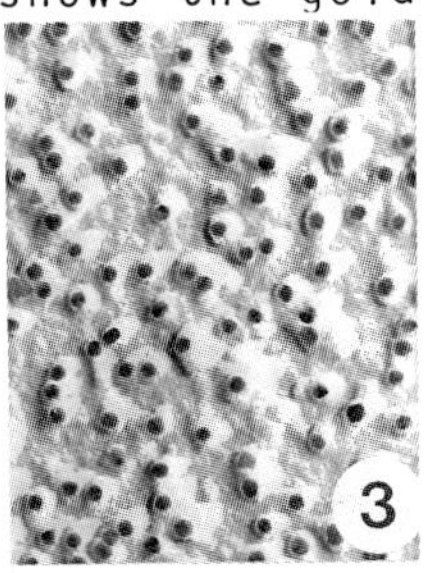

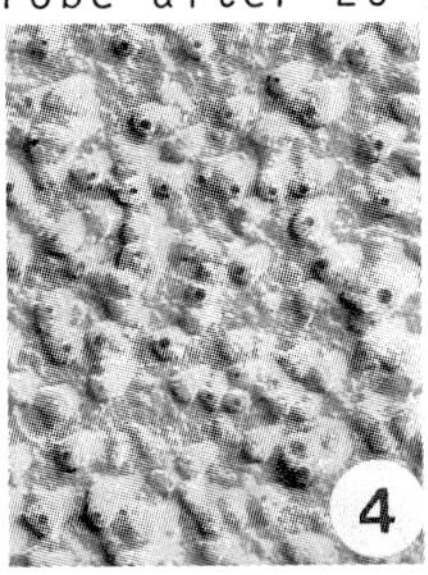

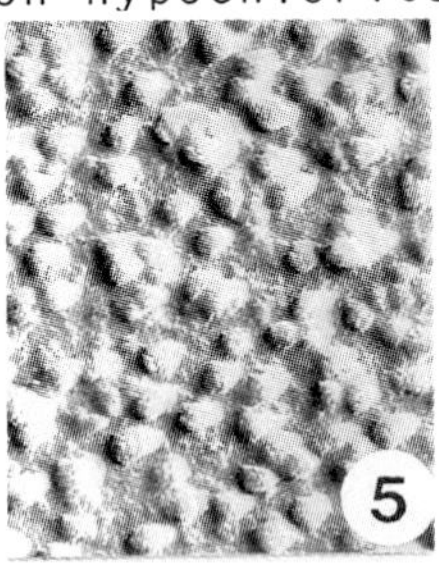

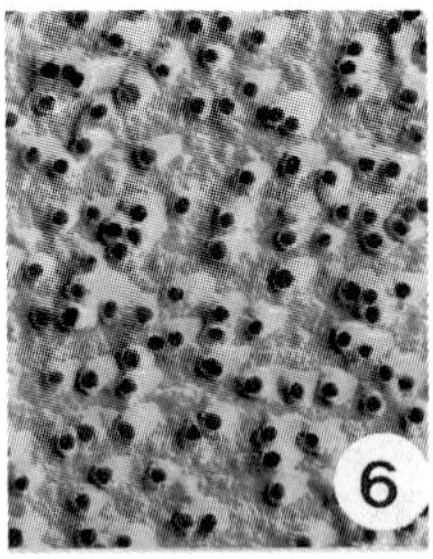

Pt/C replicas of gold probes after 1min (fig.3), 2,5h (fig.4) and 5h (fig.5) on chromic acid; after 20h on hypochlorite (fig.6) x80.000.

1. K. Mannweiler et al., J. Microsc. 126, 145-149 (1982)
2. H. Hohenberg et al., Science of Biol. Spec. Prep., 235 (1985)
3. G. Rutter et al., Eur.J.Cell Biol. 39, 443-448 (1982)
4. W. Bohn et al., Virology 149, 91-106 (1985)

Financially supported in part by Gemeinnützige Hertie-Stiftung Frankfurt/a. Main. The Heinrich-Pette-Institut is financially supported by Freie u. Hansestadt Hamburg, and by Ministerium für Jugend, Familie, Frauen und Gesundheit, Bonn.

Silver enhancement of single and double immuno-gold labelled cells imaged with backscattered electrons

E. Namork[1] and H.E. Heier[2]
Departments of Methodology[1] and Immunology[2]
National Institute of Public Health, N-0462, Oslo 4, Norway

ABSTRACT: Silver enhancement of single and double immuno-labeled red blood cells was carried out to extend the use of gold probes for cell surface labeling, visualized in the backscatter electron imaging (BEI) mode. Individual particles of colloidal gold down to 5nm were detected after enhancement, thereby providing increased labeling efficiency.

1. INTRODUCTION

The backscatter electron imaging mode has proved most successful in detecting colloidal gold probes labeling cell surface antigens. The most commonly used probe size is 40nm which is easily visualized in this mode. Probe sizes as small as possible should be used, however, to avoid steric hindrance. The smaller probes have also a narrower size distribution which is essential for safe discrimination of probes used in double labeling. However, because of decreasing backscatter electron signal with decreasing probe volume, particles smaller than 20nm sparsely labeling cells, may be overlooked. Silver enhancement, which is based on reduction of silver ions at the surface of a heavy metal, was tested to enable smaller gold probe sizes to be visualized in scanning electron microscopy.

2. MATERIALS AND METHODS

Red blood cells, sensitized with a mixture of two monoclonal antibodies (IgG and IgM), were incubated with a mixture of goat anti-mouse IgG and IgM conjugated with 15nm and 40nm colloidal gold particles, respectively (Janssen Pharmaceutica) or goat anti-mouse IgG-20nm and IgM-5nm (E.Y. Laboratories, Inc., U.S.A.). Labeling was also carried out with the four probes separately.After labeling, the cells were sedimented and fixed onto chips of mica. The specimens were then washed thoroughly in distilled water before silver enhancement.

The silver solution was prepared according to Danscher (Histochem., 1981, 71: 1-16). It vas found that the freshly prepared solutions, except the citrate buffer, could be frozen in aliquotes to make up 10 ml of developing solution: 6 ml, 25% gum arabic, 1.5 ml hydroquinone and 1.5 ml silver lactate. 1 ml aliquotes of the buffer were kept at 4°C. Thawing, mixing of solutions and developing were carried out in a darkroom under red safelight. After silver enhancement for the appropriate times, the specimens were washed, dehydrated and critical point dried. Finally, the cells, labeled and enhanced, were coated with a thin layer of carbon before mounting on stubs for electron microscopy.

Electron microscopy was performed on a JSM-840 equipped with a lanthanum hexaboride emitter, operated at 15 KeV in the BEI mode. A probe current of $1x10^{-10}$ Amp and a working distance of 7-8mm were used.

3. RESULTS

Erythrocytes labeled simultaneously and separately with 15 and 40nm probes were silver developed for 6 min and the probes measured to 25nm ± 2 and 49nm ± 3, respectively (Fig. 1a and b).

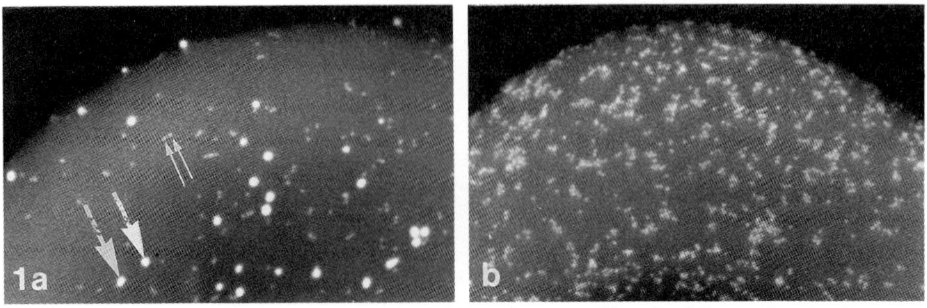

To reach a convenient size for visualization, cells double and single labeled with 5 and 20nm were enhanced for 20 min and the probes measured to 24nm ± 3 and 46nm ± 4, respectively (Fig. 2a and b). As seen from the micrographs (25.000x),the gold labels still appear spherical after enhance-

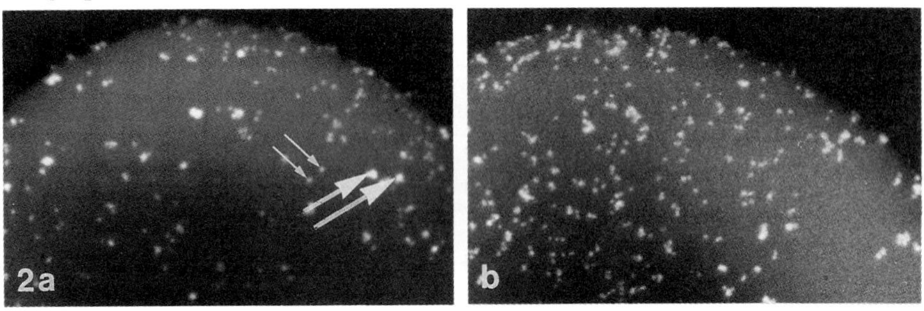

ment. Individual particles of all probes are visualized with good contrast and the two probes in the double labelings can be discerned. The single labeled cells are also easily recognized at low magnification which is an advantage in quantitating labeling density of a large number of cells.

4. DISCUSSION

The most important factor influencing the specificity of the silver precipitation is the use of a protective colloid in the developing solution. The colloid gum arabic slows down the reduction of silver ions ensuring an even growth of the gold particles which is essential in double labeling experiments, where single particles are to be visualized and discriminated. The results indicate that the growth rate varies for each probe size. Because of the big differences in surface area of the varying probe sizes, the smallest probe will have the largest percentage increase after enhancement. The results show that both the smallest probes in the double labelings are half the size of their counterparts after enhancement. The smallest probe chosen should therefore be less than half the size of the larger one, to safely discriminate between the two.

5. CONCLUSION

Silver enhancement can be used to visualize individual gold particles of any size down to 5nm, labeling one or simultaneously two antigenic sites on cell surfaces. The method thereby enables cells to be labeled with small probes for increased labeling efficiency, visualized in a SEM equipped with a conventional solid state BEI detector.

Simultaneous immunoelectron microscopic localization of DNA and histones on ultrathin sections

M Thiry

Laboratoire de Biologie cellulaire et tissulaire, Université de Liège, 4020-Liège, Belgique.

ABSTRACT: An ultrastructural approach for simultaneous demonstration of DNA and histones on the same section of Lowicryl-embedded cells has been developopped by applying sequentially a recent method using a monoclonal anti-bromodeoxyuridine antibody detecting bromodeoxyuridine previously incorporated into DNA and an indirect immunogold technique using anti-histone antisera. In this procedure, immunoglobulin gold complexes of two different sizes have been employed. Under these conditions, small and large gold particles are essentially found over the condensed chromatin.

The general distribution of DNA within cells has been recently revealed by means of immunoelectron microscopic approaches using either a monoclonal antibody directed against double and single stranded DNA (Scheer et al, 1987; Thiry et al, 1988) or a monoclonal anti-bromodeoxyuridine antibody detecting bromodeoxyuridine previously incorporated into DNA (Thiry and Dombrowicz, 1988).

On the other hand, using different anti-histone antisera, the cellular distribution of three histones (H_{2B}, H_3, H_4) has been determined at the ultrastructural level (Thiry and Muller, 1988).

In the present work, in order to compare and corrolate the spatial distribution of DNA with those of histones, a double immunogold staining procedure on ultrathin sections of Lowicryl-embedded cells has been developped.

This technique involves the sequential application of two distinct antibody-immunoglobulin gold procedures. A first indirect immunolabelling method using a immunoglobulin coupled with large gold particles was applied by floating the ultrathin sections. Once the first labelling is performed, the sections were mounted on collodion-coated gold grids and dried. The second face of the sections was then labelled using a second antibody and an immunoglobulin gold complex formed with small gold particles.

This double immunogold staining procedure allows to avoid co-labelling due to artifactual cross-reactions between reagents since the labelling of each of the faces of the sections is independently performed. Further, in this procedure, the first reaction can not hinder the second in some extent. This is particularly advantageous when the two studied antigens occur close together as suspected in the present work.

In the present work, the recent method using an anti-bromodeoxyuridine monoclonal antibody and the immunogold labelling technique using antisera raised against purified histones of chick erythrocytes have been sequentially applied to demonstrate the ultrastructural localization of DNA and histones respectively.

Under these conditions, small and large particles can be visualized on the same cellular structures. Specifically, the condensed chromatin associated with the nuclear envelope (Fig. 1) and with the nucleolus is labelled by gold particles of two different sizes.

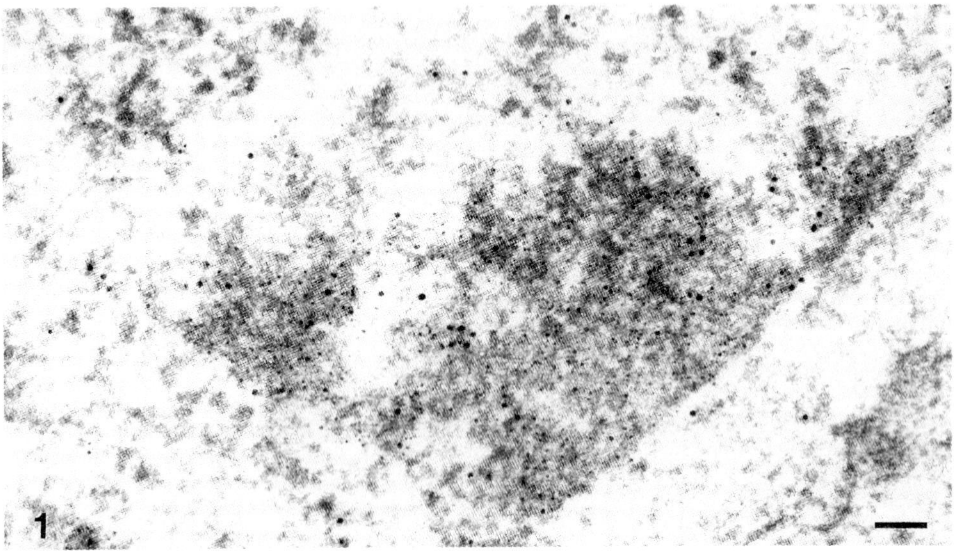

Fig. 1. Simultaneous localisation of DNA (as revealed by the small gold particles) and H_4 histone (as revealed by the large gold particles) over the condensed chromatin associated with the nuclear envelope of an Ehrlich tumour cell. Bar = 0.1 μm.

These results demonstrate the possibility to simultaneously localize DNA and histones on the same section. This approach should provide an important tool for studying the functional organization of chromatin.

Acknowledgements: The author is grateful to Dr Muller who provided anti-histone antisera, to Miss Skivée for the technical assistance, to the "Fonds de la Recherche Scientifique Médicale" (grant n°3.4512.86) and to the "Action de Recherche concertée" (grant n°85/9080) for their financial support.

Scheer U, Messner K, Hazan R, Raska I, Hansmann P, Falk H, Spiess E and Franke W 1987 Eur. J. Cell Biol. 43 358

Thiry M and Dombrowicz D 1988 Biol. Cell in press

Thiry M and Muller S 1988 in preparation

Thiry M, Scheer U and Goessens G 1988 submitted for publication

High sensitivity detection of DNA synthesis by immunolocalization of bromodeoxyuridine

R. Rizzoli, N.M. Maraldi, A. Galanzi, N. Zini, M. Falconi, M. Vitale, G. Mazzotti.

Inst. of Anatomia Umana, University of Bologna and Inst. of Citomorfologia C.N.R. c/o Inst. Codivilla-Putti, Bologna, Italy

ABSTRACT: Bromodeoxyuridine can be detected in newly-synthesized DNA by immunocytochemical method; after short time of incorporation the labelling with colloidal gold shows characteristic localizations related to the distribution of the replication units.

1. INTRODUCTION

DNA synthesis at the electron microscope level is usually monitored by ^{3}H thymidine incorporation autoradiography, that gives a resolution of about 100 nm (Maraldi 1976). Recently, for studies of cell kinetics in flow cytometry, bromodeoxyuridine (BrdUrd) has been used as a synthetic analogue of thymidine detectable by specific monoclonal antibodies (Dolbeare 1983). We have evaluated the possibility to identify BrdUrd at the electron microscope level by an immunocytochemical method performed on thin sections.

2. MATERIALS AND METHODS

Different cell lines were grown in the presence of different concentrations of BrdUrd or of 10 μCi/ml of ^{3}H thymidine for times ranging from 5 min to 24 hr. For BrdUrd detection it has been employed a post-embedding method (De Mey 1983) using a first monoclonal antibody against BrdUrd and a secondary antibody conjugated with 15 nm gold particles. Electron microscope autoradiograms, coated with monolayers of Ilford L4 emulsion, were developed with gold latensification-phenidon technique.

3. RESULTS AND DISCUSSION

The number of the positive cells is quite similar in immunostaining and autoradiography. The cells that incorporated BrdUrd for longer times show a labelling diffuse all over the nucleus, more intense on the heterochromatin, while the cells which incorporated the substance for a shorter time (from 5 min to 4 hr) show three different types of gold deposition: gold particles are mainly localized over the euchromatin (Fig. 1); the particles are exclusively associated to the heterochromatin (Fig. 2); the labelling is found at the edge between eu-heterochromatin, both around the nucleolus and close to the nuclear envelope (Fig.3). The labelling detects therefore the sites of DNA synthesis that can be related to the number of the replication units, simultaneously activated in the different moments of the S phase. The high resolution of this technique, about 40 nm, should also allow to detect the relationships between newly-synthesized DNA and the nuclear matrix in the different phases of the cell cycle.

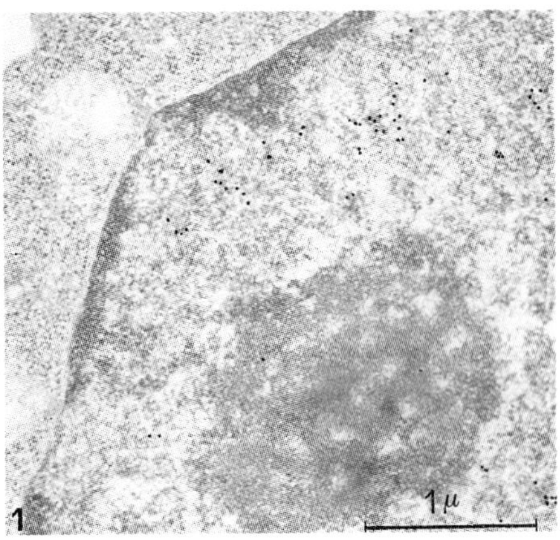

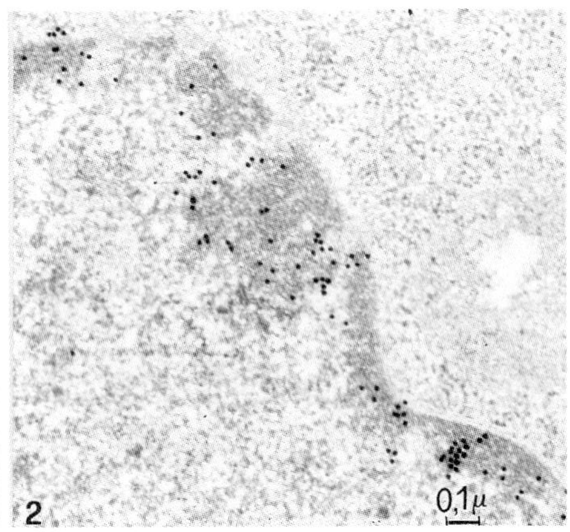

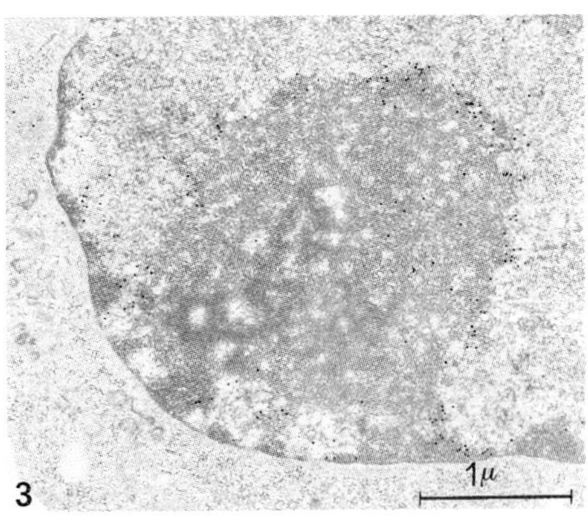

(Figs. 1-3)
Immunogold localization of BrdUrd incorporated in Friend Leukemia cell nuclei (Description in the text).

4. REFERENCES

De Mey J 1983 Immunochemistry: Practical applications in pathology and Biology (Bristol:Wright) pp 82

Dolbeare F, Gratzner H, Pallavicini M, Gray J, 1983 Proc. Natl. Acad. Sci. 80 5573

Maraldi N M 1982 Cell Growth (New York:Plenum Publ. Corp.) pp 113 -132

Inst. Phys. Conf. Ser. No. 93: Volume 3, Chapter 21
Paper presented at EUREM 88, York, England, 1988

Immunocytochemical detection of an immunogenic protein of *Coxiella burnetii* by monoclonal antibodies

T F McCaul[1], C E Snyder, Jr[1], and J C Williams[1,2].

USAMRIID, Rickettsial Dis. Lab, Airborne Dis. Div., Fort Detrick, Frederick, MD 21701, USA[1], and NIAID, NIH, Bethesda, MD 20204, USA[2].

ABSTRACT: Ultrastructural localization of a 29.5 kDa immunogenic protein was carried out on thin sections of Coxiella burnetii. Immunolabelling of the protein was enhanced after exposure of sections to a solubilization buffer containing sodium dodecyl sulphate (SDS) and sodium deoxycholate (DOC). Epitopes of the 29.5 kDa protein, recognized by monoclonal antibodies, were differentially expressed among the cell variants.

1. INTRODUCTION

Coxiella burnetii, an obligate phagolysosomal bacterium, has a developmental cycle involving two distinct cell types, namely large cell and small variants (McCaul and Williams 1981). A recently isolated immunogenic 29.5 kDa cell wall protein (Williams and Stewart 1984) could only be extracted at 100°C in a solubilization buffer containing detergents. This unusual property of extraction aroused our interest in determining the location of the 29.5 kDa protein by post-embedding immunolabelling techniques.

2. MATERIALS AND METHODS

C. burnetii (9MIC7 strain), purified from infected yolk sac material of hen eggs by Renografin gradient centrifugation, was fixed for 3 h in 1.25% glutaraldehyde in 0.066M Na-cacodylate buffer, pH = 6.8; pre-embedded in 2% Difco Agar; and then rinsed in the same buffer prior to dehydration in a graded methanol series. Polymerization was carried out in L R Gold resin (-25°C for 18 h under UV light). Immunolabelling was performed on 60-100 nm thin sections with 6 different MAbs (1C3, 4E8, 4B2, 4D6, 6D6-1, 6D6-2) reactive with the 29.5 kDa protein, as 1° antibody, and goat anti-mouse IgM-colloidal gold (15 nm) (Janssen), as 2° antibody. Some sections were treated with a solubilization buffer (100 mM Tris, 10 mM Na-EDTA, 1% SDS, 1% DOC, 1 mM phenylmethylsulphonyl fluoride and 0.02% NaN_3, pH 8.0) at 100°C for 30 min before labelling. The sections were then stained with 0.5% uranyl acetate in 50% ethanol (30 sec) and lead citrate (30 sec) and examined with a Joel 100B transmission electron microscope (TEM) operated at 80 kv.

3. RESULTS AND DISCUSSION

All MAbs against the 29.5 kDa protein revealed similar labelling. The epitopes were found on both the outer membrane and the peptidoglycan area. The label was predominant in the large cell variant (Figs 1,2). Little or no label was found on the small cell variant. More gold labels were observed in the cell wall of the large cell variant after treatment of thin sections with the detergents (Fig 2). The detergent treatment did not significantly enhance labelling of the small cell variant (Fig 2). The 29.5 kDa protein appeared to be differentially expressed among the cell variants. The inaccessibility of the monoclonal antibodies to the 29.5 kDa protein epitopes in the small cell variant implies that the protein is either masked by another component or not yet synthesized.

4. REFERENCES

McCaul T F and Williams J C 1981 J. Bact. **147** 1063

Williams J C and Stewart S 1984 Microbiology-1984 pp 257-262. The work was done while T F McCaul held a National Research Council - USAMRIID Research Associateship.

All figures are based on the reactivity of MAb #4E8 against 29.5 kDa protein.

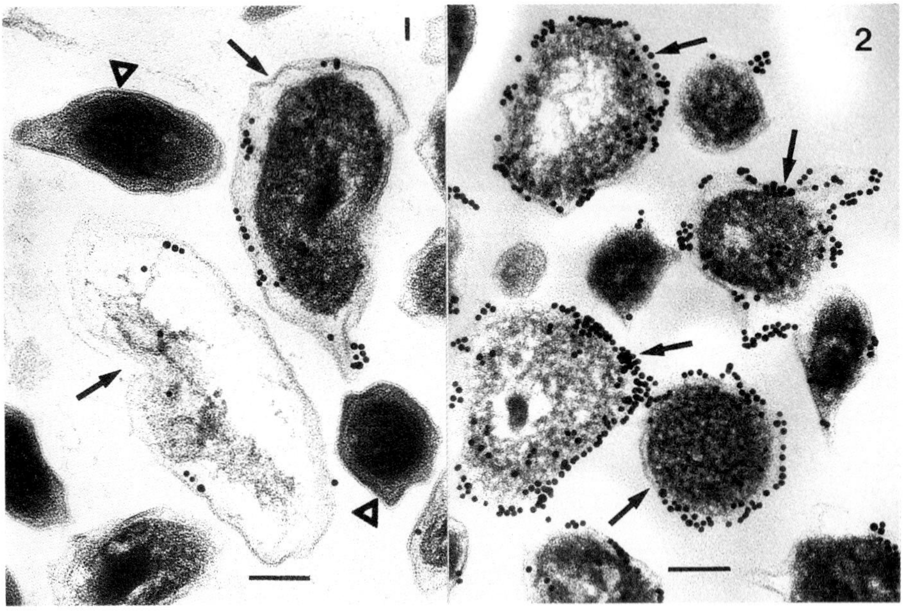

Fig.1. Thin section of Coxiella burnetii embedded in L R Gold resin. Note label on large cell variant (arrows). No label on small cell variant (∇). Bar = 100 nm

Fig.2. L R Gold embedded C. burnetii. Treatment of thin sections with detergents prior to immunolabelling. Note increase in the number of gold labels compared to Fig 1. Immunolabelling on a) inner face of outer membrane and b) peptidoglycan area of the large cell variant (arrows). Bar = 100 nm

X-ray microanalysis of frozen-dried cryosections: influence of cryotransfer and specimen cooling

T von Zglinicki(1) B Uhrik(2) K Zierold(3)

(1) Institute of Pathology, Charité, Berlin 1040, GDR
(2) Centre of Physiological Sciences, 833 06 Bratislava, CSSR
(3) Max-Planck-Institut für Systemphysiologie, 4600 Dortmund 1, FRG

ABSTRACT: S losses occur due to radiation damage in rat liver cryosections measured at ambient temperature. K seems to be the ion most sensitive against rehydration artefacts and small losses occur during fast transfer of frozen-dried sections through the room air.

1. INTRODUCTION

The use of continouos specimen cooling is commonly believed to improve the reliability of X-ray microanalytical results from biological frozen-dried cryosections. If specimen cooling during transfer and analysis is not possible, results may suffer from rehydration artefacts and radiation damage.

2. MATERIAL AND METHODS

Rat liver specimens were obtained by a cryobioptical method (von Zglinicki et al.1986) and were sectioned in a REICHERT FC 4D ultramicrotome at 110...130 K and at a nominal section thickness of 0.1 μm. Sections from the same specimen were either cryotransferred at below 150 K, freeze-dried within the electron microscope and measured at below 90 K or freeze dried in a vacuum evaporator, carbon-coated, transferred through the room air within 2 min and measured at ambient temperature. These sections will be referred to as cryoprepared liver (CPL) and air-transferred liver (ATL) sections, respectively.

In order to test the influence of the exposure to the room air separately, in a second series of experiments sections of isolated rat hepatocytes were cryotransferred into the electron microscope. Sections were freeze-dried in the microscope and elemental concentrations in clearly identified compartments were measured twice at low temperature. The first measurement was performed immediately(cryoprepared hepatocytes - CPH), whilst the second was done after the grids were warmed to room temperature, exposed to the room air for 2 min, and cooled down again to liquid nitrogen temperature (air-exposed hepatocytes - AEH).

Concentrations of Na, Mg, P, S, Cl, K, and Ca were estimated after subtraction of grid and foil background by using the QUANTEM software of the LINK 860 microanalysis system.

No morphological differences are found between CPL and ATL sections (von Zglinicki & Uhrik 1988). However, intracellular structures in AEH sections appear more coarse than those in CPH sections. Obviously, carbon-coating of ATL sections prevents morphological alterations due to rehydration during transfer.

ATL sections as compared to CPL sections display significant S losses in the cytoplasm and mitochondria, Na loss in the cytoplasm and K loss in mitochondria. Neither for the other elements analysed in these compartments nor for any of the elements in areas of rough endoplasmic reticulum significant differences are found. In sections of isolated hepatocytes mitochondria, cytoplasm and nuclei have been measured. The sole significant difference found is a loss of K from AEH as compared to CPH sections.

No significant differences in any of the elements measured have been observed, if measurements in CPH sections were repeated directly without exposure to the room air. However, if sections were stored for more than about 20 h even under high vacuum in the microscope column, gross disturbances of the ionic distributions result.

It is concluded, that S loss in ATL sections is due to radiation damage at ambient temperature (Rick et al. 1982, Zierold & Steinbrecht 1987), whilst K losses in the order of 10...20% might occur in certain compartments of rat liver cells due to rehydration even within times as short as 2min.

4. REFERENCES

Rick R, Doerge A, Thurau K 1982 J. Microsc. 125 239

von Zglinicki T, Bimmler M, Purz H J 1986 J. Microsc. 141 79

von Zglinicki T, Uhrik B 1988 J. Microsc. in press

Zierold K, Steinbrecht R A 1987 Cryotechniques in Biological Electron Microscopy ed RA Steinbrecht, K Zierold (Berlin, Heidelberg: Springer) pp 272 - 82

Inst. Phys. Conf. Ser. No. 93: Volume 3, Chapter 22
Paper presented at EUREM 88, York, England, 1988

Improved low-temperature embedding in lowicryl K4M in an agar UVF cold cabinet

Audrey M. Glauert and Robert D. Young

Strangeways Research Laboratory, Worts Causeway, Cambridge CB1 4RN, UK

The introduction of the Lowicryl resins (Carlemalm *et al* 1982) has enabled biological specimens to be embedded at low temperature, with a resulting minimization of the denaturation of proteins, enhanced preservation of antigenic and enzymatic activities, and decreased extraction of lipids. In consequence the Lowicryls have become increasingly popular, particularly for immunocytochemical studies on thin sections.

The Lowicryls are acrylate- and methacrylate-based resins and share the unfortunate property of other methacrylates in that polymerization is an exothermic reaction. The extent of the possible temperature rise within Lowicryl K4M has been described by Ashford *et al* (1986), who found that rises as high as 35 K occurred after as little as 30 min after the UV lamp had been switched on to initiate polymerization. In addition, there was another rapid rise in the temperature of the resin, sometimes to 25 K above the ambient temperature of the room, as soon as the samples were removed from the cold environment of the apparatus.

Temperature rises of this magnitude will have a deleterious effect on the specimen and to avoid this we have devised a simple apparatus in which the resin temperature can be completely controlled, both within the cold cabinet and after removal. Experiments were performed within an Agar UVF 35 low-temperature polymerization cabinet (Figure 1)(Agar Scientific, Stansted). Two UV lamps are mounted in the lid and the embedding blocks are supported at a variable distance from the lamps by a set of plastic-covered wire baskets (Figure 2). The cabinet is easily filled with dry nitrogen gas and it was found that all problems with condensation could be avoided if the cabinet was filled with nitrogen before the cooling unit was switched on.

In the initial experiments we used similar conditions to Ashford *et al* (1986) for flat embedding of up to 15 ml of Lowicryl K4M in an Al dish, with the cabinet set at 238 K. The dish rested directly on a foil-covered Perspex sheet and was irradiated with

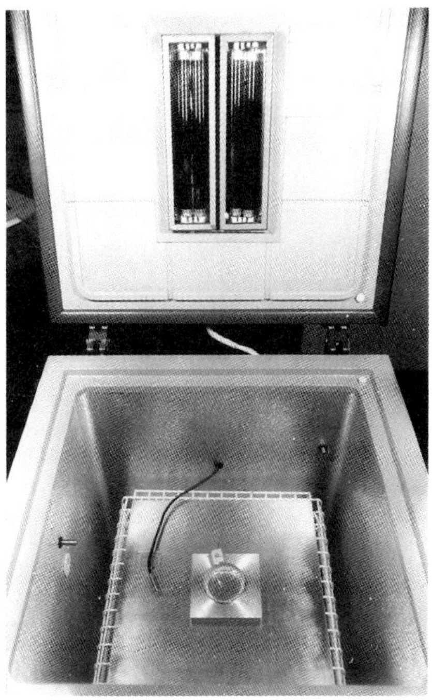

Fig. 1. The Agar UVF low-temperature cabinet for UV polymerization of Lowicryl

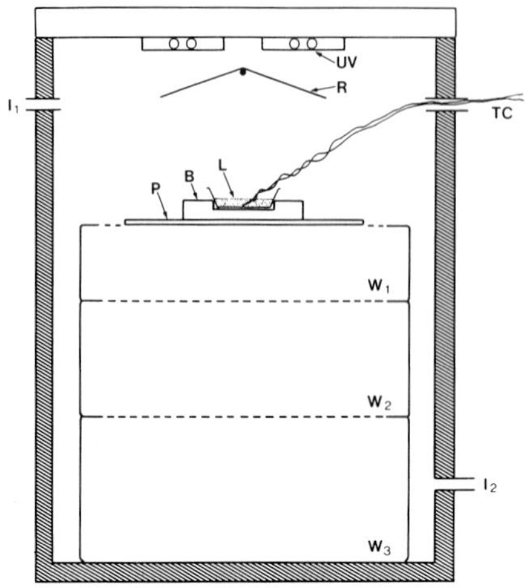

Fig. 2. Polymerization of Lowicryl (L). B, Al block; R, UV reflector; P, Perspex sheet; W_1, W_2, W_3, basket supports; I_1, I_2, outlet and inlet for nitrogen gas; TC, thermocouple

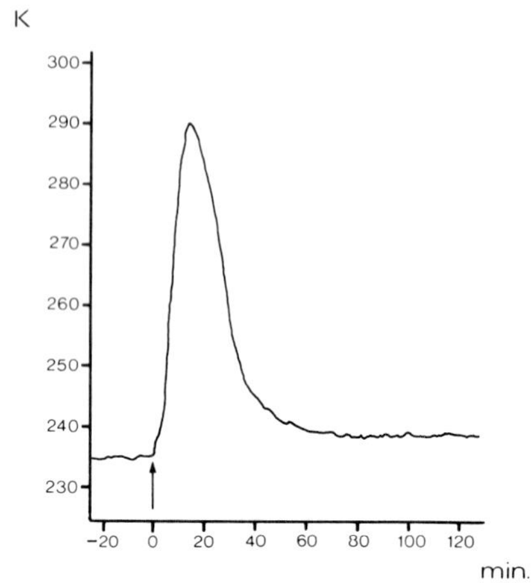

Fig. 3. Thermocouple trace of the temperature of 15 ml of Lowicryl K4M under direct UV irradiation. There is a rise of 53 K within 15 min of switching on the UV lamps (at arrow)

UV light from a distance of 40 mm. Thermocouple measurements showed that the temperature of the resin rose by 53 K within 15 min (Figure 3). There was also a rapid rise in temperature to 7.5 K above ambient on removing the dish from the cabinet. The use of indirect UV and an increase in the lamp distance to 200 mm reduced but did not suppress temperature rises in the cabinet. It was clear that the heat evolved during polymerization of the Lowicryl was not being removed. Consequently we designed a simple heat 'sink' in the form of an aluminium block with a circular cavity in which the Al dish would fit tightly (Figure 4). A small amount of ethanol was added to the cavity to ensure good thermal contact with the block. With indirect UV irradiation from a distance of 200 mm, the temperature rise of the Lowicryl within the cabinet was completely suppressed, even with the cabinet set at 248 K. In addition, it was easy to lift the hardened resin out of the cabinet while the dish was still within the Al block and then there was no rapid rise in temperature at this stage either.

We recommend the use of this simple heat 'sink' for routine use to ensure that all temperature rises are prevented during polymerization. Preliminary experiments indicate that a similar control of temperature can be obtained with Lowicryl in capsules by using an ethanol bath as a heat 'sink' (A M Glauert and R D Young, manuscript in preparation).

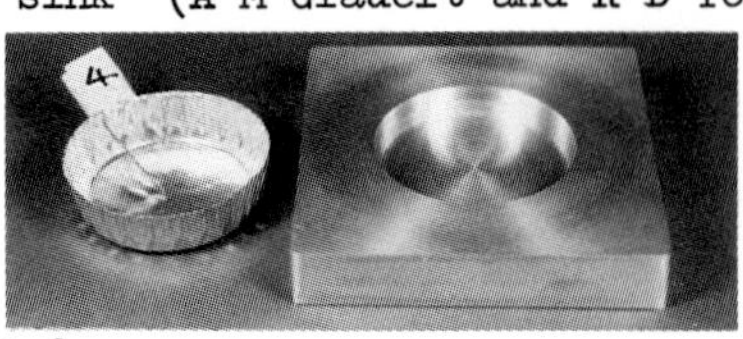

Fig. 4. An Al dish containing resin and an Al block with a circular cavity for use as a heat 'sink' during polymerization of Lowicryl K4M by UV irradiation in the cold

References

Ashford A E, Allaway W G, Gubler F, Lennon A, and Sleegers J 1986 *J. Microscopy* 144, 107–126

Carlemalm E, Garavito R M and Villiger W 1982 *J. Microscopy* 126, 123–143

Intracellular and luminal ion concentrations in sea turtle salt glands: x-ray microanalysis of frozen-hydrated bulk samples

A T Marshall

Department of Zoology, La Trobe University, Bundoora (Melbourne), Victoria 3083, Australia.

ABSTRACT: The principal cells of the secretory tubules had high concentrations of Cl^- (81 mmol l^{-1}) and high Na^+ concentrations in secreting cells. This is consistent with active Cl^- transport via a Na/2Cl/K cotransport system as postulated for elasmobranch rectal gland. High luminal K^+ concentrations may indicate apical K^+ transport. Luminal Na^+ and Cl^- concentrations were much lower than in the final secretion which may indicate ductal modification of the primary secretion.

1. INTRODUCTION

Green sea turtles (Chelonia mydas) have lachrymal salt glands. The NaCl rich fluid excreted from these glands can be hyperosmotic to the blood (330 mosmols kg^{-1}) and to seawater (1000 mosmols kg^{-1}), reaching on occassion osmotic concentrations in excess of 2000 mosmols kg^{-1}. As with other vertebrate salt glands there is much interest in how the cells of the secretory epithelium elaborate such a concentrated solution of NaCl. The determination of intracellular and luminal ion concentrations may indicate whether these glands fit the model for chloride - transporting epithelia in general and the structurally and functionally similar elasmobranch rectal gland in particular (Epstein and Silva, 1985).

2. METHODS

Lachrymal salt glands from hatchling turtles which were actively secreting (S) and not obviously secreting (NS) were frozen by pressing into solid nitrogen. Frozen glands were fractured at low temperature (88 K) under vacuum and coated with Be. Analysis of the frozen-hydrated bulk samples (100 K) was carried out in a Jeol JSM 35 with Edax 9100 x-ray spectrometers by methods reviewed in Marshall (1987, 1988). Briefly, element concentrations from apical, nuclear and basal regions of cells and from lumina of secretory tubules were obtained from characteristic x-ray intensities referenced to x-ray intensities from inorganic standards and corrected for absorption and atomic number matrix effects. The latter corrections being obtained from the Phi-Rho-Zed ionisation function. Cell water content was calculated from measured oxygen concentrations.

X-ray intensities were subjected to statistical analysis (ANOVA).

3. RESULTS

In the principal cells of the secretory tubules consistent differences were observed for some elements between basal, nuclear and apical regions. Of particular interest was a lower apical Cl concentration than in nuclear and basal regions. Since the differences were not statistically significant, however, the data were pooled and transformed into concentrations as shown in Table 1. Luminal concentrations are shown in Table 2.

Table 1. X-ray microanalysis of secretory tubule principal cells

	H2O	Na	K	Cl	Mg	Ca
	%		m mol l^{-1}		m mol kg^{-1}	
NS (N=3, n=94)	88	13±1	174±2	81±2	10±1	4±0.5
S (N=5, n=149)	88	34±1*	160±3	81±2	13±1*	3±0.5

Mean ± SE, N = number of animals, n = analyses, * p 0.05
NS - not secreting, S - secreting

Table 2. Composition of luminal fluids in secretory tubules

	Na	Cl	K	Mg	Ca
			mmol l^{-1}		
NS (N=3, n=4)	114±13	174±16	44±12	6±2	2±2
S (N=5, n=5)	122±16	167±30	38±5	10±2	5±2

There were high Cl concentrations in both types of glands and a high Na concentration in actively secreting glands.

Within the lumina it is striking that Na and Cl concentrations do not approach those in the secreted fluid from the gland (800-1000 mmol l^{-1}).

4. DISCUSSION

The high intracellular Cl concentrations, higher than could reasonably be expected from an electrochemical equilibrium distribution, suggest some form of active Cl^- uptake. This would be consistent with a Na/2Cl/K cotransport system as postulated for the functionally and structurally similar elasmobranch rectal gland. The elevated intracellular Na concentration in secreting glands could be a consequence of increased Na^+ transport via the Na/2Cl/K cotransporter. Relatively high luminal K concentrations suggest that K^+ passes across the apical cell membrane.

The measured luminal Na^+ and Cl^- concentrations did not approach the concentration of these ions in the fluid which emerges from the gland. Calculated osmotic pressures show that the luminal fluids are close to being isosmotic with the cells or slightly hyposmotic. This may indicate that the initial secretion is isosmotic to the blood plasma and that it is modified during its passage through the duct system.

5. REFERENCES

Epstein H F and Silva P 1985 Anns New York Acad. Sciences **456** 187
Marshall A T (1987) In: Cryotechniques in Biological Electron Microscopy. R A Steinbrecht and K Zierold eds (Springer Verlag) pp 240-257
Marshall A T 1988 J. Electron Microsc. Techniques (in press)

Inst. Phys. Conf. Ser. No. 93: Volume 3, Chapter 22
Paper presented at EUREM 88, York, England, 1988

Analytical electron microscopy in biology

K Zierold
Max-Planck-Institut für Systemphysiologie, Rheinlanddamm 201,
4600 Dortmund, FRG

ABSTRACT: Analytical electron microscopy can identify, localize, and quantify mass, water, elements and in some instances also molecules in biological cells. The most information with regard to sensitivity and spatial resolution is provided by scanning transmission electron microscopy, X-ray microanalysis, and electron energy loss spectroscopy of cryosections. However, the reliability of the obtained data depends critically on the adequate specimen preparation which has to be adapted to the biological problem to be investigated and to the functional properties of the cells.

1. INTRODUCTION

Electron microscopy not only extends our knowledge of the structural organization of matter by 3 orders of magnitude beyond that achievable by light microscopy (from 0.3 μm to 0.3 nm), it may additionally provide information on the elemental and chemical composition. The measurement of the various electrons and X-rays generated by the interaction of the electron beam with atoms and molecules of the specimen is called analytical electron microscopy (AEM). Details concerning the physical basis and instrumentation of AEM are described e.g. by Hren et al (1981). Applications of AEM to biological specimens have been reviewed by Moreton (1981), Hall and Gupta (1983), Morgan (1985), Hall (1986), Jeanguillaume (1987), LeFurgey et al (1988). This paper discusses first the electrons and X-rays relevant for biological AEM, and then concentrates on the adequate specimen preparation necessary for the measurement of intracellular biomolecules, elements, mass, and water.

2. ELECTRONS AND X-RAYS

The different electrons and X-rays used as signals for biological AEM are illustrated in Fig. 1. The primary electron beam interacts with the atoms of the specimen: The secondary electrons emitted from atoms close to the surface are used in scanning electron microscopy (SEM) to image surface topography. The intensity of the backscattered electrons increases with the atomic number (Z) of the specimen atoms thus giving a rough estimate of the distribution of high Z elements. Deceleration of primary electrons in the specimen causes continuum X-radiation which comprises a measure of the mass thickness. Furthermore, the primary electrons excite atoms to emit X-rays of definite energy according to Moseley's law proportional to Z^2. The measurement of the generated energy spectrum is called energy dispersive X-ray microanalysis (XMA). This method allows to measure simultaneously the

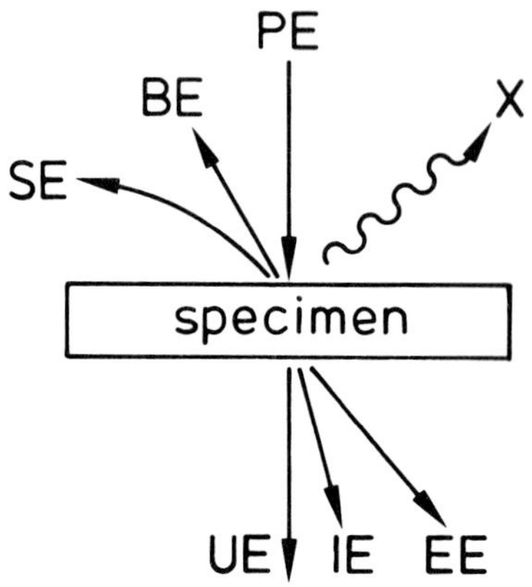

Fig. 1
Scheme of electron specimen interaction relevant for analytical electron microscopy in biology:
BE backscattered electrons
EE elastically scattered electrons
IE inelastically scattered electrons
PE primary electrons
SE secondary electrons
UE unscattered electrons
X X-rays.

content of all elements with Z higher than 10. The detection of lighter elements requires special detectors or a wavelength dispersive crystal spectrometer. If the specimen is thin enough, electrons penetrate either unscattered without any interaction, or scattered by atoms. The elastically scattered electrons are mainly responsible for the image obtained in transmission electron microscopy (TEM). In scanning transmission electron microscopy (STEM) these electrons determine the darkfield intensity which is a measure for the mass thickness of sufficiently thin specimen. The inelastically scattered electrons have lost energy at defined energy edges depending on the electron shell configuration of the atoms and molecules. The measurement of this effect is called electron energy loss spectroscopy (EELS). By this method in particular low Z elements and in some cases also organic molecules can be localized.

3. BIOMOLECULES

Water and organic molecules, in particular biomolecules such as nucleic acids, proteins, lipids, and carbohydrates are the main constituents of cells. As in conventional (not cryo) electron microscopes, water is incompatible with vacuum, preparation techniques for cells have concentrated on the stabilization of biomolecules by chemical fixatives such as aldehydes. The chemically fixed specimens are usually processed for TEM by embedding in plastic followed by ultramicrotomy. Specific molecules, for example enzymes, are localized by cytochemistry. The basic idea is to induce within the cell a chemical reaction of the molecule of interest with a substance to generate an electron dense reaction product which can be identified by TEM. Problems of this method are lack of specificity and stoichiometry of the expected chemical reactions. XMA can be a useful tool in identifying several electron dense reaction products by their element content (Sumner 1984). Higher specificity in localization is achieved by immunolabelling techniques (for an overview see e.g. Polak and Varndell 1984): Antigenic sites of biomolecules react with specific antibodies which are marked by colloidal gold particles of a size down to 5 nm in diameter. The gold-labelled molecules are localized by TEM either in Lowicryl^R plastic sections (Carlemalm et al 1986), cryofractures, or cryosections from chemically fixed cells (Boonstra et al 1987). Immunogold-labelled molecules on cell surfaces were localized by imaging of backscattered electrons in high resolution SEM (Walther and Müller 1986).

Isolated molecules can be identified without any labelling by EELS, as demonstrated for six nucleic acids by Isaacson and Crewe (1975). Leapman and Ornberg (1988) have shown that various biomolecules can be distinguished according to their elemental composition, as determined by EELS. However, this method has not yet been applied for identification of molecules within cells. The quantitative evaluation of energy loss spectra requires that they are caused by single scattering. Therefore, biological specimens for EELS of 100 kV electrons have to be thinner than 30 nm. Thus, single molecules, thin films, or very thin plastic sections are feasable samples, whereas cryosections are too thick and require deconvolution calculations to correct energy loss spectra for multiple scattering (Leapman and Ornberg 1988).

4. ELEMENTS

The stabilizing effect of chemical fixatives on biomolecules does not hold for single elements, in particular ions. Diffusible ions are washed out and even initially bound ions and elements may be dissolved and redistributed. Therefore, AEM of elements requires cryofixation without chemical pretreatment. Cryofixation techniques were reviewed extensively by e.g. Plattner and Bachmann (1982), Robards and Sleytr (1985), Sitte et al (1987). The main problem of cryofixation is that ice crystal growth during freezing causes segregation between water and organic material. For AEM the size of the ice crystal segregation compartments has to be smaller than the lateral analytical resolution of 2-5 μm in bulk specimens and less than 30 nm in 100 nm thick cryosections.

Frozen-hydrated specimens do not only allow to localize diffusible ions in aqueous compartments, the comparison of X-ray spectra obtained from the same place before and after freeze-drying provides physiologically important wet weight element concentrations (Gupta and Hall 1981). However, the analytical resolution of this method is limited to 500 nm in 1 μm thick sections and 3 μm in bulk specimens. AEM of frozen-hydrated 100 nm thick cryosections is hampered by poor contrast and mass loss due to radiation damage (Zierold 1984, 1986a). The problem of low contrast may be overcome by EELS imaging, but mass loss makes reproducible XMA in specimens thinner than 300 nm impossible. The mass loss can be neglected in sections thicker than 1 μm.

Freeze-drying improves considerably contrast and stability against radiation damage in cryosections. The element distribution appears to be well preserved, except for aqueous compartments with more than 90% water content where ions are precipitated randomly. The detection limit of energy dispersive XMA is about 10 mMol/kg dry weight for elements with Z higher than 12. For Na und Mg the detection limits are 30 and 20 mMol/kg dry weight, respectively. In 100 nm thick cryosections the lateral analytical resolution of XMA in STEM is less than 30 nm. Thus, 500 atoms can be detected, if they are concentrated within a volume of 100 x 50 x 50 nm^3 (Zierold 1986a).

The lateral analytical resolution can be improved by EELS. Recently, Shuman and Somlyo (1987) have reported the detection of 3 Ca atoms dispersed in organic matrix on carbon film by EELS. Unfortunately, cryosections are too thick for EELS. Therefore, 30 nm thick sections are prepared from cryofixed, freeze-dried, and plastic-embedded cells. Plastic embedding may alter ion gradients between intracellular compartments and element ratios within cells (Roos and Barnard 1985, Wroblewski and Wroblewski 1986). Therefore, this technique can be recommended only for the localization of elements tightly associated with structural components.

The use of freeze-substitution for analytical purposes is discussed controversely in the literature (see e.g. Harvey 1982, Wroblewski and Wroblewski 1986, Zierold and Steinbrecht 1987). Polar substitution media (methanol, acetone) should be avoided, as they may remove ions. Apolar substitution media (for example diethylether) could be tolerated for the analysis of bound elements which are not dissolved by the substitution process. Data obtained from freeze-substituted specimens should be considered with care, unless they are corroborated by AEM of cryosections.

5. MASS/WATER

Since electron microscopy images essentially the mass distribution of the specimen (Zeitler and Bahr 1962), mainly STEM can be used to measure the mass of biomolecules (see e.g. Engel 1982, Leapman et al. 1984, Wall and Hainfeld 1986). Halloran et al (1978) and Reichelt and Engel (1985) determined the mass distribution in plastic sections by STEM. Linders et al (1982) have compared the backscattered electron signal, the continuum X-radiation, and the transmitted electron signal for mass measurements and concluded that the transmission electron signal is the most reliable one in sufficiently thin sections. Egerton and Cheng (1987) measured the local thickness by EELS.

AEM is a useful tool to measure the intracellular distribution of water. The local water content is an important parameter to convert dry weight element concentrations obtained from freeze-dried cryosections by XMA to the physiologically more significant wet weight concentrations. This can be done for example by comparing X-ray spectra of cryosections in the frozen-hydrated and freeze-dried state (Gupta and Hall 1981). A similar method by EELS was reported by Leapman and Ornberg (1988). The disadvantage of these methods is that the contrast in the frozen-hydrated state is often too poor to identify intracellular compartments. Provided that cells consist essentially of water and organic material, the water content can also be determined by comparing the darkfield intensities of the 100 nm thick freeze-dried cryosection and of the film supporting the section (Zierold 1986b).

6. CONCLUSION AND OUTLOOK

The main preparation procedures for AEM of biological cells are summarized in Fig. 2. However, it should be kept in mind that biological cells are not static objects. The distribution of molecules, elements, and water in cells depends critically on their functional properties. Thus, the most challenging task for AEM in biology is the localization of substances in definite physiological states. This was demonstrated, for example, by the elucidation of calcium shifts in muscle (Wendt-Gallitelli and Wolburg 1984, Somlyo et al 1985). Thus, AEM of biological cells requires cryofixation methods adapted to the particular cell and to the physiological problem to be investigated.

REFERENCES

Boonstra J, van Maurik P and Verkleij A J 1987 in: Cryotechniques in Biological Electron Microscopy eds R A Steinbrecht, K Zierold (Berlin-Heidelberg: Springer-Verlag) pp 217-30

Carlemalm E, Villiger W, Acetarin J D and Kellenberger E 1985 in: The Science of Biological Specimen Preparation for Microscopy and Microana-

Fig. 2:
Methods to identify, localize and quantify biomolecules, elements, mass, and water in cells by analytical electron microscopy

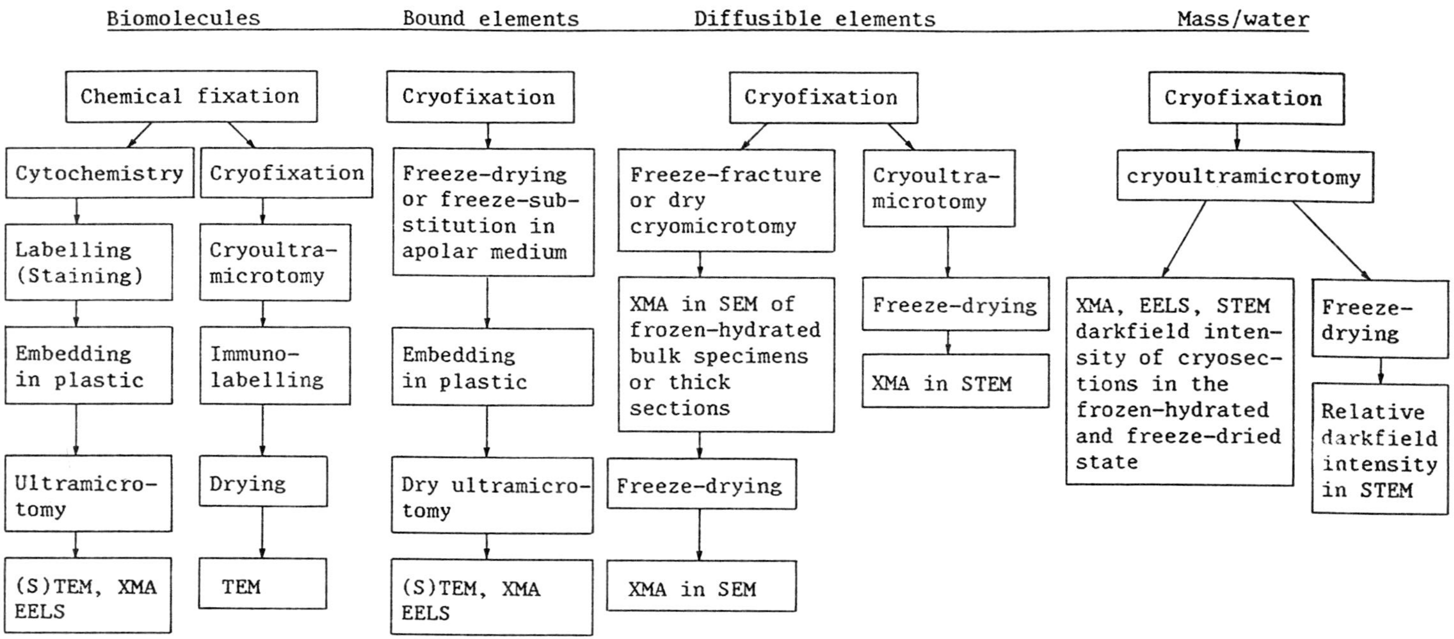

Abbreviations used:

EELS electron energy loss spectroscopy
SEM scanning electron microscopy
(S)TEM (scanning) transmission electron microscopy
XMA X-ray microanalysis

lysis 1985 eds M Müller, R P Becker, A Boyde, J J Wolosewick (AMF O'Hare, IL: SEM Inc.) pp 147-54
Egerton R F and Cheng S C 1987 Ultramicroscopy 21 231-44
Engel A 1982 Micron 13 425-36
Gupta B L and Hall T A 1981 Tissue Cell 13 623-43
Hall T A 1986 Micron & Microsc. Acta 17 91-100
Hall T A and Gupta B L 1983 Q. Rev. Biophys. 16 279-339
Halloran B P, Kirk R G and Spurr A R 1978 Ultramicroscopy 3 175-84
Harvey D M R 1982 J. Microsc. 127 209-21
Hren J J, Goldstein J I and Joy D C 1979 Introduction to Analytical Electron Microscopy (New York-London: Plenum Press)
Isaacson M S and Crewe A V 1975 Annu. Rev. Biophys. Bioeng. IV 165-84
Jeanguillaume C 1987 Scan. Microsc. 1 437-50
Leapman R D, Fiori C E and Swyt C R 1984 J. Microsc. 133 239-53
Leapman R D and Ornberg R L 1988 Ultramicroscopy 24 251-68
LeFurgey A, Bond M and Ingram P 1988 Ultramicroscopy 24 185-220
Linders P W J, Stols A L H, van de Vorstenbosch R A and Stadhouders A M 1982 Scan. Electron Microsc.1982/IV 1603-15
Moreton R B 1981 Biol. Rev. 56 409-61
Morgan A J 1985 X-ray Microanalysis in Electron Microscopy for Biologists (Oxford: Royal Microscopical Society)
Plattner H and Bachmann L 1982 Int. Rev. Cytol. 79 237-304
Polak J M and Varndell I M eds 1984 Immunolabelling for Electron Microscopy (Amsterdam-New York-Oxford: Elsevier)
Reichelt R and Engel A 1985 J. Microsc. Spectrosc. Electron. 10 491-98
Robards A W and Sleytr U B 1985 in: Practical Methods in Electron Microscopy ed A M Glauert (Amsterdam-New York-Oxford; Elsevier)
Roos N and Barnard T 1985 Ultramicroscopy 17 335-44
Shuman H and Somlyo A P (1987) Ultramicroscopy 21 23-32
Sitte H, Edelmann L and Neumann K 1987 in: Cryotechniques in Biological Electron Microscopy eds R A Steinbrecht, K Zierold (Berlin-Heidelberg: Springer-Verlag) 87-113
Somlyo A V, McClellan G, Gonzalez-Serratos H and Somlyo A P 1985 J. Biol. Chem. 260 6801-07
Sumner A T 1984 Scan. Electron Microsc. 1984/II 905-17
Wall J S and Hainfeld J F 1986 Ann. Rev. Biophys. Biophys. Chem. 15 355-76
Walther P and Müller M 1985 in: The Science of Biological Specimen Preparation for Microscopy and Microanalysis 1985 eds M Müller, R P Becker, A Boyde, J J Wolosewick (AMF O'Hare, IL: SEM Inc.) 195-201
Wendt-Gallitelli M F and Wolburg H 1984 J. Electron Microsc. Tech. 1 151-74
Wróblewski J and Wróblewski R 1986 J. Microsc. 142 351-62
Zeitler E and Bahr G F 1962 J. Appl. Phys. 33 847-53
Zierold K 1984 in: Electron Microscopy 1984, Vol 2 eds A Csanády, P Röhlich, D Szabo (Budapest: The Program Committee of the 8th Eur. Congr. on Electron Microscopy) 1397-1406
Zierold K 1986a in: The Science of Biological Specimen Preparation 1985 eds M Müller, R P Becker, A Boyde, J J Wolosewick (AMF O'Hare, IL: SEM Inc.) 119-27
Zierold K 1986b Scan. Electron Microsc. II 713-24
Zierold K and Steinbrecht R A 1987 in: Cryotechniques in Biological Electron Microscopy eds. R A Steinbrecht, K Zierold (Berlin-Heidelberg: Springer-Verlag) 272-82

Biological applications of EELS

C. Colliex

Laboratoire de Physique des Solides, Bât. 510, Université Paris-Sud, 91405 Orsay, France.

1. INTRODUCTION

International E.M. meetings offer at regular intervals unique opportunities for surveying the advances which have occured in a given field. For the present subject : "EELS in biology", the previous reference is the Hamburg conference, in the proceedings of which several reviews were devoted to rather close topics, i.e. by Somlyo et al. (1982), or by Egerton (1982). Meanwhile, important contributions have been published, such as those by Shuman et al., Colliex, Leapman and Ottensmeyer gathered in the special issue of Annals of the New-York Academy of Sciences (1986). Without being exhaustive, this introductory list of references must also quote the paper by Jeanguillaume (1987) which presents from a biologist's point of view an extensive compilation of all life sciences applications performed with EELS both on sections and on isolated molecules or organelles, and the very recent and comprehensive paper by Leapman and Ornberg (1988).

In the present contribution, I have selected, as a physicist, some major recent developments which bear characteristic marks for the present and future of EELS in biology. As in any selection of highlights, there is by human nature a subjective factor and I have decided to emphasize a few recent works (with the labels : quantification and imaging). Advances of great potential interest for biological specimens have been made in two main directions : improvements in concentration detection limits down to the range of elemental compositions of current interest in physiological studies (i.e. 1 to 100 mmol/kg or $\simeq$ 10 to 10^3 ppm) - Shuman and Somlyo (1987) - and improvements in spatial resolution and ultimate sensitivity down to single atom identification - Colliex and Mory (1988) -. The first achievement has been made possible by a combination of progress in detection efficiency and in spectral data processing techniques, because it involves high counting rates over a large number of energy loss channels. The second is the result of improved digital image acquisition techniques together with refined processing methods specifically adapted to sequences of energy filtered images : in this case, a moderate number of energy channels is involved but a large number of image pixels is necessary to visualize and identify the individual Tb and U subnanometer clusters.

II. PROGRESS IN SPECTROSCOPY

In order to obtain local elemental or chemical information from EELS spectroscopy, the first solution is to record and process spectra acquired from a selected area or with a point probe fixed or rastered over a given specimen feature.

Spatially resolved concentrations of Na, Cl, K, Ca, P, Fe between 10 and 1000 mmol/kg have been currently measured on thin biological samples with EDX techniques. With conventional EELS techniques (sequential spectrum acquisition and use of background stripping method with a power law model), most studies with standards exhibit a detectable concentration limit in the range 0.1 to 1 % at., with improved values when the edge of interest lies at lower energy losses than that of the matrix (i.e. Cl in C as compared to Ca in C).

A great jump forward has been made by Shuman et al. (1986), Shuman and Somlyo (1987). In their work concerning Ca detection with the L_{23} edge at 350 eV, in an organic matrix with a dominant C-K edge at 285 eV, typical signal/background ratios lie below 1/1000. A simple argument shows that total count numbers in excess of 10^7 are required if the detection is solely limited by statistical noise considerations. This can only be obtained within a reasonable experiment duration, when using a parallel detection system. A number of satisfactory devices (Shuman and Kruit (1985), Krivanek et al. (1987), Strauss et al. (1987), Egerton and Crozier (1987)), have been recently described. Quantification is made complex as a consequence of the non-uniformity of the detector response. Element to element gain fluctuations of 1 % seem quite common with the present state of technology. Specific methods have consequently been developed in order to average the detector spatial response by measuring a sequence of repositioned spectra.

The identification and measurement of very faint signals over intense background with eventual extended oscillations requires high counting rates over a large number of energy loss channels, and the appropriate software for processing these data. There has been a burst of new programs for EELS quantitative analysis. Difference or derivative techniques (e.g.top hat filters) have been implemented to reveal the presence of edges (Zaluzec (1985)). Different fitting processes applying both to the pre- and post- edge regions have been proposed as improvements with respect to the normal background stripping procedure (Steele et al. (1985), Bauer and Scholz (1987)). Algorithms have been developed for a more careful estimate of errors and uncertainties in background extrapolations (Trebbia (1988)). Finally, inspired by EDX quantitative software, new quantitation schemes are presently being introduced in EELS analysis. The basic idea is to simulate the experimental spectrum with a family of reference edges - instead of lines for EDX - acquired on standards ; the finite thickness influence is taken into account, by convoluting with the low energy loss spectrum (Shuman and Somlyo (1987), Leapman and Swyt (1988)). Concentrations are determined as the weighting factors of the different edges in the best least squares fit.

A few results are gathered on a graph (minimum concentration C_0, spatial resolution d, primary dose D) together with curves of equal number N of detected atoms, respectively 1 and 10 atoms, and equal signal to noise ratio SNR - see figure 1.

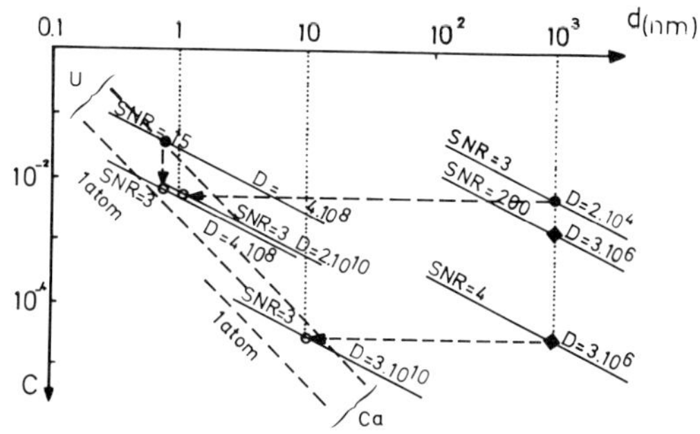

Fig. 1 Summary of achieved performance in terms of concentration and spatial resolution. Dose units are in e^-/nm^2; ● correspond to experiments achieved in Orsay with sequential detection (on Ca in araldite at 10^3 nm and on U on C at 0.7 nm; ◆ correspond to experiments by Shuman and al. on Ca in PVP ; o are reasonable extrapolations; —— are curves of equal SNR; - - - are curves of equal number of detected atoms (1 and 10 respectively) for the two investigated situations (U and Ca). See also Colliex (1985) and Castaing (1987) for discussion.

III. PROGRESS IN IMAGING

Two developments in instrumentation have contributed to the recent rapid expansion of energy filtering techniques on biological specimens. The first is the commercial availability of energy filters of the Castaing and Henry design, on a CTEM column - Zeiss EM 902. This has followed the impressive work of Ottensmeyer and collaborators (see for instance Ottensmeyer (1984)) who clearly demonstrated the impact of electron spectroscopic imaging for elemental mapping of various elements (P, Ca, Cl...) in sections or on particles (nucleosomes, ribosomal subunits...). The other factor of importance is the efficient coupling of digital recording and processing systems on STEM instruments, which have been specifically designed to handle families of images recorded through different detection channels (annular dark field, energy loss bright field...) - Mory et al. (1988). As a result, it has been possible to imagine new combinations of signals in order to enhance specific image contrasts, or to improve the quantitative character of elemental mapping, at an unprecedented level of spatial resolution.

With a filtering device, one defines a given energy window of variable width in the energy loss spectrum and records an image with this particular class of transmitted electrons. It is well known that, for a thin section (with $t/\Lambda_i \lesssim 1$), the zero loss image exhibits higher contrast and improved resolution, where Λ_i is the inelastic mean free path. On the other hand for thick biological specimens (i.e. with $3 \lesssim t/\Lambda_i \lesssim 6$), there have been several hints that the use of higher energy losses around the maximum ΔE_p of the energy loss distribution could produce better images. This great potential has now been demonstrated qualitatively with a filtering Zeiss microscope on 0.1 to 1.0 micrometer thick sections of embedded biological materials (Bauer et al. (1987). The limitation of the energy bandwith to < 20 eV clearly reduces the effects of the chromatic aberration of the objective lens, and the use of the maximum of the inelastic intensity distribution enhances the signal. With a dedicated STEM we have performed a quantitative study of the image characteristics (intensity, contrast, signal to noise ratio) as a function of the energy loss value used (Colliex et al. (1988)). Figure 2 shows a typical EELS spectrum for a ≃ 0.4 µm thick section of a chromosome together with the inelastic filtered image recorded in optimum conditions (i.e. with maximum SNR) ; it corresponds to an energy loss value slightly below that for the maximum of intensity. These results, together with observations of contrast reversal, are in good agreement with a proposed simple model of inelastic scattering in thick biological sections.

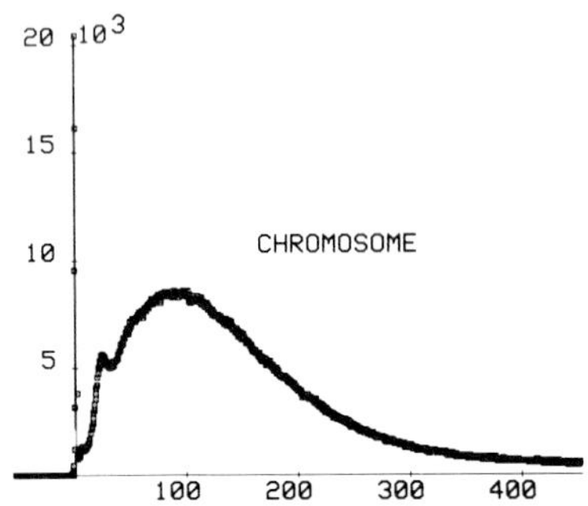

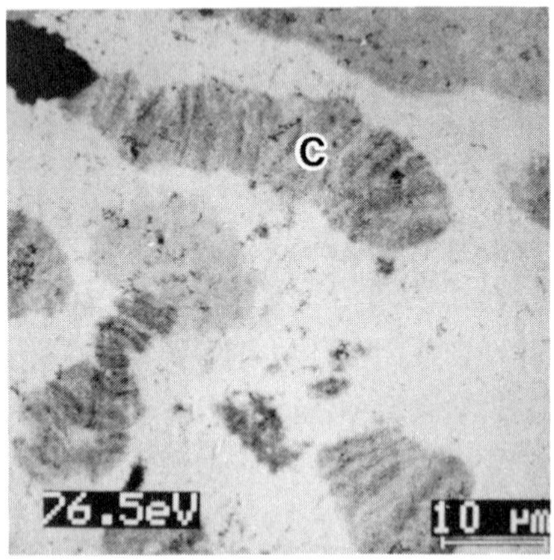

Fig. 2 Section of a salivary gland nucleus (≃ 0.4 µm thick).EELS spectrum recorded on the chromosome (C on the image) and inelastic BF image (with ΔE = 76.5 eV) corresponding to the best SNR. Specimen courtesy of A. and D. Olins.

In STEM, the availability of several simultaneous signals is also full of promise for ratio techniques which enhance the contrast of unstained sections but are only strictly valid for very thin specimens. Jeanguillaume and Tencé (1987) propose a new method, consisting of the mixing of three signals - unscattered, annular dark field and inelastic - which is really thickness independent and can provide an absolute measurement of the Λ_i/Λ_e ratio and of the collection efficiencies of the different signals.

Finally, great progress has also been made in the field of elemental mapping, through a combination of several energy filtered images connected to a core loss signal. The problem lies in the difficulty of estimating the background below the characteristic signal for each image pixel. From time acquisition considerations, the dwell time per pixel must remain short and consequently the numbers of counts involved are weak. Developments in processing image sequences for elemental mapping have greatly reduced the risks of extrapolation errors and of false positive and negative values in the resultant image (Bonnet et al. (1988)). The present level of performance of the Orsay STEM for elemental mapping is illustrated with a couple of examples. Fig. 3 corresponds to a typical nitrogen map in a biological section obtained from a sequence of 5 images at energy losses around the nitrogen K-edge at 400 eV. The image before the edge at 386 eV is representative of a carbon image with contrast reversal near the features of interest. Fig. 4 shows an uranium map of a subnanometer cluster used as a staining agent. With such test specimens, a subnanometer spatial resolution has been demonstrated on chemical maps with an energy loss of 100 eV and a detection sensitivity of one single U atom for a primary dose of 4×10^8 e^-/nm^2 (Colliex and Mory (1988)). These representative data are also inserted in the general graph of fig. 1. They are of great importance in the context of selective biochemical tagging of active sites.

As a conclusion, I wish to point out the wealth of extra information carried in the novel three dimensional spectrum images, provided by parallel detection EELS systems, as compared to conventional two dimensional micrographs. In spite of their enormous size in terms of memory content, it is now possible to imagine many original applications for imaging low elemental concentrations in complex biological structures (Jeanguillaume (1988)). Some of these developments may benefit from the use of multivariate statistical analysis as extended to sequences of energy filtered images by Bonnet and Hannequin (1988).

Acknowledgements

This review has been made possible through the daily work done at the Orsay STEM with the support of the CNRS multidisciplinary unit 120041. Thanks are due to all my colleagues who contributed to its development at various stages : N. Bonnet, C. Jeanguillaume, C. Mory, M. Tencé and P. Trebbia. It was a pleasure to collaborate on various aspects of the present work with E. Delain, E. Larras, A. Olins, D. Olins, C. Quintana, D. Ugarte.

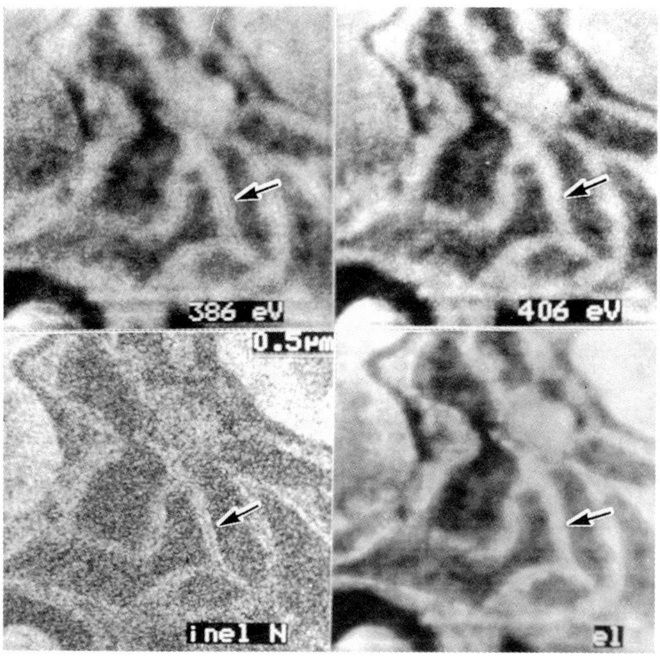

Fig. 3 Nitrogen characteristic map in a section of epithelial cell of a mouse thyroid gland. One image before the N-K edge at 386 eV and one after it at 406 eV have been selected from the sequence of five energy filtered images. The result of the process is inel N and the adf is shown for comparison. One recognizes the rough endoplasmic reticulum network. Specimen courtesy of E. Larras.

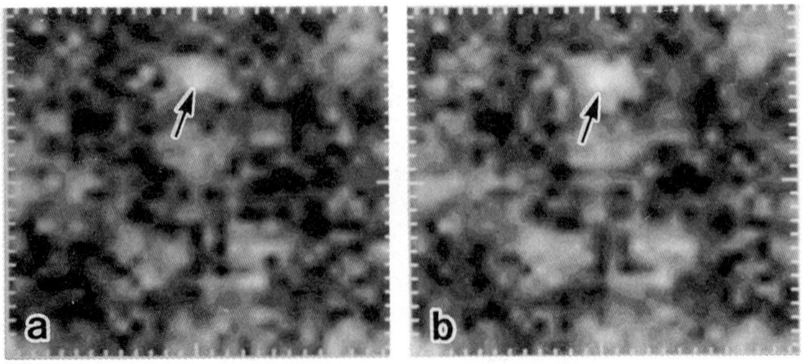

Fig. 4 Duplicated U maps of small cluster. Image size : 6.4×6.4 nm. The marked cluster corresponds to about 8 atoms with a SNR 15 and a dose of $4{\times}10^8 e/nm^2$*. Single atom identification is achievable with the same dose - see Fig. 1. Specimen courtesy of E. Delain.*

Réferences

Bauer HD and Scholz W 1987 Ultramicroscopy 23 109.
Bauer R et al. 1987 Optik 77 171.
Bonnet N et al 1988, Proc. 6th Pfefferkorn Conf. Niagara, SEM, in press.
Bonnet N and Hannequin P 1988 Proc. Aussois Workshop, in press.
Castaing R 1987 Analusis 15 517.
Colliex C 1985 Ultramicroscopy 18 131.
Colliex C 1986 Ann NY Acad. Sciences 483 311.
Colliex C and Mory C 1988 Proc. EMSA, in press.
Colliex C et al 1988 J. Microscopy, in press.
Egerton R 1982 Proc. 10th Int. Conf. E.M. Hamburg, I, 151.
Egerton R and Crozier 1987 J. Microscopy 148 157.
Jeanguillaume C 1987 Scanning Microscopy I 437.
Jeanguillaume C and Tencé M 1987 Ultramicroscopy 23 67.
Jeanguillaume C 1988, Proc Aussois Workshop, in press.
Krivanek OL et al 1987 Ultramicroscopy 22 103.
Leapman RD 1986 Ann. NY Acad Sciences 483 326.
Leapman RD and Ornberg RL 1988 Ultramicroscopy 24 251.
Leapman RD and Swyt CR 1988 Proc. Aussois Workshop, in press.
Mory C et al 1988, Proc. 6th Pfefferkorn Conf Niaga SEM in press.
Ottensmeyer FP 1986 Ann NY Acad Sciences 483 334.
Shuman H and Kruit P 1985 Rev Sci Instrum 56 231.
Shuman H et al 1986 Ann NY Acad Sciences 483 295.
Shuman H and Somlyo AP 1987 Ultramicroscopy 21 23.
Somlyo AP et al 1982 Proc. 10th Int Conf E.M. Hamburg I 143.
Steele JD et al 1985 Ultramicroscopy 17 273.
Strauss MG et al 1987 Ultramicroscopy 22 117.
Trebbia P 1988 Ultramicroscopy 24 399.
Zaluzec N 1985 Ultramicroscopy 18 185.

Inst. Phys. Conf. Ser. No. 93: Volume 3, Chapter 22
Paper presented at EUREM 88, York, England, 1988

Organic standards for quantitation of element composition in epiphyseal growth plates under vitamin D deficiency

R H BARCKHAUS*, J ALTHOFF*, H P BERTRAM**, F GREINKE, H J HÖHLING*

* Institute of Medical Physics, D-44oo Münster, FRG
**Institute of Pharmacology and Toxicology, D-44oo Münster, FRG

ABSTRACT: The preparation of organic Agar-Agar standards is described. These standards are used for quantitation of Na, Mg, P, S, Cl, K, and Ca in semithin sections of shock-frozen and freeze-dried epiphyseal growth plates of piglets suffering from Vitamin D Deficiency Rickets, type I. The organic standards are compared with internal standards, which are based on an average elemental composition of the bone mineral.

1. INTRODUCTION

One of the most important suppositions for accurate quantitative X-ray microanalysis of ultra- and semithin sections of biological tissue is a series of good standards. The choice of standards is therefore determined by specimen examined.

2. MATERIALS AND METHODS

The standards were prepared using Agar-Agar as organic matrix. The salts used were $NaH_2PO_4 \cdot H_2O$, $MgCl \cdot 6H_2O$, $MgSO_4 \cdot 7H_2O$, $CaCl_2 \cdot 2H_2O$ and KCl. - Agar-Agar solutions with salt concentrations of 25, 5o, 1oo and 2oo mmol/kg were used. Parallel to the biological samples 2 mm^3-blocks of the Agar-Agar standards were shock-frozen at 63 K, freeze-dried at 193 K, and then embedded under vacuum in Spurr's chlorine-free medium. For the determination of water-contents the standard blocks were lyophilized and afterwards prepared for chemical analyses (atomic absorption spectrophotometry, potentiometry, photometry, complexiometry). - The fully mineralized regions of trabecula bone which had been analyzed by quantitative microprobe and by chemical analyses (Althoff et al. 1982) were taken as internal standards with Ca = 3o% w/w and P = 15% w/w (Barckhaus et al. 1981). - For electron microscopical microprobe analyses 2 mm^3-blocks of proximal ulna growth plates were rapidly dissected, shock-frozen in liquid nitrogen near its melting point, freeze-dried and infiltrated with Spurr's chlorine-free medium. - Dry sections of the organic standards and of the tissue were mounted between folding grids, carbon coated and analyzed.

3. RESULTS

The organic Agar-Agar standards are easy to prepare and to handle. In scanning transmission electron microscopical micrograph filamentous structures are visible in a homogeneous groundsubstance. The X-ray spectra showed no contaminations. An example for the course of the graphical curves is given for Cl and Ca (Fig. 1). The concentrations of these elements in

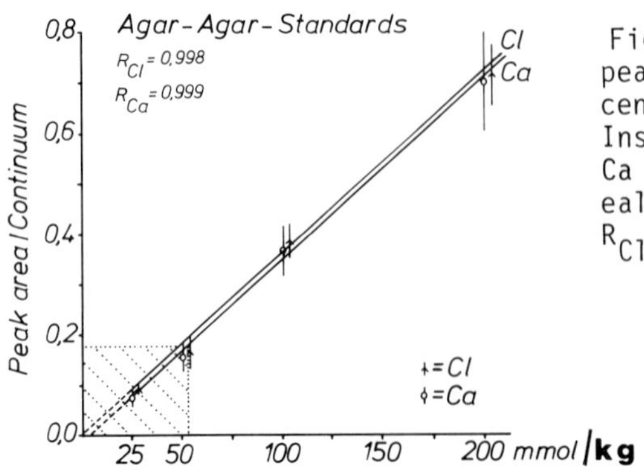

Fig. 1. Graphical presentation of peak-to-continuum ratios versus concentrations for chlorine and calcium. Insert: Concentration area of Cl and Ca in cartilage areas of the epiphyseal growth plate.
R_{Cl}, R_{Ca}: Correlation coefficients for Cl and Ca

the cartilage of epiphyseal growth plates are about 53 mmol/kg dry weight. The comparison of the internal standards with the organic standards reveals a sufficient agreement (Table 1).

	Control		Rickets	
Ca_{ic}	Int. St.	Org. St.	Int. St.	Org. St
GZ	1o,1 mmol/kg	13,3 mmol/kg	9,7 mmol/kg	12,7 mmol/kg
PZ	25,1 mmol/kg	28,5 mmol/kg	11,2 mmol/kg	14,6 mmol/kg
DZ	38,1 mmol/kg	41,2 mmol/kg	23,1 mmol/kg	26,5 mmol/kg
Ca_{ec}				
GZ	27,5 mmol/kg	31,o mmol/kg	4,1 mmol/kg	7,5 mmol/kg
LP	21,2 mmol/kg	22,6 mmol/kg	16,o mmol/kg	19,4 mmol/kg
HZ	3o,2 mmol/kg	33,4 mmol/kg	39,5 mmol/kg	42,8 mmol/kg
DZ	49,5 mmol/kg	52,8 mmol/kg	42,1 mmol/kg	45,6 mmol/kg

Table 1. Comparison of internal and organic standards. Top: Contents of intracellular Ca (Ca_{ic}); at the bottom: Contents of extracellular Ca (Ca_{ec}). Int. St: Internal Standard; Org. St.: Organic Standard; GZ: Germinative-, PZ: Proliferative-, LP: Lower Palisades-, HZ: Hypertrophic-, DZ: Degenerative-Zone. For the mmolar concentrations of the Int. St. a water-content of about 85% is assumed. The data are given in mmoles per kg wet weight.

4. REFERENCES

Althoff J, Quint P, Krefting E R, Höhling H J 1982 Histochemistry 74 541

Barckhaus R H, Krefting E R, Althoff J, Quint P, Höhling H J 1981 Cell Tissue Res 217 661

On the validity of block freeze-drying and vacuum resin embedding of cryoquenched tissues for quantitative intracellular x-ray microanalysis

H.Y. Elder, D.L. Bovell, J.D. Pediani, S.M. Wilson, S.A. McWilliams & D.McEwan Jenkinson*

Insititute of Physiology, University of Glasgow, Glasgow G12 8QQ and *Moredun Research Institute, Edinburgh EH17 7JH

ABSTRACT: Block freeze-drying, resin embedding and dry sectioning offers fidelity of intracellular elemental retention in biological specimens, better targetting of small tissue components, robust sections resistant to beam induced mass loss and can yield quantitative elemental data provided appropriate standards and corrections are applied.

1. INTRODUCTION

It is well established that rapid cryoquenching is the only way in which cellular inorganic elements can be retained in their *in vivo* locations and concentrations. Several options are available after cryofixation for tissue preparation prior to microanalysis but location of tiny target tissue sites within larger blocks presents formidable problems for most routes. We had 3% targetting success of sweat gland profiles when cryosectioning over sixty skin biopsies of 1mm width in which the gland is only about one hundredth of the total volume. In contrast, over 90% success in locating glands has been achieved with freeze-dried tissue blocks, vacuum resin embedded in non-polar resin and sectioned dry at room temperature. Tissue blocks prepared in this way can be stored dry and sectioned repeatedly until the target tissue is located. As with freeze dried cryosectioned tissues, extracellular elemental translocation occurs in freeze-dried resin-embedded tissues and various authors have raised the question of whether resin infiltration also causes intracellular translocation (Edelmann, 1986). Elements of biological interest, particularly Cl, have been reported in measurable quantities in non-polar resins. Small amounts of Na and K are detectable in some and Epon and Spurr's resin have high concentrations of Cl. The low Cl formulation of Spurr's resin (Pallaghy, 1973), gave unevenly and incompletely cured blocks. Araldite, which has no detectable Na, almost no K and little Cl has therfore been our resin of choice.

2. METHODS

An experiment was performed to assess whether elemental translocation would be a problem in quantitative measurement of intracellular electrolyte elements. Single drops of 500mM NaCl and KH_2PO_4 in 20% wt./vol. of 400kDa dextran solutions were allowed to fall from a height of 1m onto a LN_2-cooled Cu mirror, the surface of which was kept free of condensing water vapour by its location 1cm below the rim of the dewar. Frozen flakes were then laid together and freeze-dried in contact with each other for three days at -80°C, followed by a slow warm up (approx. 1°C per 6 min.) and vacuum embedded in degassed resin (Elder, *et al.*, 1986). Blocks were cured with the flakes still in contact and dry ultrathin (100 to 150nm) sections were cut across the interface. Spectra from rows of static probes (100s live time and 0.4nA) were taken across the flakes and interface. Ice crystal artefacts were always present; spectra were taken from regions with only small ice crystal damage and the probe diameter was several times that of the ice crystals. The blocks were stored desiccated for three years and the analysis repeated.

3. RESULTS and DISCUSSION

The data in Fig 1., from a block after storage, are relative mass fractions (means ± s.e., n = 3), calculated by the continuum normalization method (Hall, 1971). The difference in elemental concentrations between edge and more central locations of the NaCl flake in particular could be due to ice crystal damage in the latter. Such flakes could not be used as standards. Uniform results and freedom from ice crystal damage are obtained with aminoplastic standards (Roos & Barnard, 1984). The measured Cl data contain a contribution from the Araldite. Therefore the Cl content of pure resin was measured and subtracted from that of the resin in the cleft between the flakes. A block with a flake containing no Cl was analysed and the ratio of the Cl signal in the embedded flake to the pure resin gave a consistent value of 77% for the flake resin content and was used to calculate the corrected Cl content of the flakes in Fig. 1. Data in Fig. 1 constitute a "worst case" situation since no diffusion barrier equivalent to the plasma membrane exists to retain intracellular salts. There is no or negligible diffusion of P, K or Na between the flakes (Fig.1.). Concern that resin wash, or water formed during resin polymerization, may cause elemental translocation (Pallaghy, 1973), seem unfounded. The experiment is not definitive but even with the derived Cl values the apparent movement is minor and, particularly for comparative studies, we consider that the block freeze-dry, resin embedding and dry sectioning route is valid for quantitative intracellular microanalysis. In several recent sweat gland studies we have obtained consistent results by this route (McWilliams, *et al.*, 1987,1988; Bovell *et al.*, 1988; Wilson, *et al.*, 1988). Comparison of resin sections analysed at -150°C and at room temperature (r.t.) showed negligible increase in beam-induced mass loss at r.t. (McWilliams, unpublished).

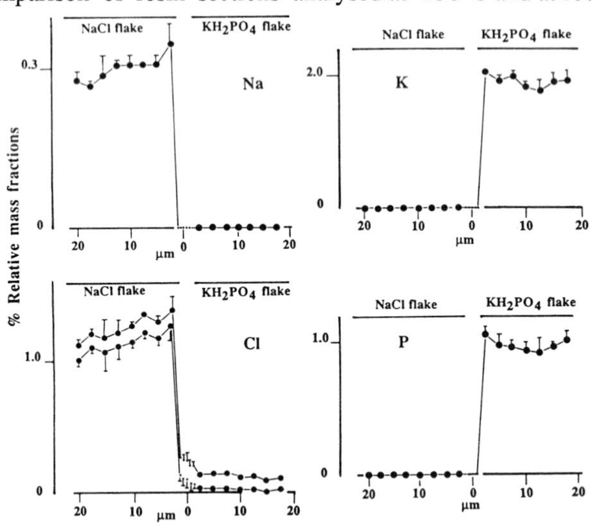

Fig.1. Concentration profiles of Na, K, P & Cl in co-embedded dextran /salt flakes and the interface between (see text). Standard errors not shown lie within the dot size. Measured and corrected Cl values are shown.

REFERENCES

Bovell, D.L., Jenkinson, D.McEwan, Elder, H.Y. & Wilson, S.M. (1988). *Res. Vet. Sci.*, in press.

Edelmann, L. (1986). *Scan. Electron. Microsc./1986/IV*, 1337-1356.

Elder, H.Y., Biddlecombe, W.H., Tetley, L., Wilson, S.M. and Jenkinson, D. McEwan. (1986). *EMSA Bull.*, **16**, 111-113.

Hall, T.A. 1971. In: *Physical techniques in biological research*. 2nd. Edn. (ed G. Oster) Vol. 1A - Optical Techniques. pp 157-275. Acad. Press, N.Y., Lond.

McWilliams, S.A., Montgomery, I., Jenkinson, D.McEwan, Elder, H.Y., Wilson, S.M. & Sutton. A.M. (1987). *Brit. J. Dermatol.*, **117**, 617-626.

McWilliams, S.A., Montgomery, I., Elder, H.Y., Jenkinson, D.McEwan & Wilson, S.M. (1988). *Tissue & Cell*, **20**, in press.

Pallaghy, C.K. (1973). *Aust. J. biol. Sci.*, **26**, 1015-1034.

Roos, N. & Barnard, T. (1984). *Ultramicroscopy*, **15**, 277-286.

Wilson, S.M., Elder, H.Y., Jenkinson, D.McEwan & McWilliams, S.A. (1988). *J. exp. Biol.*, in press.

Electron energy loss spectroscopy in ultrastructural cytochemistry

JB van Dort*, GAM Ketelaars*, T. de Jong, WC de Bruijn,

AEM unit, Clin. Pathol. Inst. Erasmus University, PO Box 1738, Rotterdam
* Hoge School West-Brabant, Lab Onderwijs. Etten Leur, The Netherlands.

ABSTRACT: Cytochemical reaction products, barium sulphate and cerium phosphate, from two lysosomal enzyme reactions present in aldehyde-fixed mouse peritoneal macrophages will be differentiated by E.E.L.S. analysis.

1. INTRODUCTION

The combination of electron energy loss spectrometer and transmission microscope (Zeiss E.M. 902) allows elemental differentiation by electron energy loss populations in relation to elastically-scattered electron populations for morphology. In one resident macrophage, localization of three enzyme reaction products were differentiated by E.E.L.S.:
(1). Peroxidase activity by di amino benzidine (DAB) visualized by Osmium tetroxide plus ferrocyanide. (2). Acid phosphatase by cerium phosphate. (3). Aryl sulphatase by barium sulphate. Reaction sequence: Peroxidase/Acid Phosphatase/Aryl Sulphatase/OsO_4 plus $K_4Fe(CN_6)$.

2. RESULTS

In Fig. 1. the lysosomal morphology is shown of an untreated ultrathin section visualized by E.E.S.I (elastically scattered electron population from which the majority of inelastically scattered electron is removed). The resident macrophage is characterized by the peroxidase/DAB/Os reaction product in nuclear envelope and R.E.R. Fine granular reaction products are present in two lysosomes. There is no differentiation between the precipitates. In Fig. 2 an E.E.L.S. image at 781- 801 eV is shown by which one of the precipitates can be identified as barium. In the lysosome containing the reaction product, the precipitate appears as bright granules against a dark cellular background. In Fig. 3 an E.E.L.S. image at 883-903 eV is shown, by which the cerium precipitate in the other lysosome is identified. In this case the bright cerium containing granules appear against the dark cellular background. The sites were barium was detected before have not yet acquired complete darkness. It has been noticed that a fine barium-containing spiculate precipitate is present dispersed through the whole cytoplasm, which remained undetected before. This precipitate is, for the time being, considered to be an artifact.
In Fig. 4. two spectra are compiled from both precipitates, showing the presence of the barium peak and the absence of the cerium peak in one, and the reverse situation in the other. The presence of osmium in the R.E.R. and the nuclear envelope could not be detected by E.E.L.S. in our instrument, in which the energy range is restricted to 2000 eV.

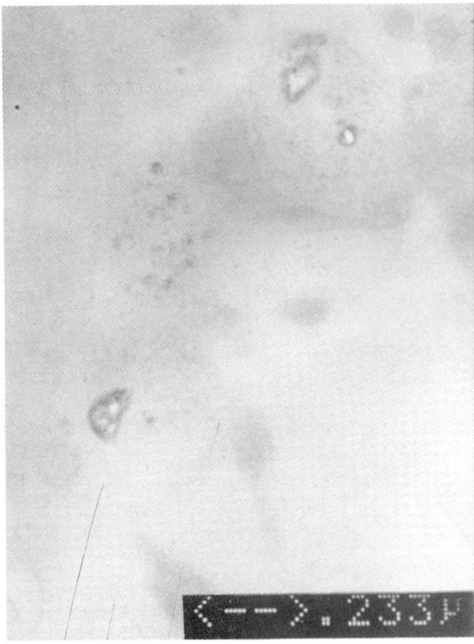

Fig. 1. Elastically scattered electron image.

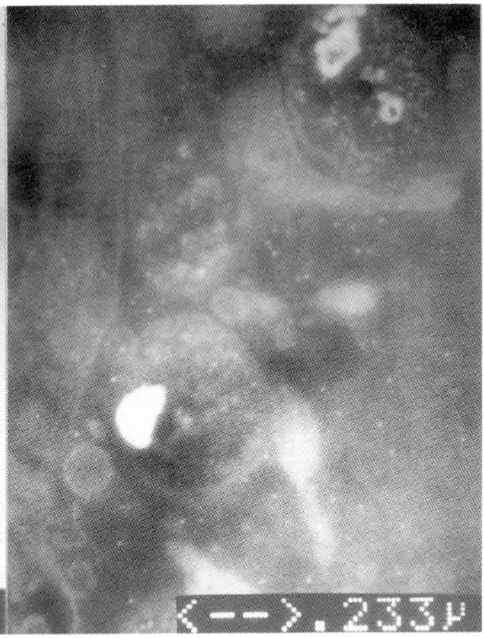

Fig. 2. Barium absorption edge image of about the same area.

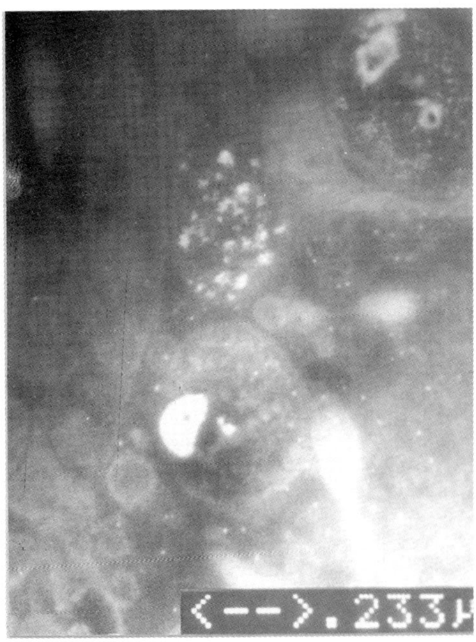

Fig. 3. Cerium absorption edge image of the same area as in Fig. 2.

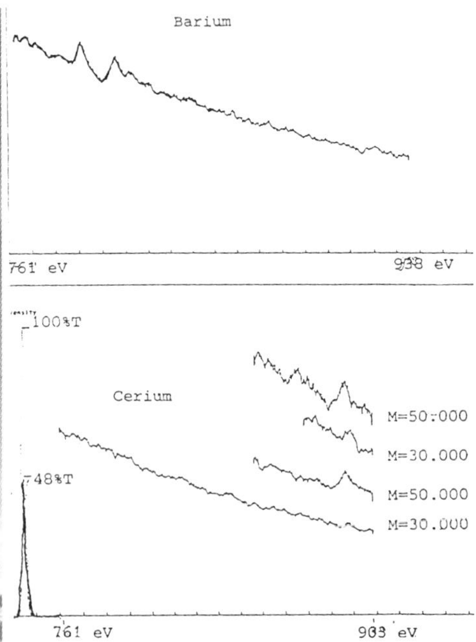

Fig. 4. Ce and Ba spectra from granular precipitates in Fig. 2-3.

Microprobe analyses on the epiphyseal growth plates and histochemistry of various intestinal enzymes in piglets suffering from vitamin D deficiency rickets, type 1

R H Barckhaus*, F Greinke*, J Harmeyer**, R Kaune**, R Gossrau***, H J Höhling*

*Institute of Medical Physics, D-4400 Münster, **Institute of Physiology, D-3000 Hannover, ***Department of Anatomy, D-1000 Berlin 33, FRG

ABSTRACT: Histochemical studies on the activities of various enterocytic enzymes were carried out with intestinal mucosal preparations and intra-/extracellular microprobe analyses on epiphyseal growth plates from piglets of the "Hannover Pig Strain" which suffer from pseudo D deficiency rickets, type I.

1. INTRODUCTION

The purpose of this study was to determine the intra-/extracellular element distribution in epiphyseal growth plates and to characterize histochemically activities of intestinal enzymes in piglets from the "Hannover Pig Strain" with vitamin D dependency rickets, type I. Results were compared with non rachitic control piglets. The rachitic piglets suffer from a hereditary defect of the renal 25-hydroxycholecalciferol-1-hydroxylase which converts $25(OH)_2D_3$ into $1,25\text{-}(OH)_2D_3$. The renal failure of $1,25\text{-}(OH)_2D_3$ production results in unphysiologically low concentrations of $1,25\text{-}(OH)_2D_3$ in plasma and symptoms of florid rickets at in age of 2.5 to 6.0 weeks. Rachitic symptoms are preceded by a drop of plasma Ca and P and a concomitant raise of alkaline phosphatase activity and results in death of the animals when remaining untreated (Kaune and Harmeyer 1987).

2. MATERIALS AND METHODS

For electronmicroscopical microprobe analyses $2mm^3$-blocks of proximal ulna growth plates were rapidly dissected, shock-frozen in liquid nitrogen near its melting point, freeze-dried and infiltrated with Spurr's chlorine-free medium. Sections of 1 µm were prepared, mounted on suitable holders and used for the microprobe analyses. Histology was examined with the scanning transmission mode of a Philips 301EM. Microscans of intra- and extracellular regions of the growth plate were selected from pictures present on the monitor. The induced x-ray spectrum was registered in the energy dispersive mode with an EDAX Si(Li)-detector system and quantified by use of organic and internal standards. For histochemistry 1o µm cryostat sections were prepared from intestinal mucosal preparations frozen in liquid nitrogen and stained for various hydrolases with corresponding methods of Lojda et al. (1979).

3. RESULTS

The microprobe analyses revealed a lower concentration of Ca, Na, K, and S in growth plates of rachitic piglets and an increase of Cl compared to controls. Phosphorus was lower in the intracellular regions of the rachitic growth plate. An intra- and extracellular reduction of Ca was present in all zones of the rachitic growth plate except for the hypertrophic and mineralizing zones in which intracellular Ca was similar to that of controls (Fig. 1). A decline of P content was measured in rachitic plates in metaphyseal direction which was mainly caused by a drop of intracellular P. This drop is interpreted as a loss of energy rich phosphates usually necessary for energy requiring transport processes. A decline of extracellular sulfur was found in all zones of the rachitic growth plate which possibly reflects a loss of sulfated proteoglycans usually capable to bind Ca. The observed changes in extra-/intracellular mineral content of the rachitic epiphyseal plate (Fig. 1) help to understand the functional disturbances in rachitic growth plate mineralization. - The histochemical studies

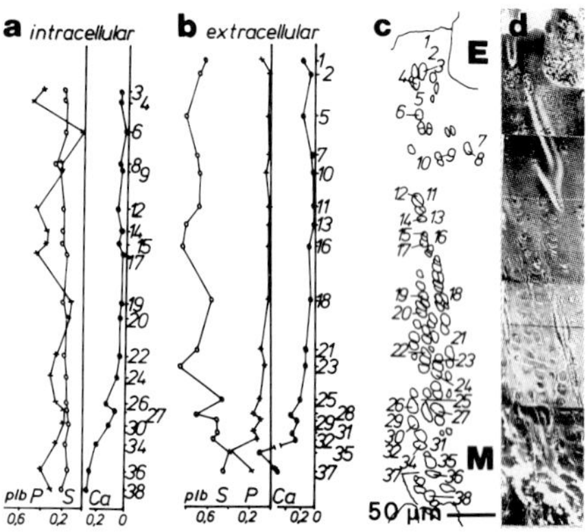

Fig. 1. Electron probe x-ray microanalyses of the intra- (a) and extracellular (b) Ca, P, and S distribution in the growth plate (c,d) of rachitic piglets. p/b: peak/background; E: epiphysis; M: metaphysis

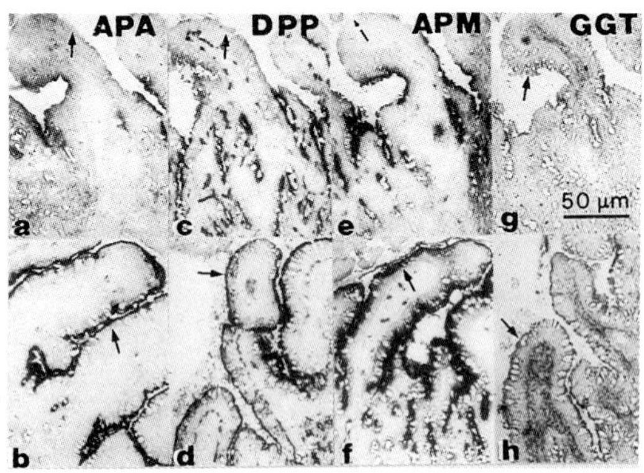

Fig. 2. Top (a,c,e,g): Acticities of microvillous hydrolases are decreased (arrows) in the intestinal enterocytes of piglets with hereditary vitamin D dependency rickets (arrows). At the bottom (b, d,f,h): Normal activities (arrows) of control animals. APA: aminopeptidase A; DPP: dipeptidylaminopeptidase IV; APM: aminopeptidase M; γGT: ß-N-acetyl-D-glucosaminidase

revealed that, compared with control animals, the activities of microvillous (aminopeptidase M and A, γ-glutamyltranspeptidase, dipeptidylaminopeptidase IV, lactase) (Fig. 2) and of lysosomal (ß-N-acetyl-D-glucosaminidase) hydrolases were decreased in the intestinal enterocytes of piglets with hereditary vitamin D dependency rickets.

4. REFERENCES

Kaune R and Harmeyer J 1987 Berl. Münch. Tierärztl. Wschr. 1oo 6

Lojda Z, Gossrau R and Schiebler T H 1979 Enzyme Histochemistry - A Laboratory Manual (Berlin Heidelberg New York: Springer)

Inst. Phys. Conf. Ser. No. 93: Volume 3, Chapter 22
Paper presented at EUREM 88, York, England, 1988

Electronprobe x-ray microanalysis of epiphyseal growth plate cartilage: calcium and potassium distribution at the mineralization front

ER Krefting

Inst. Med. Phys., Schmeddingstr. 50, D-4400 Münster, Germany (FRG)

ABSTRACT: Using a wavelength dispersive system the elemental distribution maps of Ca and K were received at the mineralization front of the epiphyseal growth plate. The tissue was prepared by cryomethods. Well preserved tissue demonstrated a high intracellular K and low Ca content and no variation in the intracellular Ca or K content in the vicinity of the mineralization front where the extracellular Ca content increases intensively. The cells seem to survive the extracellular mineralization and do not release Ca to initiate the mineralization.

It was assumed (e.g. Wuthier 1982) that the chondrocytes accumulate Ca in the columnar zone and release it in the lower hypertrophic zone to initiate the mineralization. It was demonstrated (Krefting 1987 a,b, Krefting et al. 1988) that in the columnar zone (and most likely also in the upper hypertrophic zone) a high intracellular Ca and low K content is a preparation artifact. First results concerning the Ca and K content in close vicinity to the mineralization front are given in this paper.

Growth plates of young rats were rapidly excised and frozen in liquid propane at about -190°C. 4 µm thick sections were cut in a cryostate (knife and specimen holder cooled to -45°C, chamber cooled to -35°C), transferred onto specimen holders, freeze dried and coated with carbon. The wavelength dispersive systems of a CAMECA microprobe MS46 were used to get the elemental distribution maps.

Fig. 1 is taken on a relatively well preserved tissue in the region of the mineralization front. The mineralization (demonstrated by an increase of the Ca content) starts extracellularely not in close vicinity to the cells but at sites where the distance between the cells is relatively large (arrow 1).

Even in the mineralized region (high extracellular Ca content) the intracellular Ca content remains low and that of K high (Fig. 1, arrow 2). In addition, the extracellular K content remains low. It must be concluded that the cell membrane maintains a relatively high gradient. This means that the cells were still alife in the tissue and that there was no diffusion of elements across the cell membrane in the course of the preparation. According to Shapiro and Boyde (1987) the cells in this region of the growth plate are well supplied by nearby blood vessels and it is likely that the cells do not die but survive the mineralization begin. It must be mentioned, however, that often tissue was analyzed which demonstrated a high intracellular Ca and low K content; this tissue must have been damaged during the preparation (Krefting 1987 a,b).

It must be concluded, that (1) the cells are still alife in a region with extracellular mineralization and that (2) the mineralization is not initiated by a release of intracellular Ca, as there is no indication of an intracellular Ca accumulation before the mineralization front or of a change of the intracellular Ca content across the mineralization front.

Krefting ER 1987a 11th Int Congr X-ray optics and microanalysis ed JD Brown and RH Packwood (London Canada) pp 298-302
Krefting ER 1987b Beitr elektronenmikroskop Direktabb Oberfl. 2o 239-244
Krefting ER, Höhling HJ, Felsmann M and Richter KD 1988 Histochemistry 88 321-326
Shapiro IM and Boyde A 1987 Scanning Microsc. 1 599-606
Wuthier RE 1982 Clin Orthop Rel Res 168 219-242

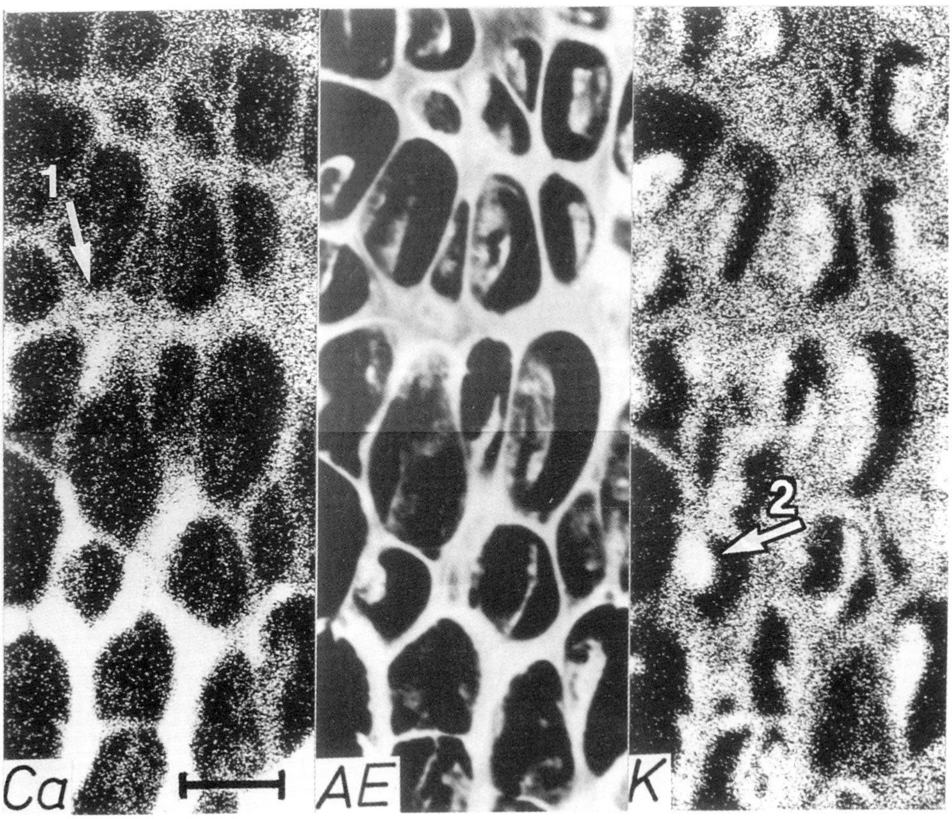

Fig. 1. Epiphyseal growth plate, rat. Mineralization front. Bar=20 µm. AE=Micrograph taken with the absorbed electrons, Ca=calcium distribution map, K-distribution map of potassium. Arrow 1: The extracellular mineralization (increase of the Ca content) begins in some distande to the cells. Arrow 2: Intracellular K content remains high and that of Ca low, even in the mineralized region.

The effect of low pH on the coniform mitochondrial rich ion transporting cell (MRITC) of Plecoptera

I R Bricknell and W T W Potts

Institute of Environmental and Biological Sciences, University of Lancaster, Bailrigg, Lancaster LA1 4YQ, U K

Plecoptera nymphs have three types of ion transporting cell in their cuticles, the coniform, caviform and bulbiform cell. The distribution of cell types is related to the pH of the water the species has to live in. Caviform and bulbiform cells are only found on nymphs resistant to low pH, while the coniform cell is found on both acid resistant and acid sensitive animals. If the coniform cell is present on acid resistant species another cell type is always present. Three relatively common British species *Perla bipunctata*, *Diura bicaudata* and *Perlodies microcephala* only have coniform MRITCS and are acid sensitive.

These large plecoptera nymphs are important in the food chain, supplying most of the protein and mineral requirements to aquatic birds such as dippers during the breeding season and both the nymphs and adults make up a large part of the diet of trout.

Animals for this work were collected from the River Lune at Hornby, Lancashire; and from the River Swale, North Yorkshire, both well buffered neutral water rivers. The nymphs were transported back to the laboratory in their natural water below 10°C to prevent final instar nymphs hatching and in the summer to prevent heat stress and death. In the laboratory they were maintained at 3-4°C in running water from their natural habitat.

The concentration of intracellular ions within the caviform cell under normal and acid conditions, was measured using low temperature SEM and X-ray microprobe analysis. The tissues containing the caviform cells were slam frozen on copper blocks cooled in liquid N_2, coated with 22 nm of Al under high vacuum in a Hexland cryochamber. Analysed in a JEOL JSM840S SEM at 15 kv with a probe current of 0.5 nA. The analysis was stopped when 80000 counts of aluminium was reached. Concentrations were calculated using standard curves obtained from 20% gelatine standards subjected to identical analysis conditions.

At pH 7 the coniform cells contained the following concentrations of ions; Sodium 30±6 mMol/l, Potassium 37±5 mMol/l, Phosphorous 57±7 mMol/l and Chloride 42±3 mMol/l. At pH 4 the concentration of intracellular ions rose to; Sodium 51±2 mMol/l, Potassium 77±1 mMol/l, Phosphorous 149±5 (Graphs 1 & 2). However after 3 days at this pH the levels of Sodium and Chloride dropped below detection while Potassium fell to 55±2 mMol/l and Phosphorous to 80±3 mMol/l (Graphs 3 & 4). This suggests that these insects have a short term response to acid episodes in the natural environment and will not survive long term acidification due to H^+ ions inhibiting the Na pump in the coniform MRITC. The high intra-cellular concentration of sodium chloride imply the presence of ion pumps at the apical membrane of the cell (see plate) in addition to the Na-K pump on the basal membrane.

X-ray microanalysis of murine gut epithelia infected with EDIM rotavirus

A J Spencer S J Haddon M P Osborne W G Starkey[1] J Collins[1]
T S Wallis[1] G J Clarke[1] K J Worton[1] & J Stephen[1]

Departments of Physiology and Microbiology[1], University of Birmingham, P.O.Box 363, Birmingham, B15 2TT.

Intracellular concentrations of elements were measured by X-ray microanalysis in gut epithelia of mice infected with EDIM rotavirus. Although infection was limited to villus tip regions in the small intestine, significant changes in concentrations of Na and Cl were observed in both villus tip and villus base cells. A new concept of secretion in diarrhoea is proposed: stimulation of villus base cells to rapid cell division is accompanied by transient accumulation of Na and Cl, the secretion of which into the lumen is the driving force for hypersecretion.

1. INTRODUCTION

Studies of Epizootic Diarrhoea of Infant Mice (EDIM) infection (an excellent model for human rotavirus gastroenteritis), have shown that an early virus-induced localised ischaemia occurs, leading to a dramatic transient shortening of villi (Osborne *et al*, 1988) and predictably to a loss of villus absorptive function. However, diarrhoea is not due to lactose intolerance (Collins *et al*, 1988). A dramatic reconstruction of villi occurs, involving rapid division of cells in villus base regions (Osborne *et al*, 1988). Here, we present evidence - obtained by X-ray microanalysis - which suggests that cell division, required to replace cells lost from upper parts of the villi, may contribute to diarrhoea.

2. METHODS

Samples of middle small intestine (MSI) from EDIM-infected neonatal mice (24-168h post infection [pi]) and age-matched controls were plunge-frozen in liquid propane. Cryosections (100-300nm thick) were mounted onto Ti grids, freeze dried and analysed in a JEOL 120CXII TEMSCAN fitted with a LaB_6 filament and a Link Systems 860 series 30mm² Si[Li] detector. Spectra were quantified using gelatin/inorganic salt standards.

3. RESULTS AND CONCLUSIONS

Concentrations of Na and Cl in cytoplasm in tip and villus base regions of MSI of EDIM rotavirus-infected mice, are plotted with age-matched controls in Figures 1 and 2.

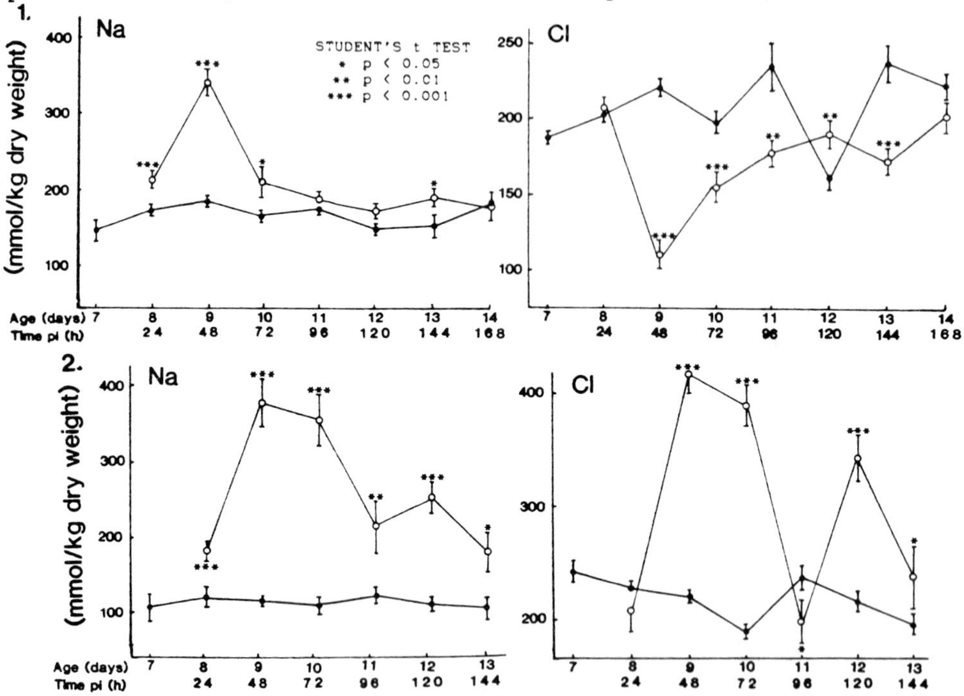

Figs. 1 and 2. Concentrations of Na and Cl in cytoplasm of villus tip cells (Fig. 1) and villus base cells (Fig. 2) of control (●-●) and EDIM-infected (○-○) mice. Mean ± SEM.

Changes in [Na] and [Cl] in villus tip cells are thought to arise from imbalancing of Na^+/H^+ and Cl^-/HCO_3^- coupled antiports due to anoxia accompanying EDIM-induced ischaemia. [Na] and [Cl] were increased in villus base cells when thymidine kinase activity was elevated (Collins *et al.*, 1988) and extensive cell proliferation occurring (Osborne *et al.*, 1988). Concentrations of Cl^- in luminal contents were increased by 3-fold at 72h pi. We propose that villus base cells, stimulated into rapid cell division by loss of mature enterocytes, accumulate high intracellular ion concentrations, which could provide an osmotically-assisted mechanism for cell volume expansion and thereby aid division. The considerable excess quantity of ions beyond that required for this process are then secreted into the lumen, to provide the osmotic basis for hypersecretion of water.

The financial support of Janssen Pharmaceutical Ltd. is gratefully acknowledged.

4. REFERENCES

Collins J, Starkey W G, Wallis T S, *et al* (1988) J. Pediatr. Gastroenterol. Nutr., in press.

Osborne M P, Haddon S J, Spencer A J, *et al* (1988) J. Pediatr. Gastroenterol. Nutr., in press.

Inst. Phys. Conf. Ser. No. 93: Volume 3, Chapter 22
Paper presented at EUREM 88, York, England, 1988

X-ray microanalysis of mouse knee meniscus

LAWTON DM: BARTLEY CJ: GARDNER DL.

Department of Histopathology, University Hospital of South Manchester
Nell Lane, Withington, M20 8LR

ABSTRACT: Dense deposits of a calcium-containing material were found in the anterior part of the knee joint meniscus in 305-day old CBA mouse. X-ray microanalysis established mean P/Ca ratios of 0.42. The results strongly suggest that mouse menisci, which do not contain bone, have a high content of hydroxyapatite, a basic calcium phosphate.

1. INTRODUCTION

The fibrocartilaginous knee joint meniscus acts as an intermediary, collagen-rich, movable 'washer' between the main femoral and tibial bearing surfaces. It is readily injured leading to impaired joint function and a susceptibility to osteoarthrosis (OA). In the course of systematic EM studies of mouse OA, we used normal CBA mice as controls and made observations on their joints by microscopy: in addition, we surveyed undecalcified material as part of a routine approach to synovial joints. The mouse knee joint is small and quickly penetrated by decalcifying agents and fixatives. As a result the well preserved components can be examined in their correct relative locations. The advantages of this approach are exemplified in this report by the demonstration of large differences anteroposteriorly in the calcification of the meniscus.

2. MATERIALS AND METHODS

Six 305-day old male CBA mice (Harlan Olac) were killed by a sharp blow to the back of the head and dislocation of the neck. Hind limbs were dissected; right knees were fixed in buffered formalin, decalcified and embedded in wax; left knees were fixed in 95% ethanol for 48 hr, dehydrated in 100% ethanol and embedded in LR White resin. Resin sections, 4 μm thick, were cut on an LKB Polycut microtome and stained with Alcoholic Toluidine Blue, Toluidine Blue, Safranin O, Picrosirius red, H & E, von Kossa and Alizarin Red. The cut surface of a block showing optimal specimen orientation was photographed with an Olympus PMT-35 macro unit (fig. 1). The same block was investigated by X-ray microanalysis in a Philips 505 scanning electron microscope fitted with an energy dispersive X-ray (EDAX) spectrometer, EDAX 9100/60 analyzer and ZAF bulk analysis programme. Thirteen sites in the joint, and pure resin outside, were selected for analysis using an accelerating voltage of 20Kv and live time of 120 sec.

3. RESULTS

Variations in the en bloc image of the meniscus (fig. 1) immediately suggested structural variability. The anterior part of the meniscus (AM), white towards its apex showed a sharp transition to grey. The posterior part of the meniscus (PM) was brown with white flecks. Light microscopy with the von Kossa technique and Alizarin Red demonstrated calcium throughout the AM and PM except for a superficial zone in which collagen and proteoglycans alone were evident (Picrosirius Red and Safranin O).

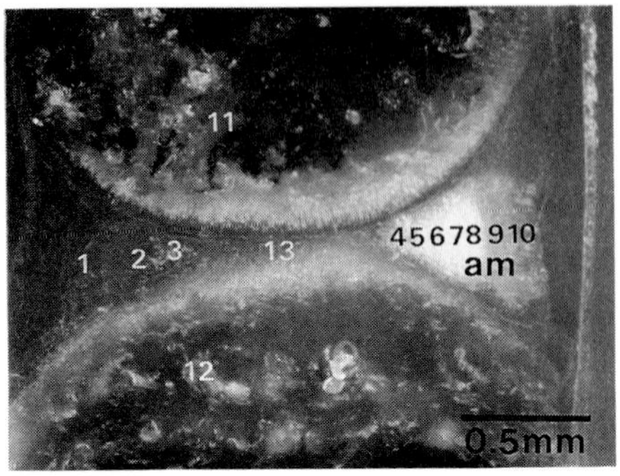

Figure 1. Macrophotograph of vertical anteroposterior face of embedded 305-day old male CBA mouse knee.

Site	CPS P	CPS Ca	K P	K Ca	P/Ca
1	31	35	0.35	0.51	0.68
2	31	32	0.36	0.49	0.75
3	31	34	0.35	0.50	0.70
4	48	85	0.25	0.58	0.44
5	38	86	0.19	0.54	0.35
6	45	93	0.23	0.61	0.37
7	69	125	0.26	0.61	0.43
8	88	161	0.26	0.61	0.42
9	73	122	0.27	0.59	0.46
10	28	47	0.26	0.55	0.46
11	81	146	0.26	0.60	0.43
12	77	121	0.28	0.58	0.49
13	0	2	0	0.24	0

Table 1. Measurements at sites enumerated in fig 1. (K=approx. Composition, wt%)

X-ray microanalysis revealed a consistent trend in levels, judged by count rates, of P and Ca across the AM. They rose from a high level at the apex to a peak density in the last of the white zone, subsiding in the grey zone to levels similar to those found in the PM. No elemental peaks were detected in pure resin. The mean P/Ca ratio for bone of the femoral condyle and tibial plateau was 0.46 ± 0.04. The P/Ca ratios for the AM and PM were 0.42 ± 0.04 and 0.71 ± 0.04 respectively.

4. DISCUSSION

These measurements of P/Ca ratio indicate that by 10 months the anterior part of the knee joint meniscus of the CBA mouse has accumulated hydroxyapatite; posteriorly the higher levels of P relative to Ca indicate a different calcium phosphate, possibly pyrophosphate (Dieppe et al, 1976). Our findings extend those of Vailas et al (1985) who found, in 120 day old rats, levels of calcium three times higher in the relatively stiff opaque anterior part of the meniscus than in the posterior part. Reports of normal human, rabbit, chicken, cow, cat, monkey, and dog present no evidence of calcification of menisci (Vailas et al, 1985). Whether meniscal calcification is a constant feature of smaller rodent or smaller animals generally requires elucidation. These differences in the meniscus have obvious implications in modelling compressive forces. It is possible to conclude that the posterior part of the rat and mouse meniscus assumes a greater role of accommodation in joint loading whilst the anterior part of the meniscus appears to be a stiff wedge.

ACKNOWLEDGEMENTS: We are most grateful to the Arthritis and Rheumatism Council for Research for financial support and to the following for expert assistance: Mr I. Brough, Materials Science Centre, UMIST, X-ray microanalysis; Mr P. Sullivan - Photography; Mrs A. Mellor - Typing.

REFERENCES

Dieppe PA: Huskisson EC: Crocker P: Willoughby DA (1976) Apatite deposition disease. Lancet 1:266-268.

Vailas AC: Zernicke RF: Matsuda J: Peller D (1985) Regional biochemical and morphological characteristics of rat knee meniscus. Comp. Biochem. Physiol. 82B:283-285.

Matrix correction procedures for the quantitative x-ray microanalysis of frozen hydrated tobacco leaf tissue

Patrick Echlin

The Botany School, University of Cambridge, Cambridge CB2 3EA, U.K.

ABSTRACT: A simple procedure to allow diffusible elements to be quantitatively analysed in a hydrated organic matrix.

1. INTRODUCTION

Quantitative x-ray microanalysis of frozen hydrated plant material is dependent on a number of factors. The problems of beam attenuation and absorption of soft x-rays by conductive coating layers, and irregular specimen geometry have been discussed in an earlier paper (Echlin & Taylor 1986). In order to calculate the correct elemental concentrations, the specimen should, if possible, remain fully hydrated throughout analysis. Pearce & Beckett (1987) showed that it is impossible to judge the degree of hydration of a sample by examining the secondary electron image. Echlin (1987) has shown that fully hydrated samples have a characteristic high background at the low energy end of the x-ray spectrum which is missing in dried or even partially dried samples. The effect of drying causes a 3-5 fold increase in the local concentration of dissolved elements.

Quantitation may be achieved by using the peak to local background ratio method either in comparison with known standards or by making the necessary corrections for the composition of the organic and aqueous matrix of those elements which cannot be measured by the x-ray spectrometer. Peripheral hydrated standards have proved impracticable and imprecise with leaf tissue. Pure element and known elemental composition standards are inaccurate when applied directly to organic material. Accurate quantitation has been achieved using the Link Systems ZAFPB quantitative programme in combination with an independent chemical analysis of leaf lamina tissue; measurement of the leaf water content at each stage of growth, and the preparation of pure element peak profiles. Details of the experimental procedures for the biological aspects of this work are in the paper by Echlin & Taylor (1986).

2. EXPERIMENTAL PROCEDURES

Mid-stalk ripe Bright (Coker 319) tobacco leaves were dried overnight at 333K and the lamina carefully separated from the main veins. The lamina was crushed to small pieces, dried at 373K for 3h and ground to a fine powder in a ball mill. An elemental analysis, using standard chemical analytical procedures, was carried out independently at the Philip Morris Research Centre, Richmond, Virginia 23261 and at Galbraith Laboratories, Knoxville, Tennessee 37921. Samples for C, H, O and N measurements were dried again at 333K and handled in a dry box containing nitrogen gas before analysis. Throughout the growing season, dry weight measurements were made repeatedly on fresh tobacco leaves at each of the five stages of growth under investigation. The water content of the leaves at the different growth stages was calculated to be: Juvenile 83.4%; Mature 77.0%; Ripe 73.1%; Senescent 79.0% and Old 85.4%. These measurements were used in combination with the chemical analysis of the dried tissue, to calculate the relative mass fractions of C, H, O, and N in frozen hydrated tissue. The results are shown in Fig.1.

	CARBON	HYDROGEN	OXYGEN	NITROGEN	ASH
Lamina DRIED	0.4476	0.0633	0.4600	0.0080	0.0211
Lamina FRESH	0.0537	0.1020	0.8219	0.0012	0.0212

Fig.1. Relative mass fractions of four elements in fresh and dried ripe tobacco leaves. (Ash = elements measured by x-ray microanalysis.)

In addition to using these figures for the analysis of the ripe lamina matrix was tested against samples from each growth stage adjusted for its water content. The results of these analyses only differed by 4% and the matrix for the dried ripe lamina was used to calculate the fresh matrix for each of the five stages of growth. The matrix corrections proved to be correct because the value obtained by x-ray microanalysis for inorganic elements in frozen hydrated samples is close to the values obtained by chemical analysis. Fig.2 shows the values for K+ in tobacco leaves at five growth stages.

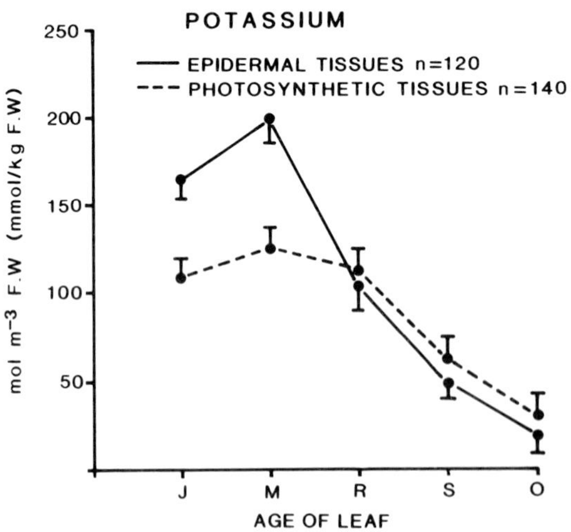

Fig.2. Changes in K+ concentration in cells from photosynthetic and non-photosynthetic tissues of tobacco leaves as a function of growth stage.

REFERENCES

Echlin P. & Taylor S.E. 1986. Journal of Microscopy 141 329-348.
Echlin P. 1987. Proc. 45th Ann. Meeting EMSA. San Francisco Press 658-659.
Pearce R.S. & Beckett A. 1987. Annals of Botany 59 191-195.

Analytical color fluorescence electron microscope

H KOIKE, T NAKANO*, T FUJIMOTO* and K OGAWA*

Akashi Beam Technology Corp., Hachioji, Tokyo, 192 Japan and *Dept. Anat., Fac. Med., Kyoto Univ., Kyoto, 606 Japan

ABRTRACT: A color fluorescence electron microscope was constructed on the basis of the cathodoluminescence technique. The chemical composition of lipid droplrts in the rat adrenal cortex, which emit intense autoluminescence, was examined by the photon counting method.

Since the electron microscope was invented, more than fifty years have already passed. During these years, several techniques of micro-area analysis based on high-resolution electron microscopy have been tested. It is no exaggeration, however, to say that no practical techniques are available for information on the chemical bonding states of constituent elements. The cathodoluminescence (CL) reflects the electronic configuration of the outer electron shell. In the case of organic compounds, including biological macromolecules, the CL reflects states of σ, π and n electrons. This suggests a possibility to obtain knowledge of the electronic structure, which is still a weak area in current analytical electron microscopy.

An analytical color fluorescence electron microscope was assembled on the basis of a high-resolution SEM (AKASHI ISI-DS130) by means of CL (Itoh et al., 1986). In order to maintain high performances at low accelerating voltages, a high-excitation objective lens was introduced. A set of three ellipsoidal mirrors, arranged at 120 degrees from each other, was placed below the upper objective lens pole-piece as a collection system for color CL images. Unfixed specimens placed on cryotips were cut with a cryomicrotome after rapid-freezing, and were observed with a cryoholder cooled by liquid nitrogen. Color images were stored in a RGB memory system to avoid the intensity reduction of CL due to electron bombardment (Fig. 1). The color fluorescence image was displayed on the color CRT simultaneously with a secondary electron image or STEM image of the same field of vision.

In the present study, the possibility of analyzing biological specimens by applying CL spectroscopy was investigated. Since such samples are generally sensitive to electron irradiation, a single-photon counting technique was employed to obtain spectra (Fig. 2). Attention was paid to lipid droplets in the rat adrenal cortex, which emit intense autoluminescence. As the rodent adrenal cortex lacks activity of 17α-hydoroxylase, it was reasonable to examine cholesterol esters, cholesterol, pregnenolone, progesterone, corticosterone and aldosterone only. The spectra of these substances in purified states were measured for comparison with the spectrum of lipid droplets (shown in the lowest line in Table 1). The result exhibits an agreement between a lipid droplet and progesterone in spectrum. Through biochemical techniques, lipid droplets are generally believed to store cholesterol in the form of esters, but this interpretation and the intracellular transport pathway of cholesterol still present some problems for discussion. Some additional investigation will also be required to know the chemical composition of lipid droplets. The results of the present study, however, revealed that intermediary metabolites of the steroid hormone can be separated by CL spectroscopy at electron-microscopic level. This suggests the potential of applying electron microscopy in a new field.

References

Itoh T, Fujimoto T, Koike H, Inoue T and Ogawa K 1986 Acta histochem. cytochem.19 pp 621-633

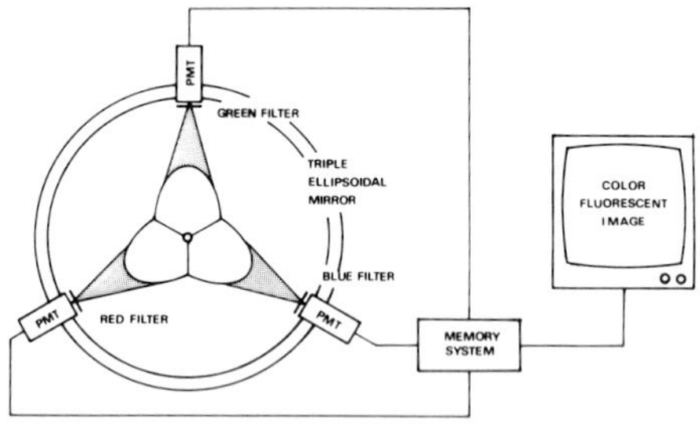

Fig. 1. Schematic diagram of the color fluorescence electron microscope.

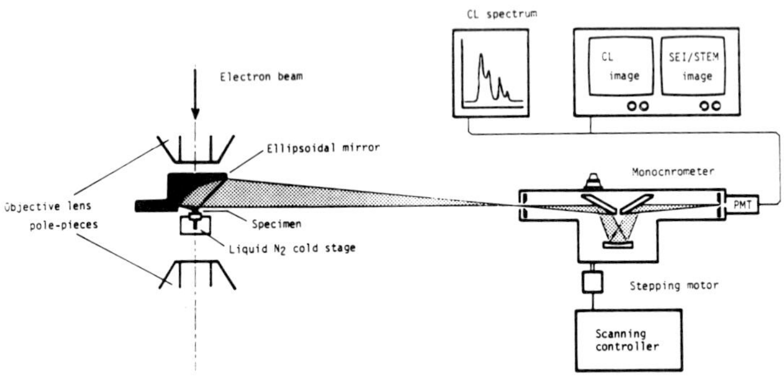

Fig. 2. Experimental arrangement of photon counting and monochromatic CL imaging systems coupled to high-excitation objective lens.

Substance	Peak wavelength (nm)					
	300	320	350	380	420	480
Intermedinary metabolites						
Cholesterol palmitate			M			H
Cholesterol				H		
Pregnenolone				H		
Progesterone		L			M	
Corticosterone					L	
Aldosterone	L					
Lipid droplets		M			M	

Intensity of peak wavelength: high H, medium M, low L.

Table 1. Comparison of the peak wavelength and intensity between intermediary metabolites in steroid biosynthesis and lipid droplets in the rat adrenal cortex.

Ultrastructure localization of chitin and cellulose using labelled colloidal gold in a ciliate cystwall and in invertebrate exoskeletons

Y. VAN DAELE, N.GRECO, Ph.COMPERE

Institute of Zoology, University of Liège, 22,Quai Van Beneden,
B-4020 LIEGE, Belgium.

Chitin and cellulose are the most widespread polysaccharides present in living organisms but their accurate localization in tissues remains problematic with the use of classical cytochemical methods.

The recent postembedding method using colloidal gold labelled with the lectin wheat germ agglutinin (Au-WGA) which is highly specific for N-acetyl-glucosamine (Horisberger 1979), was hitherto applied with success for the localization of chitin in budding yeast (Horisberger and Vonlánthen 1977; Tronchin et al. 1981) in the cyst wall of Entamoeba invadens (Arroyo-Begovich and Carabez-Trejo 1982) as well as in some structures associated to the digestive tract of invertebrates (Peters and Latka 1986).However, no information is available with regard to the validity of this method for locating chitin-protein complexes that are one of the major organic component of ciliate cyst (for example Euplotes muscicola) and of the exoskeleton in most of the protostomian organisms (such as the shore crab Carcinus maenas).

Colloidal gold labelled with 1,4 β-D glucan cellobiohydrolase (CBHI, which was kindly provided by Dr.M.Schülein, Novo, Bagsvaerd, Denmark) has been successfully used for the visualization of the enzyme adsorption at the surface of isolated cellulose microfibrils (Chanzy et al. 1984). The same complex was tentatively tested as a postembedding method to localize tunicin (cellulose-protein complex) in the test of the ascidian Halocynthia papillosa.

The two methods were applied on glutaraldehyde fixed, non osmicated material embedded in Epon 812 resin.

1.CHITIN LOCALIZATION WITH THE AU-WGA METHOD

In the ciliate Euplotes muscicola (Fig.1), only the presumed chitinous layers of the endocyst are distinctly covered with numerous gold particles. The cytoplasm, the proteic cortex of the cell and the proteic layers of the inner cyst wall remain free of particles. In the shore crab Carcinus maenas cuticle (Figs.2-3) a strong gold labelling occurs in the procuticule, where chitin is known to represent 60 to 70% of the total organic cuticular dry weight. The lipoproteic epicuticle and epidermal cells are not significantly marked. In the tunic of Halocynthia, gold particles are comparatively sparse and more scattered although some of them are associated to tunicin bundles. This peculiar distribution probably reflect that of the interfibrillar soluble chitin sulfate

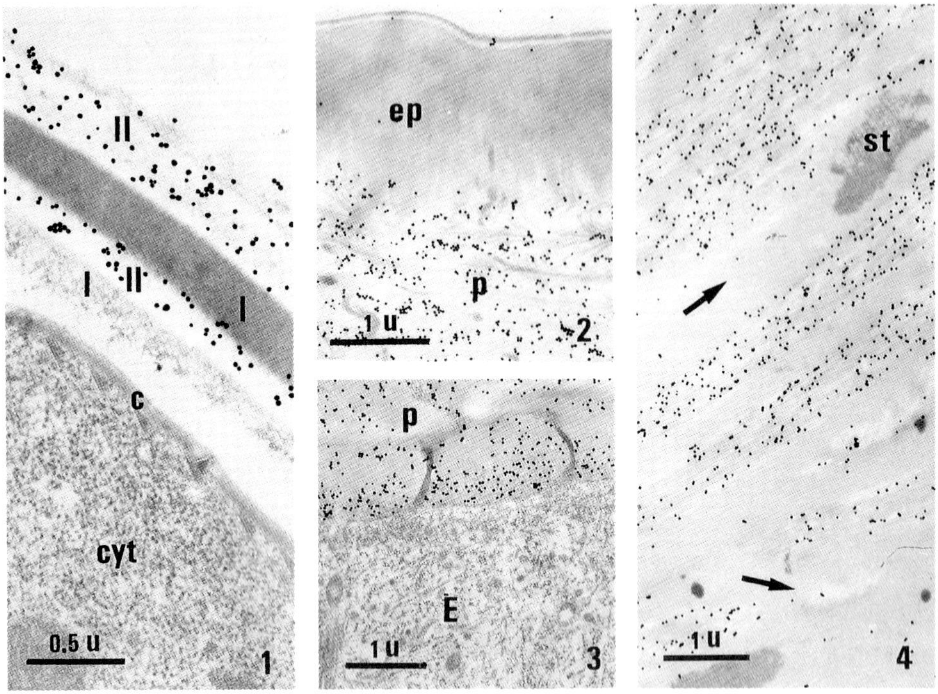

Figs.1 to 3. Au-WGA method (45 min incubation). Fig.1. Euplotes muscicola. c, proteic cortex. cyt, cytoplasm. Proteic layers (I) and chitinous layers (II) of the endocyst. Figs.2-3. Carcinus maenas. E, epidermal cell. ep, epicuticle. p, procuticle. Fig.4. Au-CBHI method (60 min incubation). Halocynthia papillosa. arrows, fundamental substance. st, striated electron dense structure.

fraction identified by Anno et al. (1974) in the same species.

2. CELLULOSE LOCALIZATION WITH THE AU-CBHI METHOD.

CBHI receptor sites were not significantly detected in the shore crab cuticle, nor in the ciliate cyst. In Halocynthia test(Fig.4) they are concentrated along the fibre bundles forming the bulk of the exoskeleton while they are lacking in striated electron dense structures (presumably collagen VI), in tunical cells, in the fundamental substance as well as in epidermal cells.

Anno K, Otsuka K and Seno N 1974 Biochim.Biophys.Acta 362 215
Arroyo-Begovich A and Carabez-Trejo A 1982 J.Parasitol 68 253
Chanzy H, Henrissat B and Vuong R 1974 FEBS Lett. 172 193
Horisberger M 1979 Biol.Cell 36 253
Horisberger M and Vonlanthen M 1977 Arch.Mikrobiol 115 1
Peters W and Latka I 1986 Histochemistry 84 155
Tronchin G, Poulain D, Herbaut J and Biguet J 1981 J.Cell Biol.26 121

This work was supported by a grant of the Belgian "Fonds de la Recherche Fondamentale Collective" (convention n° 2.4506/83).

Inst. Phys. Conf. Ser. No. 93: Volume 3, Chapter 23
Paper presented at EUREM 88, York, England, 1988

A method to visualize the results of morphometric analysis of electron-microscopic platelet sections

M J Heynen and R L Verwilghen

Dept. Haematology, university Hospital, Gasthuisberg, B3000 Leuven, Belgium.

ABSTRACT: Morphometric analysis of platelets of controls and patients with RAEB (Refractory Anaemia with Excess of Blasts as defined by the FAB group) was performed. A 2-dimensional representation of the platelet sections together with the morphometric parameters was used to visualize the differences between platelets of patients and controls.

1. INTRODUCTION

Platelet sections display a high heterogeneity due to (1) heterogeneity of circulating platelets, (2) non-random distribution of organelles in the platelet cytoplasm and (3) malignant and normal platelet populations in patients with RAEB. Due to this heterogeneity, conventional statistical procedures are not appropriate to describe the results of morphometric analysis of platelet sections. We present here a 2-dimensional plot of platelet sections together with the morphometric parameters. The utility and limitations of this method will be discussed.

2. MATERIAL AND METHODS

Area analysis of platelet sections was performed with the IBAS 1 (Kontron). the area of vacuoles of the invaginated membrane system (AAIMSPL) and of the membrane complex (AAMCPL) were expressed per area of platelet section. The area of alpha granules (AALFCYT), dense bodies (AADBCYT), mitochondria (AAMITCYT), Golgi apparatus (AAGACYT) and glycogen (AAGLYCYT) were expressed per cytoplasmic area (CYT= area of platelet section minus area of IMS vacuoles).
Each of the 492 platelet sections has been described by giving the scores of each of 7 "normalised" variables (morphometric parameters). The whole set of platelet sections can now be seen as a swarm of points in a 7-dimensional subspace. By means of Canonical Discriminant Analysis (SAS user's guide: Statistics 1982) two new variables (Canonical variables) are composed, allowing a 2-dimensional representation. In this plot the platelet sections are represented by points and the morphometric parameters by vectors. The length of the projection of this vector on the canonical variables measures the correlation between the morphometric parameter and the canonical variables.

3. RESULTS

The figure shows that (1) platelet sections of controls (+*^ :#=) and patients (BDHLMPW) are separated along the first canonical variable (Can1) and (2) Can1 is positively correlated with the morphometric parameters of alpha granules, dense bodies and membrane complex and negatively correlated with the mitochondrial parameter.

Fig.: CANONICAL DISCRIMINANT ANALYSIS OF 492 SECTIONS OF F-PRP PLATELETS FROM 6 CONTROLS (+*^:#=) and 7 RAEB PATIENTS (BDHLMPW) WITH 7 VARIABLES.

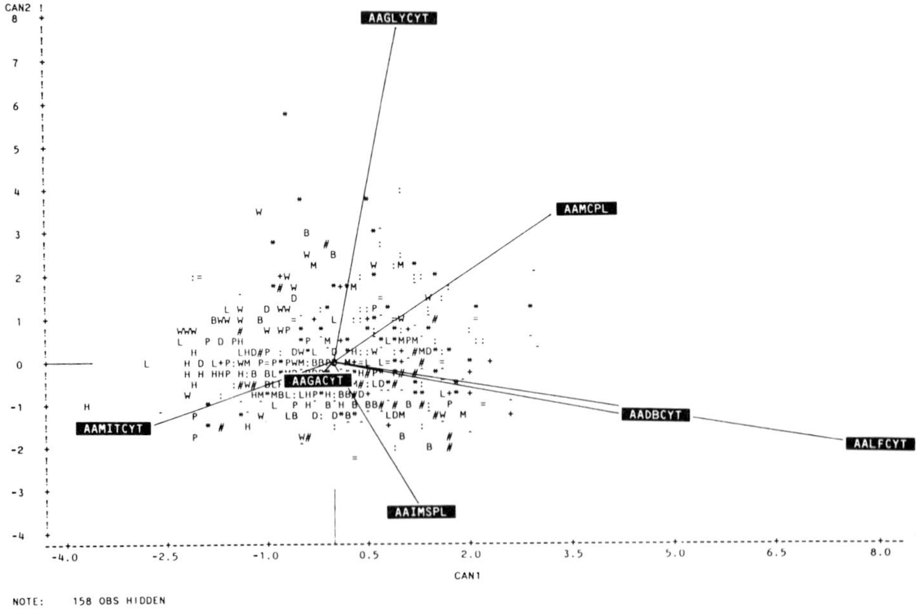

4. CONCLUSIONS

The method we present here allows us to look at single sections each with their unique set of morphometric parameters. We can see that the majority of the platelet sections of the patient score low and the majority of the control platelet sections score high on Can1. In combination with the length of the variable vectors we can conclude that platelet sections of patients with RAEB contain less alpha granules, dense bodies and membrane complex but more mitochondria than platelet sections of controls.

In this procedure we do not obtain absolute values for the volume density of organelles per platelet. But we see a trend, which becomes valuable in correlation with other data. In the study presented it was valuable to correlate the disturbed platelet function in patients with RAEB with the decreased amount of platelet-specific organelles.

Inst. Phys. Conf. Ser. No. 93: Volume 3, Chapter 24
Paper presented at EUREM 88, York, England, 1988

Immunocytochemical localization and organ specific distribution of a polypeptide growth inhibitor isolated from bovine mammary gland

F. Vogel, H. Breter, Th. Müller, F. Schneider and R. Grosse

Central Institute of Molecular Biology
Academy of Sciences of the G.D.R., 1115 Berlin-Buch, G.D.R.

Several growth factors have been recognized over the last years. Polypeptide growth inhibitors are considered to be of similar importance as growth factors in the regulation of cell proliferation. Little is known about their mode of action, their synthesis and subcellular distribution in vivo.

In order to evaluate the physiological importance of a recently purified growth inhibitor (MDGI- mammary gland derived growth inhibitor) (Boehmer et al 1987) we attempted to analyse the tissue and subcellular distribution of MDGI within the bovine mammary gland in comparison to different organs of lactating and nonlactating animals using anti MDGI-IgG. Polyclonal antibodies were raised in rabbits against MDGI and synthetic peptides corresponding to various regions of the MDGI monomer.

The distribution of MDGI has been evaluated with the immunogold technique (DeMey et al 1986) in Lowicryl K4M and cryosections using the antibodies and their appropriate controls (preimmune IgG, blocked antibody).

High concentrations of MDGI were detected in lactating epithelial cells with a preference to the basal invaginations, the apical cell parts and the decondensed euchromatin (Figure 1,a,b,c).

With the exception of fibrolasts, where no immunolabel could be detected, the connective tissue network, as endothelial cells of the blood vessels, myofibrils as well as the milk containing lumina of the aveolus exhibit a remarkable level of MDGI-related antigen.

Further studies using different organs of lactating and nonlactating animals as a target for anti MDGI antibodies reveal a strong organ specificity of MDGI for the lactating mammary gland. Over the cytoplasma and nuclei of cells from liver, pancreas and keratinocytes of the skin of both lactating and nonlactating animals only few of immunolabel were detected.

Estimating the remarkable amino acid sequence homology between MDGI and the heart fatty acid binding protein (FABP) (localized in the myofibrils of the heart muscle cells) a high level of MDGI-related antigen has been observed over the myofibrils. The nuclei of these heart muscle cells were free of label.

Howewer, like proliferation stimulating polypeptides as PDGF, EGF, NGF and insulin, which were identified in nuclear chromatin (Yeh et al 1987) it seems to be an interesting fact, that MDGI-related antigens were localized in decondensed euchromatin. As to wether these localization is of functionally relevancy to the growth inhibition of MDGI has to be a subject of further investigations.

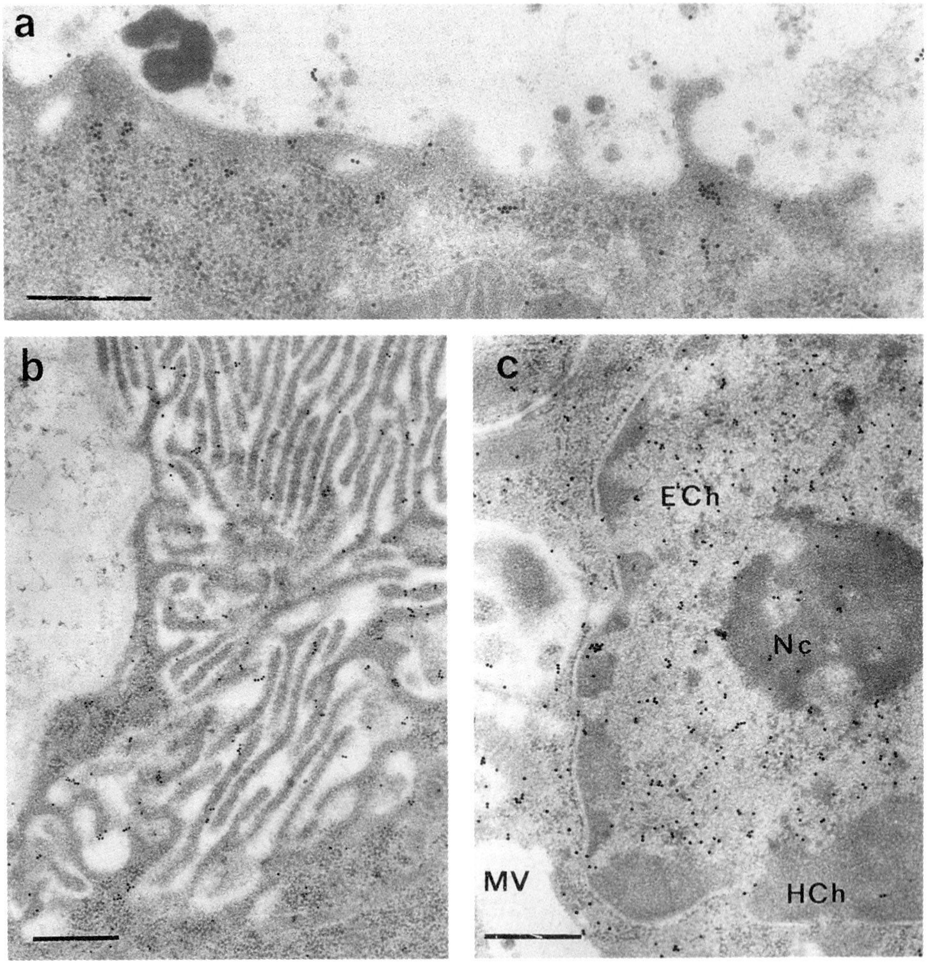

Figure 1: Localization of mammary gland derived growth inhibitor (MDGI) on (a) apical cell parts, (b) basal invaginations, and (c) nucleus of lactating epithelial cells embedded at low temperature in Lowicryl K4M utilizing the immunogold technique. Bars = 0,5 um.

References

Boehmer F-D et al 1987 J Biol Chem 262 15137
DeMey J and Moeremans M 1986 In: Tech. in Biol.Electr.Microsc. III 229 (Springer Verlag Berlin)
Yeh H-J et al 1987 Proc Natl Acad Sci USA 84 2317

Subject Index

Author Index